T0180629

Springer Proceedings in Mathematics & Statistics

Volume 47

For further volumes:
http://www.springer.com/series/10533

Springer Proceedings in Mathematics & Statistics

This book series features volumes composed of select contributions from workshops and conferences in all areas of current research in mathematics and statistics, including OR and optimization. In addition to an overall evaluation of the interest, scientific quality, and timeliness of each proposal at the hands of the publisher, individual contributions are all refereed to the high quality standards of leading journals in the field. Thus, this series provides the research community with well-edited, authoritative reports on developments in the most exciting areas of mathematical and statistical research today.

Sandra Pinelas • Michel Chipot • Zuzana Dosla
Editors

Differential and Difference Equations with Applications

Contributions from the International
Conference on Differential & Difference
Equations and Applications

 Springer

Editors
Sandra Pinelas
Departamento de Ciências Exactas
Academia Militar
Amadora, Portugal

Michel Chipot
Institut für Mathematik
University of Zürich
Zürich, Switzerland

Zuzana Dosla
Department of Mathematics
Masaryk University
Brno, Czech Republic

ISSN 2194-1009   ISSN 2194-1017 (electronic)
ISBN 978-1-4939-4179-7   ISBN 978-1-4614-7333-6 (eBook)
DOI 10.1007/978-1-4614-7333-6
Springer New York Heidelberg Dordrecht London

Mathematics Subject Classification (2010): 33C45, 34-XX, 35-XX, 37-XX, 39A10, 39A13, 39A06, 39A12, 45J05, 47-XX, 54C60, 65-XX, 76H05, 76Z05, 92D25, 93B03, 93B40

Printed on acid-free paper

Springer is part of Springer Science+Business Media (www.springer.com)

Preface

For five days from July 4 to 8, 2011, more than 230 mathematicians from 50 countries attended the International Conference on Differential and Difference Equations and Applications, held at Azores University, Ponta Delgada, Portugal.

This conference was held in honour of Professor Ravi P. Agarwal for his contributions to science and, in particular, to the mathematical community.

The scientific aim of this conference was to bring together mathematicians working in various disciplines of differential and difference equations and their applications. There were 11 plenary lectures, 21 main lectures, and 198 communications about the current research in this field. This volume contains 60 selected original papers which are connected to research lectures given at the conference. Each paper has been carefully reviewed.

We take this opportunity to thank all the participants of the conference and the contributors to these proceedings. Our special thanks belong to the Department of Mathematics, Azores University, Ponta Delgada, Portugal, for the sincere hospitality. We are also grateful to the Scientific and Organizing Committees for all the effort in the preparation of the conference.

We hope that this volume will serve researchers in all fields of differential and difference equations.

Amadora, Portugal   Sandra Pinelas
Zürich, Switzerland   Michel Chipot
Brno, Czech Republic   Zuzana Dosla

Contents

Part II   Contributions

Part I
Main Speakers

Two-Term Perturbations in Half-Linear Oscillation Theory

Ondřej Došlý and Simona Fišnarová

Abstract We consider the nonoscillatory half-linear differential equation

$$(r(t)\Phi(x'))' + c(t)\Phi(x) = 0, \quad \Phi(x) := |x|^{p-2}x, \ p > 1,$$

and we study the influence of the perturbation terms $\tilde{r}, \tilde{c}$ on oscillatory properties of the equation

$$[(r(t) + \tilde{r}(t))\Phi(x')]' + (c(t) + \tilde{c}(t))\Phi(x) = 0.$$

We prove new oscillation criteria which can be applied in situations where previously obtained criteria fail.

1 Introduction

We consider the half-linear second-order differential equation

$$L[x] := (r(t)\Phi(x'))' + c(t)\Phi(x) = 0, \quad \Phi(x) := |x|^{p-2}x, \ p > 1, \tag{1}$$

where r, c are continuous functions and $r(t) > 0$, and its "perturbation", the equation

$$\tilde{L}[x] := \left[(r(t) + \tilde{r}(t))\Phi(x')\right]' + (c(t) + \tilde{c}(t))\Phi(x) = 0 \tag{2}$$

O. Došlý (✉)
Department of Mathematics and Statistics, Masaryk University, Kotlářská 2,
CZ-611 37 Brno, Czech Republic
e-mail: dosly@math.muni.cz

S. Fišnarová
Department of Mathematics, Mendel University in Brno, Zemědělská 1,
CZ-613 00 Brno, Czech Republic
e-mail: fisnarov@mendelu.cz

S. Pinelas et al. (eds.), *Differential and Difference Equations with Applications*, Springer
Proceedings in Mathematics & Statistics 47, DOI 10.1007/978-1-4614-7333-6_1,
© Springer Science+Business Media New York 2013

with continuous functions $\tilde{r}, \tilde{c}$ and $r(t) + \tilde{r}(t) > 0$. We suppose that Eq. (1) is nonoscillatory and we investigate the influence of the perturbation terms $\tilde{r}, \tilde{c}$ on oscillatory behavior of Eq. (2).

It is a well-known fact that the oscillation theory of Eq. (1) is (almost) the same as that of the linear Sturm–Liouville differential equation

$$\left(r(t)x'\right)' + c(t)x = 0 \tag{3}$$

which is the special case $p = 2$ in Eq. (1); see [1,6] and the references given therein. The classical approach to the half-linear oscillation theory consists in regarding Eq. (1) as a perturbation of the one-term (nonoscillatory) differential equation

$$\left(r(t)\Phi(x')\right)' = 0. \tag{4}$$

The standard half-linear oscillation criteria claim, roughly speaking, that Eq. (1) is oscillatory provided the function c is "sufficiently positive with respect to r", while it is nonoscillatory, if it is not "too positive". Particular oscillation criteria along this line can be found in [6].

A more general approach is presented in [6, Sect. 5.2]. There, the equation

$$\left(r(t)\Phi(x)\right)' + \left[c(t) + \tilde{c}(t)\right]\Phi(x) = 0 \tag{5}$$

is viewed as a perturbation of Eq. (1) and the so-called principal solution of this equation plays an important role in the obtained (non)oscillation criteria. Note that in the linear oscillation theory, a two-term nonoscillatory equation (3) can be always reduced to the one-term equation $(r(t)x')' = 0$, so this approach, in contrast to the half-linear oscillation theory, brings actually nothing new.

Here we follow the idea initiated in [9] for linear equation (3) and extended to half-linear case in [3, 4]. We obtain oscillation criteria for Eq. (2) where criteria presented in the above mentioned papers fail. A typical model is the perturbed Riemann–Weber half-linear differential equation

$$\left[\left(1 + \tilde{r}(t)\right)\Phi(x')\right]' + \left[\frac{\gamma_p}{t^p} + \frac{\mu_p}{t^p \log^2 t} + \tilde{c}(t)\right]\Phi(x) = 0,$$

where γ_p, μ_p are the so-called critical constants in the half-linear Euler and Riemann–Weber equations; see [8] and also Sect. 3 of this paper.

2   Preliminaries

The principal tool we use in our paper is the Riccati technique and its modifications, which we explain at the beginning of this section. If $x(t) \neq 0$ is a solution of Eq. (1), then the Riccati variable $w = r\Phi(x'/x)$ solves the Riccati type differential equation

$$R[w] := w' + c(t) + (p-1)r^{1-q}(t)|w|^q = 0, \quad q := \frac{p}{p-1}. \tag{6}$$

If h is a positive differentiable function and we put $v = h^p(w - w_h)$, where w is a solution of Eq. (6) and $w_h = r\Phi(h'/h)$, then v is a solution of the so-called modified Riccati equation

$$R_m[v] := v' + h(t)L[h](t) + (p-1)r^{1-q}(t)h^{-q}(t)H(t,v) = 0, \qquad (7)$$

where

$$H(t,v) := |v + G(t)|^q - q\Phi^{-1}(G(t))v - |G(t)|^q, \quad G := rh\Phi(h'). \qquad (8)$$

The relationship between solutions of Eqs. (6) and (7) is described in the next lemma. The proof of statements (i) and (ii) can be found in [3]; the statement (iii) is Theorem 2.2.1 of [6].

Lemma 1. (i) *Let $h > 0$ and w be differentiable functions and let $v = h^p(w - w_h)$. Then*

$$h^p R[w] = R_m[v]. \qquad (9)$$

(ii) *The function $H(t,v)$ defined in (8) satisfies $H(t,v) \geq 0$ with the equality if and only if $v = 0$.*

(iii) *Equation (1) is nonoscillatory if and only if there exists a differentiable function w satisfying $R[w](t) \leq 0$ for large t.*

Now we give a result of [4, Theorem 3.5] which concerns nonnegativity of solutions of the modified Riccati equation (7).

Lemma 2. *Let h be a positive continuously differentiable function such that $h'(t) \neq 0$ for large t. Suppose that*

$$\int^\infty r^{-1}(t)h^{-2}(t)|h'(t)|^{2-p}\,dt = \infty, \quad h(t)L[h](t) \geq 0 \text{ for large } t, \quad \lim_{t\to\infty}|G(t)| = \infty.$$

Then all possible proper solutions (i.e., solutions which exist on some interval of the form $[T,\infty)$) of Eq. (7) are nonnegative.

The second basic method of the half-linear oscillation theory is the so-called *variational principle*, which is formulated in the next lemma.

Lemma 3. *Equation (2) is oscillatory if and only if for every $T \in \mathbb{R}$, there exists a nontrivial function $y \in W^{1,p}(T,\infty)$, with compact support in (T,∞), such that*

$$\mathscr{F}(y;T,\infty) = \int_T^\infty \left[(r(t) + \tilde{r}(t))|y'|^p - (c(t) + \tilde{c}(t)|y|^p\right]dt \leq 0. \qquad (10)$$

Next we present, for the sake of the later comparison, some known results concerning oscillation of Eq. (2). These statements are taken from [3,5]. Note that in those papers, the function h is supposed to be a positive solution of Eq. (1), so the coefficient C defined in Eq. (11) below reduces to $C = h[(\tilde{r}\Phi(h'))' + \tilde{c}\Phi(h)]$.

More precisely, Proposition 1 below is a modification of [3, Theorem 4], while Proposition 2 is a modification of [5, Theorem 4]. Here we present a more general version of the statements with arbitrary positive continuously differentiable function h. The proofs are almost the same; the only difference is in the definition of the function C.

In the rest of the paper we use the following notation, where h is a positive continuously differentiable function:

$$C := h\tilde{L}[h], \quad R := (r+\tilde{r})h^2|h'|^{p-2}, \quad \Omega := (r+\tilde{r})h\Phi(h'). \tag{11}$$

Proposition 1. *Let h be a positive continuously differentiable function such that*

$$C(t) \geq 0 \text{ for large } t, \quad \int^{\infty} C(t)\,\mathrm{d}t < \infty, \tag{12}$$

$$0 < \liminf_{t\to\infty} \Omega(t) \leq \limsup_{t\to\infty} \Omega(t) < \infty, \tag{13}$$

$$\int^{\infty} \frac{\mathrm{d}t}{R(t)} = \infty, \tag{14}$$

and

$$\int^{\infty} (r(t)+\tilde{r}(t))^{1-q}h^{-q}(t)\,\mathrm{d}t = \infty. \tag{15}$$

If there exists $\varepsilon > 0$ such that the equation

$$\left(R(t)y'\right)' + \frac{q-\varepsilon}{2}C(t)y = 0 \tag{16}$$

is oscillatory, then (2) is also oscillatory.

Remark 1. Applying the Hille–Nehari criteria to the linear equation (16), one can obtain that under conditions (12)–(15), Eq. (2) is oscillatory provided

$$\liminf_{t\to\infty} \int^t R^{-1}(s)\,\mathrm{d}s \int_t^{\infty} C(s)\,\mathrm{d}s > \frac{1}{2q}; \tag{17}$$

see [3].

Proposition 2. *Let h be the a positive continuously differentiable function satisfying conditions (15) and*

$$C(t) \geq 0 \text{ for large } t, \quad \int^{\infty} C(t)\,\mathrm{d}t = \infty, \tag{18}$$

$$\limsup_{t\to\infty} \Omega(t) < \infty. \tag{19}$$

Then Eq. (2) is oscillatory.

3 Oscillation Criteria

A typical example of perturbed equation where the previous results *do not apply* is the perturbed Riemann–Weber half-linear differential equation

$$\left[(1+\tilde{r}(t))\Phi(x')\right]' + \left[\frac{\gamma_p}{t^p} + \frac{\mu_p}{t^p\log^2 t} + \tilde{c}(t)\right]\Phi(x) = 0, \tag{20}$$

where

$$\gamma_p = \left(\frac{p-1}{p}\right)^p, \quad \mu_p = \frac{1}{2}\left(\frac{p-1}{p}\right)^{p-1}, \tag{21}$$

are the so-called critical constants in the general unperturbed Riemann–Weber equation

$$\left(\Phi(x')\right)' + \left[\frac{\gamma}{t^p} + \frac{\mu}{t^p\log^2 t}\right]\Phi(x) = 0, \tag{22}$$

where the constants $\gamma = \gamma_p$, $\mu = \mu_p$ are "separating" values between oscillation and nonoscillation in Eq. (22); see [8]. Equation (22) with these limiting values has a solution which asymptotically behaves like $h(t) = t^{\frac{p-1}{p}}\log^{\frac{1}{p}} t$. For this function, e.g., in the case when $\tilde{r}(t) = o(1)$ as $t \to \infty$, which is a typical case in applications, the limit in (19) is infinite, so criteria given in Propositions 1, 2 cannot be applied.

As main results of the paper we present criteria which can be applied also to the case when the limit (19) is infinite.

Theorem 1. *Let h be a positive continuously differentiable function satisfying conditions (12), (14), and*

$$\lim_{t\to\infty} \Omega(t) = \infty. \tag{23}$$

If there exists $\varepsilon > 0$ such that Eq. (16) is oscillatory, then Eq. (2) is also oscillatory, i.e., Eq. (2) is oscillatory provided condition (17) holds.

Proof. Suppose, by contradiction, that Eq. (2) is nonoscillatory. Then there exists a solution of the associated Riccati equation, and by Lemma 1, there exists a proper solution v of the modified Riccati equation

$$v' + C(t) + (p-1)(r(t) + \tilde{r}(t))^{1-q}h^{-q}(t)H(t,v) = 0. \tag{24}$$

Lemma 2 implies that $v(t) \geq 0$ for large t. Since $C(t)$ and $H(t,v)$ are nonnegative, it follows that $v(t)$ is nonincreasing and hence there exists a finite limit $\lim_{t\to\infty} v(t) \geq 0$. Integrating Eq. (24) and since $v(t) \geq 0$, we obtain

$$v(T) \geq \int_T^t C(s)\,\mathrm{d}s + (p-1)\int_T^t (r(s) + \tilde{r}(s))^{1-q}h^{-q}(s)H(s,v(s))\,\mathrm{d}s.$$

Consequently, since $\int^\infty C(t)\,dt < \infty$, letting $t \to \infty$, we obtain

$$\int^\infty (r(t)+\tilde{r}(t))^{1-q}h^{-q}(t)H(t,v(t))\,dt < \infty.$$

Analogously to [7] we show that $v(t) \to 0$ as $t \to \infty$. We have

$$(r(t)+\tilde{r}(t))^{1-q}h^{-q}(t)H(t,v(t)) = (r(t)+\tilde{r}(t))|h'(t)|^p F(z(t)),$$

where $F(z) = |z+1|^q - qz - 1$ and $z = v/\Omega \to 0$ as $t \to \infty$. Hence $F(z) = \frac{q(q-1)}{2}z^2 + o(z^2)$ as $z \to 0$. This means that for $\varepsilon > 0$ there exists T_1 such that for $t > T_1$

$$\frac{(q-\varepsilon)(q-1)}{2}z^2(t) < F(z(t)).$$

Hence (suppressing the integration argument t)

$$\infty > (p-1)\int^\infty (r+\tilde{r})^{1-q}h^{-q}H(t,v)\,dt = (p-1)\int^\infty (r+\tilde{r})|h'|^p F(z)\,dt$$
$$> \frac{q-\varepsilon}{2}\int^\infty (r+\tilde{r})|h'|^p z^2\,dt = \frac{q-\varepsilon}{2}\int^\infty \frac{v^2}{R}\,dt.$$

Condition $\int^\infty R^{-1}(t)\,dt = \infty$ implies that $\lim_{t\to\infty} v(t) = 0$. Now, using the Taylor polynomial of the function $H(t,v)$ at $v = 0$, we have the following estimate (see [3, Lemma 5]):

$$H(t,v) = \frac{q(q-1)}{2}|\Omega(t)|^{q-2}v^2(1+o(1)), \text{ as } v \to 0.$$

Consequently, to arbitrary $\varepsilon > 0$ there exists $T_2 > T_1$ such that

$$\frac{(q-\varepsilon)(q-1)}{2}|\Omega(t)|^{q-2}v^2(t) \le H(t,v(t)), \quad t \le T_2$$

and hence

$$\frac{q-\varepsilon}{2}\frac{v^2(t)}{R(t)} \le (p-1)(r(t)+\tilde{r}(t))^{1-q}h^{-q}(t)H(t,v(t)), \quad t \le T_2.$$

From Eq. (24) we have that v is a solution of the Riccati inequality

$$v' + C(t) + \frac{q-\varepsilon}{2}\frac{v^2}{R(t)} \le 0, \quad t \le T_2$$

and this means that Eq. (16) is nonoscillatory by Lemma 1 (iii). This is a contradiction.   □

The next statement can be regarded as a complement of Proposition 2. In contrast to that statement, we do not suppose conditions (13), (15). Here we are again motivated by the perturbed Riemann–Weber equation since in the case $h(t) = t^{\frac{p-1}{p}} \log^{1/p} t$, condition (15) (again with $\tilde{r}(t) = o(1)$ as $t \to \infty$) is not satisfied if $p < 2$.

Theorem 2. *Let h be a positive continuously differentiable function satisfying conditions (14), (18), and (23). Then Eq. (2) is oscillatory.*

Proof. The proof is based on the variational principle. Let $T \in \mathbb{R}$ be arbitrary and define for $T < t_1 < t_2 < t_3$ the function $y \in W^{1,p}(T, \infty)$ in the same way as in [5], i.e.,

$$y(t) := \begin{cases} 0 & t \in [T, t_0], \\ f(t) & t \in [t_0, t_1], \\ h(t) & t \in [t_1, t_2], \\ g(t) & t \in [t_2, t_3], \\ 0 & t \in [t_3, \infty), \end{cases}$$

where f is any continuously differentiable function satisfying $f(t_0) = 0$, $f(t_1) = h(t_1)$ and g is the solution of Eq. (1) satisfying $g(t_2) = h(t_2)$, $g(t_3) = 0$.

This is a typical construction used in the variational principle and its idea was introduced in the paper [10]. Assumptions (13), (15) are used in [5] to prove nonnegativity of a solution of the modified Riccati equation. Here we use Lemma 2 instead of these conditions, so condition (13) is replaced by condition (15) and condition (23) can be omitted. The idea of the proof is that the values $t_1 < t_2 < t_3$ can be chosen in such a way that (10) holds. Technical computations are almost the same as those in [5], so we omit them. □

Remark 2. In Theorems 1, 2, Eq. (1) actually plays no role since h is *any* differentiable function without any relationship to Eq. (1). However, in many applications, h is a solution of some known equation, typically $h(t) = t^{\frac{p-1}{p}}$ is a solution of the "critical" Euler equation

$$\left(\Phi(x')\right)' + \frac{\gamma_p}{t^p} \Phi(x) = 0$$

(see [8]), or $h(t) = t^{\frac{p-1}{p}} \log^{1/p} t$ is close to a solution of the Riemann–Weber equation (21) (see [2]). In this situation, the term $h[(r\Phi(h'))' + c\Phi(h)]$ is zero or "small" in the definition of the function C and then oscillation of (2) is really formulated in terms of the perturbation functions $\tilde{r}, \tilde{c}$.

Now we apply our results to the perturbed Riemann–Weber equation (20). We take $h(t) = t^{\frac{p-1}{p}} \log^{\frac{1}{p}} t$ and using the computations from [2] we obtain

$$\Omega(t) = (1 + \tilde{r}(t))h(t)\Phi(h'(t))$$

$$= \left(\frac{p-1}{p}\right)^{p-1} (1 + \tilde{r}(t)) \log t \left(1 + \frac{1}{(p-1)\log t}\right)^{p-1},$$

$$R(t) = (1 + \tilde{r}(t))h^2(t)|h'(t)|^{p-2}$$

$$= \left(\frac{p-1}{p}\right)^{p-2} (1 + \tilde{r}(t))t \log t \left(1 + \frac{1}{(p-1)\log t}\right)^{p-2},$$

$$C(t) = A(t) + B(t), \tag{25}$$

where

$$A(t) = h(t)\Phi(h'(t))' + h^p(t)\left(\frac{\gamma_p}{t^p} + \frac{\mu_p}{t^p \log^2 t}\right)$$

$$= t^{-1}O(\log^{-2}t),$$

$$B(t) = h(t)\left[\tilde{r}'(t)\Phi(h'(t)) + \tilde{r}(t)(\Phi(h'(t)))'\right] + \tilde{c}(t)h^p(t)$$

$$= \left(\frac{p-1}{p}\right)^{p-1} \tilde{r}'(t) \log t \left(1 + \frac{1}{(p-1)\log t}\right)^{p-1}$$

$$+ \tilde{r}(t)t^{-1}\log t\left(-\gamma_p - \frac{\mu_p}{\log^2 t} + O(\log^{-3}t)\right) + \tilde{c}(t)t^{p-1}\log t.$$

Using the above computations and Theorems 1, 2, we can formulate the following statement.

Corollary 1. *Let C be given in Eq.* (25) *and suppose that*

$$C(t) \geq 0 \text{ for large } t,$$

$$\int^{\infty} \frac{1}{(1 + \tilde{r}(t))t \log t}\, dt = \infty, \quad \lim_{t \to \infty} (1 + \tilde{r}(t)) \log t = \infty.$$

(i) *If*

$$\int^{\infty} C(t)\, dt < \infty$$

and

$$\liminf_{t \to \infty} \int^{t} \frac{ds}{(1 + \tilde{r}(s))s \log s} \int_{t}^{\infty} C(s)\, ds > \frac{1}{2}\left(\frac{p-1}{p}\right)^{p-1},$$

then (20) *is oscillatory.*

(ii) *If*

$$\int^{\infty} C(t)\, dt = \infty,$$

then (20) *is oscillatory.*

Acknowledgements Research supported by the Grants 201/11/0768 and 201/10/1032 of the Czech Grant Agency and the Research project MSM0021622409 of the Ministry of Education of the Czech Republic.

References

1. Agarwal, R.P., Grace, S.C., O'Regan, D.: Oscillation Theory of Second Order Linear, Half-Linear, Superlinear and Sublinear Differential Equations. Kluwer Academic Publishers, Dordrecht (2002)
2. Došlý, O.: Perturbations of the half-linear Euler–Weber differential equation. J. Math. Anal. Appl. **323**, 426–440 (2006)
3. Došlý, O., Fišnarová, S.: Half-linear oscillation criteria: perturbation in term involving derivative. Nonlinear Anal. Theory Methods Appl. **73**, 3756–3766 (2010)
4. Došlý, O., Fišnarová, S.: Two-parametric conditionally oscillatory half-linear differential equations. Abstr. Appl. Anal. **2011**, 16 (2011) (Article ID 182827)
5. Došlý, O., Fišnarová, S.: Variational technique and principal solution in half-linear oscillation criteria. Appl. Math. Comput. **217**, 5385–5391 (2011)
6. Došlý, O., Řehák, P.: Half-Linear Differential Equations. North-Holland Mathematical Studies, vol. 202. Elsevier, Amsterdam (2005)
7. Došlý, O., Ünal, M.: Half-linear differential equations: linearization technique and its application. J. Math. Anal. Appl. **335**, 450–460 (2007)
8. Elbert, Á., Schneider, A.: Perturbations of the half-linear Euler differential equation. Results Math. **37**, 56–83 (2000)
9. Krüger, H., Teschl, G.: Effective Prüfer angles and relative oscillation criteria. J. Differ. Equ. **245**, 3823–3848 (2008)
10. Müller-Pfeiffer, E.: Existence of conjugate points for second and fourth order differential equations. Proc. Roy. Soc. Edinburgh Sect. A **89**, 281–291 (1981)

Capillary Forces on Partially Immersed Plates

Robert Finn

1 Introduction

Writings describing the floating of objects on a liquid surface date from long prior to the starting year of the currently accepted calendar. About 350 BC Aristoteles described observations of objects that sink when fully submerged in water but which nevertheless can be made to float at the water surface. That is in striking contrast to the requirements for floating formulated by his countryman Archimedes during the following century, which specifically exclude such behavior. Two thousand years later the French physicist and priest Mariotte (1620–1684) observed and attempted [1] to explain the remarkable tendency of two floating balls either to attract or repel each other. In retrospect it cannot be surprising that the attempted explanations were at once incomplete and inconsistent; it is now generally accepted that such phenomena are closely linked with surface tension, the concept of which was initially introduced over half a century following Mariotte's decease. And an adequate description could hardly be feasible without the Calculus, which may well not have been accessible to that thinker during his lifetime.

Consistent and verifiable scientific clarification of such phenomena appears initially in two "suppléments" [2, 3] on capillarity theory that Laplace added to the tenth volume of his "mécanique céleste", and which document in strongly convincing terms the power of the magic new weapon bequeathed him by Leibniz and by Newton. Among the many contributions of that œuvre, Laplace provides quantitative calculations of the forces exerted on two parallel vertical plates, whose surfaces may consist of possibly four distinct materials, partly immersed and separated a prescribed distance in an unbounded liquid bath in the presence of a downward gravity field. Although this is not strictly a "floating" phenomenon, it

R. Finn (✉)
Mathematics Department, Stanford University, Stanford, CA 94301-2125, USA
e-mail: finn@math.stanford.edu

S. Pinelas et al. (eds.), *Differential and Difference Equations with Applications*, Springer
Proceedings in Mathematics & Statistics 47, DOI 10.1007/978-1-4614-7333-6_2,
© Springer Science+Business Media New York 2013

embodies some of the essential characteristics of that state of being; to my knowledge there is as yet no satisfactory mathematical theory that applies adequately for floating objects (although McCuan's recent paper [4] does point in a promising direction).

Laplace's second "supplément" contains general expressions for the forces on the plates in specific circumstances, and exhibits configurations for which the behavior reverses from repelling to attracting when plates of specific materials approach each other. These achievements were certainly amazing for their time and may well have intimidated the remaining scientific community. Instead of inspiring an immediate flood of further activity as one might expect, apparently no further writings on the topic appeared (beyond some expositions) during the ensuing two centuries. Then in 2010, in connection with his response to a question raised initially by Thomas Young in a letter to Poisson about 1825 and then independently in an informal email from Brian Storey in 2008, the present author approached the problem from a different point of view [5], initially without knowledge of Laplace's writings, basing his discussion on a differential-geometric identity that may not have been known to Laplace.

In a work now in preparation, Finn and Lu continue and largely complete the material of [5], with a new discussion focusing on the geometric content of the question, covering all occurring configurations, and delineating precise criteria for changes of behavior. In what follows, we outline the essential features of that work.

2 Background Material

Partly in view of continuing controversy over underlying concepts of the theory and chiefly as a courtesy to readers unfamiliar with classical surface tension theory, we begin with an outline of that material. The following remarks reflect the phenomenological approach adopted in [6].

We imagine two adjacent but distinct materials sharing a smooth surface interface S. Figure 1 illustrates the case in which S is planar, but the construction encompasses also curved surfaces. We suppose that particles interior to each material are subject to forces arising from immediately neighboring particles, which act only within infinitesimal neighborhoods, but become infinite at each point roughly as delta functions. The forces are to be isotropic interior to each material, so that at an interior point of either material, they act equally in all directions and thus will balance and cancel in an equilibrium configuration. However on the interface S, the forces may depend on direction due to the differing adjacent materials; see Fig. 1. Thus there may exist a net force acting normal to S, leading to infinitesimal compression (or expansion) at the surface, with a resultant work e (surface energy) per unit area created over S.

If the materials are both fluids, then we observe that the energy e per unit area can be identified with a force σ per unit length called *surface tension*; this basic observation was introduced in general terms by von Segner about 1750, and

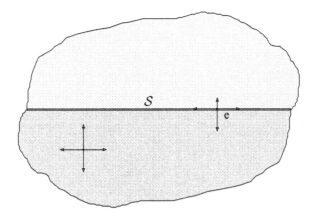

Fig. 1 Surface interface S; forces acting interior to one of the materials or on S

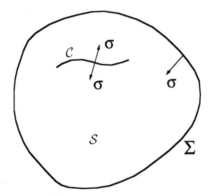

Fig. 2 Action of surface tension on a curve C interior to S or on the boundary Σ of S

developed by Thomas Young in his 1805 paper as an underlying concept for a general theory. σ is exerted on any curve C lying interior to S or on ∂S. If C is interior, then the forces on one side cancel the corresponding forces on the opposite side see Fig. 2. But if $C \in \Sigma \equiv \partial S$ then σ pulls on C and tries to contract S.

This connection is made explicit by **Young's Discovery** that if S separates two fluids, then there is a pressure jump δp across S given by

$$\delta p = 2\sigma H, \tag{1}$$

where H is the *mean curvature* of S, and also by the differential-geometric identity

$$\int_S 2\mathbf{H}\, d\mathcal{S} = \oint_\Sigma \mathbf{n}\, ds \tag{2}$$

where $\mathbf{H}$ is the mean curvature vector on S, and $\mathbf{n}$ is unit co-normal on Σ. Note that (2) equates an integral of normals to S with an integral of vectors tangential to S.

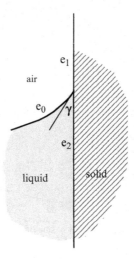

Fig. 3 Contact angle γ at a triple interface

When (2) is multiplied by σ, it identifies the net effect of *"capillary pressure"* $2\sigma|\mathbf{H}|$ distributed over $\mathcal{S}$ and acting orthogonal to $\mathcal{S}$, with the net effect of *capillary force* σ *per unit length* acting in the co-normal direction $\mathbf{n}$ (orthogonal to Σ and tangential to $\mathcal{S}$) along Σ. Note that this interpretation does not apply when one or both of the materials separated by $\mathcal{S}$ is a solid, as the δp of (1) no longer can be interpreted as a pressure change.

In a typical configuration, a fluid surface abuts on a solid *"support surface"* creating a *"triple interface"* involving one fluid/fluid interface and two fluid/solid interfaces, as indicated in Fig. 3.

Theorem of Young/Gauss *If the respective surface energy densities are* e_0, e_1, e_2 *as in Fig. 3, then the fluid/fluid interface meets $\mathcal{S}$ in an angle γ determined by*

$$\cos\gamma = \frac{e_2 - e_1}{e_0} \tag{3}$$

and thus the "contact angle" γ is a physical quantity depending only on the materials, and in no way on body forces such as gravity or on the particular geometry of the configuration.

Young originally claimed (without formal justification) that surface tensions at fluid/solid interfaces can be introduced analogously to those at fluid/fluid interfaces, and he asserted the validity of (3) with the energies replaced by tensions σ_0, σ_1, σ_2. Although that view is still commonly held (and exposited!), it is in this author's view not supportable and can lead to significant errors; see [7–12].

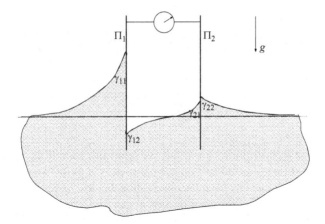

Fig. 4 Two vertical plates of conceivably four distinct materials partly immersed in an infinite fluid bath. The contact angle on each face is determined by the material of that face

3 Problem Formulation

Figure 4 shows the two vertical plates, partially immersed in an infinite liquid. The plates are assumed to be mounted on horizontal rails, constraining them to be vertical and with all faces wetted but permitting unrestricted lateral motion. The (idealized) scale measures the lateral force $\mathcal{F}$ required to hold them at prescribed distance $2a$ from each other. The fluid mass is to be connected below the plates and in mechanical equilibrium, extending as indicated to infinity, and with pressure p subject to Pascal's Law $p = -\rho g h$, with $\rho =$ density, $g =$ gravitational acceleration, and $h =$ height relative to a common rest level for the fluid at infinity. (Under reasonable hypotheses, the existence of such a level can be proved.) Across the liquid/air interface, there is a pressure jump according to (1). We assume that the pressures have been normalized so that the air pressure is zero. Using [13], the configuration can be shown to be identical in every plane parallel to that of the figure, thus reducing the problem to that of finding a planar curve $u(x)$ for the height within a representative plane. The requirement of prescribed contact angle γ forces the liquid to rise or fall between and outside the plates. Each plate may be of different material on each side; thus we are faced with eight distinct solid/fluid interfaces with (prescribed) energy coefficients e_1 to e_8. Additionally there is a (disconnected) liquid/air interface with surface tension σ. Our object is to describe the configuration quantitatively and determine the net attracting or repelling force $\mathcal{F}$ in terms of the prescribed quantities.

Variational considerations based on the principle of virtual work lead to the underlying equation of the liquid/air interface

$$(\sin \psi)_x = \kappa u \tag{4}$$

where ψ is inclination of the curve $u(x)$, and $\kappa = \rho g / \sigma$ is the *"capillarity constant."* The left side of (4) can be recognized as the planar curvature of the solution curve, thus imparting a clear geometrical substance to the problem. (Alternatively, (4) can be obtained directly from Pascal's Law of fluid pressures and from Young's Law (1).) The underlying integral identity (2) simplifies to

$$\int_{p_1}^{p_2} \mathbf{k}\, ds = \mathbf{v}_1 + \mathbf{v}_2 \tag{5}$$

Here $\mathbf{k}$ is planar curvature vector on a curve segment $\mathcal{C}$ joining the point p_1 to p_2, and $\mathbf{v}_1$, $\mathbf{v}_2$ are exterior directed unit co-normal vectors at those two endpoints of $\mathcal{C}$. The proof of (5) is immediate from the observation that $\mathbf{k} = d\mathbf{v}/ds$ when $\mathbf{v}$ is the unit tangent vector $d\mathbf{x}/ds$, with $\mathbf{x}$ the position vector on $\mathcal{C}$.

We seek solutions of (4) on the three components of the solution curve, achieving the contact angles determined from the four instances of the criterion (3) on the exposed sides of the plates. Our problem is greatly simplified by the observation that from the point of view of determining horizontal forces *the two "outer" contact angles facing the infinite domains lead in each case to the same net horizontal force σ directed toward infinity*. That is because on each "outer" segment one obtains from (1) and from (5) that the horizontal component of the net outer tension force on the plate is $\sigma \sin\gamma$. There is also a pressure force on that plate due to the lifted fluid, of $p = \rho g \int_0^u h\, dh = \frac{1}{2}\sigma\kappa u^2$, where u is rise height above "rest position" at infinity. From (4) we find, since $\tan\psi = du/dx$,

$$\sin\psi\, d\psi = \kappa u\, du \tag{6}$$

which when integrated from infinity to the plate gives

$$1 - \sin\gamma = \frac{1}{2}\kappa u^2 \tag{7}$$

Summing the net tension and pressure forces leaves only the single horizontal force σ pulling toward infinity. Thus *the horizontal effect of the outer domains is always exactly that of an undisturbed fluid pulling outward with surface force σ*.

We view (6) as an equation for determining $u(\psi)$. Using the relation $dx/d\psi = \cot\psi\, du/d\psi$ we obtain a corresponding equation for $x(\psi)$ and thus two equations in parametric form for the sought solution $u(x)$ of (4). The unique solution having inclination ψ_α at a prescribed point (x_α, u_α) is determined by

$$x = x_\alpha + \int_{\psi_\alpha}^{\psi} \frac{\cos\tau}{\kappa u}\, d\tau$$

$$u^2 = u_\alpha^2 + \frac{2}{\kappa}\left(\cos\psi_\alpha - \cos\psi\right). \tag{8}$$

If the solution extends to infinity, it is determined by the inclination ψ_α at a prescribed point x_α:

$$x = x_\alpha - \frac{1}{\sqrt{2\kappa}} \int_\psi^{\psi_\alpha} \sqrt{1 + \cos \tau} \cot \tau \, d\tau$$

$$u = \sqrt{\frac{2}{\kappa}(1 - \cos \psi)}. \tag{9}$$

There is a unique solution between the plates that meets them at prescribed angles. It can be determined by a shooting method and can be routinely calculated to arbitrary accuracy. We are interested here not in specific solutions but rather in global characterization of solution behavior, which turns out to be singular in a rather remarkable way.

4   Domains of Influence

We start by prescribing a contact angle γ_2 on the plate Π_2, $0 \leq \gamma_2 < \pi/2$. The solution given by (9) with $\psi_\alpha = (\pi/2) - \gamma_2$ determines a (unique) height $u_2 = \sqrt{\frac{2}{\kappa}(1 - \sin \gamma_2)}$, at which a solution $\mathbf{I} : u_2^{0+}(x)$ extending to negative infinity meets Π_2 in angle γ_2; see Fig. 5. Restricting attention to the interval between the plates, and assuming fluid surfaces extending to infinity on the outside of the plates, *the solution I yields zero net force between the two plates.*

Retaining the angle γ_2, let us move the contact point on Π_2 upward, a distance δ. Then by (4) the planar curvature of the new solution curve will exceed that of $\mathbf{I}$, so that it cannot contact that curve. *The new curve will be defined as a graph only within a finite interval, at the ends of which it becomes upwardly vertical, and it will have a strict minimum at a height $u_0 > 0$. If δ is small enough, the new solution will still extend to Π_1, and an attracting force $\mathcal{F}$ appears between the plates, with*

$$\mathcal{F} = \rho g u_0^2. \tag{10}$$

As we allow δ to increase, we obtain a continuum of nonintersecting solutions of (4) between the plates, all of which meet Π_2 in angle γ_2 and which meet Π_1 in successively smaller angles γ_1 and yield increasing attractive forces, until we arrive at a solution for which $\gamma_1 = 0$. This corresponds to a finite interval of initial points for (4) along Π_2. Higher points correspond to exotic materials for which one or more of the four ratios that occur in (3) exceeds unity in magnitude, and for which the fluid wets the entire plate or detaches from the plate. We do not address such configurations in the present work.

We encounter physically different behavior that can be relevant for common materials if we allow negative δ. Initially we suppose $|\delta|$ small and observe that

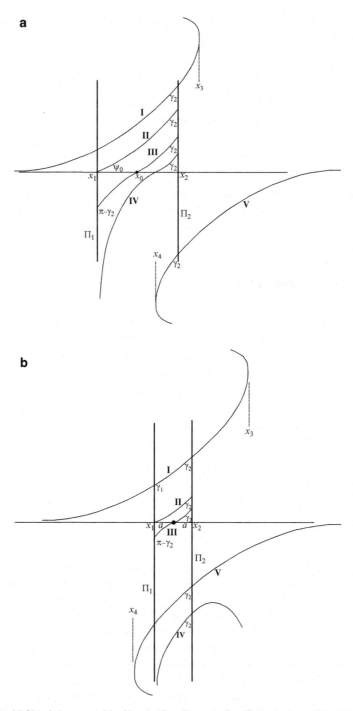

Fig. 5 (**a, b**) Sketch (not to scale) of barrier functions meeting Π_2 in angle γ_2. If $2a$ is large, then V does not extend to Π_1 and IV lies above V. If 2a is small, then IV lies below V. The barrier properties interchange at a critical separation at which IV coincides with V

the new solution curves must cross the x-axis, at an angle $\psi_0 \neq 0$. *In every such configuration, the plates repel each other, with a force*

$$\mathcal{F} = 2\sigma \left(1 - \cos \psi_0\right). \tag{11}$$

To determine the range of δ for which repelling can be expected, we must examine the global structure of the solution curves that meet Π_2 in angle γ_2. Two major variants in what can occur are illustrated in Fig. 5a, b. In preparing these figures, we have noted that the solution **I** has the character of a barrier, separating solutions with qualitatively differing behavior. We introduce four further solution curves with related barrier properties as follows:

II. The unique solution of (4) meeting Π_2 in angle γ_2 and passing through the intersection point of Π_1 with the x-axis, which it meets in an angle $\psi_0 > 0$.

III. The unique solution of (4) meeting Π_2 in angle γ_2 and Π_1 in angle $\gamma_1 = \pi - \gamma_2$. This solution crosses the x-axis at the midpoint between the plates and is symmetric relative to the crossing point.

IV. The unique solution of (4) meeting Π_2 in angle γ_2 and Π_1 in angle π.

V. The counterpart $u_2^{0-}(x)$ of **I**, meeting Π_2 in angle γ_2 and extending to $x = +\infty$. **V** is obtained from **I** via a succession of a lateral and a vertical reflection.

Restricting attention to those solutions of (4) meeting Π_2 in angle γ_2, the region bounded by the two plates, **I**, and $\max\{\mathbf{IV}, \mathbf{V}\}$ is simply covered by those solution curves. *Those curves are precisely the ones that join the two plates and yield repelling configurations. For any other solution meeting Π_2 in angle γ_2 and joining the plates, the plates will attract each other.*

Note that the net force between the plates remains uniformly bounded in all repelling configurations. Attractions become unboundedly large with decreasing plate separation. *Given an attracting configuration, if the contact angles are fixed and the plates allowed to approach each other, then the configuration remains attracting, with the force becoming unbounded as the inverse square of the plate separation.*

Repelling configurations behave differently than attracting ones. *The barrier curve **III** continues to cross the x-axis at the midpoint between the plates as $a \to 0$ and hence remains repelling; one sees easily that the net force tends to the limit $2\sigma(1-\sin\gamma_2) \neq 0$. However, **III** is the unique solution curve with that property; with decreasing a, attracting curves become rapidly more attracting, while every repelling curve distinct from **III** eventually identifies with a portion of **I** or of **V** (yielding zero net force) and then moves into the set of attracting curves, with the positive minimum (or negative maximum) u_0 becoming infinite to the order a^{-1}.*

Behavior that singular invites a closer look. We hold Π_2 fixed and move Π_1 toward it a distance δ, as indicated in Fig. 6. We focus attention on a solution curve between **I** and **III** and observe that all the angles γ_1 that appear in this range correspond to inclinations ψ_1 that also appear on the curve **I**. This latter curve remains unchanged during motions of Π_1. Thus there exists a suitable δ such that the

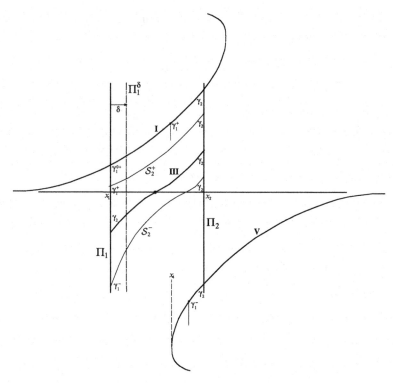

Fig. 6 Sketch of solution curves $\mathcal{S}_2^+$ and $\mathcal{S}_2^-$ in the range of repelling solutions that become attracting with decreasing plate separation. $\mathcal{S}_2^+$ shifts through the barrier **I**; $\mathcal{S}_2^-$ shifts through the barrier **V**

inclination of **I** at $x_1 + \delta$ is exactly the initial prescribed inclination $\gamma_1 - \pi/2$. Since the given solution and **I** already share the datum $\psi_2 = (\pi/2) - \gamma_2$ on Π_2, an easily proved uniqueness theorem shows that the solution determined between $x_1 + \delta$ and x_2 coincides with **I** on that interval. In other words, *when plates of materials for which the contact angles yield a solution in the range between* **I** *and* **III** *are moved toward each other, a positive separation will be attained for which no horizontal force appears.*

Bringing the plates still closer together yields attracting forces that grow rapidly, as indicated above.

Note that this behavior occurs with curves that can be arbitrarily close to **III**, and it occurs also in a corresponding "lower" neighborhood between **III** and **V**. Thus a deleted neighborhood of solutions surrounding **III** pulls apart from **III** as the plates come together, leaving **III** as the single (isolated) solution that retains its repelling property for all plate separations.

There is a still further striking behavior, which becomes evident on introducing the barrier **II** (Fig. 5), which meets Π_2 in angle γ_2 and passes through the zero level on Π_1. If we start with a solution $\mathcal{S}_2^+$ lying above **II** as in Fig. 6, and shift

Π_1 as indicated, we see that the initial contact angle on that particular solution has *increased* due to the upward convexity of $\mathcal{S}_2^+$; to maintain the same γ_1 on the plate, we must therefore move *upward* on Π_1^δ to a different solution closer to **I**; this is in agreement with the convergence to **I** just indicated. But now start with another solution, still between **I** and **III** but below **II**. That means simply that the starting point on Π_1 will be negative. In this case the sense of convexity of $\mathcal{S}_2^+$ is initially *downward*, and the identical reasoning now shows that the new curve of the family that must be chosen to keep the same contact angle will be below $\mathcal{S}_2^+$. Thus, although we have shown that the solutions converge to **I** before Π_1^δ reaches Π_2, these solutions move initially to positions *further* from **I** instead of closer to it.

The clue to clarification of this seemingly paradoxical behavior is the observation that *although the curves are moving downward, the initial points on the successive* $\{\Pi_1\}$ *are moving upward, and they eventually rise to the zero level where the convexity sense reverses.*

This theorem is illustrated in Fig. 7a, b, for the particular case $\gamma_2 = 30°$, $\gamma_1 = 145°$ (for clarity in display of details, the x-axis scale has been expanded). In the initial configuration, the initial height u_1 is negative, and Fig. 7a shows the behavior in this regime. The plate Π_1 in varying positions is matched with the corresponding solution for the given contact angles according to color. The dots (inserted by hand) are at the intersection points. One sees that as predicted the dots rise as Π_1 moves toward Π_2, reaching the zero level prior to reaching Π_2. Once at that level, the rise accelerates sharply as shown in Fig. 7b. This is as expected, for when the heights u_j are positive, the sense of curvature of the solution curves contributes to the rise instead of working against it.

As a particular consequence of this behavior, we obtain from (11) that for a "starting" curve as just described, *as the plates are brought together, there will be an initial interval in which the repelling force increases.* The force will then achieve a maximum at some intermediate position, then decrease to zero and reappear as an attracting force increasing rapidly to infinity, with successively decreasing separation of the plates.

The property of the barriers **I** and **V** to provide zero net force allows us to construct configurations, for any prescribed γ_2 in $[0, \pi/2)$, for which *the transition from repelling to attracting occurs at arbitrary prescribed separation of the plates.* It suffices to choose for γ_1 the angle with which **I** meets a plate Π_1 situated at a prescribed point $x_1 < x_2$.

The above discussion has been nominally limited to the case $0 \leq \gamma_2 < \pi/2$. If $\pi/2 < \gamma_2 \leq \pi$, we observe that a reflection in $x = x_2$ followed by a reflection in the x-axis leaves (4) invariant and changes γ_2 to $\pi - \gamma_2$. If $\gamma_2 = \pi/2$, then every nontrivial solution is attracting.

Acknowledgments I am very much indebted to Paul Concus for the computer calculations and the detailed preparation leading to Fig. 7a, b. I wish to thank the Max-Planck-Institut für Mathematik in den Naturwissenschaften, in Leipzig, for its hospitality and for the excellent working conditions that have facilitated much of this work.

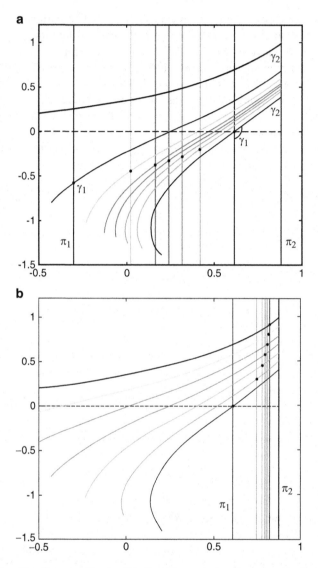

Fig. 7 (a) Transition from repelling to attracting. The contact angles γ_1, γ_2 are chosen so that the initial trajectory meets Π_1 at a height $u_1 < 0$. The successive trajectories fall but the contact points initially rise and reach the level $u = 0$. (b) Continuation of the transition. When the contact points become positive, the trajectories and the contact points both rise, the convergence accelerates, and the contact point attains the barrier **I** prior to reaching Π_2. When that happens, the solution identifies with **I**. Further approach of Π_1 to Π_2 yields rapidly rising attracting forces

References

1. Mariotte, E.: Œuvres (Pierre Vander) Leyden (1717)
2. Laplace, P.S.: Traité de mécanique céleste, Œuvres complète, vol. 4, Supplément 1, livre X, pp. 771–777. Gauthier-Villars, Paris (1805). See also the annotated English translation by N. Bowditch 1839. Chelsea, New York, 1966
3. Laplace, P.S.: Traité de mécanique céleste, Œuvres complète, vol. 4, Supplément 2, Livre X, pp. 909–945. Gauthier-Villars, Paris (1806). See also the annotated English translation by N. Bowditch 1839. Chelsea, New York, 1966
4. McCuan, J.: A variational formula for floating bodies. Pacific J. Math. **231**(1), 167–191 (2007)
5. Finn, R.: On Young's Paradox, and the attractions of immersed parallel plates. Phys. Fluids **22**, 017103 (2010)
6. Finn, R.: Equilibrium Capillary Surfaces. Grundlehren Series, vol. 284. Springer, New York (1986)
7. Finn, R.: The contact angle in capillarity. Phys. Fluids **18**, 047102 (2006)
8. Bhatnagar, R., Finn, R.: Equilibrium configurations of an infinite cylinder in an unbounded fluid. Phys. Fluids **18**, 047103 (2006)
9. Lunati, I.: Young's law and the effects of interfacial energy on the pressure at the solid–fluid interface. Phys. Fluids **19**, 118105 (2007)
10. Finn, R.: Comments related to my paper "The contact angle in capillarity". Phys. Fluids **20**, 107104 (2008)
11. Shikhmurzaev, Y.D., On Young's (1805) Equation and Finn's (2006) 'counterexample'. Phys. Lett. A **372**(2008), 704–707 (1805)
12. Finn, R., McCuan, J., Wente, H.C.: Thomas Young's surface tension diagram: its history, legacy and irreconcilabilities. J. Math. Fluid Mech. **14**(3), 445–453 (2012)
13. Finn, R., Hwang, J.-F.: On the comparison principle for capillary surfaces. J. Fac. Sci. Univ. Tokyo Sect. 1A Math **36**(1), 131–134 (1989)

Singular Problems for Integro-differential Equations in Dynamic Insurance Models

Tatiana Belkina, Nadezhda Konyukhova, and Sergey Kurochkin

Abstract A second-order linear integro-differential equation with Volterra integral operator and strong singularities at the endpoints (zero and infinity) is considered. Under limit conditions at the singular points, and some natural assumptions, the problem is a singular initial problem with limit normalizing conditions at infinity. An existence and uniqueness theorem is proved and asymptotic representations of the solution are given. A numerical algorithm for evaluating the solution is proposed; calculations and their interpretation are discussed. The main singular problem under study describes the survival (non-ruin) probability of an insurance company on infinite time interval (as a function of initial surplus) in the Cramér–Lundberg dynamic insurance model with an exponential claim size distribution and certain company's strategy at the financial market assuming investment of a fixed part of the surplus (capital) into risky assets (shares) and the rest of it into a risk-free asset (bank deposit). Accompanying "degenerate" problems are also considered that have an independent meaning in risk theory.

1 Introduction

The important problem concerning the application of financial instruments in order to reduce insurance risks has been extensively studied in recent years (see, e.g., [1, 3, 4], and references therein). In particular in [3, 4] the optimal investing strategy is studied for risky and risk-free assets in Cramér–Lundberg (C.-L.) model with budget constraint, i.e., without borrowing.

T. Belkina (✉)
Central Economics and Mathematics Institute of RAS, Nakhimovskii pr. 47, Moscow, Russia
e-mail: tbel@cemi.rssi.ru

N. Konyukhova• S. Kurochkin
Dorodnicyn Computing Centre of RAS, ul. Vavilova 40, Moscow, Russia
e-mail: nadja@ccas.ru; kuroch@ccas.ru

S. Pinelas et al. (eds.), *Differential and Difference Equations with Applications*, Springer 27
Proceedings in Mathematics & Statistics 47, DOI 10.1007/978-1-4614-7333-6_3,
© Springer Science+Business Media New York 2013

This paper complements and revises some results of [4]. The parametric singular initial problem (SIP) for an integro-differential equation (IDE) considered here is a part of the optimization problem stated and analyzed in [3, 4]: the solution of this SIP gives the survival probability corresponding to the optimal strategy when the initial surplus values are small enough. The singular problem under study is also interesting both as an independent mathematical problem and for the models in risk theory. We give more complete and rigorous analysis of this problem in comparison with [4] and add some new "degenerate" problems having independent meaning in risk theory. Some new numerical results are also discussed.

The paper is organized as follows. In Sect. 2 we set the main mathematical problem and formulate the main results concerning solvability of this problem and the solution behavior; we describe also two "degenerate" problems (when some parameters in the IDE are equal to zero) and discuss their exact solutions. In Sect. 3 we give a rather brief description of the mathematical model for which the problem in question arises (for detailed history, models' description and derivation of the IDE studied here, see [3, 4]). In Sect. 4 we describe our approach to the problem and give brief proofs of main results (for some assertions, we omit the proofs since they are given in [4]). In Sect. 5 we study an accompanying singular problem for capital stock model (the third "degenerate" problem); the results of this section are completely new. Numerical results and their interpretation are given in Sect. 6.

2 Singular Problems for IDEs and Their Solvability

2.1 Main Problem

The main singular problem under consideration has the form:

$$(b^2/2)u^2\varphi''(u) + (au+c)\varphi'(u) - \lambda\varphi(u)$$

$$+(\lambda/m)\int_0^u \varphi(u-x)\exp(-x/m)dx = 0, \qquad 0 < u < \infty, \qquad (1)$$

$$\{|\lim_{u\to+0}\varphi(u)|, |\lim_{u\to+0}\varphi'(u)|\} < \infty, \qquad \lim_{u\to+0}[c\varphi'(u) - \lambda\varphi(u)] = 0, \qquad (2)$$

$$0 \le \varphi(u) \le 1, \quad u \in \mathbf{R}_+, \qquad (3)$$

$$\lim_{u\to\infty}\varphi(u) = 1, \quad \lim_{u\to\infty}\varphi'(u) = 0. \qquad (4)$$

Here in general all the parameters a, b, c, λ, m are real positive numbers.

The second limit condition at zero is a corollary of the first one and IDE (1) itself. For this IDE, conditions (2) imply $\lim_{u\to+0}[u^2\varphi''(u)] = 0$ providing a degeneracy of the IDE (1) as $u \to +0$: any solution $\varphi(u)$ to the singular problem without initial data (1), (2) must satisfy IDE (1) up to the singular point $u = 0$.

The "truncated" problem (1)–(3) (constrained singular problem) always has the trivial solution $\varphi(u) \equiv 0$. A nontrivial solution is singled out by the additional limit conditions at infinity (4).

In what follows we use notation

$$(J_m\varphi)(u) = \frac{1}{m}\int_0^u \varphi(u-x)\exp(-x/m)dx = \frac{1}{m}\int_0^u \varphi(s)\exp(-(u-s)/m)ds, \quad (5)$$

where J_m is a Volterra integral operator and $J_m : C[0,\infty) \to C[0,\infty)$, $C[0,\infty)$ is the linear space of continuous functions defined and bounded on $\mathbf{R}_+$.

For IDE (1), the entire singular problem on $\mathbf{R}_+$ was neither posed nor studied before [4] and the present paper.

2.2 Formulation of the Main Results

The problem (1)–(4) may be rewritten in the equivalent parametrized form:

$$(b^2/2)u^2\varphi''(u) + (au+c)\varphi'(u) - \lambda[\varphi(u) - (J_m\varphi)(u)] = 0, \quad u \in \mathbf{R}_+, \quad (6)$$

$$\lim_{u\to+0}\varphi(u) = C_0, \quad \lim_{u\to+0}\varphi'(u) = \lambda C_0/c, \quad (7)$$

$$0 \le \varphi(u) \le 1, \quad u \in \mathbf{R}_+, \quad (8)$$

$$\lim_{u\to\infty}\varphi(u) = 1, \quad \lim_{u\to\infty}\varphi'(u) = 0. \quad (9)$$

Here C_0 is an unknown parameter whose value must be defined.

Lemma 1. *For IDE (6), let the values a, b, c, λ, m be fixed with $b \ne 0$, $c > 0$, $\lambda \ne 0$, $m > 0$, $a \in \mathbf{R}$. Then for any fixed $C_0 \in \mathbf{R}$ the IDE SIP (6), (7) is equivalent to the following singular Cauchy problem (SCP) for ODE:*

$$(b^2/2)u^2\varphi'''(u) + [c + (b^2+a)u + b^2u^2/(2m)]\varphi''(u)$$
$$+ (a - \lambda + c/m + au/m)\varphi'(u) = 0, \quad 0 < u < \infty, \quad (10)$$

$$\lim_{u\to+0}\varphi(u) = C_0, \quad \lim_{u\to+0}\varphi'(u) = \lambda C_0/c,$$
$$\lim_{u\to+0}\varphi''(u) = (\lambda - a - c/m)\lambda C_0/c^2. \quad (11)$$

There exists a unique solution $\varphi(u, C_0)$ to SCP (10), (11) (therefore also to the equivalent IDE SIP (6), (7)); for small u, this solution is represented by the asymptotic power series

$$\varphi(u, C_0) \sim C_0 \left[1 + \frac{\lambda}{c} \left(u + \sum_{k=2}^{\infty} D_k u^k / k \right) \right], \qquad u \sim +0, \tag{12}$$

where coefficients D_k are independent of C_0 and may be found by formal substitution of series (12) into ODE (10), namely from the recurrence relations

$$D_2 = -[(a - \lambda)/c + 1/m], \tag{13}$$

$$D_3 = -[D_2(b^2 + 2a - \lambda + c/m) + a/m]/(2c), \tag{14}$$

$$D_k = -\{D_{k-1}[(k-1)(k-2)b^2/2 + (k-1)a - \lambda + c/m]$$
$$+ D_{k-2}[(k-3)b^2/2 + a]/m\}/[c(k-1)], \quad k = 4, 5, \ldots. \tag{15}$$

Theorem 1. *For IDE* (1), *let all the parameters* a, b, c, λ, m *be fixed positive numbers and let the inequality*

$$2a/b^2 > 1 \tag{16}$$

be fulfilled. Then the following statements are valid:

1. *There exists a unique solution* $\varphi(u)$ *of the input singular linear IDE problem* (1)–(4) *and it is a smooth (infinitely differentiable) monotone nondecreasing on* $\mathbf{R}_+$ *function.*
2. *The function* $\varphi(u)$ *can be obtained as the solution* $\varphi(u, C_0)$ *of IDE SIP* (6), (7), *namely, by solving the equivalent ODE SCP* (10), (11) *where the value* $C_0 = \tilde{C}_0$ *must be chosen to satisfy conditions at infinity* (4) *(as the normalizing condition); for* $\tilde{C}_0$ *defined in this way, the restriction* $0 < \varphi(u, \tilde{C}_0) < 1$ *is valid for any finite* $u \in \mathbf{R}_+$, *i.e., for* $\varphi(u) = \varphi(u, \tilde{C}_0)$, *inequalities* (3) *are fulfilled tacitly.*
3. *If the inequality* $m(a - \lambda) + c > 0$ *is fulfilled, then the solution* $\varphi(u)$ *is concave on* $\mathbf{R}_+$; *in particular, this is true when*

$$c - \lambda m > 0. \tag{17}$$

4. *If the inequality* $m(a - \lambda) + c \le 0$ *is true, then* $\varphi(u)$ *is convex on a certain interval* $[0, \hat{u}]$ *where* $\hat{u}$ *is an inflection point,* $\hat{u} > 0$.
5. *For small* u, *due to Lemma 1 above, the solution* $\varphi(u)$ *is represented by asymptotic power series* (12)–(15) *where* $C_0 = \tilde{C}_0$, $0 < \tilde{C}_0 < 1$.
6. *For large* u, *the asymptotic representation*

$$\varphi(u) = 1 - K u^{1 - 2a/b^2}[1 + o(1)], \qquad u \to \infty, \tag{18}$$

takes place with $K = \tilde{C}_0 \tilde{K} > 0$ *where in general the value* $\tilde{K} > 0$ *(as well as the value* $\tilde{C}_0$) *cannot be determined using local analysis methods.*

2.3   The "Degenerate" Problems and Their Exact Solutions

A particular case of IDE (1) is considered "degenerate" when some of its parameters are equal to zero.

2.3.1   The First "Degenerate" Case: $a = b = 0$, $\lambda > 0$, $m > 0$, $c > \lambda m > 0$

For this case, the "degenerate" IDE problem

$$c\varphi'(u) - \lambda[\varphi(u) - (J_m\varphi)(u)] = 0, \qquad u \in \mathbf{R}_+, \tag{19}$$

$$c\varphi'(0) - \lambda\varphi(0) = 0, \qquad \lim_{u \to \infty} \varphi(u) = 1, \tag{20}$$

is equivalent to the ODE problem with one parameter:

$$c\varphi''(u) + (c/m - \lambda)\varphi'(u) = 0, \qquad u \in \mathbf{R}_+, \tag{21}$$

$$\varphi(0) = C_0, \qquad \varphi'(0) = \lambda C_0/c, \qquad \lim_{u \to \infty} \varphi(u) = 1. \tag{22}$$

Then we obtain $C_0 = \tilde{C}_0 = 1 - \lambda m/c$, $0 < \tilde{C}_0 < 1$, and

$$\varphi(u) = \varphi(u, \tilde{C}_0) = 1 - \frac{\lambda m}{c} \exp\left(-\frac{c - \lambda m}{mc} u\right), \qquad u \in \mathbf{R}_+. \tag{23}$$

If inequality (17) is not valid, i.e., $c \le \lambda m$, then there is no solution to problem (19), (20) [resp., to problem (21), (22)].

In what follows, function (23) is well known in classical C.-L. risk theory and has an independent meaning (see further Sect. 3.1).

2.3.2   The Second "Degenerate" Case: $b = 0$, $a > 0$, $c \ge 0$, $\lambda > 0$, $m > 0$

For $c > 0$, the "degenerate" IDE problem

$$(au + c)\varphi'(u) - \lambda[\varphi(u) - (J_m\varphi)(u)] = 0, \qquad u \in \mathbf{R}_+,$$

$$c\varphi'(0) - \lambda\varphi(0) = 0, \qquad \lim_{u \to \infty} \varphi(u) = 1, \tag{24}$$

is equivalent to the parametrized ODE problem:

$$(au+c)\varphi''(u) + (a - \lambda + c/m + au/m)\varphi'(u) = 0, \quad u \in \mathbf{R}_+,$$
$$\varphi(0) = C_0, \quad \varphi'(0) = \lambda C_0/c, \quad \lim_{u \to \infty} \varphi(u) = 1. \tag{25}$$

This implies $C_0 = \tilde{C}_0 = (a/\lambda)(c/a)^{\lambda/a} \left[(a/\lambda)(c/a)^{\lambda/a} + I_c(0) \right]^{-1}$, $0 < \tilde{C}_0 < 1$,

$$\varphi(u) = \varphi(u, \tilde{C}_0) = 1 - I_c(u) \left[I_c(0) + (a/\lambda)(c/a)^{\lambda/a} \right]^{-1}, \quad u \in \mathbf{R}_+, \tag{26}$$

where, taking into account the notation $\Gamma(p,z) = \int_z^\infty x^{p-1} \exp(-x) dx$, $p > 0$, for incomplete gamma-function (see, e.g., [2]), we have

$$I_c(u) = \int_u^\infty (x + c/a)^{\lambda/a - 1} \exp(-x/m) dx$$
$$= m^{\lambda/a} \exp\left(c/(am)\right) \Gamma\left(\lambda/a, u/m + c/(am)\right), \quad u \geq 0. \tag{27}$$

In particular we obtain the asymptotic representation when $u \to \infty$:

$$\varphi(u) = 1 - m \left[(a/\lambda)(c/a)^{\lambda/a} + I_c(0) \right]^{-1} u^{\lambda/a - 1} \exp(-u/m)[1 + o(1)]. \tag{28}$$

For $c = 0$, the solution to the IDE problem on $\mathbf{R}_+$,

$$u\varphi'(u) - (\lambda/a)[\varphi(u) - (J_m\varphi)(u)] = 0, \quad \lim_{u \to +0} \varphi(u) = 0, \quad \lim_{u \to \infty} \varphi(u) = 1, \tag{29}$$

can be found as a solution to the equivalent ODE problem:

$$u^2\varphi''(u) + (1 - \lambda/a + u/m)u\varphi'(u) = 0, \quad u \in \mathbf{R}_+,$$
$$\lim_{u \to +0} \varphi(u) = \lim_{u \to +0}[u\varphi'(u)] = 0, \quad \lim_{u \to \infty} \varphi(u) = 1. \tag{30}$$

This implies the same formulas (26)–(28) with $c = 0$ where $\Gamma(p) = \Gamma(p, 0)$ is the usual Euler gamma-function. In particular, using the formula

$$\varphi'(u) = [m^{\lambda/a}\Gamma(\lambda/a)]^{-1} u^{\lambda/a - 1} \exp(-u/m), \quad u \geq 0,$$

we obtain here: if $a < \lambda$ then $\varphi'(0) = 0$; if $a = \lambda$ then $\varphi'(0) = 1/m$ and $\varphi(u) = 1 - \exp(-u/m)$; if $a > \lambda$ then the function $\varphi'(u)$ is unbounded as $u \to +0$ but integrable on $\mathbf{R}_+$.

This "degenerate" case has an independent meaning in risk theory (see further Sect. 3.2).

3   Origin of the Problem: The Cramér–Lundberg Dynamic Insurance Models

3.1   The Classical C.-L. Insurance Model

Consider the classical risk process: $R_t = u + ct - \sum_{k=1}^{N_t} Z_k$, $t \geq 0$. Here R_t is the surplus of an insurance company at time t, u is the initial surplus, c is the premium rate; $\{N_t\}$ is a Poisson process with parameter λ defining, for each t, the number of claims applied on the interval $(0,t]$; $Z_1, Z_2, \ldots$ is the series of independent identically distributed random values with some distribution $F(z)$ ($F(0) = 0$, $EZ_1 = m < \infty$), describing the sequence of claims; these random values are also assumed to be independent of the process $\{N_t\}$. For this model, the positiveness condition for the net expected income ("safety loading") has the form (17).

Denote by $\tau = \inf\{t : R_t < 0\}$ the time of ruin, then $\mathbf{P}(\tau < \infty)$ is the probability of ruin at the infinite time interval.

A classical result in the C.-L. risk theory [8]: under condition (17) and assuming existence of a constant $R_L > 0$ ("the Lundberg coefficient") such that equality $\int_0^\infty [1 - F(x)] \exp(R_L x) dx = c/\lambda > 0$ holds, the probability of ruin $\xi(u)$ as a function of the initial surplus admits the estimate $\xi(u) = \mathbf{P}(\tau < \infty) \leq \exp(-R_L u)$, $u \geq 0$. Moreover, if the claims are exponentially distributed,

$$F(x) = 1 - \exp(-x/m), \quad m > 0, \quad x \geq 0, \tag{31}$$

then $R_L = (c - \lambda m)/(mc) > 0$, and the survival probability $\varphi(u) = 1 - \xi(u)$ is given by the exact formula (23), i.e., coincides with the exact solution of the first "degenerate" problem to which input singular problem (1)–(4) reduces formally as $a = b = 0$ (see Sect. 2.3.1).

For c as a bifurcation parameter, the value $c = \lambda m$ is critical: if $c \leq \lambda m$ then $\varphi(u) \equiv 0$, $u \in \mathbf{R}_+$.

3.2   The C.-L. Insurance Model with Investment into Risky Assets

Now consider the case where the surplus is invested continuously into shares with price dynamics described by geometric Brownian motion model:

$$dS_t = S_t(a\,dt + b\,dw_t), \quad t \geq 0. \tag{32}$$

Here S_t is the share price at time t, a is the expected return on shares, $0 < b$ is the volatility, $\{w_t\}$ is a standard Wiener process.

Denoting by X_t the company's surplus at time t we get $X_t = \theta_t S_t$, where θ_t is the amount of shares in the portfolio. Then the surplus dynamics meets the relation $dX_t = \theta_t dS_t + dR_t$. Taking into account (32), we obtain:

$$dX_t = aX_t dt + bX_t dw_t + dR_t, \quad t \geq 0. \tag{33}$$

In contrast with the classical model, condition (17) (the positiveness of "safety loading") is not assumed here.

For the dynamical process (33), the survival probability $\varphi(u)$ satisfies on $\mathbf{R}_+$ the following linear IDE (see, e.g., [3, 7] and references therein):

$$\lambda \int_0^u \varphi(u - z) dF(z) - \lambda \varphi(u) + (au + c)\varphi'(u) + (b^2/2)u^2\varphi''(u) = 0. \tag{34}$$

From (34), assuming exponential distribution of claims (31) we get the initial IDE (1) under study.

Assuming that there exists the solution $\varphi(u)$ of IDE (1) representing the survival probability as a function of initial surplus, the following statement (further called FKP-theorem) was obtained in [7].

Theorem 2. *Suppose $b > 0$ and the claims are distributed exponentially, i.e., (31) is valid. Then:*

1. *If inequality (16) of "robustness of shares" is fulfilled, then the asymptotic representation (18) holds with a certain constant $K > 0$.*
2. *If $2a/b^2 < 1$, then $\varphi(u) \equiv 0$, $u \in \mathbf{R}_+$.*

3.3 The C.-L. Model with Investment into a Risk-Free Asset

The model under study comprises a more general case where only a constant part α ($0 < \alpha < 1$) of the surplus is invested in shares (with the expected return μ and volatility σ) whereas remaining part $1 - \alpha$ is invested into a risk free asset (bank deposit with constant interest rate $r > 0$): the case $0 < \alpha < 1$ may be reduced to the case $\alpha = 1$ by a simple change of the parameters (shares characteristics), namely $a = \alpha\mu + (1 - \alpha)r$, $b = \alpha\sigma$.

Moreover, when the surplus is invested entirely into a risk free asset (bank deposit with constant interest rate), we obtain the second "degenerate" problem (with or without premiums) to which the input singular problem (1)–(4) reduces formally as $b = 0$. For $a > 0$, $\lambda > 0$, $m > 0$, $c \geq 0$, there exists the exact solution (26), (27) and the asymptotic representation (28) is valid (for details, see Sect. 2.3.2).

Thus when the surplus is entirely invested into a risk free asset then the survival probability is not equal to zero, for $u > 0$, even if premiums (insurance payments) are absent ($c = 0$) and has a good asymptotic behavior as $u \to \infty$.

As far as we know formulas (26)–(28) are new for risk theory.

4   On the Approach to Main Problem and Proofs of Main Results

4.1   The Singular Problem for IDE: Uniqueness of the Solution and Its Monotonic Behavior

As shown in Sect. 3, we can formulate the input singular IDE problem in the form (6), (7), (9), where operator J_m is defined by (5), C_0 is an unknown parameter whose value must be found, and, for the solution to the problem (6), (7), (9), the restrictions needed are (8).

Lemma 2. *For IDE* (6), *let the values a, b, c, λ and m be fixed with $c > 0$, $\lambda > 0$, $m > 0$ whereas a and b are any real numbers ($a, b \in \mathbf{R}$). Then the following assertions are valid:*

1. *If there exists a solution $\varphi_1(u) = \varphi_1(u, C_0)$ to problem* (6), (7), (9) *with some $C_0 > 0$, then it is a unique solution to this problem.*
2. *Such $\varphi_1(u)$ satisfies restrictions* (8), $0 < C_0 < 1$ *and $\varphi_1'(u) > 0$ for any finite $u \in \mathbf{R}_+$, i.e., $\varphi_1(u)$ is a monotone nondecreasing on $\mathbf{R}_+$ function.*

Proof.

1. Supposing the opposite, let $\varphi_2(u)$ be any other solution to problem (6), (7), (9), i.e., $\varphi_2(u) \not\equiv \varphi_1(u)$. Then two cases may occur: the first one with $\lim_{u \to +0} \varphi_2(u) = \lim_{u \to +0} \varphi_1(u)$ and the second one with $\lim_{u \to +0} \varphi_2(u) \neq \lim_{u \to +0} \varphi_1(u)$.

   For the first case, it follows that there exists a nontrivial solution $\tilde{\varphi}(u)$ of IDE (6) satisfying conditions $\lim_{u \to +0} \tilde{\varphi}(u) = \lim_{u \to \infty} \tilde{\varphi}(u) = 0$. Let $0 < \tilde{u}$ be its maximum point: $\tilde{\varphi}(\tilde{u}) = \max_{u \in [0,\infty)} \tilde{\varphi}(u) > 0$ (if $\tilde{\varphi}(u)$ takes only nonpositive values, then we consider the solution $-\tilde{\varphi}(u)$ instead). Then $\tilde{\varphi}'(\tilde{u}) = 0$, $\tilde{\varphi}''(\tilde{u}) \leq 0$. But from IDE (6) a contradiction follows:

$$(b^2/2)\tilde{u}^2 \tilde{\varphi}''(\tilde{u}) = \lambda [\tilde{\varphi}(\tilde{u}) - m^{-1} \int_0^{\tilde{u}} \tilde{\varphi}(s) \exp\left(-(\tilde{u}-s)/m\right) ds]$$

$$\geq \lambda \tilde{\varphi}(\tilde{u}) \left[1 - m^{-1} \int_0^{\tilde{u}} \exp\left(-(\tilde{u}-s)/m\right) ds\right]$$

$$= \lambda \tilde{\varphi}(\tilde{u}) \exp(-\tilde{u}/m) > 0. \tag{35}$$

   For the second case, there exists a linear combination of solutions $\hat{\varphi}(u) = c_1 \varphi_1(u) + c_2 \varphi_2(u)$ such that $\hat{\varphi}(u) \not\equiv 1$ and satisfies conditions $\lim_{u \to +0} \hat{\varphi}(u) = \lim_{u \to \infty} \hat{\varphi}(u) = 1$. If there exists a value $\hat{u} > 0$ with $\hat{\varphi}(\hat{u}) > 1$, then the first case argument is valid. Otherwise, the inequality $\hat{\varphi}(u) \leq 1$ $\forall u \in \mathbf{R}_+$ contradicts to $\lim_{u \to +0} \hat{\varphi}'(u) = \lambda/c > 0$ which follows from (7).
2. The other assertions are proved analogously.   □

4.2 SCPs for Accompanying Linear ODEs

4.2.1 Reduction of the Second-Order IDE to a Third-Order ODE

The known possibility of reducing the second-order IDE (6) to a third-order ODE is important for further exposition. First, we note that

$$(J_m\varphi)'(u) = \frac{1}{m}\left(\exp(-u/m)\int_0^u \varphi(x)\exp(x/m)dx\right)'$$
$$= [\varphi(u) - (J_m\varphi)(u)]/m. \tag{36}$$

Then differentiating IDE (6) in view of (36) gives a linear third-order IDE

$$(b^2/2)u^2\varphi'''(u) + [(b^2+a)u+c]\varphi''(u) + (a-\lambda)\varphi'(u)$$
$$+(\lambda/m)[\varphi(u)-(J_m\varphi)(u)] = 0, \quad u \in \mathbf{R}_+, \tag{37}$$

which also implies the limit condition

$$\lim_{u\to+0}[c\varphi''(u) + (a-\lambda)\varphi'(u) + (\lambda/m)\varphi(u)] = 0. \tag{38}$$

Together with the input limit condition (2), it implies the limit equality

$$\lim_{u\to+0}[c\varphi''(u) + (a-\lambda+c/m)\varphi'(u)] = 0. \tag{39}$$

In order to remove the integral term, we add IDE (37) and initial IDE (6) multiplied by $1/m$ and get the linear third-order ODE (10). Then the same limit condition (39) must be fulfilled to provide a degeneration of this ODE as $u \to +0$.

Suppose $\psi(u) = \varphi'(u)$ and rewrite ODE (10) in more canonical forms for ODEs with pole-type singularities at zero and infinity (for classification of isolated singularities of linear ODE systems and general theory of ODEs of this class, see, e.g., the monographs [5, 6, 10] complementing each other). Now, for $\psi(u)$, we have to study the following singular ODEs: for small u, we need to consider the equation

$$(b^2/2)u^3\psi''(u) + \left[c + (b^2+a)u + b^2u^2/(2m)\right]u\psi'(u)$$
$$+ \left[(a-\lambda+c/m)u + au^2/m\right]\psi(u) = 0, \quad u > 0, \tag{40}$$

and for large u, we shall consider the same equation in the form

$$(b^2/2)\psi''(u) + \left[c/u^2 + (b^2+a)/u + b^2/(2m)\right]\psi'(u)$$
$$+ \left[(a-\lambda+c/m)/u^2 + (a/m)/u\right]\psi(u) = 0, \quad u > 0. \tag{41}$$

We see that both ODE (40) and equivalent ODE (41) have irregular (strong) singularities of rank 1 as $u \to +0$ and as $u \to \infty$.

4.2.2   Singularity at Zero: Replacement of the SIP for IDE by an Equivalent SCP for ODE

Proof of Lemma 1

First, we must show that the previous transformations permit us to replace the input SIP (6), (7) for an IDE by the SCP (10), (11) for an ODE.

In the straight direction (from the IDE SIP to the ODE SCP), the statement is evident. Now let $\tilde{\varphi}(u) = \tilde{\varphi}(u, C_0)$ be a solution of ODE SCP (10), (11). We have to prove that $\tilde{\varphi}(u)$ satisfies IDE (6).

Denote the left part of IDE (6) with the function $\tilde{\varphi}(u)$ by $g(u)$. We have to prove that $g(u) \equiv 0$. Indeed, the way ODE (10) was derived means that $g(u)$ meets the first-order ODE

$$g'(u) + g(u)/m = 0, \quad 0 \le u < \infty,$$

with the general solution of the form $g(u) = \tilde{C}\exp(-u/m)$ where $\tilde{C}$ is an arbitrary constant. Since $\tilde{\varphi}(u, C_0)$ meets conditions (11), it follows from IDE (6) that $g(0) = 0$. This implies $\tilde{C} = 0$, i.e., $g(u) \equiv 0$.

The other statements of Lemma 1 follow from the results of [9] (see [4] for details).

4.2.3   SCP at Infinity and Its Two-Parameter Family of Solutions

For $\psi(u) = \varphi'(u)$, we have an SCP at infinity for the second-order ODE (41) with the conditions

$$\lim_{u \to \infty} \psi(u) = \lim_{u \to \infty} \psi'(u) = 0. \tag{42}$$

Using the known results for linear ODEs with irregular singularities, we obtain the following assertions (more complete in comparison with FKP-theorem).

Lemma 3. *For ODE (41), suppose that $b \ne 0$, $a > 0$, $m > 0$ whereas λ and c are arbitrary real numbers ($\lambda, c \in \mathbf{R}$). Then:*

1. *Any solution to ODE (41) satisfies conditions (42) so that SCP (41), (42) at infinity has a two-parameter family of solutions $\psi(u, d_1, d_2)$ where d_1 and d_2 are arbitrary constants.*
2. *For this family, the following representation holds:*

$$\psi(u, d_1, d_2) = d_1 u^{-2a/b^2}[1 + \chi_1(u)/u]$$
$$+ d_2 u^{-2}\exp(-u/m)[1 + \chi_2(u)/u]; \tag{43}$$

here the functions $\chi_j(u)$ have finite limits as $u \to \infty$ and, for large u, can be represented by asymptotic series in inverse integer powers of u,

$$\chi_j(u) \sim \sum_{k=0}^{\infty} \chi_j^{(k)}/u^k, \quad j = 1, 2, \tag{44}$$

where the coefficients $\chi_j^{(k)}$ may be found by substitution of (43), (44) in ODE (41) ($j = 1, 2, \ k \geq 0$).

3. *All solutions of the family (43) are integrable at infinity iff inequality (16) is fulfilled.*

For a detailed proof of Lemma 3, see [4].

Corollary 1. *Under the assumptions of Lemma 3, all solutions of ODE (10) have finite limits as $u \to \infty$ iff condition (16) is fulfilled.*

Summarizing all results, we obtain the proof of Theorem 1.

5 The Accompanying Singular Problem for Capital Stock Model (The Third "Degenerate" Case: c = 0, b ≠ 0, a > 0, λ > 0, m > 0)

For this case, the input singular IDE problem has the form:

$$(b^2/2)u^2\varphi''(u) + au\varphi'(u) - \lambda[\varphi(u) - (J_m\varphi)(u)] = 0, \quad u \in \mathbf{R}_+, \tag{45}$$

$$\lim_{u \to +0} \varphi(u) = \lim_{u \to +0} [u\varphi'(u)] = 0, \tag{46}$$

$$\lim_{u \to \infty} \varphi(u) = 1, \quad \lim_{u \to \infty} \varphi'(u) = 0, \tag{47}$$

and restrictions (3) are needed for the solution.

The following lemma is analogous to Lemma 2 (with a similar proof).

Lemma 4. *For IDE (45), let the values a, b, λ, and m be fixed with $\lambda > 0$, $m > 0$ whereas a and b are any real numbers ($a, b \in \mathbf{R}$). Then the following assertions are valid:*

1. *If there exists a solution $\varphi_1(u)$ to the problem (45)–(47), then it is a unique solution to this problem.*
2. *Such $\varphi_1(u)$ satisfies restrictions (3) and $\varphi_1'(u) > 0$ for any finite $u > 0$, i.e., $\varphi_1(u)$ is a monotone nondecreasing on $\mathbf{R}_+$ function.*

Analogously to the previous approach, the singular IDE problem (45)–(47) is equivalent to the following singular ODE problem:

$$(b^2/2)u^3\varphi'''(u) + \left[b^2 + a + b^2 u/(2m)\right]u^2\varphi''(u)$$
$$+ (a - \lambda + au/m)u\varphi'(u) = 0, \quad 0 < u < \infty, \tag{48}$$

$$\lim_{u \to +0} \varphi(u) = \lim_{u \to +0} \left[u\varphi'(u)\right] = \lim_{u \to +0} \left[u^2\varphi''(u)\right] = 0, \tag{49}$$

$$\lim_{u \to \infty} \varphi(u) = 1, \qquad \lim_{u \to \infty} \varphi'(u) = \lim_{u \to \infty} \varphi''(u) = 0. \tag{50}$$

First, consider SCP at regular (weak) singular point $u = 0$, i.e., SCP (48), (49) introducing notation

$$\mu_1 = 1/2 - a/b^2 + \sqrt{(1/2 - a/b^2)^2 + 2\lambda/b^2}, \tag{51}$$

$$d_1 = \mu_1 + a/b^2, \quad d_2 = \mu_1 + 2a/b^2 - 1. \tag{52}$$

The following lemma is analogous to Lemma 1.

Lemma 5. *For IDE (45), let the values a, b, λ, m be fixed with $b \neq 0$, $\lambda > 0$, $m > 0$, $a \in \mathbf{R}$. Then:*

1. *The IDE SIP (45), (46) is equivalent to the ODE SCP (48), (49).*
2. *There exists a one-parameter family of solutions $\varphi(u, P_1)$ to the ODE SCP (48), (49) (therefore also to the equivalent IDE SIP (45), (46)) and the following representation holds:*

$$\varphi(u, P_1) = P_1 \int_0^u s^{\mu_1 - 1}\eta(s)ds; \tag{53}$$

here P_1 is a parameter, $0 < \mu_1$ is defined by (51), and $\eta(u)$ is a solution to SCP

$$u^2\eta''(u) + (2d_1 + u/m)u\eta'(u) + (d_2 u/m)\eta(u) = 0, \quad u > 0, \tag{54}$$
$$\lim_{u \to +0} \eta(u) = 1, \qquad \lim_{u \to +0} \left[u\eta'(u)\right] = 0, \tag{55}$$

where d_1 and d_2 are defined by (52); there exists a unique solution $\eta(u)$ to the SCP (54), (55) and it is a holomorphic function at the point $u = 0$,

$$\eta(u) = 1 + \sum_{k=1}^{\infty} P_{k+1}u^k, \quad |u| \leq u_0, \quad u_0 > 0, \tag{56}$$

where the coefficients P_{k+1} may be found by formal substitution of series (56) into ODE (54), namely, from the recurrence relations:

$$P_2 = -d_2/(2md_1), \tag{57}$$

$$P_{k+1} = -P_k(k - 1 + d_2)/[mk(k - 1 + 2d_1)], \quad k = 2, 3, \ldots; \tag{58}$$

moreover, if $D_1 = \lim_{u \to +0} \varphi'(u, P_1)$, then $D_1 = 0$ when $a < \lambda$; $D_1 = P_1$ when $a = \lambda$; and at last $|D_1| = \infty$ when $a > \lambda$ (but $\varphi'(u, P_1)$ is integrable as $u \to +0$).

Summarizing the results and taking into account that Lemma 3 and Corollary 1 are valid for any $c \in \mathbf{R}$, we obtain

Theorem 3. *For IDE* (45), *let all the parameters a, b, λ, m be fixed positive numbers and let inequality* (16) *of "robustness of shares" be fulfilled. Then the following assertions are valid:*

1. *There exists a unique solution $\varphi(u)$ of singular linear IDE problem* (45)–(47); *it satisfies restrictions* (3) *and, for $u > 0$, is a smooth monotone nondecreasing function.*
2. *Such $\varphi(u)$ can be obtained by the formula*

$$\varphi(u) = \int_0^u s^{\mu_1 - 1} \eta(s) ds \Big/ \int_0^\infty s^{\mu_1 - 1} \eta(s) ds, \qquad u \geq 0, \tag{59}$$

where $\eta(u)$ is defined in Lemma 5.
3. *For finite $u > 0$, the solution $\varphi(u)$ is represented by a convergent series which can be obtained using formulas* (59), (56)–(58).
4. *If $a > \lambda$, then the solution $\varphi(u)$ is concave on $\mathbf{R}_+$ with $\lim_{u \to +0} \varphi'(u) = \infty$ but $\varphi'(u)$ is an integrable on $\mathbf{R}_+$ function.*
5. *If $a \leq \lambda$, then $\varphi(u)$ is convex on a certain interval $[0, \hat{u}]$ where $\hat{u}$ is an inflection point, $\hat{u} > 0$; moreover if $a < \lambda$, then $\lim_{u \to +0} \varphi'(u) = 0$ whereas if $a = \lambda$, then $\lim_{u \to +0} \varphi'(u) = 1 / \int_0^\infty \eta(s) ds > 0$.*
6. *For large u, the asymptotic representation* (18) *holds with $K > 0$ where in general the value $K > 0$ cannot be determined using local analysis methods.*

6 Numerical Examples and Their Interpretation

For the main case $c > 0$, our study shows that the input singular IDE problem (1)–(4) may be reduced to the auxiliary ODE SCP (10), (11) with the parameter C_0 to be defined, $0 < C_0 < 1$. The asymptotic expansion of the solutions at zero (12) is used to transfer the limit initial conditions (11) from the singular point $u = 0$ to a nearby regular point $u_0 > 0$; the derivatives of the solution may be evaluated by formal differentiation of the representation (12). Consequently, a regular Cauchy problem is to be solved starting from the point $u_0 > 0$. The parameter C_0 in (12) is evaluated numerically to satisfy the condition $\lim_{u \to \infty} \varphi(u) = 1$.

For the additional case $c = 0$, the singular IDE problem (45)–(47) is equivalent to the singular ODE problem (48)–(50). To solve this problem we use formula (59) and the auxiliary SCP (54), (55). The convergent power series (56)–(58) is used to transfer limit initial conditions (55) from the singular point $u = 0$ to a regular point $u_0 > 0$, and then a regular Cauchy problem is to be solved starting from this point.

Maple programming package was used as a numerical tool.

For all examples, we put $m = 1$, $\lambda = 0.09$, and for $a > 0$, $b \neq 0$, the shares are "robust": $2a/b^2 > 1$ (Figs. 1–5).

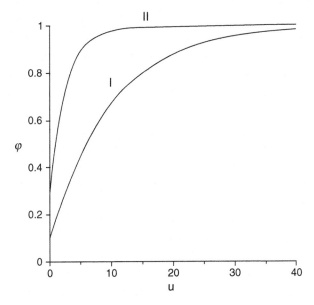

Fig. 1 The case $\mathbf{c} > \lambda\mathbf{m}$: $c = 0.1$; **I: $\mathbf{a} = \mathbf{b} = \mathbf{0}$** (the first "degenerate" case with the exact solution); $C_0 = \varphi(0) = 0.1$, $D_1 = \varphi'(+0) = 0.09$; **II: $\mathbf{a} = \mathbf{0.02}$, $\mathbf{b} = \mathbf{0.1}$**; $C_0 = 0.295$, $D_1 = 0.265$

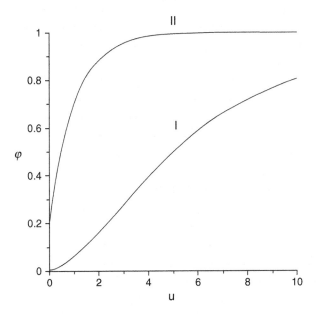

Fig. 2 The case $\mathbf{c} < \lambda\,\mathbf{m}$: $c = 0.02$, $\mathbf{b} = \mathbf{0.1}$; **I: $\mathbf{a} = \mathbf{0.02}$ ($\mathbf{m}(\lambda - \mathbf{a}) > \mathbf{c}$: $\varphi(u)$ has an inflection)**; $C_0 = 0.00527$, $D_1 = 0.0237$; **II: $\mathbf{a} = \mathbf{0.1}$ ($\mathbf{m}(\lambda - \mathbf{a}) < \mathbf{c}$: $\varphi(u)$ is concave)**; $C_0 = 0.194$, $D_1 = 0.872$

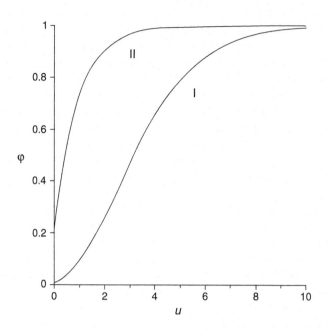

Fig. 3 The second "degenerate" case with premiums: $\mathbf{b} = \mathbf{0}$, $c = 0.02$ ($\mathbf{c} < \lambda \, \mathbf{m}$); **I:** $\mathbf{a} = \mathbf{0.02}$ ($\mathbf{m}(\lambda - \mathbf{a}) > \mathbf{c}$); $C_0 = 0.00704$, $D_1 = 0.0317$; **II:** $\mathbf{a} = \mathbf{0.1}$ ($\mathbf{m}(\lambda - \mathbf{a}) < \mathbf{c}$); $C_0 = 0.2046$, $D_1 = 0.9207$

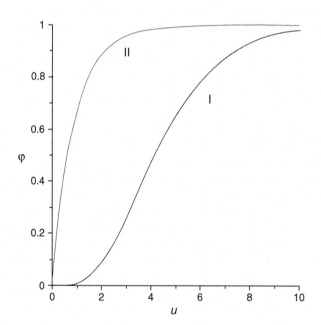

Fig. 4 The second "degenerate" case without premiums: $\mathbf{b} = \mathbf{0}, \mathbf{c} = \mathbf{0}$; **I:** $\mathbf{a} = \mathbf{0.02}$ ($\lambda > \mathbf{a}$); $\varphi(0) = \varphi'(0) = 0$; **II:** $\mathbf{a} = \mathbf{0.1}$ ($\lambda < \mathbf{a}$); $\varphi(0) = 0$, $\varphi'(+0) = \infty$

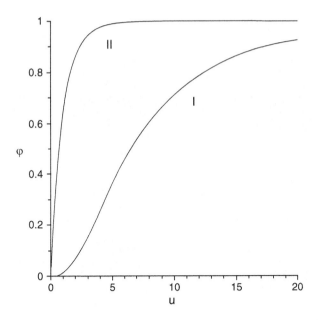

Fig. 5 The third "degenerate" case (the capital stock model): $c = 0$, $b = 0.1$; **I:** $a = 0.02$ ($\lambda > a$); $\varphi(0) = \varphi'(0) = 0$, $P_1 = 0.059587$; **II:** $a = 0.1$ ($\lambda < a$); $\varphi(0) = 0$, $\varphi'(+0) = \infty$, $P_1 = 0.861816$

7   Conclusions

The study shows that use of risky assets is not favorable for non-ruin with large initial surplus values and constant structure of the portfolio. However, the study of the cases when positiveness of the safety loading does not hold shows risky assets to be effective for small initial surplus values: while ruin is inevitable in the case without investing, the survival probability grows considerably as u grows in presence of investing even if the premiums are absent (moreover, the second derivative of the solution for small u is positive!). The study in [3, 4] of the optimal strategy for exponential distribution of claims shows that the part of risky investments should be $O(1/x)$ as present surplus x tends to infinity.

Acknowledgements This work was supported by the Russian Fund for Basic Research: Grants RFBR 10-01-00767 and RFBR 11-01-00219.

References

1. Azcue, P., Muler, N.: Optimal investment strategy to minimize the ruin probability of an insurance company under borrowing constraints. Insur. Math. Econ. **44**(1), 26–34 (2009)
2. Bateman, H., Erdélyi, A.: Higher Transcendental Functions. McGraw-Hill, New York (1953)

3. Belkina, T.A., Konyukhova, N.B., Kurkina, A.O.: Optimal investment problem in the dynamic insurance models: I. Investment strategies and the ruin probability. Survey on Appl. Ind. Math. **16**(6), 961–981 (2009) [in Russian]
4. Belkina, T.A., Konyukhova, N.B., Kurkina, A.O.: Optimal investment problem in the dynamic insurance models: II. Cramér-Lundberg model with the exponential claims. Survey on Appl. Ind. Math. **17**(1), 3–24 (2010) [in Russian]
5. Coddington, E.A., Levinson, N.: Theory of Ordinary Differential Equations. McGraw-Hill, New York (1955)
6. Fedoryuk, M.V.: Asymptotic Analysis: Linear Ordinary Differential Equations. Springer, Berlin (1993)
7. Frolova, A., Kabanov, Yu., Pergamenshchikov, S.: In the insurance business risky investments are dangerous. Finance Stochast. **6**(2), 227–235 (2002)
8. Grandell, J.: Aspects of Risk Theory. Springer, Berlin (1991)
9. Konyukhova, N.B.: Singular Cauchy problems for systems of ordinary differential equations. U.S.S.R. Comput. Maths. Math. Phys. **23**(3), 72–82 (1983)
10. Wasov, W.: Asymptotic Expansions for Ordinary Differential Equations. Dover, New York (1987)

Oscillatory Properties of Solutions
of Generalized Emden–Fowler Equations

R. Koplatadze

Abstract This work deals with the study of oscillatory properties of solutions of the equation

$$u^{(n)}(t) + p(t)\left|u(\sigma(t))\right|^{\mu(t)} \operatorname{sign} u(\sigma(t)) = 0,$$

where $p \in L_{\mathrm{loc}}(R_+; R_-)$, $\mu \in C(R_+; (0, +\infty))$, $\sigma \in C(R_+; R_+)$, and $\sigma(t) \geq t$ for $t \in R_+$. In this chapter, new sufficient (necessary and sufficient) conditions for essentially nonlinear functional differential equations to have Property **B** are established.

Keywords Oscillation • Functional differential equations • Property **B**

1 Introduction

The work with study of oscillatory properties of solutions of a differential equation of the form

$$u^{(n)}(t) + p(t)\left|u(\sigma(t))\right|^{\mu(t)} \operatorname{sign} u(\sigma(t)) = 0, \tag{1}$$

where

$$p \in L_{\mathrm{loc}}(R_+; R_-), \quad \mu \in C(R_+; (0, +\infty)), \tag{2}$$

$$\sigma \in C(R_+; R_+), \quad \text{and} \quad \sigma(t) \geq t \quad \text{for} \quad t \in R_+. \tag{3}$$

R. Koplatadze (✉)
Department of Mathematics of I. Javakhishvili, Tbilisi State University, 2,
University Street, 0186, Tbilisi, Georgia
e-mail: r_koplatadze@yahoo.com

S. Pinelas et al. (eds.), *Differential and Difference Equations with Applications*, Springer   45
Proceedings in Mathematics & Statistics 47, DOI 10.1007/978-1-4614-7333-6_4,
© Springer Science+Business Media New York 2013

Let $t_0 \in R_+$. A function $u : [t_0, +\infty) \to R$ is said to be a proper solution of Eq. (1), if it is locally absolutely continuous along with its derivatives up to the order $n - 1$ inclusive, $\sup\{|u(s)| : s \geq t\} > 0$ for $t \geq t_0$ and it satisfies Eq. (1) almost everywhere on $[t_0, +\infty)$.

A proper solution $u : [t_0, +\infty) \to R$ of the Eq. (1) is said to be oscillatory if it has a sequence of zeros tending to $+\infty$. Otherwise the solution u is said to be nonoscillatory.

Definition 1.1. We say that the Eq. (1) has Property **B** if any of its proper solutions is oscillatory or satisfies either

$$\left|u^{(i)}(t)\right| \downarrow 0, \quad \text{for} \quad t \uparrow +\infty \quad (i = 0, \ldots, n - 1) \tag{4}$$

or

$$\left|u^{(i)}(t)\right| \uparrow +\infty, \quad \text{for} \quad t \uparrow +\infty \quad (i = 0, \ldots, n - 1), \tag{5}$$

when n is even and either is oscillatory or satisfies (5), when n is odd.

In the case $\lim\limits_{t \to +\infty} \mu(t) = 1$, we call the differential equation (1) "almost linear," while $\liminf\limits_{t \to +\infty} \mu(t) \neq 1$ or $\limsup\limits_{t \to +\infty} \mu(t) \neq 1$, then we call Eq. (1) essentially nonlinear generalized Emden–Fowler-type differential equation.

Oscillatory properties of "almost linear" equations are studied well enough in [1, 3–5, 7]. In this chapter, essentially nonlinear differential equations of the type (1) are considered with one of the following conditions being satisfied:

$$\mu(t) \leq \lambda \quad \text{for} \quad t \in R_+ \quad (\lambda \in (0, 1)) \tag{6}$$

or

$$\mu(t) \geq \lambda \quad \text{for} \quad t \in R_+ \quad (\lambda \in (0, 1)). \tag{7}$$

In the present chapter, under conditions (6) and (7), sufficient (necessary and sufficient) conditions are established for the Eq. (1) to have Property **B**. Analogous results for Emden–Fowler equations are given in the paper [6].

2   Some Auxiliary Lemmas

In the sequel, $\tilde{C}_{\text{loc}}([t_0, +\infty))$ will denote the set of all functions $u : [t_0, +\infty) \to R$ absolutely continuous on any finite subinterval of $[t_0, +\infty)$ along with their derivatives of order up to and including $n - 1$.

Lemma 2.1 ([2]). *Let* $u \in \tilde{C}_{\text{loc}}^{n-1}([t_0, +\infty))$, $u(t) > 0$, $u^{(n)}(t) \geq 0$ *for* $t \geq t_0$, *and* $u^{(n)}(t) \not\equiv 0$ *in any neighborhood of* $+\infty$. *Then there exist* $t_1 \geq t_0$ *and* $\ell \in \{0, \ldots, n\}$ *such that* $l + n$ *is even and*

$$u^{(i)}(t) > 0 \quad \text{for} \quad t \geq t_1 \quad (i = 0, \ldots, \ell - 1),$$

$$(-1)^{i+\ell} u^{(i)}(t) \geq 0 \quad \text{for} \quad t \geq t_1 \quad (i = \ell, \ldots, n).$$

(8_ℓ)

Remark 2.1. If n is even and $\ell = 0$, then in (8_0) it is meant that only the second inequalities are fulfilled.

Lemma 2.2 ([2]). *Let* $u \in \tilde{C}_{\text{loc}}^{n-1}([t_0, +\infty))$ *and* (8_ℓ) *be fulfilled for some* $\ell \in \{1, \ldots, n-2\}$ *with* $l + n$ *even. Then*

$$\int_{t_0}^{+\infty} t^{n-\ell-1} u^{(n)}(t) dt < +\infty. \tag{9}$$

If, moreover,

$$\int_{t_0}^{+\infty} t^{n-\ell} u^{(n)}(t) dt = +\infty, \tag{10}$$

then there exists $t_* > t_0$ *such that*

$$\frac{u^{(i)}(t)}{t^{\ell-i}} \downarrow, \quad \frac{u^{(i)}(t)}{t^{\ell-i-1}} \uparrow \quad (i = 0, \ldots, \ell - 1), \tag{11_i}$$

$$u(t) \geq \frac{t^{\ell-1}}{\ell!} u^{(\ell-1)}(t) \quad \text{for} \quad t \geq t_*, \tag{12}$$

and

$$u^{(\ell-1)}(t) \geq \frac{t}{(n-\ell)!} \int_t^{+\infty} s^{n-\ell-1} \left| u^{(n)}(s) \right| ds +$$

$$+ \frac{1}{(n-\ell)!} \int_{t_*}^t s^{n-\ell} \left| u^{(n)}(s) \right| ds \quad \text{for} \quad t \geq t_*. \tag{13}$$

3   Necessary Conditions for the Existence of Solutions of Type (8_ℓ)

Definition 3.1. Let $t_0 \in R_+$. By $\mathbf{U}_{\ell, t_0}$ we denote the set of all proper solutions of the equation (1) satisfying the condition (8_ℓ).

Theorem 3.1. *Let the conditions* (2), (3), *and* (6) *be fulfilled,* $\ell \in \{1, \ldots, n-1\}$ *with* $\ell + n$ *even, and let*

$$\int_0^{+\infty} t^{n-\ell} (\sigma(t))^{(\ell-1)\mu(t)} |p(t)| dt = +\infty, \tag{14_ℓ}$$

$$\int_0^{+\infty} t^{n-\ell-1} (\sigma(t))^{\ell\mu(t)} |p(t)| dt = +\infty. \tag{15_ℓ}$$

If, moreover, $\mathbf{U}_{\ell,t_0} \neq \varnothing$ for some $t_0 \in R_+$, then for any $k \in N$ and $\delta \in [0,\lambda]$, and $\sigma_* \in C([t_0,+\infty))$ such that

$$t \leq \sigma_*(t) \leq \sigma(t) \quad \text{for} \quad t \geq t_0, \tag{16}$$

we have

$$\int_0^{+\infty} t^{n-\ell-1+\lambda-\delta} (\sigma_*(t))^{\mu(t)-\lambda} (\sigma(t))^{(\ell-1)\mu(t)} \big(\rho_{\ell,k}(\sigma_*(t))\big)^\delta dt < +\infty, \tag{17}$$

where

$$\rho_{\ell,1}(t) = \left(\frac{1-\lambda}{\ell!(n-1)!} \int_0^t \int_s^{+\infty} \xi^{n-\ell-1+\mu(\xi)-\lambda} (\sigma(\xi))^{(\ell-1)\mu(\xi)} \right.$$

$$\left. \times |p(\xi)| d\xi\, ds \right)^{\frac{1}{1-\lambda}}, \tag{18$_\ell$}$$

$$\rho_{\ell,i}(t) = \frac{1}{\ell!(n-\ell)!} \int_0^t \int_s^{+\infty} \xi^{n-\ell-1} (\sigma(\xi))^{(\ell-1)\mu(\xi)}$$

$$\times \big(\rho_{\ell,i-1}(\sigma(\xi))\big)^{\mu(\xi)} |p(\xi)| d\xi\, ds \quad (i=2,\dots,k). \tag{19$_\ell$}$$

Proof. Let $t_0 \in R_+$, $\ell \in \{1,\dots,n-1\}$ and $\mathbf{U}_{\ell,t_0} \neq \varnothing$. By definition of the set $\mathbf{U}_{\ell,t_0}$, the Eq. (1) has a proper solution $u \in \mathbf{U}_{\ell,t_0}$ satisfying the condition (8_ℓ). By Eqs. (1), (8_ℓ), and (14_ℓ), it is clear that the condition (10) holds. Thus by Lemma 2.2 the conditions (11_i)–(13) are fulfilled and

$$u^{(\ell-1)}(t) \geq \frac{t}{(n-\ell)!} \int_t^{+\infty} s^{n-\ell-1} \big(u(\sigma(s))\big)^{\mu(s)} |p(s)| ds$$

$$+ \frac{1}{(n-\ell)!} \int_{t_*}^t s^{n-\ell} \big(u(\sigma(s))\big)^{\mu(s)} |p(s)| ds \text{ for } t \geq t_*. \tag{20}$$

By (12) from (20), we get

$$u^{(\ell-1)}(t) \geq \frac{t}{(n-\ell)!} \int_t^{+\infty} s^{n-\ell-1} \big(u(\sigma(s))\big)^{\mu(s)} |p(s)| ds$$

$$- \frac{1}{(n-\ell)!} \int_{t_*}^t s\, d \int_s^{+\infty} \xi^{n-\ell-1} \big(u(\sigma(\xi))\big)^{\mu(\xi)} |p(\xi)| d\xi$$

$$\geq \frac{1}{(n-\ell)!} \int_{t_*}^t \int_s^{+\infty} \xi^{n-\ell-1} \big(u(\sigma(\xi))\big)^{\mu(\xi)} |p(\xi)| d\xi\, ds$$

$$\geq \frac{1}{\ell!(n-\ell)!} \int_{t_*}^t \int_s^{+\infty} \xi^{n-\ell-1} (\sigma(\xi))^{(\ell-1)\mu(\xi)}$$

$$\times \big(u^{(\ell-1)}(\sigma(\xi))\big)^{\mu(\xi)} |p(\xi)| d\xi\, ds. \tag{21}$$

Therefore, by Eqs. (3) and $(11_{\ell-1})$, from (21) we have

$$u^{(\ell-1)}(t) \geq \frac{1}{\ell!(n-\ell)!} \int_{t_*}^t \int_s^{+\infty} \xi^{n-\ell-1+\mu(\xi)}(\sigma(\xi))^{(\ell-1)\mu(\xi)}$$
$$\times \left(\frac{u^{(\ell-1)}(\xi)}{\xi}\right)^{\mu(\xi)} |p(\xi)| d\xi \, ds \text{ for } t \geq t_*. \tag{22}$$

On the other hand, by Eqs. $(11_{\ell-1})$ and (15_ℓ) it is obvious that

$$\frac{u^{\ell-1}(t)}{t} \downarrow 0 \quad \text{for} \quad t \uparrow +\infty. \tag{23}$$

According to (23), without loss of generality, we can assume that $u^{(\ell-1)}(t)/t \leq 1$ for $t \geq t_*$. Since $0 < \mu(t) \leq \lambda < 1$, from (22) we have

$$u^{(\ell-1)}(t) \geq \frac{1}{\ell!(n-\ell)!} \int_{t_*}^t \int_s^{+\infty} \xi^{n-\ell-1-\lambda+\mu(\xi)}(\sigma(\xi))^{(\ell-1)\mu(\xi)}$$
$$\times \left(u^{(\ell-1)}(\xi)\right)^\lambda |p(\xi)| d\xi \, ds. \tag{24}$$

By $(2.4_{\ell-1})$, it is obvious that

$$x'(t) \geq \frac{(u^{(\ell-1)}(t))^\lambda}{\ell!(n-\ell)!} \int_t^{+\infty} \xi^{n-\ell-1-\lambda+\mu(\xi)}(\sigma(\xi))^{(\ell-1)\mu(\xi)} |p(\xi)| d\xi, \tag{25}$$

where

$$x(t) = \frac{1}{\ell!(n-\ell)!} \int_{t_*}^t \int_s^{+\infty} \xi^{n-\ell-1-\lambda+\mu(\xi)}(\sigma(\xi))^{(\ell-1)\mu(\xi)}$$
$$\times \left(u^{(\ell-1)}(\xi)\right)^\lambda |p(\xi)| d\xi \, ds. \tag{26}$$

Thus, according to Eqs. (24) and (26), from (25) we get

$$x'(t) \geq \frac{x^\lambda(t)}{\ell!(n-\ell)!} \int_t^{+\infty} \xi^{n-\ell-1-\lambda+\mu(\xi)}(\sigma(\xi))^{(\ell-1)\mu(\xi)} |p(\xi)| d\xi \text{ for } t \geq t_*.$$

Therefore,

$$x(t) \geq \left(\frac{1-\lambda}{\ell!(n-\ell)!} \int_{t_*}^t \int_s^{+\infty} \xi^{n-\ell-1-\lambda+\mu(\xi)}(\sigma(\xi))^{(\ell-1)\mu(\xi)} |p(\xi)| d\xi \, ds\right)^{\frac{1}{1-\lambda}}$$

$$\text{for } t \geq t_*.$$

Hence, according to Eqs. (24) and (26), we have

$$u^{(\ell-1)}(t) \geq \rho_{t_*,\ell,1}(t) \quad \text{for} \quad t \geq t_*, \tag{27}$$

where

$$\rho_{t_*,\ell,1}(t) = \left(\frac{1-\lambda}{\ell!(n-\ell)!} \int_{t_*}^t \int_s^{+\infty} \xi^{n-\ell-1-\lambda+\mu(\xi)} (\sigma(\xi))^{(\ell-1)\mu(\xi)} \right.$$

$$\left. \times |p(\xi)| d\xi \, ds \right)^{\frac{1}{1-\lambda}}. \tag{28}$$

Thus by Eqs. (20), (26), and (27), we get

$$u^{(\ell-1)}(t) \geq \rho_{t_*,\ell,k}(t) \quad \text{for} \quad t \geq t_*, \tag{29}$$

where

$$\rho_{t_*,\ell,k}(t) = \frac{1}{\ell!(n-\ell)!} \int_{t_*}^t \int_s^{+\infty} \xi^{n-\ell-1} (\sigma(\xi))^{(\ell-1)\mu(\xi)}$$

$$\times \left(\rho_{t_*,\ell,k-1}(\sigma(\xi)) \right)^{\mu(\xi)} |p(\xi)| d\xi \, ds \quad (k=2,3,\dots). \tag{30}$$

On the other hand, by Eqs. (3), (8$_\ell$), (12), and (16) from (20), we have

$$u^{(\ell-1)}(t) \geq \frac{t}{\ell!(n-\ell)!} \int_t^{+\infty} s^{n-\ell-1} (\sigma(s))^{(\ell-1)\mu(s)} \left(u^{(\ell-1)}(\sigma(s)) \right)^{\mu(s)} |p(s)| ds$$

$$\geq \frac{t}{\ell!(n-\ell)!} \int_t^{+\infty} s^{n-\ell-1} (\sigma(s))^{(\ell-1)\mu(s)} \left(u^{(\ell-1)}(\sigma_*(s)) \right)^{\mu(s)} |p(s)| ds$$

$$= \frac{t}{\ell!(n-\ell)!} \int_t^{+\infty} s^{n-\ell-1} (\sigma(s))^{(\ell-1)\mu(s)} (\sigma_*(s))^{\mu(s)}$$

$$\times \left(\frac{u^{(\ell-1)}(\sigma_*(s))}{\sigma_*(s)} \right)^{\mu(s)} |p(s)| ds.$$

By Eqs. (11$_{\ell-1}$) and (16), for any $\delta \in [0,\lambda]$,

$$u^{(\ell-1)}(t) \geq \frac{t}{\ell!(n-\ell)!} \int_t^{+\infty} s^{n-\ell-1} (\sigma(s))^{(\ell-1)\mu(s)} (\sigma_*(s))^{\mu(s)-\lambda}$$

$$\times \left(u^{(\ell-1)}(\sigma_*(s)) \right)^\delta \left(u^{\ell-1}(s) \right)^{\lambda-\delta} |p(s)| ds.$$

Therefore, according to (29), we have

$$u^{(\ell-1)}(t) \geq \frac{t}{\ell!(n-\ell)!} \int_t^{+\infty} s^{n-\ell-1} (\sigma(s))^{(\ell-1)\mu(s)} (\sigma_*(s))^{\mu(s)-\lambda} \times$$

$$\times \left(\rho_{t_*,\ell,k}(\sigma_*(s)) \right)^\delta \left(u^{(\ell-1)}(s) \right)^{\lambda-\delta} |p(s)| ds \tag{31}$$

$$\text{for} \quad t \geq t_* \quad (k=1,2,\dots).$$

If $\delta = \lambda$, then from (31)

$$
\int_{t_*}^{+\infty} s^{n-\ell-1}(\sigma(s))^{(\ell-1)\mu(s)}(\sigma_*(s))^{\mu(s)-\lambda} \left(\rho_{t_*,\ell,k}(\sigma_*(s)) \right)^{\lambda} |p(s)| ds
$$

$$
\leq \ell!(n-\ell)! \frac{u^{(\ell-1)}(t_*)}{t_*} \leq \ell!(n-\ell)!. \tag{32}
$$

Let $\delta \in [0,\lambda)$. Then from (31)

$$
\left(u^{(\ell-1)}(t) \right)^{\lambda-\delta} \geq \frac{t^{\lambda-\delta}}{(\ell!(n-\ell)!)^{\lambda-\delta}} \left(\int_t^{+\infty} s^{n-\ell-1}(\sigma(s))^{(\ell-1)\mu(s)}(\sigma_*(s))^{\mu(s)-\lambda} \right.
$$

$$
\left. \times \left(\rho_{t_*,\ell,k}(\sigma_*(s)) \right)^{\delta} \left(u^{(\ell-1)}(s) \right)^{\lambda-\delta} |p(s)| ds \right)^{\lambda-\delta}
$$

$$
\text{for } t \geq t_* \ (k = 1,2,\dots).
$$

Thus we have

$$
\frac{\varphi(t)}{\left(\int_t^{+\infty} \varphi(s)ds \right)^{\lambda-\delta}} \geq \frac{1}{(\ell!(n-\ell)!)^{\lambda-\delta}} t^{n-\ell-1+\lambda-\delta}(\sigma(t))^{(\ell-1)\mu(t)}(\sigma_*(t))^{\mu(t)-\lambda}
$$

$$
\times \left(\rho_{t_*,\ell,k}(\sigma_*(t)) \right)^{\delta} |p(t)| \text{ for } t \geq t_* \ (k = 1,2,\dots),
$$

where

$$
\varphi(t) = t^{n-\ell-1}(\sigma(t))^{(\ell-1)\mu(t)}(\sigma_*(t))^{\mu(t)-\lambda} \left(\rho_{t_*,\ell,k}(\sigma_*(t)) \right)^{\delta} \left(u^{(\ell-1)}(t) \right)^{\lambda-\delta} |p(t)|.
$$

From the last inequality, we get

$$
-\int_{y(t_*)}^{y(t)} \frac{ds}{s^{\lambda-\delta}} \geq \frac{1}{(\ell!(n-\ell)!)^{\lambda-\delta}} \int_{t_*}^{t} s^{n-\ell-1+\lambda-\delta}(\sigma(s))^{(\ell-1)\mu(s)}
$$

$$
\times (\sigma_*(s))^{\mu(s)-\lambda} \left(\rho_{t_*,\ell,k}(\sigma_*(s)) \right)^{\delta} |p(s)| ds,
$$

where

$$
y(t) = \int_t^{+\infty} \varphi(s)ds. \tag{33}
$$

Therefore,

$$
\int_{t_*}^{t} s^{n-\ell-1+\lambda-\delta}(\sigma(s))^{(\ell-1)\mu(s)}(\sigma_*(s))^{\mu(s)-\lambda} \left(\rho_{t_*,\ell,k}(\sigma_*(s)) \right)^{\delta} |p(s)| ds
$$

$$
\leq (\ell!(n-\ell)!)^{\lambda-\delta} \int_0^{y(t_*)} \frac{ds}{s^{\lambda-\delta}}. \tag{34}
$$

By (33), without loss of generality we can assume that $y(t_*) \leq 1$. Thus from (34) we have

$$\int_{t_*}^t s^{n-\ell-1+\lambda-\delta} (\sigma(s))^{(\ell-1)\mu(s)} (\sigma_*(s))^{\mu(s)-\lambda} \left(\rho_{t_*,\ell,k}(\sigma_*(s))\right)^\delta |p(s)| ds$$

$$\leq (\ell!(n-\ell)!)^{\lambda-\delta} \int_0^1 \frac{ds}{s^{\lambda-\delta}} = \frac{(\ell!(n-\ell)!)^{\lambda-\delta}}{1-\lambda+\delta} \quad \text{for } t \geq t_*.$$

Passing to limit in the latter inequality, we obtain

$$\int_{t_*}^{+\infty} s^{n-\ell-1+\lambda-\delta} (\sigma(s))^{(\ell-1)\mu(s)} (\sigma_*(s))^{\mu(s)-\lambda}$$

$$\times \left(\rho_{t_*,\ell,k}(\sigma_*(s))\right)^\delta |p(s)| ds < +\infty. \tag{35}$$

Therefore, since

$$\lim_{t \to +\infty} \frac{\rho_{\ell,k}(t)}{\rho_{t_*,\ell,k}(t)} = 1 \quad (k = 1, 2, \ldots),$$

by Eqs. (32) and (35), it is obvious that for any $\delta \in [0, \lambda]$ and $k \in N$, (17) holds, which proves the validity of the theorem. $\square$

Analogously we can prove

Theorem 3.2. *Let the conditions* (2), (3), (7), (14_ℓ), *and* (15_ℓ) *be fulfilled,* $\ell \in \{1, \ldots, n-1\}$ *with* $\ell + n$ *even and* $\mathbf{U}_{\ell,t_0} \neq \varnothing$ *for some* $t_0 \in R_+$. *Then for any* $k \in N$ *and* $\delta \in [0, \lambda]$,

$$\int_0^{+\infty} t^{n-\ell-1+\delta} (\sigma(t))^{(\ell-1)\mu(t)} \left(\tilde{\rho}_{\ell,k}(\sigma(t))\right)^{\mu(t)-\delta} |p(t)| dt < +\infty, \tag{36}$$

where

$$\tilde{\rho}_{\ell,1}(t) = \left(\frac{1-\lambda}{\ell!(n-\ell)!} \int_0^t \int_s^{+\infty} \xi^{n-\ell-1} (\sigma(\xi))^{(\ell-1)\mu(\xi)} |p(\xi)| d\xi\, ds \right)^{\frac{1}{1-\lambda}}, \tag{37_ℓ}$$

$$\tilde{\rho}_{\ell,i}(t) = \frac{1}{\ell!(n-\ell)!} \int_0^t \int_s^{+\infty} \xi^{n-\ell-1} (\sigma(\xi))^{(\ell-1)\mu(\xi)}$$

$$\times \left(\tilde{\rho}_{\ell,i-1}(\sigma(\xi))\right)^{\mu(\xi)} |p(\xi)| d\xi\, ds, \quad i = 2, 3, \ldots \tag{38_ℓ}$$

4 Sufficient Conditions for Nonexistence of Solutions of the Type (8_ℓ)

Theorem 4.1. *Let the conditions* (2), (5), (6), (14_ℓ), *and* (15_ℓ) *be fulfilled, where* $\ell \in \{1,\ldots,n-1\}$ *with* $\ell + n$ *even, and there exist* $\delta \in [0,\lambda]$, $k \in N$, *and* $\sigma_* \in C(R_+)$ *satisfying the condition* (16) *such that*

$$\int_0^{+\infty} t^{n-\ell-1+\lambda-\delta}(\sigma_*(t))^{\mu(t)-\lambda}(\sigma(t))^{(\ell-1)\mu(t)}\left(\rho_{\ell,k}(\sigma_*(t))\right)^\delta |p(t)|\,dt = +\infty \quad (39_\ell)$$

holds. Then for any $t_0 \in R_+$ *we have* $\mathbf{U}_{\ell,t_0} = \varnothing$, *where* $\rho_{\ell,k}$ *is defined by Eqs.* (18_ℓ) *and* (19_ℓ).

Proof. Assume the contrary. Let there exist $t_0 \in R_+$ such that $\mathbf{U}_{\ell,t_0} \neq \varnothing$ (see Definition 3.1). Then the Eq. (1) has a proper solution $u : [t_0, +\infty) \to R$ satisfying the condition (8_ℓ). Since the conditions of Theorem 3.1 are fulfilled, for any $\delta \in [0,\lambda]$, $k \in N$, and $\sigma_* \in C([t_0,+\infty))$ satisfying the condition (16), the condition (17) holds, which contradicts Eq. (39_ℓ). The obtained contradiction proves the validity of the theorem. $\qquad\square$

Using Theorem 3.2, analogously we can prove

Theorem 4.2. *Let the conditions* (2), (3), (7), (14_ℓ), *and* (15_ℓ) *be fulfilled, where* $\ell \in \{1,\ldots,n-1\}$ *with* $\ell + n$ *even, and let there exist* $\delta \in [0,\lambda]$ *and* $k \in N$ *such that*

$$\int_0^{+\infty} t^{n-\ell-1+\delta}(\sigma(t))^{(\ell-1)\mu(t)}\left(\tilde{\rho}_{\ell,k}(\sigma(t))\right)^{\mu(t)-\delta}|p(t)|\,dt = +\infty. \quad (40_\ell)$$

Then for any $t_0 \in R_+$ *we have* $\mathbf{U}_{\ell,t_0} = \varnothing$, *where* $\tilde{\rho}_{\ell,k}$ *is defined by Eqs.* (37_ℓ) *and* (38_ℓ).

Theorem 4.3. *Let the conditions* (2), (3), (7), (15_ℓ) *and let for some* $\gamma \in (0,1)$ *the condition*

$$\liminf_{t\to+\infty} t^\gamma \int_t^{+\infty} s^{n-\ell-1}(\sigma(s))^{(\ell-1)\mu(s)}|p(s)|\,ds > 0 \quad (41_\ell)$$

be fulfilled, where $\ell \in \{1,\ldots,n-1\}$ *with* $\ell + n$ *even. If, moreover, there exist* $\alpha > 1$ *such that*

$$\liminf_{t\to+\infty} \frac{\sigma(t)}{t^\alpha} > 0 \quad (42)$$

and either

$$\alpha\lambda \geq 1 \quad (43)$$

or, if $\alpha\lambda < 1$, for some $\varepsilon > 0$

$$\int_0^{+\infty} t^{n-\ell-1+\mu(t)\left(\frac{\alpha(1-\gamma)}{1-\alpha\lambda}-\varepsilon\right)} (\sigma(t))^{(\ell-1)\mu(t)} |p(t)| dt = +\infty, \qquad (44_\ell)$$

then for any $t_0 \in R_+$ we have $U_{\ell,t_0} = \varnothing$.

Proof. It sufficient to show that the condition (40_ℓ) is satisfied for $\delta = 0$ and for some $k \in N$. Indeed, according to Eqs. (41_ℓ) and (4), there exist $\alpha > 1$, $c > 0$, $\gamma \in (0,1)$, and $t_1 \in [t_0, +\infty)$ such that

$$t^\gamma \int_t^{+\infty} s^{n-\ell-1} (\sigma(s))^{(\ell-1)\mu(s)} |p(s)| ds \geq c \quad \text{for } t \geq t_1 \qquad (45)$$

and

$$\sigma(t) \geq ct^\alpha \quad \text{for } t \geq t_1. \qquad (46)$$

Choose $\varepsilon > 0$, $k_0 \in N$, and $c_* \in (1, +\infty)$ such that

$$(1-\gamma)(k_0 - 1) \geq \frac{1}{\lambda} \quad \text{when } \alpha\lambda \geq 1, \qquad (47)$$

$$1 + \alpha\lambda + \cdots + (\alpha\lambda)^{k_0-2} \geq \frac{1}{1-\alpha\lambda} - \frac{\varepsilon}{\alpha(1-\gamma)} \quad \text{when } \alpha\lambda < 1 \qquad (48)$$

and

$$c_*^{\lambda^i} \left(\frac{c}{2\ell!(n-\ell)!(1-\gamma)(1+\alpha\lambda+\cdots+(\alpha\lambda)^{i-2})} \right)^{1+\lambda+\cdots+\lambda^{i-2}} \geq 1 \qquad (49)$$

$$(i = 1,\ldots,k).$$

According to Eqs. (45) and (37_ℓ), it is obvious that $\lim_{t\to+\infty} \tilde{\rho}_{\ell,1}(t) = +\infty$. Therefore, without loss of generality, we can assume that $\tilde{\rho}_{\ell,1}(t) \geq c_*$ for $t \geq t_1$. Thus, by (45) from (38_ℓ), we get

$$\tilde{\rho}_{\ell,2}(t) \geq \frac{c_*^\lambda}{\ell!(n-\ell)!} \int_{t_1}^t \int_s^{+\infty} \xi^{n-\ell-1} (\sigma(\xi))^{(\ell-1)\mu(\xi)} |p(\xi)| d\xi \, ds$$

$$\geq \frac{c_*^\lambda c}{\ell!(n-\ell)!} \int_{t_1}^t s^{-\gamma} ds = \frac{c_*^\lambda c}{\ell!(n-\ell)!(1-\gamma)} (t^{1-\gamma} - t_1^{1-\gamma}).$$

Choose $t_2 > t_1$ such that

$$\tilde{\rho}_{\ell,2}(t) \geq \frac{c_*^\lambda c t^{1-\gamma}}{2\ell!(n-\ell)!(1-\gamma)} \quad \text{for } t \geq t_2.$$

Then, by Eqs. (7), (45), (46), and (49) from (38_ℓ), we have

$$\tilde{\rho}_{\ell,3}(t) \geq c_*^{\lambda^2}\left(\frac{c}{2\ell!(n-\ell)!(1-\gamma)(1+\alpha\lambda)}\right)^{1+\lambda} t^{(1-\gamma)(1+\alpha\lambda)} \quad \text{for} \quad t \geq t_3,$$

where $t_3 > t_2$ is a sufficiently large number. Therefore, for $k_0 \in N$ there exists $t_{k_0} \in R_+$ such that

$$\tilde{\rho}_{\ell,k_0}(t) \geq c_*^{\lambda^{k_0-1}}\left(\frac{c}{2\ell!(n-\ell)!(1-\gamma)(1+\alpha\lambda+\cdots+(\alpha\lambda)^{k_0-2})}\right)^{1+\lambda+\cdots+\lambda^{k_0-2}}$$

$$\times t^{(1-\gamma)(1+\alpha\lambda+\cdots+(\alpha\lambda)^{k_0-2})} \quad \text{for} \quad t \geq t_{k_0}. \tag{50}$$

Assume that (43) is fulfilled. Then, according to Eqs. (7), (41_ℓ), (47), and (50), it is obvious that, if $\delta = 0$ for $k = k_0$, (40_ℓ) holds. In the case, where (43) holds, the validity of the theorem has been already proved.

Assume now that $\alpha\lambda < 1$ and for some $\varepsilon > 0$ (44_ℓ) is fulfilled. Then, by (48) from (50) we have

$$\left(\tilde{\rho}_{\ell,k_0}(\sigma(t))\right)^{\mu(t)} \geq c_1 t^{\mu}(t)\left(\frac{\alpha(1-\gamma)}{1-\alpha\lambda}-\varepsilon\right) \quad \text{for} \quad t \geq t_{k_0},$$

where $c_1 > 0$. Consequently, according to (44_ℓ), it is obvious that (40_ℓ) holds with if $\delta = 0$ and $k = k_0$. The proof of the theorem is complete. $\square$

5 Differential Equations with Property B

Theorem 5.1. *Let the conditions (2), (3), and (6) be fulfilled and for any $\ell \in \{1,\ldots,n-1\}$ with $\ell+n$ even Eqs. (14_ℓ) and (15_ℓ) hold. Let, moreover, there exist $\delta \in [0,\lambda]$, $k \in N$, and $\sigma_* \in C(R_+)$ satisfying the condition (16) such that Eq. (39_ℓ) holds and*

$$\int_0^{+\infty} t^{n-1}|p(t)|dt = +\infty \tag{51}$$

*when n is even. Then Eq. (1) has Property **B**, where $\rho_{\ell,k}$ is defined by Eqs. (18_ℓ) and (19_ℓ).*

Proof. Let the Eq. (1) have a proper nonoscillatory solution $u : [t_0, +\infty) \to (0, +\infty)$ (the case $u(t) < 0$ is similar). Then by Eqs. (2), (3), and Lemma 2.1, there exists $\ell \in \{0, 1, \ldots, n\}$ such that $\ell+n$ is even and the condition (8_ℓ) holds. Since the conditions of Theorem 4.1 are fulfilled for any $\ell \in \{1,\ldots,n-1\}$ with $\ell+n$ even, we have $\ell \notin \{1,\ldots,n-1\}$. As while proving Theorem 5.1, it can be shown that Eqs. (5) [(4)] holds if $\ell = n$ (if n is even and $\ell = 0$). Let $\ell = n$. Show that (5) be fulfilled. Indeed, by

(2.1_n), there exists $c \in (0,1)$ and $t_1 > t_0$ such that $u(t) \geq ct^{n-1}$ for $t \geq t_1$. According to (3.2_{n-1}), from Eq. (1) we have

$$u^{(n-1)}(t) \geq c^{\lambda} \int_{t_1}^{t} (\sigma(s))^{(n-1)\mu(s)} |p(s)| ds \to +\infty \quad \text{for} \quad t \to +\infty.$$

Therefore, (5) holds.

Let n is even and $\ell = 0$. Show that the condition (4) holds. If that is not the case, then there exists $c \in (0,1)$ such that $u(t) \geq c$ for sufficiently large t. According to Eqs. (2.1_0) and (6), from Eq. (1) we have

$$\sum_{i=0}^{n-1} (n-i-1)! t_1^i |u^{(i)}(t_1)| \geq \int_{t_1}^{t} s^{n-1} c^{\mu(s)} |p(s)| ds \geq c^{\lambda} \int_{t_1}^{t} s^{n-1} |p(s)| ds, \quad (52)$$

where t_1 is a sufficiently large number. The inequality (52) contradicts the condition (51). Therefore, the Eq. (1) has Property **B**. □

Theorem 5.2. *Let the conditions* (2), (3), *and* (7) *be fulfilled and for any* $\ell \in \{1,\ldots,n-1\}$ *with* $\ell + n$ *even, Eqs.* (14_ℓ) *and* (15_ℓ) *hold and*

$$\limsup_{t \to +\infty} \mu(t) < +\infty. \quad (53)$$

Let, moreover, there exist $\delta \in [0,\lambda]$ *and* $k \in N$ *such that* (40_ℓ) *hold. Then Eq.* (1) *has Property* **B**, *where* $\tilde{\rho}_{\ell,k}$ *is defined by Eqs.* (37_ℓ) *and* (38_ℓ).

Proof. The proof of the theorem is analogous to that of Theorem 5.1. We have just to use Theorem 4.2 instead of Theorem 4.1, and change λ by $\mu = \sup\{\mu(t) : t \in R_+\}$ in the inequality (52). □

Theorem 5.3. *Let the conditions* (2), (3), (6), (51), *and*

$$\liminf_{t \to +\infty} \frac{(\sigma(t))^{\mu(t)}}{t} > 0 \quad (54)$$

be fulfilled. If, moreover, there exist $\delta \in [0,\lambda]$, $k \in N$, *and* $\sigma_* \in C(R_+)$ *satisfying the condition* (16) *such that for odd n (for even n)* (4.1_1) $[(4.1_2)]$ *holds, then the Eq.* (1) *has Property* **B**, *where* $\rho_{1,k}$ $(\rho_{2,k})$ *is defined by* (3.5_1) *and* (3.6_1) $[(3.5_2)$ *and* $(3.6_2)]$.

Proof. To prove the theorem, it suffices to show that the conditions of Theorem 5.1 are fulfilled. Indeed, according to Eqs. (4.1_1) and (54) [Eqs. (4.1_2) and (54)], it is obvious that Eq. (39_ℓ) holds for any $\ell \in \{1,\ldots,n-1\}$ with $\ell + n$ even. Thus according to Eqs. (51) and (54), all the conditions of Theorem 5.1 are fulfilled, which proves the validity of the theorem. □

Using Theorem 5.2, analogously we can prove

Theorem 5.4. *Let the conditions* (2), (3), (7), (51), (53), *and* (54) *be fulfilled. Let, moreover, there exist* $\delta \in [0,\lambda]$ *and* $k \in N$ *such that for odd (for even n)* (4.2_1)

$[(4.2_2)]$ *holds. Then the Eq.* (1) *has Property* **B**, *where* $\tilde{\rho}_{1,k}$ $(\tilde{\rho}_{2,k})$ *is defined by* (3.24_1) *and* (3.25_1) $[(3.24_2)$ *and* $(3.25_2)]$.

Corollary 5.1. *Let the conditions* (2), (3), (6), (51), *and* (54) *be fulfilled. If, moreover,*

$$\int_0^{+\infty} t^{n-2+\mu(t)} |p(t)| dt = +\infty \tag{55}$$

for odd n, and

$$\int_0^{+\infty} t^{n-3+\mu(t)} (\sigma(t))^{\mu(t)} |p(t)| dt = +\infty \tag{56}$$

for even n, then the Eq. (1) *has Property* **B**.

Proof. It suffices to note that by Eqs. (6), (51), (54)–(56), all the conditions of the Theorem 5.3 are fulfilled with $\sigma_*(t) \equiv t$ and $\delta = 0$. $\qquad\square$

Corollary 5.2. *Let the conditions* (2), (3), (6), (51), *and* (54) *be fulfilled. Let, moreover, for some $k \in N$*

$$\int_0^{+\infty} t^{n-2} (\sigma(t))^{\mu(t)-\lambda} (\rho_{1,k}(\sigma(t)))^\lambda |p(t)| dt = +\infty \tag{57}$$

hold when n is odd and

$$\int_0^{+\infty} t^{n-3} (\sigma(t))^{2\mu(t)-\lambda} (\rho_{2,k}(\sigma(t)))^\lambda |p(t)| dt = +\infty \tag{58}$$

hold when n is even. Then the Eq. (1) *has Property* **B**, *where* $\rho_{1,k}$ $(\rho_{2,k})$ *is defined by Eqs.* (3.5_1) *and* (3.6_1) $[(3.5_2)$ *and* $(3.6_2)]$.

Proof. It suffices to note that by Eqs. (6), (54), (57), and (58), all the conditions of Theorem 5.3 are fulfilled with $\sigma_*(t) = \sigma(t)$ and $\delta = \lambda$. $\qquad\square$

Corollary 5.3. *Let the conditions* (2), (3), (7), (51), *and* (54) *be fulfilled. Let, moreover, for some $k \in N$*

$$\int_0^{+\infty} t^{n-2} (\tilde{\rho}_{1,k}(\sigma(t)))^{\mu(t)} |p(t)| dt = +\infty \tag{59}$$

hold when n is odd, and

$$\int_0^{+\infty} t^{n-3} (\sigma(t))^{\mu(t)} (\tilde{\rho}_{2,k}(\sigma(t)))^{\mu(t)} |p(t)| dt = +\infty \tag{60}$$

hold when n is even. Then the Eq. (1) *has Property* **B**, *where* $\tilde{\rho}_{1,k}$ $(\tilde{\rho}_{2,k})$ *is defined by Eqs.* (3.24_1) *and* (3.25_1) $[(3.24_2)$ *and* $(3.25_2)]$.

Proof. It suffices to note that by Eqs. (7), (54), (59), and (60), all the conditions of Theorem 54 are fulfilled with $\delta = 0$. $\qquad\square$

Theorem 5.5. *Let the conditions* (2), (3), (7), (51), *and* (54) *be fulfilled and* (41_1) *[(41_2)] hold for odd n (for even n). If, moreover, there exists* $\alpha \in (1, +\infty)$ *such that* (42) *holds, then for the Eq.* (1) *to have Property* **B***, it is sufficient that at least one of the conditions* (43) *or, if* $\alpha\lambda < 1$, (44_1) *[(44_2)] holds, for odd n (for even n).*

Proof. According to Eqs. (4.3_1), (54), and (4.6_1) [(4.3_2), (54), and (4.6_2)], it is obvious that for any $\ell \in \{1, \ldots, n-2\}$ with $\ell + n$ even, Eqs. (14_ℓ), (15_ℓ), and (44_ℓ) hold. Assume that the Eq. (1) has a nonoscillatory solution $u : [t_0, +\infty) \to (0, +\infty)$ satisfying to condition (8_ℓ). Then by Theorem 4.3, $\ell \notin \{1, \ldots, n-2\}$. Therefore, $\ell = n$, or n is even and $\ell = 0$. In this case by Eqs. (15_{n-1}) [(51)], it is obvious that condition (5) [(4)] holds. Therefore, the Eq. (1) has Property **B**.   □

Theorem 5.6. *Let the conditions* (2), (3), (6), (15_{n-1}), *and*

$$\limsup_{t \to +\infty} \frac{(\sigma(t))^{\mu(t)}}{t} < +\infty \tag{61}$$

be fulfilled. If, moreover, there exist $\delta \in [0, \lambda]$, $k \in N$, *and* $\sigma_* \in C(R_+)$ *satisfying the condition* (16) *such that* (39_{n-1}) *holds, then the Eq.* (1) *has Property* **B***.*

Proof. By virtue of Eqs. (2), (3), (6), (3.2_{n-1}), (4.1_{n-2}), and (61), the conditions of the Theorem 5.1 are obviously satisfied. Therefore, according to that theorem, the Eq. (1) has Property **B**.   □

The validity of the Theorem 5.7 below is proved similarly.

Theorem 5.7. *Let the conditions* (2), (3), (7), (15_{n-1}), *and* (61) *be fulfilled. If, moreover, there exist* $\delta \in [0, \lambda]$ *and* $k \in N$ *such that* (40_{n-2}) *holds, then the Eq.* (1) *has Property* **B***, where* $\tilde{\rho}_{n-1,k}$ *is defined by Eqs.* (37_{n-1}) *and* (38_{n-1}).

Corollary 5.4. *Let the conditions* (2), (3), (6), (15_{n-1}), *and* (61) *be fulfilled and*

$$\int_0^{+\infty} t(\sigma(t))^{(n-2)\mu(t)-\lambda} \rho_{n-2,1}^{\lambda}(\sigma(t))|p(t)|dt = +\infty; \tag{62}$$

then the Eq. (1) *has Property* **B***, where* $\rho_{n-2,1}$ *is defined by* (3.5_{n-2}).

Proof. It suffices to note that by Eqs. (2), (3), (6), (15_{n-1}), (61), and (62), all the conditions of the Theorem 5.6 are fulfilled with $k = 1$, $\delta = \lambda$, and $\sigma_*(t) = \sigma(t)$.   □

Corollary 5.4′. *Let the conditions* (2), (3), (6), (15_{n-1}), *and* (61) *be fulfilled and*

$$\liminf_{t \to +\infty} t^{\lambda} \int_t^{+\infty} \xi^{1+\mu(\xi)-\lambda}(\sigma(\xi))^{(n-3)\mu(\xi)}|p(\xi)|d\xi > 0. \tag{63}$$

Then the condition

$$\int_0^{+\infty} t(\sigma(t))^{(n-2)\mu(t)}|p(t)|dt = +\infty \tag{64}$$

suffices the Eq. (1) to have Property **B**.

Proof. By Eqs. (63) and (18_{n-2}), there exist $t_0 \in R_+$ and $c > 0$ such that

$$\rho_{n-2,1}(\sigma(t)) \geq c\,\sigma(t) \quad \text{for} \quad t \geq t_0.$$

Hence, according to (64) the condition (62) holds, which proves the validity of the corollary. □

Corollary 5.5. *Let the conditions* (2), (3), (6), (15_{n-1}), *and* (61) *be fulfilled and*

$$\int_0^{+\infty} t^{1+\mu(t)}(\sigma(t))^{(n-3)\mu(t)}|p(t)|dt = +\infty. \tag{65}$$

Then the Eq. (1) has Property **B**.

Proof. According to Theorem 5.6, it suffices to note that by (65) the condition (4.1_{n-2}) holds with $\delta = 0$ and $\sigma_*(t) \equiv t$. □

Corollary 5.6. *Let the conditions* (2), (3), (7), (15_{n-1}), *and* (61) *be fulfilled and*

$$\int_0^{+\infty} t^{1+\lambda}(\sigma(t))^{(n-3)\mu(t)}|p(t)|dt = +\infty. \tag{66}$$

Then the Eq. (1) has Property **B**.

Proof. According to the Theorem 5.7, it suffices to note that by (66), the condition (40_{n-1}) holds with $\delta = \lambda$. □

Corollary 5.7. *Let the conditions* (2), (3), (7), (15_{n-1}), *and* (61) *be fulfilled and*

$$\int_0^{+\infty} t(\sigma(t))^{(n-3)\mu(t)}(\tilde{\rho}_{n-2,1}(\sigma(t)))^{\mu(t)}|p(t)|dt = +\infty. \tag{67}$$

Then the Eq. (1) has Property **B**, *where $\tilde{\rho}_{n-2,1}$ is defined by* (37_{n-1}).

Proof. It is suffices to note that by condition (67), the condition (40_{n-1}) holds with $\delta = 0$. □

Analogously, to Corollary 5.4′ we can prove

Corollary 5.7′. *Let the conditions* (2), (3), (7), (15_{n-1}), *and* (61) *be fulfilled and*

$$\liminf_{t \to +\infty} t^{\lambda} \int_t^{+\infty} \xi(\sigma(\xi))^{(n-3)\mu(\xi)}|p(\xi)|d\xi > 0. \tag{68}$$

Then the condition (64) *suffices the Eq. (1) to have Property* **B**.

Theorem 5.8. *Let the conditions* (2), (3), (7), (15_{n-1}), *and* (61) *be fulfilled. If, moreover, there exists $\alpha \in (1, +\infty)$ such that the condition* (42) *holds, then for the*

Eq. (1) *to have Property* **B**, *it is sufficient that at least one of the conditions* (43) *or if* $\alpha\lambda < 1$ (44_{n-2}) *holds.*

Proof. According the Theorem 4.3, it is sufficient to note that by Eqs. (61) and (44_{n-2}) for any $\ell \in \{1,\ldots,n-1\}$ with $\ell + n$ is even, (44_ℓ) holds.   □

6   Necessary and Sufficient Conditions

Theorem 6.1. *Let n be even, the conditions* (2), (3), *and* (6) *be fulfilled, and*

$$\liminf_{t\to+\infty} \frac{\sigma(t)}{t^{\frac{2-\mu(t)}{\mu(t)}}} > 0. \tag{69}$$

Then the condition (51) *is necessary and sufficient for the Eq.* (1) *to have Property* **B**.

Proof. Necessity. Assume that the Eq. (1) has Property **B** and

$$\int_0^{+\infty} t^{n-1}|p(t)|dt < +\infty. \tag{70}$$

According to (70), by Lemma 4.1 [2] there exists $c \neq 0$ such that the equation has a proper solution $u : [t_0, \infty) \to R$ satisfying the condition $\lim_{t\to+\infty} u(t) = c$. But this contradicts the fact that the Eq. (1) has Property **B**.

Sufficiency. According to Eqs. (6) and (69), it is obvious that the condition (54) holds. On the other hand, by Eqs. (51) and (69), the condition (56) holds. Thus, since n is even, all the conditions of Corollary 5.1 are fulfilled, i.e., the Eq. (1) to have Property **B**.   □

Corollary 6.1. *Let n be even, the conditions* (2), (3), *and* (6) *be fulfilled, and*

$$\lim_{t\to+\infty} \mu(t) = \lambda \ (\lambda \in (0,1)), \quad \liminf_{t\to+\infty} t^{\mu(t)-\lambda} > 0, \quad \liminf_{t\to+\infty} \frac{\sigma(t)}{t^{\frac{2-\lambda}{\lambda}}} > 0. \tag{71}$$

Then the condition (51) *is necessary and sufficient for the Eq.* (1) *to have Property* **B**.

Remark 6.1. The condition (71) defines a set the functions σ for which the condition (51) is necessary and sufficient. It turns out that the number $\frac{2-\lambda}{\lambda}$ is optimal. Indeed, let $\varepsilon > 0$, $\lambda \in \left(\frac{1}{1+\varepsilon}, 1\right)$ and $\alpha \in (1,2)$. Consider the differential equation (1) with

$$p(t) = \alpha(\alpha-1)\cdots(\alpha-n+1)t^{-n+\alpha(1-\mu(t)(\frac{2-\lambda}{\lambda}-\varepsilon))},$$

$$\sigma(t) = t^{\frac{2-\lambda}{\lambda}}, \quad t \geq 1, \quad \lim_{t\to+\infty} \mu(t) = \lambda.$$

It is obvious that the condition (51) is fulfilled and

$$\liminf_{t\to+\infty} \frac{\sigma(t)}{t^{\frac{2-\lambda}{\lambda}}} = 0 \quad \text{and} \quad \liminf_{t\to+\infty} \frac{\sigma(t)}{t^{\frac{2-\lambda}{\lambda}-\varepsilon}} > 0.$$

On the other hand, for even n, $u(t) = t^\alpha$ is a solution of Eq. (1). Therefore, when n is even, the Eq. (1) does not have Property **B**.

Theorem 6.2. *Let the conditions* (2), (3), *and* (6) *be fulfilled and*

$$\limsup_{t\to+\infty} \frac{\sigma(t)}{t^{\frac{1+\mu(t)}{2\mu(t)}}} < +\infty. \tag{72}$$

Then the condition (3.2_{n-1}) *is necessary and sufficient for the Eq.* (1) *to have Property **B**.*

Proof. Necessity. Assume that the Eq. (1) has Property **B** and

$$\int_0^{+\infty} (\sigma(t))^{(n-1)\mu(t)} |p(t)| dt < +\infty. \tag{73}$$

According to (73), by Lemma 4.1 [2] there exists $c \neq 0$ such that the Eq. (1) has a proper solution $u : [t_0, \infty) \to R$ satisfying the condition $\lim_{t\to+\infty} u^{(n-1)}(t) = c$. But this contradicts the fact that the Eq. (1) has Property **B**.

Sufficiency. By Eqs. (6) and (72), it is obvious that (61) holds. On the other hand, by Eqs. (3.2_{n-1}) and (72), the condition (65) holds. Thus, all conditions of Corollary 5.5 are fulfilled, i.e., the Eq. (1) has Property **B**. □

Analogously to Theorem 6.2, using the Corollaries 5.4′ and 5.6, we can prove Theorems 6.3–6.4.

Theorem 6.3. *Let the conditions* (2), (3), (6), (61), *and* (63) *be fulfilled. Then the condition* (15_{n-1}) *is necessary and sufficient for the Eq.* (1) *to have Property **B**.*

Theorem 6.4. *Let the conditions* (2), (3), (7), *and* (61) *be fulfilled. Then the condition* (15_{n-1}) *is necessary and sufficient for the Eq.* (1) *to have Property **B**.*

Acknowledgements The work was supported by Sh. Rustaveli National Science Foundation (Georgia). Grant No. GNSF/ST09-81-3-101.

References

1. Graef, J., Koplatadze, R., Kvinikadze, G.: Nonlinear functional differential equations with Properties A and B. J. Math. Anal. Appl. **306**, 136–160 (2005)
2. Koplatadze, R.: On oscillatory properties of solutions of functional differential equations. Mem. Differ. Equ. Math. Phys. **3**, 3–179 (1994)
3. Koplatadze, R.: On oscillatory properties of solutions of generalized Emden-Fowler type differential equations. Proc. A. Razmadze Math. Inst. **145**, 117–121 (2007)

4. Koplatadze, R.: Quasi-linear functional differential equations with Property A. J. Math. Anal. Appl. **330**, 483–510 (2007)
5. Koplatadze, R.: On Asymptotic behavior of solutions of almost linear and essentially nonlinear differential equations. Nonlinear Anal. Theory Methods Appl. **71**(12), 396–400 (2009)
6. Koplatadze, R.: On asymptotic behaviour of solutions of n-th order Emden-Fowler differential equations with advanced argument. Czeh. Math. J. **60**(135), 817–833 (2010)
7. Koplatadze, R., Litsyn, E.: Oscillation criteria for higher order "almost linear" functional differential equations. Funct. Differ. Equ. **16**(3), 387–434 (2009)

On Conforming Tetrahedralisations of Prismatic Partitions

Sergey Korotov and Michal Křížek

Abstract We present an algorithm for conform (face-to-face) subdividing prismatic partitions into tetrahedra. This algorithm can be used in the finite element calculations and analysis.

Keywords Finite element method • Prismatic element • Tetrahedral mesh • Linear elements

Mathematical Subject Classification: 65N50, 51M20

1 Introduction

Tetrahedral, prismatic, pyramidal, or block elements are usually used in finite element approximations of various engineering three-dimensional problems. Therefore, a natural question arises which of these elements are the most suitable for a particular problem in a given domain (cf. [1, 3, 8, 9]). The use of several types of elements enables us to compare the influence of the space discretisation on the finite element solution.

S. Korotov (✉)
BCAM – Basque Center for Applied Mathematics, Alameda de Mazarredo 14, E-48009 Bilbao, Basque Country, Spain

IKERBASQUE, Basque Foundation for Science, E–48011 Bilbao, Spain
e-mail: korotov@bcamath.org

M. Křížek
Institute of Mathematics, Academy of Sciences, Žitná 25, CZ–115 67 Prague 1, Czech Republic
e-mail: krizek@math.cas.cz

S. Pinelas et al. (eds.), *Differential and Difference Equations with Applications*, Springer Proceedings in Mathematics & Statistics 47, DOI 10.1007/978-1-4614-7333-6_5, © Springer Science+Business Media New York 2013

However, among various types of elements, tetrahedral elements are the most popular for many reasons. For instance, if block trilinear elements are employed, the discrete maximum principle need not be satisfied (see [5, 7]). But if each block element is divided into six nonobtuse tetrahedra such that all of them contain the spatial diagonal, then the stiffness matrix associated to linear elements has the same size, and the discrete maximum principle is fulfilled for a large class of nonlinear elliptic problems (see [4]). Moreover, the stiffness matrix associated with linear finite elements has less nonzero diagonals than that one for trilinear block or triangular prism elements. Another important reason for the use of tetrahedral elements is their flexibility to describe complicated boundaries.

In this work we show how to conformly (face-to-face) subdivide the given prismatic partition into tetrahedra as local subdivisions of prisms into tetrahedra cannot be done independently from each other in order to get a face-to-face tetrahedralisation.

2 Subdivision of Prismatic Partitions into Tetrahedra

By *prism* (or more precisely *triangular prism*) we shall mean a prism with two parallel triangular faces and three rectangular faces.

In what follows we consider only bounded polyhedra $\overline{\Omega} \subset R^3$ which can be decomposed into prisms. Let $\mathscr{P}_h$ be a face-to-face partition of one such polyhedron into prisms (see Fig. 1). Here h stands for the usual discretisation parameter, i.e., the maximum diameter of all prisms from $\mathscr{P}_h$.

Throughout this paper we will always consider only face-to-face partitions, and therefore, the notion "face to face" will be sometimes omitted. In Sect. 2, we assume that a partition $\mathscr{P}_h$ into it is evident from the definition of a prism that any partition $\mathscr{P}_h$ into prisms consists of parallel layers of prisms. In the following we may suppose these layers to be horizontal and we call the bottom plane of a layer its base and the triangular face of a prism we call the base triangle of the prism. A more general situation is treated in Sect. 3.

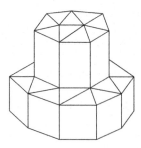

Fig. 1 Partition into prisms

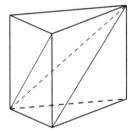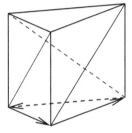

Fig. 2 Two subdivisions of rectangular faces of a prism

We shall consider such tetrahedralisations of $\mathscr{P}_h$ that the triangular faces of the prisms are not cut. Hence, the different layers of prisms can be subdivided independently into tetrahedra, and these altogether provide us with a conforming tetrahedral mesh over $\overline{\Omega}$.

We shall subdivide each prism into three tetrahedra as marked in Fig. 2 (left). We see that its rectangular faces are divided by diagonals into triangles and these diagonals determine three tetrahedra in the subdivision. However, these diagonals cannot be chosen arbitrarily. In Fig. 2 (right) we observe a division of three rectangular faces of a prism which does not correspond to any partition of the prism into tetrahedra. Therefore, we have to divide rectangular faces in the whole partition carefully.

In the next theorem we show how to practically construct from a given prismatic partition $\mathscr{P}_h$ a face-to-face tetrahedralisation, thus avoiding the situation illustrated in Fig. 2 (right) (or its mirror image) when dividing rectangular faces by diagonals.

Theorem 2.1. *For any conforming partition into prisms there exists a face-to-face subdivision into tetrahedra.*

Proof. From the beginning of this section we know that any partition of $\mathscr{P}_h$ into prisms consists of parallel layers which can be tetrahedralised independently (see Fig. 1). Consider one of such layers supposed to be horizontal and let $\mathscr{T}_h$ be the triangulation of its base corresponding to the partition $\mathscr{P}_h$. Take an arbitrary vector $\vec{v} \neq 0$ in the plane containing the triangulation $\mathscr{T}_h$ (for instance $\vec{v} = (1,0,0)$). Now we define the orientation $\vec{e}$ of each edge e of the triangulation $\mathscr{T}_h$ such that

$$(\vec{v}, \vec{e}) \geq 0. \tag{1}$$

If an edge e is perpendicular to $\vec{v}$, we may take an arbitrary orientation of $\vec{e}$. In this way we get the planar digraph $G_h = (N, E)$, where N is the set of nodes and E is the set of the directed edges.

It is clear that G_h does not contain a directed circle of which edges form a triangle of $\mathscr{T}_h$ (cf. Fig. 2). Indeed, if, on the contrary, $\vec{e}_1, \vec{e}_2, \vec{e}_3$ form a circle then $\vec{e}_1 + \vec{e}_2 + \vec{e}_3 = \vec{0}$. Taking the scalar product of both sides by $\vec{v}$ and using (1), we get that $\vec{v}$ is perpendicular to the triangle with side vectors $\vec{e}_1, \vec{e}_2, \vec{e}_3$, which is a contradiction.

□

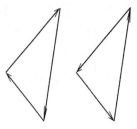

Fig. 3 Non-allowed edge orientations

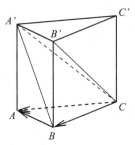

Fig. 4 A possible orientation of edges of a triangular base that leads to a partition into three tetrahedra

The algorithm can be summarised as follows:

1. Orient all edges in the triangulation of the base according to formula (1).
2. Subdivide all vertical rectangular faces in the direction defined by the orientation of edges of the base as indicated in Fig. 4.

Remark 2.1. Non-allowed edge orientations are sketched in Fig. 3. In Fig. 4 we see a partition of the prism $ABCA'B'C'$ into three tetrahedra different from that in the left part of Fig. 2.

We demonstrate now that the family of tetrahedral meshes generated as above is regular whenever the family of original prismatic meshes is regular as well.

Definition 2.1. A family of partitions $\mathscr{F} = \{\mathscr{T}_h\}_{h\to 0}$ of a polyhedron $\overline{\Omega}$ into convex elements is said to be *regular* (*strongly regular*) if there exists a constant $\kappa > 0$ such that for any partition $\mathscr{T}_h \in \mathscr{F}$ and any element $T \in \mathscr{T}_h$ we have

$$\kappa h_T^3 \leq \text{meas } T \quad (\kappa h^3 \leq \text{meas } T), \tag{2}$$

where $h_T = \text{diam } T$.

Remark 2.2. The above definition is equivalent to the inscribed ball condition (see, e.g., [2, Sect. 16]) which is more complicated. Note also that the regularity of a family of partitions into prisms is equivalent to Zlámal's minimum angle condition [2, p. 128] applied to triangular bases of all prisms provided the height of all prisms is proportional to h.

Theorem 2.2. *If a family of partitions $\{\mathcal{P}_h\}_{h\to 0}$ of a polyhedron $\overline{\Omega}$ into prisms is regular (strongly regular), then the associated family of partitions $\{\mathcal{T}_h\}_{h\to 0}$ into tetrahedra is also regular (strongly regular).*

Proof. It is evident that the volume of each of the three tetrahedra from Fig. 2 (left) is equal to the one third of the volume of the prism. Therefore, if inequalities (2) hold for prisms, then the same relations hold also for tetrahedra with another constant $\kappa' = \kappa/3$.   □

Remark 2.3. Assume that a family of partitions of $\overline{\Omega}$ into prisms is regular and that Ω has Lipschitz boundary. Then by Theorem 2.2 the optimal interpolation properties of tetrahedral elements in Sobolev space norms are satisfied.

3 Polyhedral Domains that Have No Lipschitz Boundary

In practice we meet sometimes polyhedral domains which do not have Lipschitz boundary in the sense of Nečas [10, p. 17].

Let us point out that domains with Lipschitz boundaries often stand as an important assumption in a number of useful theorems, such as various imbedding and density theorems, trace theorem, and Poincaré-Friedrichs' theorem. Many authors then apply these theorems for polyhedral domains assuming (incorrectly) that any polyhedral domain has Lipschitz boundary.

In the left part of Fig. 5 we observe a polyhedral domain whose boundary is not Lipschitz (see [6, p. 48]) near points marked by black dots. In any neighbourhood of these points the boundary is not a graph of a function. It can be expressed only by means of a multivalued function in any coordinate system (whereas any Lipschitz function is one-valued). Note that the polyhedral domain of Fig. 5 satisfies the classical external (and also internal) cone condition.

The right part of Fig. 5 shows its partition into rectangular blocks and their subdivision into triangular prisms. We see that this partition is somewhat different from that described in the previous Sect. 2, and the subdivision into tetrahedra has to be done carefully on the intersection of the two bars (see Theorem 2.1).

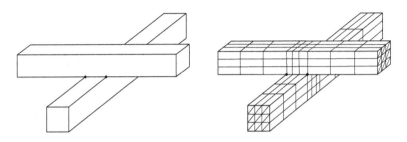

Fig. 5 Polyhedral domain whose boundary is not Lipschitz and its partition into triangular prisms

Acknowledgements This work was supported by Grant MTM2011-24766 of the MICINN, Spain, and the Grant no. IAA 100190803 of the Grant Agency of the Academy of Sciences of the Czech Republic. The authors are indebted to A. and Z. Horváth for fruitful discussions.

References

1. Apel, T., Düvelmeyer, N.: Transformation of hexahedral finite element meshes into tetrahedral meshes according to quality criteria. Preprint SFB393/03-09. Tech. Univ. Chemnitz, pp. 1–12 (2003)
2. Ciarlet, P.G.: Basic error estimates for elliptic problems. In: Ciarlet, P., Lions, G.J.L. (eds.) Handbook of Numerical Analysis, vol. II. North-Holland, Amsterdam (1991)
3. Hannukainen, A., Korotov, S., Vejchodský, T.: Discrete maximum principle for FE solutions of the diffusion-reaction problem on prismatic meshes. J. Comput. Appl. Math. **226**, 275–287 (2009)
4. Karátson, J., Korotov, S., Křížek, M.: On discrete maximum principles for nonlinear elliptic problems. Math. Comp. Simulation **76**, 99–108 (2007)
5. Korotov, S., Vejchodský, T.: A comparison of simplicial and block finite elements. In: Kreiss, G., et al. (eds.) Proceedings Eighth European Conference on Numerical Mathematics and Advanced Applications (ENUMATH2009), Uppsala, Sweden, pp. 531–540. Springer, Heidelberg (2010)
6. Křížek, M.: An equilibrium finite element method in three-dimensional elasticity. Apl. Mat. **27**, 46–75 (1982) See also www.dml.cz
7. Křížek, M., Lin, Q.: On diagonal dominance of stiffness matrices in 3D. East-West J. Numer. Math. **3**, 59–69 (1995)
8. Křížek, M., Neittaanmäki, P.: Finite element approximation of variational problems and applications. In: Pitman Monographs and Surveys in Pure and Applied Mathematics, vol. 50, Longman Scientific & Technical, Harlow (1990)
9. Liu, L., Davies, K.B., Yuan, K., Křížek, M.: On symmetric pyramidal finite elements. Dyn. Contin. Discrete Impuls. Syst. Ser. B Appl. Algorithms **11**, 213–227 (2004)
10. Nečas, J., Hlaváček, I.: Mathematical Theory of Elastic and Elasto-Plastic Bodies: An Introduction. Elsevier, Amsterdam (1981)

Swarm Dynamics and Positive Dynamical Systems

Ulrich Krause

Abstract In this paper we model swarm dynamics by a nonautonomous linear system with row-stochastic matrices. It is shown that all birds approach the same velocity if their local interaction has the property that its intensity decays not too fast and its structure becomes not too loose. This mirrors in particular specific flight regimes. The results obtained employ tools from positive discrete dynamical systems.

1 Introduction

It is fascinating to watch swarms of starlings or formations of geese up in the sky. How are birds able to coordinate each other in huge flocks or in orderly formations? (For principles of organization among birds as well as for its biological roots see the review [14]. For recent empirical research on swarm formations see [1, 15]). Those questions were asked already 2000 years ago and are investigated today in the field of swarm dynamics by means of mathematical models. But swarm dynamics is not only about birds. In recent years researchers across different disciplines—including biologists, computer scientists, engineers, physicists, and mathematicians—developed models of astonishing similarity to understand beside swarms of birds or schools of fish also opinion formation or traffic jams among people as well as distributed computing in networks or self-organizing groups of robots. (See references [3, 5–7, 17, 18] on birds, [9, 13] on opinion formation, and [2, 11, 16] on robots.) Taken abstractly, the system under consideration consists of autonomous agents without any central direction but using simple rules of interaction which are local in the sense that each agent interacts only with a small

U. Krause
University of Bremen, Bremen, Germany
e-mail: krause@math.uni-bremen.de

S. Pinelas et al. (eds.), *Differential and Difference Equations with Applications*, Springer Proceedings in Mathematics & Statistics 47, DOI 10.1007/978-1-4614-7333-6_6, © Springer Science+Business Media New York 2013

set of "neighbors." One main question on the dynamics of such a multi-agent system is to find conditions under which the system is able to approach a stable pattern, in particular a consensus or agreement among agents on a certain issue like velocity or opinions.

In this paper Sect. 2 sets out a simple model of swarm dynamics in discrete time driven by the simple rule that each agent (bird) at each time acts by taking a convex combination of the actions of a few agents. Equivalently, the model can be described as a nonautonomous linear system with row-stochastic matrices. Section 3 provides results from positive dynamical systems which will be used later on to analyze to swarm model set out. The main result used guarantees the convergence to consensus if the intensity of interaction decays not "too fast" and the structure of interaction becomes not "too loose." Broadly speaking, a positive dynamical system in discrete time and finite dimensions is about nonnegative matrices and the stability behavior of infinite products of those. (See the monographs [8, 19].) More general, a positive dynamical system in discrete time deals with linear or nonlinear self-mappings of some convex cone in a Banach space and the stability of (inhomogeneous) iterations of such mappings. Using tools from Sect. 3, in Sect. 4 a first result is obtained on the swarm formation by local interaction. This result allows to consider a modified local version of Cucker–Smale flocking which has been much studied in recent years. Section 5 presents our second result on swarm formation which allows for different regimes of local interaction. Of particular interest here is how the changing flight formations of birds have to look like for getting still a swarm altogether. Thereby a useful tool from positive dynamical systems is so-called Sarymsakov matrices and their properties.

The author thanks a referee for helpful comments.

2 A Model of Swarm Dynamics

Consider a number of n birds moving in three-dimensional space and let $x^i(t)$ be the position of bird i, for $i = 1,\ldots,n$, at time t. Assuming discrete time, that is, $t \in \mathbb{N} = \{0,1,2,\ldots\}$, the velocity $v^i(t) \in \mathbb{R}^3$ of bird i at time is given by $v^i(t) = x^i(t+1) - x^i(t)$. As a central feature of a swarm we consider the phenomenon that the velocities of the birds come close to each other. How do the birds achieve this? What kind of self-organization leads to a "consensus on velocities"?

The computer scientist Reynolds [17] detected by computer simulations that artificial birds which he called "boids" show the flocking behavior of natural birds provided three rules of interaction are satisfied. One of his by now famous rules is that of "alignment" which for a boid means to "steer towards the average heading of local flockmates." This rule we model as taking a convex combination of velocities, that is

$$v^i(t+1) \in \text{conv } \{v^1(t),\ldots,v^n(t)\} \tag{1}$$

for each bird $i \in \{1, \ldots, n\}$ and each $t \in \mathbb{N}$, where conv M denotes the convex hull of a set M. Denoting the coefficients for the convex combination in (1) by $a_{ij}(t)$ for $j = 1, \ldots, n$, we can express rule (1) by $v^i(t+1) = \sum_{j=1}^{n} a_{ij}(t)v^j(t)$ where $0 \le a_{ij}(t)$ and $\sum_{j=1}^{n} a_{ij}(t) = 1$ for all i. Thus (1) is equivalent to

$$v(t+1) = A(t)v(t) \quad for \quad t = 0, 1, 2, \ldots \tag{2}$$

with $v(t)$ being the column $v(t) = (v^1(t), \ldots, v^n(t))'$ and $A(t)$ the row-stochastic matrix $(a_{ij}(t))_{1 \le i, j \le n}$. Dealing with birds, $\mathbb{R}^3$ is the natural space for the states $v(t)$, but considering other types of agents, a state space $\mathbb{R}^d$ with $d \ge 1$ can be appropriate. The model (2), or (1), makes still sense. (Actually, for (1) to make sense the state space is only required to be a so-called mixture space, that is, a space which allows to form convex combinations of an abstract kind.) The "local flockmates" in Reynold's rule means that the convex combination in (1) involves only a subset $\emptyset \subsetneq S(i,t) \subseteq \{1, \ldots, n\}$. Equivalently, for the ith row of $A(t)$, one has $a_{ij}(t) > 0$ if and only if $j \in S(i,t)$. The set $S(i,t)$ can be interpreted as the set of birds with which bird i interacts at time t, for example, in that bird i "sees" at time t all birds in $S(i,t)$. The central feature of velocities coming closer to each other we model more pointed as velocities approaching a "consensus" in the course of time, that is

$$\lim_{t \to \infty} v^i(t) = v_* \quad \text{for all} \quad i \in \{1, \ldots, n\} \tag{3}$$

for some $v_* \in \mathbb{R}^3$ and with respect to Euclidean topology. The model given by (2) is a positive dynamical system in discrete time which is nonautonomous or time variant. In what follows we will supply conditions for property (3) to hold in terms of the structure and the intensity of interaction embodied in the matrices $A(t)$. For doing this we turn in the next section to tools from positive dynamical systems used in opinion dynamics to address a consensus problem like (3).

3 Positive Dynamical Systems: Convergence to Consensus

Let a nonautonomous positive dynamical system

$$x^i(t+1) = \sum_{j=1}^{n} a_{ij}(t)x^j(t), \ 1 \le i \le n, t \in \mathbb{N} \tag{4}$$

be given with $x^i(t) \in \mathbb{R}^d, 0 \le a_{ij}(t), \sum_{j=1}^{n} a_{ij}(t) = 1$.

Consider the following assumptions for system (4). For an infinite time sequence $t_0 < t_1 < t_2 < \ldots$ in $\mathbb{N}$ such that for some $r, N \in \mathbb{N} \setminus \{0\}$

$$1 \le t_{m+1} - t_m \le r \quad \text{for all} \quad m \ge N,$$

the following assumptions do hold:

(i) There exists a nonincreasing function $\alpha : \mathbb{R}_+ \to]0,1]$ such that for $A(t) = (a_{ij}(t))$

$$A(t) \geq \alpha(t)I \quad \text{for all} \quad t \geq t_N$$

(I the $n \times n$-identity matrix, $\mathbb{R}_+ = \{r \in \mathbb{R} \mid r \geq 0\},]0,1] = \{r \in \mathbb{R} \mid 0 < r \leq 1\}$).
(ii) There exist finitely many nonnegative $n \times n$-matrices $A_i, 1 \leq i \leq k$, and a mapping $\sigma : \mathbb{N} \to \{1,\ldots,k\}$ such that

$$A(t_m) \geq \alpha(t_m)A_{\sigma(m)} \quad \text{for all} \quad m \in \mathbb{N}.$$

The following result is proved in [13].

Theorem 1. *Suppose assumptions (i) and (ii) are satisfied and there exists $p \in \mathbb{N} \setminus \{0\}$ such that*
the intensity of interaction *decays not "too fast," that is,*

$$\int_1^\infty \alpha(t)^{pr} dt = \infty,$$

and the structure of interaction *becomes not "too loose," that is, for each $m \in \mathbb{N}$ the product $A_{\sigma(m+p)}A_{\sigma(m+p-1)} \cdots A_{\sigma(m+1)}$ is scrambling.*
Then system (4) approaches consensus, that is,

$$\lim_{t \to \infty} x^i(t) = x_* \quad \text{for all} \quad 1 \leq i \leq n \quad \text{with consensus} \quad x^* \in \mathbb{R}^d.$$

Furthermore, for two vectors of initial conditions $x(0), y(0)$ leading to consensus x_* and y_*, respectively, one has that

$$\|x_* - y_*\| \leq \max\{\|x^i(0) - y^j(0)\| \mid 1 \leq i, j \leq n\}$$

for any norm $\| \cdot \|$ on $\mathbb{R}^d$.

Thereby, a nonnegative matrix A is called *scrambling* if for any two rows i and j there exists a column k such that $a_{ik} > 0$ and $a_{jk} > 0$ or, equivalently, AA' is a (strictly) positive matrix (cf. [8,19]). For the special case where $t_m = m$ for all $m \in \mathbb{N}$ and $k = 1$, from Theorem 1 one obtains the following corollary (see also [12]).

Corollary 1. *Suppose for system (4) that $A(t) \geq \alpha(t)A$ for all $t \in \mathbb{N}$ and a nonincreasing function $\alpha : \mathbb{R}_+ \to]0,1]$. If for some $p \in \mathbb{N} \setminus \{0\}$ one has that $\int_1^\infty \alpha(t)^p dt = \infty$ and A is a nonnegative matrix with positive diagonal for which the power A^p is scrambling, then system (4) approaches consensus and the sensitivity property as in Theorem 1 holds.*

For system (4) to approach a consensus the assumptions on the intensity and structure of interaction are by no means necessary. There exist, however, examples for which one of the assumptions is not satisfied and system (4) does not approach a consensus (see [13]). The reader who wants to know more about the convergence to consensus in opinion dynamics may consult [9].

4 Swarm Formation by Local Interaction

We now come back to the swarm model (2) of Sect. 2 given by $v(t+1) = A(t)v(t)$ for $t \in \mathbb{N}$. Let $S(i,t) = \{j \mid a_{ij}(t) > 0\}$ be the set of *birds seen by i at time t*. From Corollary 1 of the previous section we obtain the following result on swarm formation.

Theorem 2. *Suppose each bird i sees always a set $S(i)$ of birds (including itself), that is, $i \in S(i) \subseteq S(i,t)$ for all t and all i. Assume further of any two birds i and j one sees the other or both see a third one, that is, $S(i) \cap S(j) \neq \emptyset$ for all $i \leq i, j \leq n$. For the minimal interaction of birds at $t \in \mathbb{R}_+$, that is,*

$$\alpha(t) = \min\{a_{ij}(s) \mid s \in \mathbb{N}, s \leq t, 1 \leq i \leq n, j \in S(i)\},$$

assume that it is not weak in the sense that $\int_1^\infty \alpha(t)dt = \infty$.
Then all birds approach a common velocity $v_\star$, that is $\lim_{t\to\infty} v^i(t) = v_\star$ for all $1 \leq i \leq n$.

Furthermore, if the initial velocities of all birds are close to each other, the corresponding common velocities are close, too.

Proof. Define $A = (a_{ij})$ by $a_{ij} = 1$ for $j \in S(i)$ and $a_{ij} = 0$ otherwise. A has a positive diagonal because of $i \in S(i)$ and A is scrambled because of $S(i) \cap (S(j) \neq \emptyset$. By the definition of $\alpha(t)$ we have that $A(t) \geq \alpha(t)A$ for all t. Obviously, $\alpha(t)$ is not increasing in t and, by definition of $S(i)$, we have that $\alpha(t) \in]0, 1]$. The assumptions of Corollary 1 being satisfied for $p = 1$ the conclusion of Theorem 2 follows. □

We relate Theorem 2 to different models from the literature. In [20] a model of self-driven particles in the plane is investigated by computer simulations. The particles are driven with a constant absolute velocity and average the direction of motion with particles in some neighborhood. Numerical evidence is given that under a small noise all particles line up in the same direction. This model has been analytically treated in [10, 11] where, in a more general context, conditions are supplied to guarantee convergence to a consensus. (For an even more general approach see [16].) In these models the structure of interaction is assumed to be symmetric that is $a_{ij}(t) > 0$ iff $a_{ji}(t) > 0$, and the intensity of interaction is assumed to be strictly positive bounded from below. Starting from [20], recently F. Cucker and S. Smale developed a model which received much attention and to which we now turn in more detail. (See the original articles [5, 6] as well as the subsequent

articles [3, 4, 7, 18].) The discrete time version of so-called *Cucker–Smale flocking*
is given, beside

$$v^i(t) = x^i(t+1) - x^i(t), \text{ by}$$

$$v^i(t+1) - v^i(t) = \sum_{i \neq j = 1}^{n} f_{ij}(x(t))(v^j(t) - v^i(t)) \tag{5}$$

with intensities for $j \neq i$

$$f_{ij}(x) = \frac{H}{(1 + \|x^i - x^j\|^2)^\beta} \quad \text{with constants} \quad H > 0, \beta \geq 0.$$

In this model the intensity of interaction is not bounded from below by a positive
constant and it may decay to zero.

Furthermore, each bird changes his velocity by a weighted sum of the differences
of its velocity with those of other birds. The weights decrease in a particular manner
with the distance between birds. Obviously, in this model interaction among birds
is symmetric and the interaction is global in that $f_{ij}(x) > 0$ for all $i \neq j$. We shall
modify model (5) by restricting on the one hand constant H to $H \leq \frac{1}{n}$ but allowing
on the other hand local interaction, that is,

$$v^i(t+1) - v^i(t) = \sum_{j \in N(i)} f_{ij}(x(t))(v^j(t) - v^i(t)), \tag{6}$$

where $N(i)$ is a nonempty subset of $\{1, \ldots, n\}$ which does not contain i. From
Theorem 2 we obtain the following result.

Corollary 2 (modified Cucker–Smale flocking). *Assume for model (6) that $i \in N(j)$ or $j \in N(i)$ or $N(i) \cap N(j) \neq \emptyset$ for any two birds i and j. Then $\lim_{t \to \infty} v^i(t) = v_\star$
for all $1 \leq i \leq n$ in case $\beta \leq \frac{1}{2}$ or the swarm remains bounded, that is, $\|x^i(t)\| \leq c$
for some constant $c > 0$ and all i, all t.*

Proof. Define a matrix $A(t) = (a_{ij}(t))$ by

$$a_{ij}(t) = \begin{cases} f_{ij}(x(t)) & \text{for } j \in N(i) \\ 1 - \sum_{k \in N(i)} f_{ik}(x(t)) & \text{for } j = i \\ 0 & \text{otherwise.} \end{cases}$$

From the definition of $f_{ij}(x)$ together with assumption $H \leq \frac{1}{n}$, we have that

$$\sum_{k \neq i} f_{jk}(x) \leq (n-1)H \leq \frac{n-1}{n} < 1 \quad \text{and, hence,}$$

$$a_{ii}(t) = 1 - \sum_{k \in N(i)} f_{ik}(x(t)) \geq 1 - \sum_{k \neq i} f_{ik}(x(t)) \geq 1 - \frac{n-1}{n} = \frac{1}{n}.$$

Thus, $A(t)$ is a nonnegative matrix such that

$$\sum_{j=1}^{n} a_{ij}(t) = \sum_{j \in N(i)} f_{ij}(t) + a_{ii}(t) = 1,$$

that is, $A(t)$ is row stochastic for all t.

Furthermore, from (6) it follows

$$v^i(t+1) = \sum_{j \in N(i)} f_{ij}(x(t))v^j(t) + (1 - \sum_{j \in N(i)} f_{ij}(x(t)))v^i(t)$$

and, hence,

$$v^i(t+1) = \sum_{j \in N(i)} a_{ij}(t)v^j(t) + a_{ii}(t)v^i(t) = \sum_{j=1}^{n} a_{ij}(t)v^j(t).$$

Thus system (6) is a swarm model of type (2).

Define $S(i) = N(i) \cup \{i\}$. To apply Theorem 2 it remains to show for $t \in \mathbb{R}_+$ and

$$\alpha(t) = \min\{a_{ij}(s) \mid s \in \mathbb{N}, s \le t, 1 \le i \le n, j \in N(i) \cup \{i\}\}$$

that $\int_1^\infty \alpha(t)dt = \infty$.

In case, the swarm remains bounded, for $i \ne j$

$$f_{ij}(s) \ge \frac{H}{(1 + (2c)^2)^\beta} \qquad \text{for all } s \text{ and, because of} \qquad a_{ii}(s) \ge \frac{1}{n},$$

it follows

$$\alpha(t) \ge \min\left\{\frac{H}{(1 + (2c)^2)^\beta}, \frac{1}{n}\right\}.$$

Therefore $\int_1^\infty \alpha(t)dt = \infty$ for a bounded swarm.

For the case $\beta \le \frac{1}{2}$, from $v(t+1) = A(t)v(t)$ with $A(t)$ row stochastic we have that $\|v^i(t+1)\| \le \max_{1 \le j \le n} \|v^j(t)\|$ for all i. For the max-norm therefore $\|v(t+1)\| \le \|v(t)\|$ for all t. From $x^i(t+1) - x^i(t) = v^i(t)$ it follows that

$$x^i(t) = x^i(0) + \sum_{s=0}^{t-1} v^i(s) \text{ and, hence, } \|x^i(t)\| \le \|x^i(0)\| + \|v(0)\|t.$$

Therefore, $\|x^i(t) - x^j(t)\| \le \|x^i(0)\| + \|x^j(0)\| + (\|v(0)\| + \|v(0)\|)t$, and there exist t_0 and $c > 0$ such that

$$1 + \|x^i(s) - x^j(s)\|^2 \le ct^2 \qquad \text{for all} \qquad t \ge t_0, s.$$

Thus, $\alpha(t) \geq \min\{\frac{H}{(ct^2)^\beta}, \frac{1}{n}\}$ for $t \geq t_0$. Since $\int_{t_0}^\infty \frac{dt}{t^{2\beta}} = \infty$ for $\beta \leq \frac{1}{2}$ this yields $\int_1^\infty \alpha(t)dt = \infty$ which proves the corollary. $\Box$

Corollary 2 shows, provided $H \leq \frac{1}{n}$, that for Cucker–Smale flocking the global interaction can be weakened to local interaction. We may simply chose, for example, in Corollary 2 $N(i) = \{k\}$ for some k and all i. This means that all birds adjust to some leading bird k. In that case interaction is neither symmetric nor global. There are other nonsymmetric, non-global cases which correspond to certain flight formations. In [5, 6] Cucker and Smale use different methods avoiding the assumption $H \leq \frac{1}{n}$ and which enable them to show for model (5) convergence of velocities, and positions too, for $\beta > \frac{1}{2}$ provided certain relations between initial velocities and positions are met. For continuous time Cucker–Smale flocking for $H \leq \frac{1}{n}$ but without local interaction is investigated in [3] and a result as in Corollary 2 for $\beta \leq \frac{1}{2}$ is proven.

5   Swarm Formation and Flight Regimes for Time-Dependent Local Interaction

In Theorem 2 of the previous section, we assumed that the set $S(i)$ of birds seen by bird i does not change with time. Now we want to admit that these sets which mirror the local interaction of the birds can depend on time. To do so we have to employ instead of Corollary 1 the stronger Theorem 1. Swarm formation then is possible in different flight regimes which we describe by using the Sarymsakov matrices introduced by Hartfiel [8].

Let A be a nonnegative $n \times n$-matrix, M a subset of $\{1,\ldots,n\}$ and $F(M) = \{1 \leq j \leq n \mid a_{ij} > 0$ for some $i \in M\}$.

Definition 1 ([8]). A is a *Sarymsakov matrix* (or *S*-matrix, for short) if for any two non-empty subsets M and M' of $\{1,\ldots,n\}$ from $M \cap M' = \emptyset$ and $F(M) \cap F(M') = \emptyset$ it follows that

$$|M \cup M'| < |F(M) \cup F(M')|,$$

where $|\cdot|$ denotes the number of elements of a finite set.

A weaker notion than that of an *S*-matrix is the following one which is easier to interpret and which in the context of swarms turns out to be equivalent to an *S*-matrix.

Definition 2. A nonempty subset M of $\{1,\ldots,n\}$ is *saturated* if $F(M) \subseteq M$ and a $n \times n$-matrix A is *coherent* if any two saturated subsets have a nonempty intersection.

The following lemma presents relationships among the concepts introduced.

Lemma 1. *Consider the following properties for a nonnegative $n \times n$-matrix A:*

1. *Scrambling.*
2. *S-matrix.*
3. *Some power is scrambling.*
4. *Coherent.*

(i) *The following implications hold:*

$$1 \Rightarrow (2) \Rightarrow (3) \Rightarrow (4).$$

(ii) *None of these implications can be reversed (not even for row-stochastic matrices).*

(iii) *If A has a positive diagonal, then the properties (2), (3), and (4) are equivalent.*

Proof.

(i) Suppose A is scrambling and M, M' are two nonempty subsets of $\{1, \ldots, n\}$. To $i \in M, j \in M'$ there exists k such that $a_{ik} > 0$ and $a_{jk} > 0$. Thus, $k \in F(M) \cap F(M')$. This proves $(1) \Rightarrow (2)$. The implication $(2) \Rightarrow (3)$ is a special case of the following theorem of Hartfiel (Theorem 4.8 in [8]): The product of $n - 1$ Sarymsakov matrices (of order n) is scrambling. To see $(3) \Rightarrow (4)$ suppose A^p is scrambling and M, M' are two saturated subsets of $\{1, \ldots, n\}$. For $i \in M, j \in M'$ there exists k such that $a_{ik}^{(p)} > 0$ and $a_{jk}^{(p)} > 0$. Therefore there exist sequences $(i_1, \ldots, i_{p-1})$ and $(j_1, \ldots, j_{p-1})$ such that

$$a_{ii_1} \cdot a_{i_1 i_2} \ldots a_{i_{p-1},k} > 0 \quad \text{and} \quad a_{jj_1} \cdot a_{j_1 j_2} \ldots a_{j_{p-1},k} > 0.$$

Since M is saturated, it follows inductively that $i_1 \in M, i_2 \in M, \ldots, i_{p-1} \in M, k \in M$. Similarly, $k \in M'$ and, hence, $k \in M \cap M'$. This shows that A is coherent.

(ii) To see that the implication $(1) \Rightarrow (2)$ cannot be reversed in general, consider the row-stochastic matrix

$$A = \begin{bmatrix} 1 & 0 & 0 \\ \frac{1}{2} & \frac{1}{2} & 0 \\ 0 & \frac{1}{2} & \frac{1}{2} \end{bmatrix}.$$

A is not scrambled but it is an S-matrix. For nonempty sets M, M', one has $F(M) \cap F(M') \neq \emptyset$ with the exception $M = \{1\}, M' = \{3\}$. For the latter case $F(M) = M$ and $F(M') = \{2, 3\}$ and, hence, $|F(M) \cup F(M')| = |\{1, 2, 3,\}| > |M \cup M'|$.

Concerning the implication (2) ⇒ (3), consider

$$A = \begin{bmatrix} 1 & 0 & 0 \\ 1 & 0 & 0 \\ 0 & 1 & 0 \end{bmatrix}. \quad \text{Since} \quad A^2 = \begin{bmatrix} 1 & 0 & 0 \\ 1 & 0 & 0 \\ 1 & 0 & 0 \end{bmatrix}$$

this power is scrambling. But A is not an S-matrix since for $M = \{3\}, M' = \{1,2\}$ one has that $F(M) \cap F(M') = \emptyset$ but $|F(M) \cup F(M')| = |\{1,2\}| < |M \cup M'|$.

Finally, concerning (3) ⇒ (4) let $A = \begin{bmatrix} 0 & 1 \\ 1 & 0 \end{bmatrix}$. Since $M = \{1,2\}$ is the only saturated set matrix A must be coherent. But, of course, no power of A is scrambling.

(iii) Let A be nonnegative with a positive diagonal. We show that (4) implies (2). Let M, M' be nonempty subsets of $\{1,\ldots,n\}$ such that $M \cap M' = \emptyset$ and $F(M) \cap F(M') = \emptyset$. Because of the positive diagonal $M \subseteq F(M)$ and $M' \subseteq F(M')$. Suppose we had equality, $M = F(M)$ and $M' = F(M')$. Then M and M' are saturated and must have by (4) a nonempty intersection—which is a contradiction. Therefore, $M \cup M' \subsetneq F(M) \cup F(M')$ which implies that $|M \cup M'| < |F(M) \cup F(M')|$ Thus, A must be an S-matrix. □

Remark 1. In (iii) not all four properties need be equivalent as the first counter-example in (ii) shows. Furthermore, the equivalences in (iii) need not hold in case some but not all elements in the diagonal are positive. This is illustrated by

$$A = \begin{bmatrix} 1 & 0 & 0 \\ 1 & 0 & 0 \\ 0 & \frac{1}{2} & \frac{1}{2} \end{bmatrix}$$

which is coherent but not an S-matrix.

For the swarm model given by (2) consider now time-dependent local interaction as described by the set $S(i,t) = \{1 \leq j \leq n \mid a_{ij}(t) > 0\}$ of birds seen by bird i at time t.

Definition 3. The birds are said to form *a coherent flight regime at t* if the matrix $A(t)$ is coherent. Equivalently, if M is a (nonempty) subset of birds which contains with each bird also all birds seen by it and M' is another set of this kind, then $M \cap M' \neq \emptyset$.

From Theorem 1 and Lemma 1 we obtain the following result on time-dependent local interaction in swarm formation.

Theorem 3. *Suppose $i \in S(i,t)$ for all i, all t and suppose there is a time sequence $t_0 < t_1 < \ldots$ in $\mathbb{N}$ with $r = \sup_m(t_{m+1} - t_m) < \infty$ such that the flight regime at each t_m is coherent. Assume further for the minimal interaction $\alpha(t)$ of birds at $t \in \mathbb{R}_+$ defined as the minimum of $\min\{a_{ij}(t_m) \mid t_m \leq t, 1 \leq i \leq n, j \in S(i,t_m)\}$ and $\min\{a_{ii}(s) \mid 1 \leq i \leq n, s \in \mathbb{N} \setminus \{t_0, t_1, \ldots\}, s \leq t\}$ that it is not weak in the sense that $\int_1^\infty \alpha(t)^{(n-1)r} dt = \infty$.*

Then all birds approach a common velocity v_*, that is, $\lim_{t \to \infty} v^i(t) = v_*$ for all $1 \le i \le n$. Furthermore, if the initial velocities of all birds are close to each other, the corresponding common velocities are close, too.

Proof. By definition of $\alpha(t)$ for $t \in \mathbb{R}_+$ the function $\alpha(\cdot)$ is not increasing and $\alpha(t) \in \,]0,1]$ by definition of $S(i,s)$ and assumption $i \in S(i,s)$ for $s \in \mathbb{N}$. Define for $m \in \mathbb{N}$ a nonnegative matrix $A_{\sigma(m)}$ by $(A_{\sigma(m)})_{ij} = 1$ for $j \in S(i,t_m)$ and $(A_{\sigma(m)})_{ij} = 0$ otherwise. Since the flight regime at t_m is assumed to be coherent and $i \in S(i,t_m)$, the matrix $A_{\sigma(m)}$ is coherent and has a positive diagonal. There are only finitely many different matrices $A_{\sigma(m)}$ for $m \in \mathbb{N}$. By Lemma 1 (iii) each of these matrices is an S-matrix. By Hartfiel's theorem (Theorem 4.8 in [8]) the product of $(n-1)$ S-matrices of order n is scrambling. Thus, for $p = n - 1$ and any $m \in \mathbb{N}$ the product $A_{\sigma(m+p)} \cdots A_{\sigma(m+1)}$ is a scrambling matrix. Finally, by definition of $\alpha(t)$ we have that $A(t) \ge \alpha(t)I$ for all $t \in \mathbb{N}$ and $A(t_m) \ge \alpha(t_m)A_{\sigma(m)}$ for all $m \in \mathbb{N}$ and $\int_1^\infty \alpha(t)^{pr}dt = \infty$ for $p = n - 1$ by assumption. All the assumptions being satisfied Theorem 1 yields the conclusion wanted. $\square$

It is a particular feature of Theorem 3 that for the birds to approach a common velocity coordination as a coherent flight regime is required only from time to time at periods t_m. In between the birds need only to coordinate in the weak sense that the matrix of the flight regime has a positive diagonal. A second interesting feature of Theorem 3 is that the coherent flight regimes at special periods t_m can change with time as well as being of a quite diverse form at different time steps. To analyze the latter in more detail we make the following definition.

Definition 4. For $t \in \mathbb{N}$ given let $C(t)$, the *core at t*, be the intersection of all subsets of $\{1,\ldots,n\}$ which are saturated with respect to $F(M) = \{j \mid a_{ij}(t) > 0 \text{ for some } i \in M\}$. Furthermore, a *sight chain at t from i to j* is a sequence $(i_1,\ldots,i_m)$ in $\mathbb{N}$ such that $i_{k+1} \in S(i_k,t)$ for all $1 \le k \le m - 1$ and $i = i_1, j = i_m$.

Using these concepts we can characterize a coherent flight regime as follows.

Proposition 1. *Let $t \in \mathbb{N}$ and $A(t)$ with positive diagonal. Then a flight regime is coherent (at t) if and only if from each bird there is a sight chain (at t) to a bird in the core (at t).*

Proof. Suppose first, the flight regime is coherent. Since intersections of saturated sets are nonempty and saturated again, as can be easily seen, it follows that $C(t)$ is nonempty and saturated. From Lemma 1 (iii) it follows that some power $A(t)^p$ is scrambling. Thus, for i, j given, there exist sight chains from i to k and j to k for some k. Since $C(t) \neq \emptyset$ there is some $j \in C(t)$ and because $C(t)$ is saturated we must have that $k \in C(t)$. That is, there is a sight chain from i to $k \in C(t)$. Conversely, if there exist from each bird a sight chain to a bird in the core then $C(t) \neq \emptyset$. Thus, for any two saturated sets M and M', we have $\emptyset \neq C(t) \subseteq M \cap M'$ which means that the flight regime at t must be coherent. $\square$

In the following we shall use Proposition 1 to discuss various possibilities for the coherent flight regimes assumed in Theorem 3. For this discussion we leave aside the other assumptions, made in Theorem 3 on the minimal interaction among the birds.

The latter would require to go more deeply into the relationship between positions and velocities as connected by $v(t) = x(t+1) - x(t)$. (We did it for Cucker–Smale flocking in Corollary 2 where, however, the flight regimes did not change with time.)

Consider first the case that the core is minimal, that is, it consists of a single bird, say, 1. Examples are *leadership*, that is, $j \in S(1,t)$ for all $j \neq i$, or *line formation*, that is $i + 1 \in S(i,t)$ for $1 \leq i \leq n - 1$. Another example is the famous *V-formation* of certain kinds of geese consisting of two line formations with 1 at the top. In general, with 1 as the core, the flight regime is an *echelon*, a treelike, hierarchical composition of *V*-formations. It need, however, not to be actually a (reversed) tree since cycles of seeing among the birds may occur. For example, the flight regime may consist of two cycles of the same or of different orientation.

The core can consist instead of a single bird also of a group of different birds. For example of a group of birds forming a *cycle* (with respect to seeing each other). Connected to this cycle could be the other birds in a treelike manner, including some further cycles. (Whereas cycles seem not so relevant for birds, in particular for geese, they seem to be important for fish schools.)

The types of flight regimes discussed may show up one after the other at different periods t_m during the flight of the swarm. For example, considering leadership the leader can change from t_m to t_{m+1}. With a different leader the swarm can change speed and direction of velocity. That is, the swarm can move around and may make even a turn. This applies also to other types of flight regimes, an echelon may move around still remaining an echelon. It is possible, however, an echelon becomes a line formation in the course of time.

It should be mentioned that the structure of the swarm in terms of sight chains does not imply necessarily a similar structure in terms of positions in space. Such a correspondence, however, is possible, especially if the structure of sight is treelike without cycles. Furthermore, "sight" is only a substitute for any kind of communication, for example acoustic communication, among the birds. The above discussion of flight regimes is driven by the question how birds coordinate to approach a common velocity. There are many other reasons relevant for swarm formation, in particular from biology or aerodynamics which we did not address in this paper. (For such issues as well as for questions of factual coordination see the profound survey [14]).)

References

1. Ballerini, M., et al.: Interaction ruling animal collective behavior depends on topological rather than metric distance: evidence from a field study. Proc. Nat. Acad. Sci. USA **105**, 1232–1237 (2008)
2. Blondel, V.D., Hendrickx, J.M., Olshevsky, A., Tsitsiklis, J.N.: Convergence in multiagent coordination, consensus, and flocking, pp. 2996–3000 In: Proceedings of 44th IEEE Conference on Decision and Control (2005)
3. Cañizo, J.A., Carillo, J.A., Rosado, J.: Collective behavior of animals: swarming and complex patterns. ARBOR (2011, to appear) doi: 10.3989/arbor

4. Chazelle, B.: The total s-energy of a multiagent system. SIAM J. Contr. Optim. **49**, 1680–1706 (2011)
5. Cucker, F., Smale, S.: Emerging behavior in flocks. IEEE Trans. Aut. Cont. **52**, 852–862 (2007)
6. Cucker, F., Smale, S.: On the mathematics of emergence. Japan. J. Math. **2**, 197–227 (2007)
7. Ha, S.-Y., Liu, J.-G.: A simple proof of the Cucker-Smale flocking dynamics and mean-field limit. Comm. Math. Sci. **7**, 297–325 (2009)
8. Hartfiel, D.J.: Nonhomogeneous Matrix Products. World Scientific Publishing Co., Singapore (2002)
9. Hegselmann, R., Krause, U.: Opinion dynamics and bounded confidence: models, analysis, and simulation. J. Art. Soc. Soc. Sim. **5** (2002) http://jasss.soc.surrey.ac.uk/5/3/2.html
10. Hendrickx, J.M., Blondel, V.D.: Convergence of different linear and non-linear Vicsek models, pp. 1229–1240 In: Proceedings of 17th International Symposium on Mathematical Theoryof Networks and Systems (2006)
11. Jadbabaie, A., Lin, J., Morse, A.S.: Coordination of groups of mobile autonomous agents using nearest neighbor rules. IEEE Trans. Aut. Cont. **48**, 988–1001 (2003)
12. Krause, U.: Positive particle interaction. Lect. Notes Contr. Inf. Sci. **294**, 199–206 (2003)
13. Krause, U.: Convergence of the multidimensional agreement algorithm when communication fades away. Lect. Notes Contr. Inf. Sci. **341**, 217–222 (2006)
14. Lebar Bajec, I., Heppner, F.H.: Organized flight in birds. Animal Behav. **78**, 777–789 (2009)
15. Lukeman, R., Li, Y.-X., Edelstein-Keshet, L.: Inferring individual rules from collective behavior. Proc. Nat. Acad. Sci. (early edition) 1–5 (2010). www.pnas.org/cgi/doi/10.1073/pnas.1001763107
16. Moreau, L.: Stability of multi-agent systems with time-dependent communication links. IEEE Trans. Aut. Cont. **50**, 169–182 (2005)
17. Reynolds, C.W.: Flocks, herds, and schools: a distributed behavioral model. Comput. Graph. **21**, 25–34 (1987)
18. Shen, J.: Cucker-Smale flocking under hierarchical leadership. SIAM J. Appl. Math. **68**, 694–719 (2008)
19. Seneta, E.: Non-Negative Matrices and Markov Chains, 2nd edn. Springer, New York (1981)
20. Vicsek, T., Czirók, A., Ben-Jacob, E., Cohen, I., Shochet, O.: Novel type of phase transition in a system of self-driven particles. Phys. Rev. Lett. **75**, 1226–1229 (1995)

Periodic Solutions of Differential and Difference Systems with Pendulum-Type Nonlinearities: Variational Approaches

Jean Mawhin

Abstract We survey some recent results on the use of variational methods in proving the existence and multiplicity of periodic solutions of systems of differential equations of the type

$$(\phi(q'))' = \nabla_q F(t,q) + h(t)$$

or systems of difference equations of the type

$$\Delta(\phi(\Delta q(n-1))) = \nabla_q F(n,q) + h(n) \quad (n \in \mathbb{Z})$$

when ϕ belongs to a class of suitable homeomorphisms between an open ball and the whole space and F is periodic in the components of q.

Keywords Periodic equations • Difference systems • Pendulum nonlinearities • Variational methods

1 Introduction

For the periodic problem associated to classical systems of forced pendulum-type

$$q'' = \nabla_q F(t,q) + h(t), \quad q(0) = q(T), \quad q'(0) = q'(T), \tag{1}$$

with $F : [0,T] \times \mathbb{R}^N \to \mathbb{R}$ and $\nabla_q F : [0,T] \times \mathbb{R}^N \to \mathbb{R}^N$ continuous, F periodic in each variable q_j, and $h \in L^1([0,T], \mathbb{R}^N)$, the existence of at least one solution was

J. Mawhin (✉)
Institut de Recherche en Mathématique et Physique, Université Catholique de Louvain, chemin du cyclotron, 2, B-1348 Louvain-la-Neuve, Belgium
e-mail: jean.mawhin@uclouvain.be

S. Pinelas et al. (eds.), *Differential and Difference Equations with Applications*, Springer Proceedings in Mathematics & Statistics 47, DOI 10.1007/978-1-4614-7333-6_7,
© Springer Science+Business Media New York 2013

proved through the direct method of the calculus of variations in [16,17], when h has mean value zero. The results are easily extended to systems involving p-Laplacians, as shown in [10].

When q'' is replaced by a "relativistic" differential operator of the type

$$\left(\frac{q'}{\sqrt{1 - |q'|^2}} \right)',$$

the scalar case was first considered in [3] and the case of a system was studied in [4]. The existence of at least one solution was proved under the same conditions. The result is proved in [4] for the more general problem

$$\left(\phi(q') \right)' = \nabla_q F(t,q) + h(t), \quad q(0) = q(T), \quad q'(0) = q'(T), \tag{2}$$

when ϕ belongs to a suitable class of homeomorphisms between the open ball $B(a) \subset \mathbb{R}^N$ of center 0 and radius $a > 0$ and $\mathbb{R}^N$. We describe in Sect. 3 a slightly different approach given in [1], based upon Szulkin's critical point theory for C^1 perturbations of lower semicontinuous convex functionals [22] and a new reduction of the corresponding variational inequality to the differential system.

Using Lusternik-Schnirelmann theory in Hilbert manifolds [18] or variants of it, Chang [5], Rabinowitz [20], and the author [8] have independently obtained results which imply that problem (1) has at least $N + 1$ geometrically distinct solutions for every $h \in L^2$ with mean value zero (see also [17]). Notice that because of the periodicity property of F, if $q(t)$ is a solution of (1), the same is true for $(q_1(t) + j_1\omega_1, q_2(t) + j_2\omega_2, \ldots, q_N(t) + j_n\omega_N)$ for any $(j_1, j_2, \ldots, j_N) \in \mathbb{Z}^N$, and hence two solutions q and $\hat{q}$ of (1) are called *geometrically distinct* if

$$q \not\equiv \hat{q} \ (\text{mod } \omega_j e_j, j = 1, 2, \ldots, N).$$

This result is an extension of an earlier one of the author and Willem [16] who proved, under the same conditions, the existence of at least two geometrically distinct solutions, using the variant of the mountain pass lemma introduced in [15] to treat the special case where $N = 1$, and in particular the *forced pendulum problem*

$$q'' + \mu \sin q = h(t), \quad q(0) = q(T), \quad q'(0) = q'(T).$$

See [9] for a survey of this problem. Very recently, Bereanu and Torres [2] have extended the mountain pass approach of [15] to obtain the existence of at least two geometrically distinct solutions for problem (2) with $N = 1$. It is not clear if their approach is applicable to system (2) with $N \geq 2$, and, would it be the case, the existence of two solutions only would be insured.

We describe in Sect. 4 some results of [11], showing that when $h \in L^{1+\varepsilon}$ has mean value zero, problem (2) has at least $N + 1$ geometrically distinct solutions. To do this, we reduce problem (2) to an equivalent *Hamiltonian* system. The advantage of the Hamiltonian formulation with respect to the Lagrangian one used in [3, 4] and presented in Sect. 4 is that the Hamiltonian action functional is defined on the whole space, so that the Hamiltonian system is trivially its Euler-Lagrange equation. The price to pay in the Hamiltonian formalism is that the Hamiltonian action functional is indefinite, excluding the obtention of existence results by minimization and of multiplicity results through classical Lusternik-Schnirelmann category. We use instead an abstract multiplicity result of Szulkin [23] for some compact perturbations of an indefinite self-adjoint operator. Although its final result is stated in terms of the classical cuplength of a finite-dimensional manifold, the underlying technique in Szulkin's paper is a more sophisticated concept of relative category. Complete details for the results of Sects. 3 and 4 can be found in the mentioned papers and in the survey [14].

The use of variational methods in the study of T-periodic solutions of systems of second-order *difference* equations having a variational structure has been introduced in 2003 by Guo and Yu in [6], and many variants and generalizations have been given since, none of them dealing with periodic nonlinearities. See [12, 13] for references. In [12], given a positive integer T, the existence of T-periodic solutions of systems of difference equations of the form

$$\Delta \phi[\Delta u(n-1)] = \nabla_u F[n, u(n)] + h(n) \quad (n \in \mathbb{Z}) \tag{3}$$

i.e., of $\mathbb{Z}$-sequences $(u(n))_{n \in \mathbb{Z}}$ such that $u(n+T) = u(n)$ for all $n \in \mathbb{Z}$ and verifying (3), has been proven by minimization of an associated real function, when F and h verify conditions analog to the ones of the differential systems. Here $\Delta u(n) := u(n+1) - u(n)$ is the usual forward difference operator. By some reduction to a Hamiltonian form and the use of Szulkin's results [23], it has been shown in [13] that system (3) has indeed at least $N+1$ geometrically distinct T-periodic solutions. Those results are described in Sect. 5.

2 Some Notations and Preliminary Results

In $\mathbb{R}^N$, we denote the usual inner product by $\langle \cdot, \cdot \rangle$ and corresponding Euclidian norm by $|\cdot|$. We denote the usual norm in $L^p := L^p(0, T; \mathbb{R}^N)$ $(1 \le p \le \infty)$ by $|\cdot|_p$, set $C := C([0, T], \mathbb{R}^N)$, $C^1 = C^1([0, T], \mathbb{R}^N)$, and $W^{1,\infty} := W^{1,\infty}([0, T], \mathbb{R}^N)$. The usual norm $|\cdot|_\infty$ is considered on C and $|v|_{1,\infty} = |v|_\infty + |v'|_\infty$ is the norm on $W^{1,\infty}$ and C^1. Each $v \in L^1$ can be written $v(t) = \bar{v} + \tilde{v}(t)$, with

$$\bar{v} := T^{-1} \int_0^T v(t)\, dt, \quad \int_0^T \tilde{v}(t)\, dt = 0.$$

So, $\bar{v}$ is the mean value of v. One has, for any $v \in L^\infty$,

$$|\tilde{v}|_\infty \le \frac{T}{4} |v'|_\infty. \tag{4}$$

For any integer $T > 0$, H_T denotes the space of T-periodic $\mathbb{Z}$-sequences in $\mathbb{R}^N$, i.e., of mappings $u : \mathbb{Z} \to \mathbb{R}^N$ such that $u(n+T) = u(n)$ for all $n \in \mathbb{Z}$. H_T is isomorphic to $(\mathbb{R}^N)^T = \mathbb{R}^{NT}$ and can be endowed with the inner product and corresponding norm

$$(u|v) := \sum_{j=1}^{T} \langle u(j), v(j) \rangle, \quad \|u\| = \left(\sum_{j=1}^{T} |u(j)|^2 \right)^{1/2}.$$

Equivalent norms are $|u|_1 = \sum_{j=1}^{T} |u(j)|$, $|u|_\infty = \max_{1 \le j \le T} |u(j)|$. For any $u = (u(n))_{n \in \mathbb{Z}} \in H_T$, we set $\bar{u} = T^{-1} \sum_{j=1}^{T} u(j) \in \mathbb{R}^N$, identify $\bar{u}$ with the corresponding constant T-periodic sequence, and set

$$\tilde{u} := u - \bar{u}, \quad \overline{H}_T := \{ u \in H_T : u = \bar{u} \}, \quad \tilde{H}_T := \{ u \in H_T : \bar{u} = 0 \}.$$

Given $a > 0$, a homeomorphism $\phi : B(a) \to \mathbb{R}^N$ is said to be of *class R* if $\phi(0) = 0$, $\phi = \nabla \Phi$, with $\Phi : \overline{B(a)} \to \mathbb{R}$ of class C^1 on $B(a)$, continuous, strictly convex on $\overline{B(a)}$, and such that $\Phi(0) = 0$. If $\Phi^* : \mathbb{R}^N \to \mathbb{R}$ is the Legendre-Fenchel transform of Φ defined by

$$\Phi^*(v) = \langle \phi^{-1}(v), v \rangle - \Phi[\phi^{-1}(v)] = \sup_{u \in \overline{B(a)}} \{ \langle u, v \rangle - \Phi(u) \},$$

then Φ^* is also strictly convex, and, with $d := \max_{u \in \overline{B(a)}} \Phi(u)$,

$$a|v| - d \le \Phi^*(v) \le a|v| \quad (v \in \mathbb{R}^N), \tag{5}$$

so that Φ^* is coercive on $\mathbb{R}^n$. Finally, Φ^* is of class C^1, $\phi^{-1} = \nabla \Phi^*$, so that

$$v = \nabla \Phi(u) = \phi(u), \quad u \in B(a) \quad \Leftrightarrow \quad u = \phi^{-1}(v) = \nabla \Phi^*(v), \quad v \in \mathbb{R}^n.$$

3 Lagrangian Variational Approach for Periodic Solutions of Differential Systems

A function $F : [0, T] \times \mathbb{R}^N \to \mathbb{R}$ is said to be of *class P* if F is continuous, $F(t, 0) = 0$ for a.e. $t \in [0, T]$ (without loss of generality), $\nabla_q F : [0, T] \times \mathbb{R}^N \to \mathbb{R}^N$ exists and is continuous, and there are some $\omega_1 > 0, \dots, \omega_N > 0$ such that

$$F(t, q_1, \dots, q_N) = F(t, q_1 + \omega_1, \dots, q_N + \omega_N)$$

for all $(t, q) \in [0, T] \times \mathbb{R}^N$. We consider the existence of solutions for the periodic problem

$$(\phi(q'))' = \nabla_q F(t, q) + h(t), \quad q(0) = q(T), \quad q'(0) = q'(T), \tag{6}$$

where $h \in L^1$, ϕ is of class R and F is of class P. A *solution* of (6) is a function $u \in C^1$, such that $|u'|_\infty < a$, $\phi(u')$ is differentiable a.e. and (6) is satisfied a.e.

The following variational setting for dealing with equations or systems of type (6) was introduced in [1]. Define the convex subset K of $W^{1,\infty}$ by

$$K := \{v \in W^{1,\infty} : |v'|_\infty \le a, \quad v(0) = v(T)\},$$

and the function $\Psi : C \to (-\infty, +\infty]$ by

$$\Psi(v) = \begin{cases} \varphi(v), & \text{if } v \in K, \\ +\infty, & \text{otherwise,} \end{cases}$$

with $\varphi : K \to \mathbb{R}$ given by

$$\varphi(v) = \int_0^T \Phi(v'(t))\, dt, \quad v \in K.$$

Obviously, Ψ is proper and convex. It is proved in [1] that Ψ is lower semicontinuous on C and that K is closed in C. Next, define $\mathcal{G} : C \to \mathbb{R}$ by

$$\mathcal{G}(u) = \int_0^T [F(t, u(t)) + \langle h(t), u(t) \rangle]\, dt, \quad u \in C.$$

A standard reasoning shows that $\mathcal{G}$ is of class C^1 on C and its derivative is given by

$$\mathcal{G}'(u)(v) = \int_0^T \langle \nabla F(t, u(t)) + h(t), v(t) \rangle\, dt, \quad u, v \in C.$$

Define the functional $I : C \to (-\infty, +\infty]$ by $I = \Psi + \mathcal{G}$. As sum of a proper convex lower semicontinuous function and of a C^1 function, I has the structure required by Szulkin's critical point theory [22]. Accordingly, a function $q \in C$ is a *critical point* of I if $q \in K$ and satisfies the inequality

$$\Psi(v) - \Psi(q) + \mathcal{G}'(q)(v - q) \ge 0, \quad \forall v \in C,$$

or, equivalently

$$\int_0^T [\Phi(v'(t)) - \Phi(q'(t)) + \langle \nabla F(t, q(t)) + h(t), v(t) - q(t) \rangle\, dt \ge 0, \quad \forall v \in K.$$

The following simple result is given in [22].

Lemma 1. *Each local minimum of I is a critical point of I.*

The following elementary lemma, proved in [1] by direct computation, is useful is relating the critical points of I to the solutions of (6).

Lemma 2. *For every $f \in L^1$, the problem*

$$(\phi(q'))' = \overline{q} + f(t), \quad q(0) = q(T), \quad q'(0) = q'(T)$$

has a unique solution q_f, also unique solution of the variational inequality

$$\int_0^T [\Phi(v'(t)) - \Phi(q'(t)) + \langle \overline{q} + f(t), v(t) - q(t) \rangle] \, dt \geq 0, \quad \forall v \in K,$$

and unique minimum over K of the strictly convex functional J defined on K by

$$J(q) = \int_0^T \left[\Phi(q'(t)) + \frac{|\overline{q}|^2}{2} + \langle f(t), q(t) \rangle \right] dt.$$

The idea of proof of the result below first occurred in [3].

Proposition 1. *If q is a critical point of I, then q is a solution of problem (6).*

Proof. For q a critical point of I, let

$$f_q := \nabla F(\cdot, q) + h - \overline{q} \in L^1.$$

Because of Lemma 2, the problem

$$(\phi(w'))' = \overline{w} + f_q(t), \quad w(0) = w(T), \quad w'(0) = w'(T).$$

has a unique solution $q^\#$, which is also the unique solution of the variational inequality

$$\int_0^T [\Phi(v'(t)) - \Phi(q^{\#'}(t)) + \langle \overline{q^\#} + f_u(t), v(t) - q^\#(t) \rangle] \, dt \geq 0, \quad \forall v \in K.$$

Since q is a critical point of I, we have

$$\int_0^T [\Phi(v'(t)) - \Phi(q'(t)) + \langle \overline{q} + f_q(t), v(t) - q(t) \rangle] \, dt \geq 0, \quad \forall v \in K.$$

It follows by uniqueness that $q = q^\#$, and q solves problem (6).   □

The following result was first proved in the scalar case in [3] and in the vector case in [4] using slightly different arguments.

Theorem 1. *If ϕ is of class R and F of class P, then, for any $h \in L^1$ with $\bar{h} = 0$, problem (6) has at least one solution which minimizes I on C (or K).*

Proof. Let

$$\rho := N^{1/2} \max_{1 \leq j \leq N} \omega_j, \qquad (7)$$

so that $[0, \omega_1] \times \cdots \times [0, \omega_N] \subset \overline{B}(\rho)$. Due to the periodicity of $F(t, \cdot)$ and because of $\bar{h} = 0$, it holds

$$I(v + j_1 \omega_1 e_1 + \cdots + j_N \omega_N e_N) = I(v)$$

for all $v \in K$ and $(j_1, \ldots, j_N) \in \mathbb{Z}^N$. Then, with $\hat{K}_\rho := \{u \in K : |\bar{u}| \leq \rho\}$,

$$\inf_{\hat{K}_\rho} I = \inf_K I = \inf_C I,$$

and it suffices to prove that there is some $q \in \hat{K}_\rho$ such that

$$I(q) = \inf_{\hat{K}_\rho} I. \qquad (8)$$

If $v \in \hat{K}_\rho$, we obtain, using (4),

$$|v|_\infty \leq |\bar{v}| + |\tilde{v}|_\infty \leq \rho + \frac{Ta}{4}.$$

This, together with $|v'|_\infty \leq a$, shows that $\hat{K}_\rho$ is bounded in $W^{1,\infty}$ and, by the compactness of the embedding $W^{1,\infty} \subset C$, the set $\hat{K}_\rho$ is relatively compact in C. Let $\{q_n\} \subset \hat{K}_\rho$ be a minimizing sequence for I. Passing to a subsequence if necessary, we may assume that $\{q_n\}$ converges uniformly to some $q \in \hat{K}_\rho$. From the lower semicontinuity of Ψ and the continuity of $\mathcal{F}$ on C, we obtain

$$I(q) \leq \liminf_{n \to \infty} I(q_n) = \lim_{n \to \infty} I(q_n) = \inf_{\hat{K}_\rho} I,$$

showing that (8) holds true. By Lemma 1 and Proposition 1, q is a solution of (6). $\qquad \square$

For the use in examples, let us introduce the continuous mapping $S : [0, T] \times \mathbb{R}^N \to \mathbb{R}^N$ by

$$S(t, u) := (\mu_1(t) \sin u_1, \mu_2(t) \sin u_2, \ldots, \mu_N(t) \sin u_N) \quad (\mu_j \in \mathbb{R}, \; j = 1, \ldots, N), \quad (9)$$

where $\mu_j : [0,T] \to \mathbb{R}$ is continuous $(j = 1,\ldots,N)$, so that

$$S(t,u) = \nabla c(t,u) \text{ with } c(t,u) := \sum_{j=1}^{N} \mu_j(t)(1 - \cos u_j).$$

Example 1. For any $T > 0$ and any $h \in L^1$ such that $\overline{h} = 0$, the problem

$$\left(\frac{q'}{\sqrt{1 - |q'|^2}}\right)' + S(t,q) = h(t), \quad u(0) = u(T), \quad u'(0) = u'(T)$$

has at least one solution.

In particular, in the scalar case, the forced relativistic pendulum problem

$$\left(\frac{q'}{\sqrt{1 - q'^2}}\right)' + \mu \sin q = h(t), \quad q(0) = q(T), \quad q'(0) = q'(T)$$

has at least one solution for any $\mu \in \mathbb{R}$ and $T > 0$ when $\overline{h} = 0$.

4   Hamiltonian Variational Approach for Periodic Solutions of Differential Systems

The change of variables $\nabla \Phi(q') = p$, equivalent to $q' = \nabla \Phi^*(p)$, transforms problem (6) into the equivalent one:

$$p' = \nabla_q F(t,q) + h(t), \quad q' = \nabla \Phi^*(p), \quad p(0) = p(T), \quad q(0) = q(T). \quad (10)$$

With the Hamiltonian function $H : [0,T] \times \mathbb{R}^N \times \mathbb{R}^N \to \mathbb{R}$ defined by

$$H(t,p,q) = \Phi^*(p) - F(t,q) - \langle h(t), q \rangle,$$

where Φ^* is the Legendre-Fenchel transform of Φ defined in Sect. 2, problem (10) takes the Hamiltonian form

$$p' = -\nabla_q H(t,p,q), \quad q' = \nabla_p H(t,p,q), \quad p(0) = p(T), \quad q(0) = q(T).$$

Formally, system (10) is the Euler-Lagrange equation associated to the (action) functional $\mathcal{A}$ defined on a suitable space of T-periodic functions (p,q) by

$$\mathcal{A}(p,q) = \int_0^T \left[-\langle p(t), q'(t) \rangle + \Phi^*(p(t)) - F(t,q(t)) - \langle h(t), q(t) \rangle \right] dt.$$

If

$$J = \begin{pmatrix} 0 & I \\ -I & 0 \end{pmatrix},$$

denotes the $2N \times 2N$ symplectic matrix, define (see, e.g., [19]) the space $H_\#^{1/2} :=$ $H_\#^{1/2}(0,T;\mathbb{R}^{2N})$ as the space of functions $z = (p,q) \in L^2(0,T;\mathbb{R}^{2N})$ with Fourier series $z(t) = \sum_{k \in \mathbb{Z}} e^{k\omega tJ} z_k$ ($\omega = 2\pi/T$), such that $z_k \in \mathbb{R}^{2N}$ ($k \in \mathbb{Z}$) and

$$|z|_{1/2}^2 := \sum_{k \in \mathbb{Z}} (1 + |k|)|z_k|^2 < +\infty.$$

With the corresponding inner product

$$(z|w) := \sum_{k \in \mathbb{Z}} (1 + |k|)\langle z_k, w_k \rangle,$$

$H_\#^{1/2}$ is a Hilbert space such that $H_\#^1(0,T;\mathbb{R}^{2N}) \subset H_\#^{1/2} \subset L^s(0,T;\mathbb{R}^{2N})$ for any $s \in [1,+\infty)$.

By easy computations based on Fourier series and Cauchy-Schwarz inequality, for (p,q), (u,v) smooth, the symmetric bilinear form

$$b[(p,q),(u,v)] := -\int_0^T [\langle p'(t), v(t) \rangle + \langle u'(t), q(t) \rangle] \, dt$$

generating the quadratic form $(p,q) \mapsto \int_0^T [-2\langle p(t), q'(t) \rangle] \, dt$ satisfies an inequality of the form

$$|b[(p,q),u,v)]| \le C|(p,q)|_{1/2}|(u,v)|_{1/2}.$$

Hence it can be extended to $H_\#^{1/2}$ as a continuous bilinear form, still noted b, and the linear self-adjoint operator $A : H_\#^{1/2} \to H_\#^{1/2}$ defined through Riesz's representation theorem by the relation

$$(A(p,q)|(u,v)) = b[(p,q),(u,v)] \quad ((p,q),(u,v) \in H_\#^{1/2}) \qquad (11)$$

is continuous. In terms of Fourier series, with $w = (u,v)$,

$$(A(p,q)|(u,v)) = 2\pi \sum_{k \in \mathbb{Z}} k\langle z_k, w_k \rangle$$

and hence

$$(A(p,q)|(p,q)) = 2\pi \sum_{k \in \mathbb{Z}} k|z_k|^2.$$

It is easily seen that the spectrum of A is made of the eigenvalues $\lambda_k = 2\pi \frac{k}{1+|k|}$ ($k \in \mathbb{Z}$), each of multiplicity $2N$, and of the elements $-2\pi, 2\pi$ in the essential spectrum. Therefore, it is standard to show that $H_\#^{1/2} = H^- \oplus H^0 \oplus H^+$ (orthogonal sum with respect to $(\cdot|\cdot)$ and to L^2), with $H^0 = \ker A \simeq \mathbb{R}^{2N}$, and, for $z^- \in H^-$, $z^+ \in H^+$,

$$(Az^- | z^-) \leq -\pi |z^-|_{1/2}^2, \quad (Az^+ | z^+) \geq \pi |z^+|_{1/2}^2.$$

Furthermore the subspaces H^- and H^+ are invariant for A.

Using estimate (5), it is well known [19] that the assumptions on ϕ and F imply that $\mathcal{A}$ is of class C^1 on $H_\#^{1/2}$ and that any critical point $(\hat{p}, \hat{q})$ of the functional

$$\mathcal{A}(p,q) = -\frac{1}{2}(A(p,q)|(p,q)) + \int_0^T [\Phi^*(p(t)) - F(t,q(t)) - \langle h(t), q(t)\rangle] \, dt$$

is a (Carathéodory) solution of (10) (see, e.g., [19]).

Let E be a real Hilbert space with inner product $(\cdot|\cdot)$ and norm $\|\cdot\|$ and V^d a compact d-dimensional C^2-manifold without boundary. Let $L : E \to E$ be a bounded linear self-adjoint operator to which there corresponds an orthogonal decomposition $E = E^- \oplus E^0 \oplus E^+$ into invariant subspaces, with $E^0 = \ker L$, and $\varepsilon > 0$ such that

$$(Lx^+|x^+) \geq \varepsilon \|x^+\|^2 \quad (x^+ \in E^+), \quad (Lx^-|x^-) \leq -\varepsilon \|x^-\|^2 \quad (x^- \in E^-).$$

The following result is due to Szulkin [23].

Lemma 3. *Let $\Psi \in C^1(E \times V^d, \mathbb{R})$ be given by $\Psi(x,v) = \frac{1}{2}(Lx|x) - \psi(x,v)$, where ψ' is compact. Suppose that $\psi'(E \times V^d)$ is a bounded set, E^0 is finite dimensional and, if $\dim E^0 > 0$, $\psi(x^0, v) \to -\infty$ (or $\psi(x^0, v) \to +\infty$) as $\|x^0\| \to \infty$, $x^0 \in E^0$. Then Φ has at least cuplength $(V^d) + 1$ critical points.*

The *cuplength* of X is the greatest number of elements of nonzero degree in the cohomology $H^*(X)$ of X with nonzero cup product, i.e., the largest integer m for which there exists $\alpha_j \in H^{k_j}(X)$, $1 \leq j \leq m$, such that $k_1, \ldots, k_m \geq 1$ and $\alpha_1 \cup \ldots \cup \alpha_m \neq 0$ in $H^{k_1 + \cdots + k_m}(X)$. For the n-dimensional torus $\mathbb{T}^n$, cuplength$(\mathbb{T}^n) = n$ [21].

Lemma 3 applied to a suitable reformulation of $\mathcal{A}$ gives our multiplicity theorem.

Theorem 2. *If ϕ is of class R and F of class P, then, for every $h \in L^s$ ($s > 1$) with $\bar{h} = 0$, problem (2) has at least $N + 1$ geometrically distinct solutions.*

Proof. The fact that F is of class P and $\bar{h} = 0$ implies that, for any $(j_1, \ldots, j_N) \in \mathbb{Z}^N$,

$$\mathcal{A}(p, q_1 + j_1 \omega_1 e_1, \ldots, q_N + j_N \omega_N e_N) = \mathcal{A}(p,q).$$

To each critical point $(\hat{p}, \hat{q})$ of $\mathcal{A}$ on $H_\#^{1/2}$ corresponds the orbit

$$(\hat{p}, \hat{q}_1 + j_1 \omega_1 1, \ldots, \hat{q}_N + j_N \omega_N e_N) \quad ((j_1, \ldots, j_N) \in \mathbb{Z}^N)$$

of critical points, which can be considered as a single critical point lying on the manifold $E \times \mathbb{T}^N$, with $\mathbb{T}^N$ the N-torus $\mathbb{R}^N / (\omega_1 \mathbb{Z}, \ldots, \omega_N \mathbb{Z})$, and

$$E = \{(p,q) \in H_\#^{1/2} : \bar{q} = 0\}.$$

Denoting by $L : E \to E$ the restriction to E of A given in (11), we have $E = H^- \oplus E^0 \oplus H^+$, where $E^0 \simeq \mathbb{R}^N = \{(p,0) \in \mathbb{R}^{2N} : p \in \mathbb{R}^N\} = \ker L$. Hence, $\mathcal{A}$ has the equivalent expression

$$\frac{1}{2}(L(p,\tilde{q})|(p,\tilde{q}) + \int_0^T [\Phi^*(p(t)) - F(t,\bar{q}+\tilde{q}(t)) - \langle h(t),\tilde{q}(t)\rangle] \, dt,$$

namely,

$$\Psi(x,v) = \frac{1}{2}\langle L(p,\tilde{q}),(p,\tilde{q})\rangle - \psi(p,\tilde{q};\bar{q})$$

requested by Szulkin's lemma with $x = (p,\tilde{q})$, $v = \bar{q}$, considered as an element of $\mathbb{T}^N$, and

$$\psi(p,\tilde{q};\bar{q}) := \int_0^T [F(t,\bar{q}+\tilde{q}(t)) + \langle h(t),\tilde{q}(t)\rangle - \Phi^*(p(t))] \, dt.$$

Therefore, for any $v, \tilde{w}, \overline{w}$, we have

$$(\psi'(p,\tilde{q};\bar{q})|(v,\tilde{w};\overline{w}))$$

$$= \int_0^T [\langle \nabla_q F(t,\bar{q}+\tilde{q}(t)),\overline{w}+\tilde{w}\rangle + \langle h(t),\tilde{w}(t)\rangle - \langle \phi^{-1}(p(t)),v(t)\rangle] \, dt.$$

Because $\nabla_q F(t,\cdot)$ and ϕ^{-1} have a bounded range, ψ' has a bounded range, and ψ' is compact by the compact embedding of $H_\#^{1/2}$ in L^s for any $s \geq 1$. On the other hand, because of (5) and the fact that any $(p,\tilde{q}) \in E^0$ has the form $(p^0,0)$ with $p^0 \in \mathbb{R}^N$, we have, for $|p^0| \to \infty$,

$$\psi(p^0,\bar{q}) = \int_0^T [-F(t,\bar{q}) - \Phi^*(p^0)] \, dt = -T[\overline{F(\cdot,\bar{q})} + T\Phi^*(p^0)] \to -\infty.$$

All the assumptions of Lemma 3 being satisfied, Ψ has at least cuplength $(\mathbb{T}^N) + 1 = N + 1$ critical points, i.e., $\mathcal{A}$ has at least $N + 1$ geometrically distinct critical points.
□

Example 2. For any $h \in L^s$ ($s > 1$) such that $\bar{h} = 0$ and S defined in (9), the problem

$$\left(\frac{q'}{\sqrt{1-|q'|^2}}\right)' + S(t,q) = h(t), \quad q(0) = q(T), \quad q'(0) = q'(T)$$

has at least $N + 1$ geometrically distinct solutions.

In particular, for any $\mu \in \mathbb{R}$, $T > 0$ and $h \in L^s$ $(s > 1)$ such that $\overline{h} = 0$, the forced relativistic pendulum problem

$$\left(\frac{q'}{\sqrt{1 - q'^2}} \right)' + \mu \sin q = h(t), \quad q(0) = q(T), \quad q'(0) = q'(T)$$

has at least two geometrically distinct solutions.

Remark 1. A similar Hamiltonian approach has been recently used by Manásevich and Ward [7] to give an alternative proof to the result of Brezis and the author on the relativistic forced pendulum [3]. The existence of the corresponding critical point for the associated Hamiltonian action is obtained using Rabinowitz' saddle point theorem [19].

5 Periodic Solutions of Systems of Difference Equations

In this section the function $F : \mathbb{Z} \times \mathbb{R}^N \to \mathbb{R}$ is said to be of *class P* if F is continuous, $F(n,0) = 0$ for $n \in \mathbb{Z}$, $\nabla_q F : \mathbb{Z} \times \mathbb{R}^N \to \mathbb{R}^N$ exists and is continuous, and there are some $\omega_1 > 0, \ldots, \omega_N > 0$ such that

$$F(n, q_1, \ldots, q_N) = F(n, q_1 + \omega_1, \ldots, q_N + \omega_N)$$

for all $(n,q) \in \mathbb{Z} \times \mathbb{R}^N$. For ϕ of class R and $h \in H_T$, let us consider the problem of the existence of T-periodic solutions of the system of difference equations, with $\Delta u(n) = u(n+1) - u(n)$,

$$\Delta \phi[\Delta q(n-1)] = \nabla_q F[n, q(n)] + h(n) \quad (n \in \mathbb{Z}). \tag{12}$$

When those assumptions hold, the real function

$$\mathcal{I}(q) := \sum_{i=1}^{T} \{ \Phi[\Delta q(i)] + F[i, q(i)] + \langle h(i), q(i) \rangle \}$$

is well defined on the closed convex subset $K \subset H_T$ defined by $K := \{ u \in H_T : |\Delta u|_\infty \leq a \}$. We first state an easy necessary and sufficient condition for the existence of a minimum to $\mathcal{I}$.

Proposition 2. *If ϕ is of class R and F of class P, $\mathcal{I}$ has a minimum over K if and only if it has a minimizing subsequence (u_k) in K such that $(\overline{u_k})$ is bounded.*

Any minimizer q of $\mathcal{I}$ on K satisfies a *variational inequality.*

Lemma 4. *If q minimizes $\mathcal{I}$ over K, then, for all $v \in K$,*

$$\sum_{i=1}^{T} \left\{ \Phi[\Delta v(i)] - \Phi[\Delta q(i)] + \langle \nabla_q F[i, q(i)] + h(i), v(i) - q(i) \rangle \right\} \geq 0.$$

In order to show that any minimizer of $\mathcal{I}$ on K is indeed a solution of (12), we need the following elementary lemma where, when appropriate, we identify $\mathbb{R}^N$ with the subspace of H_T made of constant sequences.

Lemma 5. *If ϕ is of class R, then, for any $e \in H_T$, the system$v$*

$$\Delta \phi[\Delta q(n-1)] = \overline{q} + e(n) \quad (n \in \mathbb{Z})$$

has a unique T-periodic solution $\hat{q}_e$, which is also the unique solution of the variational inequality

$$\sum_{i=1}^{T} \left\{ \Phi[\Delta v(i)] - \Phi[\Delta \hat{q}_e(i)] + \langle \overline{\hat{q}_e} + e(i), v(i) - \hat{q}_e(i) \rangle \right\} \geq 0 \quad \forall v \in H_T.$$

We can relate, like in the case of differential systems, the minimizers of $\mathcal{I}$ on K to the T-periodic solutions of system (12).

Proposition 3. *If ϕ is of class R and F of class P, any minimizer of $\mathcal{I}$ on K is a T-periodic solution of* (12).

The existence of a minimum of $\mathcal{I}$ can be proved when $\overline{h} = 0$, giving the following existence result.

Theorem 3. *If ϕ is of class R and F of class P, then, for any $h \in H_T$ such that $\overline{h} = 0$, system* (12) *has at least one T-periodic solutions which minimizes $\mathcal{I}$ over H_T.*

Proof. As F is of class P and $\overline{h} = 0$, we have, for all $(j_1, \ldots, j_N) \in \mathbb{Z}^N$,

$$\mathcal{I}(q_1 + j_1 \omega_1 e_1, \ldots, q_N + j_N \omega_N e_N) = \mathcal{I}(q_1, \ldots, q_n),$$

and hence $\inf_K \mathcal{I} = \inf_{\hat{K}_\rho} \mathcal{I}$, where $\hat{K}_\rho := \{u \in K : |\overline{u}| \leq \rho\}$ and ρ is given in (7). As $\hat{K}_\rho$ is closed and bounded in $H_T \simeq \mathbb{R}^{NT}$, Weierstrass' theorem implies the existence of some $q \in \hat{K}_\rho$ such that $\mathcal{I}(q) = \inf_{\hat{K}_\rho} \mathcal{I}$, so that q is a minimizer of $\mathcal{I}$ on K. The result follows from Proposition 3. $\qquad\qquad\square$

To obtain the multiplicity result, we again write system (12) as a system of first-order difference equations having a Hamiltonian structure through the change of variables

$$\nabla \Phi[\Delta q(n-1)] = p(n) \quad (n \in \mathbb{Z})$$

equivalent to

$$\Delta q(n-1) = \nabla \Phi^*[p(n)] \quad (n \in \mathbb{Z}),$$

with Φ^* the Legendre-Fenchel transform of Φ. System (12) is equivalent to

$$\Delta p(n) = \nabla_q F[n,q(n)] + h(n), \quad \Delta q(n) = \nabla \Phi^*[p(n+1)], \quad (n \in \mathbb{Z}). \quad (13)$$

which has the Hamiltonian form for the Hamiltonian function $H : \mathbb{Z} \times \mathbb{R}^N \times \mathbb{R}^N \to \mathbb{R}$ given by

$$H(n,u,v) = \Phi^*(u) - F(n,v) - \langle h(n), v \rangle$$

If we define the Hamiltonian action $\mathcal{A} : H_T \times H_T \to \mathbb{R}$ by

$$\mathcal{A}(p,q) = -\sum_{i=1}^{T} \{ \langle \Delta p(i), q(i) \rangle + \Phi^*[p(i)] - F[i,q(i)] - \langle h(i), q(i) \rangle \},$$

it is standard to show that the critical points of $\mathcal{A}$ on $H_T \times H_T$ correspond to the T-periodic solutions of (13) and hence to the T-periodic solutions of (12).

Now the quadratic form $\mathcal{Q}$ defined on $H_T \times H_T$ by

$$\mathcal{Q}(p,q) = -2\sum_{i=1}^{T} \langle \Delta p(i), q(i) \rangle = -2(\Delta p | q)$$

vanishes on $\overline{H}_T \times \overline{H}_T$ and is indefinite, as shown easily, and the bilinear form $b : (H_T \times H_T) \times (H_T \times H_T) \to \mathbb{R}$ defined by

$$b[(p,q),(u,v)] = (\Delta p | v) - (\Delta u | q)$$

is symmetric, so that

$$(A(p,q)|(u,v)) = -(\Delta p | v) - (\Delta u | q) \quad ((p,q),(u,v) \in H_T \times H_T),$$

for some self-adjoint operator $A : H_T \times H_T \to H_T \times H_T$ and

$$\frac{1}{2}(A(p,q)|(p,q)) = -(\Delta p | q) = \frac{1}{2}\mathcal{Q}(p,q).$$

The eigenvalues of A coincide with the eigenvalues of b and, using their Weber-Poincaré-Fischer variational characterization, satisfy the following conditions.

Lemma 6. *The eigenvalues of B (or b) can be written in a sequence*

$$-\lambda_{NT} \leq -\lambda_{NT-1} \leq \ldots \leq -\lambda_1 < \lambda_0 = 0 < \lambda_1 \leq \ldots \leq \lambda_{NT-1} \leq \lambda_{NT}.$$

The eigenspace of $\lambda_0 = 0$ is $\overline{H}_T \times \overline{H}_T$ and $H_T \times H_T$ can be decomposed in an orthogonal direct sum

$$H_T \times H_T = H^- \oplus H^0 \oplus H^+$$

with $H^0 = \overline{H}_T \times \overline{H}_T$ and

$$\mathcal{Q}(p^-,q^-) \leq -\lambda_1(\|p^-\|^2 + \|q^-\|^2) \quad ((p^-,q^-) \in H^-)$$

$$\mathcal{Q}(p^+,q^+) \geq \lambda_1(\|p^+\|^2 + \|q^+\|^2) \quad ((p^+,q^+) \in H^+).$$

Finally, the spaces H^-, H^0, H^+ are invariant for B.

Notice that

$$\mathcal{A}(p,q) = \frac{1}{2}(A(p,q)|(p,q)) - \sum_{i=1}^{T}[\Phi^*[p(i+1)] - F[i,q(i)] - \langle h(i),q(i)\rangle]$$

We apply Proposition 3 to a suitable reformulation of $\mathcal{A}$ to obtain the multiplicity theorem corresponding to Theorem 2 in the differential case. The proof is similar and details can be found in [13].

Theorem 4. *Assume that ϕ is of class R and F of class P. Then, for every $h \in H_T$ such that $\overline{h} = 0$, problem (12) has at least $N+1$ geometrically distinct solutions.*

Example 3. For any positive integer T and $h \in H_T$ such that $\overline{h} = 0$, the system

$$\Delta\phi_R[\Delta q(n-1)] + S[n,q(n)] = h(n) \quad (n \in \mathbb{Z})$$

has at least $N+1$ geometrically distinct T-periodic solutions. *In particular, for any $\mu \in \mathbb{R}$, positive integer T and $h \in H_T$ such that $\overline{h} = 0$, the equation*

$$\Delta\phi[\Delta q(n-1)] + \mu\sin q(n) = h(n) \quad (n \in \mathbb{ZZ})$$

has at least two geometrically distinct T-periodic solutions.

References

1. Bereanu, C., Jebelean, P., Mawhin, J.: Variational methods for nonlinear perturbations of singular ϕ-Laplacians. Rend. Lincei Mat. Appl. **22**, 89–111 (2011)
2. Bereanu, C., Torres, P.: Existence of at least two periodic solutions of the forced relativistic pendulum. Proc. Am. Math. Soc. **140**, 2713–2719 (2012)
3. Brezis, H., Mawhin, J.: Periodic solutions of the forced relativistic pendulum. Differ. Int. Equ. **23**, 801–810 (2010)
4. Brezis, H., Mawhin, J.: Periodic solutions of Lagrangian systems of relativistic oscillators. Commun. Appl. Anal. **15**, 235–250 (2011)
5. Chang, K.C.: On the periodic nonlinearity and the multiplicity of solutions. Nonlinear Anal. **13**, 527–537 (1989)
6. Guo, Z.M., Yu, J.S.: The existence of periodic and subharmonic solutions of subquadratic second order difference equations. J. London Math. Soc. **68**(2), 419–430 (2003)

7. Manásevich, R., Ward Jr, J.R.: On a result of Brezis and Mawhin. Proc. Amer. Math. Soc. **140**, 531–539 (2012)
8. Mawhin, J.: Forced second order conservative systems with periodic nonlinearity. Ann. Inst. Henri-Poincaré Anal. Non Linéaire **5**, 415–434 (1989)
9. Mawhin, J.: Global results for the forced pendulum equations. In: Cañada, A., Drábek, P., Fonda, A. (eds.) Handbook on Differential Equations. Ordinary Differential Equations, vol. 1, pp. 533–589. Elsevier, Amsterdam (2004)
10. Mawhin, J.: Periodic solutions of the forced pendulum: classical vs relativistic. Le Matematiche **65**, 97–107 (2010)
11. Mawhin, J.: Multiplicity of solutions of variational systems involving ϕ-Laplacians with singular ϕ and periodic nonlinearities. Discrete Contin. Dyn. Syst. **32**, 4015–4026 (2012)
12. Mawhin, J.: Periodic solutions of second order nonlinear difference systems with ϕ-Laplacian: a variational approach. Nonlinear Anal. **75**, 4672–4687 (2012)
13. Mawhin, J.: Periodic solutions of second order Lagrangian difference systems with bounded or singular ϕ-Laplacian and periodic potential Discrete Contin. Dyn. Syst. S **6**, 1065–1076 (2013)
14. Mawhin, J.: Stability and bifurcation theory for non-Autonomous differential equations. CIME (Cetraro, 2011), Jonhson R., Pera M.P. (eds.) Lect. Notes in Math, vol. 2065, pp. 103–184. Springer, Berlin (2013)
15. Mawhin, J., Willem, M.: Multiple solutions of the periodic boundary value problem for some forced pendulum-type equations. J. Differ. Equ. **52**, 264–287 (1984)
16. Mawhin, J., Willem, M.: Variational methods and boundary value problems for vector second order differential equations and applications to the pendulum equation. In: Vinti, C. (ed.) Nonlinear Analysis and Optimisation (Bologna, 1982), pp. 181–192, LNM 1107. Springer, Berlin (1984)
17. Mawhin, J., Willem, M.: Critical Point Theory and Hamiltonian Systems. Springer, New York (1989)
18. Palais, R.S.: Ljusternik-Schnirelmann theory on Banach manifolds. Topology **5**, 115–132 (1966)
19. Rabinowitz, P.: Minimax Methods in Critical Point Theory with Applications to Differential Equations. CBMS Regional Conference No. 65. American Mathematical Society, Providence (1986)
20. Rabinowitz, P.: On a class of functionals invariant under a Z_n action. Trans. Am. Math. Soc. **310**, 303–311 (1988)
21. Schwartz, J.T.: Nonlinear Functional Analysis. Gordon and Breach, New York (1969)
22. Szulkin, A.: Minimax principles for lower semicontinuous functions and applications to nonlinear boundary value problems. Ann. Inst. H. Poincaré Anal. Non Linéaire **3**, 77–109 (1986)
23. Szulkin, A.: A relative category and applications to critical point theory for strongly indefinite functionals. Nonlinear Anal. **15**, 725–739 (1990)

Two-Dimensional Differential Systems with Asymmetric Principal Part

Felix Sadyrbaev

Abstract We consider the Sturm–Liouville nonlinear boundary value problem

$$\begin{cases} x' = f(t,y) + u(t,x,y), \\ y' = -g(t,x) + v(t,x,y), \end{cases}$$

$$x(0)\cos\alpha - y(0)\sin\alpha = 0,$$
$$x(1)\cos\beta - y(1)\sin\beta = 0,$$

assuming that the limits $\lim_{y\to\pm\infty}\frac{f(t,y)}{y} = f_{\pm}$, $\lim_{x\to\pm\infty}\frac{g(t,x)}{x} = g_{\pm}$ exist. Nonlinearities u and v are bounded. The system includes various cases of asymmetric equations (such as the Fučík one). Two classes of multiplicity results are discussed. The first one is that of A. Perov–M. Krasnosel'skii; the second one has originated from the works by L. Jackson–K. Schrader and H. Knobloch.

1 Introduction

The goal of this paper is to describe and discuss two approaches for investigation of multiplicity of solutions for nonlinear boundary value problems of the type

$$\begin{cases} x' = f(t,y) + u(t,x,y), \\ y' = -g(t,x) + v(t,x,y), \end{cases} \tag{1}$$

F. Sadyrbaev (✉)
Institute of Mathematics and Computer Science, University of Latvia,
Raynis boul. 29, Riga, Latvia
e-mail: felix@latnet.lv

S. Pinelas et al. (eds.), *Differential and Difference Equations with Applications*, Springer
Proceedings in Mathematics & Statistics 47, DOI 10.1007/978-1-4614-7333-6_8,
© Springer Science+Business Media New York 2013

where f and g are principal terms and u and v are bounded nonlinearities. Functions f and g may be asymmetric functions in the meaning that $f_+ \neq f_-$ and $g_+ \neq g_-$ (see the notation in Sect. 2).

This system is considered together with the Sturm–Liouville boundary conditions

$$\begin{aligned} x(0)\cos\alpha - y(0)\sin\alpha &= 0, \\ x(1)\cos\beta - y(1)\sin\beta &= 0, \end{aligned} \tag{2}$$

where $0 \leq \alpha < \pi, 0 < \beta \leq \pi$.

Remark 1. The interval $[0,1]$ may be changed to $[a,b]$ and the homogeneous boundary conditions may be replaced by nonhomogeneous ones.

The following equations are included in the scheme (1), (2):

1. the quasi-linear problem

$$\begin{cases} x' = a(t)y + u(t,x,y), \\ y' = -b(t)x + v(t,x,y); \end{cases} \tag{3}$$

2. the Fučík type equation

$$x'' = -\lambda x^+ + \mu x^- + v(t,x,x'), \tag{4}$$

or the equivalent system

$$\begin{cases} x' = y, \\ y' = -\lambda x^+ + \mu x^- + v(t,x,y); \end{cases} \tag{5}$$

3. the equation with asymmetric principal part

$$x'' = -g(x) + v(t,x,x'), \tag{6}$$

where $\displaystyle \lim_{x \to \pm\infty} \frac{g(x)}{x} = g_\pm$, in general $g_+ \neq g_-$.

2 Perov's Approach

Perov's approach [5, Chap. 15] can be explained in the following way.
    Consider the problem

$$x'' = -g(x), \quad x(0) = 0, \ x(\pi) = 0.$$

Suppose that $g(0) = 0$ and

$$g(x) \sim k^2 x, \quad x \sim 0,$$
$$g(x) \sim m^2 x, \quad x \sim \infty,$$

where $m \neq k$.

The problem has then at least $2|k - m|$ nontrivial solutions. This can be confirmed by analyzing the (x, x')-plane by the phase plane method. Indeed, consider the Cauchy problems

$$x'' = -g(x), \quad x(0) = 0, \quad x'(0) = \gamma.$$

Introduce polar coordinates as $x = \rho \sin \varphi$, $x' = \rho \cos \varphi$. The above initial conditions in polar coordinates are (for $x'(0) > 0$ and $x'(0) < 0$, respectively)

$$\varphi_0(0) = 0, \; \rho(0) = \gamma > 0 \quad \text{or} \quad \varphi_\pi(0) = \pi, \; \rho(0) = \gamma > 0.$$

Then $\varphi_0(1; \gamma)$ takes at least $|m - k|$ values of the form πi (i is an integer) as γ varies from 0 to $+\infty$ thus producing at least $|m - k|$ solutions to the problem. Similarly, another at least $|m - k|$ solutions can be obtained analyzing the behavior of $\varphi_\pi(1; \gamma)$:

(A1) Functions f, g, u, v are continuous and continuously differentiable in phase variables x, y.

(A2) The uniform in t limits

$$\lim_{y \to +\infty} \frac{f(t,y)}{y} = f_+, \quad \lim_{y \to -\infty} \frac{f(t,y)}{y} = f_-,$$

$$\lim_{x \to +\infty} \frac{g(t,x)}{x} = g_+, \quad \lim_{x \to -\infty} \frac{g(t,x)}{x} = g_-$$

exist, where $f_\pm$ and $g_\pm$ are nonnegative.

(A3) u and v are bounded nonlinearities.

(A4) There exists a particular solution (the trivial one, for simple notation) of the problem (1), (2).

Before to state the result we need to introduce some definitions.

Consider the linear system

$$\begin{cases} x' = a(t)x + b(t)y, \\ y' = c(t)x + d(t)y. \end{cases} \tag{7}$$

Introduce polar coordinates

$$x = \rho \sin \varphi, \quad y = \rho \cos \varphi.$$

The increase of $\varphi(t)$ corresponds to clock-wise rotation in the phase plane.

Definition 1. System (7) is k-oscillatory (k is a nonnegative integer) with respect to the boundary conditions (2) if the angular function $\varphi(t)$ with the initial condition $\varphi(0) = \alpha$ satisfies the inequalities

$$\beta + k\pi < \varphi(1) < \beta + (k+1)\pi. \tag{8}$$

Consider the asymmetric system

$$\begin{cases} x' = f_+(t)y^+ - f_-(t)y^-, \\ y' = -g_+(t)x^+ + g_-(t)x^-, \end{cases} \tag{9}$$

which is a limiting system for

$$\begin{cases} x' = f(t,y) + u(t,x,y), \\ y' = -g(t,x) + v(t,x,y). \end{cases} \tag{10}$$

Introduce polar coordinates for (9) as

$$x = r\sin\Theta, \quad y = r\cos\Theta.$$

Let Θ_0 be defined by the initial condition $\Theta_0(0) = \alpha$ and Θ_π be defined by the condition $\Theta_\pi(0) = \alpha + \pi$.

Definition 2. Asymmetric system (9) is called (m,n)-oscillatory (m and n are positive integers) with respect to the boundary conditions (2) if the angular functions $\Theta_0(t)$ and $\Theta_\pi(t)$ satisfy the inequalities

$$\begin{aligned} \beta + m\pi < \Theta_0(1) < \beta + (m+1)\pi \\ \beta + (n+1)\pi < \Theta_\pi(1) < \beta + (n+2)\pi. \end{aligned} \tag{11}$$

Lemma 1. *Let m and n be the numbers in* (11).
  Then $|m - n| \leq 1$.

Proof. Lemma 3.2 in [10] or, alternatively, proof is essentially that of Proposition 2.1 in [7].   □

Remark 2. The examples exist for any possible case in Lemma 1. For instance, $m = 1$ and $n = 0$ for the problem

$$x'' = -(k\pi)^2 x^+, \quad x(0) = 0, \; x(1) = 0,$$

where $1 < k < 2$. The equation above is equivalent to the system

$$\begin{cases} x' = y, \\ y' = -(k\pi)^2 x^+. \end{cases}$$

Remark 3. The right sides in both systems (7) and (9) are positive homogeneous, and differential equations for the angular functions $\varphi(t)$ and $\Theta(t)$ do not depend on ρ and r, respectively.

Theorem 1. *Suppose the conditions (A1) to (A4) hold. Assume that*

1. The variational system around the trivial solution

$$\begin{cases} z' = f_y(t,0)w + u_x(t,0,0)z + u_y(t,0,0)w, \\ w' = -g_x(t,0)z + v_x(t,0,0)z + v_y(t,0,0)w \end{cases} \tag{12}$$

is k-oscillatory (in the sense of Definition 1*).*
2. The limiting system (9) is (m,n)-oscillatory (Definition 2).

Then the number N of nontrivial solutions to the problem (1), (2) satisfies the estimate

$$N \geq |m-k| + |k-n|. \tag{13}$$

Proof. Theorem 4.1 in [10] or, alternatively, the proof of Theorem 3.1 in [7] can be adapted for the case under consideration. □

3 Jackson–Schrader's Approach

The approach described in the previous section is based on comparison of the behaviors of nonlinearities in the right sides of a system around some specific solution (the trivial one, for instance), and at infinity the approach we start to describe now is based on the study of types of solutions to BVP.

We restrict ourselves to the boundary conditions

$$x(0) = 0, \quad x(1) = 0. \tag{14}$$

It is well known that the problem

$$x'' = f(t,x,x'), \quad x(0) = 0, \quad x(1) = 0 \tag{15}$$

is solvable if a continuous nonlinearity f is bounded. A solution $x(t)$ is a C^2-function defined in the interval $[0,1]$.

In case the right side f is an unbounded function, some specific conditions should be imposed in order to guarantee the existence of a solution to the problem. One of the suitable set of conditions in this case is the existence of the so-called lower and upper functions α and β. Namely, the problem

$$x'' = f(t,x), \quad x(0) = 0, \quad x(1) = 0 \tag{16}$$

is solvable if there exist regularly ordered upper and lower functions β and α (i.e., $\beta''(t) \leq f(t, \beta(t))$, $\alpha''(t) \geq f(t, \alpha(t))$, $\alpha \leq \beta$ for $t \in [0,1]$, $\alpha(0) \leq 0 \leq \beta(0)$, $\alpha(1) \leq 0 \leq \beta(1)$).

Similar type results are valid for $f = f(t, x, x')$. Then additional Nagumo–Bernstein type conditions are needed to ensure the solvability of the problem (one may consult the book [1] and the article [2] for the related discussion).

It was observed by Jackson and Schrader [3] (also Knobloch [4] for different boundary conditions) that in the above conditions more can be said about the expected solution. Namely, a solution $\xi(t)$ exists with the specific property that it can be approximated by a monotone sequence of solutions $\{\xi_n\}$ of auxiliary boundary problems of the Dirichlet type.

As a result, the respective equation of variations

$$y'' = f_x(t, \xi(t), \xi'(t))y + f_{x'}(t, \xi(t), \xi'(t))y', \tag{17}$$

is disconjugate in the interval $(0, 1)$, that is, a solution of the Cauchy problem (17),

$$y(0) = 0, \quad y'(0) = 1 \tag{18}$$

has not zeros in $(0, 1)$.

This observation gave motivation for the following studies. Consider a quasilinear boundary value problem

$$x'' + \lambda x = F(t, x, x'), \quad x(0) = 0, \quad x(1) = 0, \tag{19}$$

where F is bounded (and Lipschitzian, for technical reasons).

Definition 3. The linear part $x'' + \lambda x$ is i-nonresonant with respect to the boundary conditions $x(0) = 0$, $x(1) = 0$ if a solution of the Cauchy problem $x'' + \lambda x = 0$, $x(0) = 0$, $x'(0) = 1$ has exactly i zeros in $(0, 1)$ and $x(1) \neq 0$.

Definition 4. A solution $\xi(t)$ of the BVP (19) is an i-type solution if the difference $u(t; \gamma) = x(t; \gamma) - \xi(t)$ has exactly i zeros in $(0, 1)$ and $u(1; \gamma) \neq 0$ for small (in modulus) γ, where $x(t; \gamma)$ is a solution of

$$x'' + \lambda x = F(t, x, x'), \quad x(0) = 0, \quad x'(0) = \xi'(0) + \gamma.$$

It was shown in [9] that if the linear part is i-nonresonant, then an i-type solution $\xi(t)$ exists. Then a solution $y(t)$ of the respective equation of variations (if F has partial derivatives)

$$y'' + \lambda y = F_x(t, \xi(t), \xi'(t))y + F_{x'}(t, \xi(t), \xi'(t))y', \quad y(0) = 0, y'(0) = 1$$

has not zeros in $(0, 1)$. The case $y(1) = 0$ is not excluded, however, and the respective examples exist.

This knowledge can be used to detect multiple solutions in BVPs. For instance, in the work [8] the multiplicity of solutions for the problem

$$x'' = -q(t)|x|^p \text{sign}x, \quad x(0) = 0, \quad x(1) = 0$$

was studied using the quasi-linearization process. Problems of the type

$$x'' + k^2 x = \text{trunc}\{k^2 x - q(t)|x|^p \text{sign}x\}, \quad x(0) = 0, \quad x(1) = 0$$

were considered with different k, where the right side in the above line is the appropriate truncation of the function $k^2 x - q(t)|x|^p \text{sign}x$. The multiplicity results were established.

3.1 Equations with Asymmetrical Principal Part

Consider the problem with a piecewise linear left side (asymmetrical principal part) in equation

$$x'' + \lambda x^+ - \mu x^- = f(t,x,x'), \quad x(0) = 0, \quad x(1) = 0, \tag{20}$$

where $x^+ = \max\{x,0\}$, $x^- = \max\{-x,0\}$, λ and μ are nonnegative parameters.
    The respective homogeneous equation

$$x'' = -\lambda x^+ + \mu x^- \tag{21}$$

is not linear, but it possesses the positive homogeneity property, that is, for $\alpha \geq 0$ the product $\alpha x(t)$ is a solution if $x(t)$ does. The sum of two solutions $x_1(t)$ and $x_2(t)$ need not to be a solution.
    The natural question arises as is the problem (20) solvable if a continuous function f is bounded.
    To answer this question consider the problem

$$x'' = -\lambda x^+ + \mu x^-, \quad x(0) = 0, \quad x(1) = 0. \tag{22}$$

A set Σ_F of all nonnegative (λ, μ) such that the problem (22) has a nontrivial solution is called the Fučík spectrum.
    It appears that the condition $(\lambda, \mu) \notin \Sigma_F$ is not sufficient for solvability of the problem (20) even for bounded f.

Example 1.

$$x'' + \lambda x^+ = -1, \quad x(0) = 0, \, x(1) = 0,$$

where $\lambda = 4\pi^2$, $\mu = 0$. A solution to the BVP does not exist.

The solvability, however, can be proved for "good" regions as was claimed in [6].

To describe "good" regions recall that branches of the Fučík spectrum are given by the relations:

$$F_0^+ = \left\{ (\lambda, \mu) : \frac{\pi}{\sqrt{\lambda}} = 1, \quad \mu \geq 0 \right\},$$

$$F_0^- = \left\{ (\lambda, \mu) : \lambda \geq 0, \frac{\pi}{\sqrt{\mu}} = 1 \right\},$$

$$F_{2i-1}^+ = \left\{ (\lambda; \mu) : i \frac{\pi}{\sqrt{\lambda}} + i \frac{\pi}{\sqrt{\mu}} = 1 \right\},$$

$$F_{2i-1}^- = \left\{ (\lambda; \mu) : i \frac{\pi}{\sqrt{\mu}} + i \frac{\pi}{\sqrt{\lambda}} = 1 \right\},$$

$$F_{2i}^+ = \left\{ (\lambda; \mu) : (i+1) \frac{\pi}{\sqrt{\lambda}} + i \frac{\pi}{\sqrt{\mu}} = 1 \right\},$$

$$F_{2i}^- = \left\{ (\lambda; \mu) : (i+1) \frac{\pi}{\sqrt{\mu}} + i \frac{\pi}{\sqrt{\lambda}} = 1 \right\}.$$

The Fučík spectrum is depicted in Fig. 1.

Let D_i be a part of "good" region where solutions of the initial value problems

$$x'' = -\lambda x^+ + \mu x^-, \quad x(0) = 0, \quad x'(0) = \pm 1$$

have exactly i zeros in $(0,1)$. If (λ, μ) is in a "good" region, these two solutions are of opposite signs at $t = 1$, and this is important for solvability of the boundary value problem (20).

A set D_0 is a square below F_0^- and to the left of F_0^+. A set D_1 is a region bounded by F_0^-, F_0^+, and $F_1^{\pm}$. A set D_2 is a region bounded by $F_1^{\pm}$ and $\min\{F_2^+, F_2^-\}$ and so on. A union of these regions is depicted in Fig. 2.

3.2   Solvability and Properties of a Solution

Consider the problem (20) with (λ, μ) belonging to the "good" region.

Definition 5. A principal part $x'' + \lambda x^+ - \mu x^-$ is k-nonresonant with respect to boundary conditions $x(0) = 0$, $x(1) = 0$ if solutions of the Cauchy problems

$$x'' + \lambda x^+ - \mu x^- = 0, \quad x(0) = 0, \, x'(0) = 1$$
$$x'' + \lambda x^+ - \mu x^- = 0, \quad x(0) = 0, \, x'(0) = -1$$

have exactly k zeros in $(0,1)$ and $x(1) \neq 0$.

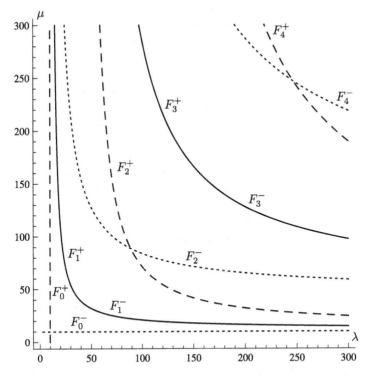

Fig. 1 The first branches of the Fučík spectrum. If $(\lambda,\mu) \in F_k^+$ (resp., $(\lambda,\mu) \in F_k^-$), then the respective solution $x(t)$ of the BVP (22) has exactly k zeros in the interval $(0,1)$ and $x'(0) > 0$ (resp., $x'(0) < 0$)

Definition 6. A solution $\xi(t)$ of the BVP (20) is called m_+-type solution (resp., m_--type solution) if the difference $u(t;\gamma) = x(t;\gamma) - \xi(t)$ has exactly m zeros in $(0,1)$ and $u(1;\gamma) \neq 0$ for small positive (resp., negative) γ, where $x(t;\gamma)$ is a solution of

$$x'' + \lambda x^+ - \mu x^- = f(t,x,x'), \quad x(0) = 0, \quad x'(0) = \xi'(0) + \gamma.$$

It is possible that $m_+ \neq m_-$.

Example 2. Consider the equation

$$x'' + (2\pi - \varepsilon)^2 x^+ - (2\pi - \varepsilon)^2 x^- = f(x),$$

where

$$f(x) = \begin{cases} p^2, & x > 1, \\ p^2 x, & 0 \leq x \leq 1, \\ 0, & x < 0. \end{cases}$$

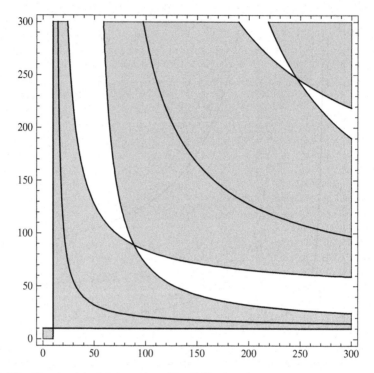

Fig. 2 "Good" region (*shaded*) is a union of D_k. If (λ, μ) is in some of D_k, then the problem (20) with a bounded f is solvable

The solution $\xi(t) \equiv 0$ is 0_+ and 1_- solution of the problem (in the sense of Definition 6), since neighboring solutions satisfy the asymmetric equation

$$x'' + [(2\pi - \varepsilon)^2 - p^2]x^+ - (2\pi - \varepsilon)^2 x^- = 0,$$

where p and ε are chosen so that

$$0 < (2\pi - \varepsilon)^2 - p^2 < \pi^2, \quad \pi^2 < (2\pi - \varepsilon)^2 < (2\pi)^2.$$

Let us state now the result concerning the problem (20). We assume that $f(t, x, x')$ in (20) is continuous and satisfies the Lipschitz conditions with respect to x and x'. Therefore solutions of the equation in (20) continuously depend on the initial data.

Theorem 2. *Boundary value problem* (20) *with k-nonresonant principal part has* k_+-*type solution and* k_--*type solution if a nonlinearity f is bounded.*

Scheme of the proof.

Step 1.   Let S be a set of all solutions of BVP. We wish to prove that the set $SI = \{\gamma \in R : x \in S, x'(0) = \gamma\}$ is compact.
First, we prove that it is bounded.
For this, consider equation

$$x'' = -\lambda x^+ + \mu x^- + f(t,x,x') \tag{23}$$

written in the form

$$\begin{cases} \dfrac{dx}{dt} = y, \\[2mm] \dfrac{dy}{dt} = -q(x) + f(t,x,y), \end{cases} \tag{24}$$

where $q(x) = \lambda x^+ - \mu x^-$. We cannot use the standard Green's function approach since the principal part in (23) is not linear. We use polar coordinates instead.

Introduce polar coordinates $(\rho(t), \theta(t))$ as

$$x(t) = \rho(t) \sin \theta(t), \quad x'(t) = \rho(t) \cos \theta(t).$$

The expression for $\theta(t)$ is

$$\begin{aligned} \frac{d\theta}{dt} &= \frac{1}{\rho} [\rho \cos^2 \theta + q(\rho \sin \theta) \sin \theta + f(t, \rho \sin \theta, \rho \cos \theta) \sin \theta] \\ &= \cos^2 \theta + q(\sin \theta) \sin \theta + \frac{1}{\rho} f(t, \rho \sin \theta, \rho \cos \theta) \sin \theta. \end{aligned} \tag{25}$$

The term

$$\frac{1}{\rho} f(t, \rho \sin \theta, \rho \cos \theta) \sin \theta \tag{26}$$

is negligibly small if $\rho(t)$ stays in a complement of the circle of sufficiently large radius for any $t \in [0,1]$. This is the case for solutions of (23) which satisfy the initial conditions

$$x(0) = 0, \quad x'(0) = \pm\gamma, \tag{27}$$

if $\gamma \to +\infty$. This is a consequence of the following result.

Lemma 2. *For solutions of the problems* (23), (27) *a function* $m(\Delta)$ *exists such that*

$$m(\gamma) \to +\infty \ as \ \gamma \to +\infty$$

and $\rho(t) \geq m(\gamma)$ *for any* $t \in [0,1]$.

Lemma 2 follows from Lemma 15.1 in [5] since all solutions of (23) are extendable to the interval $[0,1]$. The latter follows from quasi-linearity of the principal part in (20) and boundedness of f.

Therefore if a set SI is not bounded, then solutions of the BVP have arbitrarily large (in modulus) values $\gamma = x'(0)$. The respective $\rho(t)$ are arbitrarily large also and the expression (26) is arbitrarily small. The angular functions $\theta(t)$ are close then to the angular functions $\varphi(t)$ of solutions of the system

$$\begin{cases} \dfrac{dx}{dt} = y, \\ \dfrac{dy}{dt} = -q(x), \end{cases} \tag{28}$$

which is equivalent to the equation

$$x'' = -\lambda x^+ + \mu x^-.$$

The expression for $\varphi(t)$ is

$$\frac{d\varphi}{dt} = \cos^2 \varphi + q(\sin \varphi) \sin \varphi. \tag{29}$$

Due to assumption on the principal part in (20), if $\varphi(0) = 0$, then $\varphi(1) \neq \pi n$, where n is an integer.

Hence a set SI is bounded. By continuous dependence of solutions of (20) on the initial data, it is also closed, so compact in R.

Step 2. Let $\gamma_{\max}$ be maximal element in SI. The respective solution $\xi_{max}(t)$ is k_+-type solution. Suppose this is not true, say, the number k of zeros of the difference $u(t;\gamma) = x(t;\gamma) - \xi_{max}(t)$ is less than k_+. Then, increasing γ to $+\infty$ and taking into account that $x(t;\gamma)$ (and the difference $u(t;\gamma)$ also) has exactly k_+ zeros in $(0,1)$ and $u(1;\gamma) \neq 0$ for large γ, one concludes that the number of zeros of the difference $x(t;\gamma) - \xi_{max}(t)$ in the interval $(0,1)$ increases as $\gamma \to +\infty$. Then, due to continuous dependence of zeros on γ, there exists $\gamma_* > \xi'_{max}(0)$ such that $x(1;\gamma_*) - \xi_{max}(1) = 0$. Therefore $x(1;\gamma_*) = 0$ and $x(t;\gamma_*)$ is also a solution of the BVP (20). Then a contradiction with maximality of a solution $\xi_{max}(t)$ is obtained. Similarly the case $k > k_+$ can be treated.

Step 3. Let $\gamma_{\min}$ be minimal element in SI. The respective solution $\xi_{min}(t)$ is k_--type solution. The proof can be conducted in the same manner as that above.

Corollary 1. *If a solution of BVP* (20) *is unique, then it is k_+-type solution and at the same time k_--type solution.*

Corollary 2. *If problem* (20) *has a solution which is m_+-type solution and at the same time n_--type solution, where $m \neq n$, then problem has more solutions.*

4 Conclusion

The system (1) behaves like the variational system (12) in a neighborhood of the trivial solution and like asymmetric system (9) for large values of $x^2 + y^2$. This difference can be measured in terms of the angular function for solutions of the system (1). There are multiple solutions to the problem (1), (2) if oscillatory properties of the variational system and the limiting asymmetric system significantly differ.

Quasi-linear problem (20) is solvable if the coefficients λ and μ in the principal part are properly selected. In this case, a solution $\xi(t)$ exists which reflects the oscillatory properties of the principal part. If the problem is known to have a solution with properties different of those of the principal part, then there are multiple solutions to the problem.

References

1. Bernfeld, S., Lakshmikantham, V.: An Introduction to Nonlinear Boundary Value Problems. Academic, New York (1974)
2. Cabada, A.: An overview of the lower and upper solutions method with nonlinear boundary value conditions. Boundary Value Problems (2011) doi: 10.1155/2011/893753
3. Jackson, L.K., Schrader, K.W.: Comparison theorems for nonlinear differential equations. J. Diff. Equations. **3**, 248–255 (1967)
4. Knobloch, H.W.: Second order differential inequalities and a nonlinear boundary value problem. J. Diff. Equations. **5**, 55–71 (1969)
5. Krasnosel'skii, M.A., Perov, A.I., Povolockii, A.I., Zabreiko P.P.: Plane Vector Fields. Academic, New York (1966)
6. Kufner, A., Fučík, S.: Nonlinear Differential Equations. Nauka, Moscow (1988) (Russian translation of: Kufner, A., Fučik, S., Nonlinear Differential Equations, Elsevier, Amsterdam-Oxford-New York, 1980)
7. Sadyrbaev, F.: Multiplicity of solutions for two-point boundary value problems with asymptotically asymmetric nonlinearities. Nonlinear Analysis: TMA. **27**, 999–1012 (1996)
8. Sadyrbaev, F., Yermachenko, I.: Quasilinearization and multiple solutions of the Emden-Fowler type equation. Mathematical Modelling and Analysis. **10**, 41–50 (2005) doi: 10.1080/13926292.2005.9637269
9. Sadyrbaev, F., Yermachenko, I.: Types of solutions and multiplicity results for two-point nonlinear boundary value problems. Nonlinear Analysis:TMA. **63**, e1725–e1735 (2005)
10. Sadyrbaev, F., Zayakina, O.: In: Klokov, Yu. (ed.) Acta Universitatis Latviensis, Mathematics. Differential Equations, vol. 605, p. 30–47. University of Latvia, Riga (1997)

Hyperbolicity Radius of Time-Invariant Linear Systems

T.S. Doan, A. Kalauch, and S. Siegmund

Abstract Hyperbolicity of linear systems of difference and differential equations is a robust property. We provide a quantity to measure the maximal size of perturbations under which hyperbolicity is preserved. This so-called hyperbolicity radius is calculated by two methods, using the transfer operator and the input–output operator.

Keywords Hyperbolicoty radius • Transfer operator • input-output operator

1 Introduction

Hyperbolicity is an important notion in the qualitative theory of dynamical systems generated by difference equations $y_{n+1} = f(y_n)$ or differential equations $\dot{y}(t) = f(y(t))$. Roughly speaking, a constant solution y^* is said to be hyperbolic if for the according linearization $x_{n+1} = Df(y^*)x_n$ or $\dot{x}(t) = Df(y^*)x(t)$, respectively, there exists a decomposition of the state space into a stable subspace and an unstable subspace. In this paper we focus on linear systems $x_{n+1} = Ax_n$ and $\dot{x}(t) = Ax(t)$.

It is well known that hyperbolicity is robust, i.e., if one adds a small perturbation to a given hyperbolic system, then the perturbed system is still hyperbolic (see, e.g., Coppel [2]). We go one step further and discuss a measure of robustness of hyperbolicity, namely, the hyperbolicity radius. It is a quantity which measures the distance of a given hyperbolic system to the set of all nonhyperbolic systems.

T.S. Doan
Institute of Mathematics, Vietnam Academy of Science and Technology,
18 Hoang Quoc, Viet Hanoi, Vietnam
e-mail: dtson@math.ac.vn

A. Kalauch • S. Siegmund (✉)
Department for Mathematics, Institute for Analysis, TU Dresden, 01062 Dresden, Germany
e-mail: anke.kalauch@tu-dresden.de; stefan.siegmund@tu-dresden.de

S. Pinelas et al. (eds.), *Differential and Difference Equations with Applications*, Springer   113
Proceedings in Mathematics & Statistics 47, DOI 10.1007/978-1-4614-7333-6_9,
© Springer Science+Business Media New York 2013

This radius was first discussed in [9] under the name "dichotomy radius". More precisely, using the characterization of hyperbolicity by admissibility, the authors provide a lower bound for the hyperbolicity radius based on the norm of the input–output operator for nonautonomous difference equations with a special type of perturbation. In this paper we show that this lower bound is in fact equal to the hyperbolicity radius for time-invariant systems.

In case that there are no unstable directions, i.e., the unstable subspace is trivial, hyperbolicity reduces to stability. In this setting, the stability radius is introduced as the distance of a given stable system to the set of all unstable systems. In contrast to the hyperbolicity radius, the stability radius is a well-investigated notion, see, e.g., [1, 3–6]. The stability radius can be calculated by means of the transfer operator. An extension of this result to the hyperbolicity radius is given in this paper.

The paper is organized as follows. In Sect. 2 hyperbolicity is characterized for systems of time-invariant linear difference and differential equations, and the corresponding input–output operators are discussed. In Sect. 3, the hyperbolicity radii are introduced and calculated by the transfer operator and by means of the input–output operator, respectively.

To fix notation, let $\mathbb{K} = \mathbb{R}$ or $\mathbb{K} = \mathbb{C}$ and $L^2(\mathbb{Z}, \mathbb{K}^d)$ and $L^2(\mathbb{R}, \mathbb{K}^d)$ denote the set of functions $h\colon \mathbb{Z} \to \mathbb{K}^d$ and measurable functions $g\colon \mathbb{R} \to \mathbb{K}^d$, respectively, such that

$$\sum_{k=-\infty}^{\infty} \|h_k\|^2 < \infty, \quad \int_{\mathbb{R}} \|g(t)\|^2 \, dt < \infty,$$

where we use the Euclidean norm in $\mathbb{K}^d$. For a matrix $A \in \mathbb{C}^{d \times d}$, let $\sigma(A)$ denote the spectrum of A. Define $\mathbb{S}^1 := \{z \in \mathbb{C}^d : |z| = 1\}$.

2 Preliminaries

2.1 Systems of Time-Invariant Linear Difference Equations

For a matrix $A \in \mathbb{R}^{d \times d}$, the associated system of time-invariant linear difference equations is of the form

$$x_{n+1} = A x_n. \tag{1}$$

System (1) is said to be *hyperbolic* if there exist an according invariant decomposition

$$\mathbb{R}^d = S \oplus U$$

(i.e., subspaces S and U such that $AS \subset S$, $AU \subset U$) and positive constants K, α such that

$$\|A^n v\| \leq Ke^{-\alpha n}\|v\| \qquad \text{for all } v \in S, \, n \in \mathbb{N},$$

$$\|A^n v\| \geq \frac{1}{K}e^{\alpha n}\|v\| \qquad \text{for all } v \in U, \, n \in \mathbb{N}.$$

In the following theorem, several characterizations of hyperbolicity are collected.

Theorem 1. *For system* (1), *the following statements are equivalent:*

(i) System (1) *is hyperbolic.*

(ii) The spectrum of A satisfies $\sigma(A) \cap \mathbb{S}^1 = \emptyset$.

(iii) For each $s = \{s_k\}_{k \in \mathbb{Z}} \in L^2(\mathbb{Z}, \mathbb{C}^d)$ *there exists a unique* $x = \{x_k\}_{k \in \mathbb{Z}} \in L^2(\mathbb{Z}, \mathbb{C}^d)$ *satisfying*

$$x_{n+1} = Ax_n + s_n \qquad \text{for all } n \in \mathbb{Z}. \tag{2}$$

we say that $L^2(\mathbb{Z}, \mathbb{C}^d)$ *is admissible.*

Proof. See, e.g., [7]. □

The rest of this subsection is devoted to introduce the notion of input–output operator associated with a hyperbolic system. Suppose that system (1) is hyperbolic. Let T be an invertible transformation in $\mathbb{R}^{d \times d}$ such that

$$A = T \begin{pmatrix} A_1 & 0 \\ 0 & A_2 \end{pmatrix} T^{-1}, \text{ where } A_i \in \mathbb{C}^{d_i \times d_i}, \, i = 1, 2,$$

which satisfies that

$$|\lambda_1| > 1 \quad \text{for all } \lambda_1 \in \sigma(A_1), \qquad |\lambda_2| < 1 \quad \text{for all } \lambda_2 \in \sigma(A_2).$$

Define $Q_1 : \mathbb{C}^d \to \mathbb{C}^{d_1}$ and $Q_2 : \mathbb{C}^d \to \mathbb{C}^{d_2}$ by

$$Q_1(x_1, \ldots, x_d)^T = (x_1, \ldots, x_{d_1})^T, \qquad Q_2(x_1, \ldots, x_d)^T = (x_{d_1+1}, \ldots, x_d)^T.$$

Lemma 1. *Let* $B \in \mathbb{C}^{d \times m}$, $C \in \mathbb{C}^{n \times d}$. *Then, for each input* $u \in L^2(\mathbb{Z}, \mathbb{C}^m)$ *there exists a unique output* $y \in L^2(\mathbb{Z}, \mathbb{C}^n)$ *satisfying the system*

$$x_{n+1} = Ax_n + Bu_n,$$
$$y_{n+1} = Cx_n. \tag{3}$$

This output is given by

$$y_n = CT \begin{pmatrix} -\sum_{k=n}^{\infty} A_1^{n-1-k} Q_1 T^{-1} Bu_k \\ \sum_{k=-\infty}^{n-1} A_2^{n-1-k} Q_2 T^{-1} Bu_k \end{pmatrix} \tag{4}$$

and satisfies that there exists $M > 0$ *with*

$$\|y\|_2 \le M \|u\|_2 \quad \text{for all } u \in L^2(\mathbb{Z}, \mathbb{C}^m).$$

Proof. According to Theorem 1(iii), (3) has a unique solution in $L^2(\mathbb{Z}, \mathbb{C}^n)$. On the other hand, by (4) it can be easily seen that y is a solution of (3). Hence, it remains to show that $y \in L^2(\mathbb{Z}, \mathbb{C}^n)$. Indeed, since $|\lambda| < 1$ for all $\lambda \in \sigma(A_2)$ and $|\lambda| > 1$ for all $\lambda \in \sigma(A_1)$ it follows that there exist $K, \alpha > 0$ such that

$$\|A_2^m\|, \|A_1^{-m}\| \le K e^{-\alpha m} \qquad \text{for all } m \in \mathbb{N}. \tag{5}$$

By the definition of y we have

$$\|y_n\|^2 \le \|CT\|^2 \|T^{-1}B\|^2 \left(\sum_{k=n}^{\infty} \|A_1^{n-1-k}\|^2 \|u_k\|^2 + \sum_{k=-\infty}^{n-1} \|A_2^{n-1-k}\|^2 \|u_k\|^2 \right),$$

which together with (5) implies that

$$\|y_n\|^2 \le K \|CT\|^2 \|T^{-1}B\|^2 \left(\sum_{k=n}^{\infty} e^{-\alpha(k+1-n)} \|u_k\|^2 + \sum_{k=-\infty}^{n-1} e^{-\alpha(n-1-k)} \|u_k\|^2 \right).$$

Therefore,

$$\sum_{n \in \mathbb{Z}} \|y_n\|^2 \le K \|CT\|^2 \|T^{-1}B\|^2 \sum_{n \in \mathbb{Z}} \left(\sum_{k=-\infty}^{n} e^{-\alpha(n+1-k)} + \sum_{k=n+1}^{\infty} e^{-\alpha(k-1-n)} \right) \|u_n\|^2.$$

Hence,

$$\|y\|^2 \le K \frac{e^\alpha + 1}{e^\alpha - 1} \|CT\|^2 \|T^{-1}B\|^2 \|u\|^2,$$

which completes the proof.   □

For given matrices $B \in \mathbb{C}^{d \times m}$, $C \in \mathbb{C}^{n \times d}$, the operator $L : L^2(\mathbb{Z}, \mathbb{C}^m) \to L^2(\mathbb{Z}, \mathbb{C}^n)$ defined by

$$(Lu)_n := CT \begin{pmatrix} -\sum_{k=n}^{\infty} A_1^{n-1-k} Q_1 T^{-1} B u_k \\ \sum_{k=-\infty}^{n-1} A_2^{n-1-k} Q_2 T^{-1} B u_k \end{pmatrix}$$

is called the *input–output operator* associated to the hyperbolic system (1). According to Lemma 1, the operator L is well defined and bounded.

2.2 Systems of Time-Invariant Linear Differential Equations

For a matrix $A \in \mathbb{R}^{d \times d}$, the associated system of time-invariant linear differential equations is of the form

$$\dot{x}(t) = Ax(t). \tag{6}$$

Systems (6) is said to be *hyperbolic* if there exists an according invariant decomposition

$$\mathbb{R}^d = S \oplus U$$

(i.e., subspaces S and U such that $e^{tA}S \subset S$, $e^{tA}U \subset U$ for all $t \geq 0$) and positive constants K, α with

$$\|e^{tA}v\| \leq K e^{-\alpha t} \|v\| \qquad \text{for all } v \in S, t \geq 0,$$

$$\|e^{tA}v\| \geq \frac{1}{K} e^{\alpha t} \|v\| \qquad \text{for all } v \in U, t \geq 0.$$

In the following theorem, several characterizations of hyperbolicity are formulated.

Theorem 2. *For system (6), the following statements are equivalent:*

(i) *System (6) is hyperbolic.*
(ii) *The spectrum of A satisfies $\sigma(A) \cap i\mathbb{R} = \emptyset$.*
(iii) *For each $s \in L^2(\mathbb{R}, \mathbb{C}^d)$ there exists a unique solution $x \in L^2(\mathbb{R}, \mathbb{C}^d)$ satisfying*

$$\dot{x}(t) = Ax(t) + s(t) \qquad \text{for all } t \in \mathbb{R}. \tag{7}$$

Proof. See, e.g., [8]. □

Similar to the above section, the *input–output operator* of the hyperbolic system (6) with respect to the structure matrices $B \in \mathbb{C}^{d \times m}$ and $C \in \mathbb{C}^{n \times d}$ is the operator $L: L^2(\mathbb{R}, \mathbb{C}^m) \to L^2(\mathbb{R}, \mathbb{C}^n)$ defined by

$$Lu(t) := CT \begin{pmatrix} -\int_t^\infty e^{A_1(t-s)} Q_1 T^{-1} Bu(s)\, ds \\ \int_{-\infty}^t e^{A_2(t-s)} Q_2 T^{-1} Bu(s)\, ds \end{pmatrix},$$

where T is an invertible matrix which transforms matrix A into the form

$$T^{-1}AT = \begin{pmatrix} A_1 & 0 \\ 0 & A_2 \end{pmatrix} T^{-1} \quad \text{with} \quad A_i \in \mathbb{C}^{d_i \times d_i}, i = 1, 2,$$

satisfying that

$$\operatorname{Re}\lambda_1 > 0 \quad \text{for all } \lambda_1 \in \sigma(A_1), \qquad \operatorname{Re}\lambda_2 < 0 \quad \text{for all } \lambda_2 \in \sigma(A_2),$$

and $Q_1 \colon \mathbb{C}^d \to \mathbb{C}^{d_1}$ and $Q_2 \colon \mathbb{C}^d \to \mathbb{C}^{d_2}$ are defined by

$$Q_1(x_1,\dots,x_d)^T = (x_1,\dots,x_{d_1})^T, \qquad Q_2(x_1,\dots,x_d)^T = (x_{d_1+1},\dots,x_d)^T.$$

Analogously to Lemma 1, some fundamental properties of the input–output operator are given as follows.

Lemma 2. Let $B \in \mathbb{C}^{d\times m}$, $C \in \mathbb{C}^{n\times d}$. Then, for each $u \in L^2(\mathbb{R},\mathbb{C}^m)$ the unique solution $y \in L^2(\mathbb{Z},\mathbb{C}^n)$ satisfying the system

$$
\begin{aligned}
\dot{x}(t) &= Ax(t) + Bu(t), \\
y(t) &= Cx(t),
\end{aligned}
\tag{8}
$$

is $y = Lu$. Moreover, L is a bounded linear operator from $L^2(\mathbb{R},\mathbb{C}^m)$ to $L^2(\mathbb{R},\mathbb{C}^n)$.

Proof. Similar to the proof of Lemma 1.   □

3 Hyperbolicity Radius

According to Theorems 1 and 2, hyperbolicity is persistent under a small perturbation. In the following, we introduce a quantity which measures the distance between a given hyperbolic system to the set of all nonhyperbolic systems. For this purpose, we discuss first the model of perturbed systems. Let $B \in \mathbb{C}^{d\times m}$, $C \in \mathbb{C}^{n\times d}$, and $\Delta \in \mathbb{C}^{m\times n}$ be given matrices. The corresponding *structured perturbations* of (1) and (6), respectively, are given by

$$x_{n+1} = [A + B\Delta C]x_n, \tag{9}$$

and

$$\dot{x}(t) = [A + B\Delta C]x(t), \tag{10}$$

where $\Delta \in \mathbb{K}^{m\times n}$ is arbitrary.

Definition 1 (Hyperbolicity radius). Let $B \in \mathbb{C}^{d\times m}$, $C \in \mathbb{C}^{n\times d}$. Suppose that (1)/(6) is hyperbolic. The number

$$d_{\mathbb{K}}(A;B,C) := \inf\{\|\Delta\| \colon \text{system (9) / system (10) is not hyperbolic}\}$$

is called the hyperbolicity radius of (1)/(6) with respect to the structure matrices B and C, respectively.

The rest of this section is devoted to establish a characterization of hyperbolicity radius based on the norm of the input–output operator and the transfer matrix.

3.1 Systems of Time-Invariant Linear Difference Equations

Consider a hyperbolic system of time-invariant linear difference equations

$$x_{n+1} = Ax_n$$

and let $B \in \mathbb{C}^{d \times m}, C \in \mathbb{C}^{n \times d}$. We provide two methods to calculate the hyperbolicity radius.

3.1.1 Characterization of the Hyperbolicity Radius by Transfer Operator

The operator defined by

$$G(s) = C(sI - A)^{-1}B$$

is called *transfer operator* associated with (1).

Theorem 3. *The hyperbolicity radius can be computed via the transfer operator by means of*

$$d_{\mathbb{C}}(A; B, C) = \left[\max_{\theta \in [0, 2\pi]} \| G(e^{i\theta}) \| \right]^{-1}.$$

Proof. Let $\Delta \in \mathbb{C}^{m \times n}$ satisfy that $A + B\Delta C$ is not hyperbolic. Hence, there exist $0 \neq x \in \mathbb{C}^d$ and $\theta \in [0, 2\pi]$ such that

$$(A + B\Delta C)x = e^{i\theta}x,$$

which implies that

$$G(e^{i\theta})\Delta Cx = Cx.$$

As a consequence, we get

$$d_{\mathbb{C}}(A; B, C) \geq \left[\max_{\theta \in [0, 2\pi]} \| G(e^{i\theta}) \| \right]^{-1}.$$

To show the converse implication we need to construct a matrix $\Delta \in \mathbb{C}^{m \times n}$ such that

$$\| \Delta \| \leq \left[\max_{\theta \in [0, 2\pi]} \| G(e^{i\theta}) \| \right]^{-1}, \qquad \sigma(A + B\Delta C) \cap \mathbb{S}^1 \neq \emptyset. \tag{11}$$

For this purpose, we choose and fix an arbitrary $\theta \in [0, 2\pi]$. The singular value decomposition of the matrix $G(e^{i\theta})$ is given as follows

$$G(e^{i\theta}) = C(e^{i\theta}I - A)^{-1}B = \sum_{i=1}^{n} s_i u_i v_i^*,$$

where $u_i \in \mathbb{C}^{n \times 1}, v_i \in \mathbb{C}^{m \times 1}, \|u_i\| = \|v_i\| = 1$ and $s_1 > s_2 > \cdots > s_n$ are the singular values of $G(e^{i\theta})$. Define $\Delta = s_1^{-1}v_1 u_1^*$. Since $s_1 = \|G(e^{i\theta})\|$ it follows that

$$\|\Delta\| = \|G(e^{i\theta})\|^{-1}.$$

On the other hand, an elementary computation yields that

$$e^{i\theta} \in \sigma(A + B\Delta C).$$

Since θ can be chosen arbitrarily, (11) is proved and the proof is complete.   □

3.1.2 Characterization of Hyperbolicity Radius by Input–Output Operator

Theorem 4. *The hyperbolicity radius can be computed via the input–output operator by means of*

$$d_{\mathbb{C}}(A; B, C) = \frac{1}{\|L\|}.$$

Proof. We first show that

$$d_{\mathbb{C}}(A; B, C) \geq \frac{1}{\|L\|}. \tag{12}$$

For this purpose, we choose and fix an arbitrary $\Delta \in \mathbb{C}^{d \times d}$ with $\|\Delta\| \leq \frac{1}{\|L\|}$ and consider the system

$$x_{n+1} = [A + B\Delta C]x_n. \tag{13}$$

According to Theorem 1, proving the hyperbolicity of (13) is equivalent to show that the space $L^2(\mathbb{Z}, \mathbb{C}^d)$ is admissible. Choose and fix $\{s_k\}_{k \in \mathbb{Z}} \in L^2(\mathbb{Z}, \mathbb{C}^d)$ and consider the equation

$$x_{n+1} = [A + B\Delta C]x_n + s_n, \qquad \{x_k\}_{k \in \mathbb{Z}} \in L^2(\mathbb{Z}, \mathbb{C}^d).$$

This is equivalent to

$$x_{n+1} - Ax_n = B\Delta Cx_n + s_n \qquad \text{for all } n \in \mathbb{Z}. \tag{14}$$

We define $R: L^2(\mathbb{Z}, \mathbb{C}^d) \to L^2(\mathbb{Z}, \mathbb{C}^d)$ by

$$(Ru)_n = u_{n+1} - Au_n \qquad \text{for all } n \in \mathbb{Z}.$$

Clearly, R is a bounded linear operator and the inverse operator of R is determined by

$$(R^{-1}u)_n = T \begin{pmatrix} -\sum_{k=n}^{\infty} A_1^{n-1-k} Q_1 T^{-1} u_k \\ \sum_{k=-\infty}^{n-1} A_2^{n-1-k} Q_2 T^{-1} u_k \end{pmatrix}.$$

For each matrix $M \in \mathbb{C}^{p \times q}$, we define the function $M: L^2(\mathbb{Z}, \mathbb{C}^q) \to L^2(\mathbb{Z}, \mathbb{C}^p)$ by

$$(Mu)_n = Mu_n \qquad \text{for all } n \in \mathbb{Z}, \ u \in L^2(\mathbb{Z}, \mathbb{C}^q).$$

Equation (14) becomes

$$R(I - R^{-1}B\Delta C)x = s.$$

Note that $L = CR^{-1}B$ and hence, using the fact that $\|\Delta\| \|L\| \leq 1$, we obtain

$$(I - R^{-1}B\Delta C)^{-1} = I + R^{-1}B\Delta C + \sum_{k=1}^{\infty} R^{-1}B\Delta(L\Delta)^k C.$$

As a consequence, (14) has a unique solution and (12) is thus proved. In view of Theorem 1, it remains to prove that

$$\max_{\theta \in [0,2\pi]} \|G(e^{i\theta})\| \geq \|L\|. \tag{15}$$

For a fixed $\{u_k\}_{k \in \mathbb{Z}} \in L^2(\mathbb{Z}, \mathbb{C}^m)$, we define the function $f: [0, 2\pi] \to \mathbb{C}^n$ by

$$f(\theta) = \sum_{n \in \mathbb{Z}} e^{in\theta} (Lu)_n.$$

By Parseval's theorem, we get

$$\|Lu\|^2 = \sum_{n \in \mathbb{Z}} \|(Lu)_n\|^2 = \frac{1}{2\pi} \int_0^{2\pi} \|f(\theta)\|^2 \, d\theta. \tag{16}$$

An explicit form of $G(e^{i\theta})$ is given by

$$G(e^{i\theta}) = CT \begin{pmatrix} -\sum_{k=1}^{\infty} e^{i(k-1)\theta} A_1^{-k} \\ \sum_{k=0}^{\infty} e^{-i(k+1)\theta} A_2^k \end{pmatrix} T^{-1} B.$$

As a consequence, one has

$$f(\theta) = \sum_{n \in \mathbb{Z}} G(e^{-i\theta}) e^{in\theta} u_n,$$

which together with (16) implies that

$$\|Lu\|^2 \leq \frac{1}{2\pi} \int_0^{2\pi} \|G(e^{-i\theta})\|^2 \| \sum_{n \in \mathbb{Z}} e^{in\theta} u_n \|^2 \, d\theta$$

$$\leq \max_{\theta \in [0,2\pi]} \|G(e^{i\theta})\|^2 \|u\|^2,$$

which proves (15). $\qquad\qquad\qquad\qquad\qquad\qquad\qquad\qquad\qquad\square$

3.2 Systems of Time-Invariant Linear Differential Equations

Consider a hyperbolic system of time-invariant linear differential equations

$$\dot{x} = Ax$$

and let $B \in \mathbb{C}^{d \times m}$, $C \in \mathbb{C}^{n \times d}$. We again establish two methods to calculate the hyperbolicity radius.

3.2.1 Characterization of Hyperbolicity Radius by Transfer Operator

The operator defined by

$$G(s) = C(sI - A)^{-1} B$$

is called *transfer operator* associated with (6).

Theorem 5. *The hyperbolicity radius can be computed via the transfer operator by means of*

$$d_{\mathbb{C}}(A; B, C) = \left[\max_{\omega \in \mathbb{R}} \|G(i\omega)\| \right]^{-1}.$$

Proof. Let $\Delta \in \mathbb{C}^{m \times n}$ satisfy that $A + B\Delta C$ is not hyperbolic. Hence, there exists $0 \neq x \in \mathbb{C}^d$ and $\omega \in \mathbb{R}$ such that

$$(A + B\Delta C)x = i\omega x,$$

which implies that

$$G(i\omega)\Delta C x = C x.$$

As a consequence, we get

$$d_{\mathbb{C}}(A; B, C) \geq \left[\max_{\omega \in \mathbb{R}} \|G(i\omega)\| \right]^{-1}.$$

To show the converse implication we need to construct a matrix $\Delta \in \mathbb{C}^{m \times n}$ such that

$$\|\Delta\| \leq \left[\max_{\omega \in \mathbb{R}} \|G(i\omega)\| \right]^{-1}, \qquad \sigma(A + B\Delta C) \cap i\mathbb{R} \neq \emptyset. \tag{17}$$

For this purpose, we choose and fix an arbitrary $\omega \in \mathbb{R}$. The singular value decomposition of the matrix $G(i\omega)$ is given by

$$G(i\omega) = C(i\omega I - A)^{-1}B = \sum_{i=1}^{n} s_i u_i v_i^*,$$

where $u_i \in \mathbb{C}^{n \times 1}$, $v_i \in \mathbb{C}^{m \times 1}$, $\|u_i\| = \|v_i\| = 1$, and $s_1 > s_2 > \cdots > s_n$ are the singular values of $G(i\omega)$. Define $\Delta = s_1^{-1} v_1 u_1^*$. Since $s_1 = \|G(i\omega)\|$, it follows that

$$\|\Delta\| = \|G(i\omega)\|^{-1}.$$

On the other hand, an elementary computation yields

$$i\omega \in \sigma(A + B\Delta C).$$

Since ω can be chosen arbitrarily, (17) is proved. $\qquad\qquad\qquad\qquad\qquad\qquad$ □

3.2.2 Characterization of Hyperbolicity Radius by Input–Output Operator

Theorem 6. *The hyperbolicity radius can be computed via the input–output operator by means of*

$$d_{\mathbb{C}}(A; B, C) = \frac{1}{\|L\|}.$$

Proof. We first show that

$$d_{\mathbb{C}}(A;B,C) \geq \frac{1}{\|L\|}. \tag{18}$$

For this purpose, we choose and fix an arbitrary $\Delta \in \mathbb{C}^{d \times d}$ with $\|\Delta\| \leq \frac{1}{\|L\|}$ and consider the system

$$\dot{x}(t) = [A + B\Delta C]x(t). \tag{19}$$

According to Theorem 2, proving the hyperbolicity of (19) is equivalent to show that the space $L^2(\mathbb{R}, \mathbb{C}^d)$ is admissible. Choose and fix $s \in L^2(\mathbb{R}, \mathbb{C}^d)$ and consider the equation

$$\dot{x}(t) = [A + B\Delta C]x(t) + s(t).$$

This is equivalent to

$$\dot{x}(t) = Ax(t) + B\Delta Cx(t) + s(t) \qquad \text{for all } t \in \mathbb{R}. \tag{20}$$

Let T be an invertible matrix which transforms matrix A into the following form

$$T^{-1}AT = \begin{pmatrix} A_1 & 0 \\ 0 & A_2 \end{pmatrix} T^{-1} \quad \text{with} \quad A_i \in \mathbb{C}^{d_i \times d_i}, \; i = 1, 2,$$

which satisfies that

$$\text{Re}\lambda_1 > 0 \quad \text{for all } \lambda_1 \in \sigma(A_1), \qquad \text{Re}\lambda_2 < 0 \quad \text{for all } \lambda_2 \in \sigma(A_2).$$

We define $R: L^2(\mathbb{R}, \mathbb{C}^d) \to L^2(\mathbb{R}, \mathbb{C}^d)$ by

$$Ru(t) = T \begin{pmatrix} -\int_t^\infty e^{A_1(t-s)} Q_1 T^{-1} u(s) \, ds \\ \int_{-\infty}^t e^{A_2(t-s)} Q_2 T^{-1} u(s) \, ds \end{pmatrix}.$$

Clearly, $L = CRB$, and a solution $x \in L^2(\mathbb{R}, \mathbb{R}^d)$ of (20) satisfies the equation

$$x = R(B\Delta Cx + s).$$

Using the fact that $\|\Delta\|\|L\| < 1$, we obtain that

$$x = (I - RB\Delta C)^{-1} Rs$$

$$= \left[I + \sum_{k=0}^\infty RB\Delta(L\Delta)^k C \right] Rs,$$

which proves that (20) has a unique solution in $L^2(\mathbb{R}, \mathbb{R}^d)$, and (18) is shown. In view of Theorem 5, it remains to prove that

$$\max_{\omega \in \mathbb{R}} \|G(i\omega)\| \geq \|L\|. \tag{21}$$

For a fixed $u \in L^2(\mathbb{R}, \mathbb{C}^m)$, let $\hat{u}$ denote its Fourier–Plancherel transformation, i.e.,

$$\hat{u}(\omega) := \int_{-\infty}^{\infty} u(x) e^{-ix\omega} \, dx.$$

By Parseval's theorem we get

$$\|Lu\|^2 = \frac{1}{2\pi} \int_{-\infty}^{\infty} \|\widehat{Lu}(\omega)\|^2 \, d\omega. \tag{22}$$

On the other hand,

$$\widehat{Lu}(\omega) = G(i\omega)\hat{u}(\omega),$$

which together with (22) implies that

$$\|Lu\|^2 = \frac{1}{2\pi} \int_{-\infty}^{\infty} G(i\omega)\hat{u}(\omega) \, d\omega.$$

Hence, (21) is shown and the proof is complete. $\qquad \square$

References

1. Clark, S., Latushkin, Y., Montgomery-Smith, S., Randolph, T.: Stability radius and internal versus external stability in Banach spaces: an evolution semigroup approach. SIAM J. Control Optim. **38**(6), 1757–1793 (2000)
2. Coppel, W.A.: Dichotomies in Stability Theory. In: Lecture Notes in Mathematics, vol. 629. Springer, Berlin (1978)
3. Doan, T.S., Kalauch, A., Siegmund, S., Wirth, F.: Stability radii for positive linear time-invariant systems on time scales. Syst. Contr. Lett. **59**(3–4), 173–179 (2010)
4. Jacob, B.: A formula for the stability radius of time-varying systems. J. Differ. Equ. **142**(1), 167–187 (1998)
5. Hinrichsen, D., Ilchmann, A., Pritchard, A.J.: Robustness of stability of time-varying linear systems. J. Differ. Equ. **82**(2), 219–250 (1989)
6. Hinrichsen, D., Pritchard, A.J.: Stability radius for structured perturbations and the algebraic Riccati equation. Syst. Contr. Lett. **8**(2), 105–113 (1986)
7. Pham, H.A.N., Naito, T.: New characterizations of exponential dichotomy and exponential stability of linear difference equations. J. Differ. Equ. Appl. **11**(10), 909–918 (2005)
8. Preda, P., Pogan, A., Preda, C.: (L_p, L_q)–admissibility and exponential dichotomy of evolutionary processes on the half-line. Int. Equ. Oper. Theory **49**(3), 405–418 (2004)
9. Sasu, A.L.: Exponential dichotomy and dichotomy radius for difference equations. J. Math. Anal. Appl. **344**(2), 906–920 (2008)

Oscillation Criteria for Delay and Advanced Difference Equations with Variable Arguments

I.P. Stavroulakis

Abstract Consider the first-order delay difference equation

$$\Delta x(n) + p(n)x(\tau(n)) = 0, \ n \geq 0,$$

and the first-order advanced difference equation

$$\nabla x(n) - p(n)x(\mu(n)) = 0, \ \ n \geq 1, \ \ [\Delta x(n) - p(n)x(\nu(n)) = 0, \ \ n \geq 0],$$

where Δ denotes the forward difference operator $\Delta x(n) = x(n+1) - x(n)$, ∇ denotes the backward difference operator $\nabla x(n) = x(n) - x(n-1)$, $\{p(n)\}$ is a sequence of nonnegative real numbers, $\{\tau(n)\}$ is a sequence of positive integers such that $\tau(n) \leq n-1$, for all $n \geq 0$, and $\{\mu(n)\}$ $[\{\nu(n)\}]$ is a sequence of positive integers such that

$$\mu(n) \geq n+1 \text{ for all } n \geq 1, \ \ [\nu(n) \geq n+2 \text{ for all } n \geq 0].$$

The state of the art on the oscillation of all solutions to these equations is presented. Examples illustrating the results are given.

Keywords Oscillation • Delay • Advanced difference equations

Mathematical Subject Classification 2010: Primary 39A21, 39A10; Secondary 39A12.

I.P. Stavroulakis (✉)
Department of Mathematics, University of Ioannina, 451 10 Ioannina, Greece
e-mail: ipstav@cc.uoi.gr

S. Pinelas et al. (eds.), *Differential and Difference Equations with Applications*, Springer
Proceedings in Mathematics & Statistics 47, DOI 10.1007/978-1-4614-7333-6_10,
© Springer Science+Business Media New York 2013

1　Introduction

The problem of establishing sufficient conditions for the oscillation of all solutions of the first-order delay difference equation with constant arguments

$$\Delta x(n) + p(n)x(n-k) = 0, \quad n \geq 0 \tag{E}$$

has been the subject of many investigations. See, for example, [2, 8, 9, 12–17, 23, 26, 29, 34, 37, 38, 40, 43–45, 49–60, 64, 66, 67] and the references cited therein. Recently this problem was also investigated for the delay difference and the advanced difference equation with variable arguments of the form

$$\Delta x(n) + p(n)x(\tau(n)) = 0, \quad n \geq 0 \tag{E_1}$$

and

$$\nabla x(n) - p(n)x(\mu(n)) = 0, n \geq 1, \quad [\Delta x(n) - p(n)x(v(n)) = 0, n \geq 0], \tag{E_2}$$

where Δ denotes the forward difference operator $\Delta x(n) = x(n+1) - x(n)$, ∇ denotes the backward difference operator $\nabla x(n) = x(n) - x(n-1)$, $\{p(n)\}$ is a sequence of nonnegative real numbers, $\{\tau(n)\}$ is a sequence of positive integers such that $\tau(n) \leq n-1$, for all $n \geq 0$, and $\{\mu(n)\}$ $[\{v(n)\}]$ is a sequence of positive integers such that

$$\mu(n) \geq n+1 \text{ for all } n \geq 1, \quad [v(n) \geq n+2 \text{ for all } n \geq 0].$$

See, for example, [1, 3–7, 10, 18, 26, 41–43, 65], and the references cited therein.

Strong interest in the Eqs. (E_1) and (E_2) is motivated by the fact that they represent the discrete analogues of the delay

$$x'(t) + p(t)x(\tau(t)) = 0, \quad t \geq t_0, \tag{E_1}'$$

and the advanced differential equation

$$x'(t) - p(t)x(\mu(t)) = 0, \quad t \geq t_0, \tag{E_2}'$$

where $p, \tau, \mu \in C([t_0, \infty), \mathbb{R}^+)$, $\mathbb{R}^+ = [0, \infty)$, $\tau(t), \mu(t)$ are nondecreasing $\tau(t) < t$ and $\mu(t) > t$ for $t \geq t_0$ (see, for example, [11, 19–27, 30–33, 35, 36, 39, 46–48, 61–63]).

By a solution of Eq. (E_1) we mean a sequence $x(n)$ which is defined for $n \geq \min \{\tau(n) : n \geq 0\}$ and which satisfies Eq. (E_1) for all $n \geq 0$. By a solution of Eq. (E_2), we mean a sequence of real numbers $\{x(n)\}$ which is defined for $n \geq 0$ and satisfies Eq. (E_2) for all $n \geq 1$ $[n \geq 0]$. (The definition of a solution to Eq. (E) is given analogously.)

As usual, a solution $\{x(n)\}$ is said to be *oscillatory* if for every positive integer n_0 there exist $n_1, n_2 \geq n_0$ such that $x(n_1)x(n_2) \leq 0$. In other words, a solution $\{x(n)\}$ is *oscillatory* if it is neither eventually positive nor eventually negative. Otherwise, the solution is called *nonoscillatory*.

In this paper our purpose is to present the state of the art on the oscillation of all solutions to the above Eqs. $(E),(E_1)$, and (E_2), especially in the case where

$$0 < \liminf_{n \to \infty} \sum_{i=n-k}^{n-1} p(i) \leq \left(\frac{k}{k+1}\right)^{k+1} \quad \text{and} \quad \limsup_{n \to \infty} \sum_{i=n-k}^{n} p(i) < 1$$

for Eq. (E),

$$0 < \liminf_{n \to \infty} \sum_{i=\tau(n)}^{n-1} p(i) \leq \frac{1}{e} \quad \text{and} \quad \limsup_{n \to \infty} \sum_{i=\tau(n)}^{n} p(i) < 1$$

for Eq. (E_1), and

$$\limsup_{n \to \infty} \sum_{i=n}^{\mu(n)} p(i) \left[\limsup_{n \to \infty} \sum_{i=n}^{v(n)-1} p(i)\right] < 1$$

for Eq. (E_2).

2   Oscillation Criteria for Eq. (E)

In this section we study the delay difference equation with constant argument

$$\Delta x(n) + p(n)x(n-k) = 0, \quad n = 0,1,2,..., \tag{E}$$

where $\Delta x(n) = x(n+1) - x(n)$, $\{p(n)\}$ is a sequence of nonnegative real numbers, and k is a positive integer.

In 1981, Domshlak [12] was the first who studied this problem in the case where $k = 1$. Then, in 1989, Erbe and Zhang [23] established that all solutions of Eq. (E) are oscillatory if

$$\liminf_{n \to \infty} p(n) > \frac{k^k}{(k+1)^{k+1}} \tag{1}$$

or

$$\limsup_{n \to \infty} \sum_{i=n-k}^{n} p(i) > 1. \tag{C_1}$$

In the same year, 1989, Ladas et al. [37] proved that a sufficient condition for all solutions of Eq. (E) to be oscillatory is that

$$\liminf_{n \to \infty} \sum_{i=n-k}^{n-1} p(i) > \left(\frac{k}{k+1}\right)^{k+1}. \tag{C_2}$$

Therefore they improved the condition (1) by replacing the $p(n)$ of Eq. (1) by the arithmetic mean of $p(n-k), \dots, p(n-1)$ in Eq. (C_2).

Concerning the constant $\frac{k^k}{(k+1)^{k+1}}$ in Eq. (1) it should be emphasized that, as it is shown in [23], if

$$\sup p(n) < \frac{k^k}{(k+1)^{k+1}},$$

then Eq. (E) has a nonoscillatory solution.

In 1990, Ladas [34] conjectured that Eq. (E) has a nonoscillatory solution if

$$\sum_{i=n-k}^{n-1} p(i) < \left(\frac{k}{k+1}\right)^{k+1}$$

holds eventually. However, a counterexample to this conjecture was given in 1994, by Yu et al. [64].

It is interesting to establish sufficient oscillation conditions for the Eq. (E) in the case where neither Eqs. (C_1) nor (C_2) is satisfied.

In 1995, the following oscillation criterion was established by Stavroulakis [51]:

Theorem 2.1 ([51]). *Assume that*

$$\alpha_0 := \liminf_{n \to \infty} \sum_{i=n-k}^{n-1} p(i) \leq \left(\frac{k}{k+1}\right)^{k+1}$$

and

$$\limsup_{n \to \infty} p(n) > 1 - \frac{\alpha_0^2}{4}, \tag{2}$$

then all solutions of Eq. (E) oscillate.

In 2004, the same author [52] improved the condition (2) as follows:

Theorem 2.2 ([52]). *If $0 < \alpha_0 \leq \left(\frac{k}{k+1}\right)^{k+1}$, then either one of the conditions*

$$\limsup_{n \to \infty} \sum_{i=n-k}^{n-1} p(i) > 1 - \frac{\alpha_0^2}{4} \tag{C_3}$$

or

$$\limsup_{n\to\infty} \sum_{i=n-k}^{n-1} p(i) > 1 - \alpha_0^k, \tag{3}$$

implies that all solutions of Eq. (E) oscillate.

In 2006, Chatzarakis and Stavroulakis [2] established the following:

Theorem 2.3 ([2]). *If* $0 < \alpha_0 \le \left(\frac{k}{k+1}\right)^{k+1}$ *and*

$$\limsup_{n\to\infty} \sum_{i=n-k}^{n-1} p(i) > 1 - \frac{\alpha_0^2}{2(2-\alpha_0)}, \tag{4}$$

then all solutions of Eq. (E) oscillate.

Also, Chen and Yu [8] obtained the following oscillation condition:

$$\limsup_{n\to\infty} \sum_{i=n-k}^{n} p(i) > 1 - \frac{1 - \alpha_0 - \sqrt{1 - 2\alpha_0 - \alpha_0^2}}{2}. \tag{C_6}$$

3   Oscillation Criteria for Eq. (E_1)

In this section we study the delay difference equation with variable argument

$$\Delta x(n) + p(n)x(\tau(n)) = 0, \ n = 0, 1, 2, \ldots, \tag{E_1}$$

where $\Delta x(n) = x(n+1) - x(n)$, $\{p(n)\}$ is a sequence of nonnegative real numbers, and $\{\tau(n)\}$ is a nondecreasing sequence of integers such that $\tau(n) \le n - 1$ for all $n \ge 0$ and $\lim_{n\to\infty} \tau(n) = \infty$.

In 2008, Chatzarakis et al. [4] investigated for the first time the oscillatory behavior of Eq. (E_1) in the case of a general delay argument $\tau(n)$ and derived the following theorem:

Theorem 3.1 ([4]). *If*

$$\limsup_{n\to\infty} \sum_{i=\tau(n)}^{n} p(i) > 1, \tag{D_1}$$

then all solutions of Eq. (E_1) oscillate.

This result generalizes the oscillation criterion Eq. (C_1). Also in the same year Chatzarakis et al. [5] extended the oscillation criterion Eq. (C_2) to the general case of Eq. (E_1). More precisely, the following theorem has been established in [5]:

Theorem 3.2 ([5]). *Assume that*

$$\limsup_{n\to\infty} \sum_{i=\tau(n)}^{n-1} p(i) < +\infty \tag{5}$$

and

$$\alpha := \liminf_{n\to\infty} \sum_{i=\tau(n)}^{n-1} p(i) > \frac{1}{e}. \tag{D_2}$$

Then all solutions of Eq. (E_1) oscillate.

Remark 3.1. It should be mentioned that in the case of the delay differential equation

$$x'(t) + p(t)x(\tau(t)) = 0, \quad t \geq t_0, \tag{E_1}'$$

it has been proved (see [30, 36]) that either one of the conditions

$$\limsup_{n\to\infty} \int_{\tau(t)}^{t} p(s)ds > 1 \tag{D_1}'$$

or

$$\liminf_{n\to\infty} \int_{\tau(t)}^{t} p(s)ds > \frac{1}{e} \tag{D_2}'$$

implies that all solutions of Eq. $(E_1)'$ oscillate. Therefore, the conditions (D_1) and (D_2) are the discrete analogues of the conditions $(D_1)'$ and $(D_2)'$ and also the analogues of the conditions (C_1) and (C_2) in the case of a general delay argument $\tau(n)$.

Remark 3.2 ([5]). Note that the condition (5) is not a limitation since, if Eq. (D_1) holds, then all solutions of Eq. (E_1) oscillate.

Remark 3.3 ([5]). The condition (D_2) is optimal for Eq. (E_1) under the assumption that $\lim_{n\to+\infty} (n - \tau(n)) = \infty$, since in this case the set of natural numbers increases infinitely in the interval $[\tau(n), n-1]$ for $n \to \infty$.

Now, we are going to present an example to show that the condition (D_2) is optimal, in the sense that it cannot be replaced by the non-strong inequality.

Example 3.1 ([5]). Consider Eq. (E_1), where

$$\tau(n) = [\beta n], p(n) = \left(n^{-\lambda} - (n+1)^{-\lambda}\right)([\beta n])^{\lambda}, \beta \in (0,1), \lambda = -\ln^{-1}\beta \qquad (6)$$

and $[\beta n]$ denotes the integer part of βn.

It is obvious that

$$n^{1+\lambda}\left(n^{-\lambda} - (n+1)^{-\lambda}\right) \to \lambda \ for \ n \to \infty.$$

Therefore

$$n\left(n^{-\lambda} - (n+1)^{-\lambda}\right)([\beta n])^{\lambda} \to \frac{\lambda}{e} \ for \ n \to \infty. \qquad (7)$$

Hence, in view of Eqs. (6) and (7), we have

$$\liminf_{n\to\infty} \sum_{i=\tau(n)}^{n-1} p(i) = \frac{\lambda}{e} \liminf_{n\to\infty} \sum_{i=[\beta n]}^{n-1} \frac{e}{\lambda}i\left(i^{-\lambda} - (i+1)^{-\lambda}\right)([\beta i])^{\lambda} \cdot \frac{1}{i}$$

$$= \frac{\lambda}{e} \liminf_{n\to\infty} \sum_{i=[\beta n]}^{n-1} \frac{1}{i} = \frac{\lambda}{e} \ln\frac{1}{\beta} = \frac{1}{e}$$

or

$$\liminf_{n\to\infty} \sum_{i=\tau(n)}^{n-1} p(i) = \frac{1}{e}. \qquad (8)$$

Observe that all the conditions of Theorem 3.2 are satisfied except the condition (D_2). In this case it is not guaranteed that all solutions of Eq. (E_1) oscillate. Indeed, it is easy to see that the function $u = n^{-\lambda}$ is a positive solution of Eq. (E_1).

As it has been mentioned above, it is an interesting problem to find new sufficient conditions for the oscillation of all solutions of the delay difference equation (E_1), in the case where neither Eqs. (D_1) nor (D_2) is satisfied.

In 2008, Chatzarakis et al. [4] derived the following theorem:

Theorem 3.3 ([4]). *Assume that* $0 < \alpha \le \frac{1}{e}$. *Then we have*

(I) *If*

$$\limsup_{n\to\infty} \sum_{j=\tau(n)}^{n} p(j) > 1 - \left(1 - \sqrt{1-\alpha}\right)^2, \qquad (9)$$

then all solutions of Eq. (E_1) oscillate.

(II) *If in addition,*

$$p(n) \geq 1 - \sqrt{1-\alpha} \text{ for all large } n, \tag{10}$$

and

$$\limsup_{n \to \infty} \sum_{j=\tau(n)}^{n} p(j) > 1 - \alpha \frac{1 - \sqrt{1-\alpha}}{\sqrt{1-\alpha}}, \tag{11}$$

then all solutions of Eq. (E_1) oscillate.

In 2008 and 2009, the above result was improved in [6,7] as follows:

Theorem 3.4 ([6]).

(I) *If $0 < \alpha \leq \frac{1}{e}$ and*

$$\limsup_{n \to \infty} \sum_{j=\tau(n)}^{n} p(j) > 1 - \frac{1}{2}\left(1 - \alpha - \sqrt{1-2\alpha}\right), \tag{12}$$

then all solutions of Eq. (E_1) oscillate.
(II) *If $0 < \alpha \leq 6 - 4\sqrt{2}$ and in addition,*

$$p(n) \geq \frac{\alpha}{2} \text{ for all large } n, \tag{13}$$

and

$$\limsup_{n \to \infty} \sum_{j=\tau(n)}^{n} p(j) > 1 - \frac{1}{4}\left(2 - 3\alpha - \sqrt{4 - 12\alpha + \alpha^2}\right), \tag{14}$$

then all solutions of Eq. (E_1) are oscillatory.

Theorem 3.5 ([7]). *Assume that $0 < \alpha \leq \sqrt{2} - 1$, and*

$$\limsup_{n \to \infty} \sum_{j=\tau(n)}^{n} p(j) > 1 - \frac{1}{2}\left(1 - \alpha - \sqrt{1 - 2\alpha - \alpha^2}\right), \tag{C_6}'$$

then all solutions of Eq. (E_1) oscillate.

Remark 3.4. In the case where the sequence $\{\tau(n)\}$ is not assumed to be nondecreasing, define (cf. [4–7])

$$\sigma(n) = \max\{\tau(s) : 0 \leq s \leq n, s \in \mathbb{N}\}.$$

Clearly, the sequence of integers $\{\sigma(n)\}$ is nondecreasing. In this case, Theorems 3.1–3.5 can be formulated in a more general form. More precisely in the conditions (D_1), (D_2), (11), (12), (14), and $(C_6)'$ the term $\tau(n)$ is replaced by $\sigma(n)$.

Remark 3.5. Observe the following:

(i) When $0 < \alpha \leq \frac{1}{e}$, it is easy to verify that

$$\frac{1-\alpha-\sqrt{1-2\alpha-\alpha^2}}{2} > \alpha\frac{1-\sqrt{1-\alpha}}{\sqrt{1-\alpha}} > \frac{1-\alpha-\sqrt{1-2\alpha}}{2} > (1-\sqrt{1-\alpha})^2,$$

and therefore, the condition $(C_6)'$ is weaker than the conditions (11), (12), and (9).

(ii) When $0 < \alpha \leq 6 - 4\sqrt{2}$, it is easy to show that

$$\frac{1}{4}\left(2-3\alpha-\sqrt{4-12\alpha+\alpha^2}\right) > \frac{1}{2}\left(1-\alpha-\sqrt{1-2\alpha-\alpha^2}\right),$$

and therefore in this case and when Eq. (13) holds, inequality (14) improves the inequality $(C_6)'$ and especially, when $\alpha = 6 - 4\sqrt{2} \simeq 0.3431457$, the lower bound in $(C_6)'$ is 0.8929094, while in Eq. (14) is 0.7573593.

Example 3.1 ([6]). Consider the equation

$$\Delta x(n) + p(n)x(n-2) = 0,$$

where

$$p(3n) = \frac{1,474}{10,000}, \quad p(3n+1) = \frac{1,488}{10,000}, \quad p(3n+2) = \frac{6,715}{10,000}, \quad n = 0,1,2,\ldots.$$

Here $\tau(n) = n - 2$ and it is easy to see that

$$\alpha_0 = \liminf_{n\to\infty} \sum_{j=n-2}^{n-1} p(j) = \frac{1,474}{10,000} + \frac{1,488}{10,000} = 0.2962 < \left(\frac{2}{3}\right)^3 \simeq 0.2962963,$$

and

$$\limsup_{n\to\infty} \sum_{j=n-2}^{n} p(j) = \frac{1,474}{10,000} + \frac{1,488}{10,000} + \frac{6,715}{10,000} = 0.9677.$$

Observe that

$$0.9677 > 1 - \frac{1}{2}\left(1-\alpha_0-\sqrt{1-2\alpha_0}\right) \simeq 0.967317794,$$

that is, condition (12) of Theorem 3.4 is satisfied, and therefore all solutions oscillate. Also, condition $(C_6)'$ is satisfied. Observe, however, that

$$0.9677 < 1, \quad \alpha_0 = 0.2962 < \left(\frac{2}{3}\right)^3 \simeq 0.2962963,$$

$$0.9677 < 1 - \left(1 - \sqrt{1 - \alpha_0}\right)^2 \simeq 0.974055774,$$

and therefore none of the conditions (D_1), (D_2), and (9) are satisfied.

If, on the other hand, in the above equation

$$p(3n) = p(3n+1) = \frac{1,481}{10,000}, \quad p(3n+2) = \frac{6,138}{10,000}, \quad n = 0,1,2,\ldots,$$

it is easy to see that

$$\alpha_0 = \liminf_{n \to \infty} \sum_{j=n-2}^{n-1} p(j) = \frac{1,481}{10,000} + \frac{1,481}{10,000} = 0.2962 < \left(\frac{2}{3}\right)^3 \simeq 0.2962963,$$

and

$$\limsup_{n \to \infty} \sum_{j=n-2}^{n} p(j) = \frac{1,481}{10,000} + \frac{1,481}{10,000} + \frac{6,138}{10,000} = 0.91.$$

Furthermore, it is clear that $p(n) \geq \frac{\alpha_0}{2}$ for all large n. In this case

$$0.91 > 1 - \frac{1}{4}\left(2 - 3\alpha_0 - \sqrt{4 - 12\alpha_0 + \alpha_0^2}\right) \simeq 0.904724375,$$

that is, condition (14) of Theorem 3.4 is satisfied, and therefore all solutions oscillate. Observe, however, that

$$0.91 < 1, \quad \alpha_0 = 0.2962 < \left(\frac{2}{3}\right)^3 \simeq 0.2962963,$$

$$0.91 < 1 - \left(1 - \sqrt{1 - \alpha_0}\right)^2 \simeq 0.974055774,$$

$$0.91 < 1 - \frac{1}{2}\left(1 - \alpha_0 - \sqrt{1 - 2\alpha_0}\right) \simeq 0.934635588,$$

$$0.91 < 1 - \frac{1}{2}\left(1 - \alpha_0 - \sqrt{1 - 2\alpha_0 - \alpha_0^2}\right) \simeq 0.930883291,$$

and therefore, none of the conditions (D_1), (D_2), (9), (12), and $(C_6)'$ are satisfied.

4   Oscillation Criteria for Eq. (E_2)

In this section, we study the advanced difference equation with variable argument

$$\nabla x(n) - p(n)x(\mu(n)) = 0, \, n \geq 1, \quad [\Delta x(n) - p(n)x(v(n)) = 0, \, n \geq 0], \qquad (E_2)$$

where ∇ denotes the backward difference operator $\nabla x(n) = x(n) - x(n-1)$, Δ denotes the forward difference operator $\Delta x(n) = x(n+1) - x(n)$, $\{p(n)\}$ is a sequence of nonnegative real numbers, and $\{\mu(n)\}$ $[\{v(n)\}]$ is a sequence of positive integers such that

$$\mu(n) \geq n+1 \text{ for all } n \geq 1, \quad [v(n) \geq n+2 \text{ for all } n \geq 0].$$

In the special case where $\mu(n) = n + k$, $[v(n) = n + \sigma]$ the advanced difference equations (E_2) take the form

$$\nabla x(n) - p(n)x(n+k) = 0, \, n \geq 1, \quad [\Delta x(n) - p(n)x(n+\sigma) = 0, \, n \geq 0], \qquad (E')$$

where k is a positive integer greater or equal to one and σ is a positive integer greater or equal to two.

In [26], the advanced difference equation with constant argument

$$\Delta x(n) - p(n)x(n+\sigma) = 0, \quad n \geq 0$$

is studied and proved that if

$$\limsup_{n \to \infty} \sum_{i=n}^{n+\sigma-1} p(i) > 1, \qquad (C_1')$$

or

$$\liminf_{n \to \infty} \sum_{i=n+1}^{n+\sigma-1} p(i) > \left(\frac{\sigma-1}{\sigma}\right)^\sigma, \qquad (C_2')$$

then all solutions oscillate.

Very recently, Chatzarakis and Stavroulakis [3] investigated for the first time the oscillatory behavior of Eq. (E_2) with variable argument and established the following theorems:

Theorem 4.1 ([3]). *Assume that the sequence* $\{\mu(n)\}$ $[\{v(n)\}]$ *is non-decreasing. If*

$$\limsup_{n\to\infty} \sum_{i=n}^{\mu(n)} p(i) \left[\limsup_{n\to\infty} \sum_{i=n}^{v(n)-1} p(i) \right] > 1, \qquad (A_1)$$

then all solutions of Eq. (E_2) oscillate.

Theorem 4.2 ([3]). *Assume that the sequence* $\{\mu(n)\}$ $[\{v(n)\}]$ *is nondecreasing, and*

$$\liminf_{n\to\infty} \sum_{i=n+1}^{\mu(n)} p(i) \left[\liminf_{n\to\infty} \sum_{i=n+1}^{v(n)-1} p(i) \right] = \alpha.$$

If $0 < \alpha \le 1$, *and*

$$\limsup_{n\to\infty} \sum_{i=n}^{\mu(n)} p(i) \left[\limsup_{n\to\infty} \sum_{i=n}^{v(n)-1} p(i) \right] > 1 - \left(1 - \sqrt{1-\alpha}\right)^2, \qquad (15)$$

then all solutions of Eq. (E_2) oscillate.

If $0 < \alpha < (3\sqrt{5}-5)/2$, $p(n) \ge 1 - \sqrt{1-\alpha}$ *for all large n, and*

$$\limsup_{n\to\infty} \sum_{i=n}^{\mu(n)} p(i) \left[\limsup_{n\to\infty} \sum_{i=n}^{v(n)-1} p(i) \right] > 1 - \alpha \left(\frac{1}{3\sqrt{1-\alpha}+\alpha-2} - 1 \right), \qquad (16)$$

then all solutions of Eq. (E_2) oscillate.

In the special case of the advanced difference equation

$$\nabla x(n) - p(n)x(n+k) = 0, \, n \ge 1, \quad [\Delta x(n) - p(n)x(n+\sigma) = 0, \, n \ge 0], \qquad (E')$$

Theorems 4.1 and 4.2 lead to the following corollaries:

Corollary 4.1 ([3]). *Assume that*

$$\limsup_{n\to\infty} \sum_{i=n}^{n+k} p(i) \left[\limsup_{n\to\infty} \sum_{i=n}^{n+\sigma-1} p(i) \right] > 1, \qquad (A_1)'$$

then all solutions of Eq. (E') oscillate.

Corollary 4.2 ([3]). *Assume that*

$$\liminf_{n\to\infty} \sum_{i=n+1}^{n+k} p(i) \left[\liminf_{n\to\infty} \sum_{i=n+1}^{n+\sigma-1} p(i) \right] = \alpha_0.$$

If $0 < \alpha_0 \leq 1/2$, *and*

$$\limsup_{n \to \infty} \sum_{i=n}^{n+k} p(i) \left[\limsup_{n \to \infty} \sum_{i=n}^{n+\sigma-1} p(i) \right] > 1 - \left(1 - \sqrt{1 - \alpha_0} \right)^2, \qquad (4.1)'$$

then all solutions of Eq. (E') *oscillate.*

If $0 < \alpha_0 < (3\sqrt{5} - 5)/2$, $p(n) \geq 1 - \sqrt{1 - \alpha_0}$ *for all large n*, *and*

$$\limsup_{n \to \infty} \sum_{i=n}^{n+k} p(i) \left[\limsup_{n \to \infty} \sum_{i=n}^{n+\sigma-1} p(i) \right] > 1 - \alpha_0 \left(\frac{1}{3\sqrt{1 - \alpha_0} + \alpha_0 - 2} - 1 \right), \qquad (4.2)'$$

then all solutions of Eq. (E') *oscillate.*

Remark 4.1. In the case where the sequence $\{\mu(n)\}$ $[v(n)]$ is not assumed to be nondecreasing, define (cf. [3])

$$\xi(n) = \max\{\mu(s) : 1 \leq s \leq n, s \in \mathbb{N}\}, \quad [\rho(n) = \max\{v(s) : 1 \leq s \leq n, s \in \mathbb{N}\}].$$

Clearly, the sequence of integers $\{\xi(n)\}$ $[\{\rho(n)\}]$ is nondecreasing. In this case, Theorems 4.1 and 4.2 can be formulated in a more general form. More precisely, in the conditions (A_1), (15), and (16), the term $\mu(n)$ $[v(n)]$ is replaced by $\xi(n)$ $[\rho(n)]$.

Remark 4.2. When $\alpha \to 0$, then the conditions (15) and (16) of Theorem 4.2 reduce to

$$\limsup_{n \to \infty} \sum_{i=n}^{\mu(n)} p(i) \left[\limsup_{n \to \infty} \sum_{i=n}^{v(n)-1} p(i) \right] > 1,$$

that is, to the condition (A_1). However, when $0 < \alpha < (3\sqrt{5} - 5)/2$ and $p(n) \geq 1 - \sqrt{1 - \alpha}$, then we have

$$\alpha \left[\frac{1}{3\sqrt{1 - \alpha} + \alpha - 2} - 1 \right] > \left(1 - \sqrt{1 - \alpha} \right)^2$$

which means that the condition (16) is weaker than the condition (15).

Example 4.1 ([3]). Consider the advanced difference equation

$$\nabla x(n) - p(n)x(n + 1 + [\beta n]) = 0, \quad n \geq 1, \qquad (17)$$

where

$$p(n) = \begin{cases} \frac{c}{i}, & \text{if } i \neq r^n \\ \frac{96e - 100}{100e}, & \text{if } i = r^n \end{cases}, \quad n = 1, 2, \ldots,$$

and

$$c = \frac{1/e}{\ln(1+\beta)}, \quad r = 2 + \left[\frac{1}{\beta}\right], \quad \beta \in (0,1).$$

Equation (17) is of type (E_2) with $\mu(n) = n+1+[\beta n]$. Here, $\{p(n)\}$ is a sequence of positive real numbers, and $\{\mu(n)\}$ is a nondecreasing sequence of positive integers.

We will first show that

$$\lim_{n \to \infty} \sum_{i=n+1}^{n+1+[\beta n]} \frac{c}{i} = \frac{1}{e}. \tag{18}$$

Since $\frac{c}{i}$ is nonincreasing, and taking into account the fact that

$$\int_{b-1}^{b} f(x)dx \geq f(b) \geq \int_{b}^{b+1} f(x)dx,$$

where $f(x)$ is a nonincreasing positive function, we have

$$\sum_{i=n+1}^{n+1+[\beta n]} \frac{c}{i} \geq c \sum_{i=n+1}^{n+1+[\beta n]} \int_{i}^{i+1} \frac{ds}{s} = c \int_{n+1}^{n+2+[\beta n]} \frac{ds}{s} = c \ln \frac{n+2+[\beta n]}{n+1}$$

and

$$\sum_{i=n+1}^{n+1+[\beta n]} \frac{c}{i} \leq c \sum_{i=n+1}^{n+1+[\beta n]} \int_{i-1}^{i} \frac{ds}{s} = c \int_{n}^{n+1+[\beta n]} \frac{ds}{s} = c \ln \frac{n+1+[\beta n]}{n}.$$

It is easy to see that

$$\lim_{n \to \infty} \left(c \ln \frac{n+2+[\beta n]}{n+1} \right) = \lim_{n \to \infty} \left(c \ln \frac{n+1+[\beta n]}{n} \right) = c \ln(1+\beta) = \frac{1}{e}.$$

From the above it is clear that Eq. (18) holds. In particular, it follows that

$$\lim_{n \to \infty} \sum_{i=r^n+1}^{r^n+1+[\beta r^n]} \frac{c}{i} = \frac{1}{e}. \tag{19}$$

Observe that

$$r^n < r^n + 1 < r^n + 1 + [\beta r^n] < r^{n+1} \quad \text{for large } n. \tag{20}$$

Indeed, for any integer $n \geq 0$, we have $[\beta r^n] \leq \beta r^n$ and, since $\frac{\beta r^n}{r^n - 1} \to \beta < 1$, as $n \to \infty$, it holds that $[\beta r^n] < r^n - 1$, for all large n. Hence, in view of $r > 1$, we have

$$1 + [\beta r^n] < 1 + r^n - 1 = r^n < r^n(r - 1) = r^{n+1} - r^n$$

or

$$r^n + 1 + [\beta r^n] < r^{n+1} \quad \text{forlarge} n,$$

which proves Eq. (20). Thus, we get

$$\sum_{i=r^n+1}^{r^n+1+[\beta r^n]} p(i) = \sum_{i=r^n+1}^{r^n+1+[\beta r^n]} \frac{c}{i} \quad \text{foralllarge} n$$

and, because of Eq. (19),

$$\lim_{n\to\infty} \sum_{i=r^n+1}^{r^n+1+[\beta r^n]} p(i) = \frac{1}{e}. \tag{21}$$

Furthermore, since $\frac{96e-100}{100e} \geq \frac{c}{i}$ for all large i, we obtain

$$\sum_{i=n+1}^{n+1+[\beta n]} p(i) \geq \sum_{i=n+1}^{n+1+[\beta n]} \frac{c}{i} \quad \text{foralllarge} n,$$

which, by virtue of Eq. (18), gives

$$\liminf_{n\to\infty} \sum_{i=n+1}^{n+1+[\beta n]} p(i) \geq \frac{1}{e}. \tag{22}$$

From Eqs. (21) and (22) it follows that

$$\alpha = \liminf_{n\to\infty} \sum_{i=n+1}^{n+1+[\beta n]} p(i) = \frac{1}{e}. \tag{23}$$

Next, we shall prove that

$$\limsup_{n\to\infty} \sum_{i=n}^{n+1+[\beta n]} p(i) = \frac{1}{e} + \frac{96e - 100}{100e} = \frac{96}{100}. \tag{24}$$

Observe that

$$\sum_{i=r^n}^{r^n+1+[\beta r^n]} p(i) = \frac{96e - 100}{100e} + \sum_{i=r^n+1}^{r^n+1+[\beta r^n]} p(i) \quad \text{foralllarge} n,$$

and so, because of Eq. (21),

$$\lim_{n \to \infty} \sum_{i=r^n}^{r^n + 1 + [\beta r^n]} p(i) = \frac{96e - 100}{100e} + \frac{1}{e} = \frac{96}{100}. \qquad (25)$$

Furthermore, we see that

$$\lim_{n \to \infty} \left(\frac{\ln(n + 1 + [\beta n])}{\ln r} - \frac{\ln(n+1)}{\ln r} \right) = \lim_{n \to \infty} \left(\frac{\ln \frac{n+1+[\beta n]}{n+1}}{\ln r} \right) = \frac{\ln(1 + \beta)}{\ln r} < 1,$$

which implies that

$$\frac{\ln(n + 1 + [\beta n])}{\ln r} - \frac{\ln(n+1)}{\ln r} < 1 \text{ for sufficiently large } n.$$

Hence, for each large n, there exists at most one integer n^* so that

$$\frac{\ln(n+1)}{\ln r} \le n^* \le \frac{\ln(n + 1 + [\beta n])}{\ln r}$$

or

$$\ln(n+1) \le n^* \ln r \le \ln(n + 1 + [\beta n]),$$

i.e., such that

$$n + 1 \le r^{n^*} \le n + 1 + [\beta n].$$

By taking into account this fact, we obtain

$$\sum_{i=n}^{n+1+[\beta n]} p(i) \le \sum_{i=n}^{n+1+[\beta n]} \frac{c}{i} + \frac{96e - 100}{100e} = \frac{c}{n} + \sum_{i=n+1}^{n+1+[\beta n]} \frac{c}{i} + \frac{96e - 100}{100e}$$

for all large n. Combining the last inequality with Eq. (18), we have

$$\limsup_{n \to \infty} \sum_{i=n}^{n+1+[\beta n]} p(i) \le \frac{1}{e} + \frac{96e - 100}{100e} = \frac{96}{100}. \qquad (26)$$

From Eqs. (25) and (26) we conclude that Eq. (24) is always valid. Thus,

$$\limsup_{n \to \infty} \sum_{i=n}^{n+1+[\beta n]} p(i) = \frac{96}{100} = 0.96 > 1 - \left(1 - \sqrt{1 - \alpha}\right)^2 \simeq 0.957999636,$$

that is, condition (15) of Theorem 4.2 is satisfied, and therefore all solutions of Eq. (17) oscillate.

Example 4.2 ([3]). Consider the equation

$$\nabla x(n) - p(n)x(n+1) = 0, \quad n \ge 1, \tag{27}$$

where

$$p(2n-1) = \frac{1,474}{10,000}, \; p(2n) = \frac{8,396}{10,000}, \quad n \ge 1.$$

Equation (27) is of the type (E') with $k = 1$ $[\sigma = 2]$. We have

$$\alpha_0 = \liminf_{n \to \infty} \sum_{i=n+1}^{n+1} p(i) = \frac{1,474}{10,000} = 0.1474$$

and

$$\limsup_{n \to \infty} \sum_{i=n}^{n+1} p(i) = \frac{1,474}{10,000} + \frac{8,396}{10,000} = 0.987.$$

Furthermore, since $1 - \sqrt{1 - \alpha_0} \simeq 0.076636582$, we have $p(n) > 1 - \sqrt{1 - \alpha_0}$ for every $n \ge 1$. Observe that

$$0.987 > 1 - \alpha_0 \left[\frac{1}{3\sqrt{1 - \alpha_0} + \alpha_0 - 2} - 1 \right] \simeq 0.986744342,$$

that is, condition $(4.2)'$ of Corollary 4.2 is satisfied, and therefore all solutions of Eq. (27) oscillate. Observe, however, that

$$0.9677 < 1,$$

$$\alpha_0 = 0.2962 < \left(\frac{2}{3} \right)^2 = 0.2962963,$$

$$0.9677 < 1 - \left(1 - \sqrt{1 - \alpha_0} \right)^2 \simeq 0.974055774,$$

and therefore none of the conditions $(A_1)'$, (C_1'), (C_2') (since $\sigma = 2$), and $(4.1)'$ is satisfied.

Acknowledgements The author would like to thank the referee for some useful comments.

References

1. Berezansky, L., Braverman, E., Pinelas, S.: On nonoscillation of mixed advanced-delay differential equations with positive and negative coefficients. Comput. Math. Appl. **58**, 766–775 (2009)
2. Chatzarakis, G.E., Stavroulakis, I.P.: Oscillations of first order linear delay difference equations. Aust. J. Math. Anal. Appl.**3**(1), 11 (2006) (Art.14)
3. Chatzarakis, G.E., Stavroulakis, I.P.: Oscillations of difference equations with general advanced argument. Cent. Eur. J. Math. **10**(2), 807–823 (2012). DOI: 102478/s 11533-011-0137-5
4. Chatzarakis, G.E., Koplatadze, R., Stavroulakis, I.P.: Oscillation criteria of first order linear difference equations with delay argument. Nonlinear Anal. **68**, 994–1005 (2008)
5. Chatzarakis, G.E., Koplatadze, R., Stavroulakis, I.P.: Optimal oscillation criteria for first order difference equations with delay argument. Pacific J. Math. **235**, 15–33 (2008)
6. Chatzarakis, G.E., Philos, Ch.G., Stavroulakis, I.P.: On the oscillation of the solutions to linear difference equations with variable delay. Electron. J. Differ. Equ. **2008**(50), 1–15 (2008)
7. Chatzarakis, G.E., Philos, Ch.G., Stavroulakis, I.P.: An oscillation criterion for linear difference equations with general delay argument. Port. Math. **66**, 513–533 (2009)
8. Chen, M.P., Yu, Y.S.: Oscillations of delay difference equations with variable coefficients. In: Elaydi, S.N., et al. (eds.) Proceedings of First International Conference on Difference Equations, pp. 105–114. Gordon and Breach (1995)
9. Cheng, S.S. Zhang, G.: "Virus" in several discrete oscillation theorems. Appl. Math. Lett. **13**, 9–13 (2000)
10. Dannan, F.M., Elaydi, S.N.: Asymptotic stability of linear difference equations of advanced type. J. Comput. Anal. Appl. **6**, 173–187 (2004)
11. Diblik, J.: Positive and oscillating solutions of differential equations with delay in critical case. J. Comput. Appl. Math. **88**, 185–2002 (1998)
12. Domshlak, Y.: Discrete version of Sturmian Comparison Theorem for non-symmetric equations. Dokl. Azerb. Acad. Sci., **37**, 12–15 (1981) (Russian)
13. Domshlak, Y.: Sturmian comparison method in oscillation study for discrete difference equations, I. J. Differ. Int. Equ. **7**, 571–582 (1994)
14. Domshlak, Y.: Delay-difference equations with periodic coefficients: sharp results in oscillation theory. Math. Inequal. Appl. **1**, 403–422 (1998)
15. Domshlak, Y.: Riccati difference equations with almost periodic coefficients in the critical state. Dyn. Syst. Appl. **8**, 389–399 (1999)
16. Domshlak, Y.: What should be a discrete version of the Chanturia-Koplatadze Lemma? Funct. Differ. Equ. **6**, 299–304 (1999)
17. Domshlak, Y.: The Riccati difference equations near "extremal" critical states. J. Differ. Equ. Appl. **6**, 387–416 (2000)
18. Driver, R.D.: Can the future influence the present? Phys. Rev. D **19**(3), 1098–1107 (1979)
19. Elbert, A., Stavroulakis, I.P.: Oscillations of first order differential equations with deviating arguments. In: Recent Trends in Differential Equations, pp. 163–178. World Sci. Ser. Appl. Anal. 1, World Science Publishing Co., Singapore (1992)
20. Elbert, A., Stavroulakis, I.P.: Oscillation and non-oscillation criteria for delay differential equations. Proc. Am. Math. Soc.textbf123, 1503–1510 (1995)
21. Elsgolts, L.E.: Introduction to the Theory of Differential Equations with Deviating Arguments. (Translated from the Russian by R. J. McLaughlin) Holden-Day, Inc., San Francisco (1966)
22. Erbe, L.H., Kong, Q., Zhang, B.G.: Oscillation Theory for Functional Differential Equations. Dekker, New York (1995)
23. Erbe, L.H., Zhang, B.G.: Oscillation of discrete analogues of delay equations. Differ. Int. Equ. **2**, 300–309 (1989)
24. Fukagai, N., Kusano, T.: Oscillation theory of first order functional differential equations with deviating arguments. Ann. Mat. Pura Appl. **136**, 95–117 (1984)

25. Gopalsamy, K.: Stability and Oscillations in Delay Differential Equations of Population Dynamics. Kluwer Academic Publishers, Dordrecht (1992)

26. Gyori, I., Ladas, G.: Oscillation Theory of Delay Differential Equations with Applications. Clarendon Press, Oxford (1991)

27. Hale, J.K.: Theory of Functional Differential Equations. Springer, New York (1997)

28. Hoag, J.T., Driver, R.D.: A delayed-advanced model for the electrodynamics two-body problem. Nonlinear Anal. **15**, 165–184 (1990)

29. Jaroš, J., Stavroulakis, I.P.: Necessary and sufficient conditions for oscillations of difference equations with several delays. Utilitas Math. **45**, 187–195 (1994)

30. Koplatadze, R.G., Chanturija, T.A.: On the oscillatory and monotonic solutions of first order differential equations with deviating arguments. Differentsial'nye Uravneniya **18**, 1463–1465 (1982)

31. Kulenovic, M.R., Grammatikopoulos, M.K.: Some comparison and oscillation results for first-order differential equations and inequalities with a deviating argument. J. Math. Anal. Appl. **131**, 67–84 (1988)

32. Koplatadze, R.G., Kvinikadze, G.: On the oscillation of solutions of first order delay differential inequalities and equations. Georgian Math. J. **1**, 675–685 (1994)

33. Kusano, T.: On even-order functional-differential equations with advanced and retarded arguments. J. Differ. Equ. **45**, 75–84 (1982)

34. Ladas, G.: Recent developments in the oscillation of delay difference equations. In: International Conference on Differential Equations, Stability and Control. Dekker, New York (1990)

35. Ladas, G., Stavroulakis, I.P.: Oscillations caused by several retarded and advanced arguments. J. Differ. Equ. **44**, 134–152 (1982)

36. Ladas, G., Laskhmikantham, V., Papadakis, J.S.: Oscillations of higher-order retarded differential equations generated by retarded arguments. In: Delay and Functional Differential Equations and Their Applications, pp. 219–231. Academic, New York (1972)

37. Ladas, G., Philos, Ch.G., Sficas, Y.G.: Sharp conditions for the oscillation of delay difference equations. J. Appl. Math. Simul. **2**, 101–112 (1989)

38. Ladas, G., Pakula, L., Wang, Z.C.: Necessary and sufficient conditions for the oscillation of difference equations. PanAmerican Math. J. **2**, 17–26 (1992)

39. Ladde, G.S., Lakshmikantham, V., Zhang, B.G.: Oscillation Theory of Differential Equations with Deviating Arguments. Dekker, New York (1987)

40. Lalli, B., Zhang, B.G.: Oscillation of difference equations. Colloq. Math. **65**, 25–32 (1993)

41. Li, X., Zhu, D.: Oscillation and nonoscillation of advanced differential equations with variable coefficients. J. Math. Anal. Appl. **269**, 462–488 (2002)

42. Li, X., Zhu, D.: Oscillation of advanced difference equations with variable coefficients. Ann. Differ. Equ. **18**, 254–263 (2002)

43. Lin, Y.Z., Cheng, S.S.: Complete characterizations of a class of oscillatory difference equations. J. Differ. Equ. Appl. **2**, 301–313 (1996)

44. Luo, Z., Shen, J.H.: New results for oscillation of delay difference equations. Comput. Math. Appl. **41**, 553–561 (2001)

45. Luo, Z., Shen, J.H.: New oscillation criteria for delay difference equations. J. Math. Anal. Appl. **264**, 85–95 (2001)

46. Myshkis, A.D.: Linear homogeneous differential equations of first order with deviating arguments. Uspekhi Mat. Nauk, **5**, 160–162 (1950) (Russian)

47. Onose, H.: Oscillatory properties of the first-order differential inequalities with deviating argument. Funkcial. Ekvac. **26**, 189–195 (1983)

48. Sficas, Y.G., Stavroulakis, I.P.: Oscillation criteria for first-order delay equations. Bull. London Math. Soc. **35**, 239–246 (2003)

49. Shen, J.H., Luo, Z.: Some oscillation criteria for difference equations. Comput. Math. Appl. **40**, 713–719 (2000)

50. Shen, J.H., Stavroulakis, I.P.: Oscillation criteria for delay difference equations. Electron. J. Differ. Equ. **2001**(10), 1–15 (2001)

51. Stavroulakis, I.P.: Oscillations of delay difference equations. Comput. Math. Appl. **29**, 83–88 (1995)
52. Stavroulakis, I.P.: Oscillation criteria for first order delay difference equations. Mediterr. J. Math. **1**, 231–240 (2004)
53. Tang, X.H.: Oscillations of delay difference equations with variable coefficients, (Chinese). J. Central So. Univ. Technology, **29**, 287–288 (1998)
54. Tang, X.H., Cheng, S.S.: An oscillation criterion for linear difference equations with oscillating coefficients. J. Comput. Appl. Math. **132**, 319–329 (2001)
55. Tang, X.H., Yu, J.S.: A further result on the oscillation of delay difference equations. Comput. Math. Appl. **38**, 229–237 (1999)
56. Tang, X.H., Yu, J.S.: Oscillation of delay difference equations. Comput. Math. Appl. **37**, 11–20 (1999)
57. Tang, X.H., Yu, J.S.: Oscillations of delay difference equations in a critical state. Appl. Math. Lett. **13**, 9–15 (2000)
58. Tang, X.H., Yu, J.S.: Oscillation of delay difference equations. Hokkaido Math. J. **29**, 213–228 (2000)
59. Tang, X.H., Yu, J.S.: New oscillation criteria for delay difference equations. Comput. Math. Appl. **42**, 1319–1330 (2001)
60. Tian, C.J., Xie, S.L., Cheng, S.S.: Measures for oscillatory sequences. Comput. Math. Appl. **36** (10–12),149–161 (1998)
61. Wang, Z.C., Stavroulakis, I.P., Qian, X.Z.: A Survey on the oscillation of solutions of first order linear differential equations with deviating arguments. Appl. Math. E-Notes **2**, 171–191 (2002)
62. Yan, W., Yan, J.: Comparison and oscillation results for delay difference equations with oscillating coefficients. Int. J. Math. Math. Sci. **19**, 171–176 (1996)
63. Yu, J.S., Tang, X.H.: Comparison theorems in delay differential equations in a critical state and application. Proc. London Math. Soc. **63**,188–204 (2001)
64. Yu, J.S., Zhang, B.G., Wang, Z.C.: Oscillation of delay difference equations. Applicable Anal. **53**, 117–124 (1994)
65. Zhang, B.G.: Oscillation of solutions of the first-order advanced type differential equations. Sci. Explor. **2**, 79–82 (1982)
66. Zhang, B.G., Zhou, Y.: The semicycles of solutions of delay difference equations. Comput. Math. Appl. **38**, 31–38 (1999)
67. Zhang, B.G., Zhou, Y.: Comparison theorems and oscillation criteria for difference equations. J. Math. Anal. Appl. **247**, 397–409 (2000)

Positive Solutions of Nonlinear Equations with Explicit Dependence on the Independent Variable

J.R.L. Webb

Abstract We discuss positive solutions of integral equations for problems that arise from nonlinear boundary value problems. The boundary conditions can be either of local or nonlocal type. We concentrate on the case where the nonlinear term $f(t,u)$ depends explicitly on t and this dependence is crucial. We give new fixed-point index results using a comparison theorem for a class of linear operators related to the u_0-positive operators of Krasnosel'skiĭ. These are used to establish new results on existence and nonexistence of positive solutions under some conditions which can be sharp.

Keywords Positive solutions · Nonlocal boundary conditions · Fixed point index

1 Introduction

We investigate positive solutions of integral equations of the form

$$u(t) = \int_0^1 G(t,s) f(s, u(s)) \, \mathrm{d}s,$$

that arise from nonlinear boundary value problems (BVPs), for example,

$$u''(t) + f(t, u(t)) = 0, \text{ or } u^{(4)}(t) = f(t, u(t)), \ t \in (0,1),$$

subject to various boundary conditions (BCs) of local or nonlocal type. We are interested in the case where f depends explicitly on t in a crucial way, especially when it is not of the well-studied form $g(t)h(u)$.

J.R.L. Webb (✉)
School of Mathematics and Statistics, University of Glasgow, Glasgow G12 8QW, UK
e-mail: Jeffrey.Webb@glasgow.ac.uk

S. Pinelas et al. (eds.), *Differential and Difference Equations with Applications*, Springer Proceedings in Mathematics & Statistics 47, DOI 10.1007/978-1-4614-7333-6_11, © Springer Science+Business Media New York 2013

Our work covers a general situation and can be applied to many different BVPs. For example, for the above second-order differential equation, it includes BCs of separated type, possibly with one or two nonlocal terms,

$$au(0) - bu'(0) = \alpha[u], \quad cu(1) + du'(1) = \beta[u], \tag{1}$$

with a, b, c, d nonnegative, $ac + bc + ad > 0$, and $\alpha[u]$, $\beta[u]$ are linear functionals on $C[0, 1]$ that are given by

$$\alpha[u] = \int_0^1 u(s) \, dA(s), \quad \beta[u] = \int_0^1 u(s) \, dB(s), \tag{2}$$

involving Riemann–Stieltjes integrals with A, B functions of bounded variation, that is dA, dB can be sign changing measures, provided some positivity hypotheses of integral type hold; see, for example, [29, 30]. This includes the *local* case when α, β are identically 0. It treats multipoint and integral boundary conditions in a single framework. The work can also be applied to equations of higher order with local or nonlocal boundary conditions (see, for example, [30]) and to fractional differential equations (see, for example, [17]).

Let $C[0, 1]$ denote the Banach space of continuous functions defined on $[0, 1]$ endowed with the standard norm $\|u\| = \max\{u(t) : t \in [0, 1]\}$. Positive solutions of such BVPs can be obtained by finding fixed points, in a suitable cone, of the nonlinear integral operator

$$Nu(t) = \int_0^1 G(t, s) f(s, u(s)) \, ds, \tag{3}$$

where the kernel G is the Green's function for the problem. Under mild conditions this defines a compact map N in $C[0, 1]$, and, when $G \geq 0$, and $f \geq 0$, some fixed-point theory, such as the theory of fixed-point index in a sub-cone K of the cone $P = \{u \in C[0, 1] : u(t) \geq 0\}$ of nonnegative functions, can be applied to N.

A condition that is often satisfied and has proved useful in discussing positive fixed points of N is the following one:

(C) There exist a nonnegative measurable function Φ with $\Phi(s) > 0$ for a.e. $s \in (0, 1)$ and a continuous function $c \in P \setminus \{0\}$ such that

$$c(t)\Phi(s) \leq G(t, s) \leq \Phi(s), \quad \text{for } 0 \leq t, s \leq 1. \tag{4}$$

This condition was introduced in a slightly different form in [19]. For G continuous, one possibility is to take $\Phi(s) = \max_{t \in [0,1]} G(t, s)$, then the task is to determine a good function c; taking c as large as possible leads to weaker conditions.

Many papers that use various fixed-point theories have discussed the case when the nonlinear term is of the form $f(t, u) = g(t)h(u)$, where g satisfies an integrability condition but can have pointwise singularities, for example, [19, 20, 27, 28, 31, 32].

In such a case it is convenient to incorporate the term g into the kernel of the integral operator, that is, replace $G(t,s)$ by $G(t,s)g(s)$. When the nonlinearity is of the form $g(t)f(t,u)$, most previous works, for example, [4, 7, 16, 18, 23, 27, 30, 31], have concentrated on using inequalities to reduce to the case when f does not depend explicitly on t by considering hypotheses involving $\overline{f}(u) := \sup_{t \in [0,1]} f(t,u)$ and $\underline{f}(u) := \inf_{t \in [0,1]} f(t,u)$.

Some fixed-point index results, which can be used to prove existence of positive fixed points of N, have been obtained by using the related linear operator

$$Lu(t) := \int_0^1 G(t,s)u(s)\,\mathrm{d}s, \tag{5}$$

which is typically compact and has spectral radius $r(L) > 0$, and comparing the behavior of $\overline{f}(u)/u$ and $\underline{f}(u)/u$, for u near 0, and for u near ∞, with $\mu(L) := 1/r(L)$. By the Krein–Rutman theorem, $r(L)$ is an eigenvalue of L with an eigenvector in P. Then, $\mu(L)$ is called the principal characteristic value of L and is often called the "principal eigenvalue" of the corresponding BVP.

Results which employ the "eigenvalue" and give existence of at least one positive solution, and also multiplicity results, are known; see, for example, [7,27,29,31,32].

An example of an existence theorem where the conditions are sharp is as follows: there is at least one positive solution if "the nonlinearity crosses the eigenvalue," that is,

$$\text{either } \limsup_{u \to 0+} \overline{f}(u)/u < \mu(L) \text{ and } \liminf_{u \to \infty} \underline{f}(u)/u > \mu(L),$$

$$\text{or } \liminf_{u \to 0+} \underline{f}(u)/u > \mu(L) \text{ and } \limsup_{u \to \infty} \overline{f}(u)/u < \mu(L). \tag{6}$$

For local boundary conditions of separated type see, for example, [3]; for some multipoint problems see [31, 32]; and for some quite general situations see [27, 29]. In this case conditions are also known, in terms of $\underline{f}$ and $\overline{f}$, which give an arbitrary finite number of positive solutions under suitable conditions on f; see, for example, [11, 27, 29].

In a recent paper [24] we gave a new fixed-point index result which depends on the spectral radius of another related linear operator which applies on intervals bounded away from 0 and ∞. This enabled us to give a new result, related to the behavior of $\underline{f}, \overline{f}$, on the existence of two positive solutions under conditions depending only on spectral radii of linear operators.

In this paper we shall extend the results of [24] to the situation where f depends explicitly on t when this dependence is crucial.

The main contribution to the case when f depends explicitly on t, and is not simply a multiplicative factor, has been by Lan [10, 12–14, 17]. However, there has been a lack of examples for this case.

Our method uses comparison results for a class of linear operators called u_0 positive relative to two cones, introduced in [24], which is closely related to a class studied by Krasnosel'skiĭ, [8, 9]. We also give a new nonexistence result using similar comparison arguments.

In the special case when a stronger positivity condition than (C) holds, we proved in [25, 26] new existence results for *multiple* positive solutions under conditions which depend solely on the principal characteristic value of the associated linear operator, but this does not seem possible in the case studied here when only (C) holds.

To use the new results requires being able to calculate $r(L)$. The explicit t dependence usually prevents simple calculations directly from the differential equation, so some numerical method is required to obtain explicit constants. I work with the integral equation and use a C program, that runs on a desktop pc, written for me by my colleague Prof. K.A. Lindsay.

These new existence results are *complementary* to those of Lan; there are cases when one type of result is applicable but not the other.

We give two examples to illustrate our new results. We obtain explicit constants which show that our results can give improvements of Lan's result and, because we also have a nonexistence result, also can give a sharp conclusion.

2   Positive Linear Operators

A subset K of a Banach space X is called a cone if K is closed, and $x, y \in K$ and $\alpha \geq 0$ imply that $x + y \in K$ and $\alpha x \in K$, and $K \cap (-K) = \{0\}$. We always suppose that $K \neq \{0\}$. A cone defines a partial order by $x \preceq_K y \iff y - x \in K$. A cone is called *normal* if there exists $\gamma > 0$ such that for all $0 \leq x \leq y$ it follows that $\|x\| \leq \gamma \|y\|$. A cone is said to be *reproducing* if $X = K - K$ and to be *total* if $X = \overline{K - K}$.

For example, in the space $C[0, 1]$ of real-valued continuous functions on $[0, 1]$, endowed with the usual supremum norm, $\|u\| := \sup\{|u(t)| : t \in [0, 1]\}$, the cone of nonnegative functions $P := \{u \in C[0, 1] : u(t) \geq 0\}$ is well known (and easily shown) to be reproducing and normal with normality constant $\gamma = 1$.

In a recent paper [24] we have given a modification of the notion u_0-positive linear operator due to Krasnosel'skiĭ [8, 9]. We suppose that we have two cones in X, $K_0 \subset K_1$ and we let $\preceq$ denote the partial order defined by the larger cone K_1, that is, $x \preceq y \iff y - x \in K_1$. We say that L is *positive* if $L(K_1) \subset K_1$,

Definition 2.1. A positive bounded linear operator $L : X \to X$ is said to be u_0 *positive relative to the cones* (K_0, K_1), if there exists $u_0 \in K_1 \setminus \{0\}$, such that for every $u \in K_0 \setminus \{0\}$ there are constants $k_2(u) \geq k_1(u) > 0$ such that

$$k_1(u)u_0 \preceq Lu \preceq k_2(u)u_0.$$

This is stronger than requiring that L is positive and is satisfied if L is u_0-positive on K_1 according to [6, 9] (take $K_0 = K_1$ in the definition above). The idea is to utilize the extra properties satisfied by elements of the smaller cone but only use the weaker ordering of the larger cone.

In [24] we proved the following comparison theorem which is similar to one of Keener and Travis [6], which was a sharpening of some results of Krasnosel'skiĭ, see [8, Sect. 2.5.5].

Theorem 2.1 ([24]). *Let $K_0 \subset K_1$ be cones in a Banach space X, and let $\preceq$ denote the partial order of K_1. Suppose that L_1, L_2 are bounded linear operators and that at least one is u_0-positive relative to (K_0, K_1). If there exist*

$$u_1 \in K_0 \setminus \{0\}, \ \lambda_1 > 0, \ such\,that \ \lambda_1 u_1 \preceq L_1 u_1, \quad and$$

$$u_2 \in K_0 \setminus \{0\}, \ \lambda_2 > 0, \ such\,that \ \lambda_2 u_2 \succeq L_2 u_2, \tag{7}$$

and $L_1 u_j \preceq L_2 u_j$ for $j = 1, 2$, then $\lambda_1 \leq \lambda_2$.

We shall use this only for one compact linear operator L taking $L_1 = L_2 = L$ to obtain one part of the following result. Let $r(L)$ denote the spectral radius of a linear operator L.

Theorem 2.2. *Let $K_0 \subset K_1$ be cones in a Banach space X and let L be a compact positive linear operator:*

(i) *If there exist $u_1 \in K_1 \setminus \{0\}$ and $\lambda_1 > 0$ such that $\lambda_1 u_1 \preceq L u_1$, then $r(L) \geq \lambda_1$.*
(ii) *Suppose that K_1 is a total cone, L is u_0-positive relative to (K_0, K_1), and $L(K_1) \subset K_0$. If there exist $u_2 \in K_1 \setminus \{0\}$ and $\lambda_2 > 0$ such that $\lambda_2 u_2 \succeq L u_2$, then $r(L) \leq \lambda_2$.*

Proof.

(i) By Theorem 2.5 of Krasnosel'skiĭ [8] (a new short proof of this using fixed-point index theory is given in [25]), L has an eigenfunction in K_1 with an eigenvalue $\lambda \geq \lambda_1$, hence $r(L) \geq \lambda \geq \lambda_1$. In fact the result (i) holds without the compactness assumption by a known elementary argument, repeated in [25], but L may have no real eigenvalues in that case.
(ii) We may suppose that $r(L) > 0$. As K_1 is a total cone and $L : K_1 \to K_1$ is compact, $r(L)$ is an eigenvalue of L with eigenfunction $\varphi \in K_1 \setminus \{0\}$ by the Krein–Rutman theorem (see, for example [2]), that is, $r(L)\varphi = L\varphi$. Since $L(K_1) \subset K_0$ it follows that $\varphi \in K_0$. We may also assume $u_2 \in K_0$ by replacing it by $L u_2$ if necessary. By the comparison Theorem 2.1, we obtain $\lambda_2 \geq r(L)$. $\qquad\square$

3   Integral Equations and New Fixed-Point Index Results

Studying positive solutions of a BVP can be done by finding fixed points, in some sub-cone K of the cone $P = \{u \in C[0, 1] : u(t) \geq 0\}$, of the nonlinear integral operator

$$Nu(t) = \int_0^1 G(t,s)f(s,u(s))\,ds. \tag{8}$$

If the nonlinearity is of the form $g(t)f(t,u)$ with a possibly singular term g (usually integrable), then we may replace the kernel (Green's function) G by $\tilde{G} = Gg$, so in the theory we only need to consider the form (8).

Under mild conditions this defines a compact map N in the space $C[0,1]$, and, when $G \geq 0$ and $f \geq 0$, the theory of fixed-point index can often be applied to N to prove existence of multiple positive solutions for the integral equation

$$u(t) = \int_0^1 G(t,s)f(s,u(s))\,ds.$$

The rather weak conditions that we now impose on G, f, g are similar to ones in the papers [27, 29, 31].

(C_1) The kernel $G \geq 0$ is measurable, and for every $\tau \in [0,1]$ we have

$$\lim_{t \to \tau} |G(t,s) - G(\tau,s)| = 0 \text{ for almost every} \quad (\text{a.e.}) \quad s \in [0,1].$$

(C_2) There exist a nonnegative function $\Phi \in L^1$ with $\Phi(s) > 0$ for a.e. $s \in (0,1)$ and $c \in P \setminus \{0\}$ such that

$$c(t)\Phi(s) \leq G(t,s) \leq \Phi(s), \text{ for } 0 \leq t,s \leq 1. \tag{9}$$

For a subinterval $J = [t_0,t_1]$ of $[0,1]$, let $c_J := \min\{c(t) : t \in J\}$; since $c \in P \setminus \{0\}$, there exist intervals J with $c_J > 0$.

(C_3) The nonlinearity $f : [0,1] \times [0,\infty) \to [0,\infty)$ satisfies Carathéodory conditions, that is, $f(\cdot,u)$ is measurable for each fixed $u \geq 0$ and $f(t,\cdot)$ is continuous for a.e. $t \in [0,1]$, and for each $r > 0$, there exists ϕ^r such that

$$f(t,u) \leq \phi^r(t) \text{ for all } u \in [0,r] \text{ and a.e. } t \in [0,1], \text{ where } \Phi\phi^r \in L^1.$$

Clearly, (C_1) is satisfied if G is continuous. A precursor of condition (C_2) was used in [19]. The condition (C_2) is frequently satisfied by ordinary differential equations with both local and nonlocal boundary conditions, with the function c positive on $(0,1)$; see, for example, [29] for a quite general situation.

For a subinterval $J = [t_0,t_1] \subseteq [0,1]$ such that $c_J := \min\{c(t) : t \in J\} > 0$, we define cones K_c, K_J by

$$K_c := \{u \in P : u(t) \geq c(t)\|u\|, t \in [0,1]\}, \tag{10}$$

$$K_J := \{u \in P : u(t) \geq c_J\|u\|, t \in J\}. \tag{11}$$

It is clear that $K_c \subset K_J$. When we consider the cone K_J, we will always suppose that $c_J > 0$. These cones, especially the second, have been studied by many authors in the study of existence of multiple positive solutions of BVPs. For the first cone we mention [1, 15, 16]; for the second see, for example, [5, 27, 29, 31].

Let $f(t, u)$ satisfy Carathéodory conditions. For $0 < r < R$ define functions by

$$F_{r,R}(t) := \inf_{u \in [r,R]} f(t,u)/u, \quad F^{r,R}(t) := \sup_{u \in [r,R]} f(t,u)/u. \tag{12}$$

We define corresponding linear operators by

$$L^{r,R}u(t) := \int_0^1 G(t,s)F^{r,R}(s)u(s)\,ds,$$

$$L^J_{r,R}u(t) := \int_{t_0}^{t_1} G(t,s)F_{r,R}(s)u(s)\,ds. \tag{13}$$

Since $F_{r,R}(t) \leq F^{r,R}(t) \leq \sup_{u \in [r,R]} f(t,u)/r \leq \phi^R(t)/r$ for a.e t, and $G(t,s) \leq \Phi(s)$, using condition (C_3) shows that these operators are well defined on $C[0,1]$. Standard arguments show that they are compact; see, for example, [21, Proposition V.3.1].

We define $F^{R,\infty}(t)$ and $L^{R,\infty}$ in the obvious way. We also want to consider the case $r = 0$, that is, we want to define $L^{0,\rho}$ and $L_{0,\rho}$ for $\rho > 0$. Let $F^{0,\rho}(t) := \sup_{u \in (0,\rho]} f(t,u)/u$, and suppose that $\Phi F^{0,\rho} \in L^1$. Then we define

$$L^{0,\rho}u(t) := \int_0^1 G(t,s)F^{0,\rho}(s)u(s)\,ds. \tag{14}$$

Similarly we define $L_{0,\rho}u(t)$ when $F_{0,\rho}(t) := \inf_{u \in (0,\rho]} f(t,u)/u$.

Lemma 3.1. *For $0 \leq r < R$ the compact linear operators $L^J_{r,R}$ and $L^{r,R}$ map P into K_c.*

Proof. The known argument is essentially the same for each operator; we include the proof for one of them to illustrate the argument. For $u \in P$ we have

$$L^J_{r,R}u(t) \leq \int_{t_0}^{t_1} \Phi(s)F_{r,R}(s)u(s)\,ds,$$

hence $\|L^J_{r,R}u\| \leq \int_{t_0}^{t_1} \Phi(s)F_{r,R}(s)u(s)\,ds$; and also

$$L^J_{r,R}u(t) \geq c(t) \int_{t_0}^{t_1} \Phi(s)F_{r,R}(s)u(s)\,ds \geq c(t)\|L^J_{r,R}u\|.$$

$\square$

Lemma 3.2. *Suppose that $c_J > 0$ and that $\int_J \Phi(s)F_{r,R}(s)\,ds > 0$. Then $r(L^J_{r,R}) > 0$, and hence by the Krein–Rutman theorem, $r(L^J_{r,R})$ is an eigenvalue of $L^J_{r,R}$ with an eigenfunction in K_c. Similarly $r(L^{r,R})$ is an eigenvalue of $L^{r,R}$ with an eigenfunction in K_c.*

Proof. Let $\hat{1}$ denote the constant function with value 1. Then we have

$$L_{r,R}^J \hat{1} \succeq \left(c_J \int_{t_0}^{t_1} \Phi(s) F_{r,R}(s) \, ds \right) \hat{1},$$

so by Theorem 2.5 of Krasnosel'skiĭ [8], $L_{r,R}^J$ has an eigenfunction in P with an eigenvalue $\lambda \geq c_J \int_{t_0}^{t_1} \Phi(s) F_{r,R}(s) \, ds > 0$. Hence $r(L_{r,R}^J) > 0$ and, since P is a total cone, the Krein–Rutman theorem then says that $r(L_{r,R}^J))$ is an eigenvalue of $L_{r,R}^J$ with an eigenfunction in P. As $L_{r,R}^J$ maps P into K_c, the eigenfunction is in K_c. Similarly we have

$$L^{r,R} c(t) = \int_0^1 G(t,s) F^{r,R}(s) c(s) \, ds \geq \int_0^1 c(t) \Phi(s) F^{r,R}(s) c(s) \, ds$$

so $L^{r,R}$ has an eigenvalue $\lambda \geq \int_0^1 \Phi(s) F^{r,R}(s) c(s) \, ds > 0$.   □

One reason for considering the operator $L_{r,R}^J u(t) = \int_{t_0}^{t_1} G(t,s) F_{r,R}(s) u(s) \, ds$ rather than the "natural" operator $L_{r,R} u(t) = \int_0^1 G(t,s) F_{r,R}(s) u(s) \, ds$ is the following result:

Theorem 3.1. *Suppose that $c_J > 0$ and $\int_J \Phi(s) F_{r,R}(s) \, ds > 0$. Then the linear operator $L_{r,R}^J$ is u_0-positive relative to (K_c, P).*

Proof. Let $u \in K_c \setminus \{0\}$. Then we have

$$L_{r,R}^J u(t) = \int_J G(t,s) F_{r,R}(s) u(s) \, ds \leq \left(\int_J G(t,s) F_{r,R}(s) \, ds \right) \|u\|,$$

and, since $u \in K_c \subset K_J$,

$$L_{r,R}^J u(t) = \int_J G(t,s) F_{r,R}(s) u(s) \, ds \geq \left(\int_J G(t,s) F_{r,R}(s) \, ds \right) c_J \|u\|.$$

Let $u_0(t) := \int_J G(t,s) F_{r,R}(s) \, ds$, then $u_0(t) \geq c_J \int_J \Phi(s) F_{r,R}(s) \, ds > 0$ so $u_0 \in P \setminus \{0\}$.   □

Remark 3.1. It is not clear whether or not $L_{r,R}$ is u_0-positive relative to (K_c, P) without some extra hypothesis. In some special cases, G satisfies the stronger positivity condition

$$c_0 \Phi(s) \leq G(t,s) \leq \Phi(s), \quad \text{for } 0 \leq t, s \leq 1, \tag{15}$$

for a constant $c_0 > 0$. We may then take $J = [0,1]$ and then $L_{r,R}^J = L_{r,R}$ is u_0-positive relative to (K_c, P), and $L^{r,R}$ is also u_0-positive relative to (K_c, P); in this case, see [26]. Stronger results than we can prove in the present paper are possible for this restricted case, as shown in [25, 26].

For $\rho > 0$ define the open (relative to K_c) set $U_\rho = \{u \in K_c : \|u\| < \rho\}$. Note that, if $u \in \partial U_\rho$ (the boundary relative to K_c), then $0 \leq u(t) \leq \rho$ for all $t \in [0,1]$ and $c_J \rho \leq u(t) \leq \rho$ for $t \in J$.

We now give our new fixed-point index results. In the case when essentially there is no explicit t dependence, with hypotheses similar to $(C_1),(C_2),(C_3)$, these results were first proved in [31], with rather different arguments.

The first new result gives conditions so that the index equals one. It is typically applied for ρ small which corresponds to the behavior of $f(t,u)/u$ for u near 0.

Theorem 3.2. *Suppose there exists $\rho > 0$ such that $r(L^{0,\rho}) < 1$. Then*

$$i_{K_c}(N, U_\rho) = 1. \tag{16}$$

Proof. We claim that $Nu \neq \sigma u$ for all $u \in \partial U_\rho$ and all $\sigma \geq 1$, from which it follows by standard properties of fixed-point index that $i_{K_c}(N, U_\rho) = 1$; see, for example, [2, 5]. If there exist $\sigma \geq 1$ and $u \in \partial U_\rho$ with $\sigma u = Nu$ then $0 \leq u(s) \leq \rho$ for all $s \in [0,1]$ and $\|u\| = \rho$, therefore we have

$$u(t) = \frac{1}{\sigma} Nu(t) \leq \int_0^1 G(t,s) f(s, u(s)) \, ds \leq \int_0^1 G(t,s) F^{0,\rho}(s) u(s) \, ds,$$

that is, $u \preceq L^{0,\rho} u$. By Theorem 2.2 (i) this implies $r(L^{0,\rho}) \geq 1$, which contradiction proves the claim. $\qquad\square$

We next give a new result which depends on the behavior of $f(t,u)/u$ for u very large.

Theorem 3.3. *Suppose there exists $R > 0$ such that $r(L^{R,\infty}) < 1$. Then there exists $R_1 > R$ such that*

$$i_{K_c}(N, U_{R_1}) = 1. \tag{17}$$

Proof. We note that, since $r(L^{R,\infty}) < 1$ and $L^{R,\infty}$ maps P to P, $I - L^{R,\infty}$ is invertible and maps P into P. We have $f(t,u) \leq F^{R,\infty}(t)u$ for all $u \geq R$, hence

$$f(t,u) \leq F^{R,\infty}(t)u + \phi^R(t) \text{ for all } u \geq 0.$$

Let $w_R(t) := \int_0^1 G(t,s)\phi^R(s) \, ds$ and take $R_1 > \|(I - L^{R,\infty})^{-1}(w_R)\|$. We claim that $\sigma u \neq Nu$ for all $\sigma \geq 1$ and all $u \in \partial U_{R_1}$. In fact, if not, there exists u with $\|u\| = R_1$ such that

$$u(t) \leq \int_0^1 G(t,s) f(s, u(s)) \, ds \leq \int_0^1 G(t,s) \left(F^{R,\infty}(s)u + \phi^R(s) \right) ds,$$

that is, $u \preceq L^{R,\infty} u + w_R$, hence $u \preceq (I - L^{R,\infty})^{-1} w_R$. As is well known, P is a normal cone with normality constant 1, so this implies that $\|u\| \leq \|(I - L^{R,\infty})^{-1} w_R\|$ which contradicts the choice of R_1. $\qquad\square$

Our third new result gives conditions so that the fixed-point index is zero. An advantage of this result is that it depends on the behavior of f on an interval (in the u variable) that can be bounded away from both zero and infinity.

Theorem 3.4. *Suppose there exist a subinterval J with $c_J > 0$ and a constant $\rho > 0$ such that $r(L^J_{c_J\rho,\rho}) > 1$. Then*

$$i_{K_c}(N, U_\rho) = 0. \tag{18}$$

Proof. We take $e \in K_c \setminus \{0\}$ and we will show that $u \neq Nu + \sigma e$ for all $u \in \partial U_\rho$ and all $\sigma \geq 0$, from which it follows by standard properties of fixed-point index that $i_{K_c}(N, U_\rho) = 0$. If not, there exist $\sigma \geq 0$ and $u \in \partial U_\rho$ such that $u = Nu + \sigma e$. Then $c_J\rho \leq u(s) \leq \rho$ for $s \in J$, and we have

$$u(t) = Nu(t) + \sigma e(t) \geq \int_0^1 G(t,s) f(s, u(s)) \, ds \geq \int_J G(t,s) F_{c_J\rho,\rho}(s) u(s) \, ds,$$

that is,

$$u \succeq L^J_{c_J\rho,\rho} u.$$

By Theorem 2.2 (*ii*) this implies that $r(L^J_{c_J\rho,\rho}) \leq 1$, which contradiction proves the result. $\qquad\square$

We now state, in a changed notation from the original papers, the results of Lan [10, 12, 13] that correspond to these ones. One of these results makes use of the (relatively) open set $\Omega_\rho := \{u \in K_c : \min_J u(t) < c_J\rho\}$ introduced by Lan in [11]. Note that, for $u \in \overline{\Omega}_\rho$, $\|u\| \leq \rho$, and if $u \in \partial\Omega_\rho$ (the boundary relative to K_c), then $\min_J u(t) = c_J\rho$ and $c_J\rho \leq u(t) \leq \rho$ for $t \in J$. We remark that we could use this open set in place of U_ρ in Theorem 3.4 above, the same proof is valid.

Theorem 3.5.

(a) *Suppose there exists $\rho > 0$ such that $f(t,u) \leq \rho f^\rho(t)$ for $0 \leq u \leq \rho$, where $\sup_{t \in [0,1]} \int_0^1 G(t,s) f^\rho(s) \, ds < 1$. Then $i_{K_c}(N, U_\rho) = 1$.*

(b) *Suppose there exist a subinterval J with $c_J > 0$ and $\rho > 0$ such that $f(t,u) \geq c_J\rho f_\rho(t)$ for $c_J\rho \leq u \leq \rho$, and $t \in J$, where $\inf_{t \in J} \int_J G(t,s) f_\rho(s) \, ds > 1$.*

Then $i_{K_c}(N, \Omega_\rho) = 0$.

The original proofs of these results are in [10, 12, 13]. For completeness we give the proofs here, using our notation, to illustrate the arguments.

Proof.

(a) If there exist $\sigma \geq 1$ and $u \in \partial U_\rho$ with $\sigma u = Nu$ then $0 \leq u(s) \leq \rho$ for all $s \in [0,1]$ and $\|u\| = \rho$, therefore we have

$$u(t) = \frac{1}{\sigma} Nu(t) \leq \int_0^1 G(t,s) f(s,u(s)) \, ds \leq \int_0^1 G(t,s) \rho f^\rho(s) \, ds.$$

Taking the supremum for $t \in [0,1]$ gives $\rho \leq \rho \sup_{t \in [0,1]} \int_0^1 G(t,s) f^\rho(s) \, ds < \rho$, which contradiction proves the claim.

(b) Let $e \in K_c \setminus \{0\}$ and suppose that there exist $\sigma \geq 0$ and $u \in \partial \Omega_\rho$ such that $u = Nu + \sigma e$. Then $c_J \rho \leq u(s) \leq \rho$ for $s \in J$, and for each $t \in [0,1]$ we have

$$u(t) = \int_0^1 G(t,s) f(s,u(s)) \, ds + \sigma e(t) \geq \int_J G(t,s) c_J \rho f_\rho(s) \, ds.$$

Taking the infimum over J gives

$$c_J \rho \geq c_J \rho \inf_{t \in J} \int_J G(t,s) f_\rho(s) \, ds > c_J \rho,$$

a contradiction.   □

Remark 3.2.

(i) Note that, since ρ is at our disposal, the assumptions in (b) are equivalent to the following: there exists $\rho > 0$ such that $f(t,u) \geq \rho \tilde{f}_\rho(t)$ for $\rho \leq u \leq \rho/c_J$ where $\inf_{t \in J} \int_J G(t,s) \tilde{f}_\rho(s) \, ds > 1$. This is a more useful form when a hypothesis is made about the behavior of $f(t,u)/u$ for all large u, and we consider J approaching $[0,1]$.

(ii) In the result (a) we can clearly take $f^\rho(t) = \sup_{u \in [0,\rho]} f(t,u)/\rho$. Since, for each t, $f(t,u)/\rho \leq f(t,u)/u$ for all $u \in (0,\rho]$, it would then follow that $f^\rho(t) \leq F^{0,\rho}(t)$. Let $L_1 u(t) := \int_0^1 G(t,s) f^\rho(s) u(s) \, ds$, and $L_2 u(t) := \int_0^1 G(t,s) F^{0,\rho}(s) u(s) \, ds$. Then Theorem 3.5 (a) requires $\|L_1\| < 1$, while Theorem 3.2 requires $r(L_2) < 1$. In some cases, $f^\rho = F^{0,\rho}$, and then Theorem 3.2 is an improvement of Theorem 3.5 (a); in some other cases Theorem 3.5 (a) can be applied, but Theorem 3.2 cannot be applied. Similar comments apply to Theorem 3.4 and Theorem 3.5 (b). This is why our new results complement those of Lan.

4   Existence and Nonexistence of Positive Solutions

We have all the ingredients for an existence result for positive solutions. Let (C_1)–(C_3) hold and f satisfy the conditions of the previous section. The first result gives one positive solution.

Theorem 4.1. *Let J be a subinterval of $[0,1]$ such that $c_J > 0$ and let $0 < \rho_1 < c_J\rho_2 < \rho_2$. Suppose that*

$$r(L^{0,\rho_1}) < 1 \text{ and } r(L^J_{c_J\rho_2,\rho_2}) > 1.$$

Then the integral operator N has at least one fixed point in $K_c \setminus \{0\}$.

Proof. By Theorem 3.2, $i_{K_c}(N, U_{\rho_1}) = 1$ and by Theorem 3.4, $i_{K_c}(N, U_{\rho_2}) = 0$. By the additivity property of index this gives $i_{K_c}(N, U_{\rho_2} \setminus \overline{U}_{\rho_1}) = -1$, and by the solution property, N has a fixed point in $U_{\rho_2} \setminus \overline{U}_{\rho_1}$, hence is nontrivial. □

Remark 4.1. Since $i_{K_c}(N, U_{\rho_1}) = 1$, there is a fixed point of N in U_{ρ_1}; this is typically the zero solution, whose existence is usually obvious.

Because of the overlap of the intervals, it is not possible to have the conditions

$$r(L^J_{c_J\rho_1,\rho_1}) > 1 \text{ and } r(L^{0,\rho_2}) < 1,$$

so there is not a second set of conditions of this type for one positive solution.

We can obtain a result that does not involve J which gives one positive solution in either of two cases. This is similar to the original result in [31] when there was no relevant explicit t dependence. To do this we employ a result of Nussbaum, Lemma 2 of [22], which says that if L_n are compact linear operators and $L_n \to L$ in the operator norm, then $r(L_n) \to r(L)$. In particular, when $c(t) > 0$ for all $t \in (0,1)$ we may take $J = [t_0, t_1]$ to be an arbitrary subinterval of $(0,1)$, and letting $t_0 \to 0$ and $t_1 \to 1$, we have $r(L^J_{c_J\rho,\rho}) \to r(L_{0,\rho})$ and $r(L^J_{R,R/c_J}) \to r(L_{R,\infty})$.

Corollary 4.1. *Suppose that $c(t) > 0$ for all $t \in (0,1)$. Let $0 < \rho < R$ and suppose that*

$$\text{either } (i) \ r(L^{0,\rho}) < 1 \text{ and } r(L_{R,\infty}) > 1,$$

$$\text{or } (ii) \ r(L_{0,\rho}) > 1 \text{ and } r(L^{R,\infty}) < 1.$$

Then the integral operator N has at least one fixed point in $K_c \setminus \{0\}$.

Proof. Case (i) is similar to Theorem 4.1, so we only show (ii). By Theorem 3.4 and the above remarks $i_{K_c}(N, U_\rho) = 0$ and by Theorem 3.3, there is $R_1 > R$ sufficiently large with $i_{K_c}(N, U_{R_1}) = 1$. By the additivity and existence properties of index, the result follows. □

The next result gives two positive solutions.

Theorem 4.2. *Let J be a subinterval of $[0,1]$ such that $c_J > 0$ and let $0 < \rho_1 < c_J\rho_2 < \rho_2 < \rho_3$ and suppose that*

$$r(L^{0,\rho_1}) < 1, \ r(L^J_{c_J\rho_2,\rho_2}) > 1, \text{ and } r(L^{\rho_3,\infty}) < 1.$$

Then the integral operator N has at least two fixed points in $K_c \setminus \{0\}$.

Proof. By the additivity property we obtain

$$i_{K_c}(N, U_{\rho_2} \setminus \overline{U}_{\rho_1}) = -1, \quad i_{K_c}(N, U_{\rho_3} \setminus \overline{U}_{\rho_2}) = 1,$$

so N has a fixed point in $U_{\rho_3} \setminus \overline{U}_{\rho_2}$ and another in the disjoint set $U_{\rho_2} \setminus \overline{U}_{\rho_1}$. □

Remark 4.2.

(a) There is a third solution in U_{ρ_1}, typically zero.
(b) It is not possible to have the conditions

$$r(L^J_{c_J\rho_1,\rho_1}) > 1, \quad \text{and} \quad r(L^{0,\rho_2}) < 1,$$

so there is not another theorem for the existence of two positive solutions. However, in the special case when the stronger positivity condition (15) holds, it is possible to have multiple positive solutions; see [25, 26] for details. This is a big advantage but only can be proved in that restricted case,

(c) It is clear that we do not need to choose $F_{r,R}(t) = \inf_{u \in [r,R]} f(t,u)/u$, and $F^{r,R}(t) = \sup_{u \in [r,R]} f(t,u)/u$, lower and upper bounds, respectively, suffice, but choosing equalities give the most precise results.

With our comparison theorem we can also prove nonexistence results.

Theorem 4.3.

(i) *Suppose that* $f(t,u) \leq \overline{F}(t)u$ *for all* $u > 0$, *where* $\Phi\overline{F} \in L^1$. *Let* $\overline{L}u(t) := \int_0^1 G(t,s)\overline{F}(s)u(s)\,\mathrm{d}s$ *and suppose that* $r(\overline{L}) < 1$. *Then* N *has no nonzero fixed point in* K_c.

(ii) *Suppose that* $c(t) > 0$ *for* $t \in (0,1)$ *and that* $f(t,u) \geq \underline{F}(t)u$ *for all* $u > 0$, *where* $\Phi\underline{F} \in L^1$. *Let* $\underline{L}u(t) := \int_0^1 G(t,s)\underline{F}(s)u(s)\,\mathrm{d}s$ *and suppose that* $r(\underline{L}) > 1$. *Then* N *has no nonzero fixed point in* K_c.

Proof. In case (i) suppose that $u \in K_c \setminus \{0\}$ is a fixed point of N. Then

$$u = Nu \preceq \overline{L}u,$$

and by Theorem 2.2 (i) this would imply $r(\overline{L}) \geq 1$, which contradiction proves the result. In case (ii) we define $\underline{L}^J u(t) := \int_J G(t,s)\underline{F}(s)u(s)\,\mathrm{d}s$, and, using Nussbaum's result mentioned above, we select $J \subset (0,1)$ so that $r(\underline{L}^J) > 1$. By the proof of Theorem 3.1, $\underline{L}^J$ is u_0-positive relative to (K_c, P). Now if $u \in K_c \setminus \{0\}$ is a fixed point of N, then

$$u = Nu \succeq \underline{L}u \geq \underline{L}^J u,$$

and by Theorem 2.2 (ii) this gives a contradiction. □

5   Illustrative Examples

We now give two examples to illustrate our results. The first example has a simple but standard type of nonlinearity. We obtain a sharp result, whereas the results of Lan give a less precise conclusion.

Example 5.1.   Consider the BVP

$$u''(t) + \lambda f(t, u(t)) = 0, \ t \in (0, 1), \text{ with } f(t, u) = (1 - t)u^2 + tu, \qquad (19)$$

where λ is a positive parameter, with boundary conditions

$$u(0) = 0, \ u(1) = 0. \qquad (20)$$

We will show that there is $\lambda^* \approx 18.956$ such that the problem has at least one (strictly) positive solution for each $\lambda < \lambda^*$ and there is no positive solution for $\lambda > \lambda^*$. (Of course there is also the zero solution.) Thus, our result can be sharp.

The Green's function for this problem is well known to be

$$G(t, s) = \begin{cases} s(1 - t), & \text{if } s \leq t, \\ t(1 - s), & \text{if } s > t. \end{cases}$$

Hence $\Phi(s) = s(1-s)$ and $c(t) = \min\{t, 1-t\}$. Let $Nu(t) := \lambda \int_0^1 G(t, s) f(s, u(s)) \, ds$. We have $f(t, u)/u = (1-t)u + t$ and therefore $F^{0,\rho}(t) = (1-t)\rho + t$ and $F_{r,R}(t) = (1-t)r + t$. Then we have

$$L^{0,\rho} u(t) := \int_0^1 G(t, s)((1-s)\rho + s)u(s) \, ds.$$

Let $L^0 u(t) := \int_0^1 G(t, s)su(s) \, ds$. By the result of Nussbaum [22] mentioned above, for an arbitrary $\varepsilon > 0$, ρ can be chosen so small that $r(L^{0,\rho}) - r(L^0) < \varepsilon$. Thus, by Theorem 3.2, we may choose ρ sufficiently small so that $i_{K_c}(N, U_\rho) = 1$ provided $\lambda r(L^0) < 1$, that is, $\lambda < 1/r(L^0)$. By a numerical calculation, $1/r(L^0) \approx 18.956$. For any $\lambda > 0$, taking r, R sufficiently large makes $r(L_{r,R})$ as large as we wish. Hence for every $\lambda \in (0, 1/r(L^0))$, there is at least one positive solution by Corollary 4.1. By the nonexistence result, Theorem 4.3, since $f(t, u) \geq tu$ for all $u > 0$, no positive solution exists if $\lambda > 1/r(L^0)$.

Now we compare our conclusion with the one that can be obtained using Lan's result, Theorem 3.5 above. We have $f^\rho(t) = \sup_{u \in [0,\rho]} f(t, u)/\rho = (1 - t)\rho + t$. Theorem 3.5 gives $i_{K_c}(N, U_\rho) = 1$ if $\lambda \max_{t \in [0,1]} \int_0^1 G(t, s)((1-s)\rho + s) \, ds < 1$. By choosing ρ sufficiently small, it suffices to have $\lambda \max_{t \in [0,1]} \int_0^1 G(t, s)s \, ds < 1$. By a Maple calculation this requires $\lambda < 9\sqrt{3} \approx 15.588$. For each λ in $(0, 9\sqrt{3})$, choosing R sufficiently large gives a zero index by part (b) of Theorem 3.5, and thus, Lan's result gives existence of a positive solution for every $\lambda \in (0, 9\sqrt{3})$. Lan does not give a nonexistence result so does not have an upper bound on the possible allowed λ. Thus in this example, we have a better result, and our constant is sharp. We note that in this example

$$\overline{f}(u) = \begin{cases} u, & \text{if } u \le 1, \\ u^2 & \text{if } u \ge 1, \end{cases} \quad \underline{f}(u) = \begin{cases} u^2, & \text{if } u \le 1, \\ u & \text{if } u \ge 1. \end{cases}$$

Thus,

$$\lim_{u \to 0+} \overline{f}(u)/u = 1, \quad \lim_{u \to \infty} \underline{f}(u)/u = 1,$$

$$\lim_{u \to 0+} \underline{f}(u)/u = 0, \quad \lim_{u \to \infty} \overline{f}(u)/u = \infty,$$

and the results that use $\overline{f}, \underline{f}$ do not apply: there is no "crossing of the eigenvalue," as is required in (6) in the Introduction.

Remark 5.1. For the problem with $f(u) = u^2 + u$, with no explicit t dependence, the corresponding result is as follows: Let $Lu(t) := \int_0^1 G(t,s)u(s)\,ds$. By similar arguments to the above, there is no positive solution for $\lambda > 1/r(L)$, and for every $\lambda \in (0, 1/r(L))$, there is at least one positive solution. In this case, it is well known that $1/r(L) = \pi^2$. Since, *very roughly*, $u^2 + u$ is approximately double the "average" of our $f(t, u)$, the constant we found of 18.956 makes sense.

In the second example we have a result for existence of two positive solutions.

Example 5.2. Consider the BVP

$$u''(t) + \lambda f(t, u(t)) = 0, \ t \in (0, 1), \text{ with } f(t, u) = (1 - t)h(u) + tu, \quad (21)$$

where λ is a positive parameter and

$$h(u) := \begin{cases} 0, & \text{if } 0 \le u \le 1, \\ 8(u - 1)(9 - u)/7, & \text{if } 1 < u < 9, \\ 0, & \text{if } u \ge 9, \end{cases}$$

with boundary conditions

$$u(0) = 0, \ u(1) = 0.$$

We have the same Green's function as in Example 5.1. We will show that there are constants $\lambda^* \approx 3.482$, $\lambda^{**} \approx 11.85$ and $\lambda^{***} \approx 18.956$ such that the problem has no positive solution for $\lambda < \lambda^*$, has at least two positive solutions for each $\lambda^{**} < \lambda < \lambda^{***}$, and has no positive solution for $\lambda > \lambda^{***}$.

Firstly we have $F(t,u) \geq t$ for all $u \geq 0$, so by Theorem 4.3, there is no positive solution if $\lambda r(\underline{L}) > 1$ where $\underline{L}u(t) := \int_0^1 G(t,s)su(s)\,ds$, that is, if $\lambda > 1/r(\underline{L}) = \lambda^{***} \approx 18.956$ (as in the previous example).

We also have $h(u) \leq 32u/7$ so $F(t,u) \leq 32/7 - 25t/7$ for all $u \geq 0$ so there is no positive solution for $\lambda r(\overline{L}) < 1$ where $\overline{L}u(t) := \int_0^1 G(t,s)(32/7 - 25s/7)u(s)\,ds$, that is if $\lambda < 1/r(\overline{L}) = \lambda^* \approx 3.482$, by a numerical calculation.

For existence we note that $F(t,u) = t$ for $u \in [0,1]$ and for $u \geq 9$ so two conditions in Theorem 4.2 are satisfied for $\lambda < 1/r(\underline{L})$.

We now choose $J = [1/4, 3/4]$ so that $c_J = 1/4$ and choose $\rho_2 = 8$ so we are now working on the interval $[2,8]$. As $h(u) \geq u$ for $u \in [2,8]$ we have $F_{c_J\rho_2,\rho_2}^J(t) \geq 1$. By a numerical calculation we find $1/r(L_{2,8}^J) \approx 11.85$ so there are at least two positive solutions for $1/r(\underline{L}) = \lambda^{***} \geq \lambda > \lambda^{**} = 1/r(L_{2,8}^J) \approx 11.85$. We cannot get high accuracy for λ^{**} here because the numerical program used was only written for the continuous case; here the cutoff introduces discontinuities. Using the comparison Theorem 2.1, we can prove the definite bounds $11.81 \leq 1/r(L_{2,8}^J) \leq 11.89$.

In this example we easily see that, for $\overline{f}(u) := \sup_{t \in [0,1]} f(t,u)$ and $\underline{f}(u) := \inf_{t \in [0,1]} f(t,u)$,

$$\overline{f}(u) = \begin{cases} u, & \text{if } u \leq 9/8, \\ h(u), & \text{if } 9/8 \leq u \leq 8, \\ u, & \text{if } u > 8. \end{cases} \qquad \underline{f}(u) = \begin{cases} h(u), & \text{if } u \leq 9/8, \\ u, & \text{if } 9/8 \leq u \leq 8, \\ h(u), & \text{if } u > 8. \end{cases}$$

Let $Lu(t) := \int_0^1 G(t,s)u(s)\,ds$ and $\tilde{L}u(t) := \int_{1/4}^{3/4} G(t,s)u(s)\,ds$. To get two solutions using (D_1') of Theorem 5.12 of [24], we would need $\lambda < 1/r(L) = \pi^2$ and $\lambda > 1/r(\tilde{L}) \approx 11.85$, an impossibility. Thus, those results do not apply.

References

1. Anuradha, V., Hai, D.D., Shivaji, R.: Existence results for superlinear semipositone BVP's. Proc. Am. Math. Soc. **124**(3), 757–763 (1996)
2. Deimling, K.: Nonlinear Functional Analysis. Springer, Berlin (1985) (Reprinted: Dover Publications 2010, ISBN 13: 9780486474410)
3. Erbe, L.: Eigenvalue criteria for existence of positive solutions to nonlinear boundary value problems. Math. Comput. Model. **32**, 529–539 (2000)
4. Graef, J.R., Kong, L.: Existence results for nonlinear periodic boundary-value problems. Proc. Edinb. Math. Soc. **52**, 79–95 (2009)
5. Guo, D., Lakshmikantham, V.: Nonlinear Problems in Abstract Cones. Academic, San Diego (1988)
6. Keener, M.S., Travis, C.C.: Positive cones and focal points for a class of nth order differential equations. Trans. Am. Math. Soc. **237**, 331–351 (1978)
7. Kong, L., Wong, J.S.W.: Positive solutions for higher order multi-point boundary value problems with nonhomogeneous boundary conditions. J. Math. Anal. Appl. **367**, 588–611 (2010)

8. Krasnosel'skiĭ, M.A.: Positive Solutions of Operator Equations. P. Noordhoff Ltd, Groningen (1964)
9. Krasnosel'skiĭ, M.A., Zabreĭko, P.P.: Geometrical Methods of Nonlinear Analysis. Springer, Berlin (1984)
10. Lan, K.Q.: Multiple positive solutions of Hammerstein integral equations with singularities. Differ. Equ. Dynam. Syst. **8**, 175–192 (2000)
11. Lan, K.Q.: Multiple positive solutions of semilinear differential equations with singularities. J. Lond. Math. Soc. **63**(2), 690–704 (2001)
12. Lan, K.Q.: Multiple positive solutions of conjugate boundary value problems with singularities. Appl. Math. Comput. **147**, 461–474 (2004)
13. Lan, K.Q.: Multiple positive solutions of Hammerstein integral equations and applications to periodic boundary value problems. Appl. Math. Comput. **154**, 531–542 (2004)
14. Lan, K.Q.: Multiple eigenvalues for singular Hammerstein integral equations with applications to boundary value problems. J. Comput. Appl. Math. **189**, 109–119 (2006)
15. Lan, K.Q.: Positive solutions of semi-positone Hammerstein integral equations and applications. Commun. Pure Appl. Anal. **6** (2), 441–451 (2007)
16. Lan, K.Q.: Eigenvalues of semi-positone Hammerstein integral equations and applications to boundary value problems. Nonlinear Anal. **71**, 5979–5993 (2009)
17. Lan, K.Q., Lin, W.: Multiple positive solutions of systems of Hammerstein integral equations with applications to fractional differential equations. J. Lond. Math. Soc. **83**(2), 449–469 (2011)
18. Liu, B., Liu, L., Wu, Y.: Positive solutions for a singular second-order three-point boundary value problem. Appl. Math. Comput. **196**, 532–541 (2008)
19. Lan, K.Q., Webb, J.R.L.: Positive solutions of semilinear differential equations with singularities. J. Differ. Equ. **148**, 407–421 (1998)
20. Ma, R., Ren, L.: Positive solutions for nonlinear m-point boundary value problems of Dirichlet type via fixed-point index theory. Appl. Math. Lett. **16**, 863–869 (2003)
21. Martin, R.H.: Nonlinear Operators and Differential Equations in Banach Spaces. Wiley, New York (1976)
22. Nussbaum, R.D.: Periodic solutions of some nonlinear integral equations. In: Dynamical Systems, pp. 221–249. Academic, New York (1977) (Proc. Internat. Sympos., Univ. Florida, Gainesville, Fla., 1976)
23. Sun, Y., Liu, L., Zhang, J., Agarwal, R.P.: Positive solutions of singular three-point boundary value problems for second-order differential equations. J. Comput. Appl. Math. **230**, 738–750 (2009)
24. Webb, J.R.L.: Solutions of nonlinear equations in cones and positive linear operators. J. Lond. Math. Soc. **82**(2), 420–436 (2010)
25. Webb, J.R.L.: A class of positive linear operators and applications to nonlinear boundary value problems. Topol. Methods Nonlinear Anal. **39**, 221–242 (2012)
26. Webb, J.R.L.: Positive solutions of nonlinear equations via comparison with linear operators. Discrete Contin. Dyn. Syst. **33**, 5507–5519 (2013)
27. Webb, J.R.L., Infante, G.: Positive solutions of nonlocal boundary value problems: a unified approach. J. Lond. Math. Soc. **74**(2), 673–693 (2006)
28. Webb, J.R.L., Infante, G.: Positive solutions of nonlocal boundary value problems involving integral conditions. NoDEA Nonlinear Differ. Equ. Appl. **15**, 45–67 (2008)
29. Webb, J.R.L., Infante, G.: Nonlocal boundary value problems of arbitrary order. J. Lond. Math. Soc. **79**(2), 238–258 (2009)
30. Webb, J.R.L., Infante, G., Franco, D.: Positive solutions of nonlinear fourth-order boundary-value problems with local and non-local boundary conditions. Proc. Roy. Soc. Edinburgh Sect. A **138**, 427–446 (2008)
31. Webb, J.R.L., Lan, K.Q.: Eigenvalue criteria for existence of multiple positive solutions of nonlinear boundary value problems of local and nonlocal type. Topol. Methods Nonlinear Anal. **27**, 91–116 (2006)
32. Zhang, G., Sun, J.: Positive solutions of m-point boundary value problems. J. Math. Anal. Appl. **291**, 406–418 (2004)

Existence Results for a System of Third-Order Right Focal Boundary Value Problems

Patricia J.Y. Wong

Abstract We consider the following system of third-order three-point generalized right focal boundary value problems

$$u_i'''(t) = f_i(t, u_1(t), u_2(t), \ldots, u_n(t)), \quad t \in [a, b]$$

$$u_i(a) = u_i'(t_i) = 0, \qquad \gamma_i u_i(b) + \delta_i u_i''(b) = 0,$$

where $i = 1, 2, \ldots, n$, $\gamma_i \geq 0$, $\delta_i > 0$ and $\frac{1}{2}(a+b) < t_i < b$. By using a variety of tools like Leray–Schauder alternative and Krasnosel'skii's fixed point theorem, we offer several criteria for the existence of *fixed-sign* solutions of the system. A solution $u = (u_1, u_2, \ldots, u_n)$ is said to be of *fixed sign* if for each $1 \leq i \leq n$, $\theta_i u_i(t) \geq 0$ for $t \in [a, b]$ where $\theta_i \in \{-1, 1\}$ is fixed. We also consider a related eigenvalue problem

$$u_i'''(t) = \lambda f_i(t, u_1(t), u_2(t), \ldots, u_n(t)), \quad t \in [a, b]$$

$$u_i(a) = u_i'(t^*) = 0, \qquad \gamma_i u_i(b) + \delta_i u_i''(b) = 0,$$

where $i = 1, 2, \ldots, n$, $\lambda > 0$, $\gamma_i \geq 0$, $\delta_i > 0$ and $\frac{1}{2}(a+b) < t^* < b$. Criteria will be established so that the above system has a *fixed-sign* solution for values of λ that form an interval (bounded or unbounded). Explicit intervals for such λ will also be presented. We include some examples to illustrate the results obtained.

Keywords Fixed-sign solutions • Positive solutions • Right focal boundary value problems • Eigenvalues

P.J.Y. Wong (✉)
School of Electrical and Electronic Engineering, Nanyang Technological University,
50 Nanyang Avenue, Singapore 639798, Singapore
e-mail: ejywong@ntu.edu.sg

S. Pinelas et al. (eds.), *Differential and Difference Equations with Applications*, Springer
Proceedings in Mathematics & Statistics 47, DOI 10.1007/978-1-4614-7333-6_12,
© Springer Science+Business Media New York 2013

1 Introduction

In this paper we shall consider a *system* of third-order differential equations subject to generalized right focal boundary conditions. To be precise, the system is

$$u_i'''(t) = f_i(t, u_1(t), u_2(t), \ldots, u_n(t)), \quad t \in [a, b]$$

$$u_i(a) = u_i'(t_i) = 0, \qquad \gamma_i u_i(b) + \delta_i u_i''(b) = 0,$$

(1)

where $i = 1, 2, \ldots, n$, and t_i, γ_i, δ_i are fixed numbers with

$$\frac{1}{2}(a+b) < t_i < b, \quad \gamma_i \geq 0, \quad \delta_i > 0, \quad \eta_i \equiv 2\delta_i + \gamma_i(b-a)(b+a-2t_i) > 0.$$

A solution $u = (u_1, u_2, \ldots, u_n)$ of (1) will be sought in $(C[a,b])^n = C[a,b] \times C[a,b] \times \cdots \times C[a,b]$ (n times). We say that u is a solution of *fixed sign* if for each $1 \leq i \leq n$, we have $\theta_i u_i(t) \geq 0$ for $t \in [a,b]$ where $\theta_i \in \{1, -1\}$ is fixed. In particular, if we choose $\theta_i = 1$, $1 \leq i \leq n$, then our fixed-sign solution u becomes a *positive* solution. We remark that in many practical problems, it is only meaningful to have *positive* solutions. Nonetheless our definition of *fixed-sign* solution is more general and gives extra *flexibility*. By using a variety of tools like Leray–Schauder alternative and Krasnosel'skii's fixed point theorem, we shall offer several criteria for the existence of *fixed-sign* solutions of the system (1).

We also consider an eigenvalue problem related to (1), namely,

$$u_i'''(t) = \lambda f_i(t, u_1(t), u_2(t), \ldots, u_n(t)), \quad t \in [a, b]$$

$$u_i(a) = u_i'(t^*) = 0, \qquad \gamma_i u_i(b) + \delta_i u_i''(b) = 0,$$

(2)

where $i = 1, 2, \ldots, n$ and $\lambda > 0$. The numbers t^*, γ_i, δ_i are fixed constants with

$$\frac{1}{2}(a+b) < t^* < b, \quad \gamma_i \geq 0, \quad \delta_i > 0, \quad \eta_i \equiv 2\delta_i + \gamma_i(b-a)(b+a-2t^*) > 0.$$

If, for a particular λ, the system (2) has a fixed-sign solution $u = (u_1, u_2, \ldots, u_n)$, then λ is called an *eigenvalue* and u a corresponding *eigenfunction* of the system. Let E be the set of eigenvalues, i.e.,

$$E = \{\lambda \mid \lambda > 0 \text{ such that the system (2) has a fixed-sign solution}\}.$$

We shall establish criteria for E to contain an interval and to be an interval (which may either be bounded or unbounded). Upper and lower bounds for an eigenvalue λ are also established. In addition explicit subintervals of E are obtained.

Right focal boundary value problems have attracted much interests in recent years. Existence of positive solutions to the two-point right focal boundary value problem

$$(-1)^{3-k}y'''(t) = f(t,y(t)), \ t \in [0,1]$$

$$y^{(j)}(0) = 0, \ 0 \le j \le k-1; \quad y^{(j)}(1) = 0, \ k \le j \le 2,$$

where $k \in \{1,2\}$, has been well discussed in the literature [1, 4]. The related discrete problem can be found in [11, 18–20]. Work on a three-point right focal problem, a special case of (1) when $n = 1$, $\delta_1 = 1$, $\gamma_1 = 0$, is available in [6, 8]. Further, Anderson [7] has considered (1) when $n = 1$ and developed the Green's function for the boundary value problem. In our present work, we generalize the problem considered in [7] to a *system* of boundary value problems, with very *general* nonlinear terms f_i, and we seek *fixed-sign* solutions of the system—this presents a much more robust model for many nonlinear phenomena.

The paper is outlined as follows. Section 2 deals with the existence of fixed-sign solutions of (1). In Sect. 3, the eigenvalue problem (2) is presented. Throughout, we also include some examples to illustrate the results obtained. This chapter is based on the work [21, 22]. Other work on fixed-sign solutions of third-order right focal boundary value problems can be found in [23–26], while work on positive solutions of boundary value problems is abundant, a sample includes [2–5, 9, 10, 12–15, 17] and the references cited therein.

2   Existence of Fixed-Sign Solutions

In this section, we shall obtain the existence of one or two fixed-sign solutions of (1) by using *Leray–Schauder alternative* and *Krasnosel'skii's fixed point theorem in a cone*, which are stated as follows:

Theorem 2.1 ([4]). *Let B be a Banach space with $D \subseteq B$ closed and convex. Assume U is a relatively open subset of D with $0 \in U$ and $S : \overline{U} \to D$ is a continuous and compact map. Then either*

(a)   *S has a fixed point in $\overline{U}$ or*
(b)   *There exists $u \in \partial U$ and $\lambda \in (0,1)$ such that $u = \lambda Su$.*

Theorem 2.2 ([16]). *Let $B = (B, \| \cdot \|)$ be a Banach space, and let $C \subset B$ be a cone in B. Assume Ω_1, Ω_2 are open subsets of B with $0 \in \Omega_1$, $\overline{\Omega}_1 \subset \Omega_2$, and let $S : C \cap (\overline{\Omega}_2 \backslash \Omega_1) \to C$ be a completely continuous operator such that either*

(a)   *$\|Su\| \le \|u\|$, $u \in C \cap \partial \Omega_1$, and $\|Su\| \ge \|u\|$, $u \in C \cap \partial \Omega_2$ or*
(b)   *$\|Su\| \ge \|u\|$, $u \in C \cap \partial \Omega_1$, and $\|Su\| \le \|u\|$, $u \in C \cap \partial \Omega_2$.*

Then S has a fixed point in $C \cap (\overline{\Omega}_2 \backslash \Omega_1)$.

The next lemma gives the Green's function of a related boundary value problem and its properties. This lemma plays a very important role in obtaining the subsequent results.

Lemma 2.1 ([7]). *Let $g_i(t,s)$ be the Green's function of the boundary value problem*

$$\begin{cases} y'''(t) = 0, \ t \in [a,b] \\ y(a) = y'(t_i) = 0, \qquad \gamma_i y(b) + \delta_i y''(b) = 0. \end{cases}$$

We have for $t,s \in [a,b]$,

$$g_i(t,s) = \begin{cases} s \in [a,t_i] : \begin{cases} v_1(t,s), \ t \le s \\ v_2(t,s), \ t \ge s \end{cases} \\ s \in [t_i,b] : \begin{cases} v_3(t,s), \ t \le s \\ v_4(t,s), \ t \ge s \end{cases} \end{cases} \tag{3}$$

where

$$v_1(t,s) = \frac{t-a}{2}(2s-t-a) + \frac{\gamma_i(t-a)}{2\eta_i}(s-a)^2(2t_i-a-t),$$

$$v_2(t,s) = \frac{(s-a)^2}{2\eta_i}[\eta_i + \gamma_i(t-a)(2t_i-a-t)],$$

$$v_3(t,s) = \frac{t-a}{2\eta_i}(2t_i-a-t)[2\delta_i + \gamma_i(b-s)^2],$$

$$v_4(t,s) = \frac{t-a}{2\eta_i}(2t_i-a-t)[2\delta_i + \gamma_i(b-s)^2] + \frac{(t-s)^2}{2}.$$

Moreover, $g_i(t,s)$ has the following properties:

$$g_i(t,s) \ge 0, \ t,s \in [a,b]; \qquad g_i(t,s) > 0, \ t,s \in (a,b] \tag{4}$$

$$g_i(t,s) \le g_i(t_i,s), \ t,s \in [a,b] \tag{5}$$

$$\begin{cases} \text{for a fixed } h_i \in (0,b-t_i), \\ g_i(t,s) \ge M_i g_i(t_i,s), \ t \in [t_i-h_i,t_i+h_i], \ s \in [a,b] \\ \text{where } M_i = \dfrac{(t_i-a+h_i)(t_i-a-h_i)}{(t_i-a)^2}. \end{cases} \tag{6}$$

Throughout, we shall denote $u = (u_1, u_2, \ldots, u_n)$. Let the Banach space $B = (C[a,b])^n$ be equipped with the norm

$$\|u\| = \max_{1 \le i \le n} \sup_{t \in [a,b]} |u_i(t)| = \max_{1 \le i \le n} |u_i|_0,$$

where we let $|u_i|_0 = \sup_{t \in [a,b]} |u_i(t)|$, $1 \le i \le n$.

Define the operator $S : (C[a,b])^n \to (C[a,b])^n$ by

$$Su(t) = (S_1 u(t), S_2 u(t), \ldots, S_n u(t)), \ t \in [a,b], \tag{7}$$

where

$$S_i u(t) = \int_a^b g_i(t,s) f_i(s, u(s)) ds, \ t \in [a,b], \ 1 \le i \le n. \tag{8}$$

Clearly, a fixed point of the operator S is a solution of the system (1).

Using Theorem 2.1, we obtain an existence criterion for a *general* solution (need not be of fixed sign).

Theorem 2.3. *Let* $f_i : [a,b] \times \mathbb{R}^n \to \mathbb{R}$, $1 \le i \le n$ *be continuous. Suppose there exists a constant* ρ, *independent of* λ, *such that* $\|u\| \ne \rho$ *for any solution* $u \in (C[a,b])^n$ *of the system*

$$u_i(t) = \lambda \int_a^b g_i(t,s) f_i(s, u(s)) ds, \ t \in [a,b], \ 1 \le i \le n, \tag{9_λ}$$

where $0 < \lambda < 1$. *Then,* (1) *has at least one solution* $u^* \in (C[a,b])^n$ *such that* $\|u^*\| \le \rho$.

Proof. Clearly, a solution of (9_λ) is a fixed point of the equation $u = \lambda Su$ where S is defined in (7), (8). Using the Arzelà–Ascoli theorem, we see that S is continuous and completely continuous. Now, in the context of Theorem 2.1, let $U = \{u \in B \mid \|u\| < \rho\}$. Since $\|u\| \ne \rho$, where u is any solution of (9_λ) we cannot have conclusion (b) of Theorem 2.1; hence, conclusion (a) of Theorem 2.1 must hold, i.e., the system (1) has a solution $u^* \in \overline{U}$ with $\|u^*\| \le \rho$. □

We shall now present the existence results of fixed-sign solutions. Let $\theta_i \in \{1, -1\}$, $1 \le i \le n$ be fixed, and also fix the numbers h_i and M_i [see (6)]. Define the sets $\tilde{K}$ and K as follows:

$$\tilde{K} = \{u \in B \mid \theta_i u_i(t) \ge 0, \ t \in [a,b], \ 1 \le i \le n\},$$

$$K = \{u \in \tilde{K} \mid \theta_j u_j(t) > 0 \text{ for some } j \in \{1, 2, \cdots, n\} \text{ and some } t \in [a,b]\} = \tilde{K} \backslash \{0\}.$$

Applying Theorem 2.3, we get the existence of a fixed-sign solution as follows.

Theorem 2.4. *Let the following hold:*

(C1) *For each $1 \leq i \leq n$, f_i is continuous on $[a,b] \times \tilde{K}$ with $\theta_i f_i(t,u) \geq 0$ for $(t,u) \in [a,b] \times \tilde{K}$ and $\theta_i f_i(t,u) > 0$ for $(t,u) \in [a,b] \times K$.*

(C2) *For each $1 \leq i \leq n$,*

$$\theta_i f_i(t,u) \leq q_i(t) w_{i1}(|u_1|) w_{i2}(|u_2|) \cdots w_{in}(|u_n|), \quad (t,u) \in [a,b] \times \tilde{K},$$

where q_i, w_{ij}, $1 \leq j \leq n$ are continuous, $w_{ij} : [0,\infty) \to [0,\infty)$ are nondecreasing, and $q_i : [a,b] \to [0,\infty)$.

(C3) *There exists $\alpha > 0$ such that for each $1 \leq i \leq n$,*

$$\alpha > d_i w_{i1}(\alpha) w_{i2}(\alpha) \cdots w_{in}(\alpha),$$

where $d_i = \sup_{t \in [a,b]} \int_a^b g_i(t,s) q_i(s) ds$.

Then, (1) has a fixed-sign solution $u^* \in (C[a,b])^n$ such that $\|u^*\| < \alpha$, i.e., $0 \leq \theta_i u_i^*(t) < \alpha$, $t \in [a,b]$, $1 \leq i \leq n$.

Proof. To apply Theorem 2.3, we consider the system

$$u_i(t) = \int_a^b g_i(t,s) \hat{f}_i(s,u(s)) ds, \quad t \in [a,b], \ 1 \leq i \leq n, \tag{10}$$

where $\hat{f}_i : [a,b] \times \mathbb{R}^n \to \mathbb{R}$ is defined by

$$\hat{f}_i(t,u_1,u_2,\ldots,u_n) = f_i(t,\theta_1|u_1|,\theta_2|u_2|,\ldots,\theta_n|u_n|), \quad 1 \leq i \leq n.$$

Noting $(\theta_1|u_1|,\theta_2|u_2|,\ldots,\theta_n|u_n|) \in \tilde{K}$, by (C1) we see that the function $\hat{f}_i$ is well defined and is continuous. We shall show that (10) has a solution. For this, we consider the system

$$u_i(t) = \lambda \int_a^b g_i(t,s) \hat{f}_i(s,u(s)) ds, \quad t \in [a,b], \ 1 \leq i \leq n, \tag{11_λ}$$

where $0 < \lambda < 1$. Let $u \in (C[a,b])^n$ be any solution of (11_λ) We can verify that $\|u\| \neq \alpha$, then it follows from Theorem 2.3 that (10) has a solution. Moreover, this solution is of fixed sign and is also a solution of (1). $\qquad\square$

Theorem 2.4 provides the existence of a fixed-sign solution which may be trivial. Our next result guarantees the existence of a *nontrivial* fixed-sign solution.

Theorem 2.5. *Let (C1)–(C3) hold. Moreover, suppose*

(C4) *For each $1 \leq j \leq n$ and some $i \in \{1,2,\ldots,n\}$ (i depends on j),*

$$\theta_i f_i(t,u) \geq \tau_{ij}(t) w_{ij}(|u_j|), \quad (t,u) \in [t_j - h_j, t_j + h_j] \times K$$

where $\tau_{ij} : [t_j - h_j, t_j + h_j] \to (0,\infty)$ is continuous.

(C5) *There exists $\beta > 0$ such that for each $1 \leq j \leq n$, the following holds for some $i \in \{1, 2, \ldots, n\}$ (i depends on j and is the same i as in (C4)):*

$$\beta \leq w_{ij}(M_j\beta) \cdot \int_{t_j-h_j}^{t_j+h_j} g_i(\sigma_{ij}, s)\tau_{ij}(s)ds$$

where $\sigma_{ij} \in [a,b]$ is defined by

$$\int_{t_j-h_j}^{t_j+h_j} g_i(\sigma_{ij}, s)\tau_{ij}(s)ds = \sup_{t \in [a,b]} \int_{t_j-h_j}^{t_j+h_j} g_i(t, s)\tau_{ij}(s)ds.$$

Then, (1) has a fixed-sign solution $u^* \in (C[a,b])^n$ such that

(a) $\alpha < \|u^*\| \leq \beta$ and $\min_{t \in [t_j-h_j, t_j+h_j]} \theta_j u_j^*(t) > M_j\alpha$ for some $j \in \{1, 2, \ldots, n\}$, if $\alpha < \beta$.

(b) $\beta \leq \|u^*\| < \alpha$ and $\min_{t \in [t_j-h_j, t_j+h_j]} \theta_j u_j^*(t) \geq M_j\beta$ for some $j \in \{1, 2, \ldots, n\}$, if $\alpha > \beta$.

Proof. The proof uses Theorem 2.2 with

$$C = \{u \in B \mid \text{for each } 1 \leq i \leq n, \ \theta_i u_i(t) \geq 0 \text{ for } t \in [a,b],$$

$$\text{and } \min_{t \in [t_i-h_i, t_i+h_i]} \theta_i u_i(t) \geq M_i |u_i|_0\},$$

$\Omega_1 = \{u \in B \mid \|u\| < \alpha\}$ and $\Omega_2 = \{u \in B \mid \|u\| < \beta\}$, if $\alpha < \beta$.   □

Using Theorems 2.4 and 2.5(a), we obtain the existence of *two* fixed-sign solutions.

Theorem 2.6. *Let (C1)–(C5) hold with $\alpha < \beta$. Then, (1) has (at least) two fixed-sign solutions $u^1, u^2 \in (C[a,b])^n$ such that*

$$0 \leq \|u^1\| < \alpha < \|u^2\| \leq \beta$$

and

$$\min_{t \in [t_j-h_j, t_j+h_j]} \theta_j u_j^2(t) > M_j\alpha \text{ for some } j \in \{1, 2, \ldots, n\}.$$

In Theorem 2.6 it is possible to have $\|u^1\| = 0$. Applying Theorem 2.5 twice, we get the next result that guarantees the existence of *two nontrivial* fixed-sign solutions.

Theorem 2.7. *Let (C1)–(C5) and (C5)$|_{\beta=\tilde{\beta}}$ hold, where $0 < \tilde{\beta} < \alpha < \beta$. Then, (1) has (at least) two fixed-sign solutions $u^1, u^2 \in (C[a,b])^n$ such that*

$$0 < \tilde{\beta} \leq \|u^1\| < \alpha < \|u^2\| \leq \beta,$$

$$\min_{t\in[t_k-h_k,t_k+h_k]} \theta_k u_k^1(t) \geq M_k \tilde{\beta} \quad and \quad \min_{t\in[t_j-h_j,t_j+h_j]} \theta_j u_j^2(t) > M_j \alpha$$

for some $j,k \in \{1,2,\ldots,n\}$.

Using Theorems 2.4 and/or 2.5 repeatedly, we can generalize Theorems 2.6 and 2.7 and obtain the existence of *multiple* fixed-sign solutions of (1).

Theorem 2.8. *Assume (C1), (C2), and (C4) hold. Let (C3) be satisfied for* $\alpha = \alpha_\ell$, $\ell = 1,2,\ldots,k$, *and (C5) be satisfied for* $\beta = \beta_\ell$, $\ell = 1,2,\ldots,m$:

(a) *If* $m = k+1$ *and* $0 < \beta_1 < \alpha_1 < \cdots < \beta_k < \alpha_k < \beta_{k+1}$, *then (1) has (at least)* $2k$ *fixed-sign solutions* $u^1,\ldots,u^{2k} \in (C[a,b])^n$ *such that*

$$0 < \beta_1 \leq \|u^1\| < \alpha_1 < \|u^2\| \leq \beta_2 \leq \cdots < \alpha_k < \|u^{2k}\| \leq \beta_{k+1}.$$

(b) *If* $m = k$ *and* $0 < \beta_1 < \alpha_1 < \cdots < \beta_k < \alpha_k$, *then (1) has (at least)* $2k-1$ *fixed-sign solutions* $u^1,\ldots,u^{2k-1} \in (C[a,b])^n$ *such that*

$$0 < \beta_1 \leq \|u^1\| < \alpha_1 < \|u^2\| \leq \beta_2 \leq \cdots \leq \beta_k \leq \|u^{2k-1}\| < \alpha_k.$$

(c) *If* $k = m+1$ *and* $0 < \alpha_1 < \beta_1 < \cdots < \alpha_m < \beta_m < \alpha_{m+1}$, *then (1) has (at least)* $2m+1$ *fixed-sign solutions* $u^0,\ldots,u^{2m} \in (C[a,b])^n$ *such that*

$$0 \leq \|u^0\| < \alpha_1 < \|u^1\| \leq \beta_1 \leq \|u^2\| < \alpha_2 < \cdots \leq \beta_m \leq \|u^{2m}\| < \alpha_{m+1}.$$

(d) *If* $k = m$ *and* $0 < \alpha_1 < \beta_1 < \cdots < \alpha_k < \beta_k$, *then (1) has (at least)* $2k$ *fixed-sign solutions* $u^0,\ldots,u^{2k-1} \in (C[a,b])^n$ *such that*

$$0 \leq \|u^0\| < \alpha_1 < \|u^1\| \leq \beta_1 \leq \|u^2\| < \alpha_2 < \cdots < \alpha_k < \|u^{2k-1}\| \leq \beta_k.$$

Example 2.1. Consider the system of boundary value problems

$$\begin{cases} u_1'''(t) = \exp\left(|u_1|^{1/7} + |u_2|^{1/9}\right), & t \in [0,1] \\ u_2'''(t) = \exp\left(|u_1|^{1/6} + |u_2|\right), & t \in [0,1] \\ u_i(0) = u_i'(0.55) = 0, & u_i(1) + 0.5u_i''(1) = 0, \quad i = 1,2. \end{cases} \tag{12}$$

Here, $n = 2$, $a = 0$, $t_1 = t_2 = 0.55$, $b = 1$, $\gamma_1 = \gamma_2 = 1$, $\delta_1 = \delta_2 = 0.5$, $f_1(t,u) = \exp\left(|u_1|^{1/7} + |u_2|^{1/9}\right)$, and $f_2(t,u) = \exp\left(|u_1|^{1/6} + |u_2|\right)$. Fix $\theta_1 = \theta_2 = 1$. Clearly, (C1) holds. In (C2), let $q_1 = q_2 = 1$ and

$$w_{11}(u_1) = \exp\left(|u_1|^{1/7}\right), \ w_{12}(u_2) = \exp\left(|u_2|^{1/9}\right),$$

$$w_{21}(u_1) = \exp\left(|u_1|^{1/6}\right), \ w_{22}(u_2) = \exp\left(|u_2|\right).$$

Next, using the expression of $g_i(t,s)$ in (3), we compute that $d_1 = d_2 = \frac{101761}{864000}$. Therefore, condition (C3) reduces to

$$\alpha > \frac{101761}{864000} \exp\left(\alpha^{1/7} + \alpha^{1/9}\right) \quad \text{and} \quad \alpha > \frac{101761}{864000} \exp\left(\alpha^{1/6} + \alpha\right).$$

By direct computation, the above inequalities are satisfied if $\alpha \in [0.8307, 1.4421]$. Hence, (C3) holds for any $\alpha \in [0.8307, 1.4421]$.

In condition (C4), pick $\tau_{ij} = 1$ for $i, j \in \{1, 2\}$. Finally, since $\lim_{z \to \infty} \frac{z}{w_{ij}(z)} = 0$, $i, j \in \{1, 2\}$, it is easy to choose $\beta > \alpha$ such that (C5) is fulfilled.

By Theorem 2.6, the system (12) has two nonnegative solutions $u^1, u^2 \in (C[0,1])^2$ such that (from (12), it is clear that $\|u^1\| \neq 0$)

$$\begin{cases} 0 < \|u^1\| < \alpha < \|u^2\| \leq \beta, \\ \min_{t \in [0.55-h, 0.55+h]} u_j^2(t) > \dfrac{(0.55+h)(0.55-h)\alpha}{(0.55)^2} \quad \text{for some } j \in \{1,2\} \end{cases} \tag{13}$$

where h can be any number in $(0, 0.45)$. Since α can be any number in $[0.8307, 1.4421]$, we further conclude from (13) that

$$\begin{cases} 0 < \|u^1\| < 0.8307 \quad \text{and} \quad \|u^2\| > 1.4421, \\ \min_{t \in [0.55-h, 0.55+h]} u_j^2(t) > \dfrac{(0.55+h)(0.55-h)(1.4421)}{(0.55)^2} \quad \text{for some } j \in \{1,2\} \end{cases} \tag{14}$$

where h can be any number in $(0, 0.45)$.

3 Eigenvalue Problem

In this section, we shall consider the eigenvalue problem (2). To begin, let $g_i(t,s)$ be the Green's function of the boundary value problem

$$\begin{cases} y'''(t) = 0, \ t \in [a, b] \\ y(a) = y'(t^*) = 0, \qquad \gamma_i y(b) + \delta_i y''(b) = 0. \end{cases}$$

Clearly, the properties of g_i can be obtained from Lemma 2.1 with t_i replaced by t^*. In particular, (5) and (6) give

$$g_i(t,s) \leq g_i(t^*, s), \ t, s \in [a, b] \tag{15}$$

$$\begin{cases} \text{for a fixed } h \in (0, b - t^*), \\[2mm] g_i(t,s) \geq M g_i(t^*, s), \quad t \in [t^* - h, t^* + h], \ s \in [a,b] \\[2mm] \text{where } M = \dfrac{(t^* - a + h)(t^* - a - h)}{(t^* - a)^2}. \end{cases} \tag{16}$$

With the same Banach space and norm as in Sect. 2, we define the operator S : $(C[a,b])^n \to (C[a,b])^n$ by (7) and

$$S_i u(t) = \lambda \int_a^b g_i(t,s) f_i(s, u(s)) ds, \quad t \in [a,b], \ 1 \leq i \leq n. \tag{17}$$

Clearly, a fixed point of the operator S is a solution of the system (2).

Let $\theta_i \in \{1, -1\}$, $1 \leq i \leq n$ be fixed, and also fix the numbers h and M [see (16)]. Define the sets $\tilde{K}$ and K as in Sect. 2.

Lemma 3.1. *Let the following hold:*

(H1) *For each $1 \leq i \leq n$, assume that $f_i : [a,b] \times \mathbb{R}^n \to \mathbb{R}$ is an L^1-Carathéodory function, i.e., (i) the map $t \to f_i(t,u)$ is measurable for all $u \in \mathbb{R}^n$; (ii) the map $u \to f_i(t,u)$ is continuous for almost all $t \in [a,b]$; and (iii) for any $r > 0$, there exists $\mu_r \in L^1[a,b]$ such that $|u| \leq r$ implies that $|f_i(t,u)| \leq \mu_r(t)$ for almost all $t \in [a,b]$.*

Then, the operator S defined in (7), (17) is continuous and completely continuous.

The next result ensures S maps a cone C into itself. A fixed point of S in this cone will be a *fixed-sign* solution of (2).

Lemma 3.2. *Let the following hold:*

(H2) *For each $1 \leq i \leq n$, assume that*

$$\theta_i f_i(t,u) \geq 0, \ u \in \tilde{K}, \ a.e. \ t \in (a,b) \quad \text{and} \quad \theta_i f_i(t,u) > 0, \ u \in K, \ a.e. \ t \in (a,b).$$

(H3) *For each $1 \leq i \leq n$, there exist continuous functions p_i, a_i, b_i with $p_i : \tilde{K} \to [0,\infty)$ and $a_i, b_i : (a,b) \to [0,\infty)$ such that*

$$a_i(t) p_i(u) \leq \theta_i f_i(t,u) \leq b_i(t) p_i(u), \ u \in \tilde{K}, \ a.e. \ t \in (a,b).$$

(H4) *For each $1 \leq i \leq n$, the function a_i is not identically zero on any nondegenerate subinterval of (a,b), and there exists a number $0 < \rho_i \leq 1$ such that*

$$a_i(t) \geq \rho_i b_i(t), \ a.e. \ t \in (a,b).$$

Then, the operator S maps the cone C into itself, where

$$C = \{u \in B \mid \text{for each } 1 \leq i \leq n, \ \theta_i u_i(t) \geq 0 \text{ for } t \in [a,b],$$

$$\text{and } \min_{t \in [t^* - h, t^* + h]} \theta_i u_i(t) \geq M \rho_i |u_i|_0 \}.$$

(Note that $C \subseteq \tilde{K}$. A fixed point of S obtained in C or $\tilde{K}$ will be a fixed-sign solution of (2).)

Our first three results provide conditions for E to contain an interval and for E to be an interval.

Theorem 3.1. *Let (H1)–(H4) hold and let $g_i(t_i,s)b_i(s) \in L^1[a,b]$, $1 \leq i \leq n$. Then, there exists $c > 0$ such that the interval $(0,c] \subseteq E$.*

Proof. Let $R > 0$ be given. Define

$$c = R \left\{ \left[\max_{1 \leq k \leq n} \sup_{\substack{|u_j| \leq R \\ 1 \leq j \leq n}} p_k(u_1, u_2, \ldots, u_n) \right] \int_a^b g_i(t_i,s)b_i(s)ds \right\}^{-1} . \tag{18}$$

Let $\lambda \in (0,c]$. Denote $C(R) = \{u \in C \mid \|u\| \leq R\}$. We shall prove that $S(C(R)) \subseteq C(R)$. To begin, let $u \in C(R)$. By Lemma 3.2, we have $Su \in C$. Thus, it remains to show that $\|Su\| \leq R$. Using (18), we get for $t \in [a,b]$ and $1 \leq i \leq n$

$$\begin{aligned}
|S_i u(t)| &= \theta_i S_i u(t) \\
&\leq \lambda \int_a^b g_i(t_i,s)b_i(s)p_i(u(s))ds \\
&\leq \lambda \left[\sup_{\substack{|u_j| \leq R \\ 1 \leq j \leq n}} p_i(u_1, u_2, \cdots, u_n) \right] \int_a^b g_i(t_i,s)b_i(s)ds \\
&\leq \lambda \left[\max_{1 \leq k \leq n} \sup_{\substack{|u_j| \leq R \\ 1 \leq j \leq n}} p_k(u_1, u_2, \cdots, u_n) \right] \int_a^b g_i(t_i,s)b_i(s)ds \\
&\leq c \left[\max_{1 \leq k \leq n} \sup_{\substack{|u_j| \leq R \\ 1 \leq j \leq n}} p_k(u_1, u_2, \cdots, u_n) \right] \int_a^b g_i(t_i,s)b_i(s)ds = R.
\end{aligned}$$

It follows immediately that

$$\|Su\| \leq R.$$

Thus, we have shown that $S(C(R)) \subseteq C(R)$. Also, from Lemma 3.1 the operator S is continuous and completely continuous. Schauder's fixed point theorem guarantees that S has a fixed point in $C(R)$. Clearly, this fixed point is a fixed-sign solution of (2), and therefore λ is an eigenvalue of (2). Since $\lambda \in (0,c]$ is arbitrary, we have proved that the interval $(0,c] \subseteq E$. $\qquad \square$

Lemma 3.3. *Let (H1) and (H2) hold. Moreover, suppose*

(H5)   *For each $1 \le i, j \le n$, if $|u_j| \le |v_j|$, then for a.e. $t \in (a,b)$,*

$$\theta_i f_i(t, u_1, \ldots, u_{j-1}, u_j, u_{j+1}, \ldots, u_n) \le \theta_i f_i(t, u_1, \ldots, u_{j-1}, v_j, u_{j+1}, \ldots, u_n).$$

Let $\lambda^* \in E$. Then, for any $\lambda \in (0, \lambda^*)$, we have $\lambda \in E$, i.e., $(0, \lambda^*] \subseteq E$.

Proof. Let $u^* = (u_1^*, u_2^*, \ldots, u_n^*)$ be the eigenfunction corresponding to the eigenvalue λ^*. Thus, we have

$$u_i^*(t) = S_i u^*(t) = \lambda^* \int_a^b g_i(t,s) f_i(s, u^*(s)) ds, \ t \in [a,b], \ 1 \le i \le n. \tag{19}$$

Define

$$K^* = \left\{ u \in \tilde{K} \ \middle| \ \text{for each } 1 \le i \le n, \ \theta_i u_i(t) \le \theta_i u_i^*(t), \ t \in [a,b] \right\}.$$

For $u \in K^*$ and $\lambda \in (0, \lambda^*)$, applying (H2) and (H5) yields

$$0 \le \theta_i S_i u(t) = \theta_i \left[\lambda \int_a^b g_i(t,s) f_i(s, u(s)) ds \right]$$

$$\le \theta_i \left[\lambda^* \int_a^b g_i(t,s) f_i(s, u^*(s)) ds \right] = \theta_i S_i u^*(t), \ t \in [a,b], \ 1 \le i \le n,$$

where the last equality follows from (19). This immediately implies that the operator S maps K^* into K^*. Moreover, from Lemma 3.1 the operator S is continuous and completely continuous. Schauder's fixed point theorem guarantees that S has a fixed point in K^*, which is a fixed-sign solution of (2). Hence, λ is an eigenvalue, i.e., $\lambda \in E$. $\square$

Using Lemma 3.3, we obtain the following:

Theorem 3.2. *Let (H1), (H2), and (H5) hold. If $E \ne \emptyset$, then E is an interval.*

The next result provides upper and lower bounds for an eigenvalue λ.

Theorem 3.3. *Let (H2)–(H4) hold and let $g_i(t_i, s) b_i(s) \in L^1[a,b]$, $1 \le i \le n$. Moreover, suppose*

(H6)   *For each $1 \le i, j \le n$, if $|u_j| \le |v_j|$, then*

$$p_i(u_1, \ldots, u_{j-1}, u_j, u_{j+1}, \ldots, u_n) \le p_i(u_1, \ldots, u_{j-1}, v_j, u_{j+1}, \ldots, u_n).$$

Let λ be an eigenvalue of (2) and $u \in C$ be a corresponding eigenfunction with $q_i = |u_i|_0$, $1 \le i \le n$. Then, for each $1 \le i \le n$, we have

$$\lambda \ge \frac{q_i}{p_i(q_1, q_2, \ldots, q_n)} \left[\int_a^b g_i(t^*, s) b_i(s) ds \right]^{-1} \tag{20}$$

and

$$\lambda \leq \frac{q_i}{p_i(\theta_1 M \rho_1 q_1, \theta_2 M \rho_2 q_2, \ldots, \theta_n M \rho_n q_n)} \left[\int_{t^*-h}^{t^*+h} g_i(t^*,s) a_i(s) ds \right]^{-1}. \quad (21)$$

We are now ready to establish criteria for E to be a bounded/unbounded interval.

Theorem 3.4. *Let (H1)–(H6) hold and let* $g_i(t_i,s) b_i(s) \in L^1[a,b]$, $1 \leq i \leq n$. *For each* $1 \leq i \leq n$, *define*

$$F_i^B = \left\{ p : \tilde{K} \to [0,\infty) \ \middle| \ \frac{|u_i|}{p(u_1,u_2,\ldots,u_n)} \text{ is bounded for } u \in \mathbb{R}^n \right\},$$

$$F_i^0 = \left\{ p : \tilde{K} \to [0,\infty) \ \middle| \ \lim_{\min_{1 \leq j \leq n} |u_j| \to \infty} \frac{|u_i|}{p(u_1,u_2,\ldots,u_n)} = 0 \right\},$$

$$F_i^\infty = \left\{ p : \tilde{K} \to [0,\infty) \ \middle| \ \lim_{\min_{1 \leq j \leq n} |u_j| \to \infty} \frac{|u_i|}{p(u_1,u_2,\ldots,u_n)} = \infty \right\}.$$

(a) *If* $p_i \in F_i^B$ *for each* $1 \leq i \leq n$, *then* $E = (0,c)$ *or* $(0,c]$ *for some* $c \in (0,\infty)$.
(b) *If* $p_i \in F_i^0$ *for each* $1 \leq i \leq n$, *then* $E = (0,c]$ *for some* $c \in (0,\infty)$.
(c) *If* $p_i \in F_i^\infty$ *for each* $1 \leq i \leq n$, *then* $E = (0,\infty)$.

Proof.

(a) This is immediate from (21) and Theorems 3.1 and 3.2.
(b) Since $F_i^0 \subseteq F_i^B$, $1 \leq i \leq n$, it follows from Case (a) that $E = (0,c)$ or $(0,c]$ for some $c \in (0,\infty)$. In particular, $c = \sup E$. It can be shown that $c = \sup E \in E$, and this completes the proof for Case (b).
(c) Let $\lambda > 0$ be fixed. Choose $\varepsilon > 0$ so that

$$\lambda \max_{1 \leq i \leq n} \int_a^b g_i(t^*,s) b_i(s) ds \leq \frac{1}{\varepsilon}.$$

By definition, if $p_i \in F_i^\infty$, $1 \leq i \leq n$, then there exists $R = R(\varepsilon) > 0$ such that the following holds for each $1 \leq i \leq n$:

$$p_i(u_1,u_2,\ldots,u_n) < \varepsilon |u_i|, \ |u_j| \geq R, \ 1 \leq j \leq n.$$

We can prove that $S(C(R)) \subseteq C(R)$. From Lemma 3.1 the operator S is continuous and completely continuous. Schauder's fixed point theorem guarantees that S has a fixed point in $C(R)$. Clearly, this fixed point is a fixed-sign solution of (2), and therefore λ is an eigenvalue of (2). Since $\lambda > 0$ is arbitrary, we have proved that $E = (0,\infty)$. □

In the next few theorems, we establish explicit subintervals of E. For each p_i, $1 \leq i \leq n$ introduced in (H3), we define

$$\overline{p}_{0,i} = \limsup_{\max_{1 \le j \le n} |u_j| \to 0} \frac{p_i(u_1, u_2, \ldots, u_n)}{|u_i|},$$

$$\underline{p}_{0,i} = \liminf_{\max_{1 \le j \le n} |u_j| \to 0} \frac{p_i(u_1, u_2, \ldots, u_n)}{|u_i|},$$

$$\overline{p}_{\infty,i} = \limsup_{\min_{1 \le j \le n} |u_j| \to \infty} \frac{p_i(u_1, u_2, \ldots, u_n)}{|u_i|},$$

$$\underline{p}_{\infty,i} = \liminf_{|u_i| \to \infty} \frac{p_i(u_1, u_2, \ldots, u_n)}{|u_i|}.$$

It is assumed that $\underline{p}_{\infty,i}$ yields a number (which can be infinite). Using Theorem 2.2, we obtain the next two results.

Theorem 3.5. *Let (H1)–(H4) hold and let* $g_i(t^*, s)b_i(s) \in L^1[a,b]$, $1 \le i \le n$. *If* λ *satisfies*

$$\sigma_{1,i} < \lambda < \sigma_{2,i}, \ 1 \le i \le n,$$

where

$$\sigma_{1,i} = \left[\underline{p}_{\infty,i} \, M \rho_i \int_{t^*-h}^{t^*+h} g_i(t^*, s) a_i(s) ds \right]^{-1}$$

and

$$\sigma_{2,i} = \left[\overline{p}_{0,i} \int_a^b g_i(t^*, s) b_i(s) ds \right]^{-1},$$

then $\lambda \in E$. *Hence,* $(\sigma_{1,i}, \sigma_{2,i}) \subseteq E$, $1 \le i \le n$.

Theorem 3.6. *Let (H1)–(H4) hold and let* $g_i(t^*, s)b_i(s) \in L^1[a,b]$, $1 \le i \le n$. *If* λ *satisfies*

$$\sigma_{3,i} < \lambda < \sigma_{4,i}, \ 1 \le i \le n,$$

where

$$\sigma_{3,i} = \left[\underline{p}_{0,i} \, M \rho_i \int_{t^*-h}^{t^*+h} g_i(t^*, s) a_i(s) ds \right]^{-1}$$

and

$$\sigma_{4,i} = \left[\overline{p}_{\infty,i} \int_a^b g_i(t^*, s) b_i(s) ds \right]^{-1},$$

then $\lambda \in E$. *Hence,* $(\sigma_{3,i}, \sigma_{4,i}) \subseteq E$, $1 \le i \le n$.

Combining Lemma 3.3, Theorems 3.5 and 3.6, we get the next result.

Theorem 3.7. *Let (H1)–(H5) hold and let $g_i(t^*,s)b_i(s) \in L^1[a,b]$, $1 \le i \le n$. Then,*

$$\left(0, \max_{1 \le i \le n} \sigma_{2,i}\right) \subseteq E \quad and \quad \left(0, \max_{1 \le i \le n} \sigma_{4,i}\right) \subseteq E.$$

Remark 3.1. For a fixed $i \in \{1,2,\ldots,n\}$, if p_i is *superlinear* (i.e., $\overline{p}_{0,i}=0$ and $\underline{p}_{\infty,i}=\infty$) or *sublinear* (i.e., $\underline{p}_{0,i} = \infty$ and $\overline{p}_{\infty,i} = 0$), then we conclude from Theorems 3.5 and 3.6 that $E = (0,\infty)$, i.e., (2) has a fixed-sign solution for any $\lambda > 0$. We remark that superlinearity and sublinearity conditions have also been discussed for various boundary value problems in the literature for the single equation case ($n = 1$); see, for example, [2–5, 12, 13, 19] and the references cited therein.

Example 3.1. Consider the system of boundary value problems

$$\begin{cases} u_1'''(t) = \lambda \, \dfrac{(|u_1(t)|+1)^2}{|u_2(t)|+1} \left[t(t-2)\left(t-\dfrac{517}{360}\right)+1\right]^{-1}, & t \in [0,1] \\[4mm] u_2'''(t) = \lambda \, \dfrac{3(|u_2(t)|+1)^2}{|u_1(t)|+1} \left[3t(t-2)\left(t-\dfrac{517}{360}\right)+3\right]^{-1}, & t \in [0,1] \\[4mm] u_i(0) = u_i'(0.55) = 0, \qquad u_i(1)+0.5u_i''(1) = 0, & i = 1,2. \end{cases} \quad (22)$$

In this example, $n = 2$, $[a,b] = [0,1]$, $t_i = 0.55$, $\gamma_i = 1$, $\delta_i = 0.5$, $i = 1,2$,

$$f_1(t,u_1(t),u_2(t)) = \frac{(|u_1(t)|+1)^2}{|u_2(t)|+1}\left[t(t-2)\left(t-\frac{517}{360}\right)+1\right]^{-1}$$

and

$$f_2(t,u_1(t),u_2(t)) = \frac{3(|u_2(t)|+1)^2}{|u_1(t)|+1}\left[3t(t-2)\left(t-\frac{517}{360}\right)+3\right]^{-1}.$$

Clearly, (H1) is satisfied. Fix $\theta_1 = \theta_2 = 1$. Then, (H2) is fulfilled. Next, choose

$$p_1(u_1,u_2) = \frac{(|u_1|+1)^2}{|u_2|+1}, \qquad p_2(u_1,u_2) = \frac{3(|u_2|+1)^2}{|u_1|+1},$$

$$a_1(t) = b_1(t) = \left[t(t-2)\left(t-\frac{517}{360}\right)+1\right]^{-1},$$

$$a_2(t) = b_2(t) = \left[3t(t-2)\left(t-\frac{517}{360}\right)+3\right]^{-1}.$$

Then, (H3) and (H4) (with $\rho_1 = \rho_2 = 1$) are satisfied. Moreover, we have $g_i(t_i,s)b_i(s) \in L^1[0,1]$, $i = 1,2$.

By Theorem 3.1, there exists $c > 0$ such that

$$(o, c] \subseteq E. \tag{23}$$

On the other hand, it is easy to see that

$$\underline{p}_{0,1} = \infty, \quad \overline{p}_{\infty,1} = 1, \quad \underline{p}_{0,2} = \infty, \quad \text{and} \quad \overline{p}_{\infty,2} = 3.$$

Using expression (3), by direct computation we get

$$\sigma_{3,1} = \sigma_{3,2} = 0 \quad \text{and} \quad \sigma_{4,1} = \sigma_{4,2} = 13.7382. \tag{24}$$

It follows from Theorem 3.6 that

$$(0, 13.7382) \subseteq E. \tag{25}$$

In view of (25), we see that (23) holds for any positive c less than 13.7382. Moreover, as an example, when $\lambda = 6 \in (0, 13.7382) \subseteq E$, the system (22) has a positive solution $u = (u_1, u_2)$ given by

$$u_1(t) = u_2(t) = t(t - 2)\left(t - \frac{517}{360}\right). \tag{26}$$

References

1. Agarwal, R.P.: Focal Boundary Value Problems for Differential and Difference Equations. Kluwer, Dordrecht (1998)
2. Agarwal, R.P., Bohner, M., Wong, P.J.Y.: Positive solutions and eigenvalues of conjugate boundary value problems. Proc. Edinb. Math. Soc.(series 2) **42**, 349–374 (1999)
3. Agarwal, R.P., Henderson, J., Wong, P.J.Y.: On superlinear and sublinear (n, p) boundary value problems for higher order difference equations. Nonlinear World **4**, 101–115 (1997)
4. Agarwal, R.P., O'Regan, D., Wong, P.J.Y.: Positive Solutions of Differential, Difference and Integral Equations. Kluwer, Dordrecht (1999)
5. Agarwal, R.P., Wong, P.J.Y.: Advanced Topics in Difference Equations. Kluwer, Dordrecht (1997)
6. Anderson, D.: Multiple positive solutions for a three point boundary value problem. Math. Comput. Model. **27**, 49–57(1998)
7. Anderson, D.: Green's function for a third-order generalized right focal problem. J. Math. Anal. Appl. **288**, 1–14 (2003)
8. Anderson, D., Davis, J.: Multiple solutions and eigenvalues for third order right focal boundary value problems. J. Math. Anal. Appl. **267**, 135–157 (2002)
9. Baxley, J.V., Carroll, P.T.: Nonlinear boundary value problems with multiple positive solutions. Discrete Contin. Dyn. Syst. Suppl. 83–90 (2003)
10. Baxley, J.V., Houmand, C.R.: Nonlinear higher order boundary value problems with multiple positive solutions. J. Math. Anal. Appl. **286**, 682–691 (2003)
11. Davis, J.M., Henderson, J., Prasad, K.R., Yin, W.: Eigenvalue intervals for nonlinear right focal problems. Appl. Anal. **74**, 215–231 (2000)
12. Eloe, P.W., Henderson, J.: Positive solutions and nonlinear $(k, n - k)$ conjugate eigenvalue problems. Diff. Eqns. Dyn. Sys. **6**, 309–317(1998)

13. Erbe, L.H., Wang, H.: On the existence of positive solutions of ordinary differential equations. Proc. Am. Math. Soc. **120**, 743–748(1994)
14. Graef, J.R., Henderson, J.: Double solutions of boundary value problems for $2m$th-order differential equations and difference equations. Comput. Math. Appl. **45**, 873–885 (2003)
15. Graef, J.R., Qian, C., Yang, B.: A three point boundary value problem for nonlinear fourth order differential equations. J. Math. Anal. Appl. **287**, 217–233 (2003)
16. Krasnosel'skii, M.A.: Positive Solutions of Operator Equations. Noordhoff, Groningen (1964)
17. Lian, W., Wong, F., Yeh, C.: On the existence of positive solutions of nonlinear second order differential equations. Proc. Am. Math. Soc. **124**, 1117–1126 (1996)
18. Wong, P.J.Y.: Two-point right focal eigenvalue problems for difference equations. Dynam. Systems Appl. **7**, 345–364 (1998)
19. Wong, P.J.Y.: Positive solutions of difference equations with two-point right focal boundary conditions. J. Math. Anal. Appl. **224**, 34–58 (1998)
20. Wong, P.J.Y., Agarwal, R.P.: Existence of multiple positive solutions of discrete two-point right focal boundary value problems. J. Difference Equ. Appl. **5**, 517–540 (1999)
21. Wong, P.J.Y.: Contant-sign solutions for a system of generalized right focal problems. Nonlinear Anal. **63**, 2153–2163 (2005)
22. Wong, P.J.Y.: Eigenvalue characterization for a system of generalized right focal problems. Dynam. Systems Appl. **15**, 173–191 (2006)
23. Wong, P.J.Y.: Multiple fixed-sign solutions for a system of generalized right focal problems with deviating arguments. J. Math. Anal. Appl. **323**, 100–118 (2006)
24. Wong, P.J.Y.: Triple fixed-sign solutions for a system of third-order generalized right focal boundary value problems. In: Proceedings of the Conference on Differential and Difference Equations and Applications, 1139–1148, USA (2006)
25. Wong, P.J.Y.: On the existence of fixed-sign solutions for a system of generalized right focal problems with deviating arguments. Discrete Contin. Dyn. Syst. Suppl, 1042–1051 (2007)
26. Wong, P.J.Y.: Eigenvalues of a system of generalized right focal problems with deviating arguments. J. Comput. Appl. Math. **218**, 459–472 (2008)

Forced Oscillation of Second-Order Impulsive Differential Equations with Mixed Nonlinearities

A. Özbekler, and A. Zafer

Abstract In this paper we give new oscillation criteria for a class of second-order mixed nonlinear impulsive differential equations having fixed moments of impulse actions. The method is based on the existence of a nonprincipal solution of a related second-order linear homogeneous equation.

Keywords Oscillation • Mixed nonlinear • Fixed moments • Impulse • Non-principal

1 Introduction

Impulsive differential equations are of particular interest in many areas such as biology, physics, chemistry, control theory, and medicine as they model the real processes better than differential equations. Because of the lack of smoothness property of the solutions the theory of impulsive differential equations is much richer than that of differential equations without impulse. However, due to difficulties caused by impulsive perturbations, the theory is not well developed in comparison with that of non-impulsive differential equations. For basic theory of impulsive differential equations, we refer in particular to [3, 21]. Concerning the oscillation theory for second-order impulsive differential equations, we refer in particular to [2, 6, 13, 16–18] and the references therein.

A. Özbekler
Department of Mathematics, Atilim University 06836, Incek, Ankara, Turkey
e-mail: aozbekler@gmail.com

A. Zafer (✉)
Department of Mathematics, Middle East Technical University 06531, Ankara, Turkey
e-mail: zafer@metu.edu.tr

S. Pinelas et al. (eds.), *Differential and Difference Equations with Applications*, Springer
Proceedings in Mathematics & Statistics 47, DOI 10.1007/978-1-4614-7333-6_13,
© Springer Science+Business Media New York 2013

In this paper we derive oscillation criteria for solutions of nonlinear impulsive equations of the form

$$(r(t)x')' + \sum_{k=1}^{n} q_k(t)\phi_k(x) = f(t), \; t \neq \theta_i;$$
$$\Delta r(t)x' + \sum_{k=1}^{n} q_{i,k}\phi_k(x) = f_i, \qquad t = \theta_i, \tag{1}$$

where $\Delta g(t)$ denotes the difference $g(t^+) - g(t^-)$ with $g(t^\pm) = \lim_{\tau \to t^\pm} g(\tau)$.

Our main purpose in this paper is to obtain an oscillation for Eq. (1) by making use of nonprincipal solutions of a related second-order impulsive differential equation

$$(r(t)z')' + Q(t)z = 0, \; t \neq \theta_i;$$
$$\Delta r(t)z' + Q_i z = 0, \qquad t = \theta_i, \tag{2}$$

where

$$Q(t) = \sum_{k=1}^{n} q_k(t) \text{ and } Q_i = \sum_{k=1}^{n} q_{i,k}.$$

The existence of principal and nonprincipal solutions of Eq. (2) was shown by present authors in [18] when the equation is nonoscillatory or equivalently has a positive solution.

Theorem 1. *If Eq. (2) has a positive solution on $[a, \infty)$, then it has two linearly independent solutions u and v such that*

$$\lim_{t \to \infty} \frac{u(t)}{v(t)} = 0;$$

$$\int_a^\infty \frac{1}{r(t)u^2(t)} dt = \infty, \quad \int_a^\infty \frac{1}{r(t)v^2(t)} dt < \infty;$$

$$\frac{r(t)v'(t)}{v(t)} > \frac{r(t)u'(t)}{u(t)} \quad \text{for } t \text{ sufficiently large.}$$

The functions u and v are called principal and nonprincipal solutions of Eq. (2), respectively.

The use of principal and nonprincipal solutions in connection with oscillation and asymptotic behavior of second-order differential equations could be found in [1, 5, 8, 10, 14, 22, 26, 27]. For some extensions to Hamiltonian systems, half-linear differential equations, dynamic equations, and impulsive differential equations, see also [4, 7, 8, 18, 19, 28]. In [18], we obtained oscillation criteria for nonhomogeneous linear impulsive equations by making use of a nonprincipal solution of the corresponding homogeneous equation.

In the case when $n = 2$ and the impulses are absent, Eq. (1) reduces to

$$(r(t)x')' + p(t)F(x) - q(t)G(x) = f(t), \quad t \geq t_0. \tag{3}$$

Recently, the oscillation criteria obtained by Wong [27] have been extended to Eq. (3) in [15]. The arguments there are based on the existence of a positive solution of the related linear equation

$$(r(t)x')' + [p(t) - q(t)]x = 0 \tag{4}$$

when r, p, q, and f are continuous functions with $r > 0$, $p \geq 0$, and $q \geq 0$ on $[t_0, \infty)$. Moreover, F and $G \in C(\mathbb{R}, \mathbb{R})$ satisfy

(A_1) $xF(x) > 0$ and $xG(x) > 0$ for $x \neq 0$.

(A_2) $\lim\limits_{|x| \to \infty} x^{-1}F(x) > 1$, $\lim\limits_{|x| \to 0} x^{-1}F(x) < 1$, $\lim\limits_{|x| \to \infty} x^{-1}G(x) < 1$, $\lim\limits_{|x| \to 0} x^{-1}G(x) > 1$.

The result is as follows.

Theorem 2. *Suppose that Eq. (4) is nonoscillatory and let $z(t)$ be a positive solution of it satisfying*

$$\int_a^\infty \frac{1}{r(s)z^2(s)} \, ds < \infty \tag{5}$$

for some a sufficiently large, i.e., a nonprincipal solution. Let

$$\beta_0 = \max_{x \geq 0}\{x - F(x)\}, \quad \alpha_0 = -\min_{x \leq 0}\{x - F(x)\};$$
$$\delta_0 = \max_{x \leq 0}\{x - G(x)\}, \quad \gamma_0 = -\min_{x \geq 0}\{x - G(x)\}.$$

If

$$\overline{\lim_{t \to \infty}} \{\mathcal{H}_0(t) - \mathcal{N}_2(t)\} = -\underline{\lim_{t \to \infty}} \{\mathcal{H}_0(t) + \mathcal{N}_1(t)\} = \infty,$$

where

$$\mathcal{H}_0(t) := \int_a^t \frac{1}{r(s)z^2(s)} \left(\int_a^s z(\tau)f(\tau)d\tau \right) ds, \tag{6}$$

$$\mathcal{N}_1(t) := \int_a^t \frac{1}{r(s)z^2(s)} \left(\int_a^s [\beta_0 p(\tau) + \gamma_0 q(\tau)]z(\tau)d\tau \right) ds,$$

and

$$\mathcal{N}_2(t) := \int_a^t \frac{1}{r(s)z^2(s)} \left(\int_a^s \left[\alpha_0 p(\tau) + \delta_0 q(\tau) \right] z(\tau) d\tau \right) ds,$$

then Eq. (3) is oscillatory.

Theorem 2 is applicable to equations of the form

$$(r(t)x')' + p(t)|x|^{\beta-1}x - q(t)|x|^{\gamma-1}x = f(t),$$

when

$$0 < \gamma < 1 < \beta.$$

In the limiting case $\beta \to 1^+$ and $\gamma \to 1^-$, the following theorem of Wong [27] is recovered.

Theorem 3. *Let z be a positive solution of $(r(t)x')' + \tilde{q}(t)x = 0$ satisfying Eq. (5) and $\mathcal{H}_0$ be as defined in Eq. (6). If*

$$\overline{\lim_{t \to \infty}} \mathcal{H}_0(t) = -\underline{\lim_{t \to \infty}} \mathcal{H}_0(t) = \infty,$$

then

$$(r(t)z')' + \tilde{q}(t)z = f(t)$$

is oscillatory.

More results concerning the oscillation of the Emden–Fowler equation

$$(r(t)x')' + q(t)|x|^{\alpha-1}x = f(t)$$

could be found in [9, 11, 12, 20, 23–25, 27].

Denote by $\text{PLC}[t_0, \infty)$, $t_0 \in \mathbb{R}$ is fixed, the set of functions $h : [t_0, \infty) \to \mathbb{R}$ such that h is continuous on each interval (θ_i, θ_{i+1}), $h(\theta_i^{\pm})$ exist, and $h(\theta_i) = h(\theta_i^-)$ $i \in \mathbb{N}$.

Definition 1. By a solution of Eq. (1) on an interval $[t_*, \infty)$, $t_* \geq t_0$, we mean a function $x \in C[t_*, \infty)$ with x', $(rx')' \in \text{PLC}[t_*, \infty)$ that satisfies Eq. (1) for $\geq t_*$.

Definition 2. A nontrivial solution x of Eq. (1) is called oscillatory if it has arbitrarily large zeros; otherwise, it is called nonoscillatory. Equation (1) is called oscillatory (nonoscillatory) if all of its solutions are oscillatory (nonoscillatory).

We recall that according to the Sturm's separation theorem [16] every solution of Eq. (2) is oscillatory (nonoscillatory) if there is one oscillatory (nonoscillatory) solution of the equation.

2 Main Results

Let Eq. (2) be nonoscillatory and $z(t)$ its positive nonprincipal solution satisfying

$$\int_a^\infty \frac{ds}{r(s)z^2(s)} < \infty. \tag{7}$$

Note that the existence of such a solution is guaranteed by Theorem 1.

With regard to impulsive Eqs. (1) and (2), we assume throughout this work that

(i) $r, q_k, q, f \in \mathrm{PLC}[t_0, \infty)$; $r(t) > 0$ and

$$q_k(t) \begin{cases} \leq 0, \ k = 1, 2, \ldots, m \\ \geq 0, \ k = m+1, m+2, \ldots, n. \end{cases} \tag{8}$$

(ii) $\{\theta_i\}$ is a strictly increasing unbounded sequence of real numbers, $\theta_i \geq t_0$; $\{f_i\}$, $\{q_i\}$ and $\{q_{i,k}\}$ are real sequences and

$$q_{i,k} \begin{cases} \leq 0, \ k = 1, 2, \ldots, m \\ \geq 0, \ k = m+1, m+2, \ldots, n. \end{cases} \tag{9}$$

(iii) (C_1) $s\phi_k(s) > 0$ for $s \neq 0$, $k = 1, 2, \ldots, n$;
       (C_2)

$$\lim_{|s| \to \infty} s^{-1}\phi_k(s) \begin{cases} < 1, \ k = 1, 2, \ldots, m \\ > 1, \ k = m+1, m+2, \ldots, n. \end{cases}$$
$$\lim_{|s| \to 0} s^{-1}\phi_k(s) \begin{cases} > 1, \ k = 1, 2, \ldots, m \\ < 1, \ k = m+1, m+2, \ldots, n. \end{cases}$$

Using (C_1) and (C_2), it is easy to find positive constants $\rho_j(k)$, $j = 0, 1, 2, 3$, such that

$$\begin{aligned} \max_{x \leq 0} \Phi_k(x) = \rho_0(k), \quad \min_{x \geq 0} \Phi_k(x) = -\rho_1(k), \quad k = 1, 2, \ldots, m; \\ \max_{x \geq 0} \Phi_k(x) = \rho_2(k), \quad \min_{x \leq 0} \Phi_k(x) = -\rho_3(k), \quad k = m+1, m+2, \ldots, n, \end{aligned} \tag{10}$$

where

$$\Phi_k(x) = x - \phi_k(x).$$

In what follows, we define

$$\mathscr{S}_1(t) := \sum_{k=1}^m \rho_1(k) G_k(t) - \sum_{k=m+1}^n \rho_2(k) \mathscr{G}_k(t) \tag{11}$$

and

$$\mathscr{S}_2(t) := \sum_{k=1}^{m} \rho_0(k) G_k(t) - \sum_{k=m+1}^{n} \rho_3(k) \mathscr{G}_k(t), \tag{12}$$

where

$$G_k(t) := \int_a^t \frac{1}{r(s)z^2(s)} \left(\int_a^s z(\tau) q_k(\tau) d\tau + \sum_{a \le \theta_i < s} z(\theta_i) q_{i,k} \right) ds.$$

Denote

$$\mathscr{H}(t) := \int_a^t \frac{1}{r(s)z^2(s)} \left(\int_a^s z(\tau) f(\tau) d\tau + \sum_{a \le \theta_i < s} z(\theta_i) f_i \right) ds. \tag{13}$$

Theorem 4. *Suppose that Eq. (2) is nonoscillatory and let $z(t)$ be a positive solution of it satisfying Eq. (7), i.e., a nonprincipal solution. If*

$$\varlimsup_{t \to \infty} \{\mathscr{H}(t) + \mathscr{S}_2(t)\} = -\varlimsup_{t \to \infty} \{\mathscr{H}(t) - \mathscr{S}_1(t)\} = \infty, \tag{14}$$

where $\mathscr{S}_1$, $\mathscr{S}_1$, and $\mathscr{H}$ are given, respectively, by Eqs. (11), (12), and (13), then Eq. (1) is oscillatory.

Proof. Let $x(t)$ be a solution of Eq. (1). The change of variable $x = z(t)w$ transforms Eq. (1) into

$$(r(t)z^2w')' = \left\{ f(t) + \sum_{k=1}^{n} q_k(t) \Phi_k(x) \right\} z, \quad t \ne \theta_i; \tag{15}$$

$$\Delta w = 0, \quad \Delta r(t) z^2 w' = \left\{ f_i + \sum_{k=1}^{n} q_{i,k} \Phi_k(x) \right\} z, \quad t = \theta_i. \tag{16}$$

Since $z(t)$ is a solution of Eq. (2), we can express $w(t)$ by integration of Eq. (15) and using Eq. (16) as follows:

$$w(t) = c_1 + c_2 \int_a^t \frac{ds}{r(s)z^2(s)} + \int_a^t \frac{1}{r(s)z^2(s)} \left(\int_a^s z(\tau) \sum_{k=1}^{n} q_k(\tau) \Phi_k(x(\tau)) d\tau \right.$$

$$\left. + \sum_{a \le \theta_i < s} z(\theta_i) \sum_{k=1}^{n} q_{i,k} \Phi_k(x(\theta_i)) \right) ds + \mathscr{H}(t), \tag{17}$$

where $c_1 = w(a)$ and $c_2 = r(a)z^2(a)w'(a)$ are constants.

It is not difficult to see from Eqs. (8), (9), (10), and (17) that if $x(t) > 0$ on $[a, \infty)$, then

$$w(t) \leq c_1 + c_2 \int_a^t \frac{ds}{r(s)z^2(s)} + \mathcal{H}(t) - \mathcal{S}_1(t) \tag{18}$$

and that if $x(t) < 0$ on $[a, \infty)$, then

$$w(t) \geq c_1 + c_2 \int_a^t \frac{ds}{r(s)z^2(s)} + \mathcal{H}(t) + \mathcal{S}_2(t). \tag{19}$$

Note that Eqs. (7), (14), (18), and (19) imply that

$$\overline{\lim_{t \to \infty}} w(t) = -\lim_{t \to \infty} w(t) = +\infty. \tag{20}$$

Because $z(t)$ is positive, Eq. (20) implies that $x(t)$ has no definite sign on $[a, \infty)$, namely, it is oscillatory.

Remark 1. If we pick $q_j(t) = 0$ for all $j = 2, 3, \ldots, n-1$ and $q_{i,k} \equiv f_i \equiv 0$, then Eq. (1) reduces to Eq. (3) and hence, we recover [15, Theorem 2.1].

When $\phi_k(x) = |x|^{\alpha_k - 1} x$, $0 < \alpha_1 < \cdots < \alpha_m < 1 < \alpha_{m+1} < \cdots < \alpha_n$, then (iii) is satisfied with

$$\begin{aligned}
\rho_0(k) &= \rho_1(k) = (1 - \alpha_k)\alpha_k^{\alpha_k/(1-\alpha_k)} > 0, \ k = 1, 2, \ldots, m; \\
\rho_2(k) &= \rho_3(k) = (\alpha_k - 1)\alpha_k^{\alpha_k/(1-\alpha_k)} > 0, \ k = m+1, m+2, \ldots, n,
\end{aligned}$$

and we obtain the following oscillation criterion for equation

$$\begin{aligned}
(r(t)x')' + \sum_{k=1}^n q_k(t)|x|^{\alpha_k - 1} x &= f(t), \ t \neq \theta_i; \\
\Delta r(t)x' + \sum_{k=1}^n q_{i,k}|x|^{\alpha_k - 1} x &= f_i, \qquad t = \theta_i.
\end{aligned} \tag{21}$$

Theorem 5. *Suppose that Eq. (2) is nonoscillatory and let $z(t)$ be a positive solution of it satisfying Eq. (7), i.e., a nonprincipal solution. If*

$$\overline{\lim_{t \to \infty}}\{\mathcal{H}(t) + \mathcal{S}_0(t)\} = -\lim_{t \to \infty}\{\mathcal{H}(t) - \mathcal{S}_0(t)\} = \infty, \tag{22}$$

where

$$\mathcal{S}_0(t) := \sum_{k=1}^n (1 - \alpha_k)\alpha_k^{\alpha_k/(1-\alpha_k)}\mathcal{G}_k(t),$$

then Eq. (21) is oscillatory.

Remark 2. In view of

$$\lim_{\beta \to 1^{\pm}} \beta^{\beta/(1-\beta)} = 1/e,$$

we have

$$\lim_{\substack{\alpha_k \to 1^- \\ (k=1,2,\dots,m)}} (1 - \alpha_k)\alpha_k^{\alpha_k/(1-\alpha_k)} = \lim_{\substack{\alpha_k \to 1^+ \\ (k=m+1,m+2,\dots,n)}} (1 - \alpha_k)\alpha_k^{\alpha_k/(1-\alpha_k)} = 0,$$

and hence,

$$\mathscr{S}_0(t) \to 0 \text{ as } \alpha_k \to 1^- \ (k = 1, 2, \dots, m)$$

and

$$\mathscr{S}_0(t) \to 0 \text{ as } \alpha_k \to 1^+ \ (k = m+1, m+2, \dots, n).$$

Using these facts, we recover [18, Theorem 3.1] for equation

$$\begin{aligned}
(r(t)z')' + Q(t)z &= f(t), \ t \neq \theta_i; \\
\Delta r(t)z' + Q_i z &= f_i, \qquad t = \theta_i.
\end{aligned} \tag{23}$$

It is clear that two special cases of Eq. (21) are Emden–Fowler-type superlinear impulsive equation

$$\begin{aligned}
(r(t)x')' + \sum_{k=m+1}^{n} q_k(t)|x|^{\alpha_k-1}x &= f(t), \ t \neq \theta_i; \\
\Delta r(t)x' + \sum_{k=m+1}^{n} q_{i,k}|x|^{\alpha_k-1}x &= f_i, \qquad t = \theta_i,
\end{aligned} \tag{24}$$

and Emden–Fowler-type sublinear impulsive equation

$$\begin{aligned}
(r(t)x')' + \sum_{k=1}^{m} q_k(t)|x|^{\alpha_k-1}x &= f(t), \ t \neq \theta_i; \\
\Delta r(t)x' + \sum_{k=1}^{m} q_{i,k}|x|^{\alpha_k-1}x &= f_i, \qquad t = \theta_i.
\end{aligned} \tag{25}$$

Corollary 1. *Suppose that Eq. (2) with $q_k(t) \equiv 0$ and $q_{i,k} \equiv 0$, $k = 1, 2, \dots, m$, is nonoscillatory and let $z(t)$ be a positive solution of it satisfying Eq. (7), i.e., a nonprincipal solution. If*

$$\overline{\lim_{t \to \infty}}\{\mathscr{H}(t) + \mathscr{S}_{01}(t)\} = -\underline{\lim_{t \to \infty}}\{\mathscr{H}(t) - \mathscr{S}_{01}(t)\} = \infty, \tag{26}$$

where

$$\mathscr{S}_{01}(t) := \sum_{k=m+1}^{n} (1 - \alpha_k) \alpha_k^{\alpha_k/(1-\alpha_k)} \mathscr{G}_k(t),$$

then Eq. (24) is oscillatory.

Corollary 2. *Suppose that Eq. (2) with $q_k(t) \equiv 0$ and $q_{i,k} \equiv 0$, $k = m+1, m+2, \ldots, n$, is nonoscillatory and let $z(t)$ be a positive solution of it satisfying Eq. (7), i.e., a nonprincipal solution. If*

$$\overline{\lim_{t \to \infty}} \{\mathscr{H}(t) + \mathscr{S}_{02}(t)\} = -\lim_{t \to \infty} \{\mathscr{H}(t) - \mathscr{S}_{02}(t)\} = \infty, \tag{27}$$

where

$$\mathscr{S}_{02}(t) := \sum_{k=1}^{m} (1 - \alpha_k) \alpha_k^{\alpha_k/(1-\alpha_k)} \mathscr{G}_k(t),$$

then Eq. (25) is oscillatory.

Finally we give an example to illustrate one of our results. Due to impulses, the computation becomes quite tedious.

Example 1. Consider the Emden–Fowler-type sublinear impulsive equation

$$\begin{aligned} x'' - |x|^{\alpha-1}x &= 1, \quad t \neq \ln(i+1), \ t \geq 0, \ i \in \mathbb{N}; \\ \Delta x' - \frac{12(i+1)}{i(4i+5)} |x|^{\alpha-1}x &= (-1)^i i^4, \quad t = \ln(i+1), \end{aligned} \tag{28}$$

where $\alpha \in (0,1)$.

The related unforced impulsive equation

$$\begin{aligned} z'' - z &= 0, \quad t \neq \ln(i+1), \ t \geq 0, \ i \in \mathbb{N}; \\ \Delta z' - \frac{12(i+1)}{i(4i+5)} z &= 0, \quad t = \ln(i+1) \end{aligned} \tag{29}$$

is nonoscillatory with a nonoscillatory solution

$$z(t) = z_i(t) = ie^t - \frac{i(i+1)(2i+1)}{6} e^{-t}, \quad t \in (\ln i, \ln(i+1)], \quad i \in \mathbb{N}.$$

It is easy to show that

$$\int_{\ln 2}^{\infty} \frac{dt}{z^2(t)} = \sum_{k=2}^{\infty} \int_{\ln k}^{\ln(k+1)} \frac{dt}{z_k^2(t)} = \sum_{k=2}^{\infty} \frac{18(2k+1)}{k^2(k^2-1)(4k+1)(4k+5)} \cong 0.0750772,$$

i.e., Eq. (7) is satisfied.

Let $\{\ln(i_s + 1)\}$ be a subsequence of $\{\ln(i + 1)\}$ so that $s \in (\ln(i_s), \ln(i_s + 1)]$. Then

$$\int_{\ln 2}^{s} z(\tau)(-1)d\tau = -\sum_{k=2}^{i_s} \int_{\ln k}^{\ln(k+1)} z_k(\tau)d\tau = -\frac{1}{6}\sum_{k=2}^{i_s}(4k - 1) = -\frac{1}{6}(i_s - 1)(2i_s - 3)$$

and

$$\sum_{\ln 2 \leq \ln(k+1) < s} z(\ln(k + 1))\frac{12(k + 1)}{k(4k + 5)} = 2\sum_{k=1}^{i_s - 1}(k + 1) = (i_s - 1)(i_s + 2).$$

Define

$$\Psi_i(s) := -\frac{1}{6}(i_s - 1)(2i_s - 3) + (i_s - 1)(i_s + 2) = \frac{1}{6}(i_s - 1)(4i_s + 15),$$

then for $n \in \mathbb{N}$ sufficiently large, we have

$$\mathscr{G}_{1,n} = \int_{\ln 2}^{\ln n} \frac{1}{z^2(s)}\left\{-\int_{\ln 2}^{s} z(\tau)d\tau + 12\sum_{\ln 2 \leq \ln(k+1) < s} \frac{(k + 1)z(\ln(k + 1))}{k(4k + 5)}\right\}ds$$

$$= \sum_{k=2}^{n-1} \int_{\ln k}^{\ln(k+1)} \frac{\Psi_k(s)}{z_k^2(s)}ds$$

$$= \frac{1}{6}\sum_{k=2}^{n-1}\left\{(k - 1)(4k + 15)\int_{\ln k}^{\ln(k+1)} \frac{ds}{z_k^2(s)}\right\}$$

$$= \sum_{k=2}^{n-1} v_k,$$

where

$$v_k = \frac{3(2k + 1)(4k + 15)}{k^2(k + 1)(4k + 1)(4k + 5)}.$$

On the other hand

$$\int_{\ln 2}^{s} z(\tau)d\tau = \sum_{k=2}^{i_s} \int_{\ln k}^{\ln(k+1)} z_k(\tau)d\tau = \frac{1}{6}\sum_{k=2}^{i_s}(4k - 1) = \frac{1}{6}(i_s - 1)(2i_s - 3)$$

and

$$\sum_{\ln 2 \le \ln(k+1) < s} (-1)^k k^4 z(\ln(k+1))$$

$$= \frac{1}{6} \sum_{k=1}^{i_s - 1} (-1)^k k^5 (4k+5)$$

$$= \frac{(-1)^{i_s+1}}{24} (8i_s^6 - 14i_s^5 - 25i_s^4 + 40i_s^3 + 25i_s^2 - 24i_s - 5 + 5(-1)^{i_s}).$$

Define

$$\Phi_i(s) := \frac{(-1)^{i_s+1}}{24} (8i_s^6 - 14i_s^5 - 25i_s^4 + 40i_s^3 + 25i_s^2 - 24i_s - 5 + 5(-1)^{i_s})$$

$$+ \frac{1}{6}(i_s - 1)(2i_s - 3).$$

For $n \in \mathbb{N}$ sufficiently large, we get

$$\mathscr{H}_n = \int_{\ln 2}^{\ln n} \frac{1}{z^2(s)} \left\{ \int_{\ln 2}^{s} z(\tau) d\tau + \sum_{\ln 2 \le \ln(k+1) < s} (-1)^k k^4 z(\ln(k+1)) \right\} ds$$

$$= \sum_{k=2}^{n-1} \int_{\ln k}^{\ln(k+1)} \frac{\Phi_k(s)}{z_k^2(s)} ds$$

$$= \sum_{k=2}^{n-1} \left\{ \Phi_k(s) \int_{\ln k}^{\ln(k+1)} \frac{ds}{z_k^2(s)} \right\}$$

$$= \sum_{k=2}^{n-1} \eta_k + \sum_{k=2}^{n-1} (-1)^{k+1} \xi_k,$$

where

$$\eta_k = \frac{3(2k+1)(2k-3)}{k^2(k+1)(4k+1)(4k+5)}$$

and

$$\xi_k = \frac{3(2k+1)(8k^6 - 14k^5 - 25k^4 + 40k^3 + 25k^2 - 24k - 5 + 5(-1)^k)}{k^2(k^2-1)(4k+1)(4k+5)}.$$

Since

$$\sum_{k=2}^{\infty} v_k \cong 0.389354$$

and

$$\sum_{k=2}^{\infty} \eta_k \cong 0.0343956,$$

we have

$$\overline{\lim_{t \to \infty}} \{ \mathscr{H}(t) + \mathscr{S}_{02}(t) \} = \overline{\lim_{t \to \infty}} \{ \mathscr{H}_n + (1 - \alpha) \alpha^{\alpha/(1-\alpha)} \mathscr{G}_{1,n} \}$$

$$= \sigma_+ + \overline{\lim_{t \to \infty}} \sum_{k=2}^{n-1} (-1)^{k+1} \xi_k$$

$$= \infty.$$

Similarly,

$$\lim_{t \to \infty} \{ \mathscr{H}(t) - \mathscr{S}_{02}(t) \} = \lim_{t \to \infty} \{ \mathscr{H}_n - (1 - \alpha) \alpha^{\alpha/(1-\alpha)} \mathscr{G}_{1,n} \}$$

$$= \sigma_- + \lim_{t \to \infty} \sum_{k=2}^{n-1} (-1)^{k+1} \xi_k$$

$$= -\infty,$$

where $\sigma_{\pm} = 0.0343956 \pm 0.389354 \times (1 - \alpha) \alpha^{\alpha/(1-\alpha)}$.

Since the conditions of Corollary 2 are satisfied, we may conclude that every solution of Eq. (28) is oscillatory for any choice of $\alpha \in (0,1)$.

References

1. Agarwal, R.P., Grace, S.R., O'Regan, D.: Oscillation Theory for Second Order Linear, Half-Linear, Superlinear and Sublinear Dynamic Equations. Kluwer Academic Publishers, Dordrecht (2002)
2. Bainov, D.D., Domshlak, Y.I., Simeonov, P.S.: Sturmian comparison theory for impulsive differential inequalities and equations. Arch. Math. (Basel) **67**, 35–49 (1996)
3. Bainov, D.D., Simeonov, P.S.: Impulsive Differential Equations: Periodic Solutions and Applications. Longman Scientific and Technical, Essex (1993)
4. Cecchi, M., Došlá, Z., Marini, M.: Half-linear equations and characteristic properties of the principal solutions. J. Differ. Equ. **208**, 96–507 (2005)
5. Chen, S.: Asymptotic integrations of nonoscillatory second order differential equations. Tran. Am. Math. Soc. **327**, 853–865 (1991)
6. Chen, Y.S., Feng, W.Z.: Oscillations of second order nonlinear ode with impulses. J. Math. Anal. Appl. **210**, 150–169 (1997)
7. Došlý, O.: Principal solutions and transformations of linear Hamiltonian systems (English summary). Arch. Math. (Brno) **28**, 96–120 (1992)
8. Došlý, O., Řehák, P.: Half-Linear Differential Equations. Elsevier Ltd., Heidelberg (2005)
9. El-Sayed, M.A.: An oscillation criterion for a forced-second order linear differential equation. Proc. Am. Math. Soc. **118**, 813–817 (1993)

10. Hartman, P.: Ordinary Differential Equations. SIAM, Philadelphia (2002)
11. Kartsatos, A.G.: Maintenance of oscillations under the effect of a periodic forcing term. Proc. Am. Math. Soc. **33**, 377–383 (1972)
12. Keener, M.S.: Solutions of a linear nonhomogenous second order differential equations. Appl. Anal. **1**, 57–63 (1971)
13. Luo, J.: Second-order quasilinear oscillation with impulses. Comput. Math. Appl. **46**, 279–291 (2003)
14. Morse, M., Leighton, W.: Singular quadratic functionals. Trans. Am. Math. Soc. **40**, 252–286 (1936)
15. Özbekler, A., Wong, J.S.W., Zafer, A.: Forced oscillation of second-order nonlinear differential equations with positive and negative coefficients. Appl. Math. Lett. **24**, 1225–1230 (2011)
16. Özbekler, A., Zafer, A.: A Sturmian comparison theory for linear and half-linear impulsive differential equations. Nonlinear Anal. **63**, 289–297 (2005)
17. Özbekler, A., Zafer, A.: A Picone's formula for linear non-selfadjoint impulsive differential equations. J. Math. Anal. Appl. **319**, 410–423 (2006)
18. Özbekler, A., Zafer, A.: Principal and non-principal solutions of impulsive differential equations with applications. Appl. Math. Comput. **216**, 1158–1168 (2010)
19. Özbekler, A., Zafer, A.: Second order oscillation of mixed nonlinear dynamic equations with several positive and negative coefficients. In: Discrete Continuous Dynamical System Series B 2011, Dynamical Systems and Differential Equations. Proceedings of the 8th AIMS International Conference, suppl. (Accepted)
20. Rainkin, S.M.: Oscillation theorems for second-order nonhomogenous linear differential equations. J. Math. Anal. Appl. **53**, 550–553 (1976)
21. Samoilenko, A.M., Perestyuk, N.A.: Impulsive Differential Equations. World Scientific, Singapore (1995)
22. Šimša, J.: Asymptotic integration of second order ordinary differential equations. Proc. Am. Mat. Soc. **101**, 96–100 (1987)
23. Skidmore, A., Bowers, J.J.: Oscillatory behaviour of solutions of $y'' + p(x)y = f(x)$. J. Math. Anal. Appl. **49**, 317–323 (1975)
24. Skidmore, A., Leighton, W.: On the differential equation $y'' + p(x)y = f(x)$. J. Math. Anal. Appl. **43**, 46–55 (1973)
25. Teufel, H.: Forced second order nonlinear oscillations. J. Math. Anal. Appl. **40**, 148–152 (1972)
26. Trench, W.F.: Linear perturbation of a nonoscillatory second order differential equation. Proc. Am. Mat. Soc. **97**, 423–428 (1986)
27. Wong, J.S.W.: Oscillation criteria for forced second-order linear differential equation. J. Math. Anal. Appl. **231**, 235–240 (1999)
28. Zafer, A.: On oscillation and nonoscillation of second order dynamic equations. Appl. Math. Lett. **22**, 136–141 (2009)

From the Poincaré–Birkhoff Fixed Point Theorem to Linked Twist Maps: Some Applications to Planar Hamiltonian Systems

Anna Pascoletti and Fabio Zanolin

Abstract We present some results about fixed points and periodic points for planar maps which are motivated by the analysis of the twist maps occurring in the Poincaré–Birkhoff fixed point theorem and in the study of geometric configurations associated to the linked twist maps arising in some problems of chaotic fluid mixing. Applications are given to the existence and multiplicity of periodic solutions for some planar Hamiltonian systems and, in particular, to the second-order nonlinear equation $\ddot{x} + f(t,x) = 0$.

Keywords Twist maps • Planar Hamiltonian systems • Fixed points • Periodic solutions

Mathematical Subject Classification: 34C25, 37E40

1 Introduction

The study of the existence and multiplicity of periodic solutions for Hamiltonian systems represents a classical area of research which has been widely investigated. In this paper we focus our attention to the case of *nonautonomous planar* Hamiltonian systems of the form

$$\dot{x} = \frac{\partial H}{\partial y}(t,x,y), \qquad \dot{y} = -\frac{\partial H}{\partial x}(t,x,y). \tag{1}$$

A. Pascoletti • F. Zanolin (✉)
Department of Mathematics and Computer Science, University of Udine,
via delle Scienze 206, 33100 Udine, Italy
e-mail: anna.pascoletti@uniud.it; fabio.zanolin@uniud.it

S. Pinelas et al. (eds.), *Differential and Difference Equations with Applications*, Springer
Proceedings in Mathematics & Statistics 47, DOI 10.1007/978-1-4614-7333-6_14,
© Springer Science+Business Media New York 2013

Such kind of equations are relevant not only for their intrinsic interest from the point of view of the applications but also because they represent a common ground where several different techniques, ranging from nonlinear analysis (for instance, critical point theory) to the theory of dynamical systems, can compete in order to produce new results.

Here and in what follows we suppose that $H : \mathbb{R} \times \mathbb{R}^2 \to \mathbb{R}$ is a continuous function which is T-periodic in its first variable, that is,

$$H(t+T,x,y) = H(t,x,y), \quad \forall t,x,y,$$

and sufficiently smooth with respect to x and y in order to guarantee the uniqueness of the solutions for the initial value problems associated to (1). Some discontinuities in the t-variable may be allowed, provided that solutions are considered in the Carathéodory sense [12]. For instance, in the frame of (1) we can study (in the phase plane) the periodically perturbed scalar nonlinear second-order ODEs

$$\ddot{x} + f(x) = p(t) \tag{2}$$

or

$$\ddot{x} + p(t)f(x) = 0, \tag{3}$$

with $f : \mathbb{R} \to \mathbb{R}$ a locally Lipschitz function and $p : \mathbb{R} \to \mathbb{R}$ a T-periodic function with $p \in L^1([0,T])$.

A classical approach for the search of periodic solutions to (1) is based on the study of the existence and multiplicity of fixed points and periodic points for the Poincaré map. Recall that the Poincaré map associated to (1) is a function which maps a point $z_0 = (x_0, y_0) \in \mathbb{R}^2$ to the point

$$\Phi(z_0) := (x(T;t_0,z_0), y(T;t_0,z_0)),$$

where $\zeta(t) = (x(t;t_0,z_0), y(t;t_0,z_0))$ is the solution of (1) satisfying the initial condition $\zeta(t_0) = z_0$. In order to have the above definition meaningful, we assume that for the given initial point z_0, the solution $\zeta(t)$ is defined for all $t \in [t_0, t_0 + T]$. Usually, the natural choice $t_0 = 0$ is made. In such a case we use the simplified notation $(x(t;z_0), y(t;z_0)) := (x(t;0,z_0), y(t;0,z_0))$.

Since we assume the uniqueness of the solutions for the Cauchy problems associated to (1), from the fundamental theory of ODEs, we know that Φ is continuous and defined on an open subset $\Omega = \text{dom}\,\Phi \subset \mathbb{R}^2$. Actually Φ is a *homeomorphism* of Ω onto $\Phi(\Omega)$ which is also *orientation-preserving* and *area-preserving* (this latter property follows from Liouville's theorem and from the fact that the right-hand side of (1) is given by a zero-divergence vector field).

An important tool to detect fixed points for area-preserving homeomorphisms of the plane is given by the Poincaré–Birkhoff fixed point theorem. Such theorem deals with the case of a planar annulus

$$A = A[a,b] := \{(x,y) \in \mathbb{R}^2 : a^2 \le x^2 + y^2 \le b^2\} \tag{4}$$

which is subject to the action of an area-preserving homeomorphism $\phi : A \to A$, leaving the boundaries $C_a := \{(x,y) \in \mathbb{R}^2 : x^2 + y^2 = a^2\}$ and $C_b := \{(x,y) \in \mathbb{R}^2 : x^2 + y^2 = b^2\}$ invariant. A crucial assumption of the theorem is the so-called twist condition which asserts that the map ϕ rotates the points of C_a and C_b in different directions. Under these conditions, Poincaré in [22] conjectured (and proved in some situations) the existence of at least two fixed points for ϕ in the interior of A. The first proof of the theorem was given by Birkhoff in [1] (see also [2, 4]). In [1] as well as in [4] the proof is performed for the existence of at least one fixed point. The existence of a second fixed point is taken as granted from a remark of Poincaré concerning the fact that ϕ has zero fixed point index on A (rephrasing Poincaré's sentence [22, p. 377] in modern language). The need of a rigorous justification for the existence of *distinct* fixed points led to further researches into this subject. A proof for the existence of at least two fixed points was proposed by Birkhoff in [3] in a more general setting. In more recent years, convincing and rigorous proofs have been given by other authors, using improvements of Birkhoff argument [5], or different approaches. The history of the "twist" theorem and its generalizations and developments is quite interesting but impossible to summarize in few lines. Generalizations of the theorem in various different directions, with the aim also of providing a more flexible tool for the study of nonautonomous equations, have been obtained. After about 100 years of studies on this topic, some controversial "proofs" of its extensions have been settled only recently. We refer the interested reader to [7] where the part of the story concerning the efforts of avoiding the condition of boundary invariance is described. In this connection, we also recommend the recent works by Martins and Ureña [16] and by Le Calvez and Wang [14] as well as the references therein.

In the applications of the Poincaré–Birkhoff theorem to planar Hamiltonian systems, usually one has to deal with annular regions homeomorphic to A having inner and outer boundaries not necessarily invariant. Recent versions of the theorem require that the inner and outer boundaries are strictly star-shaped with respect to some point. The key fact, however, is the possibility to define a suitable lifting of ϕ to a covering space of the annulus using the standard polar coordinates or some modifications of them, for instance, suitably chosen action-angle variables. In the sequel we need to consider annular regions which are not necessarily centered at the origin. To this aim, we introduce the following notation. Given a point $P \in \mathbb{R}^2$, we define

$$A(P) = A[a,b;P] := P + A[a,b]$$

whose inner and outer boundaries will be named as

$$A_i(P) = P + C_a \quad \text{and} \quad A_o(P) = P + C_b.$$

A possible way to verify the twist condition for the Poincaré map of system (1) is based on the study of some rotation numbers associated to its solutions. Such rotation numbers provide some information about the displacement of the angular coordinate.

To begin with, we describe an elementary manner to introduce some rotation numbers. We fix a point $P = (x_p, y_p)$ and consider a system of polar coordinates around P [typically, we will have $P = 0 = (0,0)$]. Suppose that for some $z_0 \in \mathbb{R}^2$, the solution $\zeta(t; z_0) = (x(t; z_0), y(t; z_0))$ satisfies

$$\zeta(t; z_0) \neq P, \quad \forall t \in [0, \tau],$$

for some $\tau > 0$. Passing to the polar coordinates

$$x = x_p + \sqrt{2\rho} \cos \theta, \quad y = y_p + \sqrt{2\rho} \sin \theta, \tag{5}$$

we obtain

$$-\dot{\theta}(t) = \frac{\dot{x}(t)(y(t) - y_p) - \dot{y}(t)(x(t) - x_p)}{(x(t) - x_p)^2 + (y(t) - y_p)^2},$$

and thus we can define the number

$$\mathrm{rot}(t, z_0, P) := \frac{1}{2\pi} \int_0^t \frac{(y(s) - y_p)\frac{\partial H}{\partial y}(s, x(s), y(s)) + (x(s) - x_p)\frac{\partial H}{\partial x}(s, x(s), y(s))}{(x(s) - x_p)^2 + (y(s) - y_p)^2} \, ds$$

for $t \in [0, \tau]$ and $(x(t), y(t)) = (x(t; z_0), y(t; z_0))$. The choice of the form (5) is not a mandatory fact, even if it could be useful in some situations, since here we have $dx \, dy = d\rho \, d\theta$.

The *rotation number* $\mathrm{rot}(t, z_0, P)$ counts the number of windings of the solution around the point P, in the clockwise sense, in the time interval $[0, t]$.

If the above rotation number is defined for $t = mT$ (for some integer $m \geq 1$) and for all the points of an annulus $A(P)$, the twist condition in the Poincaré–Birkhoff theorem for the map $\phi = \Phi^m$ can be expressed as follows:

$$\begin{cases} \mathrm{rot}(mT, z, P) > j, & \text{for } z \in A_i(P) \\ \mathrm{rot}(mT, z, P) < j, & \text{for } z \in A_o(P) \end{cases} \tag{6}$$

(or viceversa), for some $j \in \mathbb{Z}$. The existence of a fixed point for ϕ (coming from the original version of the theorem or from some of its variants) provides a point w in the interior of the annulus which is the initial point of a mT-periodic solution of (1) and such that

$$\mathrm{rot}(mT, w, P) = j. \tag{7}$$

The additional information expressed by (7) can be exploited in order to obtain multiplicity results or some precise information about the solution.

The study of twist maps is not only a crucial step in the applications of the Poincaré–Birkhoff theorem to planar Hamiltonian systems. Twist maps naturally appear in a broad number of situations (from KAM theory to the study of some geometrical configurations involving the presence of Smale's horseshoes), and thus they have been widely considered both from the theoretical point of view and for their significance in various applications, which range from celestial mechanics to fluid dynamics.

In the past decades a grown interest has been devoted to the study of the so-called *linked twist maps* (from now on abbreviated as LTMs). A typical LTM of the plane, as presented by Devaney in [8], can be described as a composition of the form

$$\Psi = \Psi_2^k \circ \Psi_1^\ell,$$

where Ψ_1 and Ψ_2 act, respectively, in a twist manner on two different annuli $A(P_1)$ and $A(P_2)$. Some authors also assume that both Ψ_1 and Ψ_2 perform on the boundaries some rotations of angles which are multiple of 2π. In this way, the maps can be extended as identities outside the annuli. This is not in contrast with (6) provided that such multiples of 2π are not the same number j appearing in (6). In our setting we do not need to extend Ψ_1 and Ψ_2 outside the respective annuli, and therefore no further requests on the twist conditions will be added.

If $A(P_1)$ and $A(P_2)$ cross each other in a proper way, then Ψ has a rich dynamics. The correct crossing of the two annuli $A[a_1,b_1;P_1]$ and $A[a_2,b_2;P_2]$ is usually described by the relations

$$\max\{b_2 - a_1, b_1 - a_2\} < \operatorname{dist}(P_1,P_2) < a_1 + a_2$$

so that LTMs can be interpreted as a class of homeomorphisms of the two-disk minus three holes [8] (see Fig. 1).

Examples of LTMs on some manifolds (like the sphere or the torus) have been considered as well (see [24, 25] and the references therein). However, if, instead of annuli of the form $A(P_i)$, we have more general annular regions on which two twist maps act, the linking conditions can be more general (see Fig. 2). LTMs in such more general setting have been recently considered in [15].

A natural way to produce a twist-type Poincaré map associated to (1) occurs when the nonautonomous system can be viewed as a perturbation of an autonomous planar system presenting a center-like structure, as

$$\dot{x} = \frac{\partial \mathcal{H}}{\partial y}(x,y), \qquad \dot{y} = -\frac{\partial \mathcal{H}}{\partial x}(x,y). \tag{8}$$

Suppose that there exists a *topological annulus* $\mathscr{A}$ (that is a compact subset of $\mathbb{R}^2$ homeomorphic to a standard annulus $A[a,b]$) which is filled by closed (periodic) orbits of system (8). Since the trajectories of (8) lie on the level lines of the Hamiltonian, we can parameterize any orbit Γ in $\mathscr{A}$ by means of the value $\mathscr{H}(\Gamma) = c$. Under mild assumptions on $\mathscr{H}$ ($\mathscr{H}$ of class C^1 with $\nabla\mathscr{H}(x,y) \neq 0$ for all $(x,y) \in \mathscr{A}$, see [13]) it is possible to prove the continuity of the function which maps c into the period τ_c of the closed orbit in $\mathscr{A}$ all level c. One can also find a compact interval $[a,b]$ such that the inner and the outer boundaries of $\mathscr{A}$ correspond

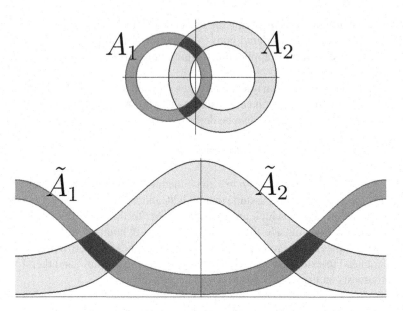

Fig. 1 Example of two standard linked annuli A_1 and A_2. For the figure we have taken $A_i = A[a_i, b_i, P_i]$ with $P_1 = (-5, 0), P_2 = (5, 0), a_1 = 6, b_1 = 8, a_2 = 6, b_2 = 10$. The sets $\tilde{A}_1$ and $\tilde{A}_2$ are the annuli A_1 and A_2 viewed from the origin (in the figure below, the scale ratio between the two axes is not respected). Since $(0, 0)$ is in the intersection of the bounded components of $\mathbb{R}^2 \setminus A_i$ (for $i = 1, 2$), using the usual polar coordinates (θ, ρ) with respect to the origin, we can lift both A_1 and A_2 as 2π-periodic strips bounded between graphs of functions $\rho = \rho(\theta)$. In this specific case, we have $\tilde{A}_i = \{(\theta, \rho) : x_{p_i} \cos\theta + (a_i^2 - x_{p_i}^2 \sin^2\theta)^{1/2} \leq \rho \leq x_{p_i} \cos\theta + (b_i^2 - x_{p_i}^2 \sin^2\theta)^{1/2}\}$, for $P_i = (x_{p_i}, 0), i = 1, 2$

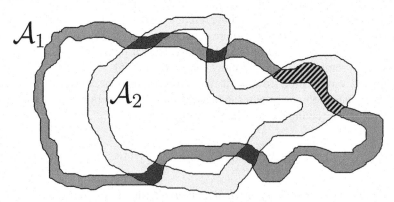

Fig. 2 Example of two linked planar topological annuli $\mathscr{A}_1$ and $\mathscr{A}_2$. Among the *five rectangular regions* which result from the intersection of the two annuli, the *four regions in darker color* are suitable for a generalized version of linked twist map theory as described in [15]), while the set painted with *zebra stripes* does not fit in that framework

to the level lines $\mathcal{H} = a$ and $\mathcal{H} = b$ (we can always enter in this situation possibly replacing $\mathcal{H}$ with $-\mathcal{H}$). In this manner, the set $\mathscr{A}$ becomes a standard annulus of the form in (4), with the level of the Hamiltonian playing the role of a radial coordinate. Angular-type coordinates can be introduced using a normalized time along the trajectories, counted from a suitable arc transversal to the annulus (such arc is obtained as a flow line of the gradient system $\dot{z} = \nabla \mathcal{H}(z)$).

If we denote by $\Phi_{\mathcal{H}}$ the Poincaré map associated to (8), for a fixed time $T > 0$, we can produce a twist condition on $\mathscr{A}$, whenever

$$\tau_a \neq \tau_b.$$

Indeed, suppose that $\tau_a < \tau_b$ and let us fix $m \geq 1$ such that the set

$$Z(m) := \left] \frac{mT}{\tau_b}, \frac{mT}{\tau_a} \right[\cap \mathbb{Z}$$

is nonempty. For each $j \in Z(m)$ we have that the points of the inner boundary $\mathscr{A}_i$ of $\mathscr{A}$ wind more than j times in the time interval $[0, mT]$. On the other hand, the points of the outer boundary $\mathscr{A}_o$ have a number of rotations strictly less than j. This simple observation guarantees that a twist condition analogous to (6) holds for $\Phi_{\mathcal{H}}^m$ relatively to $\mathscr{A}$. Such fact will imply a twist condition for Φ^m if the vector field in (1) is sufficiently close to that of (8) on $[0, mT] \times \mathscr{A}$. A recent investigation in this direction using the Poincaré–Birkhoff fixed point theorem has been performed in [11].

A possible way to produce a LTM configuration in the plane is given by a pair of planar autonomous Hamiltonian systems which periodically switch back and forth from one to the other. More precisely, let us fix $T_1, T_2 > 0$ with

$$T_1 + T_2 = T$$

and consider the systems

$$\dot{x} = \frac{\partial \mathcal{H}_1}{\partial y}(x, y), \quad \dot{y} = -\frac{\partial \mathcal{H}_1}{\partial x}(x, y), \qquad \text{for } t \in [0, T_1[\tag{9}$$

and

$$\dot{x} = \frac{\partial \mathcal{H}_2}{\partial y}(x, y), \quad \dot{y} = -\frac{\partial \mathcal{H}_2}{\partial x}(x, y), \qquad \text{for } t \in [T_1, T[, \tag{10}$$

repeating then such process in a periodic fashion (for an application to fluid mixing, see [26, Appendix B]). Allowing a discontinuity for $t \equiv 0$ and $t \equiv T_1 \pmod{T}$, the resulting system may be interpreted as a special case of (1). We enter in the generalized LTMs framework considered in [15] whenever there exist two annular regions $\mathscr{A}_1$ and $\mathscr{A}_2$ filled by periodic orbits of systems (9) and (10), respectively, and such that $\mathscr{A}_1$ and $\mathscr{A}_2$ link each other in a suitable sense (see Fig. 2). Moreover, appropriate twist conditions on each of the two annuli should be required.

The aim of this article is to briefly survey some applications of the topological methods described above and also to present some new applications to planar Hamiltonian systems, in particular to (2) and (3). Doing this, we will also present a situation which, in some sense, is intermediate between the twist maps arising in the applications of the Poincaré–Birkhoff theorem and the LTMs, namely, the so-called *bend-twist maps*. Such class of twist maps has been recently considered by Tongren Ding [10] in the analytic setting and studied for continuous functions on annular domains in [21]. We also need to introduce the following definitions.

By a *path* γ in a topological space X we mean a continuous map from an interval $[t_0, t_1]$ into X. We also set $\bar{\gamma} := \gamma([t_0, t_1])$. Usually, and without loss of generality, we take $I := [0, 1]$ as domain for the paths. An *arc* in X is the homeomorphic image of I, that is, $\bar{\gamma}$ for a one-to-one path γ in X.

Definition 1. Let X be a pathwise connected topological space and let $A, B \subset X$ be two nonempty disjoint sets. Let also $S \subset X$. We say that S *cuts the paths between A and B* if $S \cap \bar{\gamma} \neq \emptyset$, for every path $\gamma : I \to X$ such that $\gamma(0) \in A$ and $\gamma(1) \in B$.

We write $S : A \nmid B$ to express the fact that S cuts the paths between A and B.

Definition 2. Let $h : A[a, b] \to \mathscr{A} := h(A[a, b]) \subset \mathbb{R}^2$ be a homeomorphism. The set $\mathscr{A}$ is called a *topological annulus*. We decompose its boundary as $\partial \mathscr{A} = \mathscr{A}_i \cup \mathscr{A}_o$, with $\mathscr{A}_i := h(C_a)$ and $\mathscr{A}_o := h(C_b)$, the inner and outer boundaries of $\mathscr{A}$, respectively.

In the sequel, for simplifying the exposition, we present our result in the setting of maps defined on a standard annulus of the form (4). They can be easily transferred to the case of topological annuli.

2 Bend-Twist Maps

In the sequel we denote by Π the standard covering projection of $\mathbb{R} \times]0, +\infty[$ onto $\mathbb{R}^2 \setminus \{0\}$, defined by the polar coordinates $\Pi(\theta, \rho) = (\rho \cos \theta, \rho \sin \theta)$. Let $\phi : A = A[a, b] \to \mathbb{R}^2 \setminus \{0\}$ be a continuous map admitting a lifting on $\mathbb{R} \times [a, b]$ of the form

$$\tilde{\phi} : (\theta, \rho) \mapsto (\theta + \Theta(\theta, \rho), R(\theta, \rho)), \tag{11}$$

where Θ, R are continuous and real-valued functions which are 2π-periodic in the θ-variable. According to the definition of lifting, we have that $\phi \circ \Pi = \Pi \circ \tilde{\phi}$. Observe that $\Theta(\theta, \rho)$ is the same for any $(\theta, \rho) \in \Pi^{-1}(z)$. This allows to define $\Theta(z)$ as $\Theta(\theta, \rho)$ for $z = \Pi(\theta, \rho)$. The number $\Theta(z)$ is the angular displacement performed by the map ϕ on the point $z \in A$. Modulo a scaling factor, it plays the same role as the rotation number defined before. We also introduce an auxiliary function Υ giving the radial displacement

$$\Upsilon(z) := ||\phi(z)|| - ||z||.$$

By the above positions, it follows that $z \in A$ is a fixed point for ϕ if and only if

$$\exists j \in \mathbb{Z}: \quad \Theta(z) = 2j\pi, \quad \text{and} \quad \Upsilon(z) = 0. \tag{12}$$

If z_1 and z_2 solve (12) for different values of j, then they are distinct fixed points of ϕ. We say that ϕ satisfies a *twist condition* on A if there exists $j \in \mathbb{Z}$ such that

$$\Theta < 2j\pi \text{ on } A_i = C_a \quad \text{and} \quad \Theta > 2j\pi \text{ on } A_o = C_b \tag{13}$$

(or viceversa). For a twist map on A we define the set

$$\Omega_\phi^j := \{z \in A : \Theta(z) = 2j\pi\}.$$

In [10] Tongren Ding considers the case of a topological annulus $\mathscr{A}$ embedded in the plane, having as its boundaries two simple closed curves which are star-shaped with respect to the origin. It is assumed that there exists an *analytic* function $f : \mathscr{A} \to \mathscr{A}^*$, with $\mathscr{A}^* \supset \mathscr{A}$ another starlike annulus and, moreover, that f satisfies the twist condition (13). It is also observed that the set Ω_f of the points in $\mathscr{A}$ where $\Theta = 2j\pi$ contains at least a Jordan curve Γ which is not contractible in $\mathscr{A}$. The function f is called a *bend-twist map* if Υ changes its sign on the curve Γ. Then, the following theorem holds (see [10, Theorem 7.2, p. 188]):

Theorem 1. *Let $f : \mathscr{A} \to \mathscr{A}^*$ be an analytic bend-twist map. Then it has at least two distinct fixed points in $\mathscr{A}$.*

We notice that in Ding's theorem, the assumptions that f is area-preserving and leaves the annulus invariant are not needed. This represents a strong improvement of the hypotheses required for the Poincaré–Birkhoff twist theorem. On the other hand, the assumption that a given function is a bend-twist map does not seem easy to be checked in the applications. For this purpose, the following corollary (see [10, Corollary 7.3, p. 188]) provides more explicit conditions for the applicability of the abstract result:

Corollary 1. *Let $f : \mathscr{A} \to \mathscr{A}^*$ be an analytic twist map. If there are two disjoint continuous curves Γ_1 and Γ_2 in $\mathscr{A}$, each of them connecting the inner and the outer boundaries of $\mathscr{A}$ and such that $\Upsilon < 0$ on Γ_1 and $\Upsilon > 0$ on Γ_2, then f is a bend-twist map on $\mathscr{A}$, and therefore it has at least two distinct fixed points.*

Our aim is to reformulate the above results in a general topological setting in order to obtain a version of Theorem 1 and Corollary 1 for general (not necessarily analytic) maps. In order to simplify the exposition we consider the case in which the annulus is $\mathscr{A} = A[a,b] := A$. In this setting, the following result holds [20]:

Lemma 1. *Let $\phi : A \to \mathbb{R}^2 \setminus \{0\}$ be a continuous map admitting a lifting of the form (11) and satisfying the twist condition (13). Then the set Ω_ϕ^j contains a closed* *connected set C^j with the property that $C^j : A_i \nmid A_o$. Such set is essentially embedded in A, that is, the inclusion $i_{C^j} : C^j \to A$, for $i_{C^j}(x) = x$, $\forall x \in C^j$, is not homotopic to a constant map.*

Our result corresponds to [10, Lemma 7.2, p. 185] for a general ϕ. The Jordan curve $\Gamma \subset \Omega_f$ considered in [10] in the analytic case is now replaced by the essential continuum $C^j \subset \Omega^j_\phi$. Following [10] we can now give the next definition.

Definition 3. Let $\phi : A \to \mathbb{R}^2 \setminus \{0\}$ be a continuous map [admitting a lifting of the form (11)] which satisfies the twist condition (13), for some $j \in \mathbb{Z}$. We say that ϕ is a *bend-twist map* in A if Υ changes its sign on C^j.

As a consequence of this definition, the following theorem, which is a version of Theorem 1 for mappings which are not necessarily analytic, holds:

Theorem 2. *Let $\phi : A \to \mathbb{R}^2 \setminus \{0\}$ be a bend-twist map. Then it has a fixed point in* int X, *with $\Theta = 2j\pi$.*

In [21] we have provided an example of a bend-twist map having only one fixed point. To fully recover Ding's theorem, one could slightly modify the definition of bend-twist map, by observing that in the setting of Lemma 1, it is possible to prove the existence of an essentially embedded closed connected set $C^j \subset A$ with the property of being *minimal* with respect to the cutting property $C^j : A_i \nmid A_o$. For such minimal set, it holds that any real-valued continuous map, which changes its sign on it, vanishes at least twice on C^j. Clearly, from the point of view of the applications, the verification that Υ changes its sign on a minimal set (whose existence is guaranteed by Zorn's lemma) is, perhaps, of little use. For this reason, we provide the following topological version of Corollary 1 (see [21, Theorem 2.9]):

Theorem 3. *Let $\phi : A \to \mathbb{R}^2 \setminus \{0\}$ be a continuous map [admitting a lifting of the form (11)] which satisfies the twist condition (6), for some $j \in \mathbb{Z}$. If there are two disjoint arcs Γ_1 and Γ_2 in A, both connecting A_i with A_o in A and such that $\Upsilon < 0$ on Γ_1 and $\Upsilon > 0$ on Γ_2, then ϕ has at least two distinct fixed points in* int A *with $\Theta = 2j\pi$.*

Such result can be easily generalized as follows [21, Corollary 2.10]:

Theorem 4. *Let $\phi : A \to \mathbb{R}^2 \setminus \{0\}$ be a continuous map [admitting a lifting of the form (11)] which satisfies the twist condition (6), for some $j \in \mathbb{Z}$. Assume that there exist 2k disjoint arcs ($k \geq 1$), all of them connecting A_i with A_o in A. We label these arcs in a cyclic order $\Gamma_1, \Gamma_2 \ldots, \Gamma_n, \ldots \Gamma_{2k}, \Gamma_{2k+1} = \Gamma_1$ and assume that $\Upsilon < 0$ on Γ_n for n odd and $\Upsilon > 0$ on Γ_n for n even (or viceversa). Then ϕ has at least 2k distinct fixed points in* int A, *all the fixed points with $\Theta = 2j\pi$.*

Remark 1. The proofs of Theorems 3 and 4 are based on topological degree arguments. Therefore, such results are stable with respect to small continuous perturbations of the map ϕ. The assumption of area-preserving is not needed here (see Remark 3).

In the next section we investigate the different concepts discussed above regarding twist maps in the setting of the second-order nonlinear ODEs with periodic coefficients $\ddot{x} + f(t,x) = 0$ for some special cases of f. We will construct some planar annular regions obtained by the level lines of some associated autonomous Hamiltonian systems. In such a situation, the results that we have exposed for the standard annulus $A = A[a,b]$ must be translated to domains which are planar topological annuli. Due to the topological nature of our results, such extension is straightforward.

3 Applications

For the sake of brevity, we confine our applications to the case of (3). Analogous results can be obtained for (2), as well as for related equations.

Let $f : \mathbb{R} \to \mathbb{R}$ be a locally Lipschitz function and let $F(x) := \int_0^x f(s)\,ds$. We assume

$$f(0) = 0, \quad f(s)s > 0 \ \text{ for } s \neq 0, \quad F(s) \to +\infty \ \text{ for } s \to \pm\infty. \tag{14}$$

As a consequence of (14) we have that for each $\mu > 0$, the phase portrait of the first-order planar system

$$\dot{x} = y, \qquad \dot{y} = -\mu f(x) \tag{15}$$

is that of a global center at the origin. For each $c > 0$, the energy level line

$$\mathscr{E}_\mu^c := \{(x,y) : E_\mu(x,y) = c\},$$

with

$$E_\mu(x,y) := \frac{1}{2}y^2 + \mu F(x),$$

is a closed curve surrounding the origin, and it is also a periodic orbit of system (15). The fundamental period τ_μ^c of $\mathscr{E}_\mu^c$ can be expressed as

$$\tau_\mu^c = 2\int_\alpha^\beta \frac{d\xi}{\sqrt{2(c - \mu F(\xi))}} = \frac{2}{\sqrt{\mu}} \int_\alpha^\beta \frac{d\xi}{\sqrt{2(F(\beta) - F(\xi))}},$$

where the values

$$\alpha = \alpha_\mu^c < 0 < \beta = \beta_\mu^c$$

are such that

$$\mu F(\alpha) = \mu F(\beta) = c.$$

With this respect, we introduce the following notation:

$$\tau(u) := 2 \int_{u_-}^{u} \frac{d\xi}{\sqrt{2(F(\beta) - F(\xi))}}, \quad \text{for } u_- < 0 < u \text{ with } F(u_-) = F(u). \quad (16)$$

The number $\tau(u)$ is the period of the orbit of (15) for $\mu = 1$, passing through the point $(u,0)$. Hence a generic orbit of system (15) passing through the same point has period $\tau_\mu^c = \tau(u)/\sqrt{\mu}$, with $c = \mu F(u)$.

We consider now the *nonautonomous* equation (3) with a T-periodic weight function $p : \mathbb{R} \to \mathbb{R}$ which is a small perturbation of a T-periodic stepwise function $p_{B,C}$ defined on $[0, T[$ as follows:

$$p_{B,C}(t) := \begin{cases} B, \text{ for } t \in [0, T_1[\\ C, \text{ for } t \in [T_1, T_2[\end{cases}$$

with

$$T_1 + T_2 = T, \quad \text{and } 0 < B < C.$$

We perform our analysis for the equation

$$\ddot{x} + p_{B,C}(t)f(x) = 0 \quad (17)$$

and its associate first-order system in the phase plane. All our results will be extended to (3) for $\|p - p_{B,C}\|_{L^1(0,T)}$ sufficiently small (see Remarks 2, 3, 4).

For a weight function like $p_{B,C}$, it turns out that the study of (17) can be reduced to the analysis of two planar Hamiltonian systems of the form (9) and (10), defining

$$\mathscr{H}_1(x,y) := \frac{1}{2}y^2 + BF(x) = E_B(x,y), \quad \mathscr{H}_2(x,y) := \frac{1}{2}y^2 + CF(x) = E_C(x,y). \quad (18)$$

Such Hamiltonian systems correspond to (15) for $\mu = B$ and for $\mu = C$. Concerning the dynamics, we take $\mu = B$ and follow (15) for $t \in [0, T_1[$. At the time $t = T_1$, we change the parameter to the value $\mu = C$ and follow (15) on the time interval $[T_1, T[$. Due to the autonomous nature of the system, this corresponds to the study of the behavior of the solutions on the time interval $[0, T_2[$. Accordingly, the Poincaré map Φ for (17) can be split as

$$\Phi = \Phi_C \circ \Phi_B,$$

where Φ_B is the Poincaré map associated with system (15) for $\mu = B$ and for the time interval $[0, T_1]$ and, similarly, Φ_C is the Poincaré map for (15) with $\mu = C$ and $[0, T_2]$.

The results we are going to present apply to the search of T-periodic solutions, as well as to subharmonics of order m. For the sake of simplicity we consider only the case of T-periodic solutions. In order to produce a twist condition for Φ, we have to find an annulus (centered at the origin) such that (6) holds (for $m = 1$). A natural choice is an annular region defined by means of the energy level lines of (15). In our case we have energy level lines for *two* systems. Hence, a possible approach would be that of constructing an annulus and proving the twist condition for one of the two systems (for instance, for $\mu = B$) and then choose suitable assumptions ensuring that the action of the second Poincaré map will not destroy the geometry.

Suppose there are $u_0, u_1 > 0$ such that $\tau(u_0) \neq \tau(u_1)$. This is always possible if we are not in the case of an isochronous center. Just to fix a case of study, assume that

$$0 < u_0 < u_1 \quad \text{and} \quad \tau(u_0) < \tau(u_1).$$

We take as the annulus $\mathscr{A}$ the set

$$\mathscr{A} := \{(x,y) : E_B(u_0,0) \leq E_B(x,y) \leq E_B(u_1,0)\}.$$

The set $\mathscr{A}$ is a topological annulus with strictly starlike boundaries (with respect to the origin). If we describe the points of the plane by means of the modified polar coordinates given by the angle θ and with the energy level E_B playing the role of a radial coordinate, we have that $\mathscr{A} = A[E_B(u_0,0), E_B(u_1,0)]$. In order to check the twist condition, we observe that the solutions of (15) for $\mu = B$, which depart from a point of energy c, make exactly k turns around the origin (in the clockwise sense) at the time $k\tau_B^c$. Hence, (6) ($m = 1$ and $T = T_1$) is satisfied for some positive integer j if and only if

$$\sqrt{B}\frac{T_1}{\tau(u_0)} > j > \sqrt{B}\frac{T_1}{\tau(u_1)}. \tag{19}$$

This provides a twist condition for the map Φ_B. The twist condition will be satisfied for the map Φ provided that the time T_2 is not too large. Thus, we can conclude as follows:

Theorem 5. *Suppose that there exist* $0 < u_0 < u_1$ *such that* (19) *holds. Then there is* $T_2^* = T_2^*(C) > 0$ *such that* (17) *has at least two* T-*periodic solutions with* $(x(0), \dot{x}(0)) \in \text{int}\,\mathscr{A}$, *provided that* $T_2 < T_2^*$.

Proof. We give a sketch of the proof. Since Φ is an area-preserving homeomorphism of the plane with $\Phi(0) = 0$ and $\mathscr{A}$ is an annulus around the origin with strictly star-shaped boundaries, we can apply a generalized version of the Poincaré–Birkhoff theorem due to W.-Y. Ding [9], if the twist condition for Φ is satisfied.

As observed above, this is true if Φ_C does not destroy the twist condition given by Φ_B. This is guaranteed by the smallness of T_2. $\qquad\square$

Remark 2. We notice that the conclusion of Theorem 5 is true also for (3) provided that

$$\|p - p_{B,C}\|_{L^1(0,T)} < \varepsilon$$

with $\varepsilon > 0$ a suitable constant depending on the parameters of (17). For an analogous result concerning (2), see [6]. For the proof (as for the proof of Theorem 5) we need a version of the Poincaré–Birkhoff theorem in which the assumption of the invariance of the boundaries may not be satisfied. In our case, we rely on a version of the theorem for strictly star-shaped boundaries [23], [27, Corollary 1].

As a next step, we propose, for the same equation, an application of the bend-twist maps theorem. In this case we have the following:

Theorem 6. *Suppose that there exist* $0 < u_0 < u_1$ *such that* (19) *holds. Then there is* $T_2^{\#} = T_2^{\#}(C) > 0$ *such that* (17) *has at least four T-periodic solutions with* $(x(0), \dot{x}(0)) \in \text{int}\,\mathscr{A}$, *provided that* $T_2 < T_2^{\#}$.

Proof. We give a sketch of the proof. For a fixed angle ϑ, let Λ_{ϑ} be the intersection of the half-line $L_{\vartheta} := \{(\rho\cos\vartheta, \rho\sin\vartheta) : \rho > 0\}$ with the annulus. We are interested in the motion of the points of Λ_{ϑ} under the action of Φ_C, and hence we consider (15) for $\mu = C$. Just to fix the ideas, suppose that $\vartheta \in]0, \pi/2]$ (see Fig. 3). The points of Λ_{ϑ} move in the clockwise sense, and therefore they remain in the first quadrant if T_2 is sufficiently small. Now we consider the energy of the first system evaluated along the solutions of the second one. This is given by the function $E(t) :=$ $E_B(x(t), y(t))$. A differentiation yields to $\frac{dE(t)}{dt} = -(C-B)f(x(t))y(t)$. Hence there is an energy loss for E_B as long as the solution remains in the first quadrant (or in the third quadrant), while the energy increases in the second and in the fourth quadrant. We have taken Λ_{ϑ} in the first quadrant, and therefore $\Phi_C(\Lambda_{\vartheta})$ will be still in the first quadrant for T_2 sufficiently small. In this case, we conclude that $E_B(\Phi_C(z)) < E_B(z)$, for all $z \in \Lambda_{\vartheta}$. The bigger we take $\vartheta \in]0, \pi/2]$, the larger T_2 can be allowed. At this point, we recall that the annulus $\mathscr{A}$ is invariant for the solutions of (15) for $\mu = B$ and the energy E_B is constant along such solutions. Hence, if we take $\Gamma_1 := \Phi_B^{-1}(\Lambda_{\vartheta})$, we find that $E_B(\Phi(z)) < E_B(z)$ for each $z \in \Gamma_1$. Since for the annulus $\mathscr{A}$ we have taken E_B as a radial coordinate, we conclude that $\Upsilon < 0$ on Γ_1. In the same manner, we can take $\Gamma_2, \Gamma_3,$ and Γ_4 in $\mathscr{A}$ such that the assumptions of Theorem 4 are satisfied for $k = 2$.

Remark 3. We notice that the conclusion of Theorem 6 is true also for equation

$$\ddot{x} + c\dot{x} + p(t)f(x) = 0 \qquad (20)$$

provided that $\|p - p_{B,C}\|_{L^1(0,T)} < \varepsilon$ and $|c| < \varepsilon$ with $\varepsilon > 0$, a suitable constant depending on the parameters of (17).

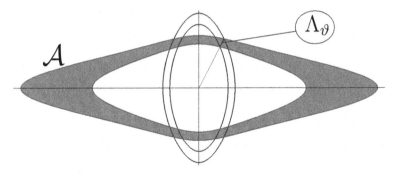

Fig. 3 As a comment to the proof of Theorem 6 we plot the phase portrait of system (15) for different values of μ. In the present example we have taken $f(x) = x/(1+x^2)$, $u_0 = 3$, $u_1 = 5$, $B = 1$, and $C = 9.4$. The scale ratio between $x-y$ axis has been slightly modified

A natural question that arises now is what happens if we take also T_2 large. As explained in the Introduction, if we are able to produce a sufficiently large twist for both the systems, then we can enter in the setting of generalized LTMs, as considered in [15, 18, 19]. In this situation, we can construct two topological annuli $\mathscr{A}_1$ and $\mathscr{A}_2$ filled by the periodic orbits of the system (15) for $\mu = B$ and $\mu = C$, respectively, which cross each other in an appropriate manner. For instance, if we can take four positive numbers

$$0 < v_0 < v_1 < u_0 < u_1$$

such that

$$CF(v_0) > BF(u_1), \tag{21}$$

then the intersection of the two annuli $\mathscr{A}_1 := A[E_B(u_0, 0), E_B(u_1, 0)]$ and $\mathscr{A}_2 := A[E_C(v_0, 0), E_C(v_1, 0)]$ consists of four rectangular regions, $\mathscr{R}_i$ for $i = 1, \ldots, 4$, like in Fig. 4. In this situation, the following result holds:

Theorem 7. *Suppose that there exist $0 < v_0 < v_1 < u_0 < u_1$, and let B, C such that (21) holds. If*

$$\tau(u_0) \neq \tau(u_1) \quad and \quad \tau(v_0) \neq \tau(v_1), \tag{22}$$

then there exist $\bar{T}_1, \bar{T}_2 > 0$ such that for each T_1, T_2 with $T_1 > \bar{T}_1$ and $T_2 > \bar{T}_2$, (17) has at least one T-periodic solution with $(x(0), \dot{x}(0)) \in \mathscr{R}_i$, for every $i = 1, \ldots, 4$. Moreover, in each of the $\mathscr{R}_i$'s, there is chaotic dynamics.

The proof can be obtained from the arguments developed in [15, 19]. The term "chaotic dynamics" is meant in the sense that inside each of the $\mathscr{R}_i$, there is a compact invariant set for the Poincaré map where Φ is *semiconjugate* to a full Bernoulli shift on $m \geq 2$ symbols. To have a larger m (and hence a larger topological

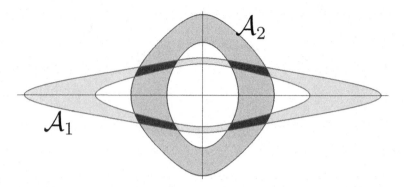

Fig. 4 Plotting of two annular regions $\mathscr{A}_1$ and $\mathscr{A}_2$ as required in the proof of Theorem 7. The regions $\mathscr{R}_i$ ($i = 1, \ldots, 4$) are painted in a *darker color*. In the present example we have taken $f(x), B, C$ as in Fig. 3 with $v_0 = 1, v_1 = 2, u_0 = 3, u_1 = 5$. The scale ratio between $x - y$ axis has been slightly modified. Since $f(x)/x = 1/(1 + x^2)$ is strictly decreasing on $]0, +\infty)$, according to Opial [17], we have that the time mapping $u \mapsto \tau(u)$ is strictly increasing on $]0, +\infty)$. Hence condition (22) is satisfied and we have chaotic dynamics in each of the four parts of $\mathscr{A}_1 \cap \mathscr{A}_2$ (for T_1 and T_2 sufficiently large)

entropy), we have just to take larger values of T. As a consequence, on each of the $\mathscr{R}_i$'s, there are infinitely many periodic points of Φ, leading to the existence of infinitely many subharmonics for (17).

Remark 4. We notice that the conclusion of Theorem 7 is true also for (20) for c small and $p(t)$ a small perturbation of $p_{B,C}(t)$.

Acknowledgements The authors thank the referee for the careful checking of the manuscript and the pertinent remarks.

References

1. Birkhoff, G.D.: Proof of Poincaré's geometric theorem. Trans. Am. Math. Soc. **14**, 14–22 (1913)
2. Birkhoff, G.D.: Dynamical systems with two degrees of freedom. Trans. Am. Math. Soc. **18**(2), 199–300 (1917)
3. Birkhoff, G.D.: An extension of Poincaré's last geometric theorem. Acta Math. **47**, 297–311 (1925)
4. Birkhoff, G.D.: Dynamical Systems. American Mathematical Society, Providence (1927)
5. Brown, M., Neumann, W.D.: Proof of the Poincaré-Birkhoff fixed point theorem. Michigan Math. J. **24**(1), 21–31 (1977)
6. Buttazzoni, P., Fonda, A.: Periodic perturbations of scalar second order differential equations. Discrete Contin. Dynam. Syst. **3**(3), 451–455 (1997)
7. Dalbono, F., Rebelo, C.: Poincaré-Birkhoff fixed point theorem and periodic solutions of asymptotically linear planar Hamiltonian systems. Rend. Sem. Mat. Univ. Politec. Torino **60**(4), 233–263 (2002)

8. Devaney, R.L.: Subshifts of finite type in linked twist mappings. Proc. Am. Math. Soc. **71**(2), 334–338 (1978)
9. Ding, W.Y.: A generalization of the Poincaré-Birkhoff theorem. Proc. Am. Math. Soc. **88**(2), 341–346 (1983)
10. Ding, T.: Approaches to the qualitative theory of ordinary differential equations. In: Peking University Series in Mathematics, vol. 3. World Scientific Publishing Co. Pte. Ltd., Hackensack (2007)
11. Fonda, A., Sabatini, M., Zanolin, F.: Periodic solutions of perturbed Hamiltonian systems in the plane by the use of the Poincaré–Birkhoff Theorem. Topol. Methods Nonlinear Anal. **40**(1), 29–52 (2012)
12. Hale, J.: Ordinary Differential Equations. Robert E. Krieger Publishing Co. Inc., Huntington (1980)
13. Henrard, M., Zanolin, F.: Bifurcation from a periodic orbit in perturbed planar Hamiltonian systems. J. Math. Anal. Appl. **277**(1), 79–103 (2003)
14. Le Calvez, P., Wang, J.: Some remarks on the Poincaré-Birkhoff theorem. Proc. Am. Math. Soc. **138**(2), 703–715 (2010)
15. Margheri, A., Rebelo, C., Zanolin, F.: Chaos in periodically perturbed planar Hamiltonian systems using linked twist maps. J. Differ. Equ. **249**(12), 3233–3257 (2010)
16. Martins, R., Ureña, A.J.: The star-shaped condition on Ding's version of the Poincaré-Birkhoff theorem. Bull. Lond. Math. Soc. **39**(5), 803–810 (2007)
17. Opial, Z.: Sur les périodes des solutions de l'équation différentielle $x'' + g(x) = 0$. Ann. Polon. Math. **10**, 49–72 (1961)
18. Pascoletti, A., Pireddu, M., Zanolin, F.: Multiple periodic solutions and complex dynamics for second order ODEs via linked twist maps. In: The 8th Colloquium on the Qualitative Theory of Differential Equations, *Proc. Colloq. Qual. Theory Differ. Equ.*, vol. 8, pp. No. 14, 32. Electron. J. Qual. Theory Differ. Equ., Szeged (2008)
19. Pascoletti, A., Zanolin, F.: Chaotic dynamics in periodically forced asymmetric ordinary differential equations. J. Math. Anal. Appl. **352**(2), 890–906 (2009)
20. Pascoletti, A., Zanolin, F.: A crossing lemma for annular regions and invariant sets. J. Math. (to appear)
21. Pascoletti, A., Zanolin, F.: A topological approach to bend-twist maps, with applications. Int. J. Differ. Equ. Art. ID 612041, 20 (2011)
22. Poincaré, H.: Sur un théorème de geométrie. Rend. Circ. Mat. Palermo **33**, 375–389 (1912)
23. Rebelo, C.: A note on the Poincaré-Birkhoff fixed point theorem and periodic solutions of planar systems. Nonlinear Anal. **29**(3), 291–311 (1997)
24. Springham, J.: Ergodic properties of linked-twist maps. Ph.D. thesis, University of Bristol (2008). http://arxiv.org/abs/0812.0899v1
25. Sturman, R., Ottino, J.M., Wiggins, S.: The mathematical foundations of mixing. In: Cambridge Monographs on Applied and Computational Mathematics, vol. 22. Cambridge University Press, Cambridge (2006)
26. Wiggins, S., Ottino, J.M.: Foundations of chaotic mixing. Philos. Trans. R. Soc. Lond. Ser. A Math. Phys. Eng. Sci. **362**(1818), 937–970 (2004)
27. Zanolin, F.: Time-maps and boundary value problems for ordinary differential equations. In: Tricomi's ideas and contemporary applied mathematics (Rome/Turin, 1997), vol. 147, pp. 281–302. Atti Convegni Lincei, Accademia Nazionale dei Lincei, Rome (1998)

Part II
Contributions

Pullback Attractors for NonAutonomous Dynamical Systems

María Anguiano, Tomás Caraballo, José Real, and José Valero

Abstract We study a nonautonomous reaction-diffusion equation with zero Dirichlet boundary condition, in an unbounded domain containing a nonautonomous forcing term taking values in the space H^{-1}, and with a continuous nonlinearity which does not ensure uniqueness of solution. Using results of the theory of set-valued nonautonomous (pullback) dynamical systems, we prove the existence of minimal pullback attractors for this problem. We ensure that the pullback attractors are connected and also establish the relation between these attractors.

Keywords Pullback attractor • Non-autonomous reaction-diffusion equation • Set-valued dynamical system • Unbounded domain

1 Introduction

The understanding of the asymptotic behavior of dynamical systems is one of the most important problems of modern mathematical physics. One way to treat this problem for systems having some dissipativity properties is to analyze the existence and structure of its global attractor. On some occasions, some phenomena are modeled by nonlinear evolutionary equations which do not take into account all the relevant information of the real systems. Instead, some neglected quantities can be modeled as an external force which, in general, becomes time dependent.

M. Anguiano (✉) • T. Caraballo • J. Real
Dpto. Ecuaciones Diferenciales y Análisis Numérico, Universidad de Sevilla,
c/Tarfia s/n, 41012 Sevilla, Spain
e-mail: anguiano@us.es; caraball@us.es; jreal@us.es

J. Valero
Dpto. Estadística y Matemática Aplicada, Universidad Miguel Hernández,
Avda. de la Universidad s/n, 03202, Elche, Spain
e-mail: jvalero@umh.es

S. Pinelas et al. (eds.), *Differential and Difference Equations with Applications*, Springer
Proceedings in Mathematics & Statistics 47, DOI 10.1007/978-1-4614-7333-6_15,
© Springer Science+Business Media New York 2013

For this reason, nonautonomous systems are of great importance and interest. The theory of pullback attractors is an important mathematical tool for studying the qualitative behavior of infinite-dimensional dynamical systems. By using this theory during the last few years, many results concerning attractors for evolution differential equations have been obtained (see [2, 3, 8, 9, 12] among others). However, these results cannot be applied to a wide class of initial-boundary problems, in which the solution may not be unique. Good examples of such systems are differential inclusions, variational inequalities, control infinite-dimensional systems, and also some partial-differential equations as the three-dimensional Navier–Stokes equations or the nonautonomous reaction-diffusion equations without uniqueness of solution. For the qualitative analysis of the above-mentioned systems, from the point of view of the theory of dynamical systems, it is necessary the theory for set-valued nonautonomous dynamical systems.

The study of reaction-diffusion equations without uniqueness of solutions in a bounded domain in the autonomous case or in the nonautonomous case under strong uniformity properties on the time-dependent terms can be found in [5, 11], among others. In the autonomous case, when the domain is unbounded, several studies on the problem can be found, for instance, in [13, 14]. In this sense, our aim is to consider a much more general problem.

Let $\Omega \subset \mathbb{R}^N$ be a nonempty open set, not necessarily bounded, and suppose that Ω satisfies the Poincaré inequality, i.e., there exists a constant $\lambda_1 > 0$ such that

$$\int_\Omega |u(x)|^2 \, dx \le \lambda_1^{-1} \int_\Omega |\nabla u(x)|^2 \, dx \quad \forall u \in H_0^1(\Omega). \tag{1}$$

Let us consider the following nonautonomous reaction-diffusion equation with zero Dirichlet boundary condition in Ω:

$$\begin{cases} \dfrac{\partial u}{\partial t} - \Delta u = f(x, u) + h(t) \text{ in } \Omega \times (\tau, +\infty), \\ u = 0 \text{ on } \partial\Omega \times (\tau, +\infty), \\ u(x, \tau) = u_\tau(x), \ x \in \Omega, \end{cases} \tag{2}$$

where $\tau \in \mathbb{R}$, $u_\tau \in L^2(\Omega)$, $h \in L_{loc}^2(\mathbb{R}; H^{-1}(\Omega))$, and $f : \Omega \times \mathbb{R} \to \mathbb{R}$ is a Carathéodory function, that is, $f(\cdot, u)$ is a measurable function for any $u \in \mathbb{R}$ and $f(x, \cdot) \in C(\mathbb{R})$ for almost every $x \in \Omega$, and satisfies that there exist constants $\alpha_1 > 0$, $\alpha_2 > 0$, and $p \ge 2$ and positive functions $C_1(x), C_2(x) \in L^1(\Omega)$ such that

$$|f(x, s)|^{\frac{p}{p-1}} \le \alpha_1 |s|^p + C_1(x) \quad \forall s \in \mathbb{R}, \text{a.e.} x \in \Omega, \tag{3}$$

$$f(x, s)s \le -\alpha_2 |s|^p + C_2(x) \ \forall s \in \mathbb{R}, \text{a.e.} x \in \Omega, \tag{4}$$

where these assumptions do not ensure uniqueness of solution of (2). Due to the nonautonomous character of the problem, in Sect. 2, we developed a theory of pullback attractors in the framework of set-valued problems. First, we recall some basic definitions for set-valued nonautonomous dynamical systems and establish a sufficient condition for the existence of pullback attractors for these systems. In Sect. 3 we prove the existence of solution of (2) and we show a sufficient condition ensuring the existence of minimal pullback attractors in $L^2(\Omega)$.

2 Theory of Set-Valued Nonautonomous Dynamical Systems

The theory of set-valued nonautonomous dynamical systems is well established as has been extensively developed over the last one and a half decades. We can find results about this theory in the work of Caraballo and Kloeden [6], among others. Most results in this section are slight modifications and generalizations of the results of this paper.

Let $X = (X, d_X)$ be a metric space, let $\mathscr{P}(X)$ denote the family of all nonempty subsets of X, and let us denote $\mathbb{R}_d^2 := \{(t,s) \in \mathbb{R}^2 : t \geq s\}$.

Definition 1. A multivalued map $U : \mathbb{R}_d^2 \times X \to \mathscr{P}(X)$ is called a **multivalued nonautonomous dynamical system** (**MNDS**) on X if $U(\tau, \tau, x) = \{x\}$ for all $\tau \in \mathbb{R}$, $x \in X$, and $U(t, \tau, x) \subset U(t, s, U(s, \tau, x))$ for all $\tau \leq s \leq t$, $x \in X$.

An **MNDS** is said to be **strict** if $U(t, \tau, x) = U(t, s, U(s, \tau, x))$ for all $\tau \leq s \leq t$, $x \in X$.

Definition 2. An MNDS U on X is said to be **upper semicontinuous** if for all $t \geq \tau$ the mapping $U(t, \tau, \cdot)$ is upper semicontinuous from X into $\mathscr{P}(X)$, i.e., for any $x_0 \in X$ and for every neighborhood $\mathscr{N}$ in X of the set $U(t, \tau, x_0)$, there exists $\delta > 0$ such that $U(t, \tau, y) \subset \mathscr{N}$ whenever $d_X(x_0, y) < \delta$.

Let $\mathscr{D}$ be a class of sets parameterized in time, $\hat{D} = \{D(t) : t \in \mathbb{R}\} \subset \mathscr{P}(X)$.

We will say that the class $\mathscr{D}$ is inclusion-closed, if $\hat{D} \in \mathscr{D}$ and $\emptyset \neq D'(t) \subset D(t)$ for all $t \in \mathbb{R}$, imply that $\hat{D}' = \{D'(t) : t \in \mathbb{R}\}$ belongs to $\mathscr{D}$.

Definition 3. We say that a family $\hat{D}_0 = \{D_0(t) : t \in \mathbb{R}\} \subset \mathscr{P}(X)$ is **pullback $\mathscr{D}$-absorbing** for the MNDS U if for every $\hat{D} \in \mathscr{D}$ and every $t \in \mathbb{R}$, there exists $\tau(t, \hat{D}) \leq t$ such that $U(t, \tau, D(\tau)) \subset D_0(t)$ for all $\tau \leq \tau(t, \hat{D})$.

Definition 4. The MNDS U is **pullback asymptotically compact** with respect to a family $\hat{B} = \{B(t) : t \in \mathbb{R}\} \subset \mathscr{P}(X)$ (or **pullback $\hat{B}$-asymptotically compact**) if for all $t \in \mathbb{R}$ and every sequence $\tau_n \leq t$ tending to $-\infty$, any sequence $y_n \in U(t, \tau_n, B(\tau_n))$ is relatively compact in X.

Definition 5. A family $\mathscr{A} = \{\mathscr{A}(t) : t \in \mathbb{R}\} \subset \mathscr{P}(X)$ is said to be a **global pullback $\mathscr{D}$-attractor** for the MNDS U if it satisfies:

1. $\mathscr{A}(t)$ is compact for any $t \in \mathbb{R}$.
2. $\mathscr{A}$ is pullback $\mathscr{D}$-attracting, i.e. $\lim_{\tau \to -\infty} dist_X(U(t, \tau, D(\tau)), \mathscr{A}(t)) = 0 \ \forall t \in \mathbb{R}$, for all $\hat{D} \in \mathscr{D}$.
3. $\mathscr{A}$ is negatively invariant, i.e., $\mathscr{A}(t) \subset U(t, \tau, \mathscr{A}(\tau))$, for any $(t, \tau) \in \mathbb{R}_d^2$.

$\mathscr{A}$ is said to be a **strict global pullback** $\mathscr{D}$**-attractor** if the invariance property in the third item is strict, i.e., $\mathscr{A}(t) = U(t, \tau, \mathscr{A}(\tau))$, for $(t, \tau) \in \mathbb{R}_d^2$.

The main tool to prove the existence of an attractor is the concept of pullback-omega-limit set.

Definition 6. For any family $\hat{B} = \{B(t) : t \in \mathbb{R}\} \subset \mathscr{P}(X)$, we define the **pullback-omega-limit set** as the t-dependent set $\Lambda(\hat{B}, t)$ given by $\Lambda(\hat{B}, t) = \bigcap_{s \leq t} \overline{\bigcup_{\tau \leq s} U(t, \tau, B(\tau))}^X$.

Now, we will establish a sufficient condition ensuring the existence of pullback attractors with respect to a general universe $\mathscr{D}$ (as in [8]). When this universe consists of bounded sets, the results have already been proved in [7]. This is our main result in this section.

Theorem 1. *Assume that* $\hat{D}_0 = \{D_0(t) : t \in \mathbb{R}\} \subset \mathscr{P}(X)$ *is pullback* $\mathscr{D}$*-absorbing for an MNDS* U*, which is also pullback* $\hat{D}_0$*-asymptotically compact. Then, the family* $\mathscr{A}_{\mathscr{D}} = \{\mathscr{A}_{\mathscr{D}}(t) : t \in \mathbb{R}\}$ *given by* $\mathscr{A}_{\mathscr{D}}(t) = \overline{\bigcup_{\hat{D} \in \mathscr{D}} \Lambda(\hat{D}, t)}^X$, $t \in \mathbb{R}$, *satisfies the following properties:*

1. *For each* $t \in \mathbb{R}$ *the set* $\mathscr{A}_{\mathscr{D}}(t)$ *is a nonempty compact subset of* X*, and* $\mathscr{A}_{\mathscr{D}}(t) \subset \Lambda(\hat{D}_0, t)$.
2. $\mathscr{A}_{\mathscr{D}}$ *is pullback* $\mathscr{D}$*-attracting and in fact is the minimal family of closed sets that pullback attracts all elements of* $\mathscr{D}$.
3. *If* $\hat{D}_0 \in \mathscr{D}$*, then* $\mathscr{A}_{\mathscr{D}}(t) = \Lambda(\hat{D}_0, t) \subset \overline{D_0(t)}^X$*, for all* $t \in \mathbb{R}$.
4. *If* U *is upper semicontinuous and with closed values,* $\mathscr{A}_{\mathscr{D}}$ *is a global pullback* $\mathscr{D}$*-attractor for* U.
5. *If* U *is upper semicontinuous, with closed and connected values, and for each* $t \in \mathbb{R}$ $\mathscr{A}_{\mathscr{D}}(t) \subset C(t)$*, where* $\hat{C} \in \mathscr{D}$ *and* $C(t)$ *is a connected subset of* X*, then* $\mathscr{A}_{\mathscr{D}}$ *is connected, i.e.,* $\mathscr{A}_{\mathscr{D}}(t)$ *is connected for any* $t \in \mathbb{R}$.
6. *If* $\hat{D}_0 \in \mathscr{D}$*, each* $D_0(t)$ *is closed, and the universe* $\mathscr{D}$ *is inclusion-closed, then* $\mathscr{A}_{\mathscr{D}} \in \mathscr{D}$*. If moreover* U *is upper semicontinuous and with closed values,* $\mathscr{A}_{\mathscr{D}}$ *is the unique global pullback* $\mathscr{D}$*-attractor belonging to* $\mathscr{D}$*. In this case, if moreover* U *is strict, then* $\mathscr{A}_{\mathscr{D}}$ *is a strict global pullback* $\mathscr{D}$*-attractor for* U.

Proof. See [1]. $\square$

We denote $\mathscr{D}_F^X$ the universe of fixed nonempty bounded subsets of X, i.e., the class of all families $\hat{D}$ of the form $\hat{D} = \{D(t) = D : t \in \mathbb{R}\}$ with D a fixed nonempty bounded subset of X. In the particular case of considering the universe $\mathscr{D}_F^X$, the corresponding minimal pullback $\mathscr{D}_F^X$-attractor for the process U is the pullback attractor defined by Crauel et al. [10, Theorem 1.1, p. 311] and will be denoted $\mathscr{A}_{\mathscr{D}_F^X}$. Then, it is easy to conclude the following result.

Corollary 1. *Assume that* $\hat{D}_0 = \{D_0(t) : t \in \mathbb{R}\} \subset \mathscr{P}(X)$ *is pullback* $\mathscr{D}$*-absorbing for an MNDS* U*, which is also pullback* $\hat{D}_0$*-asymptotically compact, upper semicontinuous, and with closed values. Then, if the universe* $\mathscr{D}$ *contains the universe*

$\mathscr{D}_F^X$, both attractors $\mathscr{A}_{\mathscr{D}_F^X}$ and $\mathscr{A}_{\mathscr{D}}$ exist, and the following relation holds:

$$\mathscr{A}_{\mathscr{D}_F^X}(t) \subset \mathscr{A}_{\mathscr{D}}(t) \text{ for all} t \in \mathbb{R}.$$

Remark 1. It can be proved (see [12]) that, under the assumptions of the preceding corollary, if, moreover, for some $T \in \mathbb{R}$ the set $\cup_{t \leq T} D_0(t)$ is a bounded subset of X, then $\mathscr{A}_{\mathscr{D}_F^X}(t) = \mathscr{A}_{\mathscr{D}}(t)$ for all $t \leq T$.

3   Existence of Pullback Attractors for (2)

The aim of this section is to show the existence of pullback attractors, which are connected, in the phase space $L^2(\Omega)$ for the problem (2) using Theorem 1.

To do this we need a theorem on the existence of solutions of problem (2), which we will see in the following subsection.

3.1   Existence of Solution

We state in this section a result on the existence of solutions of problem (2). First, we give the definition of weak solution of it.

By $|\cdot|$, $\|\cdot\| = |\nabla \cdot|$, $\|\cdot\|_*$, and $\|\cdot\|_{L^p(\Omega)}$ we denote the norms in the spaces $L^2(\Omega)$, $H_0^1(\Omega)$, $H^{-1}(\Omega)$, and $L^p(\Omega)$, respectively. We will use $(\cdot, \cdot)$ to denote the scalar product in $L^2(\Omega)$ or $[L^2(\Omega)]^N$ and $\langle \cdot, \cdot \rangle$ to denote the duality product either between $H^{-1}(\Omega)$ and $H_0^1(\Omega)$ or between $L^{p'}(\Omega)$ and $L^p(\Omega)$, where $p' = \frac{p}{p-1}$ is the conjugate exponent of p.

Definition 7.   A weak solution of (2) is a function $u \in L^p(\tau, T; L^p(\Omega)) \cap L^2(\tau, T; H_0^1(\Omega))$ for all $T > \tau$ and such that

$$(u(t), w) + \int_\tau^t (\nabla u(s), \nabla w)\, ds = (u_\tau, w) + \int_\tau^t \langle f(x, u(s)) + h(s), w \rangle\, ds \quad \forall t \geq \tau, \quad (5)$$

for all $w \in L^p(\Omega) \cap H_0^1(\Omega)$.

Theorem 2.   *Assume that Ω satisfies (1), $h \in L^2_{loc}(\mathbb{R}; H^{-1}(\Omega))$ and f is Carathéodory and satisfies (3) and (4). Then, for all $\tau \in \mathbb{R}$, $u_\tau \in L^2(\Omega)$, there exists at least one weak solution u of (2).*

Proof. See [1,4].   □

3.2 Existence of Pullback Attractors

In this section we prove our main result. First, we need a priori estimates and a continuity result.

3.2.1 A Priori Estimates and a Continuity Result

For each $\tau \in \mathbb{R}$ and $u_\tau \in L^2(\Omega)$, let us denote $S(\tau, u_\tau)$ the set of all weak solutions of (2) defined for all $t \geq \tau$. We define a multivalued map $U : \mathbb{R}_d^2 \times L^2(\Omega) \to \mathscr{P}(L^2(\Omega))$ by

$$U(t, \tau, u_\tau) = \{u(t) : u \in S(\tau, u_\tau)\}, \quad \tau \leq t, \quad u_\tau \in L^2(\Omega). \tag{6}$$

It is easy to conclude the following results, whose proofs can be found in [1, 4].

Lemma 1. *Under the assumptions of Theorem 2, the multivalued mapping U defined by (6) is a strict MNDS on $L^2(\Omega)$.*

Now, we define the universe in $\mathscr{P}(L^2(\Omega))$. We denote by $\mathscr{D}_{\lambda_1}$ the class of all families $\hat{D} = \{D(t) : t \in \mathbb{R}\} \subset \mathscr{P}(L^2(\Omega))$ such that $D(t) \subset \overline{B}_{L^2(\Omega)}(0, r_{\hat{D}}(t))$ and $\lim_{t \to -\infty} e^{\lambda_1 t} r_{\hat{D}}^2(t) = 0$.

According to the notation introduced in the last section, $\mathscr{D}_F^H$ will denote the class of all families $\hat{D}$ of the form $\hat{D} = \{D(t) = D : t \in \mathbb{R}\}$ with D a fixed nonempty bounded subset of $L^2(\Omega)$.

Remark 2. We note that $\mathscr{D}_F^H \subset \mathscr{D}_{\lambda_1}$ and both universes are inclusion-closed.

Lemma 2. *Suppose that Ω satisfies (1) and suppose that f is Carathéodory and satisfies (3) and (4). Let $h = \sum_{i=1}^{N} \frac{\partial h_i}{\partial x_i}$, with $h_i \in L_{loc}^2(\mathbb{R}; L^2(\Omega))$ for all $1 \leq i \leq N$, such that*

$$\sum_{i=1}^{N} \int_{-\infty}^{t} e^{\lambda_1 s} |h_i(s)|^2 \, ds < +\infty \ \forall t \in \mathbb{R}. \tag{7}$$

Then, the balls $B_{\lambda_1}(t) = \overline{B}_{L^2(\Omega)}(0, R_{\lambda_1}(t))$, where $R_{\lambda_1}(t)$ is the nonnegative number given for each $t \in \mathbb{R}$ by

$$R_{\lambda_1}^2(t) = 2e^{-\lambda_1 t} \sum_{i=1}^{N} \int_{-\infty}^{t} e^{\lambda_1 s} |h_i(s)|^2 \, ds + 2\lambda_1^{-1} \|C_2\|_{L^1(\Omega)} + 1, \tag{8}$$

form a family $\hat{B}_{\lambda_1} \in \mathscr{D}_{\lambda_1}$ which is pullback $\mathscr{D}_{\lambda_1}$-absorbing for the MNDS U defined by (6).

From now on, for all $m \geq 1$, we denote $\Omega_m = \Omega \cap \{x \in \mathbb{R}^N : |x|_{\mathbb{R}^N} < m\}$, where $|\cdot|_{\mathbb{R}^N}$ denotes the Euclidean norm in $\mathbb{R}^N$. We need the following results whose proofs can be found in [1,4].

Lemma 3. *Under the assumptions in Lemma 2, for any real numbers $t_1 \leq t_2$ and any $\varepsilon > 0$, there exist $T = T(t_1, t_2, \varepsilon, \hat{B}_{\lambda_1}) \leq t_1$ and $M = M(t_1, t_2, \varepsilon, \hat{B}_{\lambda_1}) \geq 1$ verifying*

$$\int_{\Omega \cap \{|x|_{\mathbb{R}^N} \geq 2m\}} u^2(x,t)\,dx \leq \varepsilon, \; \forall \tau \leq T, \; t \in [t_1, t_2], \; m \geq M,$$

for any weak solution $u \in S(\tau, u_\tau)$ where $u_\tau \in B_{\lambda_1}(\tau)$.

Lemma 4. *Under the assumptions in Lemma 2, let K be a relatively compact set in $L^2(\Omega)$. Then, for all $\tau \leq T$ and $\varepsilon > 0$ there exists $M = M(\tau, T, \varepsilon, K)$ such that*

$$\int_{\Omega \cap \{|x|_{\mathbb{R}^N} \geq 2m\}} u^2(x,t)\,dx \leq \varepsilon, \; \forall t \in [\tau, T], \quad \forall m \geq M,$$

for any $u \in S(\tau, u_\tau)$, where $u_\tau \in K$ is arbitrary.

Further, we obtain a continuity result leading to the upper semicontinuity of the MNDS U.

Proposition 1. *Under the assumptions in Lemma 2, let $\tau \in \mathbb{R}$ and $\{u_\tau^n\} \subset L^2(\Omega)$ be a sequence converging weakly in $L^2(\Omega)$ to an element $u_\tau \in L^2(\Omega)$. For each $n \geq 1$ let us fix $u_n \in S(\tau, u_\tau^n)$. Then there exists a subsequence $\{u_\mu\} \subset \{u_n\}$ satisfying that there exists $u \in S(\tau, u_\tau)$ such that*

$$u_\mu(t) \rightharpoonup u(t) \text{ weakly in } L^2(\Omega) \, \forall t \geq \tau, \tag{9}$$

$$u_\mu \rightharpoonup u \text{ weakly in } L^2(\tau, T; H_0^1(\Omega)) \; \forall T > \tau, \tag{10}$$

$$u_\mu \rightharpoonup u \text{ weakly in } L^p(\tau, T; L^p(\Omega)) \; \forall T > \tau, \tag{11}$$

$$f(x, u_\mu) \rightharpoonup f(x, u) \text{ weakly in } L^{p'}(\tau, T; L^{p'}(\Omega)) \; \forall T > \tau, \tag{12}$$

$$u_\mu|_{\Omega_m} \to u|_{\Omega_m} \text{ strongly in } L^2(\tau, T; L^2(\Omega_m)) \; \forall T > \tau, \quad \forall m \geq 1. \tag{13}$$

Finally, if the sequence $\{u_\tau^n\}$ converges strongly in $L^2(\Omega)$ to u_τ, then

$$u_\mu \to u \text{ strongly in } L^2(\tau, T; L^2(\Omega)) \; \forall T > \tau, \tag{14}$$

and

$$u_\mu(t) \to u(t) \text{ strongly in } L^2(\Omega) \, \forall t \geq \tau. \tag{15}$$

Proof. This result can be proved in much the same way as Theorem 2 and using Lemmas 3 and 4. $\qquad\square$

3.2.2 Existence of the Global Pullback Attractor

Using the previous results we can prove the following lemma, which is necessary to prove our main result.

Lemma 5. *Under the assumptions in Lemma 2, the MNDS U defined by (6) is upper semicontinuous, has closed values, and is pullback asymptotically compact with respect to the family $\hat{B}_{\lambda_1}$ defined in that lemma.*

Now, we are ready to obtain the main result.

Theorem 3. *Under the assumptions in Lemma 2, the MNDS U defined by (6) possesses a unique pullback $\mathscr{D}_{\lambda_1}$-attractor $\mathscr{A}_{\mathscr{D}_{\lambda_1}}$ belonging to $\mathscr{D}_{\lambda_1}$, which is strictly invariant and connected and is given by $\mathscr{A}_{\mathscr{D}_{\lambda_1}}(t) = \Lambda\left(\hat{B}_{\lambda_1}, t\right)$, where $\hat{B}_{\lambda_1}$ was defined in Lemma 2. Moreover, there exists the minimal pullback $\mathscr{D}_F^H$-attractor, $\mathscr{A}_{\mathscr{D}_F^H}$, which is also connected, and we have the following relation:*

$$\mathscr{A}_{\mathscr{D}_F^H}(t) \subset \mathscr{A}_{\mathscr{D}_{\lambda_1}}(t) \subset \overline{B}_{L^2(\Omega)}\left(0, R_{\lambda_1}(t)\right) \text{ for all } t \in \mathbb{R}. \tag{16}$$

Proof. As a direct consequence of the preceding results, Theorem 1 and Corollary 1, we obtain the existence of the unique pullback $\mathscr{D}_{\lambda_1}$-attractor belonging to $\mathscr{D}_{\lambda_1}$, which is strictly invariant, and the minimal pullback $\mathscr{D}_F^H$-attractor for the MNDS U defined by (6), and we also have the relation (16). Moreover, we can prove that $U(t, \tau, u_\tau)$ has connected values in $L^2(\Omega)$ (see [1] for details of the proof). On the other hand, as $\hat{B}_{\lambda_1}$ is pullback $\mathscr{D}_{\lambda_1}$-absorbing, taking into account that $\hat{B}_{\lambda_1} \in \mathscr{D}_{\lambda_1}$, thanks to the third statement of Theorem 1 we have that all conditions of the fifth statement of Theorem 1 are also satisfied. Then, we have that $\mathscr{A}_{\mathscr{D}_{\lambda_1}}$ is connected. Using similar arguments we have that $\mathscr{A}_{\mathscr{D}_F^H}$ is also connected. $\square$

Remark 3. If we also assume that $\sup_{t \leq 0} e^{-\lambda_1 t} \sum_{i=1}^N \int_{-\infty}^t e^{\lambda_1 s} |h_i(s)|^2\, ds < \infty$, then we have that $\cup_{t \leq T} \overline{B}_{L^2(\Omega)}\left(0, R_{\lambda_1}(t)\right)$ is a bounded subset of $L^2(\Omega)$. Therefore, taking into account Remark 1, we can deduce that $\mathscr{A}_{\mathscr{D}_F^H}(t) = \mathscr{A}_{\mathscr{D}_{\lambda_1}}(t)$ for all $t \leq T$.

References

1. Anguiano, M.: Atractores para EDP parabólicas no lineales y no autónomas en dominios no acotados. Ph.D. Dissertation, Universidad de Sevilla (2011)
2. Anguiano, M., Caraballo, T., Real, J.: Existence of pullback attractor for a reaction-diffusion equation in some unbounded domains with non-autonomous forcing term in H^{-1}. Int. J. Bifurcat. Chaos **20**(9), 2645–2656 (2010)
3. Anguiano, M., Caraballo, T., Real, J.: An exponential growth condition in H^2 for the pullback attractor of a non-autonomous reaction-diffusion equation. Nonlinear Anal. **72**(11), 4071–4076 (2010)

4. Anguiano, M., Caraballo, T., Real, J., Valero, J.: Pullback attractors for reaction-diffusion equations in some unbounded domains with an H^{-1}-valued non-autonomous forcing term and without uniqueness of solutions. Discret. Contin. Dyn. Syst. Ser. B **14**(2), 307–326 (2010)
5. Anguiano, M., Kloeden, P.E., Lorenz, T.: Asymptotic behaviour of nonlocal reaction-diffusion equations. Nonlinear Anal. **73**(9), 3044–3057 (2010)
6. Caraballo, T., Kloeden, P.E.: Non-autonomous attractors for integro-differential evolution equations. Discret. Contin. Dyn. Syst. Ser. S **2**(1), 17–36 (2009)
7. Caraballo, T., Langa, J.A., Melnik, V.S., Valero, J.: Pullback attractors of nonautonomous and stochastic multivalued dynamical systems. Set Valued Anal. **11**, 153–201 (2003)
8. Caraballo, T., Lukaszewicz, G., Real, J.: Pullback attractors for asymptotically compact non-autonomous dynamical systems. Nonlinear Anal. **64**, 484–498 (2006)
9. Caraballo, T., Lukaszewicz, G., Real, J.: Pullback attractors for non-autonomous 2D Navier-Stokes equations in unbounded domains. C. R. Math. Acad. Sci. Paris **342**, 263–268 (2006)
10. Crauel, H., Debussche, A., Flandoli, F.: Random attractors. J. Dynam. Differ. Equ. **9**(2), 307–341 (1997)
11. Iovane, G., Kapustyan, A.V., Valero, J.: Asymptotic behaviour of reaction-diffusion equations with non-damped impulsive effects, Nonlinear Anal. **68**, 2516–2530 (2008)
12. Marín-Rubio, P., Real, J.: On the relation between two different concepts of pullback attractors for non-autonomous dynamical systems. Nonlinear Anal. **71**, 3956–3963 (2009)
13. Morillas, F., Valero, J.: Attractors for reaction-diffusion equations in R^N with continuous nonlinearity. Asymptotic Anal. **44**, 111–130 (2005)
14. Morillas, F., Valero, J.: On the Kneser property for reaction-diffusion systems on unbounded domains. Topol. Appl. **156**, 3029–3040 (2009)

Uniform Estimates and Existence of Solutions with Prescribed Domain to Nonlinear Third-Order Differential Equation

Irina Astashova

Abstract For differential equation of the third order with power nonlinearity, uniform estimates of solutions with the same domain are obtained. The existence of solutions with prescribed domain is proved.

Consider the differential equation

$$y''' + p(x,y,y',y'')|y|^{k-1}y = 0, \quad k > 1, \tag{1}$$

the function $p(x,y_0,y_1,y_2)$ defined on $\mathbb{R} \times \mathbb{R}^3$ is continuous in x and Lipschitz continuous in y_0,y_1,y_2 with

$$0 < p_* \leq p(x,y_0,y_1,y_2) \leq p^*, \tag{2}$$

where p_*, p^* are positive constants.
   Put $\beta = \frac{k-1}{3} > 0$.

1   Uniform Estimates of Solutions

Theorem 1. *For any $k > 1$, $p_* > 0$, $p^* > p_*$, $h > 0$ there exists a constant $C > 0$ such that for any $p(x,y,y',y'')$ any solution $y(x)$ to (1) satisfying the condition $|y(x_0)| = h > 0$ at some point $x_0 \in \mathbb{R}$ cannot be extended to the interval $(x_0 - Ch^{-\beta}, x_0 + Ch^{-\beta})$.*

I. Astashova (✉)
Lomonosov Moscow State University, 1 Leninskiye Gory, GSP-1, 119991 Moscow, Russia

Moscow State University of Economics, Statistics and Informatics, 7 Nezhinskaya,
119501 Moscow, Russia
e-mail: ast@diffiety.ac.ru

S. Pinelas et al. (eds.), *Differential and Difference Equations with Applications*, Springer
Proceedings in Mathematics & Statistics 47, DOI 10.1007/978-1-4614-7333-6_16,
© Springer Science+Business Media New York 2013

Theorem 2. *For any $k > 1$, $p_* > 0$, $p^* > p_*$ there exists a constant $C > 0$ such that for any $p(x,y,y',y'')$ and any solution $y(x)$ to (1) defined on $[-a,a]$, it holds $|y(0)| \leq (\frac{C}{a})^{1/\beta}$.*

Theorem 3. *For any $k > 1$, $p_* > 0$, $p^* > p_*$ there exists a constant $C > 0$ such that for any $p(x,y,y',y'')$ and any solution $y(x)$ to (1) defined on $[a,b]$, it holds*

$$|y(x)| \leq C \min(x-a, b-x)^{-1/\beta}. \tag{3}$$

Remark 1. In [3] uniform estimates for positive solutions with the same domain to the equation

$$y^{(n)} + \sum_{j=0}^{n-1} a_j(x) \, y^{(i)} + p(x) \, |y|^{k-1} y = 0$$

with continuous functions $p(x)$ and $a_j(x)$, $n \geq 1$, $k > 1$ were obtained. In [2] similar uniform estimates for absolute values of all solutions to the equation

$$y^{(n)} + \sum_{j=0}^{n-1} a_j(x) \, y^{(i)} + p(x) \, |y|^k = 0$$

were proved. See also [4].

Remark 2. The proof of Theorems 1–3 was published in [6]. For the case $p(x, y_0, y_1, y_2) = 1$, it was published in [5].

2 The Existence of Solutions with Prescribed Domain

Consider the differential equation (1) with the same propositions about the function $p(x, y_0, y_1, y_2)$.

Definition 1. A solution $y(x)$ has a *resonance asymptote* $x = x^*$ if

$$\overline{\lim_{x \to x^*}} \, y(x) = +\infty, \qquad \underline{\lim_{x \to x^*}} \, y(x) = -\infty.$$

2.1 Main Results

Theorem 4. *Let $y(x)$ be a solution to (1) defined on $[x_0, x^*)$ with the resonance asymptote $x = x^*$. Then the position of the asymptote $x = x^*$ depends continuously on $y(x_0), y'(x_0), y''(x_0)$.*

Theorem 5. *For any finite $x_* < x^*$ there exists a non-extensible solution $y(x)$ to (1) defined on (x_*, x^*) with the vertical asymptote $x = x_*$ and the resonance asymptote $x = x^*$.*

Corollary 1. *For any $x_* \in \mathbb{R}$ there exists a Kneser solution (see the definition in [8]) of (1) with the vertical asymptote $x = x_*$ defined on the interval $(x_*, +\infty)$ and tending to 0 as $x \to +\infty$.*

Corollary 2. *For any $x^* \in \mathbb{R}$ there exists a non-extensible solution $y(x)$ of (1) with the resonance asymptote $x = x^*$ defined on the interval $(-\infty, x_*)$ and tending to 0 as $x \to -\infty$.*

Remark 3. Note that Corollaries 1 and 2 follow also from more general results from [7] on the existence of a blow-up Kneser solution to equation (1).

Theorem 6. *For any finite or infinite $x_* < x^*$ there exists a non-extensible solution $y(x)$ of (1) with domain (x_*, x^*).*

Remark 4. In [1,4] asymptotic behavior of all possible solutions to (1) is described. The similar results for (1) of the second order were published in [9].

2.2   Proofs of Main Results

2.2.1   Lemmas for Theorem 4

Lemma 1. *Suppose a solution $y(x)$ to (1) satisfies, at some point x_0, the inequalities*

$$y(x_0) \geq 0, \qquad y'(x_0) > 0, \qquad y''(x_0) \geq 0.$$

Then $y(x)$ has a local maximum at some point $x_0' > x_0$ satisfying the following estimates:

$$x_0' - x_0 \leq \left(\mu \, y'(x_0) \right)^{-\frac{k-1}{k+2}}, \tag{4}$$

$$y(x_0') > \left(\mu \, y'(x_0) \right)^{\frac{3}{k+2}}, \tag{5}$$

$$y''(x_0') < - \left(\mu \, y'(x_0) \right)^{\frac{2k+1}{k+2}}, \tag{6}$$

where $\mu > 0$ is a constant depending only on k, m, and M.

Proof. We may assume that $x_0 = 0$ and put $V = y'(0)^{\frac{1}{k+2}}$.
    First consider the case

$$y''(0) < V^{2k+1}.$$

Let $[0,x_1']$ be the longest possible segment with the inequality $y'(x) \geq \dfrac{V^{k+2}}{2}$ satisfied on it. Then we also have on this segment

$$y(x) > \frac{V^{k+2}x}{2}, \qquad y'''(x) < -mV^{k^2+2k}2^{-k}x^k,$$

$$\frac{V^{k+2}}{2} \leq y'(x) < V^{k+2} + V^{2k+1}x - \frac{mV^{k^2+2k}x^{k+2}}{2^k(k+1)(k+2)},$$

whence

$$\frac{m\left(V^{k-1}x_1'\right)^{k+2}}{2^k(k+1)(k+2)} - \left(V^{k-1}x_1'\right) - \frac{1}{2} < 0$$

and $x_1' < r_{mk}V^{-k+1}$, where $r_{mk} > 0$ is the maximum root to the equation

$$\frac{2^{-k}m}{k^2+3k+2}r^{k+2} - r - \frac{1}{2} = 0.$$

Since the derivative $y'(x)$ changes from V^{k+2} to $\dfrac{V^{k+2}}{2}$ on the segment $[0,x_1']$, there exists a point $x_1'' \in [0,x_1']$ with the inequality

$$y''(x_1'') < -\frac{V^{k+2}}{2r_{mk}V^{-k+1}} = -\frac{V^{2k+1}}{2r_{mk}}$$

holding also for $x > x_1''$, while $y(x)$ remains positive. This implies firstly that for sufficiently small μ inequality (6) holds and secondly that the derivative $y'(x)$ vanishes at some point x_0' with

$$x_0' - x_1' < \frac{V^{k+2}}{2}\cdot\frac{2r_{mk}}{V^{2k+1}} = r_{mk}V^{-k+1}.$$

This yields the inequality $x_0' < 2r_{mk}V^{-k+1}$ and for sufficiently small μ also inequality (4).

Since on the segment $[0,x_0']$ the second derivative changes from a negative value to a nonnegative one, which is less than $-\dfrac{V^{2k+1}}{2r_{mk}}$, then at some point $x''' \in [0,x_0']$, we have

$$y'''(x''') < -\frac{V^{2k+1}}{2r_{mk}x_0'} < -\frac{V^{3k}}{4r_{mk}^2},$$

whence

$$y(x_0') \geq y(x''') > V^3 \left(4M\, r_{mk}^2\right)^{-\frac{1}{k}},$$

yielding, for sufficiently small μ, inequality (5).

It remains to consider the case

$$y''(0) \geq V^{2k+1}.$$

Now by the similar methods we prove that at some point

$$x_2'' < \left(\frac{2^{2k-1}(2k+1)}{m\,y''(0)^{k-1}}\right)^{\frac{1}{2k+1}}$$

the second derivative $y''(x)$ becomes two times less than $y''(0)$ and then vanishes at some point $x_3'' < 2x_2'' < \left(\dfrac{2^{4k}(2k+1)}{m}\right)^{\frac{1}{2k+1}} V^{-k+1}$. At the same time the first derivative increases and we obtain the situation from the previous case. □

Lemma 2. *Suppose a solution $y(x)$ to (1) satisfies, at some point x_0', the inequalities*

$$y(x_0') > 0, \qquad y'(x_0') \leq 0, \qquad y''(x_0') \leq 0.$$

Then $y(x)$ vanishes at some point $x_0 > x_0'$ with the following estimates:

$$x_0 - x_0' \leq \left(\mu\, y(x_0')\right)^{-\frac{k-1}{3}}, \tag{7}$$

$$y'(x_0) < -\left(\mu\, y(x_0')\right)^{\frac{k+2}{3}}, \tag{8}$$

$$y''(x_0) < -\left(\mu\, y(x_0')\right)^{\frac{2k+1}{3}}, \tag{9}$$

where $\mu > 0$ is a constant depending only on k, m, and M.

Proof. First as in the previous lemma we obtain the properties of the point where the solution halves. Then we observe it vanishing. □

Lemma 3. *Under the conditions of Lemma 2 for any $x_1 > x_0$ with $y(x_1) = 0$, it holds*

$$|y'(x_1)| > Q\,|y'(x_0)|, \tag{10}$$

where $Q > 1$ is a constant depending only on k, m and M.

Proof. First notice that between x_0 and x_1, there exists a point x_0'', such that $y''(x_0'') = 0$ and $y'(x_0'') \geq y'(x_0) > 0$. Hence

$$
\begin{aligned}
y'(x_1)^2 - y'(x_0)^2 &\geq y'(x_1)^2 - y'(x_0'')^2 = 2\int_{x_0''}^{x_1} y'(\xi)y''(\xi)\,d\xi \\
&= 2\int_{x_0''}^{x_1} y''(\xi)\,dy(\xi) = 2y(\xi)y''(\xi)\Big|_{x_0''}^{x_1} - 2\int_{x_0''}^{x_1} y(\xi)\,dy''(\xi) \\
&= -2\int_{x_0''}^{x_1} y(\xi)\,dy''(\xi) = 2\int_{x_0''}^{x_1} p(\xi,y(\xi),y'(\xi),y''(\xi))\,|y(\xi)|^{k+1}\,d\xi.
\end{aligned}
$$

$$(11)$$

Further, between x_0'' and x_1, there exists a point x_0', such that $y'(x_0') = 0$ and, according to Lemma 1,

$$
y''(x_0') < -\left(\mu\, y'(x_0'')\right)^{\frac{2k+1}{k+2}}.
$$

Between x_0'' and x_0', there exists a point x_1'' with an intermediate value of the second derivative: $y''(x_1'') = -\dfrac{1}{2}\left(\mu\, y'(x_0'')\right)^{\frac{2k+1}{k+2}}$. Hence, between x_0'' and x_1'', there exists a point x_0''', such that

$$
y'''(x_0''') \leq -\frac{\left(\mu\, y'(x_0'')\right)^{\frac{2k+1}{k+2}}}{2(x_0' - x_0'')} < -\frac{1}{2}\left(\mu\, y'(x_0'')\right)^{\frac{3k}{k+2}},
$$

$$
y(x_0''') \geq \left(\frac{y(x_0''')}{M}\right)^{-\frac{1}{k}} > \left(\mu\, y'(x_0'')\right)^{\frac{3}{k+2}} (2M)^{-\frac{1}{k}}.
$$

Inequality sequence (11) can be continued taking into account that

$$
\left| y''(x_1'') - y''(x_0''') \right| > \frac{1}{2}\left(\mu\, y'(x_0'')\right)^{\frac{2k+1}{k+2}},
$$

and $y(x) > y(x_0''')$ for any $x \in [x_1'', x_0']$:

$$
\begin{aligned}
y'(x_1)^2 - y'(x_0'')^2 &= 2\int_{x_0''}^{x_1} p(\xi,y(\xi),y'(\xi),y''(\xi))\,|y(\xi)|^{k+1}\,d\xi \\
&\geq 2\int_{x_1''}^{x_0'} p(\xi,y(\xi),y'(\xi),y''(\xi))\,|y(\xi)|^{k+1}\,d\xi \\
&= -2\int_{x_1''}^{x_0'} y(\xi)\,dy''(\xi) \geq 2\cdot y(x_0''')\cdot \left| y''(x_1'') - y''(x_0''') \right| \\
&> 2\cdot \left(\mu\, y'(x_0'')\right)^{\frac{3}{k+2}} (2M)^{-\frac{1}{k}}\cdot \frac{1}{2}\left(\mu\, y'(x_0'')\right)^{\frac{2k+1}{k+2}} \\
&= (2M)^{-\frac{1}{k}}\left(\mu\, y'(x_0'')\right)^2.
\end{aligned}
$$

So,

$$\left| y'(x_1) \right| > \sqrt{y'(x_0'')^2 + (2M)^{-\frac{1}{k}} \left(\mu\, y'(x_0'') \right)^2}$$

$$= \left| y'(x_0'') \right| \sqrt{1 + (2M)^{-\frac{1}{k}} \mu^2} \geq \left| y'(x_0) \right| \sqrt{1 + (2M)^{-\frac{1}{k}} \mu^2}.$$

$\square$

2.2.2 Proof of Theorem 4

Proof. The three lemmas proved imply that if a solution $y(x)$ to (1) satisfies at some point x_0 the inequalities

$$y(x_0) \geq 0, \qquad y'(x_0) > 0, \qquad y''(x_0) \geq 0,$$

then there exists a point x_1 such that

$$x_1 - x_0 \leq \left(\mu'\, y'(x_0) \right)^{-\frac{k-1}{k+2}},$$

$$y(x_1) = 0, \qquad y'(x_1) < -Qy'(x_0), \qquad y''(x_1) < 0,$$

where $\mu' > 0$ and $Q > 1$ are constants depending only on k, m, and M.

At the point x_1 we obtain the initial situation mirrored relative to the axis Ox. So we can apply the same lemmas. Repeating the same procedure we obtain a sequence of segments such that the solution keeps the same sign inside each of them. The absolute value of the first derivative at the right boundary of the next segment is at least $Q > 1$ times greater than for the previous one. Hence the length L_j of the jth segment satisfies the inequality

$$L_j \leq \left(\mu'\, y'(x_0) \right)^{-\frac{k-1}{k+2}},$$

and the maximum y_j^* of $|y(x)|$ on it satisfies the inequality

$$y_j^* \geq \left(\mu\, y'(x_0) \right)^{\frac{3}{k+2}} Q^{\frac{3}{k+2}}.$$

Thus, $y_j^* \to +\infty$ as $j \to +\infty$ and

$$\sum_{j=0}^{\infty} L_j \leq \frac{\left(\mu'\, y'(x_0) \right)^{-\frac{k-1}{k+2}}}{1 - Q^{-\frac{k-1}{k+2}}} < +\infty. \tag{12}$$

So, an estimate is obtained for the distance to the vertical asymptote depending on the value of the derivative at the point with nonnegative values of $y(x)$, $y'(x)$, and $y''(x)$.

Now let $y(x)$ be a solution to (1) with a resonance asymptote at the point x^* and $\varepsilon > 0$. Consider a point $x_0 < x^*$ with positive values of $y(x)$, $y'(x)$, and $y''(x)$ and with $y'(x_0)$ sufficiently great to provide the estimate (12) be less than ε. These properties remain true under sufficiently small changes of the three values. Hence the resonance asymptote is not farther than at the point $x_0 + \varepsilon < x^* + \varepsilon$.

On the other hand, general properties of differential equations imply that if the above changes are sufficiently small, then the solution can be extended up to the point $x^* - \varepsilon$. So, the resonance asymptote cannot be closer than at the last point.

Continuity of the resonance asymptote position is proved. $\square$

2.2.3   Proof of Theorem 5

Proof. Put

$$\Delta = \{(u,v) \in [-1,1] \times [-1,1] : u < v\}.$$

Consider the map $\Gamma : \mathbb{R}^4 \to \Delta$ taking each quartet $(x_0, y_0, y_1, y_2) \in \mathbb{R}^4$ to the pair $(\tanh x_*, \tanh x^*) \in \Delta$, where x_* and x^* are the left and right boundaries (may be infinite) of the domain for the inextensible solution to (1) with initial data

$$y(x_0) = y_0, \quad y'(x_0) = y_1, \quad y''(x_0) = y_2.$$

According to Theorem 4 the map Γ is continuous. We need to prove the inclusion $\Gamma(\mathbb{R}^4) \supset \Delta \setminus \partial\Delta$.

Suppose $(u_0, v_0) \in \Delta \setminus \partial\Delta$, i. e., $-1 < u_0 = \tanh x_* < v_0 = \tanh x^* < 1$. Now we construct a loop L in $\mathbb{R}^4$ such that its image $\Gamma(L) \subset \Delta$ surrounds the point (u_0, v_0). The loop is composed of seven arcs:

$$L = L_1 \cup L_2 \cup L_3 \cup L_4 \cup L_5 \cup L_6 \cup L_7.$$

First take a point $(x_1, 0, y'_1, 0) \in \mathbb{R}^4$ with $x_1 = \dfrac{x_* + x^*}{2}$ and sufficiently great $y'_1 > 0$ providing the closure of the domain for the related solution $y_1(x)$ to be inside the interval (x_*, x^*). Notice that $y_1(x) \to -\infty$ near the left boundary of the domain for $y_1(x)$ and the point $(u_1, v_1) = \Gamma(x_1, 0, y'_1, 0)$ is located to the right of and below the point (u_0, v_0), i. e., $u_1 > u_0$, $v_1 < v_0$.

Take sufficiently great $y''_2 > 0$, to provide the solution $y_2(x)$ with initial data $(x_1, 0, y'_1, y''_2)$ to tend to $+\infty$ near the left boundary. Then there exists $y''_3 \in [0, y''_2]$ such that the solution $y_3(x)$ with initial data $(x_1, 0, y'_1, y''_3)$ is oscillatory near both boundaries.

The point $\Gamma(x_1, 0, y'_1, y''_3) = (-1, v_3)$ is located to the left of and below the point (u_0, v_0), i. e., $-1 < u_0$, $v_3 < v_0$.

Define the first arc as

$$L_1 = \{(x_1, 0, y_1', t y_3'') : 0 \leq t \leq 1\}$$

with the image $\Gamma(L_1)$ joining the points (u_1, v_1) and $(-1, v_3)$ strictly below the point (u_0, v_0) due to the choice of y_1'.

The second arc is defined as

$$L_2 = \{(x_t, y_3(x_t), y_3'(x_t), y_3''(x_t)) : 1 \leq t \leq 2\},$$
$$x_t = x_1 + (t-1)(x_2 - x_1) \quad \text{for } 1 \leq t \leq 2,$$
$$x_2 = x_* - 1.$$

The image $\Gamma(L_2)$ coincides with the point $(-1, v_3)$ since L_2 consists of initial data at various points of the same oscillatory solution $y_3(x)$.

The third arc is defined as

$$L_3 = \{(x_2, \tau_t y_3(x_2), \tau_t y_3'(x_2), \tau_t y_3''(x_2)) : 2 \leq t \leq 3\},$$
$$\tau_t = 3 - t \quad \text{for } 2 \leq t \leq 3.$$

Its image $\Gamma(L_3)$ joins the point $(-1, v_3)$ with the point $(-1, 1)$ corresponding to the trivial solution and passes strictly to the left of the point (u_0, v_0) due to the choice of $x_2 < x_*$.

Now put $x_3 = x^* + 1$ and define the next arc

$$L_4 = \{(x_2 + (t-3)(x_3 - x_2), 0, 0, 0) : 3 \leq t \leq 4\},$$

with the image $\Gamma(L_4)$ coinciding with the point $(-1, 1)$ since L_4 consists of initial data at various points of the same trivial solution.

Further, choose $y_4 < 0$ and $y_4'' < 0$ such that the solution $y_4(x)$ with initial data (x_1, y_4, y_1', y_4'') is a Kneser one. Due to the choice of y_1', the left boundary of its domain is to the right of the point x_*.

Define the fifth arc as

$$L_5 = \{(x_3, \tau_t y_4(x_3), \tau_t y_4'(x_3), \tau_t y_4''(x_3)) : 4 \leq t \leq 5\},$$
$$\tau_t = t - 4 \quad \text{for } 4 \leq t \leq 5.$$

Its image $\Gamma(L_4)$ joins the points $(-1, 1)$ and $(u_4, 1)$, passing strictly above the point (u_0, v_0) due to the choice $x_4 > x^*$.

The sixth arc is define as

$$L_6 = \{(x_t, y_4(x_t), y_4'(x_t), y_4''(x_t)) : 5 \leq t \leq 6\},$$
$$x_t = x_3 + (t-5)(x_1 - x_3) \quad \text{for } 5 \leq t \leq 6.$$

Since the arc L_6 consists of initial data at various points for the same Kneser solution $y_4(x)$, its image $\Gamma(L_6)$ coincides with the point $(u_4, 1)$ located above and to the right of the point (u_0, v_0), i. e., $u_4 > u_0$, $1 > v_0$.

Finally, define the seventh arc as

$$L_7 = \left\{ (x_1, (7-t)y_4, y_1', (7-t)y_4') : 6 \le t \le 7 \right\}.$$

Its image $\Gamma(L_7)$ joins the points $(u_4, 1)$ and (u_1, v_1) passing strictly to the right of the point (u_0, v_0) due to the choice of y_1.

The loop constracted is contractible in the space $\mathbb{R}^4$ and is mapped by Γ to another loop, which surrounds the point (u_0, v_0) and is contractible in the space $\Gamma(\mathbb{R}^4)$. This could not be possible if the point (u_0, v_0) did not belong to the image $\Gamma(\mathbb{R}^4)$. Thus, some point (x_0, y_0, y_0', y_0'') is mapped by Γ to the point (u_0, v_0), and this concludes the proof of the theorem. □

2.2.4   Proof of Corollaries and Theorem 6

Proof. The two corollaries are proved by limit considerations. A sequence of solutions with two asymptotes is constructed, the first coinciding with the given one and the second tending to the infinity needed. The sequence of the related initial data at some point is bounded. The limit of any of its subsequence provides the initial data of the solution requested. □

Since the zero solution is defined on the whole axis $(-\infty, +\infty)$, all above results may be combined as Theorem 6.

Acknowledgements The work was partially supported by the Russian Foundation for Basic Researches (Grant 11-01-00989) and by Special Program of the Ministery of Education and Science of the Russian Federation (Project 2.1.1/13250).

References

1. Astashova, I.V.: Application of dynamical systems to the study of asymptotic properties of solutions to nonlinear higher-order differential equations. J. Math. Sci. (Springer Science+Business Media) **126**(5), 1361–1391 (2005)
2. Astashova, I.V.: Uniform estimates of positive solutions to quasi-linear differential inequalities. J. Math. Sci. **143**(4), 3198–3204 (2007); Translated from Tr. Semin. Im. I. G. Petrovskogo **143**(26), 27–36 (2007)
3. Astashova, I.V.: Uniform estimates for positive solutions of quasi-linear ordinary differential equations. Izvestiya: Mathematics **72**(6),1141–1160 (2008) (translation)
4. Astashova, I.V.: Qualitative properties of solutions to quasilinear ordinary differential equations, in: Astashova I.V. (ed.) Qualitative Properties of Solutions to Differential Equations and Related Topics of Spectral Analysis: scientific edition, pp. 22-290. M.: UNITY-DANA (2012) (Russian)
5. Astashova, I.V.: Uniform estimates for solutions to the third order Emden–Fowler type autonomous differential equation. Funct. Differ. Equ. **18**, 55–64 (2011)
6. Astashova, I.V.: On uniform estimates of solutions to nonlinear third-order differential equations. Tr. Semin. Im. I. G. Petrovskogo **29**, 146–161 (2013) (Russian)

7. Kiguradze, I.: On blow-up Kneser solutions of higher order nonlinear differential equations. Differentsial'nye Uravneniya **37**(6), 735–743 (2001) (Russian); English translation: Differ. Equ. **37**(6), 768–777 (2001)
8. Kiguradze, I.T., Chanturia, T.A.: Asymptotic Properties of Solutions of Nonautonomous Ordinary Differential Equations. Kluver Academic Publishers, London (1993)
9. Kondratiev, V.A., Nikishkin, V.A.: On positive solutions of $y'' = p(x)y^k$. Some Problems of Qualitative Theory of Differential Equations and Motion Control, pp. 134–141. Saransk (1980) (Russian)

A Second-Order Difference Scheme for a Singularly Perturbed Reaction-Diffusion Problem

Basem S. Attili

Abstract We consider a singularly perturbed one-dimensional reaction-diffusion three-point boundary value problem. To approximate the solution numerically, we employ an exponentially fitted finite uniform difference scheme defined on a piecewise uniform Shishkin mesh which is second order and uniformly convergent independent of the perturbation parameter. We will present some numerical examples to show the efficiency of the proposed method.

Keywords Reaction-diffusion • Three-point BVPs • Finite difference • Singularly perturbed • Exponentially fitted scheme

1 Introduction

The problem under consideration is a singularly perturbed semilinear reaction-diffusion boundary value problem of the form

$$\varepsilon^2 y'' + \varepsilon f(x)y'(x) = g(x,y), \quad 0 < x < l, \tag{1}$$

$$\varepsilon y'(0) = \Psi(y(0)), \; y(l) = \Phi(y(l_1)), \quad 0 < l_1 < l, \tag{2}$$

where $0 < \varepsilon \ll 1$, $f(x) \geq 0$, $g(x,y)$, $\Psi(y)$, and $\Phi(y)$ are sufficiently smooth functions on their respective domains. Also, $0 < k_1 \leq \frac{\partial g}{\partial y} \leq k_2 < \infty$, $\frac{d\Psi}{dy} \geq k_3 > 0$, $\left| \frac{d\Phi}{dy} \right| \leq \kappa < 1$. This problem usually has boundary layers at the boundaries.

Singularly perturbed boundary value problems often arise in applied sciences and engineering; for example, see Nayfeh [9]. Among the examples are reaction-diffusion equations, control theory, and quantum mechanics (Shao [14] and Natesan

B.S. Attili (✉)
Department of Mathematics, Sharjah University, Sharjah, United Arab Emirates
e-mail: b.attili@sharjah.ac.ae

S. Pinelas et al. (eds.), *Differential and Difference Equations with Applications*, Springer Proceedings in Mathematics & Statistics 47, DOI 10.1007/978-1-4614-7333-6_17, © Springer Science+Business Media New York 2013

[7]). A well-known fact is that the solution of such problems displays sharp boundary or interior layers when the singular perturbation parameter ε is very small. Numerically, the presence of the perturbation parameter leads to difficulties when classical numerical techniques are used to solve such problems and convergence will not be uniform. This is due to the presence of boundary layers in these problems; see for example O'Malley [11]. Even in the case when only the approximate solution is required, finite difference schemes and finite element methods produced unsatisfactory results; see Samarski [13]. It was shown in [15] that the results of using classical methods are also unsatisfactory even when a very fine grid is used. This suggests having numerical methods where the error in the approximate solution tends to zero as $h \longrightarrow 0 (N \longrightarrow \infty)$ independently of the parameter ε; that is, uniform convergence is desired; see Attili [1, 2] and Kadalbajoo and Reddy [6]. Hence the primary objective in singular perturbation analysis of such problems is to develop asymptotic approximations to the true solution that are uniformly valid with respect to the perturbation parameter.

Fitted schemes that allow the use of the meshes with an arbitrary distribution of nodes can be found in Doolan et al. [5]. Special fitted schemes were used, and some kind of regularization of the singularity was suggested by others before applying special schemes of boundary value solvers; see Berger et al. [3], Caker and Amiraliyev [4] and Natesan et al. [8]. Three-point boundary value problem for singularly perturbed semilinear differential equations was considered by Vrabel [16]. More on the singularly perturbed problems can be found in a book by O'Malley [10] and Roos et al. [12].

The organization of this paper is as follows. We give some necessary bounds on solutions in the next section. The numerical scheme will be derived and presented in Sect. 3. Finally in Sect. 4 we give some numerical details and examples to illustrate the method.

2   Bounds on the Solution

Rewritting (1)–(2) with $f(x) = 0$ and replacing ε^2 by ε, in the form

$$\varepsilon y'' + m y(x) = g(x, y), \quad m < 0, \ 0 < x < l, y'(0) = 0, \ y(l) - y(l_1) = 0, \ 0 < l_1 < l, \tag{3}$$

we will show that the solution of the resulting nonlinear boundary value problem has an asymptotic behavior. The analysis is based on both the lower and upper solutions of the problem. The following lemma gives such uniform bounds that help in the analysis of the difference scheme which will be developed later.

Lemma 1. *Let* $f(x)$ *and* $g(x, y)$ *be sufficiently smooth. Then the solution* $y(x)$ *to problem (1)–(2) satisfies the inequality*

$$\|y(x)\|_\infty < M_0 \tag{4}$$

with $\|y(x)\|_\infty = \max_{0 \le x \le l} |y(x)|$ *where*

$$M_0 = \frac{1}{1-\kappa} \left\{ |\Phi(0)| + \kappa \left(\frac{1}{k_3} \left| \Psi(0) + \frac{1}{k_1} \|g(x,0)\| \right| \right) \right\}$$

and

$$|y'(x)| \le M \left\{ 1 + \frac{1}{\varepsilon} \left(e^{\frac{-\gamma_1 x}{\varepsilon}} + e^{\frac{-\gamma_2(L-x)}{\varepsilon}} \right) \right\}, \quad 0 \le x \le l$$

with

$$\gamma_1 = \sqrt{f^2(0) + 4k_1} + f(0) \ and \ \gamma_2 = \sqrt{f^2(l) + 4k_1} + f(l). \tag{5}$$

Proof. Rewrite problem (1)–(2) in the form

$$Ly = \varepsilon^2 y'' + \varepsilon f(x) y'(x) - r(x) y(x) = h(x), \quad 0 < x < l, \tag{6}$$

$$L_0 y = -\varepsilon y'(0) + c y(0) = \alpha, \quad y(l) - a y(l_1) = \beta, \quad 0 < l_1 < l, \tag{7}$$

where

$$h(x) = g(x,0), r(x) = \frac{\partial g}{\partial y}(x, cy), \quad 0 < c < 1$$

$$c = \frac{\partial \Psi}{\partial y}(\eta_1, y(0)), \ 0 < \eta_1 < 1, \ a = \frac{\partial \Phi}{\partial y}(\eta_2, y(0)), \ 0 < \eta_2 < 1.$$

Using the maximum principle, with L and L_0 as defined in (6) and (7) and $v(x) \in C^2[0, l]$, if $L_0 v \ge 0$, $Lv(l) \ge 0$, and $Lv \le 0$, then $v(x) \ge 0$ for all $x \in [0, l]$ and

$$|v(x)| \le \frac{1}{\kappa} |\alpha| + |v(l)| + \frac{1}{k_1} \|h(x)\|_\infty, \quad x \in [0, l]. \tag{8}$$

Applying this result on problem (6)–(7), we have

$$|y(x)| \le \frac{1}{\kappa} |\alpha| + |y(l)| + \frac{1}{k_1} \|h(x)\|_\infty$$

and

$$|y(l_1)| \le \frac{1}{\kappa} |\alpha| + |y(l)| + \frac{1}{k_1} \|h(x)\|_\infty, \quad x \in [0, l]. \tag{9}$$

From the boundary conditions, we have

$$|y(l)| \le |\beta| + \kappa |y(l_1)|. \tag{10}$$

Combining (9) and (10), we obtain

$$|y(l)| \leq \frac{1}{1-\kappa} \left\{ |\beta| + \kappa \left(\frac{1}{\kappa} |\alpha| + |y(l)| + \frac{1}{k_1} \|h(x)\|_\infty \right) \right\}. \qquad (11)$$

As a result and from (7) and (8), we get (4) the required result.   □

3 Numerical Descretization

We start by considering Δ_N to be any mesh on $[0, l]$ which may be nonuniform with $\Delta_N = \{x_0 = 0 < x_1 < x_2 < \cdots < x_{N-1} < x_N = l\}$ with step size $h_i = x_i - x_{i-1}$, $i = 1, 2, \ldots, N$ and $\|\Delta_N\|_\infty = \max_{0 \leq i \leq N} |h_i|$.

The difference scheme proposed is derived from the identity

$$\frac{1}{\alpha_i \hat{h}_i} \int_0^l (Ly)\, \phi_i(x)\, dx = 0, \quad i = 1, 2, \ldots, N-1, \qquad (12)$$

where $\phi_i(x)$ are some basis and $\hat{h}_i = \dfrac{h_i + h_{i+1}}{2}$. The basis functions are given as solutions to the following differential equations, with $f_i = f(x_i)$:

1.

$$\varepsilon \phi'' - f_i \phi' = 0; \quad x_{i-1} < x < x_i, \quad \phi(x_{i-1}) = 0, \quad \phi(x_i) = 1 \qquad (13)$$

which when solved leads to

$$\phi_{i1}(x) = \frac{e^{\frac{f_i(x-x_{i-1})}{\varepsilon}} - 1}{e^{\frac{f_i h_i}{\varepsilon}} - 1}; \ f_i \neq 0 \ \text{ and } \ \phi_{i1}(x) = \frac{x - x_{i-1}}{h_i}; \ f_i = 0. \qquad (14)$$

2.

$$\varepsilon \phi'' - f_i \phi' = 0; \quad x_i < x < x_{i+1} \quad \phi(x_i) = 1, \quad \phi(x_{i+1}) = 0 \qquad (15)$$

which when solved leads to

$$\phi_{i2}(x) = \frac{1 - e^{\frac{-f_i(x_{i+1}-x)}{\varepsilon}}}{1 - e^{\frac{-f_i h_{i+1}}{\varepsilon}}}; \ f_i \neq 0 \ \text{ and } \ \frac{x_{i+1} - x}{h_{i+1}}; \ f_i = 0. \qquad (16)$$

Hence the basis we are going to use are

$$\phi_i(x) = \begin{cases} \phi_{i1}(x); \ x_{i-1} < x < x_i \\ \phi_{i2}(x) \ \ x_i < x < x_{i+1} \\ 0 \qquad\qquad Otherwise \end{cases} \qquad (17)$$

for $i = 1, 2, \ldots, N-1$.

The coefficients α_i are given as $\alpha_i = \alpha_{i1} + \alpha_{i2}$ where

$$\alpha_{i1} = \frac{1}{\hat{h}_i} \int_{x_{i-1}}^{x_i} \phi_{i1}(x)\,dx = \begin{cases} \frac{1}{\hat{h}_i}\left(\frac{\varepsilon}{f_i} + \frac{h_i}{1-e^{\frac{f_i h_i}{\varepsilon}}}\right); & f_i \neq 0 \\ \frac{1}{\hat{h}_i}\frac{h_i}{2} & f_i = 0 \end{cases} \tag{18}$$

$$\alpha_{i2} = \frac{1}{\hat{h}_i} \int_{x_i}^{x_{i+1}} \phi_{i2}(x)\,dx = \begin{cases} \frac{1}{\hat{h}_i}\left(\frac{h_{i+1}}{1-e^{\frac{-f_i h_{i+1}}{\varepsilon}}} - \frac{\varepsilon}{f_i}\right); & f_i \neq 0 \\ \frac{1}{\hat{h}_i}\frac{h_{i+1}}{2} & f_i = 0 \end{cases}. \tag{19}$$

Then substituting in (12), using integration by parts, and rearranging, we obtain for $i = 1, 2, \ldots N - 1$:

$$\frac{-\varepsilon^2}{\alpha_i \hat{h}_i} \int_{x_{i-1}}^{x_{i+1}} \phi_i'(x)\,y'(x)dx + \frac{\varepsilon f_i}{\alpha_i \hat{h}_i} \int_{x_{i-1}}^{x_{i+1}} \phi_i(x)\,y'(x)dx - g(x_i, y_i) + \tau_i = 0. \tag{20}$$

Notice that the first two parts of (20), the integrals, need evaluation, while the last part τ_i is the truncation error and can be neglected in the proposed difference scheme as usual. This truncation error is given as

$$\tau_i = \frac{\varepsilon}{\alpha_i \hat{h}_i} \int_{x_{i-1}}^{x_{i+1}} [f_i - f(x_i)]\,\phi_i(x)\,y'(x)dx$$

$$- \frac{1}{\alpha_i \hat{h}_i} \int_{x_{i-1}}^{x_{i+1}} dx.\phi_i(x) \int_{x_{i-1}}^{x_{i+1}} \frac{d}{dx}\left(g(\sigma, y(\sigma))\right) R_i(x, \sigma)\,dx, \tag{21}$$

where $R_i(x, \sigma) = T(x - \sigma) - T(x_i - \sigma)$ and $T(x - \sigma) = \begin{cases} 1; & x \geq \sigma \\ 0; & x < \sigma \end{cases}$.

The first two integrals simplify to

$$\frac{-\varepsilon^2}{\alpha_i \hat{h}_i} \int_{x_{i-1}}^{x_{i+1}} \phi_i'(x)\,y'(x)dx + \frac{\varepsilon f_i}{\alpha_i \hat{h}_i}\bigg|_{x_{i-1}}^{x_{i+1}} \phi_i(x)\,y'(x)dx$$

$$= \varepsilon^2 \left\{ \frac{1}{\alpha_i}\left[1 + \frac{\hat{h}_i f_i}{2\varepsilon}(\alpha_{2i} - \alpha_{1i})\right]\right\} y_{xx,i} + \varepsilon f_i y_{xa,i}, \tag{22}$$

where α_{1i} and α_{2i} are as given in (18) and (19), respectively,

$$y_{xx,i} = \frac{1}{\hat{h}_i}\left[\frac{y_{i+1} - y_i}{h_{i+1}} - \frac{y_{i+1} - y_i}{h_i}\right] \text{ and } y_{xa,i} = \frac{1}{2}\left[\frac{y_{i+1} - y_i}{h_{i+1}} + \frac{y_{i+1} - y_i}{h_i}\right]. \tag{23}$$

Having the formulas for α_i, α_{1i}, and α_{2i}, then the coefficient of $y_{xx,i}$ in (22) can be written as

$$c_i = \frac{1}{\alpha_i}\left[1 + \frac{\hat{h}_i f_i}{2\varepsilon}\left(\alpha_{2i} - \alpha_{1i}\right)\right] = \begin{cases} \dfrac{\hat{h}_i f_i}{2\varepsilon}\left[\dfrac{h_{i+1}\left(e^{\frac{f_i h_i}{\varepsilon}} - 1\right) + h_i\left(1 - e^{\frac{-f_i h_i}{\varepsilon}}\right)}{h_{i+1}\left(e^{\frac{f_i h/}{\varepsilon}} - 1\right) - h_i\left(1 - e^{\frac{-f_i h_i}{\varepsilon}}\right)}\right] ; & f_i \neq 0 \\ 1; & f_i = 0 \end{cases}.$$

(24)

Combining (22)–(24), we obtain for $i = 1, 2, \ldots N - 1$

$$\varepsilon^2 c_i y_{xx,i} + + \varepsilon f_i y_{xa,i} - g\left(x_i, y_i\right) + \tau_i = 0.$$

(25)

It remains to consider the boundary conditions. For the first part, consider

$$\int_0^{x_1} (Lu)\,\phi_0 dx = 0$$

(26)

with ϕ_0, as before, the solution of the second-order differential equation:

$$\varepsilon\phi_0'' - f_0\phi_0' = 0, \quad x_0 < x < x_1 \quad \phi_0(x_0) = 1, \quad \phi_0(x_1) = 0.$$

(27)

That is, for $x_0 < x < x_1$,

$$\phi_0 = \begin{cases} \dfrac{1 - e^{\frac{-f_i(x_1 - x)}{\varepsilon}}}{1 - e^{\frac{-f_i h_1}{\varepsilon}}}; & f_i \neq 0 \\ \dfrac{x_1 - x}{h_1}; & f_i = 0 \\ 0; & Otherwise \end{cases}.$$

(28)

Again substituting into the differential equation and simplifying as we have done in (22), we obtain

$$-\varepsilon c_{0,0}\frac{y_0 - y_1}{h_1} + \Psi\left(y_0\right) + c_{0,1}g\left(x_0, y_0\right) - \tau_0 = 0,$$

(29)

where

$$c_{0,0} = \begin{cases} \dfrac{f_0 h_1}{\varepsilon\left(1 - e^{\frac{-f_0 h_1}{\varepsilon}}\right)}; & f_0 \neq 0 \\ 1; & f_0 = 0 \end{cases}, \quad c_{0,1} = \begin{cases} \dfrac{f_0 h_1}{\varepsilon\left(1 - e^{\frac{-f_0 h_1}{\varepsilon}}\right)} - \dfrac{1}{f_0}; & f_0 \neq 0 \\ \dfrac{h_1}{2\varepsilon}; & f_0 = 0 \end{cases}$$

(30)

and the truncation error

$$\tau_0 = \int_{x_0}^{x_1} [f(x_0) - f_0]\,\phi_0(x)\,y'(x)dx$$

$$-\frac{1}{\varepsilon}\int_{x_{i-1}}^{x_{i+1}} dx.\phi_0(x)_{x_0}^{x_1}\frac{d}{dx}\left(g\left(\sigma, y(\sigma)\right)\right)R_0(x, \sigma)dx, \quad \sigma\varepsilon(x_0, x_1).$$

(31)

For the second part of the boundary conditions, we have x_{N_0} the closest to l_1, $y_N - \Phi\left(y_{N_0}\right) + \tau_1 = 0$ where the truncation error $\tau_1 = \left(y\left(l_1\right) - y\left(x_{N_0}\right)\right)\phi'(\eta)$; η between $y\left(x_N\right)$ and $y\left(l_1\right)$.

Combining (25)–(27) and neglecting the truncation errors, we arrive at the difference scheme given as

$$\varepsilon^2 c_i y_{xx,i} + \varepsilon f_i y_{xa,i} - g\left(x_i, y_i\right) = 0$$

$$-\varepsilon c_{0,0}\frac{y_0 - y_1}{h_1} + \Psi\left(y_0\right) + c_{0,1}g\left(x_0, y_0\right) = 0$$

$$y_N - \Phi\left(y_{N_0}\right) = 0, \quad i = 1, 2, \ldots, N-1, \qquad (32)$$

where c_i, $c_{0,0}$, and $c_{0,1}$ are as given in (24) and (30).

4   Numerical Details and Examples

To implement the scheme given in (32), we start by subdividing the interval $[0, l]$ into three regions, namely, $[0, a_1]$, $[a_1, l - a_2]$, and $[a_2, l]$. For the mesh choice and in order to have an ε−uniform convergent scheme, we will use the Shishkin mesh. Choose $N_1 = 4N$ a positive integer and divide the intervals $[0, a_1]$ and $[a_2, l]$ into $N = \dfrac{N_1}{4}$ equal parts and the interval $[a_1, l - a_2]$ into $2N$ equal parts with a_1 and a_2 given, respectively, by

$$a_1 = \min\left\{\frac{l}{4}, \frac{\varepsilon \ln N}{\gamma_1}\right\} \quad \text{and} \quad a_2 = \min\left\{\frac{l}{4}, \frac{\varepsilon \ln N}{\gamma_2}\right\} \qquad (33)$$

with γ_1 and γ_2 as given in Lemma 1 by (10)

For numerical testing, we used the following examples:

Example 1. We solved the example

$$\varepsilon^2 y'' + \varepsilon\left(1 + \cos \pi x\right)y' - \left(1 + \sin\left(\frac{\pi x}{2}\right)\right)y$$

$$= 2\left(\varepsilon \pi\right)^2 \cos 2\pi x + \varepsilon \pi\left(1 + \cos \pi x\right)\sin 2\pi x - \left(1 + \sin\left(\frac{\pi x}{2}\right)\right)\sin^2 2\pi x$$

$$y(0) = 0, \quad y(1) - 0.5y(0.5) = 1. \qquad (34)$$

The numerical scheme converges and the results obtained are shown in Figs. 1 and 2. Figure 1 gives the solution on the interval $[0, 1]$ and Fig. 2 gives the solution on $[0.9, 1.0]$ to show the boundary layer area.

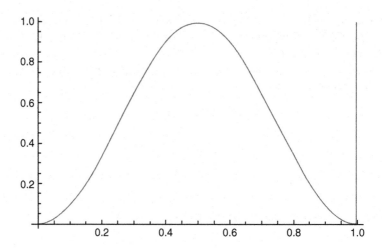

Fig. 1 Example 1, the solution on $[0,1]$

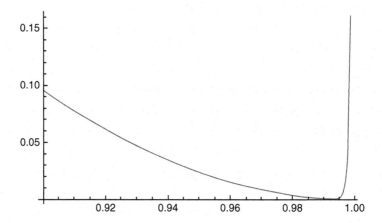

Fig. 2 Example 1 where the boundary layer is enlarged

Example 2.

$$\varepsilon^2 y'' + \varepsilon (1+x) y' - y - \arctan (y+x) = 0$$

subject to

$$-\varepsilon y'(0) + 2y(0) + \sin y(0) = 0, \quad y(1) - \cos \frac{\pi y(0.5)}{4} = 1. \qquad (35)$$

The results obtained are given in Fig. 3 where the boundary layer is clear and close to 1.

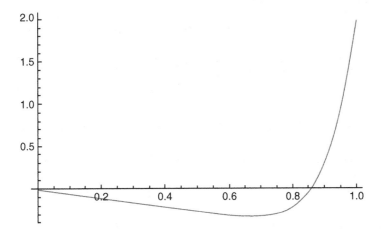

Fig. 3 Example 2 showing the boundary layer

References

1. Attili, B.S.: A numerical algorithm for some singularly perturbed boundary value problems. J. Comput. Appl. Math. **184**, 467–474 (2005)
2. Attili, B.S.: Numerical treatment of singularly perturbed two point boundary value problems exhibiting boundary layers. Commun. Nonlin. Sci. Numer. Simulat. **16**, 3504–3511 (2011)
3. Berger, A., Han, H., Kellogg R.: A priori estimates and analysis of a numerical method for a turning point problem. Math. Comp. **42**, 465–492 (1984)
4. Cakır, M., Amiraliyev, G.M.: Numerical solution of a singularly perturbed three-point boundary value problem. Int. J. Comput. Math. **84**, 1465—1481 (2007)
5. Doolan, E.P., Miller, J.H., Schilders W.A.: Uniform Numerical Methods for Problems With Initial and Boundary Layers. Boole Press, Dublin (1980)
6. Kadalbajoo, M.K., Reddy, Y.N.: A boundary value method for a class of nonlinear singular perturbation problems. Commun. Appl. Numer. Method **4**, 587–594 (1988)
7. Natesan, S., Bawa, R.: Second-order numerical scheme for singularly perturbed reaction diffusion robin problems. J. Numer. Anal. Indus. Appl. Math. **3–4**, 177–192 (2007)
8. Natesan, S., Jayakumar, J., Vigo-Aguiar, J.: Parameter uniform numerical method for singularly perturbed turning point problems exhibiting boundary layers. J. Comput. Appl. Math. **158**, 121–134 (2003)
9. Nayfeh, A.H.: Introduction to Perturbation Techniques. Wiley, New York (1993)
10. O'Malley Jr., R.E.: Introduction to Singular Perturbation. Academic, New York (1974)
11. O'Malley Jr., R.E.: Singular perturbation methods for ordinary differential equations. Applied Mathematical Sciences, vol. 89. Springer, New York (1991)
12. Roos, H.-G., Stynes, M., Tobiska L.: Robust Numerical Methods for Singularly Perturbed Differential Equations: Convection-Diffusion and Flow Problems. Springer, Berlin (2008)
13. Samarski, A.A.: Theory of Difference Schemes. Nauka, Moscow (1980)
14. Shao, S.: Asymptotic analysis and domain decomposition for a singularly perturbed reaction–convection–diffusion system with shock–interior layer interactions. Nonlinear Anal. **66**, 271–287 (2007)

15. Shishkin, G.I.: A finite difference scheme on a priori adapted meshes for a singularly perturbed parabolicconvection-diffusion equation. Numer. Math. Theor. Methods Appl. **1**, 214–234 (2008)
16. Vrabel, R., Three point boundary value problem for singularly perturbed semilinear differential equations. Electron. J. Qualit. Theo. Diff. Equ. **70**, 1–4 (2009). http://www.math.u-szeged.hu/ejqtde/

Parametric Dependence of Boundary Trace Inequalities

Giles Auchmuty

Abstract Some results about the dependence of the optimal constants in some trace inequalities for H^1-functions on a region Ω are described. These constants are shown to be the primary Steklov eigenvalue of $\mu I - \Delta$ on the region. They are related to the norm of an associated trace operator. In particular the eigenvalue is shown to be a locally Lipschitz continuous function of μ, and its inverse is a convex function of μ.

Keywords Robin eigenproblems • Steklov eigenproblems • Bases of Sobolev spaces • Comparisons of eigenvalues • Spectral representations of weak solutions

1 Introduction

This paper will derive some qualitative properties of the best constant $\delta_1(\mu)$ for the H^1-trace inequality:

$$\mathcal{A}(u,\mu) := \int_\Omega \left[|\nabla u|^2 + \mu |u|^2 \right] dx \geq \delta_1(\mu) \int_{\partial\Omega} \rho\, u^2\, d\sigma \quad \text{for all} \quad u \in H^1(\Omega). \tag{1}$$

Here Ω is a bounded region in $\mathbb{R}^N$ with a boundary $\partial\Omega$ obeying the conditions described below, ρ is a continuous probability density function on $\partial\Omega$, and $\mu \geq 0$. Our particular aim is to prove regularity and convexity/concavity properties of $\delta_1(\mu)$ as a function of μ.

G. Auchmuty (✉)
Department of Mathematics, University of Houston, Houston, TX 77204-3008, USA
e-mail: auchmuty@uh.edu

S. Pinelas et al. (eds.), *Differential and Difference Equations with Applications*, Springer Proceedings in Mathematics & Statistics 47, DOI 10.1007/978-1-4614-7333-6_18,
© Springer Science+Business Media New York 2013

Let $L^p(\Omega), H^1(\Omega)$ be the usual real Lebesgue and Sobolev spaces of functions on Ω. The norm on $L^p(\Omega)$ is denoted $\|.\|_p$. $H^1(\Omega)$ is a real Hilbert space under the standard H^1-inner product:

$$[u,v]_1 := \int_\Omega [u(x).v(x) + \nabla u(x) \cdot \nabla v(x)] \, dx. \tag{2}$$

All derivatives here will be weak derivatives, ∇u is the gradient of the function u, and the associated norm is denoted $\|u\|_{1,2}$.

We will treat these as problems posed in the space $H^1(\Omega)$ and require some regularity of the boundary $\partial\Omega$. Namely, the trace results of Auchmuty [2] should hold. These can be summarized as follows:

(B1): Ω is a bounded region in $\mathbb{R}^N$ and its boundary $\partial\Omega$ is the union of a finite number of disjoint closed Lipschitz surfaces, each surface having finite surface area.

The region Ω is said to satisfy *Rellich–Kondrachov (RK) theorem* provided the imbedding of $H^1(\Omega)$ into $L^p(\Omega)$ is compact for $1 \le p < p_S$ for $p_S = 2N/(N-2)$ when $N \ge 3$ or $p_S = \infty$ when $N = 2$.

The region Ω is said to satisfy the L^2-*compact trace theorem* provided the trace map of $H^1(\Omega)$ into $L^2(\partial\Omega, d\sigma)$ is compact. Our standard assumption will be

(B2): Ω is a region such that (B1), the RK theorem, and the L^2-compact trace theorem hold.

The notation of Evans [6] should be used for terms not defined here except that a function is said to be positive if it is ≥ 0 everywhere; strictly positive it is strictly greater than zero.

The assumption on the boundary weight function ρ will be

(B3): ρ is a continuous probability density function on the boundary $\partial\Omega$.

That is, ρ is a continuous positive function on $\partial\Omega$ with $\int_{\partial\Omega} \rho \, d\sigma = 1$. Let γ be the usual trace map of $H^1(\Omega)$ into $L^2(\partial\Omega, \rho d\sigma)$. When (1) holds and is sharp, one observes that the operator norm $\|\gamma\| = \delta_1(\mu)^{-1}$.

2 Steklov Eigenproblems

A Steklov eigenproblem is one where the eigenparameter appears solely in the boundary condition. The associated Steklov eigenfunctions provide fundamental information about the solution of Dirichlet-type boundary value problems for associated linear elliptic operators on the region. They also may be used to describe boundary trace operators and the Dirichlet to Neumann map of the problem. See Auchmuty [1–3] for descriptions of such results.

Here our interest is in the H^1-Steklov eigenproblem of finding those δ such that there are nontrivial solutions $s \in H^1(\Omega)$ satisfying

$$a(s,v;\mu) := \int_\Omega [\nabla s \cdot \nabla v + \mu s v] \, dx = \delta \int_{\partial\Omega} \rho s v \, d\sigma \quad \text{for all} \quad v \in H^1(\Omega).$$
(3)

Here $\mu \geq 0$; the case $\mu = 0$ is called the harmonic Steklov eigenproblem and has been extensively studied. See Bandle [4], Auchmuty [2], and their references for more information. For $\mu > 0$, the bilinear form $a(.,.;\mu)$ is an equivalent inner product on $H^1(\Omega)$ to the usual one. The nontrivial solutions s of (3) are called Steklov eigenfunctions of the linear operator $L_\mu := \mu I - \Delta$ associated with the Steklov eigenvalue δ.

Note that (3) is the weak form of the system:

$$L_\mu s = \mu s - \Delta s = 0 \quad \text{on } \Omega \text{ and } \quad D_\nu s = \delta \rho s \quad \text{on } \partial\Omega. \quad (4)$$

Here $D_\nu s := \nabla s \cdot \nu$ is the usual normal component of the gradient of s in the exterior normal direction.

Eigenproblems like these were studied in [1] where it was shown that they have a discrete spectrum with infinitely many positive eigenvalues $\delta_j(\mu)$ that diverge to ∞ as j increases. Each eigenvalue has finite multiplicity.

The smallest or primary, eigenvalue $\delta_1(\mu)$ may be characterized variationally in a number of different ways. First it is related to the maximal value of a quadratic boundary functional on a closed convex subset of $H^1(\Omega)$. Define C_μ to be the subset of $H^1(\Omega)$ of all functions satisfying $\mathcal{A}(u,\mu) \leq 1$ and consider the variational problem of maximizing the functional $b : H^1(\Omega) \to [0,\infty)$ defined by

$$b(u) := \int_{\partial\Omega} \rho u^2 \, d\sigma$$

and evaluating

$$\beta(\mu) := \sup_{u \in C_\mu} b(u).$$

Results about this problem may be summarized as follows.

Theorem 2.1. *Assume (B1)–(B3) hold and $\mu > 0$. Then $\beta(\mu)$ is finite, and strictly positive and there are functions $\pm s_1(\mu)$ in C_μ such that $s_1(\mu)$ maximizes b on C_μ. Inequality (1) holds with $\delta_1(\mu) := \beta(\mu)^{-1}$.*

Proof. First note that when $\mu > 0, \mathcal{A}(.,\mu)$ defined by (1) is an equivalent norm on $H^1(\Omega)$ so C_μ is a closed bounded convex set. The functional b is weakly continuous as the trace map is compact. Hence the supremum $\beta(\mu)$ is finite, strictly positive, and attained on C_μ. By homogeneity, one sees that

$$b(u) \leq \beta(\mu) \mathcal{A}(u,\mu) \quad \text{for all} \quad u \in H^1(\Omega).$$

The definition of $\delta_1(\mu)$ now yields (1). $\qquad\square$

In particular take $u \equiv 1$ in (1) to see that $\delta_1(\mu) \leq \mu|\Omega|$.

Theorem 2.2. *Assume (B1)–(B3) hold and $\mu > 0$. If $s_1(\mu)$ maximizes b on C_μ, then $s_1(\mu)$ is a Steklov eigenfunction of L_μ corresponding to the Steklov eigenvalue $\delta_1(\mu)$.*

Proof. This follows from Theorem 3.1 in [3] applied to this problem. In particular Sect. 8 of that paper proves many further results for such Steklov eigenproblems. □

When $v \equiv 1$ is substituted here, one sees that

$$\mu \int_\Omega s_1(\mu)\,dx = \delta_1(\mu) \int_{\partial\Omega} \rho\, s_1(\mu)\,d\sigma \tag{5}$$

which provides a useful formula for $\delta_1(\mu)$.

This characterization of the maximizer as a weak solution of (4) implies that each $s_1(\mu)$ is actually C^∞ on Ω. It is continuous on $\overline{\Omega}$ from Corollary 4.2 of Daners [5]. So $D_v s_1(\mu)$ is continuous on the boundary $\partial\Omega$ when (B3) holds.

Given $u \in H^1(\Omega)$, define $u_+(x) = \max(u(x),0)$ and $u_-(x) = -\min(u(x),0)$. Then u_+, u_- are in $H^1(\Omega)$, and their derivatives are given pointwise by well-known formulae. The following result helps in evaluating the primary Steklov eigenvalue of L_μ.

Theorem 2.3. *Assume (B1)–(B3) hold and $s_1(\mu)$ is a Steklov eigenfunction of L_μ corresponding to the primary Steklov eigenvalue $\delta_1(\mu)$. Then $|s_1(\mu)|$ is also a Steklov eigenfunction corresponding to $\delta_1(\mu)$. If $s_1(\mu)$ changes sign on $\overline{\Omega}$, then both $s_{1+}(\mu), s_{1-}(\mu)$ are Steklov eigenfunctions corresponding to $\delta_1(\mu)$.*

Proof. When $u \in H^1(\Omega)$ then $|u| \in H^1(\Omega)$ and $\mathcal{A}(|u|,\mu) = \mathcal{A}(u,\mu)$, $b(|u|) = b(u)$. Thus, if $s_1(\mu)$ maximizes b on C_μ, so does $|s_1(\mu)|$.

Suppose that s_1 changes sign so that both $s_{1+}(\mu), s_{1-}(\mu)$ are nonzero functions in $H^1(\Omega)$. Then using the essential disjointness of their supports, one sees that

$$\mathcal{A}(s_1,\mu) = \mathcal{A}(s_{1+}(\mu),\mu) + \mathcal{A}(s_{1-}(\mu),\mu) \geq \delta_1(\mu)\,[b(s_{1+}(\mu)) + b(s_{1-}(\mu))]$$

from the fact that (1) holds. The first and last terms here are both equal to $\delta_1(\mu)b(s_1)$, so equality must hold throughout. Thus,

$$\mathcal{A}(s_{1+}(\mu),\mu) = \delta_1(\mu)b(s_{1+}(\mu)) \quad \text{and} \quad \mathcal{A}(s_{1-}(\mu),\mu) = \delta_1(\mu)b(s_{1-}(\mu)).$$

Hence both $s_{1+}(\mu), s_{1-}(\mu)$ are eigenfunctions of (3) corresponding to $\delta_1(\mu)$. □

It would be interesting to know whether there are Steklov eigenfunctions corresponding to the eigenvalue $\delta_1(\mu)$ with $\mu > 0$ that change sign. In this case the primary eigenvalue will not be a simple eigenvalue.

3 Unconstrained Variational Principles for $\delta_1(\mu)$

Since the left-hand side of the inequality (1) is an increasing function of μ, it is obvious that the optimal constant $\delta_1(\mu)$ will be an increasing positive function when $\mu \geq 0$. Some better and different results may be proved by using an unconstrained variational characterization of the Steklov eigenvalues and eigenfunctions.
Define the functional $\mathcal{F} : H^1(\Omega) \times [0, \infty) \to \mathbb{R}$ by

$$\mathcal{F}(u, \mu) := \frac{1}{2} \int_\Omega \left[|\nabla u|^2 + \mu |u|^2 \right] dx - \left[\int_{\partial\Omega} \rho \, u^2 \, d\sigma \right]^{1/2}. \qquad (6)$$

Consider the problem of minimizing $\mathcal{F}(., \mu)$ on $H^1(\Omega)$ and finding

$$\alpha(\mu) := \inf_{u \in H^1(\Omega)} \mathcal{F}(u, \mu).$$

The essential results about this unconstrained variational principle may be summarized as follows.

Theorem 3.1. *Assume (B1)–(B3) and $\mu > 0$. Then $\alpha(\mu)$ is finite and there are minimizers $\pm \hat{u}(\mu)$ of $\mathcal{F}(., \mu)$ on $H^1(\Omega)$. They satisfy the equation*

$$\int_\Omega \left[\nabla u \cdot \nabla v + \mu u v \right] dx = \delta \int_{\partial\Omega} \rho \, u v \, d\sigma \quad \text{for all} \quad v \in H^1(\Omega) \qquad (7)$$

with $\delta = b(\hat{u}(\mu))^{-1/2}$. Thus $\hat{u}(\mu) = [\delta_1(\mu)]^{-1/2} s_1$ where s_1 is a maximizer of b on C_μ and the minimal value $\alpha(\mu) = -\frac{1}{2}\beta(\mu)$.

Proof. The functions $\mathcal{A}(., \mu)$ and b are weakly l.s.c and weakly continuous on $H^1(\Omega)$, respectively, so $\mathcal{F}(., \mu)$ is weakly l.s.c. There are constants c_0, c_1, strictly positive, such that

$$\mathcal{F}(u, \mu) \geq c_0 \|u\|_{1,2}^2 - c_1 \|u\|_{1,2} \quad \text{for all} \quad u \in H^1(\Omega)$$

with $c_0 = \min(1, \mu)$ and c_1 depending on $\|\rho\|_\infty$. Thus, $\mathcal{F}(., \mu)$ is coercive and attains a finite infimum on $H^1(\Omega)$.
The first variation of $\mathcal{F}(., \mu)$ at u is

$$\delta \mathcal{F}(u, v; \mu) = a(u, v; \mu) - b(u)^{-1/2} \int_{\partial\Omega} \rho \, u v \, d\sigma.$$

At a minimizer $\hat{u}(\mu)$, this is zero for all $v \in H^1(\Omega)$ so (7) holds with δ as indicated. Put $v = \hat{u}(\mu)$ here then $\mathcal{A}(\hat{u}(\mu), \mu) = b(\hat{u}(\mu))^{1/2} = \delta^{-1}$. Thus, the critical values of this functional are $-1/[2\delta(\mu)]$. This value is minimized at the primary Steklov eigenvalue and then $\hat{u}(\mu) = c s_1(\mu)$ with $c^2 \delta_1(\mu) = 1$. The other claims of the theorem now follow. $\square$

The proof of this result shows that the nonzero critical points of $\mathcal{F}(.,\mu)$ on $H^1(\Omega)$ are Steklov eigenfunctions of the operator L_μ. Since the minimizer of $\mathcal{F}(.,\mu)$ on $H^1(\Omega)$ corresponds to a Steklov eigenfunction s_1 of L_μ for which the corresponding Steklov eigenvalue is the least, we have the following properties of $\alpha(\mu)$ and $\delta_1(\mu)$.

Theorem 3.2. *Assume (B1)–(B3) and $\mu > 0$. Then:*

(i) $\alpha(\mu)$ *is a strictly negative and strictly increasing, concave function of μ.*
(ii) $\beta(\mu)$ *is strictly positive, strictly decreasing, and convex function on $(0,\infty)$.*
(iii) $\delta_1(\mu)$ *is a strictly positive, strictly increasing, and sublinear function of μ on $(0,\infty)$.*

$\alpha(\mu)$, $\beta(\mu)$, *and* $\delta_1(\mu)$ *are locally Lipschitz continuous functions on* $(0,\infty)$.

Proof. The functional $\mathcal{F}(u,\mu)$ is an affine function of μ for each u so $\alpha(\mu)$ is a concave functional on $(0,\infty)$. Since $\delta_1(\mu) > 0$ for all $\mu > 0$, the last part of the preceding theorem implies that $\alpha(\mu)$ is strictly negative. It is strictly increasing as each $\mathcal{F}(u,.)$ with u nonzero is strictly increasing. Finally α is locally Lipschitz as it is concave and finite on $(0,\infty)$.

From the previous theorem $\beta(\mu) = -2\alpha(\mu)$ so the results for β follow from those for part (i).

The formula $\delta_1(\mu) = -1/(2\alpha(\mu))$ yields the corresponding results for $\delta_1(\mu)$. The inequality $\delta_1(\mu) \le \mu|\Omega|$ described above shows that δ_1 is sublinear. $\square$

This result suggests that it is likely to be preferable to work with $\beta(\mu)$ rather than $\delta_1(\mu)$ in many situations since it is a convex function of μ.

It would be of interest to improve these results about the asymptotics of α and δ_1 as $\mu \to \infty$. In particular do the asymptotics of these quantities reflect geometric properties of the region Ω?

Acknowledgements This research was partially supported by NSF awards DMS 0808115 and 1108754.

References

1. Auchmuty, G.: Steklov eigenproblems and the representation of solutions of elliptic boundary value problems. Numer. Func. Anal. Opt. **25**, 321–348 (2004)
2. Auchmuty, G.: spectral characterizations of the trace spaces $H^s(\partial\Omega)$. SIAM J. Math. Anal. **38**, 894–905 (2006)
3. Auchmuty, G.: Bases and comparison results for linear elliptic eigenproblems. J. Math. Anal. Appl. **383**, 25–34 (2011)
4. Bandle, C.: Isoperimetric Inequalities and Applications. Pitman, London (1980)
5. Daners, D.: Inverse positivity for general Robin problems on Lipschitz domains. Trans. Amer. Math. Soc. **352**, 4207–4236 (2000)
6. Evans, L.C.: Partial Differential Equations. American Mathematical Society, Providence (1991)

On Properties of Third-Order Functional Differential Equations

Blanka Baculíková and Jozef Džurina

Abstract The objective of this paper is to offer sufficient conditions for all nonoscillatory solutions of the third-order functional differential equation

$$\left[a(t)\left[x'(t)\right]^{\gamma}\right]'' + p(t)x^{\beta}(\tau(t)) = 0$$

tend to zero. Our results are based on the new comparison theorems. Studied equation is in a canonical form, i.e., $\int^{\infty} a^{-1/\gamma}(s)\,ds = \infty$, and we consider both delay and advanced case of it. The results obtained essentially improve and complement earlier ones.

Keywords Third-order differential equations • Comparison theorem • Oscillation • Nonoscillation

1 Introduction

We are concerned with the oscillatory and asymptotic behavior of all solutions of the third-order functional differential equations:

$$\left[a(t)\left[x'(t)\right]^{\gamma}\right]'' + p(t)x^{\beta}(\tau(t)) = 0. \qquad (E)$$

In the sequel, we will assume $a, p \in C([t_0, \infty))$, $\tau \in C^1([t_0, \infty))$ and

(H1) γ, β are the ratios of two positive odd integers,
(H2) $a(t) > 0$, $p(t) > 0$, $\tau'(t) > 0$, $\lim\limits_{t \to \infty} \tau(t) = \infty$.

B. Baculíková (✉) • J. Džurina
Department of Mathematics, Technical University of Košice, Letná 9,
042 00 Košice, Slovakia, Europe
e-mail: blanka.baculikova@tuke.sk; jozef.dzurina@tuke.sk

S. Pinelas et al. (eds.), *Differential and Difference Equations with Applications*, Springer
Proceedings in Mathematics & Statistics 47, DOI 10.1007/978-1-4614-7333-6_19,
© Springer Science+Business Media New York 2013

Throughout the paper, we assume that (E) is in a canonical form, i.e.,

$$R(t) = \int_{t_0}^{t} a^{-1/\gamma}(s) \, ds \to \infty \quad \text{as} \quad t \to \infty.$$

By a solution of (E) we mean a function $x(t) \in C^1[T_x, \infty)$, $T_x \geq t_0$, which has the property $a(t)(x'(t))^\gamma \in C^2([T_x, \infty))$ and satisfies (E) on $[T_x, \infty)$. We consider only those solutions $x(t)$ of (E) which satisfy $\sup\{|x(t)| : t \geq T\} > 0$ for all $T \geq T_x$. We assume that (E) possesses such a solution. A solution of (E) is called oscillatory if it has arbitrarily large zeros on $[T_x, \infty)$ and otherwise it is called to be nonoscillatory. Equation (E) is said to be oscillatory if all its solutions are oscillatory.

Recently, (E) and its particular cases (see enclosed references) have been intensively studied. Various techniques were established for examination of such equations. Especially comparison theorems seem to be very effective means by the reason that they can reduce examination of third-order differential equation to that of lower-order equations. In the papers [1–4, 7, 9] the authors compared studied equation with a set of the first-order delay/advanced equation, in the sense that oscillation of these first-order equations yields desired properties of third-order equation.

In this paper we shall establish new comparison principles, we compare our third-order equation with the second-order differential inequality, and this reduction essentially simplifies investigation of the properties of our equation, forasmuch as further we deal with the second-order inequality.

Our results complement and extend earlier ones presented in [2–12].

Remark 1. All functional inequalities considered in this chapter are assumed to hold eventually; that is, they are satisfied for all t large enough.

2 Main Results

We begin with the classification of the possible nonoscillatory solutions of (E).

Lemma 1. *Let $x(t)$ be a nonoscillatory solution of (E). Then $x(t)$ satisfies, eventually, one of the following conditions:*

(C_1) $x(t)x'(t) < 0$, $x(t)\left[a(t)\left[x'(t)\right]^\gamma\right]' > 0$, $x(t)\left[a(t)\left[x'(t)\right]^\gamma\right]'' < 0$.

(C_2) $x(t)x'(t) > 0$, $x(t)\left[a(t)\left[x'(t)\right]^\gamma\right]' > 0$, $x(t)\left[a(t)\left[x'(t)\right]^\gamma\right]'' < 0$.

Proof. The proof follows immediately from the canonical form of (E).   □

To simplify formulation of our main results, we recall the following definition:

Definition 1. We say that (E) enjoys property (A) if every nonoscillatory solution satisfies (C_1).

Property (A) of (E) has been studied by various author; see enclosed references. We offer new technique for investigation property (A) of (E) based on comparison theorems.

Remark 2. It is known that condition

$$\int_{t_0}^{\infty} p(s)\,ds = \infty \tag{1}$$

implies property (A) of (E). Consequently, in the sequel, we may assume that the integral in (1) is convergent.

Now, we offer a comparison result in which we reduce property (A) of (E) to the absence of certain positive solution of the suitable second-order inequality.

Theorem 1. *If for some $c \in (0,1)$ the second-order differential inequality*

$$\left(\frac{1}{p^{1/\beta}(t)} \left(z'(t) \right)^{1/\beta} \right)' + c\frac{\tau'(t)\tau^{1/\gamma}(t)}{a^{1/\gamma}(\tau(t))} z^{1/\gamma}(\tau(t)) \leq 0 \tag{E_1}$$

has not any solution satisfying

$$z(t) > 0, \quad z'(t) < 0, \quad \left(\frac{1}{p^{1/\beta}(t)} \left(z'(t) \right)^{1/\beta} \right)' < 0, \tag{P_1}$$

then (E) has property (A).

Proof. Assume the contrary: let $x(t)$ be a nonoscillatory solution of (E), satisfying (C_2). We may assume that $x(t) > 0$ for $t \geq t_0$. Using the monotonicity of $\left[a(t) \left[x'(t) \right]^{\gamma} \right]'$, we see that

$$a(t) \left[x'(t) \right]^{\gamma} \geq \int_{t_1}^{t} \left[a(s) \left[x'(s) \right]^{\gamma} \right]'\,ds \geq \left[a(t) \left[x'(t) \right]^{\gamma} \right]' (t - t_1)$$

$$\geq c^{\gamma} t \left[a(t) \left[x'(t) \right]^{\gamma} \right]',$$

eventually, where $c \in (0,1)$ arbitrary. Then evaluating $x'(t)$ and integrating from t_1 to t, we are lead to

$$x(t) \geq c \int_{t_1}^{t} \frac{s^{1/\gamma}}{a^{1/\gamma}(s)} \left(\left[a(s) \left[x'(s) \right]^{\gamma} \right]' \right)^{1/\gamma}\,ds. \tag{2}$$

Setting to (E), we get

$$\left[a(t) \left[x'(t) \right]^{\gamma} \right]'' + c^{\beta} p(t) \left[\int_{t_1}^{\tau(t)} \frac{s^{1/\gamma}}{a^{1/\gamma}(s)} \left(\left[a(s) \left[x'(s) \right]^{\gamma} \right]' \right)^{1/\gamma}\,ds \right]^{\beta} \leq 0.$$

Integrating from t to ∞, we see that $y(t) = \left[a(t)\left[x'(t)\right]^{\gamma}\right]'$ satisfies

$$y(t) \geq c^{\beta} \int_t^{\infty} p(s) \left[\int_{t_1}^{\tau(s)} \frac{u^{1/\gamma}}{a^{1/\gamma}(u)} y^{1/\gamma}(u)\,du\right]^{\beta} ds. \tag{3}$$

Let us denote the right-hand side of (13) by $z(t)$. Then $z(t)$ satisfies (P_1), and moreover,

$$\left(\frac{1}{p^{1/\beta}(t)}\left(z'(t)\right)^{1/\beta}\right)' + c\,\frac{\tau'(t)\tau^{1/\gamma}(t)}{a^{1/\gamma}(\tau(t))}y^{1/\gamma}(\tau(t)) = 0.$$

Consequently, $z(t)$ is a solution of the differential inequality (E_1), which contradicts our assumption.   □

Now we establish, some criteria for elimination of solutions of (E_1) satisfying (P_1) to obtain sufficient conditions for property (A) of (E). We present these criteria in general form and then we adapt them for (E_1). We consider the noncanonical differential inequality:

$$\left(q(t)\left(u'(t)\right)^{\alpha}\right)' + b(t)u^{\delta}(\sigma(t)) \leq 0, \tag{E_2}$$

where

(H3) α, δ are the ratios of two positive odd integers.
(H4) $q(t) > 0$, $b(t) > 0$, $\sigma'(t) > 0$, $\lim_{t\to\infty} \sigma(t) = \infty$.

Let us denote

$$\rho(t) = \int_t^{\infty} q^{-1/\alpha}(s)\,ds.$$

Theorem 2. *Assume that $\delta > \alpha$ and $\sigma(t) \geq t$. If*

$$\int_{t_0}^{\infty} \rho^{\delta}(\sigma(s))b(s)\,ds = \infty, \tag{4}$$

then (E_2) has not any solution satisfying

$$u(t) > 0, \quad u'(t) < 0, \quad \left(q(t)\left(u'(t)\right)^{\alpha}\right)' < 0. \tag{5}$$

Proof. Let $u(t)$ be a positive solution of (E_2), such that (5) holds. An integration of (E_2) from t_1 to t leads to

$$-u'(t) \geq \frac{1}{q^{1/\alpha}(t)}\left[\int_{t_1}^{t} b(s)u^{\delta}(\sigma(s))\,ds\right]^{1/\alpha}.$$

Integrating again from $\sigma(t)$ to ∞, we obtain

$$u(\sigma(t)) \geq \int_{\sigma(t)}^{\infty} \frac{1}{q^{1/\alpha}(v)} \left[\int_{t_1}^{v} b(s) u^{\delta}(\sigma(s)) \, ds \right]^{1/\alpha} dv$$

$$\geq \int_{\sigma(t)}^{\infty} \frac{1}{q^{1/\alpha}(v)} \, dv \left[\int_{t_1}^{t} b(s) u^{\delta}(\sigma(s)) \, ds \right]^{1/\alpha}.$$

That is,

$$u^{\delta}(\sigma(t)) \geq \rho^{\delta}(\sigma(t)) \left[\int_{t_1}^{t} b(s) u^{\delta}(\sigma(s)) \, ds \right]^{\delta/\alpha}.$$

Let us denote

$$F(t) = \int_{t_1}^{t} b(s) u^{\delta}(\sigma(s)) \, ds.$$

Then

$$\frac{u^{\delta}(\sigma(t))}{F^{\delta/\alpha}(t)} b(t) \geq \rho^{\delta}(\sigma(t)) b(t).$$

Integrating from $t_2 > t_1$ to t, we have

$$\int_{t_2}^{t} \rho^{\delta}(\sigma(s)) b(s) \, ds \leq \frac{F^{1-\delta/\alpha}(t)}{1 - \delta/\alpha} - \frac{F^{1-\delta/\alpha}(t_2)}{1 - \delta/\alpha} \leq \frac{F^{1-\delta/\alpha}(t_2)}{\delta/\alpha - 1}.$$

Letting t to ∞, we get a contradiction with (4). This finishes our proof. □

Now, we are prepared to combine Theorem 1 together with Theorem 2. We set $\alpha = 1/\beta$, $\delta = 1/\gamma$, $\sigma(t) = \tau(t)$, $q(t) = p^{-1/\beta}(t)$, and $b(t) = c \dfrac{\tau'(t) \tau^{1/\gamma}(t)}{a^{1/\gamma}(\tau(t))}$. Then it is easy to check that $\rho(t) = \rho_1(t) = \displaystyle\int_{t}^{\infty} p(s) \, ds$.

Theorem 3. Let $\beta > \gamma$, $\tau(t) \geq t$. If

$$\int_{t_0}^{\infty} \left(\frac{s \rho_1(s)}{a(s)} \right)^{1/\gamma} ds = \infty, \tag{6}$$

then (E) has property (A).

Proof. By Theorem 2, the condition

$$\int_{t_0}^{\infty} \rho_1^{1/\gamma}(\tau(s)) \frac{\tau'(s) \tau^{1/\gamma}(s)}{a^{1/\gamma}(\tau(s))} \, ds = \infty,$$

or simply (6) ensures that (E_1) has not any solution satisfying (P_1). The assertion now follows from Theorem 1. □

Corollary 1. *Assume that* (E) *enjoys property* (A). *If, moreover,*

$$\int_{t_0}^{\infty} \frac{1}{a^{1/\gamma}(v)} \left(\int_{v}^{\infty} \int_{u}^{\infty} p(s)\, ds\, du \right)^{1/\gamma} dv = \infty, \tag{7}$$

then every nonoscillatory solution of (E) *tends to zero as* $t \to \infty$.

Proof. Since (E) has property (A), every nonoscillatory solution $x(t)$ satisfies (C_1). We assume that $x(t)$ is a positive and it follows from (C_1) that there exists $\lim_{t \to \infty} x(t) = \ell \geq 0$. We claim that $\ell = 0$. Assume the contrary, i.e., $\ell > 0$. Integrating twice (E) from t to ∞, we obtain

$$-a(t)\left[x'(t)\right]^{\gamma} \geq \ell^{\beta} \int_{t}^{\infty} \int_{u}^{\infty} p(s)\, ds\, du.$$

Extracting $x'(t)$ and integrating once more from t_1 to ∞, we get

$$x(t_1) \geq \ell^{\beta/\gamma} \int_{t_1}^{\infty} \frac{1}{a^{1/\gamma}(v)} \left(\int_{v}^{\infty} \int_{u}^{\infty} p(s)\, ds\, du \right)^{1/\gamma} dv,$$

which contradicts condition (7). And we conclude that $\lim_{t \to \infty} x(t) = 0$. □

Example 1. Consider the third-order nonlinear advanced differential equation

$$\left(t\left(x'(t)\right)^{3} \right)'' + \frac{a}{t^2} x^5(\lambda t) = 0, \quad t \geq 1 \tag{E_{x1}}$$

with $a > 0$ and $\lambda \geq 1$. It is easy to check that both conditions (6) and (7) are fulfilled, and then Theorem 3 implies that (E_{x1}) enjoys property (A), and moreover Corollary 1 guarantees that every nonoscillatory solution of (E_{x1}) tends to zero as $t \to \infty$. For $a = 30\lambda^5$ one solution is $x(t) = 1/t$.

The criterion for property (A) presented in Theorem 3 has a very simple form, but it does not take into account the gap between $\tau(t)$ and t. If the difference $\tau(t) - t$ is large enough, we can offer another very simple criterion for deducing property (A). We start with the following auxiliary result:

Lemma 2. *Let* $\tau(t) > t$. *Assume that* $x(t)$ *satisfies* (C_2). *Then for any constant* $k \in (0,1)$, *it holds*

$$\left| x(\tau(t)) \right| \geq k \frac{R(\tau(t))}{R(t)} \left| x(t) \right|, \tag{8}$$

eventually.

Proof. Assume that $x(t) > 0$. The monotonicity of $w(t) = a(t)[x'(t)]^\gamma$ implies that

$$x(\tau(t)) - x(t) = \int_t^{\tau(t)} x'(s)\,ds = \int_t^{\tau(t)} w^{1/\gamma}(s)a^{-1/\gamma}(s)\,ds$$

$$\geq w^{1/\gamma}(t) \int_t^{\tau(t)} a^{-1/\gamma}(s)\,ds = w^{1/\gamma}(t)\Big[R(\tau(t)) - R(t)\Big].$$

That is,

$$\frac{x(\tau(t))}{x(t)} \geq 1 + \frac{w^{1/\gamma}(t)}{x(t)}\Big[R(\tau(t)) - R(t)\Big]. \tag{9}$$

On the other hand, since $x(t) \to \infty$ as $t \to \infty$, then for any $k \in (0,1)$ there exists a t_1 large enough, such that

$$kx(t) \leq x(t) - x(t_1) = \int_{t_1}^t w^{1/\gamma}(s)a^{-1/\gamma}(s)\,ds$$

$$\leq w^{1/\gamma}(t) \int_{t_1}^t a^{-1/\gamma}(s)\,ds \leq w^{1/\gamma}(t)R(t)$$

or equivalently

$$\frac{w^{1/\gamma}(t)}{x(t)} \geq \frac{k}{R(t)}. \tag{10}$$

Using (10) in (9), we get

$$\frac{x(\tau(t))}{x(t)} \geq 1 + \frac{k}{R(t)}\Big[R(\tau(t)) - R(t)\Big] \geq k\frac{R(\tau(t))}{R(t)}.$$

This completes the proof. □

Theorem 4. *Let $\tau(t) > t$. If*

$$\int_{t_0}^\infty \frac{R^\beta(\tau(s))p(s)}{R^\beta(s)}\,ds = \infty, \tag{11}$$

then (E) has property (A).

Proof. Assume the contrary, let $x(t) > 0$ satisfies (C_2). Then $x(t) > \ell > 0$. Setting (8) into (E), we get

$$\Big[a(t)[x'(t)]^\gamma\Big]'' + k^\beta\, p(t)\frac{R^\beta(\tau(t))}{R^\beta(t)}x^\beta(t) \leq 0.$$

An integration from t_1 to t yields

$$\left[a(t_1)\left[x'(t_1)\right]^\gamma\right]' \geq (k\ell)^\beta \int_{t_1}^t \frac{R^\beta(\tau(s))p(s)}{R^\beta(s)}\,ds.$$

Letting $t \to \infty$, we are led to a contradiction. □

Example 2. Consider the third-order nonlinear advanced differential equation

$$\left(t\left(x'(t)\right)^3\right)'' + \frac{a}{t^2}x^5(t^2) = 0, \quad t \geq 1 \tag{E_{x2}}$$

with $a > 0$. It is easy to check that both conditions (11) and (7) are fulfilled. Thus, Theorem 4 together with Corollary 1 provides that every nonoscillatory solution of (E_{x2}) tends to zero as $t \to \infty$. For $a = \left(\frac{2}{7}\right)^3\left(\frac{20}{7}\right)\left(\frac{27}{7}\right)$, one such solution is $x(t) = t^{-2/7}$.

Now, we turn our attention to the delayed form of (E). Let us denote

$$p_1(t) = \frac{p\left(\tau^{-1}(t)\right)}{\tau'\left(\tau^{-1}(t)\right)}. \tag{12}$$

We are prepared to modify our previous results to cover also delay differential equations.

Theorem 5. *Let* $\tau(t) < t$. *If for some* $c \in (0,1)$ *the second-order differential inequality*

$$\left(\frac{1}{p_1^{1/\beta}(t)}\left(z'(t)\right)^{1/\beta}\right)' + c\frac{t^{1/\gamma}(t)}{a^{1/\gamma}(t)}z^{1/\gamma}(t) \leq 0 \tag{E_3}$$

has not any solution satisfying

$$z(t) > 0, \quad z'(t) < 0, \quad \left(\frac{1}{p_1^{1/\beta}(t)}\left(z'(t)\right)^{1/\beta}\right)' < 0, \tag{P_2}$$

then (E) *has property (A).*

Proof. Assume the contrary, let $x(t)$ be a positive solution of (E), satisfying (C_2). An integration of (E) from t to ∞ yields

$$\left[a(s)\left[x'(s)\right]^\gamma\right]' \geq \int_t^\infty p(s)x^\beta(\tau(s))\,ds = \int_{\tau(t)}^\infty \frac{p\left(\tau^{-1}(s)\right)}{\tau'\left(\tau^{-1}(s)\right)}x^\beta(s)\,ds$$

$$\geq \int_t^\infty \frac{p\left(\tau^{-1}(s)\right)}{\tau'\left(\tau^{-1}(s)\right)}x^\beta(s)\,ds.$$

Using (2), one can see that $y(t) = \left[a(t) \left[x'(t)\right]^\gamma\right]'$ satisfies

$$y(t) \geq c^\beta \int_t^\infty p_1(s) \left[\int_{t_1}^s \frac{u^{1/\gamma}}{a^{1/\gamma}(u)} y^{1/\gamma}(u) \, du\right]^\beta ds. \qquad (13)$$

Let us denote the right-hand side of (13) by $z(t)$. Then $z(t)$ satisfies (P_2), and moreover,

$$\left(\frac{1}{p_1^{1/\beta}(t)} \left(z'(t)\right)^{1/\beta}\right)' + c \frac{t^{1/\gamma}}{a^{1/\gamma}(t)} y^{1/\gamma}(t) = 0.$$

Therefore, $z(t)$ is a solution of the differential inequality (E_2), which contradicts our assumption. □

If we put $\alpha = 1/\beta$, $\delta = 1/\gamma$, $\sigma(t) = t$, $q(t) = p_1^{-1/\beta}(t)$, and $b(t) = c \frac{t^{1/\gamma}}{a^{1/\gamma}(t)}$, then it is easy to verify that $\rho(t) = \rho_2(t) = \int_t^\infty p_1(s) \, ds$. Combining Theorem 5 together with Theorem 2, we immediately have

Theorem 6. *Let $\beta > \gamma$, $\tau(t) < t$. If*

$$\int_{t_0}^\infty \left(\frac{s\rho_2(s)}{a(s)}\right)^{1/\gamma} ds = \infty, \qquad (14)$$

then (E) has property (A).

Example 3. Consider the third-order nonlinear delay differential equation

$$\left(t \left(x'(t)\right)^3\right)'' + \frac{a}{t^2} x^5(\lambda t) = 0, \quad t \geq 1, \qquad (E_{x3})$$

where $a > 0$ and $0 < \lambda < 1$. Since both conditions (14) and (7) hold, Theorem 6 together with Corollary 1 ensure that every nonoscillatory solution of (E_{x3}) tends to zero as $t \to \infty$. For $a = 30\lambda^5$ one solution is $x(t) = 1/t$.

3 Summary

In this paper, we have presented new comparison principles for deducing property (A) of third-order differential equation from the properties the suitable second-order delay differential inequality. Imposing additional condition, we have obtained also criteria for all nonoscillatory solutions of (E) tend to zero. Our results can be applied to both delay and advanced third-order differential equations. The criteria obtained are easy verifiable, and each of them has been precedented by suitable illustrative example.

Our method essentially simplifies the examination of the third-order equations, and what is more, it supports backward the research on the second-order delay/advanced differential equations and inequalities.

References

1. Agarwal, R.P., Shien, S.L., Yeh, C.C.: Oscillation criteria for second order retarded differential equations. Math. Comput. Modelling **26**, 1–11 (1997)
2. Agarwal, R.P., Grace, S.R., O'Regan, D.: On the oscillation of certain functional differential equations via comparison methods. J. Math. Anal. Appl. **286**, 577–600 (2003)
3. Agarwal, R.P., Grace, S. R., Smith, T.: Oscillation of certain third order functional differential equations. Adv. Math. Sci. Appl. **16**, 69–94 (2006)
4. Baculíková, B., Džurina, J.: Oscillation of third-order neutral differential equations. Math. Comput. Modelling **52**, 215–226 (2010)
5. Cecchi, M., Došlá, Z., Marini, M.: On third order differential equations with property A and B. J. Math. Anal. Appl. **231**, 509–525 (1999)
6. Džurina, J.: Asymptotic properties of third order delay differential equations. Czech. Math. J. **45**, 443–448 (1995)
7. Džurina, J., Baculíková, B. : Oscillation of third-order differential equations with mixed arguments. In: Proceedings in Mathematics (2012)
8. Erbe, L., Kong, H.Q., Zhang, B.G.: Oscillation Theory for Functional Differential Equations. Dekker, New York (1994)
9. Grace, S.R., Agarwal, R.P., Pavani, R., Thandapani, E.: On the oscillation of certain third order nonlinear functional differential equations. Appl. Math. Comp. **202**, 102–112 (2008)
10. Györi, I., Ladas, G.: Oscillation Theory of Delay with Applications. Clarendon Press, Oxford (1991)
11. Kusano, T., Naito, M.: Comparison theorems for functional differential equations with deviating arguments. J. Math. Soc. Japan **3**, 509–533 (1981)
12. Ladde, G. S., Lakshmikantham, V., Zhang, B.G.: Oscillation Theory of Differential Equations with Deviating Arguments. Dekker, New York (1987)

On the Structure of Two-Layer Cellular Neural Networks

Jung-Chao Ban, Chih-Hung Chang, and Song-Sun Lin

Abstract Let $\mathbf{Y} \subseteq \{-1,1\}^{\mathbb{Z}_{\infty} \times 2}$ be the mosaic solution space of a two-layer cellular neural network (TCNN). We decouple $\mathbf{Y}$ into two subspaces, say $Y^{(1)}$ and $Y^{(2)}$, and give a necessary and sufficient condition for the existence of factor maps between them. In such a case, $Y^{(i)}$ is a sofic shift for $i = 1, 2$. This investigation is equivalent to study the existence of factor maps between two sofic shifts. Moreover, we investigate whether $Y^{(1)}$ and $Y^{(2)}$ are topological conjugate, strongly shift equivalent, shift equivalent, or finitely equivalent via the well-developed theory in symbolic dynamical systems. This clarifies, in a TCNN, each layer's structure.

1 Introduction

Two-layer cellular neural networks (TCNNs) are large aggregates of analogue circuits presenting themselves as arrays of identical cells which are locally coupled. TCNNs have been widely applied in studying the signal propagation between

J.-C. Ban (✉)
Department of Applied Mathematics, National Dong Hwa University,
Hualien, 97401, Taiwan, R.O.C.
e-mail: jcban@mail.ndhu.edu.tw

C.-H. Chang
Department of Applied Mathematics, Feng Chia University, Taichung 40724, Taiwan, R.O.C.
e-mail: chihhung@mail.fcu.edu.tw

S.-S. Lin
Department of Applied Mathematics, National Chiao-Tung University,
Hsinchu 30050, Taiwan, R.O.C.
e-mail: sslin@math.nctu.edu.tw

S. Pinelas et al. (eds.), *Differential and Difference Equations with Applications*, Springer
Proceedings in Mathematics & Statistics 47, DOI 10.1007/978-1-4614-7333-6_20,
© Springer Science+Business Media New York 2013

neurons and in image processing, pattern recognition, and information technology [3–5, 12–14]. A one-dimensional TCNN is realized in the following form:

$$
\begin{cases}
\dfrac{dx_i^{(1)}}{dt} = -x_i^{(1)} + \sum_{|k| \le d} a_k^{(1)} y_{i+k}^{(1)} + \sum_{|k| \le d} b_k^{(1)} u_{i+k}^{(1)} + z^{(1)}, \\[4mm]
\dfrac{dx_i^{(2)}}{dt} = -x_i^{(2)} + \sum_{|k| \le d} a_k^{(2)} y_{i+k}^{(2)} + \sum_{|k| \le d} b_k^{(2)} u_{i+k}^{(2)} + z^{(2)},
\end{cases}
\tag{1}
$$

for some $d \in \mathbb{N}, i \in \mathbb{Z}$, where $u_i^{(2)} = y_i^{(1)}, u_i^{(1)} = u_i, x_i(0) = x_i^0$, and $y = f(x) = \frac{1}{2}(|x + 1| - |x - 1|)$ are the output function. For $1 \le \ell \le 2$, $A^{(\ell)} = (a_{-d}^{(\ell)}, \cdots, a_d^{(\ell)})$ is called the feedback template, $B^{(\ell)} = (b_{-d}^{(\ell)}, \cdots, b_d^{(\ell)})$ is called the controlling template, and $z^{(\ell)}$ is the threshold. The quantity $x_i^{(\ell)}$ denotes the state of a cell C_i in the ℓth layer. The stationary solutions $\bar{x} = (\bar{x}_i^{(\ell)})$ of (1) are essential for understanding the system, and their outputs $\bar{y}_i^{(\ell)} = f(\bar{x}_i^{(\ell)})$ are called *output patterns*. Among the stationary solutions, the mosaic solutions are crucial for studying the complexity of (1) [7–10]. A *mosaic solution* $(\bar{x}_i^{(\ell)})$ satisfies $|\bar{x}_i^{(\ell)}| > 1$ for all i, ℓ, and the output of a mosaic solution is called a mosaic output pattern. In a TCNN system, the "status" of each cell is taken as an input for a cell in the next layer except for those cells in the second layer. The results that can be recorded are the output of the cells in the second layer. Since the phenomena that can be observed are only the output patterns of the second layer, the second layer of (1) is called the *output layer*, while the first layer is called *hidden layer*.

Juang and Lin [7] and Ban et al. [2] investigated mosaic solutions systematically and characterized the complexity of mosaic patterns via topological entropy. In the present study, a pattern stands for a stationary solution for (1). Since the feedback and controlling templates are spatially invariant, the global pattern formation is thus completely determined by the so-called admissible local patterns. Hence, investigation of admissible local patterns is essential for studying the complexity of global patterns. The difficulty stems from the fact that the set of admissible local patterns is constrained by the differential equation (1). Suppose $\mathscr{B}$ is a basic set of admissible local patterns. The predicament is that there exists a subset of $\{-1, 1\}^{\mathbb{Z}(2d+1) \times 2}$ that cannot be realized via TCNNs. Such a constraint arises from the so-called *linear separation property*. Hsu et al. [6] demonstrated that, for one-layer CNNs without input, the parameter space can be divided into a finite number of partitions such that any two sets of parameters in the same partition admit the same basic set of admissible local patterns. This property remains true for TCNNs [1,2]. Proposition 2.1 gives a brief introduction to the procedure for determining the partitions of the parameter space of simplified TCNNs.

Suppose $\mathbf{Y}$ is the solution space of a TCNN. For $\ell = 1, 2$, let $Y^{(\ell)} = \{\cdots y_{-1}^{(\ell)} y_0^{(\ell)} y_1^{(\ell)} \cdots\}$ be the space which consists of patterns in the ℓth layer of $\mathbf{Y}$. Then $Y^{(2)}$ is called the *output space* and $Y^{(1)}$ is called the *hidden space*. There is a canonical projection $\phi^{(\ell)} : \mathbf{Y} \to Y^{(\ell)}$ for each ℓ. It is natural to ask whether there exists a relation between $Y^{(1)}$ and $Y^{(2)}$. The existence of map connecting $Y^{(1)}$ and $Y^{(2)}$ that commutes with $\phi^{(1)}$ and $\phi^{(2)}$ means the decoupling of the solution space $\mathbf{Y}$. More precisely, if there exists $\pi_{12} : Y^{(1)} \to Y^{(2)}$ such that $\pi_{12} \circ \phi^{(1)} = \phi^{(2)}$, then π_{12} enables the investigation of structures between the output space and hidden space.

Ban et al. [2] demonstrated that the output space $Y^{(2)}$ is a one-dimensional sofic shift. An analogous argument asserts that the hidden space $Y^{(1)}$ is also a sofic shift. To study the existence of $\pi_{12} : Y^{(1)} \to Y^{(2)}$ is equivalent to illustrate the existence between two sofic shifts. This elucidation gives a systematic strategy for determining whether there exists a map between $Y^{(1)}$ and $Y^{(2)}$ via well-developed theory in symbolic dynamical systems. Readers are referred to [11] for more details.

The following section analyzes simplified TCNNs. Some discussion and conclusions are given in Sect. 3.

2 Simplified Two-Layer Cellular Neural Networks

A simplified TCNN is realized as the following:

$$
\begin{cases}
\dfrac{dx_i^{(1)}}{dt} = -x_i^{(1)} + a^{(1)} y_i^{(1)} + a_r^{(1)} y_{i+1}^{(1)} + z^{(1)}, \\[2mm]
\dfrac{dx_i^{(2)}}{dt} = -x_i^{(2)} + a^{(2)} y_i^{(2)} + a_r^{(2)} y_{i+1}^{(2)} + b^{(2)} u_i^{(2)} + b_r^{(2)} u_{i+1}^{(2)} + z^{(2)}.
\end{cases}
\tag{2}
$$

Suppose $\mathbf{y} = \begin{pmatrix} \cdots y_{-1}^{(2)} y_0^{(2)} y_1^{(2)} \cdots \\ \cdots y_{-1}^{(1)} y_0^{(1)} y_1^{(1)} \cdots \end{pmatrix}$ is a mosaic pattern. For $i \in \mathbb{Z}$, $y_i^{(1)} = 1$ if and only if $x_i^{(1)} > 1$. This derives

$$
a^{(1)} + z^{(1)} - 1 > -a_r^{(1)} y_{i+1}^{(1)}.
\tag{3}
$$

Similarly, $y_i^{(1)} = -1$ if and only if $x_i^{(1)} < -1$. This implies $y_i^{(1)} = -1$ if and only if

$$
a^{(1)} - z^{(1)} - 1 > a_r^{(1)} y_{i+1}^{(1)}.
\tag{4}
$$

The same argument asserts

$$
a^{(2)} + z^{(2)} - 1 > -a_r^{(2)} y_{i+1}^{(2)} - (b^{(2)} u_i^{(2)} + b_r^{(2)} u_{i+1}^{(2)}),
\tag{5}
$$

and

$$
a^{(2)} - z^{(2)} - 1 > a_r^{(2)} y_{i+1}^{(2)} + (b^{(2)} u_i^{(2)} + b_r^{(2)} u_{i+1}^{(2)})
\tag{6}
$$

are the necessary and sufficient condition for $y_i^{(2)} = -1$ and $y_i^{(2)} = 1$, respectively. Note that the quantity $u_i^{(2)}$ in (5) and (6) satisfies $|u_i^{(2)}| = 1$ for each i. Define $\xi_1 :$ $\{-1,1\} \to \mathbb{R}$ and $\xi_2 : \{-1,1\}^{\mathbb{Z}_{3\times 1}} \to \mathbb{R}$ by $\xi_1(w) = a_r^{(1)}w$ and $\xi_2(w_1,w_2,w_3) = a_r^{(2)}w_1 + b^{(2)}w_2 + b_r^{(2)}w_3$, respectively. Let $\mathscr{B}^{(1)}, \mathscr{B}^{(2)}$ represent the basic sets of admissible local patterns of the first and second layer of (2). The set of admissible local patterns $\mathscr{B}$ of (2) is then

$$\mathscr{B} = \left\{ \boxed{\begin{array}{c} yy_r \\ uu_r \end{array}} : \boxed{\begin{array}{c} yy_r \\ uu_r \end{array}} \in \mathscr{B}^{(2)} \text{ and } \boxed{uu_r} \in \mathscr{B}^{(1)} \right\}.$$

Since we only consider mosaic patterns, $a^{(1)} + z^{(1)} - 1 = -\xi_1(y_r^{(1)})$ and $a^{(1)} + z^{(1)} - 1 = \xi_1(y_r^{(1)})$ partition $a^{(1)} - z^{(1)}$ plane into 9 regions, and $a^{(2)} + z^{(2)} - 1 > -\xi_2(y_r^{(2)}, u^{(2)}, u_r^{(2)})$ and $a^{(2)} + z^{(2)} - 1 > \xi_2(y_r^{(2)}, u^{(2)}, u_r^{(2)})$ partition $a^{(2)} - z^{(2)}$ plane into 81 regions. The "order" of lines $a^{(1)} + z^{(1)} - 1 = (-1)^\ell \xi_1(y_r^{(1)})$, $\ell = 0, 1$, comes from the sign of $a_r^{(1)}$. Thus, the parameter space $\{(a^{(1)}, a_r^{(1)}, z^{(1)})\}$ is partitioned into $2 \times 9 = 18$ regions. Similarly, the parameter space $\{(a^{(2)}, a_r^{(2)}, b^{(2)}, b_r^{(2)}, z^{(2)})\}$ is partitioned into $8 \times 6 \times 2 \times 81 = 7,776$ regions. Each region associates a basic set of admissible local patterns. This indicates the following proposition:

Proposition 2.1. *Let $\mathscr{P}^8 = \{(a^{(1)}, a_r^{(1)}, a^{(2)}, a_r^{(2)}, b^{(2)}, b_r^{(2)}, z^{(1)}, z^{(2)})\}$ be the parameter space of (2). There exists $139,968$ regions in $\mathscr{P}$ such that any two sets of templates that locate in the same region infer the same basic set of admissible local patterns. Conversely, suppose $\mathscr{B} \subseteq \{-1,1\}^{\mathbb{Z}_{2\times 2}}$ comes from a simplified TCNN. Then there exists a partition that admits $\mathscr{B}$ as its basic set of admissible local patterns.*

2.1 Ordering Matrix, Transition Matrix, and Graph

Proposition 2.1 demonstrates that each partition of the parameter space associates with a collection of local patterns that allow for generalization of global patterns. Hence the basic set of admissible local patterns plays an essential role for investigating TCNNs. This section studies the structure of admissible local patterns through defining the ordering for each pattern.

Substitute mosaic patterns -1 and 1 as symbols $-$ and $+$, respectively. Define the ordering matrix $\mathbb{X} = (x_{pq})_{1 \leq p,q \leq 4}$ as a 4×4 matrix consisting of all possible choice of 2×2 patterns. Suppose that $\mathscr{B}$ is given. The transition matrix $T \equiv T(\mathscr{B})$ is a $0 - 1$ matrix defined by $T(p,q) = 1$ if and only if $x_{pq} \in \mathscr{B}$. A directed graph G consists of a pair of two finite sets $\mathscr{V}$ and $\mathscr{E}$, where $\mathscr{V}$ is the vertex set and $\mathscr{E}$ is the edge set. For an edge $e \in \mathscr{E}$, we sometimes denote e by $e \equiv (i(e), t(e))$ for specificity. Here $i(e), t(e) \in \mathscr{V}$ are the initial and the terminal states of e. For each transition matrix there associates a directed graph $G_T = (\mathscr{V}, \mathscr{E})$. It is well known that a graph G_T can induce a shift space X_{G_T} called a shift of finite type, where X_{G_T} is defined by

$$X_{G_T} = \{\xi = (\xi_j)_{j\in\mathbb{Z}} \in \mathscr{V}^{\mathbb{Z}} : \exists e_j \in \mathscr{E} \text{ such that } i(e_j) = \xi_j, t(e_j) = \xi_{j+1} \forall j \in \mathbb{Z}\} \quad (7)$$

According to the definition, each bi-infinite sequence in X_{G_T} describes a bi-infinite walk on G_T.

Let $\mathbf{Y} = \left\{ \begin{pmatrix} y_i \\ u_i \end{pmatrix}_{i\in\mathbb{Z}} : \boxed{\begin{matrix} y_i y_{i+1} \\ u_i u_{i+1} \end{matrix}} \in \mathscr{B} \text{ for } i \in \mathbb{Z} \right\}$ be the solution space of (2). For

ease of notation, denote $\boxed{\begin{matrix} y_1 y_2 \\ u_1 u_2 \end{matrix}}$ by $y_1 y_2 \diamond u_1 u_2$ and

$$\mathbf{y} \diamond \mathbf{u} \equiv \boxed{\begin{matrix} \mathbf{y} \\ \mathbf{u} \end{matrix}} = \frac{\cdots y_{-2} y_{-1} y_0 y_1 y_2 \cdots}{\cdots u_{-2} u_{-1} u_0 u_1 u_2 \cdots}, \quad \text{where} \quad \mathbf{y} = (y_i)_{i\in\mathbb{Z}}, \mathbf{u} = (u_i)_{i\in\mathbb{Z}}.$$

Define $\phi^{(1)}, \phi^{(2)} : \mathbf{Y} \to \{-,+\}^{\mathbb{Z}}$ by $\phi^{(1)}(\mathbf{y} \diamond \mathbf{u}) = \mathbf{u}$ and $\phi^{(2)}(\mathbf{y} \diamond \mathbf{u}) = \mathbf{y}$. Set $Y^{(\ell)} = \phi^{(\ell)}(\mathbf{Y})$ for $\ell = 1, 2$. $Y^{(1)}$ is called the *hidden space*, and $Y^{(2)}$ is called the *output space*. Obviously the dynamical behavior of the output space $Y^{(2)}$ is influenced by the hidden space $Y^{(1)}$. For instance, a phenomenon which cannot be seen in one-layer CNNs is that $Y^{(1)}$ would break the symmetry of the entropy diagram of $Y^{(2)}$ [2]. This motivates the study of the relation between $Y^{(1)}$ and $Y^{(2)}$.

2.2 Labeled Graph and Symbolic Transition Matrix

Ban et al. [2] show that $Y^{(1)}, Y^{(2)}$ are sofic shifts. The difference between a shift of finite type and a sofic shift is that a shift of finite type comes from a directed graph, while a sofic shift comes from a so-called *labeled graph*.

Definition 2.2. Suppose $G = (\mathscr{V}, \mathscr{E})$ is a directed graph with vertices $\mathscr{V}$ and edges $\mathscr{E}$ and $\mathscr{A}$ is a finite alphabet. A labeled graph $\mathscr{G}$ is a pair $(G, \mathscr{L})$, and the labeling $\mathscr{L} : \mathscr{E} \to \mathscr{A}$ assigns to each edge e of G a label $\mathscr{L}(e) \in \mathscr{A}$. The underlying graph of $\mathscr{G}$ is G.

Suppose $\mathscr{G} = (G, \mathscr{L})$ is a labeled graph. The shift space $X_{\mathscr{G}}$ is called a sofic shift. Moreover, we say that $\mathscr{G}$ is *right-resolving* if $\mathscr{L}((v, w)) \neq \mathscr{L}((v, w'))$ for $v \in \mathscr{V}$ and $(v, w), (v, w') \in \mathscr{E}$.

Assume that $G_T = (\mathscr{V}, \mathscr{E})$ is the graph representation of (2). Let $\mathscr{A} = \{\alpha_0, \alpha_1, \alpha_2, \alpha_3\} = \{--, -+, +-, ++\}$. Define $\mathscr{L}^{(1)}, \mathscr{L}^{(2)} : \mathscr{E} \to \mathscr{A}$ by

$$\mathscr{L}^{(1)}(e) = \alpha_{2\tau(i(e)) + \tau(t(e))}, \quad \text{where} \quad \tau(c) := c \mod 2 \quad (8)$$

$$\mathscr{L}^{(2)}(e) = \alpha_{2[i(e)/2] + [t(e)/2]}, \quad \text{where} \quad [\cdot] \text{ is the Gauss function.} \quad (9)$$

These two labeling $\mathscr{L}^{(1)}$ and $\mathscr{L}^{(2)}$ define two labeled graphs $\mathscr{G}^{(1)} = (G_T, \mathscr{L}^{(1)})$ and $\mathscr{G}^{(2)} = (G_T, \mathscr{L}^{(2)})$, respectively. Ban et al. [2] demonstrated that $Y^{(\ell)}$ is a sofic shift and $Y^{(\ell)} = X_{\mathscr{G}^{(\ell)}}$ for $\ell = 1, 2$.

It is seen that $(\mathscr{L}^{(\ell)})_\infty$ is conjugate to $\phi^{(\ell)}$, where $(\mathscr{L}^{(\ell)})_\infty : X_{G_T} \to X_{\mathscr{G}^{(\ell)}}$ is defined by $(\mathscr{L}^{(\ell)})_\infty(\xi)_j = \mathscr{L}^{(\ell)}((\xi_j, \xi_{j+1}))$ for $j \in \mathbb{Z}$. The symbolic transition matrix $S^{(\ell)}$ of $\mathscr{G}^{(\ell)}$ is defined by

$$S^{(\ell)}(p,q) = \begin{cases} \alpha_j, & \text{if } T^{(\ell)}(p,q) = 1 \text{ and } \mathscr{L}^{(\ell)}((p,q)) = \alpha_j \text{ for some } j; \\ \varnothing, & \text{otherwise.} \end{cases} \tag{10}$$

Herein $\varnothing$ means there exists no local pattern in $\mathscr{B}$ related to its corresponding entry in the ordering matrix.

2.3 Classification of Hidden and Output Spaces

To investigate whether there is a relation between $Y^{(1)}$ and $Y^{(2)}$ turns out that topological entropy provides some evidence for the existence of the map $\pi : Y^{(1)} \to Y^{(2)}$ (or $\pi' : Y^{(2)} \to Y^{(1)}$). For the rest of this section, we assume that $h(Y^{(1)}) = h(Y^{(2)})$ unless otherwise stated, where $h(X)$ indicates the topological entropy of X.

Suppose $\mathscr{G}^{(1)}$ and $\mathscr{G}^{(2)}$ are both right-resolving. First we consider a relation between two shift spaces called *finite equivalence*.

Two shift spaces X and Y are *finitely equivalent*, denoted by $X \sim_{\mathscr{F}} Y$, if there exists a shift of finite-type W together with finite-to-one factor maps $\phi_X : W \to X$ and $\phi_Y : W \to Y$. We say that W is a common extension of X and Y, and the triple (W, ϕ_X, ϕ_Y) is a finite equivalence between X and Y.

Proposition 2.3. *If $\mathscr{G}^{(1)}$ and $\mathscr{G}^{(2)}$ are both right-resolving, then $Y^{(1)}$ and $Y^{(2)}$ are finitely equivalent.*

A graph G is essential if for every vertex v there are edges e_1, e_2 such that $i(e_1) = v$ and $t(e_2) = v$. A necessary and sufficient condition thus follows for the case where $Y^{(1)}$ and $Y^{(2)}$ are irreducible sofic shifts.

Theorem 2.4. *Suppose $Y^{(1)}, Y^{(2)}$ are irreducible sofic shifts. If G_T is an essential graph, then $Y^{(1)} \cong Y^{(2)}$ if and only if $PS^{(1)} = S^{(2)}P$, where $P = \begin{pmatrix} 1 & 0 & 0 & 0 \\ 0 & 0 & 1 & 0 \\ 0 & 1 & 0 & 0 \\ 0 & 0 & 0 & 1 \end{pmatrix}$.*

When $\mathscr{L}^{(1)}$ and $\mathscr{L}^{(2)}$ are both right-resolving, the common extension for $Y^{(1)}$ and $Y^{(2)}$ is the original space $\mathbf{Y}$. However, $\mathbf{Y}$ is no longer $Y^{(1)}, Y^{(2)}$'s common extension if either $\mathscr{G}^{(1)}$ or $\mathscr{G}^{(2)}$ is not right-resolving. For this reason, we need to construct a real common extension W.

Theorem 2.5. *Suppose either $\mathscr{G}^{(1)}$ or $\mathscr{G}^{(2)}$ is not right-resolving. There exists finite equivalence $(W, \phi_{W^{(1)}}, \phi_{W^{(2)}})$ between $Y^{(1)}$ and $Y^{(2)}$. Moreover, there exists an integral matrix F such that $FT_{G^{(1)}} = T_{G^{(2)}}F$.*

Given two integral matrices $A \in \mathbb{R}^{m \times m}, B \in \mathbb{R}^{n \times n}$. Suppose $F \in \mathbb{R}^{m \times n}$ is an integral matrix satisfies $FA = BF$. F is called factor-like if there is at most one 1s in each row of F. Let $\mathscr{G} = (G, \mathscr{L})$ be a labeled graph and let w be a word of $X_{\mathscr{G}}$. We say that w is a synchronizing word for $\mathscr{G}$ if all paths in G presenting w terminate at the same vertex.

Proposition 2.6. *Under the same assumption of Theorem 2.5, if F is factor-like, then there exists $\pi : W^{(1)} \to W^{(2)}$ which preserves topological entropy. Moreover, if all words of $Y^{(1)}$ of length N are synchronizing for some $N \in \mathbb{N}$, then there exists a factor map $\pi : Y^{(1)} \to Y^{(2)}$ which preserves entropy.*

Suppose there exists a factor map $\bar{\pi}$ from $Y^{(2)}$ to $Y^{(1)}$; it is natural to ask whether $\bar{\pi}$ is invertible. That is, is $\bar{\pi}$ actually a conjugacy? To answer this question, we introduce a definition first.

Let A and B be nonnegative integral matrices. An *elementary equivalence* from A to B is a pair (R,S) of rectangular nonnegative matrices satisfying $A = RS$ and $B = SR$. In this case we write $(R,S) : A \approx B$. A *strong shift equivalence* from A to B is a sequence of ℓ elementary equivalences:

$$(R_1, S_1) : A = A_0 \approx A_1, (R_2, S_2) : A_1 \approx A_2, \ldots, (R_\ell, S_\ell) : A_{\ell-1} \approx A_\ell = B$$

for some ℓ. In this case we say that A is strong shift equivalent to B and write $A \sim_{\mathscr{F}_{SS}} B$.

Williams classification theorem demonstrates that if A and B are nonnegative integral matrices, then X_A and X_B are conjugate if and only if A and B are strong shift equivalent. It is still hard to find a strong shift equivalence between $T_{G_{Y^{(1)}}}$ and $T_{G_{Y^{(2)}}}$. What we find instead is a weaker relation called *shift equivalence*.

Let A and B be nonnegative integral matrices. A shift equivalence from A to B is a pair (R,S) of rectangular nonnegative integral matrices satisfying

$$AR = RB, \quad SA = BS, \quad A^\ell = RS, \quad B^\ell = SR$$

for some $\ell \in \mathbb{N}$. In this case we say that A is *shift equivalent* to B and write $A \sim_{\mathscr{F}_S} B$. It follows directly that $A \sim_{\mathscr{F}_{SS}} B$ implies $A \sim_{\mathscr{F}_S} B$.

Suppose A is an $n \times n$ nonnegative integral matrix. The *eventual range* of A is defined by $\mathscr{R}_A = \bigcap_{k=1}^{\infty} \mathbb{Q}^n A^k$, where $\mathbb{Q}^n$ is the n-dimensional rational space. Let A be an $n \times n$ nonnegative integral matrix. The *dimension group* of A is $\triangle_A = \{\mathbf{v} \in \mathscr{R}_A : \mathbf{v}A^k \in \mathbb{Z}^n \text{ for some } k \geq 0\}$. The dimension group automorphism δ_A of A is the restriction of A to $\triangle_A$ such that $\delta_A(\mathbf{v}) = \mathbf{v}A$ for $\mathbf{v} \in \triangle_A$. We call $(\triangle_A, \delta_A)$ the dimension pair of A. Moreover, we define the dimension semigroup of A to be $\triangle_A^+ = \{\mathbf{v} \in \mathscr{R}_A : \mathbf{v}A^k \in (\mathbb{Z}^+)^n \text{ for some } k \geq 0\}$. We call $(\triangle_A, \triangle_A^+, \delta_A)$ the dimension triple of A.

Theorem 2.7 ([11, Theorem 7.5.8]). *Let A and B be nonnegative integral matrices.*
$A \sim_{\mathscr{F}_S} B$ *if and only if* $(\triangle_A, \triangle_A^+, \delta_A)$ *is group isomorphic to* $(\triangle_B, \triangle_B^+, \delta_B)$.

Instead of demonstrating a strong shift equivalence between two matrices, it is much easier to determine whether their dimension groups are isomorphic to one another. Furthermore, the Jordan forms $J(A), J(B)$ are necessary conditions for $A \sim_{\mathscr{F}_S} B$.

Let A be an $n \times n$ integral matrix. The invertible part $A^\times$ of A is the linear transformation obtained by restricting A to its eventual range. That is, $A^\times : \mathscr{R}_A \to \mathscr{R}_A$ is defined by $A^\times(\mathbf{v}) = \mathbf{v}A$.

Theorem 2.8 ([11, Theorem 7.4.10]). *Suppose A and B are nonnegative integral matrices. If* $A \sim_{\mathscr{F}_S} B$, *then* $J^\times(A) = J^\times(B)$, *where* $J^\times(A)$ *is the Jordan form of* $A^\times$.

3 Discussion

The existence of a factor map between two subspaces depends on whether there exists a factor map between their covering spaces. Note that a covering space of a sofic shift is a shift of finite type. In other words, to classify the subspaces of a solution space is equivalent to the classification of subshifts of finite type induced by simplified TCNNs.

The above investigation can be extended to general TCNNs and n-layer CNNs for $n \geq 2$. The illustration of general cases is ongoing.

References

1. Ban, J.C., Chang, C.H.: On the monotonicity of entropy for multi-layer cellular neural networks. Int. J. Bifurcat. Chaos Appl. Sci. Eng. **19**, 3657–3670 (2009)
2. Ban, J.C., Chang, C.H., Lin, S.S., Lin, Y.H.: Spatial complexity in multi-layer cellular neural networks. J. Differ. Equ. **246**, 552–580 (2009)
3. Chua, L.O., Roska, T.: Cellular Neural Networks and Visual Computing. Cambridge University Press, Cambridge (2002)
4. Chua, L.O., Yang, L.: Cellular neural networks: Theory. IEEE Trans. Circuits Syst. **35**, 1257–1272 (1988)
5. Crounse, K.R., Roska, T., Chua, L.O.: Image halftoning with cellular neural networks. IEEE Trans. Circuits Syst. **40**, 267–283 (1993)
6. Hsu, C.H., Juang, J., Lin, S.S., Lin, W.W.: Cellular neural networks: Local patterns for general template. Int. J. Bifurcat. Chaos Appl. Sci. Eng. **10**, 1645–1659 (2000)
7. Juang, J., Lin, S.S.: Cellular neural networks: Mosaic pattern and spatial chaos. SIAM J. Appl. Math. **60**, 891–915 (2000)
8. Ke, Y.Q., Miao, C.F.: Existence analysis of stationary solutions for rtd-based cellular neural networks. Int. J. Bifurcat. Chaos Appl. Sci. Eng. **20**, 2123–2136 (2010)
9. Li, X.: Analysis of complete stability for discrete-time cellular neural networks with piecewise linear output functions. Neural Comput. **21**, 1434–1458 (2009)

10. Lin, S.S., Shih, C.W.: Complete stability for standard cellular neural networks. Int. J. Bifurcat. Chaos Appl. Sci. Eng. **9**, 909–918 (1999)
11. Lind, D., Marcus, B.: An Introduction to Symbolic Dynamics and Coding. Cambridge University Press, Cambridge (1995)
12. Murugesh, V.: Image processing applications via time-multiplexing cellular neural network simulator with numerical integration algorithms. Int. J. Comput. Math. **87**, 840–848 (2010)
13. Peng. J., Zhang, D., Liao, X.: A digital image encryption algorithm based on hyper-chaotic cellular neural network. Fund Inform. **90**, 269–282 (2009)
14. Yang, Z., Nishio, Y., Ushida, A.: Image processing of two-layer CNNs − applications and their stability. IEICE Trans. Fundam. **E85-A**, 2052–2060 (2002)

Linear Integral Equations with Discontinuous Kernels and the Representation of Operators on Regulated Functions on Time Scales

Luciano Barbanti, Berenice Camargo Damasceno, Geraldo Nunes Silva, and Marcia Cristina Anderson Braz Federson

Abstract We present here the linear Cauchy–Stieltjes integral on regulated functions with values in Banach spaces on time scales and represent a linear operator on the space of the regulated functions by means of an appropriate kernel in the integral.

Keywords Cauchy integral • Banach spaces • Regulated functions • Time scales

1 Introduction

This paper deals with solutions of the equation

$$Ay = u, \tag{1}$$

with Y, Z Banach spaces and $y \in Y$, $u \in Z$, where A is a linear operator from Y to Z.

We are interested in the case in which Y is a function space $Y = \mathscr{F}([a,b]_{\mathbb{T}}, X)$ (X is a Banach space) containing discontinuous functions, on a time scale $\mathbb{T}$ (that is, $\mathbb{T}$ is a closed non-void subset of the real numbers $\mathbb{R}$, and $[a,b]_{\mathbb{T}} = [a,b] \cap \mathbb{T}$ in $\mathbb{R}$).

Discontinuous functions arise in a natural way when we are describing phenomenon in Mathematics, Physics, or Technology, as, for instance, in collision

L. Barbanti (✉) • B.C. Damasceno
MAT-FE/IS, UNESP, Al. Rio de Janeiro 266, 15385-000 Ilha Solteira, SP, Brasil
e-mail: barbanti@mat.feis.unesp.br; berenice@mat.feis.unesp.br

G.N. Silva
IBILCE, UNESP, R. Cristovao Colombo 2265, 15054-000 Sao Jose do Rio Preto, SP, Brasil
e-mail: gsilva@ibilce.unesp.br

M.C.A.B. Federson
ICMC, USP, Av. Trabalhador Sao-carlense 400, 13566-590 Sao Carlos, SP, Brasil
e-mail: federson@icmc.usp.br

S. Pinelas et al. (eds.), *Differential and Difference Equations with Applications*, Springer 275
Proceedings in Mathematics & Statistics 47, DOI 10.1007/978-1-4614-7333-6_21,
© Springer Science+Business Media New York 2013

theory when the displacement changes suddenly the direction or even when using the classical play operator with variable characteristics in the theory on hysteresis [1].

The following are examples of Banach spaces containing discontinuous functions:

1. The space of the regulated functions on time scales $\mathbb{T}$: We say that $f : [a,b]_\mathbb{T} \to X$ is *regulated* if there exists $f(t^+)$[respectively $f(t_-)$] whenever $\inf\{s : s > t\} = t$ [respectively $\sup\{s : s < t\} = t$]. In this case we write $f \in G([a,b]_\mathbb{T}, X)$. If moreover $f(t^-) = f(t)$ when $\sup\{s : s < t\} = t$, then we write $f \in G^-([a,b]_\mathbb{T}, X)$. Observe that both $G([a,b]_\mathbb{T}, X)$ and $G^-([a,b]_\mathbb{T}, X)$ are Banach spaces if endowed with the sup norm.

2. The spaces (on the time scale $\mathbb{T} = \mathbb{R}$) $L_p([a,b],X), 1 \le p < \infty$ of the measurable functions in the Bochner or Lebesgue sense or the space $BV([a,b],X)$ of functions of bounded variation, the Sobolev spaces, etc.

The main purpose in this work is to represent the operator A in (1), $A \in L$ $(G^-([a,b]_\mathbb{T}, X), Z))$, according the integral

$$Af(t) = \int_{[a,b]_\mathbb{T}} \mathbb{D}_s \alpha(s).f(s),$$

where the kernel α and the integral on time scales $\mathbb{T}$ itself are described below.

2 Time Scales

Settled in 1988 by Stefan Hilger [2] the Calculus on time scales was created to unify the theory between the continuous and the discrete time dynamical systems. The integral on time scales is the subject of several works in the literature [3, 4]. Here we take the notations of [3].

Considerations on the theory of integral (in the Riemann or in the generalized Riemann senses) on time scales include, among others, the Riemann delta and nabla-integral, alfa-integral, the Lebesgue and nabla-integrals, and the Henstock–Perron–Kurzweil ones [3]. The work by Mozyrska–Pawluszewicz–TorresIt is more recent (2009), [5], in which the Riemann–Stieltjes integral is considered.

Let us take a look at these integrals.

2.1 The Riemann–Stieltjes Integral on Time Scales $\mathbb{T}$

Let $\mathbb{T}$ be the time scale and consider the continuous real-valued functions f,g taking values on the interval $\mathbb{I} = [a,b]_\mathbb{T} = [a,b] \cap \mathbb{T}$ (in $\mathbb{R}$) with g strictly increasing and f bounded. Let $\wp_{[a,b]_\mathbb{T}}$ be the set of all finite divisions of the interval $[a,b]_\mathbb{T}$, that is,

$P \in \wp_{[a,b]_{\mathbb{T}}}$, $P = \{a = t_0 < t_1 < \cdots < t_n = b\}$. The upper and lower Darboux–Stieltjes sum for the P respect to f and g are

(the upper) $\quad U(P; f; g) = \sum_{j=1}^{n} \sup_{[t_{j-1}, t_j]_{\mathbb{T}}} f(t) \Delta g_j \quad$ and

(the lower) $\quad L(P; f; g) = \sum_{j=1}^{n} \inf_{[t_{j-1}, t_j]_{\mathbb{T}}} f(t) \Delta g_j$,

where $\Delta g_j = g(t_j) - g(t_{j-1})$ for $j = 1, 2, \ldots, n$.

The upper and lower Darboux–Stieltjes $\square$-integrals are, respectively, the numbers

$$(U) \int_a^b f(t) \square g(t) = \inf_{P \in \wp_{[a,b]_{\mathbb{T}}}} = U(P; f; g)$$

$$(L) \int_a^b f(t) \square g(t) = \sup_{P \in \wp_{[a,b]_{\mathbb{T}}}} = L(P; f; g).$$

If both the values are the same, we say that the Riemann–Stieltjes integral, denoted in accordance with the authors in [5], $\int_a^b f(t) \square g(t)$ is such common value. Looking at this definition we observe that there are some difficulties inherent to the Darboux sum used in it when dealing with noncontinuous functions. It is shown in the next examples.

Example 1. Let

$$\mathbb{I} = [-1, 1]_{\mathbb{T}} = \left\{ \frac{1}{2^k}; k \in \mathbb{N}^* \right\} \cup \left\{ -\frac{1}{2^k}; k \in \mathbb{N}^* \right\} \cup \{0\},$$

and define

$$g(t) = \begin{cases} 1 + t, & \text{if} \quad t \in \{\frac{1}{2^k}, k \in \mathbb{N}^*\} \cup \{0\} \\ t, & \text{otherwise} \end{cases},$$

$$f(t) = \begin{cases} 1, \text{ if} \quad t \in \{\frac{1}{2^k}, k \in \mathbb{N}^*\} \cup \{0\} \\ 0, \quad \text{otherwise} \end{cases}.$$

Then we have the following: there exist both $\int_{-1}^{0} f(t) \square g(t)$ and $\int_{0}^{1} f(t) \square g(t)$, but $\int_{-1}^{1} f(t) \square g(t)$ is not defined.

Example 2. Let

$$\mathbb{I} = [-1, 1]_{\mathbb{T}} = \{-1, 0, 1\}$$

and $f, g : \mathbb{I} \to \mathbb{R}$, with

$$g(1) = 1; g(0) = 0; g(-1) = -1; f(-1) = 1; f(0) = 0; f(1) = 4.$$

Then we have the following: the $\square$-integral of f with respect to g does not exist. In fact, the upper and lower Darboux–Stieltjes $\square$-integrals are, respectively, the numbers

$$U(P; f; g) = 5$$

$$L(P; f; g) = 0.$$

With the purpose, among others, to overcome these difficulties when discontinuous functions are considered, it is necessary to stand a greater vision about the Riemann–Stieltjes integral (reflected in a more general notion for extending it).

2.2 A Generalized Riemann–Stieltjes Integral on Time Scales $\mathbb{T}$

Here we introduce the right Cauchy–Stieltjes integral on time scales. The considerations on this type of integral when introduced in the time scales environment are appropriated for several reasons that are more and more evident with its use.

As shown in Example 1 the Riemann–Stieltjes integral may not exist if the functions f and g have discontinuities at the same point 0. In Example 2, in which $\mathbb{I}$ is discrete, it is shown that the continuity of f and g is not the issue. Moreover, both the examples show that the Riemann–Stieltjes integral does not possess, in the general, the additive property:

$$\int_a^c f(t) \square g(t) + \int_c^b f(t) \square g(t) = \int_a^b f(t) \square g(t).$$

When using the Cauchy–Stieltjes integral, we will be retrieving these fundamental properties.

Moreover by considering the right Cauchy–Stieltjes integral (see [6]), we have, roughly speaking, a great degree of compatibility among the integral, the left continuous functions in time scale $\mathbb{T}$, and the kind of filtering convergence used in the definition of the integral itself.

Consider as in Sect. 2.1 the class $\wp_{[a,b]_{\mathbb{T}}}$ of all partitions of $\mathbb{I} = [a,b]_{\mathbb{T}}$.

Let us take $f : (a,b) \longrightarrow X, g : (a,b) \longrightarrow L(X,W)$, where W is a Banach space. The (Cauchy) sum associated to

$$P = \{a = t_0, t_1, \ldots, t_n = b\} \in \wp_{[a,b]_{\mathbb{T}}}$$

of f relatively to α is defined by

$$\sigma_P(f;\alpha) = \sum_{i=0}^{n-1} [\alpha(t_{i+1}) - \alpha(t_i)] \cdot f(t_{i+1}).$$

The (right) Cauchy–Stieltjes integral $(rC - S)$ of f relatively to α, $(rC - S) \int_{[a,b]_{\mathbb{T}}} \mathbb{D}_s \alpha(s).f(s)$—or simply $\int_{[a,b]_{\mathbb{T}}} \mathbb{D}_s \alpha(s).f(s)$—is the value

$$\int_{[a,b]_{\mathbb{T}}} \mathbb{D}_s \alpha(s).f(s) = \lim_{P \in \wp_{[a,b]_{\mathbb{T}}}} \sigma_P(f;\alpha),$$

provided it exists.

Recall that the expression $\lim_{P \in \wp_{[a,b]_{\mathbb{T}}}} \sigma_P(f;\alpha) = z$ in a general topological space means that for every neighborhood V of z, there exists $P_0 \in \wp_{[a,b]_{\mathbb{T}}}$ with $\sigma_P(f;\alpha) \in V$ whenever $P \in \wp_{[a,b]_{\mathbb{T}}}$ and $P \supseteq P_0$.

2.3   Existence of the $(rC - S)$ Integral on Time Scales

To proceed let be the following definition.

Definition 1 (Semivariation). Let $\wp_{[a,b]_{\mathbb{T}}}$, be the set of the partitions of $[a,b]_{\mathbb{T}}$ and $P \in \wp_{[a,b]_{\mathbb{T}}}$ and $\alpha : [a,b]_{\mathbb{T}} \longrightarrow L(X,W)$. The function α is of bounded semivariation, and we write $\alpha \in SV([a,b]_{\mathbb{T}}, L(X,W))$ if

$$SV[\alpha] = \sup_{P \in \wp_{[a,b]_{\mathbb{T}}}} SV_P[\alpha]$$

with

$$SV_P[\alpha] = \sup \left\{ \left\| \sum_{i=1}^{|P|} [\alpha(t_i) - \alpha(t_{i-1})]x_i \right\|_W ; x_i \in X; \|x_i\| < 1 \right\}$$

and $SV[\alpha]$ is finite.

Observe that $SV[\alpha]$ is a semi-norm. It is a norm on the class of all $\alpha \in SV([a,b]_{\mathbb{T}}, L(X,W))$ with $\alpha(a) = 0$. If α is such that satisfies these conditions, we write $\alpha \in SV_0([a,b]_{\mathbb{T}}, L(X,W))$.

In the following we recall some conditions on α and f for the existence of the Cauchy–Stieltjes integral at right $(r - CS) \int_{[a,b]_{\mathbb{T}}} d_s \alpha(s).f(s)$ as well as some of its first properties.

Theorem 1 ([7], Theorem II.1(b)). *For every time scale $\mathbb{T}$, let $f \in G^-([a,b]_{\mathbb{T}}, X)$ and $\alpha \in SV([a,b]_{\mathbb{T}}, L(X,W))$. Then $\mathfrak{I}_\alpha(f) = \int_{[a,b]_{\mathbb{T}}} \mathbb{D}_s \alpha(s).f(s) \in W$, and $\mathfrak{I}_\alpha \in L(G^-([a,b],X),W)$. Furthermore, $\|\mathfrak{I}_\alpha\| \leq SV[\alpha]$, and for every $c \in [a,b]_{\mathbb{T}}$, we have*

$$\int_{[a,c]_{\mathbb{T}}} \mathbb{D}_s \alpha(s).f(s) + \int_{[c,b]_{\mathbb{T}}} \mathbb{D}_s \alpha(s).f(s) = \int_{[c,b]_{\mathbb{T}}} \mathbb{D}_s \alpha(s).f(s).$$

Remark 1. Observe that in Examples 1 and 2 in Sect. 2.1, we have $\int_{[-1,1]_{\mathbb{T}}}$ $\mathbb{D}_s\alpha(s).f(s) = 1$ and 4, respectively.

In the next section we are presenting the main results in this work.

2.4 Integral Representation for Linear Operators on $G^-([a,b],X)$

As seen in Theorem 1 above, every $\alpha \in SV([a,b]_{\mathbb{T}},L(X,W))$ defines the linear continuous operator $\mathfrak{I}_\alpha$ on the space $G^-([a,b]_{\mathbb{T}},X)$ with the use of the right Cauchy–Stieltjes integral on time scales.

The next theorem shows a kind of converse situation that all the linear-bounded operators $A \in L(G^-([a,b],X),W)$ can be represented by an operator $\mathfrak{I}_\alpha \in L(G^-([a,b],X),W)$ for a convenient α.

Theorem 2. *Let X, W be Banach spaces, and for every $Z \subset \mathbb{R}$, the characteristic function $\mathscr{X}_Z{:}\mathbb{R} \to L(X)$*

$$\mathscr{X}_Z = \begin{cases} Id \text{ if } z \in Z \\ 0 \quad \text{if } z \notin Z \end{cases}.$$

Then

$$\Lambda : SV_0([a,b]_{\mathbb{T}},L(X,W)) \longrightarrow L(G^-([a,b],X),W)$$

with $\Lambda(\alpha) = \mathfrak{I}_\alpha$ is an isometry of the first space onto the second one. Further, $\alpha(t)x = \mathfrak{I}_\alpha(\mathscr{X}_{(a,t]_{\mathbb{T}}}x)$. Moreover,

$$A(f) = \int_{[a,b]_{\mathbb{T}}} \mathbb{D}_s\alpha(s).f(s).$$

Proof. According to Theorem II.1(b) in [7], we have Λ well defined. Moreover, it is linear and continuous and we have $\|\Lambda(\alpha)\| \leq SV[\alpha]$. The operator Λ is one to one. In fact, for $\alpha \neq 0$—remembering $\alpha(a) = 0$—there are $\tau \in (a,b]_{\mathbb{T}}$ and $x \in X$ such that $\alpha(\tau)x \neq 0$. Taking $f(t) = \chi_{(a,\tau]}x \in G^-([a,b]_{\mathbb{T}},X)$ in this way we get $\mathfrak{I}_\alpha(f) \neq 0$ and so $\mathfrak{I}_\alpha$ because $\mathfrak{I}_\alpha(f) = \alpha(\tau)x$. To end the proof we only need to show that Λ is onto and that $\|\Lambda(\alpha)\| \geq SV[\alpha]$. The operator Λ is onto: given $A \in L(G^-([a,b]_{\mathbb{T}},X),W)$, let us take as definition that

$$\alpha(t)x = A(\mathscr{X}_{(a,t]_{\mathbb{T}}}x).$$

We will show that $\Lambda(\alpha) = A$. To achieve this it is sufficient to be showing that $\Lambda(\alpha)$ and A are coincident on the total set $\{\mathscr{X}_{(a,\tau]_{\mathbb{T}}}x; \tau \in (a,b]_{\mathbb{T}}\}$ in $G^-([a,b]_{\mathbb{T}},X)$. In fact,

$$\Lambda(\alpha)(\mathscr{X}_{(a,\tau]_{\mathbb{T}}}x) = \int_{[a,b]_{\mathbb{T}}} \mathbb{D}_s\alpha(s) \cdot \mathscr{X}_{(a,\tau]_{\mathbb{T}}}(s)x = \alpha(\tau)x = A(\mathscr{X}_{(a,\tau]_{\mathbb{T}}}x).$$

To end the proof, observe the definition of $SV[\alpha]$.   □

2.5   Examples

Example 1 (The Operator Evaluation at a Point). Suppose $X = W$, $t_0 \in (a,b]_{\mathbb{T}}$, and $A(f) = f(t_0)$. The mapping α is $\alpha(\tau)x = A(\mathscr{X}_{(a,t_0]_{\mathbb{T}}}x) = \mathscr{X}_{(t_0,b]_{\mathbb{T}}}x$. In this way,

$$f(t_0) = \int_{[a,b]_{\mathbb{T}}} \mathbb{D}_s\mathscr{X}_{(t_0,b]_{\mathbb{T}}}(s)f(s).$$

Example 2 (The Riemann Integral). Suppose $X = W = \mathbb{R}$ and the delta integral on time scales in the Darboux sense on $(a,b]_{\mathbb{T}}$ defined in a similar way as ([3], Definition 2.1):

$$A(f) = \int_{(a,b]_{\mathbb{T}}} f(s)ds.$$

Then,

$$\alpha(\tau)x = A(\mathscr{X}_{(a,\tau]_{\mathbb{T}}}x) = \int_{[a,b]_{\mathbb{T}}} \mathbb{D}_s\mathscr{X}_{(a,\tau]_{\mathbb{T}}}(s)x = (\tau - a)x.$$

Hence, $A(f) = \int_{[a,b]_{\mathbb{T}}} f(s)ds = \int_{[a,b]_{\mathbb{T}}} d_s(s-a) \cdot f(s)$.

Example 3 (The Dual of $G^-([a,b]_{\mathbb{T}},X)$). Take W as the field $\mathbb{R}$ or $\mathbb{C}$, X a Banach space, and X' its dual. We observe that according to Theorem 2, we can identify the dual space $G^-([a,b]_{\mathbb{T}},X)'$ with the space $SV_0([a,b]_{\mathbb{T}},X)'$.

Notice that this result allows us to improve the one obtained in [1], for the case $\mathbb{T} = \mathbb{R}$ or $\mathbb{C}$, in Hilbert spaces with the use of the Young integral.

References

1. Brokate, M., Krejc, P.: Duality in the space of regulated functions and the play operator. Mathematische Zeitschrift **245**, 667–688 (2003)
2. Hilger, S.: Analysis on measure chains - a unified approach to continuous and discrete calculus. Results Math. **18**, 18–56 (1990)
3. Guseinov, G.S.: Integration on time scales. J. Math. Anal. Appl. **285**, 107–127 (2003)
4. Agarwal, R., Bohner, M., O'Regan, D., Peterson, A.: Dynamic equations on time scales: a survey. J. Comput. Appl. Math. **141**, 1–26 (2002)

5. Mozyrska, D., Pawluszewicz, E., Torres, D.F.M.: Riemann-Stieltjes integral in time scales. The Aust. J. Math. Anal. Appl. 7(1), 1–14 (2010)
6. Hildebrandt, T.H.: Introduction to the Theory of Integration. Academic, NewYork (1963)
7. Damasceno, B.C., Barbanti, L.: Cauchy-Stieltjes integral in the space of regulated functions on time scales (2013) (Submitted)

Strong Solutions to Buoyancy-Driven Flows in Channel-Like Bounded Domains

Michal Beneš

Abstract We consider a boundary-value problem for steady flows of viscous incompressible heat-conducting fluids in channel-like bounded domains. The fluid flow is governed by balance equations for linear momentum, mass, and internal energy. The internal energy balance equation of this system takes into account the phenomena of the viscous energy dissipation and includes the adiabatic heat effects. The system of governing equations is provided by suitable mixed boundary conditions modeling the behavior of the fluid on fixed walls and open parts of the channel. Due to the fact that some uncontrolled "backward flow" can take place at the outlets of the channel, there is no control of the convective terms in balance equations for linear momentum and internal energy, and consequently, one is not able to prove energy type estimates. This makes the qualitative analysis of this problem more difficult. In this paper, the existence of the strong solution is proven by a fixed-point technique for sufficiently small external forces.

Keywords Navier-Stokes equations • Heat equation • Heat-conducting fluids • Qualitative properties • Mixed boundary conditions

1 Introduction

Let Ω be a two-dimensional bounded domain with the boundary $\partial\Omega$. Let $\partial\Omega = \overline{\Gamma}_D \cup \overline{\Gamma}_N$ be such that Γ_D and Γ_N are open, not necessarily connected; the one-dimensional measure of $\Gamma_D \cap \Gamma_N$ is zero and $\Gamma_D \neq \emptyset$ ($\Gamma_N = \bigcup_i^m \Gamma_N^{(i)}$, $\overline{\Gamma}_N^{(i)} \cap \overline{\Gamma}_N^{(j)} = \emptyset$ for $i \neq j$). In a physical sense, Ω represents a "truncated" region of an unbounded channel system occupied by a moving fluid. Γ_D will denote the "lateral" surface and

M. Beneš (✉)
Department of Mathematics, Czech Technical University in Prague, Thákurova 7,
166 29 Prague 6, Czech Republic
e-mail: benes@mat.fsv.cvut.cz

S. Pinelas et al. (eds.), *Differential and Difference Equations with Applications*, Springer   283
Proceedings in Mathematics & Statistics 47, DOI 10.1007/978-1-4614-7333-6_22,
© Springer Science+Business Media New York 2013

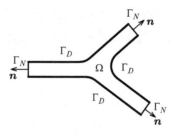

Fig. 1 Ω represents a "truncated" region of an unbounded channel system occupied by a fluid

Γ_N represents the open parts of the region Ω. It is assumed that in-/outflow pipe segments extend as straight pipes. All portions of Γ_N are taken to be flat, and the boundary Γ_N and rigid boundary Γ_D form a right angle at each point in which the boundary conditions change their type (cf. Fig. 1). Moreover, we assume that all parts of Γ_D are smooth.

We are going to study the boundary-value problem for stationary buoyancy-driven flows of viscous incompressible heat-conducting homogeneous fluids with dissipative and adiabatic heating in Ω. The strong formulation of our problem is as follows:

$$\rho(\mathbf{u}\cdot\nabla)\mathbf{u} - \nu\Delta\mathbf{u} + \nabla\pi = \rho(1-\alpha_0\theta)\mathbf{f} \qquad \text{in } \Omega, \qquad (1)$$

$$\nabla\cdot\mathbf{u} = 0 \qquad \text{in } \Omega, \qquad (2)$$

$$c_p\rho\mathbf{u}\cdot\nabla\theta - \kappa\Delta\theta - \alpha_1\nu\mathbf{e}(\mathbf{u}):\mathbf{e}(\mathbf{u}) = \rho\alpha_2\theta\mathbf{f}\cdot\mathbf{u} \qquad \text{in } \Omega, \qquad (3)$$

$$\mathbf{u} = \mathbf{0} \qquad \text{on } \Gamma_D, \qquad (4)$$

$$\theta = g \qquad \text{on } \Gamma_D, \qquad (5)$$

$$-\pi\mathbf{n} + \nu(\nabla\mathbf{u})\mathbf{n} = \mathbf{0} \qquad \text{on } \Gamma_N, \qquad (6)$$

$$\nabla\theta\cdot\mathbf{n} = 0 \qquad \text{on } \Gamma_N. \qquad (7)$$

Here $\mathbf{u} = (u_1, u_2)$, π and θ denote the unknown velocity, pressure, and temperature, respectively. Tensor $\mathbf{e}(\mathbf{u})$ denotes the symmetric part of the velocity gradient $\mathbf{e}(\mathbf{u}) = [\nabla\mathbf{u} + (\nabla\mathbf{u})^\top]/2$.

Data of the problem are as follows: $\mathbf{f}$ is a body force and g is a given function representing the distribution of the temperature θ on Γ_D. Positive constant material coefficients represent the kinematic viscosity ν, density ρ, heat conductivity κ, specific heat at constant pressure c_p, and thermal expansion coefficient of the fluid α_0. Coefficients α_1 and α_2 reflect the dissipation and adiabatic effects, respectively. Here we suppose that all functions in (1)–(7) are smooth enough. For rigorous derivation of the model we refer the readers to [5].

Due to the presence of the dissipative term $\alpha_1\nu\mathbf{e}(\mathbf{u}):\mathbf{e}(\mathbf{u})$ in the energy equation, (1)–(7) represent the elliptic system with strong nonlinearities without appropriate general existence and regularity theory (cf. [2]). Moreover, because of the "do-nothing" boundary condition (6), without any additional consideration on the in-/output region of the channel, we are not able to prove an "a priori" estimate for the convective terms in the system (see [6]). This makes the studied problem more difficult than in the frequently used case of Dirichlet boundary condition on the whole boundary. The core of the main result of this work lies in the proof of the existence of the strong solution (in the sense that the solution possesses second derivatives) for sufficiently small external force $\mathbf{f}$, nevertheless, without any additional restrictions on the distribution of the temperature θ on Γ_D described by the function g.

Remark 1. Let us note that our results can be extended to the problem if we consider the so-called "free surface" boundary condition on Γ_N and replace (6) by (see [3,4])

$$-\pi\mathbf{n} + v[\nabla\mathbf{u} + (\nabla\mathbf{u})^\top]\mathbf{n} = \mathbf{0}.$$

However, to ensure the smoothness of the solution and exclude boundary singularities near the points where the boundary conditions change their type, some additional requirements on the geometry of the domain need to be introduced. This means that Γ_N and Γ_D form an angle $\omega < \pi/4$ at each point where the boundary conditions change (see [9]).

2   Basic Notation and Some Function Spaces

Vector functions and operators acting on vector functions are denoted by boldface letters. Throughout the paper, we will always use positive constants $c, c_1, c_2, \ldots$, which are not specified and which may differ from line to line. For an arbitrary $r \in [1, +\infty]$, $L^r(\Omega)$ denotes the usual Lebesgue space equipped with the norm $\| \cdot \|_{L^r(\Omega)}$, and $W^{k,p}(\Omega)$, $k \geq 0$ (k need not to be an integer, see [8]), $p \in [1, +\infty]$, denotes the usual Sobolev space with the norm $\| \cdot \|_{W^{k,p}(\Omega)}$. For simplicity we denote shortly $\mathbf{W}^{k,p} \equiv W^{k,p}(\Omega)^2$ and $\mathbf{L}^r \equiv L^r(\Omega)^2$.

To simplify mathematical formulations we introduce the following notations:

$$a_u(\mathbf{u}, \mathbf{v}) := v \int_\Omega \nabla\mathbf{u} : \nabla\mathbf{v}\,\mathrm{d}\Omega, \tag{8}$$

$$b(\mathbf{u}, \mathbf{v}, \mathbf{w}) := \rho \int_\Omega (\mathbf{u} \cdot \nabla)\mathbf{u} \cdot \mathbf{w}\,\mathrm{d}\Omega, \tag{9}$$

$$a_\theta(\theta, \varphi) := \kappa \int_\Omega \nabla\theta \cdot \nabla\varphi\,\mathrm{d}\Omega, \tag{10}$$

$$d(\mathbf{u}, \theta, \varphi) := c_p\rho \int_\Omega \mathbf{u} \cdot \nabla\theta\,\varphi\,\mathrm{d}\Omega, \tag{11}$$

$$e(\mathbf{u}, \mathbf{v}, \varphi) := \alpha_1 v \int_\Omega \mathbf{e}(\mathbf{u}) : \mathbf{e}(\mathbf{v})\varphi\,\mathrm{d}\Omega, \tag{12}$$

$$(\mathbf{u}, \mathbf{v}) := \int_\Omega \mathbf{u} \cdot \mathbf{v}\,\mathrm{d}\Omega, \tag{13}$$

$$(\theta, \varphi)_\Omega := \int_\Omega \theta\varphi\,\mathrm{d}\Omega. \tag{14}$$

In (8)–(14) all functions $\mathbf{u}, \mathbf{v}, \mathbf{w}, \theta, \varphi$ are smooth enough, such that all integrals on the right-hand sides make sense. Let

$$\mathbf{E}_u := \left\{ \mathbf{u} \in C^\infty(\overline{\Omega})^2; \operatorname{div} \mathbf{u} = 0, \operatorname{supp} \mathbf{u} \cap \Gamma_D = \emptyset \right\} \tag{15}$$

and

$$E_\theta := \left\{ \theta \in C^\infty(\overline{\Omega}); \operatorname{supp} \theta \cap \Gamma_D = \emptyset \right\} \tag{16}$$

and $\mathbf{V}_u^{k,p}$ be the closure of $\mathbf{E}_u$ in the norm of $W^{k,p}(\Omega)^2$, $k \geq 0$ and $1 \leq p \leq \infty$. Similarly, let $V_\theta^{k,p}$ be a closure of E_θ in the norm of $W^{k,p}(\Omega)$. Then $\mathbf{V}_u^{k,p}$ and $V_\theta^{k,p}$, respectively, are Banach spaces with the norms of the spaces $W^{k,p}(\Omega)^2$ and $W^{k,p}(\Omega)$, respectively. Note that $\mathbf{V}_u^{1,2}$, $V_\theta^{1,2}$, $\mathbf{V}_u^{0,2}$ and $V_\theta^{0,2}$, respectively, are Hilbert spaces with scalar products (8), (10), (13) and (14), respectively. Further, define the spaces

$$\mathbf{D}_u := \left\{ \mathbf{u} \mid \mathbf{f} \in \mathbf{V}_u^{0,2}, \frac{1}{\nu} a_u(\mathbf{u}, \mathbf{v}) = (\mathbf{f}, \mathbf{v}) \text{ for all } \mathbf{v} \in \mathbf{V}_u^{1,2} \right\} \tag{17}$$

and

$$D_\theta := \left\{ \theta \mid h \in V_\theta^{0,2}, \frac{1}{\kappa} a_\theta(\theta, \varphi) = (h, \varphi)_\Omega \text{ for all } \varphi \in V_\theta^{1,2} \right\}, \tag{18}$$

equipped with the norms

$$\|\mathbf{u}\|_{\mathbf{D}_u} := \|\mathbf{f}\|_{\mathbf{V}_u^{0,2}} \quad \text{and} \quad \|\theta\|_{D_\theta} := \|h\|_{V_\theta^{0,2}}, \tag{19}$$

where $\mathbf{u}$ and $\mathbf{f}$ are corresponding functions via (17) and θ and h are corresponding functions via (18).

The key embeddings

$$\mathbf{D}_u \hookrightarrow \mathbf{W}^{2,2} \tag{20}$$

and

$$D_\theta \hookrightarrow W^{2,2}(\Omega) \tag{21}$$

("$\hookrightarrow$" denotes the continuous embedding) are consequences of assumptions on the domain Ω and the regularity results for the steady Stokes system in channel-like domains with "do-nothing" condition (see [1] Remark 2.2, and Corollary 2.3) and the "classical" regularity results for the stationary linear heat equation (the Poisson equation) with the mixed boundary conditions (see, for instance, [7]).

3 Formulation of the Problem and the Main Result

Now we can formulate our problem. Suppose that $\mathbf{f} \in \mathbf{L}^2$ and $g \in W^{2,2}(\Omega)$. Find a couple $[\mathbf{u}, \theta]$ such that $\mathbf{u} \in \mathbf{D}_u$, $\theta \in g + D_\theta$ and the following system

$$a_u(\mathbf{u}, \mathbf{v}) + b(\mathbf{u}, \mathbf{u}, \mathbf{v}) = (\rho(1 - \alpha_0\theta)\mathbf{f}, \mathbf{v}), \qquad (22)$$

$$a_\theta(\theta, \varphi) + d(\mathbf{u}, \theta, \varphi) - e(\mathbf{u}, \mathbf{u}, \varphi) = (\rho\alpha_2\theta\mathbf{f} \cdot \mathbf{u}, \varphi)_\Omega \qquad (23)$$

holds for every $[\mathbf{v}, \varphi] \in \mathbf{V}_u^{1,2} \times V_\theta^{1,2}$. The couple $[\mathbf{u}, \theta]$ is called the strong solution to the system (1)–(7).

Theorem 1 (Main result). *Assume* $\mathbf{f} \in \mathbf{L}^2$ *and* $g \in W^{2,2}(\Omega)$. *Let* $\|\mathbf{f}\|_{\mathbf{L}^2}$ *be "small enough." Then there exists the strong solution to the system* (1)–(7).

4 Proof of the Main Result

For arbitrary fixed $[\mathbf{u}_0, \vartheta_0] \in \mathbf{D}_u \times D_\theta$ we now consider the following problem: to find a couple $[\mathbf{u}, \vartheta] \in \mathbf{D}_u \times D_\theta$, such that

$$a_u(\mathbf{u}, \mathbf{v}) = (\rho(1 - \alpha_0\vartheta_0)\mathbf{f}, \mathbf{v})$$
$$- (\rho\alpha_0 g\mathbf{f}, \mathbf{v}) - b(\mathbf{u}_0, \mathbf{u}_0, \mathbf{v}), \qquad (24)$$

$$a_\theta(\vartheta, \varphi) + d(\mathbf{u}, \vartheta_0 + g, \varphi) - e(\mathbf{u}, \mathbf{u}, \varphi) = (\rho\alpha_2\vartheta_0\mathbf{f} \cdot \mathbf{u}_0, \varphi)_\Omega$$
$$+ (\rho\alpha_2 g\mathbf{f} \cdot \mathbf{u}_0, \varphi)_\Omega - a_\theta(g, \varphi) \quad (25)$$

for every $[\mathbf{v}, \varphi] \in \mathbf{V}_u^{1,2} \times V_\theta^{1,2}$. In what follows we prove that the map K, $K([\mathbf{u}_0, \vartheta_0]) = [\mathbf{u}, \vartheta]$, has a fixed point (denoted again by $[\mathbf{u}, \vartheta]$) in $\mathbf{D}_u \times D_\theta$. Consequently, the couple $[\mathbf{u}, \theta] := [\mathbf{u}, \vartheta + g]$ is the strong solution to the system (1)–(7).

First, let us check that the right-hand sides in (24)–(25) are well defined. For an arbitrary $[\mathbf{u}_0, \theta_0] \in \mathbf{D}_u \times D_\theta$ we have

$$\|(\rho(1 - \alpha_0\vartheta_0)\mathbf{f}, \cdot)\|_{\mathbf{V}_u^{0,2}} \le \rho\|(\mathbf{f}, \cdot)\|_{\mathbf{V}_u^{0,2}} + \rho\alpha_0\|(\vartheta_0\mathbf{f}, \cdot)\|_{\mathbf{V}_u^{0,2}}$$
$$\le \rho\|(\mathbf{f}, \cdot)\|_{\mathbf{V}_u^{0,2}} + \rho\alpha_0 c_1\|\vartheta_0\|_{D_\theta}\|(\mathbf{f}, \cdot)\|_{\mathbf{V}_u^{0,2}}, \qquad (26)$$

further

$$\|(\rho\alpha_0 g\mathbf{f}, \cdot)\|_{\mathbf{V}_u^{0,2}} \le \rho\alpha_0 c_1\|g\|_{W^{2,2}(\Omega)}\|(\mathbf{f}, \cdot)\|_{\mathbf{V}_u^{0,2}} \qquad (27)$$

and

$$\|b(\mathbf{u}_0, \mathbf{u}_0, \cdot)\|_{\mathbf{V}_u^{0,2}} \leq \rho \|\mathbf{u}_0\|_{\mathbf{V}_u^{0,\infty}} \|\mathbf{u}_0\|_{\mathbf{V}_u^{1,2}}$$

$$\leq \rho c_1 \|\mathbf{u}_0\|_{\mathbf{D}_u}^2. \tag{28}$$

The inequalities (26)–(28) together with (19)–(20) yield the estimate for the solution of (24)

$$\|\mathbf{u}\|_{\mathbf{D}_u} \leq c(\Omega, \rho, \nu, \alpha_0) \left(\|(\mathbf{f}, \cdot)\|_{\mathbf{V}_u^{0,2}} (1 + \|g\|_{W^{2,2}(\Omega)} + \|\vartheta_0\|_{D_\theta}) + \|\mathbf{u}_0\|_{\mathbf{D}_u}^2 \right). \tag{29}$$

Now with $\mathbf{u} \in \mathbf{D}_u$ in hand, $\mathbf{D}_u \hookrightarrow \mathbf{W}^{2,2}$ [cf. (20)], we deal with (25). For the right-hand side we derive the estimates

$$\|a_\theta(g, \cdot)\|_{V_\theta^{0,2}} \leq c(\Omega, \kappa) \|g\|_{W^{2,2}(\Omega)}, \tag{30}$$

$$\|(\rho \alpha_2 \vartheta_0 \mathbf{f} \cdot \mathbf{u}_0, \cdot)_\Omega\|_{V_\theta^{0,2}} \leq c(\Omega, \rho, \alpha_2) \|(\mathbf{f}, \cdot)\|_{\mathbf{V}_u^{0,2}} \|\mathbf{u}_0\|_{\mathbf{D}_u} \|\vartheta_0\|_{D_\theta} \tag{31}$$

and

$$\|(\rho \alpha_2 g \mathbf{f} \cdot \mathbf{u}_0, \cdot)_\Omega\|_{V_\theta^{0,2}} \leq c(\Omega, \rho, \alpha_2) \|(\mathbf{f}, \cdot)\|_{\mathbf{V}_u^{0,2}} \|\mathbf{u}_0\|_{\mathbf{D}_u} \|g\|_{W^{2,2}(\Omega)}. \tag{32}$$

The inequalities (30)–(32) together with (19) and (21) yield the estimate for the solution of (25)

$$\|\vartheta\|_{D_\theta} \leq c_1(\Omega, \kappa) \|g\|_{W^{2,2}(\Omega)}$$

$$+ c_2(\Omega, \kappa, \rho, \alpha_2) \|(\mathbf{f}, \cdot)\|_{\mathbf{V}_u^{0,2}} (\|g\|_{W^{2,2}(\Omega)} + \|\vartheta_0\|_{D_\theta}) \|\mathbf{u}_0\|_{\mathbf{D}_u}$$

$$+ c_3(\Omega, \kappa) \left(\|d(\mathbf{u}, g + \vartheta_0, \cdot)\|_{V_\theta^{0,2}} + \|e(\mathbf{u}, \mathbf{u}, \cdot)\|_{V_\theta^{0,2}} \right), \tag{33}$$

where the last two norms can be further estimated using (29) to obtain

$$\|d(\mathbf{u}, g + \vartheta_0, \cdot)\|_{V_\theta^{0,2}} \leq c_1(\Omega, \rho, c_p) \|\mathbf{u}\|_{\mathbf{D}_u} (\|g\|_{W^{2,2}(\Omega)} + \|\vartheta_0\|_{D_\theta})$$

$$\leq c_2(\Omega, \rho, c_p, \nu, \alpha_0) (\|g\|_{W^{2,2}(\Omega)} + \|\vartheta_0\|_{D_\theta})$$

$$\left(\|(\mathbf{f}, \cdot)\|_{\mathbf{V}_u^{0,2}} (1 + \|g\|_{W^{2,2}(\Omega)} + \|\vartheta_0\|_{D_\theta}) + \|\mathbf{u}_0\|_{\mathbf{D}_u}^2 \right) \tag{34}$$

and (similarly)

$$\|e(\mathbf{u}, \mathbf{u}, \cdot)\|_{V_\theta^{0,2}} \leq c_1(\Omega, \nu, \alpha_1) \|\mathbf{u}\|_{\mathbf{D}_u}^2$$

$$\leq c_2(\Omega, \rho, \nu, \alpha_0, \alpha_1) \left(\|(\mathbf{f}, \cdot)\|_{\mathbf{V}_u^{0,2}} (1 + \|g\|_{W^{2,2}(\Omega)} + \|\vartheta_0\|_{D_\theta}) + \|\mathbf{u}_0\|_{\mathbf{D}_u}^2 \right)^2. \tag{35}$$

Now denote by $M(R_u, R_\theta) \subset X$ the bounded set $(R_u, R_\theta > 0)$

$$M(R_u, R_\theta) := \left\{ [\mathbf{w}, v] \in \mathbf{D}_u \times D_\theta;\; \|\mathbf{w}\|_{\mathbf{D}_u} \leq R_u \text{ and } \|v\|_{D_\theta} \leq R_\theta \right\}. \tag{36}$$

We are going to show that the map K,

$$K : \mathbf{D}_u \times D_\theta \to \mathbf{D}_u \times D_\theta \qquad \text{with } K([\mathbf{u}_0, \vartheta_0]) = [\mathbf{u}, \vartheta],$$

has a fixed point in $M(R_u, R_\theta)$ for some positive numbers R_u and R_θ and provided $\|\mathbf{f}\|_{\mathbf{V}_u^{0,2}}$ is "sufficiently small". Taking R_θ sufficiently large and R_u together with $\|\mathbf{f}\|_{\mathbf{V}_u^{0,2}}$ sufficiently small, the estimates (29) and (33) imply

$$K : M(R_u, R_\theta) \to M(R_u, R_\theta). \tag{37}$$

Moreover, our aim is to prove that $\|\mathbf{f}\|_{\mathbf{V}_u^{0,2}}$, R_θ and R_u can be chosen in such a way that, in addition to (37), K realizes contraction in $M(R_u, R_\theta)$.

Let $[\mathbf{u}_0, \vartheta_0], [\bar{\mathbf{u}}_0, \bar{\vartheta}_0] \in \mathbf{D}_u \times D_\theta$ and $K([\mathbf{u}_0, \vartheta_0]) = [\mathbf{u}, \vartheta], K([\bar{\mathbf{u}}_0, \bar{\vartheta}_0]) = [\bar{\mathbf{u}}, \bar{\vartheta}]$. By noting (19) we arrive at the estimates

$$\|\mathbf{u} - \bar{\mathbf{u}}\|_{\mathbf{D}_u} \leq c(\nu) \Big(\|b(\mathbf{u}_0 - \bar{\mathbf{u}}_0, \bar{\mathbf{u}}_0, \cdot)\|_{\mathbf{V}_u^{0,2}}$$
$$+ \|b(\mathbf{u}_0, \mathbf{u}_0 - \bar{\mathbf{u}}_0, \cdot)\|_{\mathbf{V}_u^{0,2}} + \|(\rho \alpha_0(\vartheta_0 - \bar{\vartheta}_0)\mathbf{f}, \cdot)\|_{\mathbf{V}_u^{0,2}} \Big) \tag{38}$$

and

$$\|\vartheta - \bar{\vartheta}\|_{D_\theta} \leq c(\kappa) \Big(\|(\rho \alpha_2 \mathbf{f}(\vartheta_0 - \bar{\vartheta}_0) \cdot \mathbf{u}_0, \cdot)_\Omega\|_{V_\theta^{0,2}}$$
$$+ \|(\rho \alpha_2 \mathbf{f} \bar{\vartheta}_0 \cdot (\mathbf{u}_0 - \bar{\mathbf{u}}_0), \cdot)_\Omega\|_{V_\theta^{0,2}} + \|(\rho \alpha_2 \mathbf{f} g \cdot (\mathbf{u}_0 - \bar{\mathbf{u}}_0), \cdot)_\Omega\|_{V_\theta^{0,2}}$$
$$+ \|e(\mathbf{u}, \mathbf{u} - \bar{\mathbf{u}}, \cdot)\|_{V_\theta^{0,2}} + \|e(\mathbf{u} - \bar{\mathbf{u}}, \bar{\mathbf{u}}, \cdot)\|_{V_\theta^{0,2}}$$
$$+ \|d(\mathbf{u}, \bar{\vartheta}_0 - \vartheta_0, \cdot)\|_{V_\theta^{0,2}} + \|d(\bar{\mathbf{u}} - \mathbf{u}, \bar{\vartheta}_0 + g, \cdot)\|_{V_\theta^{0,2}} \Big). \tag{39}$$

Estimating all terms on the right-hand side of (38) one obtains [applying (20) and (21)]

$$\|b(\mathbf{u}_0 - \bar{\mathbf{u}}_0, \bar{\mathbf{u}}_0, \cdot)\|_{\mathbf{V}_u^{0,2}} \leq c(\Omega, \rho)\|\mathbf{u}_0 - \bar{\mathbf{u}}_0\|_{\mathbf{D}_u}\|\bar{\mathbf{u}}_0\|_{\mathbf{D}_u}, \tag{40}$$

$$\|b(\mathbf{u}_0, \mathbf{u}_0 - \bar{\mathbf{u}}_0, \cdot)\|_{\mathbf{V}_u^{0,2}} \leq c(\Omega, \rho)\|\mathbf{u}_0\|_{\mathbf{D}_u}\|\mathbf{u}_0 - \bar{\mathbf{u}}_0\|_{\mathbf{D}_u}, \tag{41}$$

$$\|(\rho \alpha_0(\vartheta_0 - \bar{\vartheta}_0)\mathbf{f}, \cdot)\|_{\mathbf{V}_u^{0,2}} \leq c(\Omega, \rho, \alpha_0)\|(\mathbf{f}, \cdot)\|_{\mathbf{V}_u^{0,2}}\|\vartheta_0 - \bar{\vartheta}_0\|_{D_\theta} \tag{42}$$

and consequently

$$\|\mathbf{u} - \bar{\mathbf{u}}\|_{\mathbf{D}_u} \leq c_1(\Omega, \nu, \rho)(\|\mathbf{u}_0\|_{\mathbf{D}_u} + \|\bar{\mathbf{u}}_0\|_{\mathbf{D}_u})\|\mathbf{u}_0 - \bar{\mathbf{u}}_0\|_{\mathbf{D}_u}$$
$$+ c_2(\Omega, \nu, \rho, \alpha_0)\|(\mathbf{f}, \cdot)\|_{\mathbf{V}_u^{0,2}}\|\vartheta_0 - \bar{\vartheta}_0\|_{D_\theta}. \tag{43}$$

Similarly, estimating all terms on the right-hand side of (39) one obtains

$$\|(\rho\alpha_2\mathbf{f}(\vartheta_0-\bar{\vartheta}_0)\cdot\mathbf{u}_0,\cdot)_\Omega\|_{V_\theta^{0,2}} \leq c(\Omega,\rho,\alpha_2)\|(\mathbf{f},\cdot)\|_{\mathbf{V}_u^{0,2}}\|\vartheta_0-\bar{\vartheta}_0\|_{D_\theta}\|\bar{\mathbf{u}}_0\|_{\mathbf{D}_u}, \quad (44)$$

$$\|(\rho\alpha_2\mathbf{f}\bar{\vartheta}_0\cdot(\mathbf{u}_0-\bar{\mathbf{u}}_0),\cdot)_\Omega\|_{V_\theta^{0,2}} \leq c(\Omega,\rho,\alpha_2)\|(\mathbf{f},\cdot)\|_{\mathbf{V}_u^{0,2}}\|\bar{\vartheta}_0\|_{D_\theta}\|\mathbf{u}_0-\bar{\mathbf{u}}_0\|_{\mathbf{D}_u}, \quad (45)$$

$$\|(\rho\alpha_2\mathbf{f}g\cdot(\mathbf{u}_0-\bar{\mathbf{u}}_0),\cdot)_\Omega\|_{V_\theta^{0,2}} \leq c(\Omega,\rho,\alpha_2)\|(\mathbf{f},\cdot)\|_{\mathbf{V}_u^{0,2}}\|g\|_{W^{2,2}(\Omega)}\|\mathbf{u}_0-\bar{\mathbf{u}}_0\|_{\mathbf{D}_u}. \quad (46)$$

For the dissipative terms, we arrive at

$$\|e(\mathbf{u},\mathbf{u}-\bar{\mathbf{u}},\cdot)\|_{V_\theta^{0,2}} \leq c(\Omega,\nu,\alpha_1)\|\mathbf{u}\|_{\mathbf{D}_u}\|\mathbf{u}-\bar{\mathbf{u}}\|_{\mathbf{D}_u}, \quad (47)$$

$$\|e(\mathbf{u}-\bar{\mathbf{u}},\bar{\mathbf{u}},\cdot)\|_{V_\theta^{0,2}} \leq c(\Omega,\nu,\alpha_1)\|\mathbf{u}-\bar{\mathbf{u}}\|_{\mathbf{D}_u}\|\bar{\mathbf{u}}\|_{\mathbf{D}_u} \quad (48)$$

and finally

$$\|d(\mathbf{u},\bar{\vartheta}_0-\vartheta_0,\cdot)\|_{V_\theta^{0,2}} \leq c(\Omega,\rho,c_p)\|\mathbf{u}\|_{\mathbf{D}_u}\|\bar{\vartheta}_0-\vartheta_0\|_{D_\theta}, \quad (49)$$

$$\|d(\bar{\mathbf{u}}-\mathbf{u},g+\bar{\vartheta}_0,\cdot)\|_{V_\theta^{0,2}} \leq c(\Omega,\rho,c_p)\|\bar{\mathbf{u}}-\mathbf{u}\|_{\mathbf{D}_u}(\|g\|_{W^{2,2}(\Omega)}+\|\bar{\vartheta}_0\|_{D_\theta}). \quad (50)$$

Applying the estimates (44)–(50) into (39), we get

$$\|\vartheta-\bar{\vartheta}\|_{D_\theta} \leq c_1(\Omega,\rho,\alpha_2,\kappa)\|(\mathbf{f},\cdot)\|_{\mathbf{V}_u^{0,2}}(\|\bar{\vartheta}_0\|_{D_\theta}+\|g\|_{W^{2,2}(\Omega)})\|\mathbf{u}_0-\bar{\mathbf{u}}_0\|_{\mathbf{D}_u}$$
$$+\left(c_2(\Omega,\rho,\alpha_2,\kappa)\|(\mathbf{f},\cdot)\|_{\mathbf{V}_u^{0,2}}\|\bar{\mathbf{u}}_0\|_{\mathbf{D}_u}+c_3(\Omega,\rho,c_p,\kappa)\|\mathbf{u}\|_{\mathbf{D}_u}\right)\|\vartheta_0-\bar{\vartheta}_0\|_{D_\theta}$$
$$+\left(c_4(\Omega,\nu,\alpha_1,\kappa)(\|\mathbf{u}\|_{\mathbf{D}_u}+\|\bar{\mathbf{u}}\|_{\mathbf{D}_u})+c_5(\Omega,\rho,c_p,\kappa)(\|g\|_{W^{2,2}(\Omega)}+\|\bar{\vartheta}_0\|_{D_\theta})\right)\|\bar{\mathbf{u}}-\mathbf{u}\|_{\mathbf{D}_u}. \quad (51)$$

Now we are ready to conclude that one can choose R_θ sufficiently large and R_u together with $\|\mathbf{f}\|_{\mathbf{V}_u^{0,2}}$ sufficiently small, such that K maps $M(R_u,R_\theta)$ into itself [as a consequence of (29) and (33)] and realizes contraction in the set $M(R_u,R_\theta)$ [as a consequence of (43) and (51)]. Hence, we have a fixed point $[\mathbf{u},\vartheta]=K([\mathbf{u},\vartheta])$, and the couple $[\mathbf{u},\theta]:=[\mathbf{u},\vartheta+g]$ is the strong solution to the system (1)–(7).

Acknowledgements This research was supported by the project GAČR P201/10/P396.

References

1. Beneš, M.: Solutions to the mixed problem of viscous incompressible flows in a channel. Arch. Math. **93**, 287–297 (2009)
2. Frehse, J.: A discontinuous solution of mildly nonlinear elliptic systems. Math. Z. **134**, 229–230 (1973)

3. Galdi, G.P., Rannacher, R., Robertson, A.M., Turek, S.: Hemodynamical Flows: Modeling, Analysis and Simulation. Oberwolfach Seminars vol. 37. Birkhauser, Basel (2008)
4. Gresho, P.M.: Incompressible fluid dynamics: some fundamental formulation issues. Ann. Rev. Fluid Mech. **23**, 413–453 (1991)
5. Kagei, Y., Ružička, M., Thäter, G.: Natural convection wit dissipative heating. Commun. Math. Phys. **214**, 287–313 (2000)
6. Kračmar, S., Neustupa, J.: A weak solvability of a steady variational inequality of the Navier-Stokes type with mixed boundary conditions. Nonlinear Anal. **47**, 4169–4180 (2001)
7. Kufner, A., Sändig, A.M.: Some Applications of Weighted Sobolev Spaces. Teubner-Texte zur Mathematik, Leipzig (1987)
8. Kufner, A., John, O., Fučík, S.: Function Spaces. Academia, Prague (1977)
9. Orlt, M., Sändig, A.M.: Regularity of viscous Navier-Stokes flows in nonsmooth domains. Lect. Notes Pure Appl. Math. **167**, 185–201 (1993)

Infinite-Horizon Variational Principles and Almost Periodic Oscillations

Joël Blot and Abdelkader Bouadi

Abstract We give variational principles for bounded, almost periodic and asymptotic almost periodic solutions to characterise the bounded, almost periodic and asymptotic almost periodic solutions of Cauchy problems of Euler–Lagrange equations in the form $L_u(t,u(t),u'(t)) + rL_{u'}(t,u(t),u'(t)) - \frac{d}{dt}L_{u'}(t,u(t),u'(t)) = b(t)$. Using an adapted weighted Sobolev space, we establish that the b which are bounded (respectively almost periodic, respectively asymptotically almost periodic) for which there exists a bounded (respectively almost periodic, respectively asymptotic almost periodic) solution to the previous equation (with an initial condition) is dense with respect to a special norm.

Keywords Almost-periodic function • Asymptotically almost periodic function • Euler–Lagrange equation • Infinite-horizon variational problems

Mathematical Subject Classification: 34C27, 49J75, 42A75.

J. Blot (✉)
Laboratoire SAMM EA4543, Université Paris 1 Panthéon-Sorbonne, centre P.M.F., 90 rue de Tolbiac, 75634 Paris cedex 13, France
e-mail: blot@univ-paris1.fr

A. Bouadi
Laboratoire SAMM EA4543, Université Paris 1 Panthéon-Sorbonne, Université Ibn Khaldoune Tiaret Algrie, centre P.M.F., 90 rue de Tolbiac, 75634 Paris cedex 13, France
e-mail: kbouadifr@yahoo.fr

S. Pinelas et al. (eds.), *Differential and Difference Equations with Applications*, Springer Proceedings in Mathematics & Statistics 47, DOI 10.1007/978-1-4614-7333-6_23,
© Springer Science+Business Media New York 2013

1 Introduction

When $r \in (0,\infty)$ and when $L : \mathbb{R}_+ \times \mathbb{R}^n \times \mathbb{R}^n \to \mathbb{R}$ is a function which admits a partial differential with respect to the second (vector) variable, denoted by L_u, and a partial differential with respect to the third (vector) variable, denoted by $L_{u'}$, we can formulate the following Euler–Lagrange differential equation:

$$L_u(t, u(t), u'(t)) = -rL_{u'}(t, u(t), u'(t)) + \frac{d}{dt}L_{u'}(t, u(t), u'(t)), \tag{1}$$

with an initial condition $u(0) = x$ (Cauchy problem). Such equations arise in various scientific fields: in the models of macroeconomic optimal growth [11, 15, 24] and in the management of fisheries or of forests [16]. They appear as first-order necessary condition of optimality for variational problems with an infinite-horizon criterion in the form

$$J(u) := \int_0^\infty e^{-rt} L(t, u(t), u'(t)) dt : \tag{2}$$

see, for instance, [7–9].

When $b : \mathbb{R} \to \mathbb{R}^n$ is a forcing term, we also consider the following equation:

$$L_u(t, u(t), u'(t)) + rL_{u'}(t, u(t), u'(t)) - \frac{d}{dt}L_{u'}(t, u(t), u'(t)) = b(t), \tag{3}$$

ever with an initial condition $u(0) = x$. The aim of this paper is to study the bounded solution of (3) when b is bounded, the almost periodic solutions of (3) when b is almost periodic, and the asymptotically almost periodic solutions of (3) when b is asymptotically almost periodic. About the almost periodic solutions, this variational formalism is different to the so-called calculus of variations in mean developed in [4–6], where the criterion uses a mean value instead an integral.

Now we briefly describe the contents of the paper. In Sect. 2 we precise the function spaces that we use, and we establish several results on the density of certain spaces into other spaces: spaces of bounded functions, spaces of almost periodic functions, spaces of asymptotically almost periodic functions and a weighted Sobolev space. In Sect. 3 we establish a variational principle for the bounded solutions of (1). In Sect. 4 we establish a variational principle for the almost periodic solutions of (1). In Sect. 5 we establish a variational principle for the asymptotically almost periodic solutions of (1). In Sect. 6 we establish a variational principle for some locally absolutely continuous solutions of (1) by using the weighted Sobolev space previously considered. In Sect. 7 we establish that, for all $x \in \mathbb{R}^n$, the set of the forcing b which are bounded (respectively almost periodic, respectively asymptotic almost periodic) for which there exists a bounded (respectively almost periodic, respectively asymptotic almost periodic) solution u such that $u(0) = x$ is dense into the space of the bounded (respectively almost periodic, respectively asymptotic almost periodic) functions with respect to an adapted norm.

2 Function Spaces

The usual inner product on $\mathbb{R}^n$ is simply denoted by a point. $x.y = \sum_{j=1}^{n} x_j y_j$ when $x = (x_1, \ldots, x_n)$ and $y = (y_1, \ldots, y_n)$, and the associated Euclidean norm is denoted as $|x| = \sqrt{x.x}$.

$C^0(\mathbb{R}_+, \mathbb{R}^n)$ is the space of the continuous functions from $\mathbb{R}_+ = [0, \infty)$ into $\mathbb{R}^n$. Setting $\mathbb{N}_* := \mathbb{N} \setminus \{0\}$, when $k \in \mathbb{N}_* \cup \{\infty\}$, $C^k(\mathbb{R}_+, \mathbb{R}^n)$ is the space of the k-times differentiable functions from $\mathbb{R}_+$ into $\mathbb{R}^n$. $BC^0(\mathbb{R}_+, \mathbb{R}^n)$ is the space of the bounded continuous functions from $\mathbb{R}_+$ into $\mathbb{R}^n$; endowed with the norm $\|u\|_\infty := \sup_{t \in \mathbb{R}_+} |u(t)|$, it is a Banach space [3] (Proposition 1, p. 25). When $k \in \mathbb{N}_*$, $BC^k(\mathbb{R}_+, \mathbb{R}^n)$ is the space of the functions $u \in BC^0(\mathbb{R}_+, \mathbb{R}^n) \cap C^k(\mathbb{R}_+, \mathbb{R}^n)$ such that $u^{(j)} = \frac{d^j u}{dt^j} \in BC^0(\mathbb{R}_+, \mathbb{R}^n)$ for all $j = 1, \ldots, k$; endowed with the norm $\|u\|_{BC^k} := \|u\|_\infty + \sum_{j=1}^{k} \|u^{(j)}\|_\infty$, it is a Banach space. $C_0^0(\mathbb{R}_+, \mathbb{R}^n)$ is the space of the functions $u \in C^0(\mathbb{R}_+, \mathbb{R}^n)$ such that $\lim_{t \to \infty} u(t) = 0$; it is a Banach subspace of $(BC^0(\mathbb{R}_+, \mathbb{R}^n), \|.\|_\infty)$. When $k \in \mathbb{N}_* \cup \{\infty\}$, $C_0^k(\mathbb{R}_+, \mathbb{R}^n)$ is the space of the functions $u \in C_0^0(\mathbb{R}_+, \mathbb{R}^n) \cap C^k(\mathbb{R}_+, \mathbb{R}^n)$ such that $u^{(j)} = \frac{d^j u}{dt^j} \in C_0^0(\mathbb{R}_+, \mathbb{R}^n)$ for all $j = 1, \ldots, k$.

When Ω is a nonempty open subset of $\mathbb{R}$, $C_c^0(\Omega, \mathbb{R}^n)$ is the space of the continuous functions from Ω into $\mathbb{R}^n$ which have a compact support. When $k \in \mathbb{N}_* \cup \{\infty\}$, $C_c^k(\Omega, \mathbb{R}^n)$ is the space of the k-times differentiable functions from Ω into $\mathbb{R}^n$ which have a compact support. $\mathscr{D}'(\Omega)$ is the space of the distributions on Ω [25]. $\mathscr{S}'(\mathbb{R}, \mathbb{R})$ denotes the space of the tempered distributions on $\mathbb{R}$ [21, 25].

When $T \in (0, \infty)$, $P_T^0(\mathbb{R}_+, \mathbb{R}^n)$ denotes the space of the T-periodic continuous functions from $\mathbb{R}_+$ into $\mathbb{R}^n$, and $P_T^1(\mathbb{R}_+, \mathbb{R}^n)$ denotes the space $P_T^0(\mathbb{R}_+, \mathbb{R}^n) \cap C^1(\mathbb{R}_+, \mathbb{R}^n)$.

$AP^0(\mathbb{R}_+, \mathbb{R}^n)$ is the space of the Bohr almost periodic functions from $\mathbb{R}_+$ into $\mathbb{R}^n$ [27, 28]; it is a Banach subspace of $(BC^0(\mathbb{R}_+, \mathbb{R}^n), \|.\|_\infty)$ [28]. When M is a $\mathbb{Z}$-submodule of $\mathbb{R}$, $AP^0(\mathbb{R}_+, \mathbb{R}^n; M)$ denotes the space of the functions which belong to $AP^0(\mathbb{R}_+, \mathbb{R}^n)$ such that their module of frequencies is included into M; it is a Banach subspace of $AP^0(\mathbb{R}_+, \mathbb{R}^n)$. When $k \in \mathbb{N}_*$, $AP^k(\mathbb{R}_+, \mathbb{R}^n; M)$ is the space of the functions $u \in AP^0(\mathbb{R}_+, \mathbb{R}^n; M) \cap C^k(\mathbb{R}_+, \mathbb{R}^n)$ such that $u^{(j)} \in AP^0(\mathbb{R}_+, \mathbb{R}^n; M)$ for all $j = 1, \ldots, k$. $AP^k(\mathbb{R}_+, \mathbb{R}^n; M)$ is a Banach subspace of $(BC^k(\mathbb{R}_+, \mathbb{R}^n), \|.\|_{BC^k})$.

$AAP^0(\mathbb{R}_+, \mathbb{R}^n)$ is the space of the $u \in BC^0(\mathbb{R}_+, \mathbb{R}^n)$ such that $u = v + w$ where $v \in AP^0(\mathbb{R}_+, \mathbb{R}^n)$ and $w \in C_0^0(\mathbb{R}_+, \mathbb{R}^n)$; it is a Banach subspace to $BC^0(\mathbb{R}_+, \mathbb{R}^n)$ [28]. The functions which belong to $AAP^0(\mathbb{R}_+, \mathbb{R}^n)$ are so-called the asymptotically almost periodic functions [27, 28]. When $k \in \mathbb{N}_*$, $AAP^k(\mathbb{R}_+, \mathbb{R}^n)$ is the space of the $u \in AAP^0(\mathbb{R}_+, \mathbb{R}^n) \cap C^k(\mathbb{R}_+, \mathbb{R}^n)$ such that their derivative $u^{(j)}$ belongs to $AAP^0(\mathbb{R}_+, \mathbb{R}^n)$ for all $j = 1, \ldots, k$; it is a Banach subspace of $BC^k(\mathbb{R}_+, \mathbb{R}^n)$.

We denote by μ the Lebesgue measure on $\mathbb{R}_+$. We consider the function $e_r : \mathbb{R}_+ \to \mathbb{R}$ defined by $e_r(t) := e^{-rt}$. Denoting by $\mathscr{B}(\mathbb{R}_+)$ the Borel σ-field of $\mathbb{R}_+$, we define the finite positive measure $\nu : \mathscr{B}(\mathbb{R}_+) \to \mathbb{R}_+$ by setting $\nu(B) := \int_B e_r d\mu$ for all $B \in \mathscr{B}(\mathbb{R}_+)$; e_r is the density of ν with respect to μ [1]. When $p \in [1, \infty)$, $L^p(\mathbb{R}_+, \mu; \mathbb{R}^n)$ (respectively $L^p(\mathbb{R}_+, \nu; \mathbb{R}^n)$) denotes the Lebesgue space of the measurable functions $u : \mathbb{R}_+ \to \mathbb{R}^n$ such that $|u|^p$ is μ integrable (respectively ν

integrable) on $\mathbb{R}_+$. The usual norm on $L^p(\mathbb{R}_+,\mu;\mathbb{R}^n)$ (respectively $L^p(\mathbb{R}_+,\nu;\mathbb{R}^n)$) is denoted by $\|.\|_{L^p(\mu)}$ (respectively $\|.\|_{L^p(\nu)}$). Note that $u \in L^1(\mathbb{R}_+,\nu;\mathbb{R}^n)$ if and only if $e_r u \in L^1(\mathbb{R}_+,\mu;\mathbb{R}^n)$ and $\|u\|_{L^1(\nu)} = \|e_r u\|_{L^1(\mu)}$. The inner product of the Hilbert space $L^2(\mathbb{R}_+,\nu;\mathbb{R}^n)$ is $(u,v)_{L^2(\nu)} = \int_{\mathbb{R}_+} e_r u.v d\mu = (\sqrt{e_r}u, \sqrt{e_r}v)_{L^2(\mu)}$.

Remark 1. We have $L^2(\mathbb{R}_+,\nu;\mathbb{R}^n) \subset L^1(\mathbb{R}_+,\nu;\mathbb{R}^n)$, and for all $u \in L^2(\mathbb{R}_+,\nu;\mathbb{R}^n)$, the following inequality holds: $\|u\|_{L^1(\nu)} \leq \frac{1}{\sqrt{r}}.\|u\|_{L^2(\nu)}$.

Since the measure of a singleton, for μ and for ν, is equal to zero, we can assimilate $L^2(\mathbb{R}_+,\nu;\mathbb{R}^n)$ and $L^2((0,\infty),\nu;\mathbb{R}^n)$, and so when $u = (u_1,\ldots,u_n) \in L^2(\mathbb{R}_+,\nu;\mathbb{R}^n)$, we can define the distributional derivative Du_j of u_j in $\mathscr{D}'((0,\infty),\mathbb{R})$, and we can define the following weighted Sobolev space:

$$H^1(\mathbb{R}_+,\nu;\mathbb{R}^n) := \{u \in L^2(\mathbb{R}_+,\nu;\mathbb{R}^n) : \forall j = 1,\ldots,n,\ Du_j \in L^2(\mathbb{R}_+,\nu;\mathbb{R})\}.$$

Endowed with the inner product $(u,v)_{H^1(\nu)} := (u,v)_{L^2(\nu)} + \sum_{j=1}^n (Du_j, Dv_j)_{L^2(\nu)}$, $H^1(\mathbb{R}_+,\nu;\mathbb{R}^n)$ is a Hilbert space. Note that we have $BC^1(\mathbb{R}_+,\mathbb{R}^n) \subset H^1(\mathbb{R}_+,\nu;\mathbb{R}^n)$. When $u \in H^1(\mathbb{R}_+,\nu;\mathbb{R}^n)$, it is easy to verify that u is locally absolutely continuous on $\mathbb{R}_+$, and so the distributional derivatives Du_j are represented by the ordinary derivatives u'_j (defined Lebesgue-almost everywhere) [25]. More deep properties of this kind of space can be found in [2, 22].

Proposition 1. *The following assertions hold:*

(i) $C_0^0(\mathbb{R}_+,\mathbb{R}^n)$ *is dense into* $L^2(\mathbb{R}_+,\nu;\mathbb{R}^n)$.
(ii) $BC^0(\mathbb{R}_+,\mathbb{R}^n)$ *is dense into* $L^2(\mathbb{R}_+,\nu;\mathbb{R}^n)$.
(iii) $AAP^0(\mathbb{R}_+,\mathbb{R}^n)$ *is dense into* $L^2(\mathbb{R}_+,\nu;\mathbb{R}^n)$.
(iv) $C_0^1(\mathbb{R}_+,\mathbb{R}^n)$ *is dense into* $H^1(\mathbb{R}_+,\nu;\mathbb{R}^n)$.
(v) $AAP^1(\mathbb{R}_+,\mathbb{R}^n)$ *is dense into* $H^1(\mathbb{R}_+,\nu;\mathbb{R}^n)$.

Proof. By using [14] (Théorème IV.12), we know that $C_c^0((0,\infty),\mathbb{R}^n)$ is dense in $L^2(\mathbb{R}_+,\mu;\mathbb{R}^n)$. For $f \in L^2(\mathbb{R}_+,\nu;\mathbb{R}^n)$ and $\varepsilon > 0$. Then, since $\sqrt{e_r}f \in L^2(\mathbb{R}_+,\mu;\mathbb{R}^n)$, there exists $\varphi \in C_c^0((0,\infty),\mathbb{R}^n)$ such that $\|\varphi - \sqrt{e_r}f\|_{L^2(\mu)} \leq \varepsilon$. Note that $\psi := \frac{1}{\sqrt{e_r}}\varphi \in C_c^0((0,\infty),\mathbb{R}^n)$, and that we have $\|\psi - f\|_{L^2(\nu)} = \|\sqrt{e_r}\psi - \sqrt{e_r}f\|_{L^2(\mu)} = \|\varphi - \sqrt{e_r}f\|_{L^2(\mu)} \leq \varepsilon$. And so we have proven that $C_c^0((0,\infty),\mathbb{R}^n)$ is dense into $L^2(\mathbb{R}_+,\nu;\mathbb{R}^n)$. Since $C_c^0((0,\infty),\mathbb{R}^n) \subset C_0^0((0,\infty),\mathbb{R}^n)$ we obtain the assertion (i).

Since $C_0^0((0,\infty),\mathbb{R}^n) \subset BC^0(\mathbb{R}_+,\mathbb{R}^n)$, the assertion (ii) is a consequence of (i). Since $C_0^0((0,\infty),\mathbb{R}^n) \subset AAP^0(\mathbb{R}_+,\mathbb{R}^n;M)$, the assertion (iii) is a consequence of (i).

By using [14] (Théorème VIII.6), we know that $C_c^1((0,\infty),\mathbb{R}^n)$ is dense into $H^1(\mathbb{R}_+,\mu;\mathbb{R}^n) := \{u \in L^2(\mathbb{R}_+,\mu;\mathbb{R}^n) : \forall j = 1,\ldots,n,\ Du_j \in L^2(\mathbb{R}_+,\mu;\mathbb{R})\}$. We fix $u \in H^1(\mathbb{R}_+,\nu;\mathbb{R}^n)$ and $\varepsilon > 0$. We set $\varepsilon_1 := \frac{\varepsilon}{\sqrt{1+(1+\frac{r}{2})^2}}$. Since $u \in H^1(\mathbb{R}_+,\nu;\mathbb{R}^n)$, we have $\sqrt{e_r}u \in L^2(\mathbb{R}_+,\mu;\mathbb{R}^n)$, and for all $j = 1,\ldots,n$, $\sqrt{e_r}Du_j \in L^2(\mathbb{R}_+,\mu;\mathbb{R})$. Note that we have $D(\sqrt{e_r}u_j) = \frac{-r}{2}\sqrt{e_r}u_j + \sqrt{e_r}Du_j \in L^2(\mathbb{R}_+,\mu;\mathbb{R})$. Then there exists $v \in C_c^1((0,\infty),\mathbb{R}^n)$ such that $\|v - \sqrt{e_r}u\|_{L^2(\mu)} \leq \varepsilon_1$ and $\|v' - (\frac{-r}{2}\sqrt{e_r}u + \sqrt{e_r}Du)\|_{L^2(\mu)} \leq \varepsilon_1$ where $Du = (Du_1,\ldots,Du_n)$. Note also that $w := \frac{1}{\sqrt{e_r}}v \in C_c^1((0,\infty),\mathbb{R}^n)$. We have

$\|w - u\|_{L^2(v)} = \|\sqrt{e_r}w - \sqrt{e_r}u\|_{L^2(\mu)} = \|v - \sqrt{e_r}u\|_{L^2(\mu)} \le \varepsilon_1$. Since $v' = \frac{-r}{2}\sqrt{e_r}w + \sqrt{e_r}w'$, we obtain $\|\|\sqrt{e_r}w' - \sqrt{e_r}Du\|_{L^2(\mu)} - \|\frac{-r}{2}(\sqrt{e_r}w - \sqrt{e_r}u)\|_{L^2(\mu)}\| \le \|\frac{-r}{2}(\sqrt{e_r}w - \sqrt{e_r}u) + \sqrt{e_r}w' - \sqrt{e_r}Du\|_{L^2(\mu)} = \|\frac{-r}{2}\sqrt{e_r}w + \sqrt{e_r}w' - (\frac{-r}{2}\sqrt{e_r}u + \sqrt{e_r}Du)\|_{L^2(\mu)} = \|v' - (\frac{-r}{2}\sqrt{e_r}u + \sqrt{e_r}Du)\|_{L^2(\mu)} \le \varepsilon_1$ that implies $\|\sqrt{e_r}w' - \sqrt{e_r}Du\|_{L^2(\mu)} \le \varepsilon_1 + \frac{r}{2}\|\sqrt{e_r}w - \sqrt{e_r}u\|_{L^2(\mu)} \le \varepsilon_1 + \frac{r}{2}\varepsilon_1 = (1 + \frac{r}{2})\varepsilon_1$. Then we obtain $\|w - u\|_{H^1(v)} \le \sqrt{\varepsilon_1^2 + (1 + \frac{r}{2})^2\varepsilon_1^2} = \varepsilon$. And so we have proven that $C_c^1((0,\infty),\mathbb{R}^n)$ is dense into $H^1(\mathbb{R}_+, v; \mathbb{R}^n)$. Since $C_c^1((0,\infty),\mathbb{R}^n) \subset C_0^1(\mathbb{R}_+, \mathbb{R}^n)$, the assertion (iv) is proven. Since $C_0^1((0,\infty),\mathbb{R}^n) \subset AAP^1(\mathbb{R}_+, \mathbb{R}^n; M)$, the assertion (iv) implies (v). $\square$

When $x \in \mathbb{R}^n$ and when $k \in \{0, 1, 2\}$, we set $AP^k(\mathbb{R}_+, \mathbb{R}^n; M, x) := \{u \in AP^k(\mathbb{R}_+, \mathbb{R}^n; M) : u(0) = x\}$, $AAP^k(\mathbb{R}_+, \mathbb{R}^n, x) := \{u \in AAP^k(\mathbb{R}_+, \mathbb{R}^n; M) : u(0) = x\}$, $BC^k(\mathbb{R}_+, \mathbb{R}^n; x) := \{u \in BC^k(\mathbb{R}_+, \mathbb{R}^n) : u(0) = x\}$ and $H^1(\mathbb{R}_+, v; \mathbb{R}^n; x) := \{u \in H^1(\mathbb{R}_+, v; \mathbb{R}^n) : u(0) = x\}$.

Proposition 2. *Let M be a $\mathbb{Z}$-submodule of $\mathbb{R}$ which is generated by at least two $\mathbb{Z}$-linearly independent real numbers, and let x be an arbitrary vector in $\mathbb{R}^n$. The two following assertions hold:*

(i) $AP^0(\mathbb{R}_+, \mathbb{R}^n; M)$ is dense into $L^2(\mathbb{R}_+, v; \mathbb{R}^n)$.
(ii) $AP^2(\mathbb{R}_+, \mathbb{R}^n, M)$ and $AP^1(\mathbb{R}_+, \mathbb{R}^n, M)$ are dense into $H^1(\mathbb{R}_+, v; \mathbb{R}^n)$.
(iii) $AP^2(\mathbb{R}_+, \mathbb{R}^n; M, x)$ is dense into $H^1(\mathbb{R}_+, v; \mathbb{R}^n; x)$.
(iv) $BC^2(\mathbb{R}_+, \mathbb{R}^n; x)$ is dense into $H^1(\mathbb{R}_+, v; \mathbb{R}^n; x)$.
(v) $AAP^2(\mathbb{R}_+, \mathbb{R}^n, x)$ is dense into $H^1(\mathbb{R}_+, v; \mathbb{R}^n; x)$.
(vi) $C_c^1((0,\infty),\mathbb{R}^n)$ is dense into $H^1(\mathbb{R}_+, v; \mathbb{R}^n; 0)$.
(vii) $H^1(\mathbb{R}_+, v; \mathbb{R}^n; 0)$ is dense into $L^2(\mathbb{R}_+, v; \mathbb{R}^n)$.

Proof. To prove the assertion (i), we will prove that the orthogonal subspace to $AP^0(\mathbb{R}_+, \mathbb{R}^n; M)$ in $L^2(\mathbb{R}_+, v; \mathbb{R}^n)$ is reduced to the singleton $\{0\}$. Let $f = (f_1, \ldots, f_n) \in L^2(\mathbb{R}_+, v; \mathbb{R}^n)$ be an orthogonal function to $AP^0(\mathbb{R}_+, \mathbb{R}^n; M)$. We fix $j \in \{1, \ldots, n\}$, then for all $\varphi \in AP^0(\mathbb{R}_+, \mathbb{R}; M)$, we have $\int_{\mathbb{R}_+} e_r f_j \varphi d\mu = 0$. Therefore, setting $e_r(t) := 0$ and $f(t) := 0$ when $t < 0$, for all $\lambda \in M$, successively by taking $\varphi(t) = \cos(-\lambda t)$ and $\varphi(t) = \sin(-\lambda t)$, we obtain $\int_{\mathbb{R}_+} e_r(t) f_j(t) e^{-i\lambda t} d\mu(t) = 0$, i.e. $\int_{\mathbb{R}} e_r(t) f_j(t) 1_{\mathbb{R}_+}(t) e^{-i\lambda t} dt = 0$, i.e. $\mathscr{F}(\tilde{e}_r \tilde{f}_j 1_{\mathbb{R}_+})(\lambda) = 0$ for all $\lambda \in M$, where $\mathscr{F}$ denotes the Fourier transform, defined by $\mathscr{F}(\phi)(\lambda) := \int_{\mathbb{R}} \phi(t) e^{-i\lambda t} dt$, and where $\tilde{e}_r$ (respectively $\tilde{f}$) is the extension of e_r (respectively f) to $\mathbb{R}$ by the value 0 on $(-\infty, 0)$. Using [13] (Proposition 1, p. TG V.1) we know that M is dense in $\mathbb{R}$, and since $\mathscr{F}(e_r f_j 1_{\mathbb{R}_+})$ is continuous [26] (Theorem 1.1, p. 2), we obtain that $\mathscr{F}(\tilde{e}_r \tilde{f}_j 1_{\mathbb{R}_+})(\lambda) = 0$ for all $\lambda \in \mathbb{R}$. Then by using the uniqueness theorem on Fourier transforms [26] (Corollary 1.22, p. 12), we obtain $\tilde{e}_r \tilde{f}_j 1_{\mathbb{R}_+} = 0$ Lebesgue-a.e. on $\mathbb{R}$. Therefore we obtain $\tilde{e}_r \tilde{f}_j = 0$ μ-a.e. on $\mathbb{R}_+$, and then $f_j = 0$ v-a.e. on $\mathbb{R}_+$. Consequently we have $f = 0$ in $L^2(\mathbb{R}_+, v; \mathbb{R}^n)$. And so the assertion (i) is proven.

To prove the assertion (ii), note that the density of $AP^2(\mathbb{R}_+, \mathbb{R}^n; M)$ implies this one of $AP^1(\mathbb{R}_+, \mathbb{R}^n; M)$. And to prove the density of $AP^2(\mathbb{R}_+, \mathbb{R}^n; M)$, we will prove that the orthogonal subspace to $AP^2(\mathbb{R}_+, \mathbb{R}^n; M)$ in $H^1(\mathbb{R}_+, v; \mathbb{R}^n)$ is reduced to the singleton $\{0\}$. Let $f \in H^1(\mathbb{R}_+, v; \mathbb{R}^n)$ be orthogonal to $AP^2(\mathbb{R}_+, \mathbb{R}^n; M)$.

Then for all $\varphi \in AP^2(\mathbb{R}_+,\mathbb{R}^n;M)$, we have $\int_{\mathbb{R}_+} e_r f.\varphi d\mu + \int_{\mathbb{R}_+} e_r f'.\varphi' d\mu = 0$. Considering the functions which have $t \mapsto \cos(-\lambda t)$ and $t \mapsto \sin(-\lambda t)$ as one of their coordinates and the value 0 for their other coordinates, these functions belong to $AP^2(\mathbb{R}_+,\mathbb{R}^n;M)$ when $\lambda \in M$, and we obtain $\int_0^\infty e^{-rt} f_j(t)e^{-i\lambda t}dt +$ $\int_0^\infty e^{-rt} f_j'(t)e^{-i\lambda t}(-i\lambda)dt = 0$, for all $j = 1,\ldots,n$, i.e.

$$\forall \lambda \in M, \int_0^\infty e^{-rt} f(t)e^{-i\lambda t}dt = i\lambda \int_0^\infty e^{-rt} f'(t)e^{-i\lambda t}. \tag{4}$$

Recall that when $\phi \in L^1(\mathbb{R},\mathbb{R}^n)$ (for the Lebesgue measure) with $\phi' \in L^1(\mathbb{R},\mathbb{R}^n)$ then [26] (Theorem 1.7, p. 4), we have

$$\forall \lambda \in \mathbb{R}, \ \mathscr{F}(\phi')(\lambda) = i\lambda \mathscr{F}(\phi)(\lambda). \tag{5}$$

Then (4) becomes

$$\forall \lambda \in M, \ \mathscr{F}(\tilde{e}_r \tilde{f})(\lambda) = i\lambda \mathscr{F}(\tilde{e}_r \tilde{f}')(\lambda). \tag{6}$$

As a consequence of (5) we obtain $\forall \lambda \in M, \ \mathscr{F}(\frac{d}{dt}(\tilde{e}_r \tilde{f}))(\lambda) = i\lambda \mathscr{F}(\tilde{e}_r \tilde{f})(\lambda)$. For μ-almost all $t \in \mathbb{R}_+$, we have $\frac{d}{dt}(e^{-rt} f(t)) = -re^{-rt} f(t) + e^{-rt} f'(t)$, so that for Lebesgue-almost all $t \in \mathbb{R}_+$, we have $\frac{d}{dt}(e^{-rt} 1_{\mathbb{R}_+}(t)f(t)) = -re^{-rt} 1_{\mathbb{R}_+}(t)f(t) +$ $e^{-rt} 1_{\mathbb{R}_+}(t)f'(t)$ that implies $\mathscr{F}(\frac{d}{dt}(\tilde{e}_r \tilde{f}))(\lambda) = -r\mathscr{F}(\tilde{e}_r \tilde{f})(\lambda) + \mathscr{F}(\tilde{e}_r \tilde{f}')(\lambda)$, for all $\lambda \in M$.Therefore we obtain $i\lambda \mathscr{F}(\tilde{e}_r \tilde{f})(\lambda) = -r\mathscr{F}(\tilde{e}_r \tilde{f})(\lambda) + \mathscr{F}(\tilde{e}_r \tilde{f}')(\lambda)$, and using (6) we have $(i\lambda)^2 \mathscr{F}(\tilde{e}_r \tilde{f}')(\lambda) = -r(i\lambda)\mathscr{F}(\tilde{e}_r \tilde{f}')(\lambda) + \mathscr{F}(\tilde{e}_r \tilde{f}')(\lambda)$, which give $[-(\lambda^2 + 1) + i(r\lambda)]\mathscr{F}(\tilde{e}_r \tilde{f}')(\lambda) = 0$, for all $\lambda \in M$. Note that $[-(\lambda^2 + 1) + i(r\lambda)] \neq 0$ for all $\lambda \in \mathbb{R}$ that implies $\mathscr{F}(\tilde{e}_r \tilde{f}') = 0$ on M, and using (6) we obtain $\mathscr{F}(\tilde{e}_r \tilde{f}) = 0$ on M. And since $\mathscr{F}(\tilde{e}_r \tilde{f})$ is continuous on $\mathbb{R}$, using the density of M into $\mathbb{R}$, we obtain $\mathscr{F}(\tilde{e}_r \tilde{f}) = 0$ on $\mathbb{R}$. Then using the uniqueness theorem we obtain $\tilde{e}_r \tilde{f} = 0$ Lebesgue-a.e. on $\mathbb{R}$, and consequently $e_r f = 0$ μ-a.e. on $\mathbb{R}_+$, and since $e_r(t) > 0$ for all t, we obtain $f = 0$ μ-a.e. on $\mathbb{R}_+$. And so we have proven that the orthogonal of $AP^2(\mathbb{R}_+,\mathbb{R}^n)$ in $H^1(\mathbb{R}_+,v,;\mathbb{R}^n)$ is reduced to zero that implies that the closure of $AP^2(\mathbb{R}_+,\mathbb{R}^n)$ into $H^1(\mathbb{R}_+,v,;\mathbb{R}^n)$ is equal to $H^1(\mathbb{R}_+,v,;\mathbb{R}^n)$. So (ii) is proven.

To prove (iii) we need to establish the following preliminary inequality:

$$|u(0)| \leq \xi \|u\|_{H_v^1}, \tag{7}$$

where $u \in H^1(\mathbb{R}_+,v;\mathbb{R}^n)$ and $\xi \in (0,\infty)$ is independent of u.

When $u \in H^1(\mathbb{R}_+,v;\mathbb{R}^n)$, u is locally absolutely continuous on $\mathbb{R}_+$, and consequently the function $e_r.u$ is also locally absolutely continuous on $\mathbb{R}_+$. Note that $(e_r u)'(t) = -r(e_r u)(t) + (e_r u')(t)$ for μ-almost all $t \in \mathbb{R}_+$. Then we have $e_r(t)u(t) = u(0) + \int_0^t (e_r u)'(s)ds = u(0) - r\int_0^t (e_r u)(s)ds + \int_0^t (e_r u')(s)ds$ that implies $u(0) = e_r(t)u(t) + r\int_0^t (e_r u)(s)ds - \int_0^t (e_r u')(s)ds$. Using the Lebesgue-dominated convergence theorem, we know that $\lim_{t \to \infty} \int_0^t (e_r u)(s)ds = \int_{\mathbb{R}_+} (e_r u)d\mu$ and

$\lim_{t \to \infty} \int_0^t (e_r u')(s) ds = \int_{\mathbb{R}_+} (e_r u') d\mu$, and consequently $\gamma := \lim_{t \to \infty} (e_r u)(t)$ exists into $\mathbb{R}^n$. If we suppose that $\gamma \neq 0$, then there exists $T \in \mathbb{R}_+$ such that $\|(e_r u)(t)\| \geq \frac{1}{2}\|\gamma\|$ for all $t \geq T$. Then we obtain $\int_{\mathbb{R}_+} \|e_r u\| d\mu = \infty$ that is impossible since $e_r u \in L^1(\mathbb{R}_+, \mu; \mathbb{R}^n)$. And so we necessarily have $\gamma = 0$ that implies $u(0) = r \int_{\mathbb{R}_+} e_r u d\mu - \int_{\mathbb{R}_+} e_r u' d\mu$, then we have $|u(0)| \leq r\|u\|_{L^1_v} + \|u'\|_{L^1_v} \leq \frac{r}{\sqrt{r}}\|u\|_{L^2_v}$ $+ \frac{1}{\sqrt{r}}\|u'\|_{L^2_v} \leq \max\{\sqrt{r}, \frac{1}{\sqrt{r}}\}(\|u\|_{L^2_v} + \|u'\|_{L^2_v}) \leq \max\{\sqrt{r}, \frac{1}{\sqrt{r}}\}\sqrt{2}\|u\|_{H^1_v}$. And so it suffices to set $\xi := \max\{\sqrt{r}, \frac{1}{\sqrt{r}}\}\sqrt{2}$ to obtain (7). Now we fix $u \in H^1(\mathbb{R}_+, v; \mathbb{R}^n; x)$, and by using (ii), there exists a sequence $(v_j)_j$ into $AP^2(\mathbb{R}_+, \mathbb{R}^n; M)$ such that $\lim_{j \to \infty} \|v_j - u\|_{H^1_v} = 0$. By inequality (5) we deduce $\lim_{j \to \infty} \|v_j(0) - u(0)\| = 0$. We set $w_j := v_j - v_j(0) + u(0)$ for all $j \in \mathbb{N}$. Then we have $w_j \in AP^2(\mathbb{R}_+, \mathbb{R}^n; M)$, and since $w_j(0) = v_j(0) - v_j(0) + u(0) = u(0) = x$, we obtain that $w_j \in AP^2(\mathbb{R}_+, \mathbb{R}^n; M, x)$. The inequalities $\|w_j - u\|_{H^1_v} \leq \|v_j - u\|_{H^1_v} + \|v_j(0) - u(0)\|_{H^1_v} \leq \|v_j - u\|_{H^1_v} + \frac{1}{r}\|v_j(0) - u(0)\|$ imply $\lim_{j \to \infty} \|v_j - u\|_{H^1_v} = 0$, and so $AP^2(\mathbb{R}_+, \mathbb{R}^n; M, x)$ is dense into $H^1(\mathbb{R}_+, v; \mathbb{R}^n; x)$, which prove (iii).

Since $AP^2(\mathbb{R}_+, \mathbb{R}^n; M, x) \subset BC^2(\mathbb{R}_+, \mathbb{R}^n, x)$, the assertion (iv) is a straightforward consequence of (iii). Taking $M = \mathbb{R}$, since $AP^2(\mathbb{R}_+, \mathbb{R}^n; M, x)$ is included in $AAP^2(\mathbb{R}_+, \mathbb{R}^n, x)$, the assertion (v) is a consequence of (iii). The proof of the assertion (vi) is similar to this one of the assertion (iv) of the previous proposition. Since $C_c^1((0,\infty), \mathbb{R}^n)$ is dense into $L^2(\mathbb{R}_+, v; \mathbb{R}^n)$, and since we have $C_c^1((0,\infty), \mathbb{R}^n) \subset H^1(\mathbb{R}_+, v; \mathbb{R}^n, 0) \subset L^2(\mathbb{R}_+, v; \mathbb{R}^n)$, (vi) is a consequence of (v). $\qquad \square$

We denote by $\overline{P_T^0}(\mathbb{R}_+, \mathbb{R}^n)$ the closure of $P_T^0(\mathbb{R}_+, \mathbb{R}^n)$ in $L^2(\mathbb{R}_+, v, \mathbb{R}^n)$.

Proposition 3. *The following assertions hold:*

(i) If $u \in \overline{P_T^0}(\mathbb{R}_+, \mathbb{R}^n)$, then $u(t + T) = u(t)$ for μ-a.e. $t \in \mathbb{R}_+$.

(ii) $\overline{P_T^0}(\mathbb{R}_+, \mathbb{R}^n) \cap C_0^0(\mathbb{R}_+, \mathbb{R}^n) = \{0\}$. Consequently $P_T^0(\mathbb{R}_+, \mathbb{R}^n)$ is not dense into $L^2(\mathbb{R}_+, v, \mathbb{R}^n)$.

Proof. Note that the operator $\Delta : u \mapsto u(. + T) - u$ is linear continuous from $L^2(\mathbb{R}_+, v; \mathbb{R}^n)$ into itself. And so $\ker\Delta$ is closed, and since $P_T^0(\mathbb{R}_+, \mathbb{R}^n) \subset \ker\Delta$, we also have $\overline{P_T^0}(\mathbb{R}_+, \mathbb{R}^n) \subset \ker\Delta$ that proves (i).

If $u \in \overline{P_T^0}(\mathbb{R}_+, \mathbb{R}^n) \cap C_0^0(\mathbb{R}_+, \mathbb{R}^n)$, then u is continuous and therefore the function $t \mapsto |u(t + T) - u(t)|$ is also continuous on $\mathbb{R}_+$, and this last function is locally Lebesgue integrable on $\mathbb{R}_+$. Moreover, on each segment $[a, b] \subset \mathbb{R}_+$, its Lebesgue integral is equal to its Riemann integral. By using (ii) we have $\int_a^b |u(t + T) - u(t)| dt = 0$ for all $a < b$ in $\mathbb{R}_+$, and then we obtain $|u(t + T) - u(t)| = 0$ for all $t \in [a, b]$ that implies $u(t + T) = u(t)$ for all $t \in \mathbb{R}_+$. Since $\lim_{t \to \infty} u(t) = 0$, we have $\forall \varepsilon > 0, \exists \tau_\varepsilon > 0, \forall t \geq \tau_\varepsilon, |u(t)| \leq \varepsilon$. We fix $\varepsilon > 0$ and we denote by k_ε the least integer number greater than τ_ε. Then, for all $t \in [k_\varepsilon T, (k_\varepsilon + 1)T]$, we have $|u(t)| \leq \varepsilon$. When $t \in [0, T]$, we have $u(t) = u(t + k_\varepsilon T)$ that implies that $|u(t)| \leq \varepsilon$. And so we obtain $\sup_{t \in [0,T]} |u(t)| \leq \varepsilon$. By doing $\varepsilon \to 0+$, we obtain $u(t) = 0$ for all $t \in [0, T]$, and then by using the periodicity, we obtain $u = 0$ that proves the assertion (ii). $\qquad \square$

3   Variational Principle for Bounded Functions

In this section we establish a variational principle in the space $BC^1(\mathbb{R}_+, \mathbb{R}^n)$. First
we recall a notion of [10]: $F \in \mathscr{U}(\mathbb{R}_+ \times \mathbb{R}^p, \mathbb{R}^q)$ if and only if $F \in C^0(\mathbb{R}_+ \times \mathbb{R}^p, \mathbb{R}^q)$,
$F(.,x) \in BC^0(\mathbb{R}_+, \mathbb{R}^q)$ for all $x \in \mathbb{R}^p$, and f satisfies the following condition:

$$\forall K \in \mathscr{P}_c(\mathbb{R}^p), \forall \varepsilon > 0, \exists \delta = \delta(K,\varepsilon) > 0, \forall x \in K, \forall y \in K,$$

$$|x - y| \le \delta \Longrightarrow \sup_{t \in \mathbb{R}_+} |F(t,y) - F(t,x)| \le \varepsilon,$$

where $\mathscr{P}_c(\mathbb{R}^p)$ denotes the set of the compact subsets of $\mathbb{R}^p$.
Adapting to our setting a result due to Blot et al. in [10] (Lemma 6, p. 710), we
obtain the following lemma:

Lemma 1. *Let $F \in \mathscr{U}(\mathbb{R}_+ \times \mathbb{R}^p, \mathbb{R}^q)$. We assume that for all $t \in \mathbb{R}_+$, the partial
mapping $F(t,.)$ is Fréchet differentiable on $\mathbb{R}^p$, and its differential F_x belongs to
$\mathscr{U}(\mathbb{R}_+ \times \mathbb{R}^p, \mathscr{L}(\mathbb{R}^p, \mathbb{R}^q))$.
Then the Nemytski operator $N_F : BC^0(\mathbb{R}_+, \mathbb{R}^p) \to BC^0(\mathbb{R}_+, \mathbb{R}^q)$, defined by $N_F(u) :=
[t \mapsto F(t, u(t))]$, is of class C^1, and for all $u, h \in BC^0(\mathbb{R}_+, \mathbb{R}^p)$, we have $DN_F(u).h =
[t \mapsto F_x(t, u(t)).h(t)]$.*

We consider the following list of conditions:

(A1) $L \in \mathscr{U}(\mathbb{R}_+ \times \mathbb{R}^n \times \mathbb{R}^n, \mathbb{R})$.
(A2) For all $t \in \mathbb{R}_+$, the partial differentials of L with respect to the second variable,
denoted by L_u, and with respect to the third variable, denoted by $L_{u'}$, exist,
and moreover $L_u \in \mathscr{U}(\mathbb{R}_+ \times \mathbb{R}^n \times \mathbb{R}^n, \mathscr{L}(\mathbb{R}^n, \mathbb{R}))$ and $L_{u'} \in \mathscr{U}(\mathbb{R}_+ \times \mathbb{R}^n \times
\mathbb{R}^n, \mathscr{L}(\mathbb{R}^n, \mathbb{R}))$.

Under these two conditions we consider the functional $J_{BC} : BC^1(\mathbb{R}_+, \mathbb{R}^n) \to \mathbb{R}$
defined by

$$J_{BC}(u) := \int_0^\infty e^{-rt} L(t, u(t), u'(t)) dt = \int_{\mathbb{R}_+} e_r L(., u, u') d\mu. \tag{8}$$

Theorem 1. *Under (A1)–(A2), J_{BC} is of class C^1 on $BC^1(\mathbb{R}_+, \mathbb{R}^n)$. Fixing $x \in \mathbb{R}^n$,
when $u \in BC^1(\mathbb{R}_+, \mathbb{R}^n, x)$, we have $DJ_{BC}(u) = 0$ on $BC^1(\mathbb{R}_+, \mathbb{R}^n, 0)$ if and only if u
is a solution of the Euler–Lagrange equation (1) on $\mathbb{R}_+$ such that $u(0) = x$.*

Proof. We consider the following mappings:

- $j : BC^1(\mathbb{R}_+, \mathbb{R}^n) \to BC^0(\mathbb{R}_+, \mathbb{R}^n) \times BC^0(\mathbb{R}_+, \mathbb{R}^n) = BC^0(\mathbb{R}_+, \mathbb{R}^{2n})$ defined by
  $j(u) := (u, u')$.
- $N_L : BC^0(\mathbb{R}_+, \mathbb{R}^{2n}) \to BC^0(\mathbb{R}_+, \mathbb{R})$ defined by $N_L(u, v)(t) := L(t, u(t), v(t))$.
- $\Lambda : BC^0(\mathbb{R}_+, \mathbb{R}) \to \mathbb{R}$ defined by $\Lambda(\varphi) := \int_0^\infty e^{-rt} \varphi(t) dt$.

Since j is linear continuous, it is of class C^1. By using Lemma 1, N_L is of class C^1. Since Λ is linear continuous, it is of class C^1. Since $J_{BC} = \Lambda \circ N_L \circ j_1$, J_{BC} is of class C^1 as a composition of three C^1 mappings. Moreover by using the chain rule and Lemma 1, we obtain $DJ_{BC}(u).h = \Lambda(DN_L(j(u))).j(h)$ that implies $DJ_{BC}(u).h = \int_0^\infty (e^{-rt}L_u(t,u(t),u'(t)).h(t) + e^{-rt}L_{u'}(t,u(t),u'(t)).h'(t))dt$. If $DJ_{BC}(u) = 0$, then, for all $h \in BC^1(\mathbb{R}_+,\mathbb{R}^n,0)$, we have $\int_0^\infty e^{-rt}L_u(t,u(t),u'(t)).h(t)dt = -\int_0^\infty e^{-rt} L_{u'}(t,u(t),u'(t)).h'(t)dt$, and consequently this equality holds for all $h \in C_c^\infty((0,\infty), \mathbb{R}^n)$, so for all $k = 1,\ldots,n$ and for all $\varphi \in C_c^\infty((0,\infty),\mathbb{R})$, we have

$$\int_0^\infty e^{-rt}L_{u_k}(t,u(t),u'(t)).\varphi(t)dt = -\int_0^\infty e^{-rt}L_{u'_k}(t,u(t),u'(t)).\varphi'(t)dt. \quad (9)$$

So by using the distributional derivative definition [25], we have $e_r.L_{u_k}(.,u,u') = D(e_rL_{u'_k}(.,u,u'))$ (equality in $\mathscr{D}'((0,\infty),\mathbb{R})$). Since $e_rL_{u_k}(.,u,u')$ and $e_rL_{u'_k}(.,u,u')$ are continuous on $(0,\infty)$, by using [25] (Théorème III, p. 54), we obtain, for all $t \in (0,\infty)$, $e^{-rt}L_{u_k}(t,u(t),u'(t)) = \frac{d}{dt}(e^{-rt}L_{u'_k}(t,u(t),u'(t)))$, and by using the continuity of $e_rL_{u_k}(.,u,u')$ and $e_rL_{u'_k}(.,u,u')$ on $\mathbb{R}_+$, we obtain that for all $t \in \mathbb{R}_+$, $e^{-rt}L_{u_k}(t,u(t),u'(t)) = \frac{d}{dt}(e^{-rt}L_{u'_k}(t,u(t),u'(t)))$. Since e_r is C^∞ on $\mathbb{R}_+$, we obtain that $e_rL_{u'_k}(.,u,u')$ is differentiable on $\mathbb{R}_+$ as a product of two differentiable functions, and we deduce (1) from (9).

Conversely, we assume that u satisfies (1). Then, for all $h \in BC^1(\mathbb{R}_+,\mathbb{R}^n)$, using (9) and the dominated convergence theorem of Lebesgue on the sequence $(1_{[0,m]}(e_rL_u(.u,u').h + e_rL_{u'}(.,u,u')))_{m \in \mathbb{N}_*}$, we have $DJ_{BC}(u).h = \int_0^\infty (e_rL_u(.,u,u').h + e_rL_{u'}(.,u,u').h')dt = \lim_{m \to \infty} \int_0^m (e_rL_u(.,u,u').h + e_rL_{u'}(.,u,u').h')dt$. Using (1), we have $\int_0^m (e_r.L_u(.,u,u').h + e_rL_{u'}(.,u,u').h')dt = \int_0^m \frac{d}{dt}(e_rL_{u'}(.,u,u').h)dt = e^{-rm}L_{u'}(m,u(m),u'(m)).h(m) - L_{u'}(0,u(0),u'(0)).h(0) = e^{-rm}L_{u'}(m,u(m),u'(m)).h(m)$, for all $m \in \mathbb{N}_*$ by using $h(0) = 0$, and since the sequence $(L_{u'}(m,u(m),u'(m)).h(m))_m$ is bounded, and $\lim_{m \to \infty} e^{-rm} = 0$, we obtain that $DJ_{BC}(u).h = 0$. $\square$

Remark 2. Theorem 1 is an improvement at the nonautonomous case of a result of Blot–Cartigny in [7].

4 Variational Principle for Almost Periodic Functions

In this section we establish a variational principle on the space $AP^1(\mathbb{R}_+,\mathbb{R}^n;M)$. First we recall that $F \in APU(\mathbb{R}_+ \times \mathbb{R}^p, \mathbb{R}^q)$ when $F \in C^0(\mathbb{R}_+ \times \mathbb{R}^p, \mathbb{R}^q)$ and when f satisfies the following condition

$$\forall K \in \mathscr{P}(\mathbb{R}^p), \forall \varepsilon > 0, \exists \ell = \ell(K,\varepsilon) > 0, \forall r \in \mathbb{R}_+, \exists \tau \in [r, r+\ell],$$

$$\forall (t,x) \in \mathbb{R}_+ \times \mathbb{R}^p, |F(t+\tau,x) - F(t,x)| \leq \varepsilon;$$

see [12, 27]. We say that $F \in APU(\mathbb{R}_+ \times \mathbb{R}^p, \mathbb{R}^q; M)$ when $F \in APU(\mathbb{R}_+ \times \mathbb{R}^p, \mathbb{R}^q)$ and when the module of frequencies of F is included into M. Adapting to our setting a result due to Blot et al. in [10] (Lemma 7, p. 710), and [19] (Theorem 4.5) on the modules of frequencies, we obtain the following lemma:

Lemma 2. Let $F \in APU(\mathbb{R}_+ \times \mathbb{R}^p, \mathbb{R}^q; M)$. We assume that for all $t \in \mathbb{R}_+$, the partial mapping $F(t, .)$ is Fréchet differentiable on $\mathbb{R}^p$ and its differential F_x belongs to $APU(\mathbb{R}_+ \times \mathbb{R}^p, \mathscr{L}(\mathbb{R}^p, \mathbb{R}^q)); M)$.
Then the Nemytski operator $N_F^1 : AP^0(\mathbb{R}_+, \mathbb{R}^p) \to AP^0(\mathbb{R}_+, \mathbb{R}^q; M)$, defined by $N_F^1(u) := [t \mapsto F(t, u(t))]$, is of class C^1, and for all $u, h \in AP^0(\mathbb{R}_+, \mathbb{R}^p; M)$, we have $DN_F^1(u).h = [t \mapsto F_x(t, u(t)).h(t)]$.

When F is autonomous, such a result is established in [4] (Proposition 3, p. 12).
   We consider the following list of conditions:

(A3) $L \in APU(\mathbb{R}_+ \times \mathbb{R}^n \times \mathbb{R}^n, \mathbb{R}; M)$.
(A4) For all $t \in \mathbb{R}_+$, the partial differentials of L with respect to the second variable, denoted by L_u, and with respect to the third variable, denoted by $L_{u'}$, exist, and moreover $L_u \in APU(\mathbb{R}_+ \times \mathbb{R}^n \times \mathbb{R}^n, \mathscr{L}(\mathbb{R}^n, \mathbb{R}); M)$ and $L_{u'} \in APU(\mathbb{R}_+ \times \mathbb{R}^n \times \mathbb{R}^n, \mathscr{L}(\mathbb{R}^n, \mathbb{R}); M)$.

Under these two conditions we consider the functional $J_{AP} : AP^1(\mathbb{R}_+, \mathbb{R}^n; M) \to \mathbb{R}$ defined by

$$J_{AP}(u) := \int_0^\infty e^{-rt} L(t, u(t), u'(t)) dt = \int_{\mathbb{R}_+} e_r L(., u, u') d\mu. \tag{10}$$

Theorem 2. Under (A3)–(A4), J_{AP} is of class C^1 on $AP^1(\mathbb{R}_+, \mathbb{R}^n; M)$. Fixing $x \in \mathbb{R}^n$, when $u \in AP^1(\mathbb{R}_+, \mathbb{R}^n; M, x)$, $DJ_{AP}(u) = 0$ on $AP^1(\mathbb{R}_+, \mathbb{R}^n; M, 0)$ if and only if u is a solution of the Euler–Lagrange equation (1) on $\mathbb{R}_+$ such that $u(0) = x$.

Proof. To prove that J_{AP} is of class C^1, we proceed like in a way which is similar to this one of the proof of Theorem 1: we split $J_{AP} = \Lambda \circ N_L \circ j_1$ where $j_1 :$ $AP^1(\mathbb{R}_+, \mathbb{R}^n) \to AP^0(\mathbb{R}_+, \mathbb{R}^n) \times AP^0(\mathbb{R}_+, \mathbb{R}^n)$ is $j_1(u) := (u, u')$, $N_L^1 : AP^0(\mathbb{R}_+, \mathbb{R}^n \times \mathbb{R}^n) \to AP^0(\mathbb{R}_+, \mathbb{R})$ is $N_L^1(u, v) := [t \mapsto L(t, u(t), v(t))]$, and $\Lambda^1 : AP^0(\mathbb{R}_+, \mathbb{R}) \to \mathbb{R}$ is $\Lambda^1(\varphi) := \int_0^\infty e^{-rt} \varphi(t) dt$. And so by using Lemma 2 and the basic results of the differential calculus, $J_{AP} = \Lambda^1 \circ N_L^1 \circ j_1$ is of class C^1. And so an analogous, we obtain the formula of the differential of J_{AP} which is analogous to this one of DJ_{BC} in proof of Theorem 1. Then, for all $\lambda \in M$, since the functions $t \mapsto \cos(-\lambda t)$ and $t \mapsto \sin(-\lambda t)$ belong to $AP^1(\mathbb{R}_+, \mathbb{R}; M)$, we obtain the following relation: $\int_0^\infty e^{-rt} L_{u_k}(t, u(t), u'(t)) e^{-i\lambda t} dt = -\int_0^\infty e^{-rt} L_{u'_k}(t, u(t), u'(t)) (-i\lambda) e^{-i\lambda t} dt$, i.e. for all $\lambda \in M$, $\int_0^\infty e^{-rt} L_{u_k}(t, u(t), u'(t)) e^{-i\lambda t} dt = i\lambda \int_0^\infty e^{-rt} L_{u'_k}(t, u(t), u'(t)) e^{-i\lambda t} dt$, that is denoting by $\tilde{L}_{u_k}$ the extension on $\mathbb{R}$ by the value 0 on $(-\infty, 0)$ of the function $t \mapsto L_{u_k}(t, u(t), u'(t))$ and by $\tilde{L}_{u'_k}$ the extension on $\mathbb{R}$ by the value 0 on $(-\infty, 0)$ of the function $t \mapsto L_{u'_k}(t, u(t), u'(t))$, for all $\lambda \in M$ and for all $k = 1, \ldots, n$, $\mathscr{F}(\tilde{e}_r \tilde{L}_{u_k})(\lambda) = i\lambda \mathscr{F}(\tilde{e}_r \tilde{L}_{u'_k})(\lambda)$. Using the continuity of $\mathscr{F}(\tilde{e}_r \tilde{L}_{u_k})$ and of $\mathscr{F}(\tilde{e}_r \tilde{L}_{u'_k})$ on $\mathbb{R}$ and the density of M into $\mathbb{R}$, the previous equality holds for all $\lambda \in \mathbb{R}$. Since we have $\mathscr{F}(D(\tilde{e}_r \tilde{L}_{u'_k})) = (i.id_{\mathbb{R}})\mathscr{F}(\tilde{e}_r \tilde{L}_{u'_k})$, where D denotes the distributional derivative in $\mathscr{S}'(\mathbb{R}, \mathbb{R})$, we have $\mathscr{F}(\tilde{e}_r \tilde{L}_{u_k}) = \mathscr{F}(D(\tilde{e}_r \tilde{L}_{u'_k}))$. By the uniqueness theorem for the Fourier transform of tempered distributions we obtain $\tilde{e}_r \tilde{L}_{u_k} = D(\tilde{e}_r \tilde{L}_{u'_k})$ (equality in $\mathscr{S}'(\mathbb{R}, \mathbb{R})$), and consequently $\tilde{e}_r \tilde{L}_{u_k} = D(\tilde{e}_r \tilde{L}_{u'_k})$ (equality in $\mathscr{D}'(\mathbb{R}, \mathbb{R})$), i.e. using the definition of the distributional derivative $\int_{\mathbb{R}} \tilde{e}_r \tilde{L}_{u_k} \varphi dt = -\int_{\mathbb{R}} \tilde{e}_r \tilde{L}_{u'_k} \varphi' dt$ for all $\varphi \in C_c^\infty(\mathbb{R}, \mathbb{R})$, we have $\int_{\mathbb{R}} \tilde{e}_r \tilde{L}_{u_k} \varphi dt = -\int_{\mathbb{R}} \tilde{e}_r \tilde{L}_{u'_k} 1_{\mathbb{R}_+} \varphi' dt$ for all $\varphi \in C_c^\infty((0, \infty), \mathbb{R})$, so we obtain the equality (9). Proceeding like in the proof of Theorem 1, u satisfies (1). The converse of the equivalence is proven like this one of assertion (i) of Theorem 1.   □

5 Variational Principle for Asymptotically Almost Periodic Functions

In this section we establish a variational principle on the space $AAP^1(\mathbb{R}_+, \mathbb{R}^n)$. First we recall, following Zaidman [28], that $F \in AAPU(\mathbb{R}_+ \times \mathbb{R}^p, \mathbb{R}^q)$ when F is continuous and when the following condition is fulfilled :

$$\forall K \in \mathscr{P}_c(\mathbb{R}^p), \forall \varepsilon > 0, \exists T = T(K, \varepsilon) > 0, \exists \ell = \ell(K, \varepsilon) > 0, \forall r \in \mathbb{R}_+,$$

$$\exists \tau \in [r, r + \ell], \forall x \in K, \forall t \geq T, |F(t + \tau, x) - F(t, x)| \leq \varepsilon.$$

Then using a result due to Blot et al. (Theorem 8.5 in [12], p. 66), we obtain the following lemma:

Lemma 3. *Let $F \in AAPU(\mathbb{R}_+ \times \mathbb{R}^p, \mathbb{R}^q; M)$. We assume that for all $t \in \mathbb{R}_+$, the partial mapping $F(t, .)$ is Fréchet differentiable on $\mathbb{R}^p$, and its differential F_x belongs to $AAPU(\mathbb{R}_+ \times \mathbb{R}^p, \mathscr{L}(\mathbb{R}^p, \mathbb{R}^q)); M)$.*
Then the Nemytski operator $N_F^2 : AAP^0(\mathbb{R}_+, \mathbb{R}^p; M) \to AAP^0(\mathbb{R}_+, \mathbb{R}^q; M)$, defined by $N_F^2(u) := [t \mapsto F(t, u(t))]$, is of class C^1, and for all $u, h \in AAP^0(\mathbb{R}_+, \mathbb{R}^p; M)$, we have $DN_F^2(u).h = [t \mapsto F_x(t, u(t)).h(t)]$.

We introduce the following conditions:

(A5) $L \in AAPU(\mathbb{R}_+ \times \mathbb{R}^n \times \mathbb{R}^n, \mathbb{R})$.
(A6) For all $t \in \mathbb{R}_+$, the partial differentials of L with respect to the second variable, denoted by L_u, and with respect to the third variable, denoted by $L_{u'}$, exist, and moreover $L_u \in AAPU(\mathbb{R}_+ \times \mathbb{R}^n \times \mathbb{R}^n, \mathscr{L}(\mathbb{R}^n, \mathbb{R}))$ and $L_{u'} \in AAPU(\mathbb{R}_+ \times \mathbb{R}^n \times \mathbb{R}^n, \mathscr{L}(\mathbb{R}^n, \mathbb{R}))$.

Under these two conditions we define the functional $J_{AAP} : AAP^1(\mathbb{R}_+, \mathbb{R}^n) \to \mathbb{R}$ by setting, for all $u \in AAP^1(\mathbb{R}_+, \mathbb{R}^n)$,

$$J_{AAP}(u) := \int_0^\infty e^{-rt} L(t, u(t), u'(t)) dt. \tag{11}$$

Theorem 3. *Under (A5)–(A6), J_{AAP} is of class C^1 on $AAP^1(\mathbb{R}_+, \mathbb{R}^n)$. Fixing $x \in \mathbb{R}^n$, when $u \in AAP^1(\mathbb{R}_+, \mathbb{R}^n, x)$, we have $DJ_{AAP}(u) = 0$ on $AAP^1(\mathbb{R}_+, \mathbb{R}^n, 0)$ if and only if u is a solution of the Euler–Lagrange equation (1) on $\mathbb{R}_+$ such that $u(0) = x$.*

Since $C_c^\infty((0, \infty), \mathbb{R}) \subset AAP^1(\mathbb{R}_+, \mathbb{R}, 0)$, the proof of this theorem is similar to this one of Theorem 1.

6 Variational Principle on a Weighted Sobolev Space

In this section we establish a variational principle on the space $H^1(\mathbb{R}_+, v; \mathbb{R}^n)$. We need some additional assumptions.

(A7) For Lebesgue-almost all $t \in \mathbb{R}_+$, $L(t, ., ., .)$ is of class C^1 on $\mathbb{R}^n \times \mathbb{R}^n$, and for all $(x, y) \in \mathbb{R}^n \times \mathbb{R}^n$, $L(., x, y)$, $L_u(., x, y)$ and $L_{u'}(., x, y)$ are Lebesgue-measurable on $\mathbb{R}_+$.

(A8) There exist $\alpha \in (0, \infty)$ and $\beta \in L^1(\mathbb{R}_+, v, \mathbb{R}_+)$ such that $|L_u(t, x, y)| \leq \alpha.(|x| + |y|) + \beta(t)$ and $|L_{u'}(t, x, y)| \leq \alpha.(|x| + |y|) + \beta(t)$ for all $(t, x, y) \in \mathbb{R}_+ \times \mathbb{R}^n \times \mathbb{R}^n$.

Note that (A7) means that L, L_u and $L_{u'}$ are Caratheodory functions. Under these conditions we define the functional $J_{H^1}(u) : H^1(\mathbb{R}_+, v, \mathbb{R}^n) \to \mathbb{R}$ by setting

$$J_H(u) := \int_0^\infty e^{-rt} L(t, u(t), u'(t)) dt, \tag{12}$$

and we can formulate the following variational principle:

Theorem 4. *Under (A7)–(A8), J_H is of class C^1 on $H^1(\mathbb{R}_+, v, \mathbb{R}^n)$. Fixing $x \in \mathbb{R}^n$, when $u \in H^1(\mathbb{R}_+, v, \mathbb{R}^n, x)$, we have $DJ_H(u) = 0$ on $H^1(\mathbb{R}_+, v, \mathbb{R}^n, 0)$ if and only if u is a locally absolutely continuous solution of the Euler–Lagrange equation (1) on $\mathbb{R}_+$ such that $u(0) = x$.*

Proof. To replace the measure of Lebesgue on a bounded subset of $\mathbb{R}^m$ by the measure v on $\mathbb{R}_+$ does not change the proof of Theorem 2.6, p. 14, in [18] (which some parts are proved in [20]) that permits us to assert that the Nemytski operator on L, $N_L^3 : L^2(\mathbb{R}_+, v, \mathbb{R}^n) \times L^2(\mathbb{R}_+, v, \mathbb{R}^n) \to L^2(\mathbb{R}_+, v, \mathbb{R})$, defined by $N_L^3(u, v)(t) := L(t, u(t), v(t))$, is of class C^1, and its differential is given by $(DN_L^3(u, v).(h, k))(t) = L_u(t, u(t), v(t)).h(t) + L_{u'}(t, u(t), v(t)).k(t)$. Since the operator $j_3 : H^1(\mathbb{R}_+, v, \mathbb{R}^n) \to L^2(\mathbb{R}_+, v, \mathbb{R}^n) \times L^2(\mathbb{R}_+, v, \mathbb{R}^n)$ defined by $j_3(u) := (u, u')$ and the functional $\Lambda^3 : L^2(\mathbb{R}_+, v, \mathbb{R}) \to \mathbb{R}$ defined by $\Lambda^3(\varphi) := \int_0^\infty e^{-rt} \varphi(t) dt$ are linear continuous, they are of class C^1. Since $J_H = \Lambda^3 \circ N_L^3 \circ j_3$, J_H is of class C^1 on $H^1(\mathbb{R}_+, v, \mathbb{R}^n)$ as a composition of C^1 mappings. Moreover by using the chain rule of the differential calculus, we obtain the following formula:

$$DJ_H(u).h = \int_0^\infty e^{-rt}(L_u(t, u(t), u'(t)).h(t) + L_{u'}(t, u(t), u'(t)).h'(t)) dt \tag{13}$$

for all $u, h \in H^1(\mathbb{R}_+, v, \mathbb{R}^n)$. For a solution $u \in H^1(\mathbb{R}_+, v, \mathbb{R}^n, x)$ of (1) such that $u(0) = x$, then $t \mapsto L_{u'}(t, u(t), u'(t))$ is locally absolutely continuous, and the equality $e^{-rt}L_u(t, u(t), u'(t)) = -rL_{u'}(t, u(t), u'(t)) + \frac{d}{dt}L_{u'}(t, u(t), u'(t)) = \frac{d}{dt}(e^{-rt}L_{u'}(t, u(t), u'(t)))$ holds for μ-a.e. $t \in \mathbb{R}_+$. Therefore for all $h \in C_c^1((0, \infty), \mathbb{R}^n)$, we obtain $\int_0^\infty e^{-rt}L_u(t, u(t), u'(t)).h(t)dt = \int_0^\infty \frac{d}{dt}(e^{-rt}.L_{u'}(t, u(t), u'(t)))dt$. Using the integration by parts for the absolutely continuous functions [23] (p. 54–55), for all $T > \sup \operatorname{supp}(h)$, we have $\int_0^T \frac{d}{dt}(e^{-rt}.L_{u'}(t, u(t), u'(t))).h(t)dt = 0 - \int_0^T e^{-rt} L_{u'}(t, u(t), u'(t)).h'(t)dt$, and by using the Lebesgue-dominated convergence theorem, we obtain $\int_0^\infty \frac{d}{dt}(e^{-rt}L_{u'}(t, u(t), u'(t))).h(t)dt = -\int_0^\infty e^{-rt}L_{u'}(t, u(t), u'(t)).h'(t)dt$, and so we obtain

$$\int_0^\infty e^{-rt}(L_u(t, u(t), u'(t)).h(t) + L_{u'}(t, u(t), u'(t)).h'(t)dt = 0,$$

i.e. $DJ_H(u).h = 0$ for all $h \in C_c^1((0, \infty), \mathbb{R}^n)$. Using Proposition 2, v, since $C_c^1((0, \infty), \mathbb{R}^n)$ is dense into $H^1(\mathbb{R}_+, v, \mathbb{R}^n, 0)$, this last relation implies $DJ_H(u).h = 0$ for all $h \in H^1(\mathbb{R}_+, v, \mathbb{R}^n, 0)$. Conversely, we assume that $DJ_H(u).h = 0$ for all $h \in H^1(\mathbb{R}_+, v, \mathbb{R}^n, 0)$, where $u \in H^1(\mathbb{R}_+, v, \mathbb{R}^n, x)$. Then using (13) we know that

$$\int_0^\infty e^{-rt}L_u(t, u(t), u'(t)).h(t)dt = -\int_0^\infty e^{-rt}L_{u'}(t, u(t), u'(t)).h'(t)dt$$

for all $h \in C_c^1((0, \infty), \mathbb{R}^n)$ and consequently for all $h \in C_c^\infty((0, \infty), \mathbb{R}^n)$. This last relation implies that $D(e_r L_{u'}(., u, u')) = e_r L_u(., u, u')$ in $\mathscr{D}'((0, \infty), \mathbb{R}^n)$. Since $e_r L_u(., u, u') \in L^1((0, \infty), \mathbb{R}^n)$ the function $e_r L_{u'}(., u, u')$ is locally absolutely continuous on $\mathbb{R}_+$, and we have $\frac{d}{dt}(e^{-rt}L_{u'}(t, u(t), u'(t))) = e^{-rt}L_u(t, u(t), u'(t))$ for μ-almost all $t \in \mathbb{R}_+$, and then u is a locally absolutely continuous solution of (1). □

7   Results on Forced Equations

When $b \in L^2(\mathbb{R}_+, v, \mathbb{R}^n)$, to treat (2) we consider the functional $\Phi_b : H^1(\mathbb{R}_+, v, \mathbb{R}^n) \to \mathbb{R}$ defined by

$$\Phi_b(u) := J_H(u) - (u, b)_{L^2_{(v)}}. \tag{14}$$

Note that introducing the function $K : \mathbb{R}_+ \times \mathbb{R}^n \times \mathbb{R}^n \to \mathbb{R}$ defined by $K(t, x, y) := L(t, x, y) - x.b(t)$, we obtain

$$\Phi_b(u) := \int_0^\infty K(t, u(t), u'(t))dt. \tag{15}$$

We consider the following conditions on L:

(A9) For all $t \in \mathbb{R}_+$, $L(t, ., .)$ is convex on $\mathbb{R}^n \times \mathbb{R}^n$.

(A10) There exist $\alpha_1 \in (0,\infty)$ and $\beta_1 \in L^1(\mathbb{R}_+, \nu, \mathbb{R})$ such that $L(t,x,y) \geq \alpha_1(|x|^2 + |y|^2) + \beta_1(t)$ for all $(t,x,y) \in \mathbb{R}_+ \times \mathbb{R}^n \times \mathbb{R}^n$.

Lemma 4. *Under (A7)–(A10), for all $b \in L^2(\mathbb{R}_+, \nu, \mathbb{R}^n)$ and for all $x \in \mathbb{R}^n$, there exists a locally absolutely continuous solution u of (2) on $\mathbb{R}_+$ such that $u(0) = x$.*

Proof. First we verify that Φ_b is of class C^1 on $H^1(\mathbb{R}_+, \nu, \mathbb{R}^n)$. Since $u \mapsto (u,b)_{L^2(\nu)}$ is linear continuous, it is of class C^1 on $H^1(\mathbb{R}_+, \nu, \mathbb{R}^n)$, and using Theorem 4, we know that J_H is of class C^1 on $H^1(\mathbb{R}_+, \nu, \mathbb{R}^n)$. Then Φ_b is of class C^1 as a difference of two C^1 functionals.

It is easy to see that the function K also satisfies the conditions (A7)–(A8), and so using (15) we can use Theorem 4 on Φ_b (instead of J_H). Then when we fix $x \in \mathbb{R}^n$, and when $u \in H^1(\mathbb{R}_+, \nu, \mathbb{R}^n, x)$, we obtain that $D\Phi_b(u) = 0$ on $H^1(\mathbb{R}_+, \nu, \mathbb{R}^n, 0)$ if and only if $K_u(t, u(t), u'(t)) + rK_{u'}(t, u(t), u'(t)) - \frac{d}{dt}K_{u'}(t, u(t), u'(t)) = 0$ that is exactly (2).

(A9) implies that Φ_b is convex on the closed affine set $H^1(\mathbb{R}_+, \nu, \mathbb{R}^n, x)$, and since Φ_b is of class C^1, using Theorem 1.2 in p. 49 of [17], we can assert that Φ_b is weakly lower semicontinuous on $H^1(\mathbb{R}_+, \nu, \mathbb{R}^n, x)$. (A10) implies that Φ_b is coercive. Then using Theorem 1.1 in p. 48 of [17], we know that there exists $u \in H^1(\mathbb{R}_+, \nu, \mathbb{R}^n, x)$ such that $\Phi_b(u) = \inf \Phi_b(H^1(\mathbb{R}_+, \nu, \mathbb{R}^n, x))$. Since $H^1(\mathbb{R}_+, \nu, \mathbb{R}^n, 0)$ is the tangent vector space of $H^1(\mathbb{R}_+, \nu, \mathbb{R}^n, x)$ with Φ_b of class C^1, we obtain that $D\Phi_b(u) = 0$ on $H^1(\mathbb{R}_+, \nu, \mathbb{R}^n, 0)$, and consequently u is a locally absolutely continuous solution of (2) on $\mathbb{R}_+$ such that $u(0) = x$. $\square$

We introduce $\|.\|_* : L^2(\mathbb{R}_+, \nu, \mathbb{R}^n) \to \mathbb{R}$ by setting

$$\|u\|_* := \sup\{|(u,h)_{L^2(\nu)}| : h \in H^1(\mathbb{R}_+, \nu, \mathbb{R}^n, 0), \|h\|_{H^1(\nu)} \leq 1\}. \qquad (16)$$

It is easy to verify that $\|.\|_*$ is a norm on $L^2(\mathbb{R}_+, \nu, \mathbb{R}^n)$; using the density of $H^1(\mathbb{R}_+, \nu, \mathbb{R}^n, 0)$ into $L^2(\mathbb{R}_+, \nu, \mathbb{R}^n)$, we have the implication $\|u\|_* = 0 \Longrightarrow u = 0$. We can verify that the inequality $\|.\|_* \leq \|.\|_{L^2(\nu)}$ holds.

We also introduce the following condition:

(A11) $L \in C^2(\mathbb{R}_+ \times \mathbb{R}^n \times \mathbb{R}^n, \mathbb{R})$.

Theorem 5. *Under (A1), (A2), and (A8)–(A11), for all $x \in \mathbb{R}^n$, the set of the b belonging to $BC^0(\mathbb{R}_+, \mathbb{R}^n)$ such that there exists $u \in BC^1(\mathbb{R}_+, \mathbb{R}^n)$ solution of (2) on $\mathbb{R}_+$ satisfying $u(0) = x$ is dense into $BC^0(\mathbb{R}_+, \mathbb{R}^n)$ with respect the norm $\|.\|_*$.*

Proof. When $x \in \mathbb{R}^n$, we introduce the operator $\mathcal{T}_x : \mathcal{D}(\mathcal{T}_x) \to L^2(\mathbb{R}_+, \nu, \mathbb{R}^n)$:

$$\mathcal{T}_x(u) := L_u(., u, u') + rL_{u'}(.u, u') - \frac{d}{dt}L_{u'}(.u, u'), \qquad (17)$$

where $\mathcal{D}(\mathcal{T}_x) \subset H^1(\mathbb{R}_+, \nu, \mathbb{R}^n, x)$ is the set of the $u \in H^1(\mathbb{R}_+, \nu, \mathbb{R}^n)$ such that $\mathcal{T}_x(u)$ is well defined.

After (A11) we have $BC^2(\mathbb{R}_+, \mathbb{R}^n, x) \subset \mathcal{D}(\mathcal{T}_x)$. Since $\mathcal{T}_x(u) := L_u(., u, u') + rL_{u'}(., u, u') - L_{u't}(., u, u') - L_{u'u}(., u, u').u' - L_{u'u'}(., u, u').u'' \in BC^0(\mathbb{R}_+, \mathbb{R}^n)$, we have

$$\mathcal{T}_x(BC^2(\mathbb{R}_+, \mathbb{R}^n, x)) \subset BC^0(\mathbb{R}_+, \mathbb{R}^n). \tag{18}$$

Note that (A7) is a consequence of (A1)–(A2), and by using Lemma 4, we know that for all $b \in L^2(\mathbb{R}_+, v, \mathbb{R}^n)$, there exists $u \in H^1(\mathbb{R}_+, v, \mathbb{R}^n, x)$ such that $\mathcal{T}_x(u) = b$, and so we have

$$\mathcal{T}_x \text{ is onto, i.e. } \mathcal{T}_x(\mathcal{D}(\mathcal{T}_x)) = L^2(\mathbb{R}_+, v, \mathbb{R}^n). \tag{19}$$

When $u \in \mathcal{D}(\mathcal{T}_x)$ and $h \in H^1(\mathbb{R}_+, v, \mathbb{R}^n, 0)$, we have, using the integration by parts,

$$(\mathcal{T}_x(u), h)_{L^2(v)}$$

$$= \int_0^\infty e^{-rt}(L_u(t, u(t), u'(t)) + rL_{u'}(t, u(t), u'(t)) - \frac{d}{dt}L_{u'}(t, u(t), u'(t))).h(t)dt$$

$$= \int_0^\infty e^{-rt}(L_u(t, u(t), u'(t)).h(t) + L_{u'}(t, u(t), u'(t)).h'(t))dt = DJ_H(u).h,$$

and consequently for all $u, v \in \mathcal{D}(\mathcal{T}_x)$, we obtain

$$\|\mathcal{T}_x(u) - \mathcal{T}_x(v)\|_*$$

$$= \sup\{|DJ_H(u).h - DJ_H(v).h| : h \in H \backslash E(\mathbb{R}_+, v, \mathbb{R}^n, 0), \|h\|_{H^1(v)} \le 1\}$$

$$\le \sup\{\|DJ_H(u) - DJ_H(v)\|_{\mathcal{L}}\|h\|_{H^1(v)} : h \in H^1(\mathbb{R}_+, v, \mathbb{R}^n, 0), \|h\|_{H^1(v)} \le 1\}$$

$$= \|DJ_H(u) - DJ_H(v)\|_{\mathcal{L}}$$

that implies, since DJ is continuous, the following assertion:

$$\mathcal{T}_x : (\mathcal{D}(\mathcal{T}_x), \|.\|_{H^1(v)}) \to (L^2(\mathbb{R}_+, v, \mathbb{R}^n), \|.\|_*) \text{ is continuous.} \tag{20}$$

Let $b \in BC^0(\mathbb{R}_+, \mathbb{R}^n)$. Since $BC^0(\mathbb{R}_+, \mathbb{R}^n) \subset L^2(\mathbb{R}_+, v, \mathbb{R}^n)$, after (18) we know that there exists $u^b \in \mathcal{D}(\mathcal{T}_x)$ such that $\mathcal{T}_x(u^b) = b$. Since $BC^2(\mathbb{R}_+, \mathbb{R}^n, x)$ is dense into $H^1(\mathbb{R}_+, v, \mathbb{R}^n, x)$, $BC^2(\mathbb{R}_+, \mathbb{R}^n, x)$ is dense into $\mathcal{D}(\mathcal{T}_x)$, and therefore there exists a sequence $(u_j)_j$ in $BC^2(\mathbb{R}_+, \mathbb{R}^n, x)$ such that $\lim_{j \to \infty} \|u_j - u^b\|_{H^1(v)} = 0$. For all $j \in \mathbb{N}$, we set $b_j := \mathcal{T}_x(u_j)$. After (18) we know that $b_j \in BC^0(\mathbb{R}_+, \mathbb{R}^n)$ and that u_j is a solution of (2) on $\mathbb{R}_+$ for b_j. Using (20) we have $\lim_{j \to \infty} b_j = \lim_{j \to \infty} \mathcal{T}_x(u_j) = \mathcal{T}_x(u^b)$. And so the proof is complete. $\qquad \square$

The proofs of the two following theorems are similar to this one of Theorem 5:

Theorem 6. *Under (A3), (A4), and (A8)–(A11), for all $x \in \mathbb{R}^n$, the set of the b belonging to $AP^0(\mathbb{R}_+, \mathbb{R}^n; M)$ such that there exists $u \in AP^1(\mathbb{R}_+, \mathbb{R}^n; M)$ solution of (1) on $\mathbb{R}_+$ satisfying $u(0) = x$ is dense into $AP^0(\mathbb{R}_+, \mathbb{R}^n; M)$ with respect the norm $\|.\|_*$.*

Theorem 7. *Under (A5), (A6), and (A8)–(A11), for all $x \in \mathbb{R}^n$, the set of the b belonging to $AAP^0(\mathbb{R}_+, \mathbb{R}^n)$ such that there exists $u \in AAP^1(\mathbb{R}_+, \mathbb{R}^n)$ solution of (1) on $\mathbb{R}_+$ satisfying $u(0) = x$ is dense into $AAP^0(\mathbb{R}_+, \mathbb{R}^n)$ with respect the norm $\|.\|_*$.*

References

1. Aliprantis, C.D., Border, K.C.: Infinite Dimensional Analysis, 2nd edn. Springer, Berlin (1999)
2. Antoci, F.: Some necessary and sufficient conditions for the compactness of the embeddings of weighted Sobolev spaces. Ricerche Math. **52**, 55–71 (2003)
3. Aubin, J.-P.: Applied Abstract Analysis. Wiley, New York (1977)
4. Blot, J.: Une approche variationnelle des orbites quasi-périodiques des systèmes hamiltoniens. Ann. Sci. Math. Québec **13**(2), 7–32 (1989)
5. Blot, J.: Almost periodically forced pendulum. Funkcial. Ekvac. **36**(2), 235–250 (1993)
6. Blot, J.: Oscillations presque-périodiques forcées d'équations d'Euler-Lagrange. Bull. Soc. Math. France **122**, 285–304 (1994)
7. Blot, J., Cartigny, P.: Bounded solutions and oscillations of convex lagrangian systems in presence of a discount rate. Z. Anal. Anwendungen **14**(4), 731–750 (1995)
8. Blot, J., Hayek, N.: Second Order necessary conditions for the infinite-horizon variational problems. Math. Oper. Res. **21**(4), 979–990 (1996)
9. Blot, J., Michel, P.: First-order necessary conditions for the infinite-horizon variational problems. J. Optim. Theory Appl. **88**(2), 339–364 (1996)
10. Blot, J., Cieutat, P., Mawhin, J.: Almost periodic oscillations of monotone second order systems. Adv. Differ. Equ. **2**(5), 693–714 (1997)
11. Blot, J., D'Onofrio, B., Violi, R.: Relative stability in concave Lagrangian systems. Int. J. Evolut. Equ. **1**(2), 1253–159 (2005)
12. Blot, J., Cieutat, P., N'Guérékata, G.M., Pennequin, D.: Superposition operators between various almost periodic function spaces and applications. Commun. Math. Anal. **6**(1), 42–70 (2009)
13. Bourbaki, N.: Éléments de mathématiques. In: Topologie Générale, Chapitres 5 à 10. Hermann, Paris (1974)
14. Brezis, H.: Analyse Fonctionnelle: Théorie et Applications. Masson, Paris (1983)
15. Carlson, D.A., Haurie, A.B., Leizarowitz, A.: Infinite Horizon Optimal Control: Deterministic and Stochastic Systems, 2nd edn. Springer, Berlin (1991)
16. Clark, C.W.: Mathematical Bioeconomics: Optimal Management of Renewable Resources, 2nd edn. Wiley Inc., Hoboken (2005)
17. Dacorogna, B.: Direct Methods in the Calculus of Variations. Springer, Berlin (1989)
18. de Figueiredo, D.G.: The Ekeland Variational Principle with Applications and Detours. Tata Institute of Fundamental Research. Springer, Berlin (1989)
19. Fink, A.M.: Almost Periodic Differential Equations. Lecture Notes in Mathematics, vol. 377, Springer, Berlin (1974)
20. Krasnoselski, M.A.: Topological Methods in the Theory of Nonlinear Integral Equations. Pergamon Press, Oxford (1964) (English edition)
21. Lang, S.: Real and Functional Analysis, 3rd edn. Springer, New York (1993)
22. Pickenhaim, S., Lykia, V., Wagner, M.: On the lower semicontinuity of functionals involving Lebesgue or improper Riemann integrals in infinite horizon optimal control problems. Contr. Cybernet. **37**, 451–468 (2008)
23. Riesz, F., Nagy, B.S.: Analyse Fonctionnelle, 6th French edn. Gauthier-Villars, Paris (1975)
24. Sargent, T.J.: Macroeconomic Theory, 2nd edn. Academic Inc., Orlando (1987)
25. Schwartz, L.: Théorie des Distributions. Hermann, Paris (1966)

26. Stein, E.M., Weiss, G.: Introduction to Fourier Analysis on Euclidean Spaces. Princeton University Press, Princeton (1971)
27. Yoshizawa, T.: Stability Theory and the Existence of Periodic Solutions and Almost Periodic Solutions. Springer Inc., New York (1975)
28. Zaidman, S.D.: Almost-Periodic Functions in Abstract Spaces. Pitman Publishing Inc., Marshfield (1985)

A Note on Elliptic Equations Involving the Critical Sobolev Exponent

Gabriele Bonanno, Giovanni Molica Bisci, and Vicenţiu Rădulescu

Abstract In this work we obtain some existence results for a class of elliptic Dirichlet problems involving the critical Sobolev exponent and containing a parameter. Through a weak lower semicontinuity result and by using a critical point theorem for differentiable functionals, the existence of a precise open interval of positive eigenvalues for which the treated problems admit at least one non-trivial weak solution is established. The attained results represent a more precise version of some contributions on the treated subject.

1 Introduction

In this we study the existence of one non-trivial weak solution for the following elliptic Dirichlet problem:

G. Bonanno (✉)
Department of Science for Engineering and Architecture (Mathematics Section)
Engineering Faculty, University of Messina, 98166 Messina, Italy
e-mail: bonanno@unime.it

G.M. Bisci
Dipartimento MECMAT, University of Reggio Calabria, Via Graziella, Feo di Vito,
89124 Reggio Calabria, Italy
e-mail: gmolica@unirc.it

V. Rădulescu
Institute of Mathematics "Simion Stoilow" of the Romanian Academy,
014700 Bucharest, Romania

Department of Mathematics, University of Craiova, 200585 Craiova, Romania
e-mail: vicentiu.radulescu@imar.ro

S. Pinelas et al. (eds.), *Differential and Difference Equations with Applications*, Springer
Proceedings in Mathematics & Statistics 47, DOI 10.1007/978-1-4614-7333-6_24,
© Springer Science+Business Media New York 2013

$$(P_{\mu,\lambda}) \quad \begin{cases} -\Delta_p u = \mu \left(\int_\Omega |u(x)|^{p^*} dx \right)^{p/p^*-1} |u|^{p^*-2} u + \lambda f(x,u) & \text{in} \quad \Omega \\ u|_{\partial\Omega} = 0, \end{cases}$$

where Ω is a bounded domain in $\mathbb{R}^N$ ($N \geq 2$) with smooth boundary $\partial\Omega$, $1 < p < N$, $\Delta_p u := \text{div}(|\nabla u|^{p-2}\nabla u)$ stands for the usual p-Laplace operator, $p^* := pN/(N-p)$ and $f : \Omega \times \mathbb{R} \to \mathbb{R}$ is a Carathéodory function satisfying the subcritical growth condition

$$|f(x,t)| \leq a_1 + a_2|t|^{q-1}, \quad \forall(x,t) \in \Omega \times \mathbb{R}, \tag{h_∞}$$

where a_1, a_2 are non-negative constants and $q \in]1, pN/(N-p)[$. Finally, λ and μ are two real parameters, respectively, positive and non-negative.

Problem $(P_{0,\lambda})$ has been extensively studied during the last few years, where the nonlinearity being a continuous function provided with certain growth properties at zero and infinity, respectively. We just mention, in the large literature on the subject, the papers [14, 15, 19]; see also the recent monograph by Kristály, Rădulescu and Varga [16] as general reference for this topic.

When $\mu \neq 0$, the classical variational approach cannot be applied due to the presence of the term

$$\left(\int_\Omega |u(x)|^{p^*} dx \right)^{p/p^*-1} |u|^{p^*-2} u.$$

Indeed, the classical Sobolev inequality ensures that the embedding of the space $W_0^{1,p}(\Omega)$ into the Lebesgue space $L^{p^*}(\Omega)$ is continuous but not compact. Due to this lack of compactness the classical methods cannot be used in order to prove the weak lower semicontinuity of the energy functional associated to $(P_{\mu,\lambda})$. In our setting we overcome this difficulty through a lower semicontinuity result obtained by Montefusco in [17]. In this paper, bearing in mind the well-known inequality

$$\left(\int_\Omega |u(x)|^{p^*} dx \right)^{1/p^*} \leq \frac{1}{S^{1/p}} \left(\int_\Omega |\nabla u(x)|^p dx \right)^{1/p}, \quad \forall u \in W_0^{1,p}(\Omega) \tag{1}$$

where S is the best constant in the Sobolev inclusion $W_0^{1,p}(\Omega) \hookrightarrow L^{p^*}(\Omega)$ (see Sect. 2), fixing $\mu \in [0, S[$ and requiring a suitable behaviour of the nonlinearity f at zero, we determine a precise open interval of positive parameters λ, for which problem $(P_{\mu,\lambda})$ admits at least one non-trivial weak solution in $W_0^{1,p}(\Omega)$.

The proof of our main results are based on a recent abstract critical point theorem proved by Bonanno and Candito in [1, Theorem 3.1, part (a)] which is substantially a refinement of the variational principle established by Ricceri in [18]; see also Bonanno and Molica Bisci [3, Theorem 2.1, part (a)].

We explicitly observe that our results are a more precise form of the contributions obtained by Faraci and Livrea in [13]. Indeed, in Theorem 3.1 of the cited paper, fixing $\mu \in [0, S[$, the authors proved the existence of a positive parameter v_μ^* such that, for every $\lambda \in]0, v_\mu^*[$, the problem $(P_{\mu,\lambda})$ admits at least one non-trivial weak solution. However, by using their approach, no concrete expression of this parameter is given.

Here, through a different strategy previously developed by Bonanno and Molica Bisci in [5], an explicit value of the parameter v_μ^* is presented. A particular case of our results (see Theorem 3 and Remark 1 below) reads as follows.

Theorem 1. *Let* $f : \mathbb{R} \to \mathbb{R}$ *be a continuous function satisfying the following subcritical growth condition:*

$$|f(t)| \leq a_1 + a_2 |t|^{p-1}, \quad \forall t \in \mathbb{R},$$

where a_1, a_2 *are non-negative constants. Furthermore, assume that*

$$\lim_{\xi \to 0^+} \frac{\int_0^\xi f(t)dt}{\xi^p} = +\infty. \tag{h_0'}$$

Then, for every $\mu \in [0, S[$ *there exists a positive number* v_μ^* *given by*

$$v_\mu^* := \frac{S - \mu}{a_2 S} \left(\frac{\omega_N}{\text{meas}(\Omega)} \right)^{p/N},$$

where ω_N *is the volume of the unit ball in* $\mathbb{R}^N$, *such that, for every* $\lambda \in]0, v_\mu^*[$, *the Dirichlet problem*

$$(\bar{P}_{\mu,\lambda}) \quad \begin{cases} -\Delta_p u = \mu \left(\int_\Omega |u(x)|^{p^*} dx \right)^{p/p^* - 1} |u|^{p^* - 2} u + \lambda f(u) & \text{in} \quad \Omega \\ u|_{\partial \Omega} = 0, \end{cases}$$

admits at least one non-trivial weak solution $u_\lambda \in W_0^{1,p}(\Omega)$. *Moreover, one has that*

$$\lim_{\lambda \to 0^+} \|u_\lambda\| = 0.$$

2 Abstract Framework and Main Results

Let Ω be a bounded domain in $\mathbb{R}^N$ with smooth boundary $\partial \Omega$ and denote by X the space $W_0^{1,p}(\Omega)$ endowed with the norm

$$\|u\| := \left(\int_\Omega |\nabla u(x)|^p dx \right)^{1/p}.$$

Fixing $q \in [1, p^*[$, from the Sobolev embedding theorem, there exists a positive constant c_q such that

$$\|u\|_{L^q(\Omega)} \leq c_q \|u\|, \quad u \in X, \tag{2}$$

and, in particular, the embedding $X \hookrightarrow L^q(\Omega)$ is compact. Moreover, the best constant that appears in inequality (1) is given by $S = 1/c^p$, where

$$c = \frac{1}{N\sqrt{\pi}} \left(\frac{N! \Gamma\left(\dfrac{N}{2} \right)}{2\Gamma\left(\dfrac{N}{p} \right) \Gamma\left(N + 1 - \dfrac{N}{p} \right)} \right)^{1/N} \eta^{1-1/p}, \tag{3}$$

and

$$\eta := \frac{N(p-1)}{N-p};$$

see, for instance, the quoted paper [20]. Let us define $F(x, \xi) := \int_0^\xi f(x,t)dt$, for every $(x, \xi) \in \Omega \times \mathbb{R}$, and consider the functional $E_{\mu,\lambda} : X \to \mathbb{R}$ given by

$$E_{\mu,\lambda}(u) := \Phi_\mu(u) - \lambda \Psi(u), \ u \in X,$$

where

$$\Phi_\mu(u) := \frac{1}{p} \int_\Omega |\nabla u(x)|^p dx - \frac{\mu}{p} \left(\int_\Omega |u(x)|^{p^*} dx \right)^{p/p^*}, \quad \Psi(u) := \int_\Omega F(x, u(x))dx.$$

Fixing $\mu \in [0, S[$, from the Sobolev inequality (1), it follows that

$$\left(\frac{S-\mu}{pS} \right) \|u\|^p \leq \Phi_\mu(u) \leq \frac{\|u\|^p}{p}, \tag{4}$$

for every $u \in X$. Furthermore, Montefusco, in [17], proved that Φ_μ is sequentially weakly lower semicontinuous for $\mu \in [0, S[$ and in this setting, since (1) holds, it is also a coercive functional. Moreover, note that $E_{\mu,\lambda} \in C^1(X, \mathbb{R})$ and a critical point $u \in X$ is a weak solution of the non-local problem

$$-\Delta_p u = \mu \left(\int_\Omega |u(x)|^{p^*} dx \right)^{p/p^*-1} |u|^{p^*-2} u + \lambda f(x,u) \quad \text{in} \quad \Omega.$$

Our main tool in order to obtain the existence of one non-trivial solution to problem $(P_{\mu,\lambda})$ is the following critical point theorem.

Theorem 2. *Let X be a reflexive real Banach space, and let $\Phi, \Psi : X \to \mathbb{R}$ be two Gâteaux differentiable functionals such that Φ is (strongly) continuous, sequentially weakly lower semicontinuous and coercive. Further, assume that Ψ is sequentially weakly upper semicontinuous. For every $r > \inf_X \Phi$, put*

$$
\varphi(r) := \inf_{u \in \Phi^{-1}(]-\infty,r[)} \frac{\left(\sup_{v \in \Phi^{-1}(]-\infty,r[)} \Psi(v) \right) - \Psi(u)}{r - \Phi(u)}.
$$

Then, for each $r > \inf_X \Phi$ and each $\lambda \in \,]0, 1/\varphi(r)[$, the restriction of $J_\lambda := \Phi - \lambda\Psi$ to $\Phi^{-1}(] - \infty, r[)$ admits a global minimum, which is a critical point (local minimum) of J_λ in X.

As pointed out in Introduction, this result is a refinement of the variational principle of Ricceri; see the quoted paper [18]. Moreover, we recall that Theorem 2 has been used in order to obtain some theoretical contributions on the existence of either three or infinitely many critical points for suitable functionals defined on reflexive Banach spaces; see [2, 3]. As consequences of the above-cited results, on the vast literature on the subject, we mention here some recent works [4, 6–12] on the existence of weak solutions for some different classes of elliptic problems.

The main result reads as follows.

Theorem 3. *Let $f : \Omega \times \mathbb{R} \to \mathbb{R}$ be a Carathéodory function with $f(x,0) \neq 0$ in Ω and satisfying condition (h_∞). Then, for every $\mu \in [0, S[$ there exists a positive number v_μ^* given by*

$$
v_\mu^* := q \sup_{\gamma > 0} \left(\frac{\gamma^{p-1}}{qa_1c_1 \left(\dfrac{pS}{S-\mu} \right)^{1/p} + a_2c_q^q \left(\dfrac{pS}{S-\mu} \right)^{q/p} \gamma^{q-1}} \right),
$$

such that, for every $\lambda \in\,]0, v_\mu^*[$, the following elliptic Dirichlet problem

$$
(P_{\mu,\lambda}) \qquad \begin{cases} -\Delta_p u = \mu \left(\displaystyle\int_\Omega |u(x)|^{p^*} dx \right)^{p/p^*-1} |u|^{p^*-2}u + \lambda f(x,u) & \text{in} \quad \Omega \\ u|_{\partial\Omega} = 0, \end{cases}
$$

admits at least one non-trivial weak solution $u_\lambda \in X$. Moreover,

$$
\lim_{\lambda \to 0^+} \|u_\lambda\| = 0
$$

and the function $\lambda \to E_{\lambda,\mu}(u_\lambda)$ is negative and decreasing in $]0, v_\mu^*[$.

Proof. Fix $\mu \in [0, S[$ and $\lambda \in]0, v_\mu^*[$. Our aim is to apply Theorem 2 with $X = W_0^{1,p}(\Omega)$; $\Phi := \Phi_\mu$ and Ψ are the functionals introduced before. Since $\mu \in [0, S[$, $\Phi : X \to \mathbb{R}$ is a continuously Gâteaux differentiable and sequentially weakly lower semicontinuous functional as well as the map $\Psi : X \to \mathbb{R}$ is continuously Gâteaux differentiable and sequentially weakly upper semicontinuous. Moreover, Φ is coercive and clearly $\inf_{u \in X} \Phi(u) = 0$. Thanks to the growth condition (h$_\infty$), one has that

$$F(x, \xi) \leq a_1 |\xi| + a_2 \frac{|\xi|^q}{q}, \tag{5}$$

for every $(x, \xi) \in \Omega \times \mathbb{R}$. Since $0 < \lambda < v_\mu^*$, there exists $\overline{\gamma} > 0$ such that

$$\lambda < v_\mu^\star(\overline{\gamma}) := \frac{q \overline{\gamma}^{p-1}}{q a_1 c_1 \left(\dfrac{pS}{S - \mu} \right)^{1/p} + a_2 c_q^q \left(\dfrac{pS}{S - \mu} \right)^{q/p} \overline{\gamma}^{q-1}}. \tag{6}$$

Now, set $r \in]0, +\infty[$ and consider the function

$$\chi(r) := \frac{\displaystyle\sup_{u \in \Phi^{-1}(]-\infty, r[)} \Psi(u)}{r}.$$

Taking into account (5) it follows that

$$\Psi(u) = \int_\Omega F(x, u(x)) dx \leq a_1 \|u\|_{L^1(\Omega)} + \frac{a_2}{q} \|u\|_{L^q(\Omega)}^q.$$

Then, due to (4), we get

$$\|u\| < \left(\frac{pSr}{S - \mu} \right)^{1/p}, \tag{7}$$

for every $u \in X$ and $\Phi(u) < r$. Now, from (2) and by using (7), for every $u \in X$ such that $\Phi(u) < r$, one has

$$\Psi(u) < c_1 a_1 \left(\frac{pS}{S - \mu} \right)^{1/p} r^{1/p} + a_2 \frac{c_q^q}{q} \left(\frac{pS}{S - \mu} \right)^{q/p} r^{q/p}.$$

Hence

$$\sup_{u \in \Phi^{-1}(]-\infty, r[)} \Psi(u) \leq c_1 a_1 \left(\frac{pS}{S - \mu} \right)^{1/p} r^{1/p} + a_2 \frac{c_q^q}{q} \left(\frac{pS}{S - \mu} \right)^{q/p} r^{q/p}.$$

Then

$$\chi(r) \leq c_1 a_1 \left(\frac{pS}{S-\mu}\right)^{1/p} r^{1/p-1} + a_2 \frac{c_q^q}{q} \left(\frac{pS}{S-\mu}\right)^{q/p} r^{q/p-1}, \qquad (8)$$

for every $r > 0$.

Hence, in particular

$$\chi(\overline{\gamma}^p) \leq c_1 a_1 \left(\frac{pS}{S-\mu}\right)^{1/p} \overline{\gamma}^{1-p} + a_2 \frac{c_q^q}{q} \left(\frac{pS}{S-\mu}\right)^{q/p} \overline{\gamma}^{q-p}. \qquad (9)$$

Now, observe that

$$\varphi(\overline{\gamma}^p) := \inf_{u \in \Phi^{-1}(]-\infty,\overline{\gamma}^p[)} \frac{\left(\sup_{v \in \Phi^{-1}(]-\infty,\overline{\gamma}^p[)} \Psi(v)\right) - \Psi(u)}{r - \Phi(u)} \leq \chi(\overline{\gamma}^p),$$

because $u_0 \in \Phi^{-1}(]-\infty,\overline{\gamma}^p[)$ and $\Phi(u_0) = \Psi(u_0) = 0$, where $u_0 \in X$ is the identically zero function. In conclusion, bearing in mind (6), the above inequality together with (9) gives

$$\varphi(\overline{\gamma}^p) \leq \chi(\overline{\gamma}^p) \leq c_1 a_1 \left(\frac{pS}{S-\mu}\right)^{1/p} \overline{\gamma}^{1-p} + a_2 \frac{c_q^q}{q} \left(\frac{pS}{S-\mu}\right)^{q/p} \overline{\gamma}^{q-p} < \frac{1}{\lambda}.$$

In other words,

$$\lambda \in \left] 0, \frac{q\overline{\gamma}^{p-1}}{qa_1 c_1 \left(\frac{pS}{S-\mu}\right)^{1/p} + a_2 c_q^q \left(\frac{pS}{S-\mu}\right)^{q/p} \overline{\gamma}^{q-1}} \right[\subseteq]0, 1/\varphi(\overline{\gamma}^p)[.$$

Thanks to Theorem 2, there exists a function $u_\lambda \in \Phi^{-1}(]-\infty,\overline{\gamma}^p[)$ such that

$$E'_{\mu,\lambda}(u_\lambda) = \Phi'(u_\lambda) - \lambda \Psi'(u_\lambda) = 0,$$

and, in particular, u_λ is a global minimum of the restriction of $E_{\mu,\lambda}$ to $\Phi^{-1}(]-\infty,\overline{\gamma}^p[)$. Further, since $f(x,0) \neq 0$ in Ω, the function u_λ cannot be trivial, that is, $u_\lambda \neq 0$. Hence, for $\mu \in [0,S[$ and for every $\lambda \in]0,v_\mu^*[$ the problem $(P_{\mu,\lambda})$ admits a non-trivial solution $u_\lambda \in X$. From now on, we argue in similar way of [13, Theorem 3.1] in order to prove that $\|u_\lambda\| \to 0$ as $\lambda \to 0^+$ and that the function $\lambda \to E_{\mu,\lambda}(u_\lambda)$ is negative and decreasing in $]0,v_\mu^*[$. The proof is complete. $\qquad \square$

Remark 1. Also when $f(x, 0) = 0$ in Ω the statements of Theorem 3 are still true if, in addition to assumption (h_∞), the function f satisfies the following hypothesis: *There are a non-empty open set $D \subseteq \Omega$ and a set $B \subseteq D$ of positive Lebesgue measure such that*

$$\limsup_{\xi \to 0^+} \frac{\inf_{x \in B} F(x, \xi)}{\xi^p} = +\infty \quad \text{and} \quad \liminf_{\xi \to 0^+} \frac{\inf_{x \in D} F(x, \xi)}{\xi^p} > -\infty. \qquad (h_0)$$

Condition (h_0) ensures that the solution, achieved by using Theorem 3, is non-trivial.

Remark 2. In conclusion, we just mention that the technical approach adopted in this manuscript has been used in different settings in order to obtain existence and multiplicity results for several kinds of differential problems, for instance, by Bonanno, Molica Bisci and Rădulescu in [8] for elliptic problems on compact Riemannian manifolds without boundary and by D'Aguì and Molica Bisci for an elliptic Neumann problem involving the p-Laplacian; see [10].

Acknowledgements V. Rădulescu acknowledges the support through Grant CNCSIS PCCE-8/2010 *"Sisteme diferenţiale în analiza neliniară şi aplicaţii"*.

References

1. Bonanno, G., Candito, P.: Non-differentiable functionals and applications to elliptic problems with discontinuous nonlinearities. J. Differ. Equ. **244**, 3031–3059 (2008)
2. Bonanno, G., Marano, S.A.: On the structure of the critical set of non-differentiable functions with a weak compactness condition. Appl. Anal. **89**, 1–10 (2010)
3. Bonanno, G., Molica Bisci, G.: Infinitely many solutions for a boundary value problem with discontinuous nonlinearities. Bound. Value Probl. **2009**, 1–20 (2009)
4. Bonanno, G., Molica Bisci, G.: Infinitely many solutions for a Dirichlet problem involving the p-Laplacian. Proc. Roy. Soc. Edinburgh Sect. A **140**, 1–16 (2009)
5. Bonanno G., Molica Bisci, G.: Three weak solutions for Dirichlet problems. J. Math. Anal. Appl. **382**, 1–8 (2011)
6. Bonanno, G., Molica Bisci, G., Rădulescu, V.: Existence of three solutions for a non-homogeneous Neumann problem through Orlicz-Sobolev spaces. Nonlinear Anal. **74**(14), 4785–4795 (2011)
7. Bonanno, G., Molica Bisci, G., Rădulescu, V.: Infinitely many solutions for a class of nonlinear eigenvalue problems in Orlicz-Sobolev spaces. C. R. Acad. Sci. Paris Ser. I **349**, 263–268 (2011)
8. Bonanno, G., Molica Bisci, G., Rădulescu, V.: Multiple solutions of generalized Yamabe equations on Riemannian manifolds and applications to Emden-Fowler problems. Nonlinear Anal. Real World Appl. **12**, 2656–2665 (2011)
9. Bonanno, G., Molica Bisci, G., Rădulescu, V.: Arbitrarily small weak solutions for a nonlinear eigenvalue problem in Orlicz-Sobolev spaces. Monatsh Math. 1–14 (2011). doi: 10.1007/s00605-010-0280-2
10. D'Aguì, G., Molica Bisci, G.: Infinitely many solutions for perturbed hemivariational inequalities. Bound. Value Probl. Art. ID 363518, 1–15 (2010)
11. D'Aguì, G., Molica Bisci, G.: Existence results for an Elliptic Dirichlet problem. Le Matematiche Fasc. I **LXVI**, 133–141 (2011)

12. D'Aguì, G., Molica Bisci, G.: Three non-zero solutions for elliptic Neumann problems. Anal. Appl. **9**(4), 1–12 (2011)
13. Faraci, F., Livrea, R.: Bifurcation theorems for nonlinear problems with lack of compactness. Ann. Polon. Math. **82**(1), 77–85 (2003)
14. Guo, Z., Webb, J.R.L.: Large and small solutions of a class of quasilinear elliptic eigenvalue problems. J. Differ. Equ. **180**, 1–50 (2002)
15. Hai, D.D.: On a class of sublinear quasilinear elliptic problems. Proc. Amer. Math. Soc. **131**, 2409–2414 (2003)
16. Kristály, A., Rădulescu, V., Varga, C: Variational Principles in Mathematical Physics, Geometry, and Economics: Qualitative Analysis of Nonlinear Equations and Unilateral Problems. Encyclopedia of Mathematics and its Applications, vol. 136. Cambridge University Press, Cambridge (2010)
17. Montefusco, E.: Lower semicontinuity of functionals via the concentration-compactness principle. J. Math. Anal. Appl. **263**, 264–276 (2001)
18. Ricceri, B.: A general variational principle and some of its applications. J. Comput. Appl. Math. **113**, 401–410 (2000)
19. Saint Raymond, J.: On the multiplicity of solutions of the equations $-\Delta u = \lambda f(u)$. J. Differ. Equ. **180**, 65–88 (2002)
20. Talenti, G.: Best constants in Sobolev inequality. Ann. Mat. Pura Appl. **110**, 353–372 (1976)

Sign-Changing Subharmonic Solutions to Unforced Equations with Singular ϕ-Laplacian

Alberto Boscaggin and Maurizio Garrione

Abstract We prove the existence of infinitely many subharmonic solutions (with a precise nodal characterization) to the equation

$$\left(\frac{u'}{\sqrt{1 - u'^2}}\right)' + g(t,u) = 0,$$

in the unforced case $g(t,0) \equiv 0$. The proof is performed via the Poincaré–Birkhoff fixed point theorem.

1 Introduction and Statement of the Main Result

In the very last years, an increasing attention has been devoted to the periodic problem associated with scalar nonlinear differential equations of the type

$$(\phi(u'))' + g(t,u) = 0, \tag{1}$$

being $\phi :\ ]-a,a[\to \mathbb{R}$ (with $0 < a < +\infty$) an increasing global homeomorphism satisfying $\phi(0) = 0$, and $g : \mathbb{R} \times \mathbb{R} \to \mathbb{R}$ a continuous function, which is T-periodic in the first variable (for a fixed period $T > 0$). Since $a < +\infty$, the differential operator $u \mapsto -(\phi(u'))'$ is called *singular* ϕ-Laplacian. As a model situation for (1), one can consider the case corresponding to the relativistic acceleration, namely, $\phi(x) = \frac{x}{\sqrt{1-x^2}}$, leading to the equation

A. Boscaggin (✉)
Dipartimento di Matematica, Università di Torino via Carlo Alberto 10, 10123 Torino, Italy
e-mail: alberto.boscaggin@unito.it

M. Garrione
Dipartimento di Scienze Matematiche, Università Politecnica delle Marche, aia Brecce Bianche
e-mail: m.garrione@univpm.it

S. Pinelas et al. (eds.), *Differential and Difference Equations with Applications*, Springer Proceedings in Mathematics & Statistics 47, DOI 10.1007/978-1-4614-7333-6_25,
© Springer Science+Business Media New York 2013

$$\left(\frac{u'}{\sqrt{1-u'^2}}\right)' + g(t,u) = 0. \tag{2}$$

Suitable assumptions on the function $g(t,x)$, ensuring the existence of one or "few" T-periodic solutions to (1) or (2), were introduced by various authors. In particular, using both variational and topological tools, the cases when $g(t,x)$ satisfies some conditions at infinity (of Landesman–Lazer or Ahmad–Lazer–Paul type, or sign assumptions; see, for instance, [1, 3]), or when $g(t,x)$ is a pendulum-like nonlinearity [2, 4, 10, 11], have been extensively studied. We refer to [12] for an exhaustive survey and a very wide bibliography about the subject.

In this brief note, we propose to consider the *unforced* case, namely, $g(t,0) \equiv 0$, adding an assumption on the behavior of $g(t,x)$ near $x = 0$. For the second-order equation $u'' + g(t,u) = 0$, by means of the Poincaré–Birkhoff fixed point theorem, this situation was shown to be the source of "many" sign-changing periodic solutions, both harmonic (i.e., T-periodic see, e.g., [15]) and subharmonic (i.e., kT-periodic for an integer $k \geq 2$, see, e.g., [6]). Yet, to the best of our knowledge, there are no results concerning (1) and (2) in this direction.

From now on, for simplicity we will limit ourselves to the investigation of (2). Possible extensions to other ϕ-operators will be discussed in Remark 2. To motivate our result, let us first consider the autonomous equation

$$\left(\frac{u'}{\sqrt{1-u'^2}}\right)' + \lambda u = 0, \tag{3}$$

being $\lambda > 0$ a real parameter. In this case, it is easily seen that the function

$$E(u,v) = \frac{1}{\sqrt{1-v^2}} + \frac{\lambda}{2}u^2 - 1, \quad u \in \mathbb{R}, |v| < 1,$$

is a first integral for (3), i.e., $E(u(t), u'(t))$ is constant whenever $u(t)$ solves (3). By a direct qualitative analysis in the phase-plane (u, u'), it turns out that all the nontrivial solutions to (3) are periodic and wind the origin in the clockwise sense. A computation of the time map (i.e., the time $T_\lambda(c)$ needed for a solution $(u(t), u'(t))$ such that $E(u(t), u'(t)) \equiv c$ to perform one revolution around the origin) yields

$$T_\lambda(c) = 4 \int_0^{\sqrt{\frac{2c}{\lambda}}} \frac{d\xi}{\sqrt{1 - \frac{4}{(2c+2-\lambda\xi^2)^2}}}, \quad c > 0.$$

It can be seen that $T_\lambda(c)$ is continuous and strictly increasing, with

$$\lim_{c \to 0^+} T_\lambda(c) = \frac{2\pi}{\sqrt{\lambda}}, \qquad \lim_{c \to +\infty} T_\lambda(c) = +\infty.$$

Therefore, the intermediate value theorem implies that for every $\tau > \frac{2\pi}{\sqrt{\lambda}}$, there exists exactly one solution—modulo translations—to (3) of minimal period τ and having exactly two zeros in the interval $[0, \tau[$.[1] As a consequence, if $k, j \geq 1$ are integer numbers such that

$$\frac{2\pi}{\sqrt{\lambda}} < \frac{kT}{j}, \tag{4}$$

there exists exactly a kT-periodic solution $u_{k,j}(t)$ to (3) having $2j$ zeros in the interval $[0, kT[$ (of course, such a solution is not of minimal period kT when $j > 1$; indeed, the definition of subharmonic solution needs some clarifications; see Remark 1). Notice that the larger is the k, the greater is the number of integers j satisfying (4).

Our goal is to extend such elementary considerations to the nonautonomous case, by the use of the Poincaré–Birkhoff fixed point theorem, in the formulation by Ding [8]. Roughly speaking, we are going to prove that for any fixed time interval $[0, kT]$, with k a positive integer, "large" solutions to (2) do not complete a full revolution around the origin. On the other hand, "small" solutions perform a finite number of turns, depending on k. Thus, a twist condition for the kth iterate of the Poincaré map associated with (2) will be fulfilled, giving the existence of subharmonic solutions with a "low" number of zeros.

Here is the precise statement of our main result. From now on, we will suppose that $g(t, x)$ is locally Lipschitz continuous in the x-variable, uniformly in t.

Theorem 1. *Assume that*

(g_0) $g(t, 0) \equiv 0$, *and*

$$\liminf_{x \to 0} \frac{g(t, x)}{x} \geq q(t), \quad uniformly \ in \ t \in [0, T], \tag{5}$$

for a suitable $q \in L^\infty(0, T)$ satisfying $\int_0^T q(t)\, dt > 0$;
(g_∞) *there exist $\alpha, \beta > 0$ such that*

$$|g(t, x)| \leq \alpha|x| + \beta, \quad for \ every \ t \in [0, T], \ x \in \mathbb{R}. \tag{6}$$

Then, there exists an integer $k^* \geq 1$ such that for every integer $k \geq k^*$, there exists another integer m_k such that for every integer j relatively prime with k and such that $1 \leq j \leq m_k$, (2) has at least two subharmonic solutions $u_{k,j}^{(1)}(t), u_{k,j}^{(2)}(t)$ of order k (not belonging to the same periodicity class), with exactly $2j$ zeros in the interval $[0, kT[$. Moreover, the following estimate for m_k holds:

---

[1] Incidentally, observe that this situation is really different from the linear problem $u'' + \lambda u = 0$; in particular, here resonance phenomena do not appear for any $\lambda > 0$ (see [1, Remark 6]).

$$m_k \geq n_k := \mathscr{E}^- \left(\frac{k}{2\pi} \frac{\int_0^T q(t)\,dt}{\sqrt{\operatorname{ess\,sup}_{[0,T]} q(t)}} \right), \tag{7}$$

where for r > 0, we denote by $\mathscr{E}^-(r)$ the greatest integer strictly *less than r.*

The result can be viewed as a variant of [6, Corollary 3.1] (see also [7, Theorem 1.1]). However, while in [6] the low angular speed of large solutions came from a sublinearity assumption on $g(t,x)$, here it is provided by the differential operator itself, so that $g(t,x)$ is allowed to grow linearly at infinity.

Remark 1. We recall that as usual in this setting, by a *subharmonic solution of order k* to (2), we mean a kT-periodic solution which is not lT-periodic for any integer $l = 1,\ldots,k-1$. Moreover, by *periodicity class* of a subharmonic $u(t)$ of order k, we mean the set $\{u(t), u(t+T), u(t+2T),\ldots,u(t+(k-1)T)\}$; since $g(\cdot,x)$ is T-periodic, such functions are subharmonic solutions of order k to (2), as well.

Referring to the discussion about the autonomous case, together with formula (4), notice that even if the minimal period of $u_{k,j}(t)$ is $\frac{kT}{j} \leq kT$, the number kT is indeed the minimal period in the class of the integer multiples of T whenever k, j are relatively prime. Notice also that in general, we cannot exclude $u_{k,j}^{(1)}(t) \equiv u_{k,j}^{(2)}(t+\sigma)$ for $\sigma \notin T\mathbb{N}$. Indeed, this is precisely what happens in the autonomous case, where for k, j as in the statement, there exists a one-parameter family of subharmonics (all the time translations of a fixed one).

Observe that the estimate (7) gives sharp information concerning the order of the subharmonics produced, so that the following corollary dealing with the existence of multiple T-periodic solutions can be stated.

Corollary 1. *Denote by* $\lambda_h := (\frac{2\pi h}{T})^2$, $h = 0,1,2,\ldots$, *the eigenvalues of the linear differential operator* $u \mapsto -u''$, *with T-periodic boundary conditions, and assume*

(g'_0) $g(t,0) \equiv 0$ *and, for suitable* $h \geq 1$ *and* $q > 0$,

$$\liminf_{x \to 0} \frac{g(t,x)}{x} \geq q > \lambda_h, \quad \text{uniformly in } t \in [0,T],$$

and (g_∞) *as in Theorem 1. Then,* (2) *has at least 2h T-periodic solutions. More precisely, for every integer j such that* $1 \leq j \leq h$, (2) *has at least two T-periodic solutions with exactly 2j zeros in the interval* $[0,T[$.

2   Proof of the Main Result and Further Remarks

As a preliminary, we briefly recall that for real numbers $\mu, \nu > 0$ and a $\mathscr{C}^1$-path $z : [t_1, t_2] \to \mathbb{R}^2$ such that $z(t) \neq 0$ for every $t \in [t_1, t_2]$, the *modified rotation number* of $z(t) = (u(t), v(t))$ around the origin is defined as

$$\text{Rot}_{\mu,\nu}\left(z(t);[t_1,t_2]\right) = \frac{\sqrt{\mu\nu}}{2\pi} \int_{t_1}^{t_2} \frac{\nu(t)u'(t) - u(t)v'(t)}{\mu u(t)^2 + \nu v(t)^2}\, dt.$$

The choice $\mu = \nu = 1$ leads to the usual notion of rotation number (which we will denote simply by Rot); the remarkable property which we will exploit (see [14]) is that for every *integer* j,

$$\text{Rot}\left(z(t);[t_1,t_2]\right) \lesseqgtr j \iff \text{Rot}_{\mu,\nu}\left(z(t);[t_1,t_2]\right) \lesseqgtr j. \tag{8}$$

We are now in a position to start the proof.

Proof (of Theorem 1). For simplicity of notation, let us set $\varphi_r(x) = \frac{x}{\sqrt{1-x^2}}$. Let us write (2) as the equivalent planar Hamiltonian system

$$u' = \varphi_r^{-1}(v), \quad v' = -g(t,u). \tag{9}$$

Notice that this leads us to work in the new plane $(u, \varphi_r(u'))$ and that the vector field in (9) is defined for every $(u,v) \in \mathbb{R}^2$. Moreover, there is uniqueness and global continuability for the solutions to the initial value problems associated with (9). As a consequence, the Poincaré map

$$\Psi : \bar{z} = (\bar{u}, \bar{v}) \in \mathbb{R}^2 \mapsto z(T;\bar{z}) = (u(T;\bar{z}), v(T;\bar{z}))$$

associated with (9) is well defined (here, by $z(t;\bar{z})$ we mean the unique vector solution to (9) such that $z(0;\bar{z}) = \bar{z}$); moreover, initial values (at time $t = 0$) of kT-periodic solutions (for $k \geq 1$ integer) to (9) correspond to fixed points of the kth iterate of Ψ. To find such fixed points by means of the Poincaré–Birkhoff theorem, we have to exhibit a gap between the rotation numbers of small and large solutions to (9).

As a first step, define k^* to be the least integer such that $n_{k^*} \geq 1$ and fix $k \geq k^*$. Setting $Q = \text{ess\,sup}_{[0,T]} q(t)$, it is possible to choose $\delta, \eta > 0$ such that

$$\frac{k\sqrt{1-\delta}}{2\pi\sqrt{Q}} \int_0^T (q(t) - \eta)\, dt > n_k;$$

since $(\varphi_r^{-1})'(0) = 1$, and in view of (5), there exists $r^* > 0$ such that

$$\varphi_r^{-1}(v)v \geq (1-\delta)v^2 \text{ and } g(t,u)u \geq (q(t)-\eta)u^2, \quad \text{for } t \in [0,T],\ u^2+v^2 \leq (r^*)^2.$$

Therefore, if $z : [0,kT] \to \mathbb{R}^2$ solves (9) with $0 < |z(t)| \leq r^*$ for every $t \in [0,kT]$, we have

$$\text{Rot}_{1,\frac{1-\delta}{Q}}\left(z(t);[0,kT]\right) = \frac{\sqrt{1-\delta}}{2\pi\sqrt{Q}} \int_0^{kT} \frac{\varphi_r^{-1}(v(t))v(t) + g(t,u(t))u(t)}{u(t)^2 + \frac{1-\delta}{Q}v(t)^2}\, dt$$

$$\geq \frac{\sqrt{1-\delta}}{2\pi\sqrt{Q}} \int_0^{kT} \frac{\frac{1-\delta}{Q}(q(t)-\eta)v(t)^2 + (q(t)-\eta)u(t)^2}{u(t)^2 + \frac{1-\delta}{Q}v(t)^2} \, dt$$

$$= \frac{k\sqrt{1-\delta}}{2\pi\sqrt{Q}} \int_0^T (q(t)-\eta) \, dt > n_k.$$

Using the fact that in view of the uniqueness, nontrivial solutions to (9) never reach the origin, it is easy to see that there exists $r \in \,]0, r^*]$ such that the previous relation holds true whenever $|z(0)| = r$.

On the other hand, fix $\varepsilon > 0$ such that

$$\frac{kT\sqrt{\alpha\varepsilon}}{2\pi} < \frac{1}{2};$$

in view of the boundedness of φ_r^{-1}, there exists $C_\varepsilon > 0$ such that

$$\varphi_r^{-1}(v)v \leq \varepsilon v^2 + C_\varepsilon, \quad \text{for } v \in \mathbb{R}.$$

Using (6), it follows that if $z : [0, kT] \to \mathbb{R}^2$ solves (9) with $z(t) \neq 0$ for every $t \in [0, kT]$, then

$$\mathrm{Rot}_{\alpha,\varepsilon}(z(t); [0, kT]) = \frac{\sqrt{\alpha\varepsilon}}{2\pi} \int_0^{kT} \frac{\varphi_r^{-1}(v(t))v(t) + g(t, u(t))u(t)}{\alpha u(t)^2 + \varepsilon v(t)^2} \, dt$$

$$\leq \frac{\sqrt{\alpha\varepsilon}}{2\pi} \left(kT + \int_0^{kT} \frac{C_\varepsilon + \beta|u(t)|}{\alpha u(t)^2 + \varepsilon v(t)^2} \, dt \right) < 1,$$

provided that $|z(t)| \geq R^*$ for a sufficiently large R^*. By standard compactness arguments (the so-called elastic property), it is possible to find $R \geq R^*$ such that the previous relation holds true whenever $|z(0)| = R$.

In view of (8), a gap between the standard rotation numbers is produced on the boundaries of the annulus $\{z \in \mathbb{R}^2 \mid r \leq |z| \leq R\}$. We can thus conclude using the Poincaré–Birkhoff fixed point theorem for Ψ^k as in [6], to find, for every integer j with $1 \leq j \leq n_k$, two kT-periodic solutions $u_{k,j}^{(i)}(t)$ ($i = 1, 2$) to (2) with $\mathrm{Rot}((u_{k,j}^{(i)}(t), \varphi_r(\frac{d}{dt}u_{k,j}^{(i)}(t))); [0, kT]) = j$. It is clear that this implies the equality $j = \mathrm{Rot}((u_{k,j}^{(i)}(t), \frac{d}{dt}u_{k,j}^{(i)}(t)); [0, kT])$, this latter being half the number of zeros of $u_{k,j}^{(i)}(t)$ in the interval $[0, kT[$. $\qquad\square$

Example 1. Theorem 1 applies, for instance, to the equation

$$\left(\frac{u'}{\sqrt{1-u'^2}} \right)' + a(t)\sin u = 0,$$

which describes the motion of a relativistic pendulum in a space with periodically varying gravity. Here, $a : \mathbb{R} \to \mathbb{R}$ is a continuous and T-periodic function, with $\int_0^T a(t)\,dt \neq 0$. Indeed, the case when $a(t)$ has positive average directly fits in the framework of Theorem 1, yielding the existence of sign-changing subharmonics $u_{k,j}(t)$. On the other hand, if the average of $a(t)$ is negative, one can perform the change of variable $x(t) = u(t) + \pi$ and apply Theorem 1, finding subharmonics $x_{k,j}(t)$ oscillating around π. Notice also that with the same arguments as in [7, Theorem 2.1], it is possible to show that the subharmonics produced take values in the intervals $]-\pi, \pi[$ in the former case and $]0, 2\pi[$ in the latter.

Remark 2. The conclusion of Theorem 1 still holds (up to slightly modifying (7)) if $\phi :]-a, a[\to \mathbb{R}$ (with $0 < a \leq +\infty$) is an increasing homeomorphism, with locally Lipschitz continuous inverse, such that $\phi(0) = 0$ and

$$\liminf_{x \to 0} \frac{\phi^{-1}(x)}{x} > 0, \qquad \limsup_{|x| \to +\infty} \frac{\phi^{-1}(x)}{x} = 0. \tag{10}$$

Observe, on one hand, that the case $a = +\infty$ is not excluded; on the other hand, notice that the second formula in (10) always holds if $a < +\infty$. For instance, $\phi(x) = x + x^3$ and $\phi(x) = \tan x$ satisfy these requirements, with $a = +\infty$ and $a = \frac{\pi}{2}$, respectively.

Remark 3. It is natural to wonder what kind of results could be proved, with similar arguments, for the mean curvature equation (see again [12] and the references therein)

$$\left(\frac{u'}{\sqrt{1 + u'^2}} \right)' + g(t, u) = 0. \tag{11}$$

Let us consider first the autonomous case

$$\left(\frac{u'}{\sqrt{1 + u'^2}} \right)' + \lambda u = 0, \tag{12}$$

which admits the first integral

$$E(u, v) = -\frac{1}{\sqrt{1 + v^2}} + \frac{\lambda}{2} u^2 + 1, \qquad u, v \in \mathbb{R}.$$

By standard phase-plane analysis, it can be seen that for $c \geq 1$, the solutions at energy level $E(u(t), u'(t)) \equiv c$ are not periodic (even more, they are not globally extendable; for this reason, it is natural to investigate the existence of "nonclassical" solutions, as in [5, 13]). On the contrary, solutions with energy $c \in]0, 1[$ are periodic, with time map given by

$$T_\lambda(c) = 4 \int_0^{\sqrt{\frac{2c}{\lambda}}} \frac{d\xi}{\sqrt{\frac{4}{(2c-2-\lambda\xi^2)^2} - 1}}.$$

It can be seen that $T_\lambda(c)$ is continuous and strictly decreasing, with

$$\lim_{c \to 0^+} T_\lambda(c) = \frac{2\pi}{\sqrt{\lambda}}, \qquad \lim_{c \to 1^-} T_\lambda(c) = \frac{\zeta}{\sqrt{\lambda}},$$

where $\zeta = 4\sqrt{2} \int_0^1 \frac{\xi^2}{\sqrt{1+\xi^2}\sqrt{1-\xi^2}} d\xi < 2\pi$. Thus, for every $\tau \in]\frac{\zeta}{\sqrt{\lambda}}, \frac{2\pi}{\sqrt{\lambda}}[$, the inter-mediate value theorem implies the existence of exactly one—up to translations—solution to (12) of minimal period τ and having exactly two zeros in the interval $[0, \tau[$. Consequently, if $k, j \geq 1$ are integer numbers such that

$$\frac{\zeta}{\sqrt{\lambda}} < \frac{kT}{j} < \frac{2\pi}{\sqrt{\lambda}}, \tag{13}$$

there exists a kT-periodic solution $u_{k,j}(t)$ to (12) having $2j$ zeros in the interval $[0, kT[$. Notice, on the one hand, that for $\lambda > 0$ fixed, a number theory argument (see the proof of [9, Theorem 2.3]) implies that there are "many" k, j relatively prime satisfying (13); on the other hand, letting $\lambda \to +\infty$, it is possible to see that multiple T-periodic solutions appear.

It could be interesting to understand if a similar result holds true for the general nonautonomous case (11).

Acknowledgements The authors wish to thank SISSA for the financial support which has given the pleasant opportunity of taking part in the *International Conference on Differential and Difference Equations and Applications* in Ponta Delgada, July 2011.

References

1. Bereanu, C., Mawhin, J.: Existence and multiplicity results for some nonlinear problems with singular ϕ-Laplacian. J. Differ. Equ. **243**, 536–557 (2007)
2. Bereanu, C., Jebelean, P., Mawhin, J.: Periodic solutions of pendulum-like perturbations of singular and bounded ϕ-Laplacians. J. Dynam. Differ. Equ. **22**, 463–471 (2010)
3. Bereanu, C., Jebelean, P., Mawhin, J.: Variational methods for nonlinear perturbations of singular ϕ-Laplacians. Atti Accad. Naz. Lincei Cl. Sci. Fis. Mat. Natur. Rend. Lincei Mat. Appl. **22**(9), 89–111 (2011)
4. Bereanu, C., Torres, P.: Existence of at least two periodic solutions of the forced relativistic pendulum. Proc. Amer. Math. Soc. **140**, 2713–2719 (2012)
5. Bonheure, D., Habets, P., Obersnel, F., Omari, P.: Classical and non-classical solutions of a prescribed curvature equation. J. Differ. Equ. **243**, 208–237 (2007)
6. Boscaggin, A.: Subharmonic solutions of planar Hamiltonian systems: a rotation number approach. Adv. Nonlinear Stud. **11**, 77–103 (2011)

7. Boscaggin, A., Zanolin, F.: Subharmonic solutions for nonlinear second order equations in presence of lower and upper solutions. Discrete Contin. Dyn. Syst. **33**, 89–110 (2013)

8. Ding, W.-Y.: Fixed points of twist mappings and periodic solutions of ordinary differential equations (Chinese). Acta Math. Sinica **25**, 227–235 (1982)

9. Ding, T., Iannacci, R., Zanolin, F.: Existence and multiplicity results for periodic solutions of semilinear Duffing equations. J. Differ. Equ. **105**, 364–409 (1993)

10. Fonda, A., Toader, R.: Periodic solutions of pendulum-like Hamiltonian systems in the plane Adv. Nonlinear Stud. **12**, 395–408 (2012)

11. Marò, S.: Periodic solutions of a forced relativistic pendulum via twist dynamics to appear on Topol. Methods Nonlinear Anal.

12. Mawhin, J.: Stability and bifurcation theory for non-autonomous differential equations (Cetraro, 2011), Lecture Notes in Math. 2065, to appear on Topol. Methods Nonlinear Anal. Springer, Berlin, (2013), 103–184

13. Obersnel, F. Omari, P.: Multiple bounded variation solutions of a periodically perturbed sinecurvature equation. Commun. Contemp. Math. **13**, 863–883 (2011)

14. Rebelo, C., Zanolin, F.: Multiplicity results for periodic solutions of second order ODEs with asymmetric nonlinearities. Trans. Amer. Math. Soc. **348**, 2349–2389 (1996)

15. Zanini, C.: Rotation numbers, eigenvalues, and the Poincaré-Birkhoff theorem. J. Math. Anal. Appl. **279**, 290–307 (2003)

Nonlinear Difference Equations with Discontinuous Right-Hand Side

Pasquale Candito and Roberto Livrea

Abstract A discrete nonlinear problem involving the $p-$Laplacian and with a discontinuous right-hand side is studied. The existence of a precise open interval of positive eigenvalues, for which it admits at least three solutions, is established. The approach adopted is fully based on the variational methods developed in [14,15] for non-smooth functions.

1 Introduction

Let $1 < p < +\infty$ and $\lambda \in \mathbb{R}^+$. Let N be a positive integer, $q_k \in \mathbb{R}_0^+$ for all $k \in [1,N]$ and let $f : [1,N] \times \mathbb{R} \to \mathbb{R}$ be a measurable locally bounded function. Denote with $[1,N]$ the discrete interval $\{1,\dots,N\}$ and with $\Delta u_{k-1} := u_k - u_{k-1}$ the forward difference operator; in this paper, we study the existence of multiple solutions for the following nonlinear discrete Dirichlet problem with discontinuous nonlinearity:

$$(D_\lambda) \begin{cases} -\Delta(|\Delta u_{k-1}|^{p-2}\Delta u_{k-1}) + q_k|u_k|^{p-2}u_k \in \lambda [f^-(k,u_k), f^+(k,u_k)], & k \in [1,N], \\ u_0 = u_{N+1} = 0, \end{cases}$$

where, for every $k \in [1,N]$,

$$f^-(k,u_k) := \lim_{\delta \to 0^+} \operatorname*{ess\,inf}_{|\zeta-u_k|<\delta} f(k,\zeta) \quad \text{and} \quad f^+(k,u_k) := \lim_{\delta \to 0^+} \operatorname*{ess\,sup}_{|\zeta-u_k|<\delta} f(x,\zeta).$$

When f is a continuous function, problem (D_λ) has been investigated by many authors; see [4,5,10,16,18]. Therefore, it is interesting to treat this type of problems

P. Candito (✉) • R. Livrea
Dipartimento MECMAT, University of Reggio Calabria, 89124 Reggio Calabria, Italy
e-mail: pasquale.candito@unirc.it; roberto.livrea@unirc.it

S. Pinelas et al. (eds.), *Differential and Difference Equations with Applications*, Springer Proceedings in Mathematics & Statistics 47, DOI 10.1007/978-1-4614-7333-6_26,
© Springer Science+Business Media New York 2013

requiring that f is only a measurable and locally bounded function. However, to the best of our knowledge, for difference inclusions there are only few papers involving the second-order difference operator. For instance, in [3], the existence of at least one solution is obtained through the set-valued mapping theory, while in [20] it has been used the same approach that we propose here to prove the existence of at least three solutions for $p = 2$ and $q_k = 0$ for every $k \in [1,N]$.

Our main results are contained in Sect. 3. In particular, the conclusions of Corollary 1 are new even for continuous nonlinearities. The approach followed is fully variational as is described in Sect. 2; see Lemma 1.

Finally, to have general references on many important questions related to difference equations and their applications to different fields of research, we cite the monographs [1, 2, 17]; while for a survey on variational methods and difference equations, see [8].

2 Preliminaries and Variational Framework

Let $(X, \|\cdot\|)$ be a real Banach space; a function $J : X \to \mathbb{R}$ is called coercive whenever

$$\lim_{\|u\| \to +\infty} J(u) = +\infty.$$

While J is said to be locally Lipschitz when for every $u \in X$, there corresponds a neighbourhood U of u and a constant $L \geq 0$ such that

$$|J(v) - J(w)| \leq L\|v - w\|,$$

for all $v, w \in U$. As usual, X^* denotes the dual space of X, and $< \cdot, \cdot >$ stands for the duality pairing between X^* and X. The generalised directional derivative of a locally Lipschitz function J at the point u along the direction v is defined as follows:

$$J^0(u;v) = \limsup_{w \to u \, t \to 0^+} \frac{J(w+tv) - J(w)}{t}.$$

Moreover, the generalised gradient of J at u is the following set:

$$\partial J(u) = \{u^* \in X^* : < u^*, v > \leq J^0(u;v) \quad \forall v \in X\}.$$

We recall that if J is continuously Gâteaux differentiable at u, then J is locally Lipschitz at u and $\partial J(u) = \{J'(u)\}$. Further, a point $u \in X$ is said to be a (generalized) critical point of the locally Lipschitz function J if $0_{X^*} \in \partial J(u)$, namely,

$$J^0(u;v) \geq 0,$$

for all $v \in X$. Clearly, if J is a continuously Gâteaux differentiable at u, then u becomes a (classical) critical point of J, that is, $J'(u) = 0_{X^*}$.

For an exhaustive overview on the non-smooth calculus, we mention the excellent monographs [15, 19]. In [9] a three-critical-point theorem for locally Lipschitz and coercive functionals has been given. Here we state a version of such result which will be the main tool in order to achieve our conclusions.

Theorem 1. *Let* $(X, \| \cdot \|)$ *be a finite dimensional real Banach space and let* $\Phi, \Psi :$ $X \to \mathbb{R}$ *two locally Lipschitz functionals. Assume that* $\Phi(0) = \Psi(0) = 0$ *and*

(i_1) *There exist* $r > 0$, $u \in X$ *with* $r < \Phi(u)$ *such that*

$$\sup_{[\Phi \leq r]} \Psi < r\frac{\Psi(u)}{\Phi(u)},$$

where $[\Phi \leq r] := \{w \in S : \Phi(w) \leq r\}$.

Put

$$\Lambda_r := \left] \frac{\Phi(u)}{\Psi(u)}, \frac{r}{\sup_{[\Phi \leq r]} \Psi} \right[,$$

assume that

(i_2) *for each* $\lambda \in \Lambda_r$, $\Phi - \lambda\Psi$ *is a coercive functional.*

Then, for each $\lambda \in \Lambda_r$, *the functional* $\Phi - \lambda\Psi$ *admits at least three critical points with at least one belonging to* $[\Phi < r]$ *and another one in* $X \setminus [\Phi \leq r]$.

Consider now the N-dimensional Banach space $S = \{u : [0, N+1] \to \mathbb{R} : u_0 = u_{N+1} = 0\}$ endowed with the norm

$$\|u\| := \left(\sum_{k=1}^{N+1} |\Delta u_{k-1}|^p + q_k |u_k|^p \right)^{1/p}, \qquad \forall u \in S,$$

and denoted by λ_1 the first eigenvalue of the problem

$$\begin{cases} -\Delta(|\Delta u_{k-1}|^{p-2}\Delta u_{k-1}) = \lambda |u_{k-1}|^{p-2} u_{k-1}, & k \in [1, N]; \\ u_0 = u_{N+1} = 0, \end{cases}$$

one has that

$$\lambda_1 = \min_{u \in S \setminus \{0\}} \frac{\sum_{k=1}^{N+1} |\Delta u_{k-1}|^p}{\sum_{k=1}^{N} |u_k|^p}.$$

In particular, for $p = 2$ one has

$$\lambda_k := 4\sin^2\left(\frac{k\pi}{2(N+1)}\right), \quad \forall k \in [1,N]; \tag{1}$$

see [10], being λ_k the eigenvalues of the matrix

$$A := \begin{pmatrix} 2 & -1 & 0 & \dots & 0 \\ -1 & 2 & -1 & \dots & 0 \\ & \dots & \dots & \dots & \\ 0 & \dots & -1 & 2 & -1 \\ 0 & \dots & 0 & -1 & 2 \end{pmatrix}_{N\times N}.$$

Let us define the following functionals:

$$\Phi(u) := \frac{\|u\|^p}{p}, \qquad \Psi(u) := \sum_{k=1}^{N} F(k,u_k), \tag{2}$$

for every $u \in S$, where $F(k,t) := \int_0^t f(k,\xi)d\xi$ for every $(k,t) \in [1,N] \times \mathbb{R}$.

It is simple to verify that Φ is continuously Gâteaux differentiable, while Ψ is locally Lipschitz continuous.

Lemma 1. *Assume that $u \in S$ is a critical point of the functional $\Phi - \lambda\Psi$, being $\lambda > 0$. Then u is a solution of problem (D_λ).*

Proof. By the definition of critical point and in view of [15, Propositions 2.3.1 and 2.3.3], one has that

$$\Phi'(u)(v) \le \lambda\Psi^0(u;v) \le \lambda\sum_{k=1}^{N} F^0((k,u_k);v_k) \tag{3}$$

for every $v \in S$. Moreover, arguing as in the proof of [8, Lemma 39], we obtain that

$$\sum_{k=1}^{N+1} |\Delta u_{k-1}|^{p-2}\Delta u_{k-1}\Delta v_{k-1} = -\sum_{k=1}^{N} \Delta(|\Delta u_{k-1}|^{p-2}\Delta u_{k-1})v_k, \tag{4}$$

for every $v \in S$. In particular, for every $\xi \in \mathbb{R}$ and $k \in [1,N]$, putting in (3) $v = \xi e_k$, where e_k are the canonical unit vectors of $\mathbb{R}^N$, and taking in mind (4), we get

$$\langle -\Delta(|\Delta u_{k-1}|^{p-2}\Delta u_{k-1}) + q_k|u_k|^{p-2}u_k, \xi\rangle_{\mathbb{R}} = \Phi'(u)(v) \le \lambda F^0((k,u_k);\xi),$$

namely,

$$-\Delta(|\Delta u_{k-1}|^{p-2}\Delta u_{k-1}) + q_k|u_k|^{p-2}u_k \in \lambda\partial F(k,u_k).$$

Finally, because it is well known that (see [15, Example 2.2.5])

$$\partial F(k,u_k) = [f^-(k,u_k), f^+(k,u_k)], \quad \forall\, k \in [1,N],$$

it follows that

$$-\Delta(|\Delta u_{k-1}|^{p-2}\Delta u_{k-1}) + q_k|u_k|^{p-2}u_k \in [f^-(k,u_k), f^+(k,u_k)], \quad \forall\, k \in [1,N].$$

Therefore, our conclusion is proved. $\qquad\square$

3 Main Results

Now, we give

Theorem 2. *Let* $f : [1,N] \times \mathbb{R} \to \mathbb{R}$ *be a measurable locally bounded function. Put*

$$Q = \sum_{k=1}^{N} q_k, \ \text{assume that:}$$

(j_1) *There exist two positive constants c and d with* $c < d$ *such that*

$$\frac{\sum_{k=1}^{N} \sup_{|\xi| \le c} F(k,\xi)}{c^p} < \frac{2^p}{(N+1)^{p-1}(2+Q)} \frac{\sum_{k=1}^{N} F(k,d)}{d^p}.$$

(j_2) $\displaystyle\limsup_{|\xi| \to +\infty} \frac{F(k,\xi)}{|\xi|^p} \le \frac{\lambda_1}{p\lambda^*},$ *for every* $k \in [1,N].$

Then, for every

$$\lambda \in \Lambda := \,]\lambda_*, \lambda^*[\, := \left]\frac{2+Q}{p}\frac{d^p}{\sum_{k=1}^{N} F(k,d)}, \frac{2^p}{p(N+1)^{p-1}}\frac{c^p}{\sum_{k=1}^{N} \sup_{|\xi| \le c} F(k,\xi)}\right[,$$

problem (D_λ) *admits at least three solutions, one with* $\|u\|_\infty \le c$ *and one with*

$$\|u\|_\infty > \frac{2c}{[2^p(N+1)^p + \|q\|_\infty N(N+1)^{p-1}]^{1/p}}.$$

Proof. We apply Theorem 1, by putting $X = S$, $r = \frac{(2c)^p}{p(N+1)^{p-1}}$, $u = v \in S$ where v is given by

$$v_k = \begin{cases} d & \text{if } k \in [1,N], \\ 0 & \text{otherwise} \end{cases}$$

and Φ and Ψ are the two locally Lipschitz functionals introduced in (2). Since $c < d$ one has $c < \left(\frac{N+1}{2}\right)^{(p-1)/p} d$ and it results that $\|v\| > (pr)^{1/p}$, that is, $\Phi(u) > r$. Moreover, taking into account both

$$\| u \|_\infty := \max_{k \in [1,N]} |u_k| \le \frac{(N+1)^{(p-1)/p}}{2} \|u\|$$

(see [11]) and condition (j_1), an easy computation ensures that

$$\frac{\sup_{[\Phi \le r]} \Psi}{r} \le \frac{\sup_{\|u\| \le (pr)^{1/p}} \sum_{k=1}^{N} F(k,u_k)}{r} \le \frac{p(N+1)^{p-1}}{2^p} \frac{\sum_{k=1}^{N} \sup_{|\xi| \le c} F(k,\xi)}{c^p}$$
$$< \frac{p}{2+Q} \frac{\sum_{k=1}^{N} F(k,d)}{d^p} = \frac{\Psi(v)}{\Phi(v)}.$$

Therefore, (i_1) holds and $\Lambda \subseteq \Lambda_r$.

From (j_2), fixed $\lambda \in \Lambda$, there exist two constants a and ρ with $a > 0$ and $\rho < \frac{\lambda_1}{p\lambda}$ such that

$$F(k,\xi) \le \rho |\xi|^p + a \quad \forall \xi \in \mathbb{R}, \ k \in [1,N].$$

Now, arguing as in [8, Theorem 46], we verify (i_2). Indeed, since

$$\Psi(u) \le \rho \sum_{k=1}^{N} |u_k|^p + aN \le \frac{\rho}{\lambda_1} \|u\|^p + aN.$$

Therefore, for every $u \in S$, we have

$$\Phi(u) - \lambda \Psi(u) \ge \left(\frac{1}{p} - \rho \frac{\lambda}{\lambda_1}\right) \|u\|^p - aN,$$

which implies that the functional $\Phi - \lambda \Psi$ is coercive.

So, the assumptions of Theorem 1 are satisfied and our conclusions follow from Lemma 1. $\square$

Remark 1. We explicitly observe that Theorem 2 gives back [8, Theorem 49] whenever f is a continuous function. In addition, if $q_k = 0$ for every $k \in [1,N]$, it furnishes a more precise version of [12, Theorem 3.1]. Moreover, we point out that (j_2) is a slight generalisation of the analogous conditions adopted in [6, 12, 20] to ensure the coercivity of the energy functional associated to the difference problem; see also [13].

Corollary 1. *Let $\alpha : [1,N] \to \mathbb{R}$ be a nonnegative function with $A := \sum_{k=1}^{N} \alpha_k > 0$ and let $g : \mathbb{R} \to \mathbb{R}$ be a locally bounded and measurable function. Assume that:*

(j_1') *There exists $d > 0$ such that $g(t) > 0$ for every $0 < |t| < d$.*

(j_1'') $\displaystyle\limsup_{t \to 0^+} \frac{g(t)}{t} = 0.$

(j_2') $\displaystyle\limsup_{|t| \to +\infty} \frac{g(t)}{t} \le 0, \quad \forall k \in [1,N].$

Then, for each $\lambda > \dfrac{2+Q}{2A} \dfrac{d^2}{\int_0^d g(t)dt}$, the problem

$$\begin{cases} -\Delta(\Delta u_{k-1}) + q_k u_k \in \lambda\, \alpha_k[g^-(u_k), g^+(u_k)], & k \in [1,N] \\ u(0) = u(N+1) = 0, \end{cases}$$

admits at least three solutions.

Proof. We see that our conclusion follows at once from Theorem 2 being $f(k,t) = \alpha_k g(t)$ for each $(k,t) \in [1,N] \times \mathbb{R}$ and $p = 2$. Indeed, putting $G(\xi) = \int_0^\xi g(t)dt$ for every $\xi \in \mathbb{R}$, by (j_1'), one has

$$G(d) > 0, \qquad \sup_{|\xi| \le c} G(\xi) = G(c), \ \forall\, c \in (0,d]. \tag{5}$$

Fix $\lambda > \dfrac{2+Q}{2A} \dfrac{d^2}{G(d)}$. By (j_1'') there exists $\delta \in (0,d)$ such that

$$g(t) < \frac{2}{\lambda A(N+1)}\, t \quad \forall t \in \,]0, \delta[.$$

Hence, it results

$$A \sup_{t \in (0,\delta)} \frac{G(t)}{t^2} \le \frac{2}{\lambda(N+1)}. \tag{6}$$

On the other hand we can observe that

$$\frac{2^p}{(N+1)^{p-1}(2+Q)} \frac{\sum_{k=1}^N F(k,d)}{d^p} = \frac{4A}{(N+1)(2+Q)} \frac{G(d)}{d^2} > \frac{2}{\lambda(N+1)}. \tag{7}$$

Taking $c \in (0, \delta)$ and combining (5)–(7), we have

$$\begin{aligned} \frac{\sum_{k=1}^N \sup_{|\xi| \le c} F(k,\xi)}{c^p} &= A \frac{\sup_{|\xi| \le c} G(\xi)}{c^2} \\ &< \frac{4A}{(N+1)(2+Q)} \frac{G(d)}{d^2} \\ &= \frac{2^p}{(N+1)^{p-1}(2+Q)} \frac{\sum_{k=1}^N F(k,d)}{d^p}, \end{aligned} \tag{8}$$

namely, assumption (j_1) holds.

Fix now a positive number ε in the interval. From the definition of maximum limit and the locally boundness of g, it is easy to verify that there exists $M > 0$ such that

$$g(t) < \varepsilon t + M \ \forall t \geq 0 \quad \text{and} \quad g(t) > \varepsilon t - M \ \forall t \leq 0.$$

Consequently, by integrating the previous conditions, we obtain that

$$G(t) \leq \frac{\varepsilon}{2} t^2 + M|t| \ \forall t \in \mathbb{R}. \tag{9}$$

Hence, for every $k \in [1, N]$, (9) implies that

$$\limsup_{|\xi| \to +\infty} \frac{F(k, \xi)}{|\xi|^p} = \alpha_k \limsup_{|\xi| \to +\infty} \frac{G(\xi)}{\xi^2} \leq A \frac{\varepsilon}{2},$$

and, by the arbitrary of $\varepsilon > 0$, it is clear that (j_2) holds too. Finally, we achieve the conclusion applying Theorem 2 and observing that condition (8) ensures that $\lambda \in \Lambda$. $\qquad \square$

Remark 2. We explicitly observe that if $g^-(0) > 0$, then the solutions ensured by Corollary 1 are all nontrivial, while if g is continuous at zero, one of such solutions could be zero. Moreover, owing to [7, Theorem 2.2], the solutions of the problem are positive whenever g is nonnegative and nontrivial, with $g^-(0) > 0$.

References

1. Agarwal, R.P.: Difference Equations and Inequalities: Theory, Methods and Applications. Dekker, New York (2000)
2. Agarwal, R.P., O'Regan, D., Wong, P.J.Y.:Positive Solutions of Differential, Difference and Integral Equations. Kluwer Academic Publishers, Dordrecht (1999)
3. Agarwal, R.P., O'Regan, D., Lakshmikantham, V.: Discrete second order inclusions. J. Difference Equ. Appl. **9**, 879–885 (2003)
4. Agarwal, R.P., Perera, K., O'Regan, D.: Multiple positive solutions of singular and nonsingular discrete problems via variational methods. Nonlinear Anal. **58**, 69–73 (2004)
5. Agarwal, R. P., Perera, K., O'Regan, D.: Multiple positive solutions of singular discrete p-Laplacian problems via variational methods. Adv. Differ. Equ. **2**, 93–99 (2005)
6. Bonanno, G., Candito, P.: Nonlinear difference equations investigated via critical methods. Nonlinear Anal. **70**, 3180–3186 (2009)
7. Bonanno, G., Candito, P.: Infinitely many solutions for a class of nonlinear discrete boundary value problems. Appl. Anal. **88**, 605–616 (2009)
8. Bonanno, G., Candito, P.: Nonlinear difference equations through variational methods. In: Gao, D.Y., Motreanu, D. (eds.) Handbook of Nonconvex Analysis and Applications, pp. 1–44. International Press of Boston Inc., Sommerville (2010)
9. Bonanno, G., Marano, S.A.: On the structure of the critical set of non-differentiable functions with a weak compactness condition. App. Anal. **89**, 1–10 (2010)
10. Bereanu C., Mawhin J.: Existence and multiplicity results for nonlinear second order difference equations with Dirichlet boundary conditions. Math. Bohem. **131**, 145–160 (2006)

11. Cabada, A., Iannizzotto A., Tersian, S.: Multiple solutions for discrete boundary value problems. J. Math. Anal. Appl. **356**, 418–428 (2009)
12. Candito, P., Giovannelli, N.: Multiple solutions for a discrete boundary value problem. Comput. Math. Appl. **56**, 959–964 (2008)
13. Candito P., Molica Bisci G.: Existence of two solutions for a second-order discrete boundary value problem. Adv. Nonlinear Stud. **11**, 443–453 (2011)
14. Chang, K.C.: Variational methods for non-differentiable functionals and their applications to partial differential equations. J. Math. Anal. Appl. **80**, 102–129 (1981)
15. Clarke, F. H.: Optimization and Nonsmooth Analysis. Classics in Applied Mathematics, vol. 5. SIAM, Philadelphia (1992)
16. Henderson, J., Thompson H. B.: Existence of multiple solutions for second order discrete boundary value problems. Comput. Math. Appl. **43**, 1239–1248 (2002)
17. Kelly, W.G., Peterson, A.C.: Difference Equations: An Introduction with Applications. Academic, San Diego (1991)
18. Liang, H., Weng, P.: Existence and multiple solutions for a second-order difference boundary value problem via critical point theory. J. Math. Anal. Appl. **326**, 511–520 (2007)
19. Motreanu, D., Rădulescu, V.: Variational and Non-variational Methods in Nonlinear Analysis and Boundary Value Problems. Kluwer, Dordrecht (2003)
20. Zhang, G., Zhang, W., Liu, S.: Multiplicity result for a discrete eigenvalue problem with discontinuous nonlinearities. J. Math. Anal. Appl. **328**, 1068–1074 (2007)

Pullback Attractors of Stochastic Lattice Dynamical Systems with a Multiplicative Noise and Non-Lipschitz Nonlinearities

Tomás Caraballo, Francisco Morillas, and José Valero

Abstract In this paper we study the asymptotic behavior of solutions of a first-order stochastic lattice dynamical system with a multiplicative noise. We do not assume any Lipschitz condition on the nonlinear term, just a continuity assumption together with growth and dissipative conditions. Using the theory of multivalued random dynamical systems we prove the existence of a pullback compact global attractor.

Keywords Stochastic lattice dynamical system • Pullback attractor • Multivalued random dynamical system • Non-uniqueness of solutions

1 Introduction

This paper is devoted to the long-term behavior of the following stochastic lattice differential equation:

$$\frac{du_i(t)}{dt} = \nu(u_{i-1} - 2u_i + u_{i+1}) - f_i(u_i) + \sum_{j=1}^{N} c_j u_i \circ \frac{dw_j(t)}{dt}, \quad i \in \mathbb{Z}, \quad (1)$$

T. Caraballo
Depto. Ecuaciones Diferenciales y Análisis Numérico, Universidad de Sevilla,
Apdo. de Correos 1160, 41080 Sevilla, Spain
e-mail: caraball@us.es

F. Morillas
Departament d'Economia Aplicada, Universitat de València, Campus dels Tarongers s/n,
46022 València, Spain
e-mail: francisco.morillas@uv.es

J. Valero
Centro de Investigación Operativa, Universidad Miguel Hernández,
Avda. de la Universidad s/n, 03202 Elche, Spain
e-mail: jvalero@umh.es

S. Pinelas et al. (eds.), *Differential and Difference Equations with Applications*, Springer Proceedings in Mathematics & Statistics 47, DOI 10.1007/978-1-4614-7333-6_27, © Springer Science+Business Media New York 2013

where $u = (u_i)_{i \in \mathbb{Z}} \in \ell^2$, $\mathbb{Z}$ denotes the integer set, v is a positive constant, f_i is a continuous function satisfying a dissipative condition, $c_j \in \mathbb{R}$, for $j = 1, \ldots, N$, and w_j are mutually independent Brownian motions, where $\circ$ denotes the Stratonovich sense in the stochastic term.

Stochastic lattice differential equations arise naturally in a wide variety of applications where the spatial structure has a discrete character, and uncertainties or random influences, called noises, are taken into account (see [3, 7] for more details on the importance of this model and for some references on it).

In Sect. 2, we introduce basic concepts concerning multivalued random dynamical systems (MRDSs) and global random attractors. In Sect. 3, we show that the stochastic lattice differential equation (1) generates a multivalued infinite dimensional random dynamical system. The existence of the global pullback attractor is stated in Sect. 4.

2 Multivalued Random Dynamical Systems

We recall now some basic definitions for set-valued non-autonomous and random dynamical systems and formulate sufficient conditions ensuring the existence of a pullback attractor for these systems.

A pair (Ω, θ) where $\theta = (\theta_t)_{t \in \mathbb{R}}$ is a flow on Ω, that is, $\theta : \mathbb{R} \times \Omega \to \Omega$, and satisfies that $\theta_0 = \mathrm{id}_\Omega$ and $\theta_{t+\tau} = \theta_t \circ \theta_\tau =: \theta_t \theta_\tau$ for $t, \tau \in \mathbb{R}$, is called a *non-autonomous perturbation*.

Let $\mathscr{P} := (\Omega, \mathscr{F}, \mathbb{P})$ be a probability space. On this probability space, we consider a measurable non-autonomous flow $\theta : (\mathbb{R} \times \Omega, \mathscr{B}(\mathbb{R}) \otimes \mathscr{F}) \to (\Omega, \mathscr{F})$.

In addition, $\mathbb{P}$ is supposed to be ergodic with respect to θ. Hence, $\mathbb{P}$ is invariant with respect to θ_t. The quadruple $(\Omega, \mathscr{F}, \mathbb{P}, \theta)$ is called a metric dynamical system.

Let $X = (X, d_X)$ be a Polish space. Let $D : \omega \to D(\omega) \in 2^X$ be a multivalued mapping. The set of multifunctions $D : \omega \to D(\omega) \in 2^X$ with closed and nonempty images is denoted by $C(X)$. Let also denote by $P_f(X)$ the set of all nonempty closed subsets of the space X. Thus, it is equivalent to write that D is in $C(X)$, or $D : \Omega \to P_f(X)$.

Let $D : \omega \to D(\omega)$ be a multivalued mapping in X over $\mathscr{P}$. Such a mapping is called a random set if the map $\omega \mapsto \inf_{y \in D(\omega)} d_X(x, y)$ is a random variable for every $x \in X$.

Definition 1. A multivalued map $G : \mathbb{R}^+ \times \Omega \times X \to P_f(X)$ is called a *multivalued non-autonomous dynamical system* (MNDS) if:

(i) $G(0, \omega, \cdot) = \mathrm{id}_X$.

(ii) $G(t + \tau, \omega, x) \subset G(t, \theta_\tau \omega, G(\tau, \omega, x))$ *(cocycle property) for all* $t, \tau \in \mathbb{R}^+$, $x \in X, \omega \in \Omega$.

It is called a *strict* MNDS if, in addition:

(iii) $G(t + \tau, \omega, x) = G(t, \theta_\tau \omega, G(\tau, \omega, x))$ *for all* $t, \tau \in \mathbb{R}^+, x \in X, \omega \in \Omega$.

An MNDS is called a *MRDS* if the multivalued mapping $(t, \omega, x) \mapsto G(t, \omega, x)$ is $\mathscr{B}(\mathbb{R}^+) \otimes \mathscr{F} \otimes \mathscr{B}(X)$ measurable.

Notice that for $V \subset X$, $G(t, \omega, V)$ is defined by $G(t, \omega, V) = \bigcup_{x_0 \in V} G(t, \omega, x_0)$.

A multivalued mapping D is said to be *negatively, strictly, or positively invariant* for the MNDS G if $D(\theta_t \omega) \subset, =, \supset G(t, \omega, D(\omega))$ for $\omega \in \Omega$, $t \in \mathbb{R}^+$.

Let $\mathscr{D}$ be the family of multivalued mappings with values in $C(X)$. We say that a family $K \in \mathscr{D}$ is *pullback $\mathscr{D}$-attracting* if, for every $D \in \mathscr{D}$,

$$\lim_{t \to +\infty} \mathrm{dist}_X (G(t, \theta_{-t} \omega, D(\theta_{-t} \omega)), K(\omega)) = 0, \text{ for all } \omega \in \Omega.$$

$B \in \mathscr{D}$ is said to be *pullback $\mathscr{D}$-absorbing* if for every $D \in \mathscr{D}$, there exists $T = T(\omega, D) > 0$ such that

$$G(t, \theta_{-t} \omega, D(\theta_{-t} \omega)) \subset B(\omega), \text{ for all } t \geq T. \tag{2}$$

Let $\mathscr{D}$ be a set of multivalued mappings in $C(X)$ satisfying the inclusion-closed property: if we suppose that $D \in \mathscr{D}$ and D' is a multivalued mapping in $C(X)$ such that $D'(\omega) \subset D(\omega)$ for $\omega \in \Omega$, then $D' \in \mathscr{D}$.

Definition 2. A family $\mathscr{A} \in \mathscr{D}$ is said to be a *global pullback $\mathscr{D}$-attractor* for the MNDS G if it satisfies:

1. $\mathscr{A}(\omega)$ is compact for any $\omega \in \Omega$.
2. $\mathscr{A}$ is pullback $\mathscr{D}$-attracting.
3. $\mathscr{A}$ is negatively invariant.

$\mathscr{A}$ is said to be a *strict global pullback $\mathscr{D}$-attractor* if the invariance in (3) is strict.

A natural modification of this definition for MRDS is the following.

Definition 3. Suppose that G is an MRDS and suppose that the properties of Definition 2 are satisfied. In addition, we suppose that A is a random set with respect to $\mathscr{P}^c$ (the completion of $\mathscr{P}$). Then $\mathscr{A}$ is called a *random global pullback $\mathscr{D}$-attractor*.

The next two general results were proved in [6].

Theorem 1. *Suppose that the MNDS $G(t, \omega, \cdot)$ is upper-semicontinuous for $t \geq 0$ and $\omega \in \Omega$. Let $K \in \mathscr{D}$ be a multivalued mapping such that the MNDS is pullback $\mathscr{D}$-asymptotically compact with respect to K, i.e., for every sequence $t_n \to +\infty$, $\omega \in \Omega$ every sequence $y_n \in G(t_n, \theta_{-t_n} \omega, K(\theta_{-t_n} \omega))$ is pre-compact. In addition, suppose that K is pullback $\mathscr{D}$-absorbing. Then, the set $\mathscr{A}$ given by*

$$\mathscr{A}(\omega) := \bigcap_{s \geq 0} \overline{\bigcup_{t \geq s} G(t, \theta_{-t} \omega, K(\theta_{-t} \omega))} \tag{3}$$

is a pullback $\mathscr{D}$-attractor. furthermore, $\mathscr{A}$ is the unique element from $\mathscr{D}$ with these properties. In addition, if G is a strict MNDS, then $\mathscr{A}$ is strictly invariant.

Theorem 2. *Let G be an MRDS. Under the assumptions in Theorem 1, let $\omega \mapsto G(t, \omega, K(\omega))$ be a random set for $t \geq 0$ with respect to $\mathscr{P}^c$. Assume also that $G(t, \omega, K(\omega))$ is closed for all $t \geq 0$ and $\omega \in \Omega$. Then the set $\mathscr{A}$ defined by (3) is a random set with respect to $\mathscr{P}^c$, so that it is a random global pullback $\mathscr{D}$-attractor.*

3 Stochastic Lattice Differential Equations

We consider a stochastic lattice differential equation

$$\frac{du_i(t)}{dt} = \nu(u_{i-1} - 2u_i + u_{i+1}) - f_i(u_i) + \sum_{j=1}^{N} c_j u_i \circ \frac{dw_j(t)}{dt}, \quad i \in \mathbb{Z}, \qquad (4)$$

where $u = (u_i)_{i \in \mathbb{Z}} \in \ell^2$, $\mathbb{Z}$ denotes the integer set, ν is a positive constant, f_i is a continuous function satisfying the assumptions below, $c_j \in \mathbb{R}$, for $j = 1, \ldots, N$, and w_j are mutually independent two-sided Brownian motions on the same probability space $(\Omega, \mathscr{F}, \mathbb{P})$. Notice that system (4) is interpreted in integral form as

$$u_i(t) = u_i(0) + \int_0^t \left(\nu(u_{i-1}(s) - 2u_i(s) + u_{i+1}(s)) - f_i(u_i(s)) \right) ds \qquad (5)$$

$$+ \int_0^t \sum_{j=1}^{N} c_j u_i(s) \circ dw_j(t), \quad i \in \mathbb{Z},$$

where the stochastic integral is understood in the sense of Stratonovich.

Assumptions on the Nonlinearity f_i: Let $f_i : \mathbb{R} \to \mathbb{R}$ satisfy the following assumptions:

(H1)   There exist $c_0 \in l^1$ and $\lambda > 0$ such that, for all $x \in \mathbb{R}$, $i \in \mathbb{Z}$, it holds that $f_i(x)x \geq \lambda x^2 - c_{0,i}$.

(H2)   There exist $c_1 \in l^2$, $c_{1,i} \geq 0$, and a continuous increasing function $C(\cdot) \geq 0$ such that, for all $x \in \mathbb{R}$, $i \in \mathbb{Z}$, it follows $|f_i(x)| \leq C(|x|)|x| + c_{1,i}$.

(H3)   The maps $f_i : \mathbb{R} \to \mathbb{R}$, $i \in \mathbb{Z}$, are continuous.

We now formulate system (4) as a stochastic differential equation in ℓ^2. Denote by $||\cdot||$ the norm in the space ℓ^2, and by $B, B^*, C_j, j = 1, \ldots, N$, and A the linear operators from ℓ^2 to ℓ^2 defined as follows. For $u = (u_i)_{i \in \mathbb{Z}} \in \ell^2$, and for each $i \in \mathbb{Z}$,

$$(Bu)_i = u_{i+1} - u_i, \ (B^*u)_i = u_{i-1} - u_i, \ (C_j u)_i = c_j u_i, \ (Au)_i = -u_{i-1} + 2u_i - u_{i+1}.$$

Then, we have that $A = BB^* = B^*B$, and $(B^*u,v) = (u,Bv)$ for all $u,v \in \ell^2$. Therefore, $(Au,u) \geq 0$ for all $u \in \ell^2$.

Let $\tilde{f}$ be the Nemytski operator associated to f_i, that is, for $u = (u_i)_{i\in\mathbb{Z}} \in \ell^2$, let $\tilde{f}(u) := (f_i(u_i))_{i\in\mathbb{Z}}$. Then, thanks to (H1)–(H2), this operator is well defined and

$$\|\tilde{f}(u)\|_{l^2}^2 = \sum_{i\in\mathbb{Z}} |f_i(u_i)|^2 \leq \sum_{i\in\mathbb{Z}} (C(|u_i|)|u_i| + c_{1,i})^2 \leq 2M(u)\|u\|_{l^2}^2 + \|c_1\|_{l^2}^2, \quad (6)$$

where $M(u) = \max_{i\in\mathbb{Z}} C(|u_i|)$.

Similar to (6), one can easily see that $\tilde{f}$ also satisfies:

$$\tilde{f}(u,u) \geq \lambda\|u\|_{l^2}^2 - \|c_0\|_{l^1}, \quad \forall u \in l^2, \quad (7)$$

and $\tilde{f} : l^2 \to l^2$ is continuous and weakly continuous (see [7] for a similar proof).

System (4) with initial value $u^0 \equiv (u_i^0)_{i\in\mathbb{Z}} \in \ell^2$ can be rewritten in ℓ^2 for $t \geq 0$ and $\omega \in \Omega$, as

$$u(t) = u^0 + \int_0^t (-\nu Au(s) - \tilde{f}(u(s)))ds + \sum_{j=1}^N \int_0^t C_j u(s) \circ dw_j(t). \quad (8)$$

To prove that this stochastic equation (8) generates a random dynamical system, we will transform it into a random differential equation in ℓ^2.

Let us consider the one-dimensional stochastic differential equation

$$dz = -\alpha z\,dt + dw(t), \quad (9)$$

for $\alpha > 0$, where $w(t)$ is a standard Brownian motion. This equation possesses a random fixed point in the sense of random dynamical systems generating a stationary solution known as the stationary Ornstein–Uhlenbeck process. More properties on this process can be found in Caraballo et al. [5].

Let us consider $\alpha = 1$ and denote by z_j^* its associated Ornstein–Uhlenbeck process corresponding to (9) with w_j instead of w.

Then for any $j = 1,\ldots,N$ we have a stationary Ornstein–Uhlenbeck process generated by a random variable $z_j^*(\omega)$ on $\bar{\Omega}_j$ defined on the metric dynamical system $(\bar{\Omega}_j, \mathscr{F}_j, \mathbb{P}_j, \theta)$. Now we set $(\Omega, \mathscr{F}, \mathbb{P}, \theta)$, where $\Omega = \bar{\Omega}_1 \times \cdots \times \bar{\Omega}_N$, $\mathscr{F} = \bigotimes_{i=1}^N \mathscr{F}_i$, $\mathbb{P} = \mathbb{P}_1 \times \mathbb{P}_2 \times \cdots \times \mathbb{P}_N$, and θ is the flow of Wiener shifts.

Notice that operator C_j generates a strongly continuous semigroup (in fact, group) of operators $S_{C_j}(t)$. More precisely, $S_{C_j}(t)$ is given by $S_{C_j}(t)u = e^{c_j t}u$, for $u \in \ell^2$, and, for simplicity, we denote $\delta(\omega) := \sum_{j=1}^N c_j z_j^*(\omega)$.

Thanks to the change of variable $v(t) = e^{-\delta(\theta_t \omega)}u(t)$, where u is a solution to (8),

$$dv(t) = e^{-\delta(\theta_t\omega)}du(t) - \sum_{j=1}^{N} c_j e^{-\delta(\theta_t\omega)}u(t) \circ dz_j^*(\theta_t\omega)$$

$$= \left(-\nu Av(t) - e^{-\delta(\theta_t\omega)}\tilde{f}(e^{\delta(\theta_t\omega)}v(t)) + \delta(\theta_t\omega)v(t)\right)dt.$$

So we can consider the following evolution equation with random coefficients:

$$\frac{dv}{dt} = -\nu Av + \delta(\theta_t\omega)v - e^{-\delta(\theta_t\omega)}\tilde{f}\left(e^{\delta(\theta_t\omega)}v\right), \tag{10}$$

and initial condition $v(0) = v^0 \in H$.

From (H1) to (H2), for every $x \in \mathbb{R}$, we easily obtain

$$e^{-\delta(\theta_t\omega)}f_i\left(e^{\delta(\theta_t\omega)}x\right)x \geq \lambda x^2 - e^{-2\delta(\theta_t\omega)}c_{0,i}. \tag{11}$$

Now we establish the following result.

Theorem 3. *Let $T > 0$ and $v^0 \in H$ be fixed. Then, for every $\omega \in \Omega$, (10) admits at least a solution $v(\cdot, \omega, v^0) \in \mathscr{C}([0,T], \ell^2)$.*

Proof. The existence of at least one local solution follows the same arguments from [7]. That this solution is global follows as in [3] (see also [8]). □

Now, we say that $u(\cdot) = u(\cdot, \omega, u^0)$ is a solution of (8) [or (4)] if $u(t) = e^{\delta(\theta_t\omega)} \times$ $\times v(t, \omega, e^{-\delta(\omega)}u^0)$ with $v(\cdot, \omega, e^{-\delta(\omega)}u^0)$ a solution of (10) with initial value u^0.

Let $\mathscr{S}\left(v^0, \omega\right)$ be the set of all solutions to (10) corresponding to the initial datum $v^0 \in \ell^2$ and $\omega \in \Omega$.

We define the multivalued map $G : \mathbb{R}^+ \times \Omega \times \ell^2 \to P\left(\ell^2\right)$ as follows:

$$G\left(t, \omega, u^0\right) = \left\{e^{\delta(\theta_t\omega)}v(t) : v \in \mathscr{S}(e^{-\delta(\omega)}u^0, \omega)\right\}. \tag{12}$$

Arguing in a standard way (see, e.g., [4]), we prove that (12) is an MNDS. Namely:

Lemma 1. *The map G defined by (12) satisfies $G(0, \omega, \cdot) = Id_{\ell^2}$ and $G(t + \tau, \omega, x) = G(t, \theta_\tau\omega, G(\tau, \omega, x))$, for all $t, \tau \in \mathbb{R}^+$, $x \in \ell^2$, $\omega \in \Omega$. Moreover, G is upper-semicontinuous and possesses closed values.*

4   Existence of Pullback Attractors

We now establish the existence of a global pullback attractor for the MNDS generated by (4). As universe $\mathscr{D}$ we consider the family of multivalued mappings D in ℓ^2 with $D(\omega) \subset B_{\ell^2}(0, \rho(\omega))$, the closed ball centered at zero and radius $\rho(\omega)$, with sub-exponential growth, i.e., $\lim_{t \to \pm\infty} \frac{\log^+ \rho(\theta_t\omega)}{t} = 0$, $\omega \in \Omega$.

$\mathscr{D}$ is called the family of sub-exponentially growing multifunctions in $C\left(l^2\right)$. The properties on $\mathscr{D}$ given in Definition 2 also hold. Our main result is the next one.

Theorem 4. *The MNDS G generated by (8) possesses a unique pullback attractor.*

To prove this theorem we use Theorem 1 and proceed in the following way. We first prove that there exists an absorbing set for G in $\mathscr{D}$, and we then prove that the asymptotic compactness holds, while the other properties follow from Lemma 1.

4.1 Existence of the Pullback Absorbing Set for the MNDS

We will construct now a pullback D-absorbing set $K(\omega) \in \mathscr{D}$.
Let $v(t) = v(t, \omega, u_0 e^{-\delta(\omega)})$ be the solution of (10) for some $u_0 \in B\left(\theta_{-t}\omega\right)$. Then,

$$||v(t)||^2 \leq e^{-2\lambda t + 2\int_0^t \delta(\theta_s\omega)ds}||v_0||^2$$

$$+ ||c_0||_{l^1} e^{-2\lambda t + 2\int_0^t \delta(\theta_s\omega)ds} \int_0^t e^{-2\delta(\theta_r\omega)+2\lambda r - 2\int_0^r \delta(\theta_s\omega)ds} dr.$$

Substituting ω by $\theta_{-t}\omega$ and u_0 by $e^{-\delta(\theta_{-t}\omega)}u_0$ in the expression of $v(\cdot)$,

$$\left\|v\left(t, \theta_{-t}\omega, e^{-\delta(\theta_{-t}\omega)}u_0\right)\right\|^2 \leq e^{-2\lambda t - 2\delta(\theta_{-t}\omega)+2\int_{-t}^0 \delta(\theta_s\omega)ds}||u_0||^2$$

$$+ ||c_0||_{l^1} \int_{-\infty}^0 e^{-2\delta(\theta_r\omega)+2\lambda r + 2\int_r^0 \delta(\theta_s\omega)ds} dr.$$

Thanks to the properties of the Ornstein–Uhlenbeck process z^*, it follows that $\int_{-\infty}^0 e^{-2\delta(\theta_r\omega)+2\lambda r + 2\int_r^0 \delta(\theta_s\omega)ds} dr < +\infty$. Noticing that for any $u_0 \in B(\theta_{-t}\omega)$, we have $u(t, \theta_{-t}\omega, u_0) = e^{\delta(\omega)}v(t, \theta_{-t}\omega, e^{-\delta(\theta_{-t}\omega)}u_0)$, then

$$||u(t, \theta_{-t}\omega, u_0)||^2 \leq e^{\delta(\omega)}e^{-2\lambda t - 2\delta(\theta_{-t}\omega)+2\int_{-t}^0 \delta(\theta_s\omega)ds}d(B(\theta_{-t}\omega))^2$$

$$+ e^{\delta(\omega)} ||c_0||_{l^1} \int_{-\infty}^0 e^{-2\delta(\theta_s\omega)+\lambda s + 2\int_s^0 \delta(\theta_r\omega)dr} ds,$$

where $d\left(B\left(\theta_{-t}\omega\right)\right)$ denote the supremum of the norm of $B\left(\theta_{-t}\omega\right)$.

Denoting by $R^2(\omega) = e^{\delta(\omega)} ||c_0||_{l^1} \int_{-\infty}^0 e^{-2\delta(\theta_s\omega)+\lambda s + 2\int_s^0 \delta(\theta_r\omega)dr} ds$, and noticing that $\lim_{t \to +\infty} e^{\delta(\omega)}e^{-2\lambda t - 2\delta(\theta_{-t}\omega)+2\int_{-t}^0 \delta(\theta_s\omega)ds}d(B(\theta_{-t}\omega))^2 = 0$, it follows that $K(\omega) = B_{\ell^2}(0, R(\omega))$ is a pullback $\mathscr{D}$-absorbing set. We will now prove that $K \in \mathscr{D}$. To this end, we have to check that $\lim_{t \to +\infty} e^{-\beta t}R(\theta_{-t}\omega) = 0$. Indeed, observe that

$$e^{-\beta t}R^2(\theta_{-t}\omega) = 2e^{-\beta t}e^{\delta(\theta_{-t}\omega)}\|c_0\|_{l^1}\int_{-\infty}^0 e^{-2\delta(\theta_{s-t}\omega)+\lambda s+2\int_s^0\delta(\theta_{r-t}\omega)dr}ds$$

$$= 2e^{-\beta t}e^{\delta(\theta_{-t}\omega)}\|c_0\|_{l^1}\int_{-\infty}^{-t} e^{-2\delta(\theta_s\omega)+\lambda(s+t)+2\int_s^{-t}\delta(\theta_r\omega)dr}ds \to 0.$$

4.2 Asymptotic Compactness

We first need some estimates on the tails of the solutions.

Lemma 2. *Let* $u^0(\omega)\in K(\omega)$ *(the set constructed in Sect. 4.1). Then for every* $\varepsilon>0$, *there exist* $T(\varepsilon,\omega) > 0$ *and* $N(\varepsilon,\omega) > 0$ *such that any solution* $u(\cdot)$ *of (4) given by* $u(t) = e^{\delta(\theta_t\omega)}v(t)$ *with* $v(\cdot) \in \mathscr{S}\left(u^0(\theta_{-t}\omega)e^{-\delta}(\omega),\theta_{-t}\omega\right)$ *satisfies*

$$\sum_{|i|\geq N(\varepsilon,\omega)}\left|u_i(t,\theta_{-t}\omega,u^0(\theta_{-t}\omega))\right|^2 \leq \varepsilon, \text{ for all } t \geq T(\varepsilon,\omega).$$

Proof. The proof follows the same lines of that one in, say [1, 3] or [2] amongst others, in the single-valued case, but arguing (in a uniform way) for all solutions of our problem associated to each initial value. See also [8] for more details. □

Theorem 5. *For* $\omega \in \Omega$ *the set* $K(\omega)$ *is asymptotically compact, i.e., every sequence* $p^n \in G(t_n,\theta_{-t_n}\omega,K(\theta_{-t_n}\omega))$, *with* $t_n \to \infty$, *has a convergent subsequence in* ℓ^2.

Proof. Consider $(t_n)_{n\in\mathbb{N}}$ with $\lim_{n\to\infty}t_n = \infty$ and $p^n \in G(t_n,\theta_{-t_n}\omega,K(\theta_{-t_n}\omega))$. Then, there exists $x^n \in K(\theta_{-t}\omega)$ such that $p^n \in G(t_n,\theta_{-t_n}\omega,x_n)$. We will show that $\{p^n\}_{n\in\mathbb{N}}$ possesses a convergent subsequence. Since $K(\omega)$ is a bounded absorbing set, for large n, $p^n \in K(\omega)$. Thus, there exists $v \in \ell^2$ and a subsequence of $\{p^n\}_{n\in\mathbb{N}}$ (still denoted by $\{p^n\}_{n\in\mathbb{N}}$) such that$\{p^n\}_{n\in\mathbb{N}} \to v$ weakly in ℓ^2. Next, we show that this weak convergence is actually strong, i.e., for each $\varepsilon > 0$ there is $N^*(\varepsilon,\omega) > 0$ such that, for $n \geq N^*(\varepsilon,\omega)$, we have $\|p^n - v\| \leq \varepsilon$.

By Lemma 2, there exist $N_1^*(\varepsilon,\omega) > 0$ and $k_1(\varepsilon,\omega) > 0$ such that for $n > N_1^*$ $\sum_{|i|\geq k_1(\varepsilon,\omega)}|p_i^n - v_i|^2 \leq \frac{1}{8}\varepsilon^2$. On the other hand, since $v \in \ell^2$, there exists $k_2(\varepsilon)$ such that $\sum_{|i|\geq k_2(\varepsilon)}|v_i|^2 \leq \frac{1}{8}\varepsilon^2$. Letting $k(\varepsilon,\omega) = \max\{k_1(\varepsilon,\omega),k_2(\varepsilon)\}$, by the previous weak convergence, we have for each $|i| \leq k(\varepsilon,\omega)$ $p_i^n \to v_i$, as $n \to \infty$, which implies that there exists $N_2^*(\varepsilon,\omega) > 0$ such that, when $n \geq N_2^*(\varepsilon,\omega)$, it follows $\sum_{|i|\leq k(\varepsilon)}|p_i^n|^2 \leq \frac{1}{2}\varepsilon^2$. Let $N^*(\varepsilon,\omega) = \max\{N_1^*(\varepsilon,\omega),N_2^*(\varepsilon,\omega)\}$. Then, from the above estimates, we obtain for $n \geq N^*(\varepsilon,\omega)$:

$$\|p^n - v\|^2 = \sum_{|i|\leq k(\varepsilon)}|p_i^n - v_i|^2 + \sum_{|i|>k(\varepsilon)}|p_i^n - v_i|^2$$

$$\leq \frac{1}{2}\varepsilon^2 + 2\sum_{|i|>K(\varepsilon)}\left(|p_i^n|^2 + |v_i|^2\right) \leq \varepsilon^2.$$

Hence, p^n converges to v strongly. The proof is complete and we have thus proved Theorem 4. □

Remark 1. Our next objective is to prove that our MNDS G is, in fact, an MRDS and possesses a random attractor, and particularly, that the previous pullback attractor is also a random pullback attractor for G. But this is the topic of our paper [8].

Acknowledgements This work has been partially supported by Ministerio de Ciencia e Innovación (Spain) MTM2008-00088, and Junta de Andalucía P07-FQM-02468 and FEDER.

References

1. Bates, P.W., Lisei, H., Lu, K.: Attractors for stochastic lattice dynamical systems. Stochas. Dynam. **6**(1), 1–21 (2006)
2. Bates, P.W., Lu, K., Wang, B.: Attractors for lattice dynamical systems. Int. J. Bifurcat. Chaos **11**, 143–153 (2001)
3. Caraballo, T., Lu, K.: Attractors for stochastic lattice dynamical systems with a multiplicative noise. Front. Math. China **3**, 317–335 (2008)
4. Caraballo, T., Langa, J.A., Valero, J.: Global attractors for multivalued random dynamical systems. Nonlinear Anal. **48**, 805–829 (2002)
5. Caraballo, T., Kloeden, P.E., Schmalfuß, B.: Exponentially stable stationary solutions for stochastic evolution equations and their perturbation. Appl. Math. Optim. **50**, 183–207 (2004)
6. Caraballo, T., Garrido-Atienza, M.J., Schmalfuß, B., Valero, J.: Non-autonomous and random attractors for delay random semilinear equations without uniqueness. Discrete Contin. Dyn. Syst. **21**, 415–443 (2008)
7. Caraballo, T., Morillas, F., Valero, J.: Random Attractors for stochastic lattice systems with non-Lipschitz nonlinearity. J. Diff. Equat. App. **17**(2), 161–184 (2011)
8. Caraballo, T., Morillas, F., Valero, J.: Attractors of stochastic lattice dynamical systems with a multiplicative noise and non-Lipschitz non-linearities, J. Differ. Equ. **253**, 667–693 (2012)

Discrete-Time Counterparts of Impulsive Hopfield Neural Networks with Leakage Delays

Haydar Akça, Valéry Covachev, and Zlatinka Covacheva

Abstract A discrete-time counterpart of a class of Hopfield neural networks with impulses and concentrated and infinite distributed delays as well as a small delay in the leakage terms is introduced. Sufficient conditions for the existence and global exponential stability of a unique equilibrium point of the discrete-time system considered are obtained.

1 Introduction

Hopfield neural networks have found applications in a broad range of disciplines [4–6] and have been studied both in the continuous- and discrete-time cases by many researchers. Moreover, there are many real-world systems and natural processes that behave in a piecewise continuous style interlaced with instantaneous and abrupt changes (impulses). Signal transmission between the neurons causes time delays. Therefore the dynamics of Hopfield neural networks with discrete or distributed delays has a fundamental concern.

H. Akça
Department of Applied Sciences and Mathematics, Abu Dhabi University,
P.O. Box 59911, Abu Dhabi, UAE
e-mail: Haydar.Akca@adu.ac.ae

V. Covachev (✉)
Department of Mathematics and Statistics, Sultan Qaboos University, 123, Muscat, Oman

Institute of Mathematics, Bulgarian Academy of Sciences, Sofia, Bulgaria
e-mail: vcovachev@hotmail.com; valery@squ.edu.om

Z. Covacheva
Middle East College Muscat, Oman

Higher College of Telecommunications and Post, Sofia, Bulgaria
e-mail: zkovacheva@hotmail.com

S. Pinelas et al. (eds.), *Differential and Difference Equations with Applications*, Springer Proceedings in Mathematics & Statistics 47, DOI 10.1007/978-1-4614-7333-6_28, © Springer Science+Business Media New York 2013

It is known from the literature on population dynamics [1] that time delays in the stabilizing negative feedback terms have a tendency to destabilize the system. Due to some theoretical and technical difficulties [3], so far there have been very few existing works with time delay in leakage (or "forgetting") terms [1,3,7,9].

Our goal in this paper is to introduce a discrete-time counterpart of a class of Hopfield neural networks with impulses and concentrated and infinite distributed delays as well as a small delay in the leakage terms, without essentially changing its stability characteristics. Note that conditions of smallness of the leakage delays have been introduced in [3,7]. We obtain sufficient conditions for the existence and global exponential stability of a unique equilibrium point of the resulting discrete-time system.

2 Impulsive Continuous-Time Hopfield Neural Network: Existence of a Unique Equilibrium

Consider an impulsive continuous-time neural network consisting of m elementary processing units (or neurons) whose state variables x_i ($i = \overline{1,m}$ which henceforth will stand for $i = 1, 2, \ldots, m$) are governed by the system

$$\frac{dx_i(t)}{dt} = -a_i x_i(t - \sigma) + \sum_{j=1}^{m} b_{ij} f_j(x_j(t)) + \sum_{j=1}^{m} c_{ij} g_j(x_j(t - \tau_{ij}))$$

$$+ \sum_{j=1}^{m} d_{ij} h_j \left(\int_0^\infty K_{ij}(s) x_j(t - s) \, ds \right) + I_i, \quad t > 0, \quad t \neq t_k, \quad (1)$$

$$\Delta x_i(t_k) = B_{ik} x_i(t_k) + \int_{t_k - \sigma}^{t_k} \psi_{ik}(s) x_i(s) \, ds + \gamma_{ik}, \quad i = \overline{1,m}, \quad k \in \mathbb{N}, \quad (2)$$

with initial values prescribed by piecewise continuous functions $x_i(s) = \phi_i(s)$ which are bounded for $s \in (-\infty, 0]$. In (1), $\sigma > 0$ denotes a delay in the stabilizing (or negative) feedback term $-a_i(x_i - \sigma)$, also called *leakage* or *forgetting term* of the unit i; $f_j(\cdot)$, $g_j(\cdot)$, $h_j(\cdot)$ denote activation functions; the parameters b_{ij}, c_{ij}, d_{ij} are real numbers that represent the weights (or strengths) of the synaptic connections between the jth unit and the ith unit; the real constant I_i represents an input signal introduced from outside the network to the ith unit; τ_{ij} are nonnegative real numbers whose presence indicates the delayed transmission of signals at time $t - \tau_{ij}$ from the jth unit to the unit i; and the delay kernels K_{ij} incorporate the fading past effects (or fading memories) of the jth unit on the ith unit. In (2), $\Delta x_i(t_k) = x_i(t_k^+) - x_i(t_k^-)$ denote impulsive state displacements at fixed instants of time t_k ($k \in \mathbb{N}$) involving integral terms whose kernels $\psi_{ik} : [t_k - \sigma, t_k] \to \mathbb{R}$ are measurable functions, essentially bounded on the respective interval. Here it is assumed that $x_i(t_k^+) = \lim\limits_{t \to t_k^+} x_i(t)$ and $x_i(t_k) = x_i(t_k^-) = \lim\limits_{t \to t_k^-} x_i(t)$, and the sequence of times $\{t_k\}_{k=1}^{\infty}$ satisfies $0 < t_1 < t_2 < \cdots < t_k \to \infty$ as $k \to \infty$ and $\Delta t_k = t_k - t_{k-1} \geq \theta$, where $\theta > 0$ denotes the minimum time interval between successive impulses. In other words, the value $\theta > 0$ means that the impulses do not occur too often.

The assumptions that accompany the impulsive network (1) and (2) are given as follows:

A$_1$. $0 < a_i < 1/\sigma$, $i = \overline{1,m}$.

A$_2$. The activation functions $f_j, g_j, h_j : \mathbb{R} \to \mathbb{R}$ are Lipschitz continuous with respective constants F_j, G_j, H_j, $j = \overline{1,m}$.

A$_3$. $a_i - F_i \sum\limits_{j=1}^{m} |b_{ji}| - G_i \sum\limits_{j=1}^{m} |c_{ji}| - H_i \sum\limits_{j=1}^{m} |d_{ji}| > 0$, $i = \overline{1,m}$.

A$_4$. The kernels $K_{ij} : [0,\infty) \to [0,\infty)$ are bounded and piecewise continuous, normalized by $\int_0^{\infty} K_{ij}(s)\,ds = 1$, and there exists a positive number μ such that $\int_0^{\infty} K_{ij}(s)\,e^{\mu s}\,ds < \infty$ for $i,j = \overline{1,m}$.

An equilibrium point of the impulsive network (1) and (2) is denoted by $x^* = (x_1^*, x_2^*, \ldots, x_m^*)^T$ whereby the components x_i^* are governed by the algebraic system

$$a_i x_i^* = \sum_{j=1}^{m} b_{ij} f_j(x_j^*) + \sum_{j=1}^{m} c_{ij} g_j(x_j^*) + \sum_{j=1}^{m} d_{ij} h_j(x_j^*) + I_i, \quad i = \overline{1,m}, \qquad (3)$$

and satisfy the linear equations

$$\left(B_{ik} + \int_{t_k-\sigma}^{t_k} \psi_{ik}(s)\,ds \right) x_i^* + \gamma_{ik} = 0, \quad k \in \mathbb{N}, \, i = \overline{1,m}. \qquad (4)$$

Lemma 1. *Let $a_i > 0$ ($i = \overline{1,m}$) and conditions A_2, A_3 be satisfied. Then system (3) has a unique solution $x^* = (x_1^*, x_2^*, \ldots, x_m^*)^T$.*

In other words, if $a_i > 0$ ($i = \overline{1,m}$) and conditions A_2–A_4 are satisfied, the system without impulses (1) has a unique equilibrium point $x^* = (x_1^*, x_2^*, \ldots, x_m^*)^T$.

Proof. In system (3) we perform the substitution $y_i = a_i x_i^*$, $i = \overline{1,m}$. Thus, we obtain the system

$$y_i = \Phi_i(y) \equiv \sum_{j=1}^{m} \left[b_{ij} f_j\left(\frac{y_j}{a_j}\right) + c_{ij} g_j\left(\frac{y_j}{a_j}\right) + d_{ij} h_j\left(\frac{y_j}{a_j}\right) \right] + I_i, \quad i = \overline{1,m}.$$

We can show that the mapping $y \mapsto \Phi(y) = (\Phi_1(y), \Phi_2(y), \ldots, \Phi_m(y))^T$ acts as a contraction in the space $\mathbb{R}^m$ equipped with the norm $\|y\| = \sum\limits_{i=1}^{m} |y_i|$. Thus, it has a unique fixed point y^*. Then $x^* = (y_1^*/a_1, y_2^*/a_2, \ldots, y_m^*/a_m)^T$ is the unique equilibrium point of system (1). $\qquad \square$

3   Formulation of a Discrete-Time Impulsive Analogue

Our goal in the present section is to introduce a discrete-time counterpart of system (1) and (2) without essentially changing its stability characteristics. The leakage terms $-a_i x_i(t - \sigma)$ in the right-hand side of (1) make it difficult to apply the semi-discretization procedure described in [2, 8]. Instead, we will discretize all terms in the right-hand side of (1).

Suppose that $\sigma < \theta$. Let the positive integer N be sufficiently large, in particular, such that

$$\left(1 + \frac{1}{N}\right) a_i \sigma < 1, \quad i = \overline{1, m}, \qquad \left(1 + \frac{1}{N}\right) \sigma < \theta. \tag{5}$$

We choose a discretization step $h = \sigma/N$ and denote by $n = \left[\frac{t}{h}\right]$ the greatest integer in t/h, $\kappa_{ij} = \left[\frac{\tau_{ij}}{h}\right]$ and, for brevity, $x_i(n) = x_i(nh)$, $n \in \mathbb{Z}$. We further replace the integral term $\int_0^\infty K_{ij}(s) x_j(t - s)\, ds$ $(i, j = \overline{1, m})$ by a sum of the form $\sum\limits_{p=1}^{\infty} \mathcal{K}_{ij}(p) x_j(n - p)$, where the discrete kernels $\mathcal{K}_{ij}(\cdot)$, $i, j = \overline{1, m}$, satisfy the following condition:

$\mathbf{A}_4'$. $\mathcal{K}_{ij}(p) \in [0, \infty)$ are bounded for $p \in \mathbb{N}$, normalized by $\sum\limits_{p=1}^{\infty} \mathcal{K}_{ij}(p) = 1$, and there exists a number $v > 1$ such that $\sum\limits_{p=1}^{\infty} \mathcal{K}_{ij}(p) v^p < \infty$.

Thus, we obtain the following discretization of the right-hand side of (1):

$$-a_i x_i(n - N) + \sum_{j=1}^{m} b_{ij} f_j(x_j(n)) + \sum_{j=1}^{m} c_{ij} g_j(x_j(n - \kappa_{ij}))$$

$$+ \sum_{j=1}^{m} d_{ij} h_j \left(\sum_{p=1}^{\infty} \mathcal{K}_{ij}(p) x_j(n - p)\right) + I_i, \quad n \in \mathbb{N}, \quad i = \overline{1, m}.$$

The negative sign of the first term makes difficult the use of Lyapunov's functionals as in [2, 8]. We eliminate this term by using for σ small enough the approximation

$$\frac{dx_i}{dt}(nh) \approx \frac{1 - Nha_i}{h} x_i(n + 1) - \frac{1 - (N + 1)ha_i}{h} x_i(n) - a_i x_i(n - N).$$

Let us recall that $Nha_i = \sigma a_i < 1$ by condition $\mathbf{A}_1$ and $(N + 1)ha_i = \left(1 + \frac{1}{N}\right) \sigma a_i < 1$ by virtue of (5). Thus, we obtain the following discrete-time analogue of the system without impulses (1):

$$(1 - Nha_i)x_i(n+1)$$

$$= (1 - (N+1)ha_i)x_i(n) + h\left(\sum_{j=1}^{m} b_{ij}f_j(x_j(n))\right)$$

$$+ \sum_{j=1}^{m} c_{ij}g_j(x_j(n - \kappa_{ij})) + \sum_{j=1}^{m} d_{ij}h_j\left(\sum_{p=1}^{\infty} \mathcal{K}_{ij}(p)x_j(n - p)\right) + I_i\right), \quad (6)$$

$n \in \mathbb{N}$, $i = \overline{1,m}$, with initial values of the form $x_i(-\ell) = \phi_i(-\ell)$ ($\ell \in \{0\} \cup \mathbb{N}$), where the sequences $\{\phi_i(-\ell)\}_{\ell=0}^{\infty}$ are bounded for all $i = \overline{1,m}$.

Next we discretize the impulse conditions (2). If we denote $n_k = \left[\frac{t_k}{h}\right]$, $k \in \mathbb{N}$, we obtain a sequence of positive integers $\{n_k\}_{k=1}^{\infty}$ satisfying $0 < n_1 < n_2 < \cdots < n_k \to \infty$ as $k \to \infty$ and $\Delta n_k = n_k - n_{k-1} \geq \left[\frac{\theta}{h}\right] - 1$. With each such integer n_k we associate two values of the solution $x(n)$, namely, $x(n_k)$ which can be regarded as the value of the solution before the impulse effect and whose components are evaluated by (6) and $x^+(n_k)$ which can be regarded as the value of the solution after the impulse effect and whose components are evaluated by the equations

$$x_i^+(n_k) - x_i(n_k) = \sum_{\ell=n_k-N}^{n_k} B_{ik\ell}x_i(\ell) + \gamma_{ik}, \quad i = \overline{1,m}, \quad k \in \mathbb{N}, \quad (7)$$

where $B_{ik\ell}$ are suitably chosen constants.

Further on we will call system (6) and (7) the discrete-time analogue of the system with impulses (1) and (2).

The components of an equilibrium point $x^* = (x_1^*, x_2^*, \ldots, x_m^*)^T$ of system (6) and (7) must satisfy the (3) and

$$\sum_{\ell=n_k-N}^{n_k} B_{ik\ell}x_i^* + \gamma_{ik} = 0. \quad (8)$$

To ensure that systems (1), (2) and (6), (7) have the same equilibrium points if any, we choose the constants $B_{ik\ell}$ so that

$$\sum_{\ell=n_k-N}^{n_k} B_{ik\ell} = B_{ik} + \int_{t_k-\sigma}^{t_k} \psi_{ik}(s)\,\mathrm{d}s, \quad i = \overline{1,m}, \quad k \in \mathbb{N}.$$

Definition 1. The equilibrium point $x^* = (x_1^*, x_2^*, \ldots, x_m^*)^T$ of system (6) and (7) is said to be *globally exponentially stable with a multiplier* ρ if there exist constants $M > 1$ and $\rho \in (0,1)$, and any other solution $x(n) = (x_1(n), x_2(n), \ldots, x_m(n))^T$ of system (6) and (7) is defined for all $n \in \mathbb{N}$ and satisfies the estimate

$$\sum_{i=1}^{m} |x_i(n) - x_i^*| \leq M\rho^n \sum_{i=1}^{m} \sup_{\ell \in \{0\} \cup \mathbb{N}} |x_i(-\ell) - x_i^*|. \quad (9)$$

4 Main Results: Sufficient Conditions for Global Exponential Stability of the Equilibrium Point

Theorem 1. *Let system* (6) *and* (7) *satisfy conditions* $\mathbf{A}_1$–$\mathbf{A}_3$, $\mathbf{A}_4'$, (5), *and the components of the unique equilibrium point* $x^* = (x_1^*, x_2^*, \ldots, x_m^*)^T$ *of system* (6) *satisfy* (8). *Then there exist constants* $M' > 1$ *and* $\lambda \in (1, v]$ *such that any other solution* $x(n) = (x_1(n), x_2(n), \ldots, x_m(n))^T$ *of system* (6) *and* (7) *is defined for all* $n \in \mathbb{N}$ *and satisfies the estimate*

$$\sum_{i=1}^{m} |x_i(n) - x_i^*| \leq M' \lambda^{-n} \prod_{k=1}^{i(1,n)} B_k' \sum_{i=1}^{m} \sup_{\ell \in \{0\} \cup \mathbb{N}} |x_i(-\ell) - x_i^*|, \tag{10}$$

$$i(1,n) = \begin{cases} 0, & n \leq n_1, \\ \max\{k \in \mathbb{N} : n_k < n\}, & n > n_1, \end{cases} \quad B_k' = B_k \left(1 + c \max_{i=\overline{1,m}} (1 - ca_i)^{-1} \right) \text{ and}$$

$$B_k = \max_{i=\overline{1,m}} \max \left\{ \left| 1 + B_{ikn_k} \right|, \max_{n_k - N \leq \ell \leq n_k - 1} |B_{ik\ell}| \right\}, \quad k \in \mathbb{N}.$$

Proof. From the conditions of the theorem it follows that system (6) and (7) has a unique equilibrium point $x^* = (x_1^*, x_2^*, \ldots, x_m^*)^T$. For any $n \in \mathbb{N} \cup \{0\}$, from (6) and (3), by virtue of condition $\mathbf{A}_2$, we derive the inequalities

$$(1 - Nha_i)|x_i(n+1) - x_i^*| \leq (1 - (N+1)ha_i)|x_i(n) - x_i^*|$$

$$+ h \sum_{j=1}^{m} \left\{ |b_{ij}| F_j |x_j(n) - x_j^*| + |c_{ij}| G_j |x_j(n - \kappa_{ij}) - x_j^*| \right.$$

$$\left. + |d_{ij}| H_j \sum_{p=1}^{\infty} \mathcal{K}_{ij}(p)|x_j(n - p) - x_j^*| \right\}, \quad i = \overline{1,m}.$$

For $\lambda \in [1, v]$, let us denote $y_i(n) = \lambda^n |x_i(n) - x_i^*|$, $n \in \mathbb{Z}$, and define a Lyapunov functional $V(\cdot)$ by

$$V(n) = \sum_{i=1}^{m} \left\{ (1 - \sigma a_i) y_i(n) + h \sum_{j=1}^{m} \left[|c_{ij}| G_j \lambda^{\kappa_{ij}+1} \sum_{\ell=n-\kappa_{ij}}^{n-1} y_j(\ell) \right. \right.$$

$$\left. \left. + |d_{ij}| H_j \sum_{p=1}^{\infty} \mathcal{K}_{ij}(p) \lambda^{p+1} \sum_{\ell=n-p}^{n-1} y_j(\ell) \right] \right\}. \tag{11}$$

It is easy to see that $V(n) \geq 0$ for $n \in \mathbb{N} \cup \{0\}$ and $V(0) < \infty$ by $\mathbf{A}_4'$. More precisely,

$$V(0) \leq M \sum_{i=1}^{m} \sup_{\ell \in \mathbb{N} \cup \{0\}} |x_i(-\ell) - x_i^*|, \tag{12}$$

where

$$
M = \max_{i=\overline{1,m}} \left\{ 1 - \sigma a_i + h \left[G_i \sum_{j=1}^{m} |c_{ji}| \lambda^{\kappa_{ji}+1} + H_i \sum_{j=1}^{m} |d_{ji}| \sum_{p=1}^{\infty} \mathcal{K}_{ji}(p) \lambda^{p+1} \right] \right\}.
$$

Further on, we obtain

$$
V(n+1) - V(n) \le -\sum_{i=1}^{m} \Psi_i(\lambda) y_i(n),
$$

where $\quad \Psi_i(\lambda) \;=\; h\lambda \left(a_i - F_i \sum_{j=1}^{m} |b_{ji}| - G_i \sum_{j=1}^{m} |c_{ji}| \lambda^{\kappa_{ji}} - H_i \sum_{j=1}^{m} |d_{ji}| \sum_{p=1}^{\infty} \mathcal{K}_{ji}(p) \lambda^{p} \right)$

$-(\lambda - 1)(1 - \sigma a_i)$. By condition $\mathbf{A}_4'$ the functions $\Psi_i(\lambda)$ $(i = \overline{1,m})$ are well defined

and continuous for $\lambda \in [1, v]$. Moreover, $\Psi_i(1) = h \left(a_i - F_i \sum_{j=1}^{m} |b_{ji}| \; - \; G_i \sum_{j=1}^{m} |c_{ji}| \right.$

$\left. -H_i \sum_{j=1}^{m} |d_{ji}| \right) > 0$, $i = \overline{1,m}$, by virtue of $\mathbf{A}_4'$ and $\mathbf{A}_3$. By continuity, for each

$i = \overline{1,m}$, there exists a number $\lambda_i \in (1, v]$ such that $\Psi_i(\lambda) \ge 0$ for $\lambda \in (1, \lambda_i]$. If
we denote $\lambda_0 = \min_{i=\overline{1,m}} \lambda_i$, then $\lambda_0 > 1$ and $\Psi_i(\lambda) \ge 0$ for $\lambda \in (1, \lambda_0]$ and $i = \overline{1,m}$.
This implies $V(n+1) \le V(n)$ for $n \neq n_k$ and $V(n_k + 1) \le V^+(n_k)$, where $V^+(n_k)$
contains $|x_i^+(n_k) - x_i^*|$ instead of $|x_i(n_k) - x_i^*|$. The above inequalities yield

$$
V(n) \le \begin{cases} V^+(n_k) & \text{for } n_k < n \le n_{k+1}, \\ V(0) & \text{for } 0 < n \le n_1. \end{cases} \tag{13}
$$

From equalities (7) and (8), we find

$$
|x_i^+(n_k) - x_i^*| \le B_k \sum_{\ell = n_k - N}^{n_k} |x_i(\ell) - x_i^*|,
$$

where the constants B_k were introduced in the statement of Theorem 1, and

$$
V^+(n_k) \le B_k' V^+(n_{k-1})
$$

for $k \ge 2$ and, similarly, $V^+(n_1) \le B_1' V(0)$.
Combining the last inequalities and (13), we derive the estimate

$$
V(n) \le \prod_{k=1}^{i(1,n)} B_k' V(0). \tag{14}
$$

Finally, from the inequalities

$$\sum_{i=1}^{m} |x_i(n) - x_i^*| \leq \max_{i=\overline{1,m}}(1 - ca_i)^{-1}\lambda^{-n}V(n),$$

(14) and (12) we deduce (10) with $M' = M \max_{i=\overline{1,m}}(1 - ca_i)^{-1}$ and any $\lambda \in (1, \lambda_0]$. $\square$

For three sets of additional assumptions, we show that inequality (10) implies global exponential stability of the equilibrium point x^* of the discrete-time system (6) and (7).

Corollary 1. *Let all conditions of Theorem 1 hold. Suppose that $B'_k \leq 1$ for all sufficiently large values of $k \in \mathbb{N}$. Then the equilibrium point x^* of the discrete-time system (6) and (7) is globally exponentially stable with multiplier $1/\lambda_0$.*

Corollary 2. *Let all conditions of Theorem 1 hold and $\limsup_{n \to \infty} \frac{i(1,n)}{n} = p < +\infty$. Let there exist a positive constant B such that $B'_k \leq B$ for all sufficiently large values of $k \in \mathbb{N}$ and $B^p < \lambda_0$. Then for any $\rho \in \left(\frac{B^p}{\lambda_0}, 1\right)$ the equilibrium point x^* of the discrete-time system (6) and (7) is globally exponentially stable with multiplier ρ.*

Corollary 3. *Let all conditions of Theorem 1 hold. Suppose that there exists a constant $\mu \in (1, \lambda_0)$ such that $B'_k \leq \mu^{n_k - n_{k-1}}$ for all sufficiently large values of $k \in \mathbb{N}$. Then the equilibrium point x^* of the discrete-time system (6) and (7) is globally exponentially stable with multiplier μ/λ_0.*

References

1. Gopalsamy, K.: Stability and Oscillations in Delay Differential Equations of Population Dynamics. Kluwer, Dordrecht (1992)
2. Gopalsamy, K.: Stability of artificial neural networks with impulses. Appl. Math. Comput. **154**, 783–813 (2004)
3. Gopalsamy, K.: Leakage delays in BAM. J. Math. Anal. Appl. **325**, 1117–1132 (2007)
4. Hopfield, J.J.: Neural networks and physical systems with emergent collective computational abilities. Proc. Natl. Acad. Sci. **79**, 2554–2558 (1982)
5. Hopfield, J.J.: Neurons with graded response have collective computational properties like those of two-state neurons. Proc. Natl. Acad. Sci. **81**, 3088–3092 (1984)
6. Hopfield, J.J., Tank, D.W.: Computing with neural circuits; a model. Science. **233**, 625–633 (1986)
7. Li, X., Fu, X., Balasubramaniam, P., Rakkiyappan, R.: Existence, uniqueness and stability analysis of recurrent neural networks with time delay in the leakage term under impulsive perturbations. Nonlinear Anal. Real World Appl. **11**, 4092–4108 (2010)
8. Mohamad, S., Gopalsamy, K.: Dynamics of a class of discrete-time neural networks and their continuous-time counterparts. Math. Comput. Simul. **53**, 1–39 (2000)
9. Peng, S.: Global attractive periodic solutions of BAM neural networks with continuously distributed delays in the leakage terms. Nonlinear Anal. Real World Appl. **11**, 2141–2151 (2010)

A Symmetric Quantum Calculus

Artur M.C. Brito da Cruz, Natália Martins, and Delfim F.M. Torres

Abstract We introduce the α,β-symmetric difference derivative and the α, β-symmetric Nörlund sum. The associated symmetric quantum calculus is developed, which can be seen as a generalization of the forward and backward h-calculus.

1 Introduction

Quantum derivatives and integrals play a leading role in the understanding of complex physical systems. The subject has been under strong development since the beginning of the twentieth century [5–8, 11]. Roughly speaking, two approaches to quantum calculus are available. The first considers the set of points of study to be the lattice $\overline{q^{\mathbb{Z}}}$ or $h\mathbb{Z}$ and is nowadays part of the more general time scale calculus [1, 3, 9]; the second uses the same formulas for the quantum derivatives but the set of study is the set $\mathbb{R}$ of real numbers [2, 4, 10]. Here we take the second perspective.

Given a function f and a positive real number h, the h-derivative of f is defined by the ratio $(f(x+h) - f(x))/h$. When $h \to 0$, one obtains the usual derivative of the function f. The symmetric h-derivative is defined by $(f(x+h) - f(x-h))/(2h)$, which coincides with the standard symmetric derivative [12] when we let $h \to 0$.

A.M.C. Brito da Cruz (✉)
Escola Superior de Tecnologia de Setúbal, Instituto Politécnico de Setúbal, Estefanilha, 2910-761 Setúbal, Portugal

Department of Mathematics, Center for Research and Development in Mathematics and Applications, University of Aveiro, 3810–193 Aveiro, Portugal
e-mail: artur.cruz@estsetubal.ips.pt

N. Martins • D.F.M. Torres
Department of Mathematics, Center for Research and Development in Mathematics and Applications, University of Aveiro, 3810–193 Aveiro, Portugal
e-mail: natalia@ua.pt; delfim@ua.pt

S. Pinelas et al. (eds.), *Differential and Difference Equations with Applications*, Springer Proceedings in Mathematics & Statistics 47, DOI 10.1007/978-1-4614-7333-6_29,
© Springer Science+Business Media New York 2013

We introduce the α, β-symmetric difference derivative and Nörlund sum and then develop the associated calculus. Such an α, β-symmetric calculus gives a generalization to (both forward and backward) quantum h-calculus.

The text is organized as follows. In Sect. 2 we recall the basic definitions of the quantum h-calculus, including the Nörlund sum, i.e., the inverse operation of the h-derivative. Our results are then given in Sect. 3: in Sect. 3.1 we define and prove the properties of the α, β-symmetric derivative, in Sect. 3.2 we define the α, β-symmetric Nörlund sum, and Sect. 3.3 is dedicated to mean value theorems for the α, β-symmetric calculus—we prove α, β-symmetric versions of Fermat's theorem for stationary points, Rolle's, Lagrange's, and Cauchy's mean value theorems.

2 Preliminaries

In what follows we denote by $|I|$ the measure of the interval I.

Definition 1. Let α and β be two positive real numbers, $I \subseteq \mathbb{R}$ be an interval with $|I| > \alpha$, and $f : I \to \mathbb{R}$. The α-forward difference operator Δ_α is defined by

$$\Delta_\alpha [f] (t) := \frac{f(t+\alpha) - f(t)}{\alpha}$$

for all $t \in I \backslash [\sup I - \alpha, \sup I]$, in case $\sup I$ is finite, or, otherwise, for all $t \in I$. Similarly, for $|I| > \beta$ the β-backward difference operator ∇_β is defined by

$$\nabla_\beta [f] (t) := \frac{f(t) - f(t-\beta)}{\beta}$$

for all $t \in I \backslash [\inf I, \inf I + \beta]$, in case $\inf I$ is finite, or, otherwise, for all $t \in I$. We call to $\Delta_\alpha [f]$ the α-forward difference derivative of f and to $\nabla_\beta [f]$ the β-backward difference derivative of f.

Definition 2. Let $I \subseteq \mathbb{R}$ be such that $a, b \in I$ with $a < b$ and $\sup I = +\infty$. For $f : I \to \mathbb{R}$ we define the Nörlund sum (the α-forward integral) of f from a to b by

$$\int_a^b f(t) \Delta_\alpha t = \int_a^{+\infty} f(t) \Delta_\alpha t - \int_b^{+\infty} f(t) \Delta_\alpha t,$$

where

$$\int_x^{+\infty} f(t) \Delta_\alpha t = \alpha \sum_{k=0}^{+\infty} f(x+k\alpha),$$

provided the series converges at $x = a$ and $x = b$. In that case, f is said to be α-forward integrable on $[a, b]$. We say that f is α-forward integrable over I if it is α-forward integrable for all $a, b \in I$.

Remark 1. If $f : I \to \mathbb{R}$ is a function such that $\sup I < +\infty$, then we can easily extend f to $\tilde{f} : \tilde{I} \to \mathbb{R}$ with $\sup \tilde{I} = +\infty$ by letting $\tilde{f}|_I = f$ and $\tilde{f}|_{\tilde{I} \backslash I} = 0$.

Remark 2. Definition 2 is valid for any two real points a, b and not only for points belonging to $\alpha\mathbb{Z}$. This is in contrast with the theory of time scales [1, 3].

Similarly, one can introduce the β-backward integral.

Definition 3. Let I be an interval of $\mathbb{R}$ such that $a, b \in I$ with $a < b$ and $\inf I = -\infty$. For $f : I \to \mathbb{R}$ we define the β-backward integral of f from a to b by

$$\int_a^b f(t) \nabla_\beta t = \int_{-\infty}^b f(t) \nabla_\beta t - \int_{-\infty}^a f(t) \nabla_\beta t,$$

where

$$\int_{-\infty}^x f(t) \nabla_\beta t = \beta \sum_{k=0}^{+\infty} f(x - k\beta),$$

provided the series converges at $x = a$ and $x = b$. In that case, f is said to be β-backward integrable on $[a, b]$. We say that f is β-backward integrable over I if it is β-backward integrable for all $a, b \in I$.

The β-backward Nörlund sum has similar results and properties as the α-forward Nörlund sum.

3 Main Results

We begin by introducing in Sect. 3.1 the α, β-symmetric derivative; in Sect. 3.2 we define the α, β-symmetric Nörlund sum, while Sect. 3.3 is dedicated to mean value theorems for the new α, β-symmetric calculus.

3.1 The α, β-Symmetric Derivative

In what follows, $\alpha, \beta \in \mathbb{R}_0^+$ with at least one of them positive and I is an interval such that $|I| > \max\{\alpha, \beta\}$. We denote by I_β^α the set

$$I_\beta^\alpha = \begin{cases} I \backslash ([\inf I, \inf I + \beta] \cup [\sup I - \alpha, \sup I]) & \text{if } \inf I \neq -\infty \wedge \sup I \neq +\infty \\ I \backslash ([\inf I, \inf I + \beta]) & \text{if } \inf I \neq -\infty \wedge \sup I = +\infty \\ I \backslash ([\sup I - \alpha, \sup I]) & \text{if } \inf I = -\infty \wedge \sup I \neq +\infty \\ I & \text{if } \inf I = -\infty \wedge \sup I = +\infty. \end{cases}$$

Definition 4. The α, β-symmetric difference derivative of $f : I \to \mathbb{R}$ is given by

$$D_{\alpha, \beta}[f](t) = \frac{f(t + \alpha) - f(t - \beta)}{\alpha + \beta}$$

for all $t \in I_\beta^\alpha$.

Remark 3. The α,β-symmetric difference operator is a generalization of both the α-forward and the β-backward difference operators. Indeed, the α-forward difference is obtained for $\alpha > 0$ and $\beta = 0$, while for $\alpha = 0$ and $\beta > 0$, we obtain the β-backward difference operator.

Remark 4. The classical symmetric derivative [12] is obtained by choosing $\beta = \alpha$ and taking the limit $\alpha \to 0$. When $\alpha = \beta = h > 0$, the α,β-symmetric difference operator is called the h-symmetric derivative.

Remark 5. If $\alpha,\beta \in \mathbb{R}^+$, then $D_{\alpha,\beta}\,[f]\,(t) = \frac{\alpha}{\alpha+\beta}\Delta_\alpha\,[f]\,(t) + \frac{\beta}{\alpha+\beta}\nabla_\beta\,[f]\,(t)$, where Δ_α and ∇_β are, respectively, the α-forward and the β-backward differences.

The symmetric difference operator has the following properties:

Theorem 1. *Let $f,g : I \to \mathbb{R}$ and $c,\lambda \in \mathbb{R}$. For all $t \in I_\beta^\alpha$, one has:*

1. $D_{\alpha,\beta}\,[c]\,(t) = 0.$
2. $D_{\alpha,\beta}\,[f+g]\,(t) = D_{\alpha,\beta}\,[f]\,(t) + D_{\alpha,\beta}\,[g]\,(t).$
3. $D_{\alpha,\beta}\,[\lambda f]\,(t) = \lambda D_{\alpha,\beta}\,[f]\,(t).$
4. $D_{\alpha,\beta}\,[fg]\,(t) = D_{\alpha,\beta}\,[f]\,(t)\,g\,(t+\alpha) + f\,(t-\beta)\,D_{\alpha,\beta}\,[g]\,(t).$
5. $D_{\alpha,\beta}\,[fg]\,(t) = D_{\alpha,\beta}\,[f]\,(t)\,g\,(t-\beta) + f\,(t+\alpha)\,D_{\alpha,\beta}\,[g]\,(t).$
6. $D_{\alpha,\beta}\left[\dfrac{f}{g}\right](t) = \dfrac{D_{\alpha,\beta}\,[f]\,(t)\,g\,(t-\beta) - f\,(t-\beta)\,D_{\alpha,\beta}\,[g]\,(t)}{g\,(t+\alpha)\,g\,(t-\beta)}$
   provided $g\,(t+\alpha)\,g\,(t-\beta) \neq 0.$
7. $D_{\alpha,\beta}\left[\dfrac{f}{g}\right](t) = \dfrac{D_{\alpha,\beta}\,[f]\,(t)\,g\,(t+\alpha) - f\,(t+\alpha)\,D_{\alpha,\beta}\,[g]\,(t)}{g\,(t+\alpha)\,g\,(t-\beta)}$
   provided $g\,(t+\alpha)\,g\,(t-\beta) \neq 0.$

Proof. Property 1 is a trivial consequence of Definition 4. Properties 2, 3, and 4 follow by direct computations:

$$D_{\alpha,\beta}\,[f+g]\,(t) = \frac{(f+g)\,(t+\alpha) - (f+g)\,(t-\beta)}{\alpha+\beta}$$

$$= \frac{f\,(t+\alpha) - f\,(t-\beta)}{\alpha+\beta} + \frac{g\,(t+\alpha) - g\,(t-\beta)}{\alpha+\beta}$$

$$= D_{\alpha,\beta}\,[f]\,(t) + D_{\alpha,\beta}\,[g]\,(t);$$

$$D_{\alpha,\beta}\,[\lambda f]\,(t) = \frac{(\lambda f)\,(t+\alpha) - (\lambda f)\,(t-\beta)}{\alpha+\beta}$$

$$= \lambda\frac{f\,(t+\alpha) - f\,(t-\beta)}{\alpha+\beta}$$

$$= \lambda D_{\alpha,\beta}\,[f]\,(t);$$

$$D_{\alpha,\beta}\,[fg]\,(t) = \frac{(fg)\,(t+\alpha) - (fg)\,(t-\beta)}{\alpha+\beta}$$

$$= \frac{f\,(t+\alpha)\,g\,(t+\alpha) - f\,(t-\beta)\,g\,(t-\beta)}{\alpha+\beta}$$

$$= \frac{f\,(t+\alpha) - f\,(t-\beta)}{\alpha+\beta}\,g\,(t+\alpha) + \frac{g\,(t+\alpha) - g\,(t-\beta)}{\alpha+\beta}\,f\,(t-\beta)$$

$$= D_{\alpha,\beta}\,[f]\,(t)\,g\,(t+\alpha) + f\,(t-\beta)\,D_{\alpha,\beta}\,[g]\,(t).$$

Equality 5 is obtained from 4 interchanging the role of f and g. To prove 6, we begin by noting that

$$D_{\alpha,\beta}\left[\frac{1}{g}\right](t) = \frac{\frac{1}{g}\,(t+\alpha) - \frac{1}{g}\,(t-\beta)}{\alpha+\beta} = \frac{\frac{1}{g(t+\alpha)} - \frac{1}{g(t-\beta)}}{\alpha+\beta}$$

$$= \frac{g\,(t-\beta) - g\,(t+\alpha)}{(\alpha+\beta)\,g\,(t+\alpha)\,g\,(t-\beta)} = -\frac{D_{\alpha,\beta}\,[g]\,(t)}{g\,(t+\alpha)\,g\,(t-\beta)}.$$

Hence,

$$D_{\alpha,\beta}\left[\frac{f}{g}\right](t) = D_{\alpha,\beta}\left[f\frac{1}{g}\right](t) = D_{\alpha,\beta}\,[f]\,(t)\,\frac{1}{g}\,(t+\alpha) + f\,(t-\beta)\,D_{\alpha,\beta}\left[\frac{1}{g}\right](t)$$

$$= \frac{D_{\alpha,\beta}\,[f]\,(t)}{g\,(t+\alpha)} - f\,(t-\beta)\,\frac{D_{\alpha,\beta}\,[g]\,(t)}{g\,(t+\alpha)\,g\,(t-\beta)}$$

$$= \frac{D_{\alpha,\beta}\,[f]\,(t)\,g\,(t-\beta) - f\,(t-\beta)\,D_{\alpha,\beta}\,[g]\,(t)}{g\,(t+\alpha)\,g\,(t-\beta)}.$$

Equality 7 follows from simple calculations:

$$D_{\alpha,\beta}\left[\frac{f}{g}\right](t) = D_{\alpha,\beta}\left[f\frac{1}{g}\right](t) = D_{\alpha,\beta}\,[f]\,(t)\,\frac{1}{g}\,(t-\beta) + f\,(t+\alpha)\,D_{\alpha,\beta}\left[\frac{1}{g}\right](t)$$

$$= \frac{D_{\alpha,\beta}\,[f]\,(t)}{g\,(t-\beta)} - f\,(t+\alpha)\,\frac{D_{\alpha,\beta}\,[g]\,(t)}{g\,(t+\alpha)\,g\,(t-\beta)}$$

$$= \frac{D_{\alpha,\beta}\,[f]\,(t)\,g\,(t+\alpha) - f\,(t+\alpha)\,D_{\alpha,\beta}\,[g]\,(t)}{g\,(t+\alpha)\,g\,(t-\beta)}. \qquad \square$$

3.2 The α,β-Symmetric Nörlund Sum

Having in mind Remark 5, we define the α,β-symmetric integral as a linear combination of the α-forward and the β-backward integrals.

Definition 5. Let $f : \mathbb{R} \to \mathbb{R}$ and $a, b \in \mathbb{R}$, $a < b$. If f is α-forward and β-backward integrable on $[a, b]$, then we define the α, β-symmetric integral of f from a to b by

$$\int_a^b f(t)\, d_{\alpha,\beta} t = \frac{\alpha}{\alpha+\beta} \int_a^b f(t)\, \Delta_\alpha t + \frac{\beta}{\alpha+\beta} \int_a^b f(t)\, \nabla_\beta t.$$

The function f is α, β-symmetric integrable if it is α, β-symmetric integrable for all $a, b \in \mathbb{R}$.

Remark 6. Note that if $\alpha \in \mathbb{R}^+$ and $\beta = 0$, then $\int_a^b f(t)\, d_{\alpha,\beta} t = \int_a^b f(t)\, \Delta_\alpha t$; if $\alpha = 0$ and $\beta \in \mathbb{R}^+$, then $\int_a^b f(t)\, d_{\alpha,\beta} t = \int_a^b f(t)\, \nabla_\beta t$.

The properties of the α, β-symmetric integral follow from the corresponding α-forward and β-backward integral properties. It should be noted, however, that the equality $D_{\alpha,\beta} \left[s \mapsto \int_a^s f(\tau)\, d_{\alpha,\beta}\tau \right](t) = f(t)$ is not always true in the α, β-symmetric calculus, despite both forward and backward integrals satisfy the corresponding fundamental theorem of calculus. Indeed, let $f(t) = \begin{cases} \frac{1}{2^t} & \text{if } t \in \mathbb{N}, \\ 0 & \text{otherwise.} \end{cases}$

Then, for a fixed $t \in \mathbb{N}$,

$$\int_0^t f(\tau) d_{1,1}\tau = \frac{1}{2} \int_0^t f(\tau)\Delta_1\tau + \frac{1}{2}\int_0^t f(\tau)\nabla_1\tau$$

$$= \frac{1}{2}\left(\sum_{k=0}^{+\infty} f(0+k) - \sum_{k=0}^{+\infty} f(t+k)\right) + \frac{1}{2}\left(\sum_{k=0}^{+\infty} f(t-k) - \sum_{k=0}^{+\infty} f(0-k)\right)$$

$$= \frac{1}{2}\left(1 + \frac{1}{2} + \cdots + \frac{1}{2^{t-1}}\right) + \frac{1}{2}\left(\frac{1}{2^t} + \frac{1}{2^{t-1}} + \cdots + \frac{1}{2}\right)$$

$$= \frac{1}{2}\frac{1-\frac{1}{2^t}}{1-\frac{1}{2}} + \frac{1}{4}\frac{1-\frac{1}{2^t}}{1-\frac{1}{2}} = \frac{3}{2}\left(1 - \frac{1}{2^t}\right)$$

and $D_{1,1}\left[s \mapsto \int_0^s f(\tau) d_{1,1}\tau \right](t) = \frac{3}{2} D_{1,1}\left[s \mapsto 1 - \frac{1}{2^s}\right](t) = -\frac{3}{2}\frac{\frac{1}{2^{t+1}} - \frac{1}{2^{t-1}}}{2} = \frac{9}{2^{t+3}}.$
Therefore, $D_{1,1}\left[s \mapsto \int_0^s f(\tau) d_{1,1}\tau \right](t) \neq f(t)$.

3.3 Mean Value Theorems

We begin by remarking that if f assumes its local maximum at t_0, then there exist $\alpha, \beta \in \mathbb{R}_0^+$ with at least one of them positive, such that $f(t_0 + \alpha) \leqslant f(t_0)$ and $f(t_0) \geqslant f(t_0 - \beta)$. If $\alpha, \beta \in \mathbb{R}^+$, this means that $\Delta_\alpha [f](t) \leqslant 0$ and $\nabla_\beta [f](t) \geqslant 0$.

Also, we have the corresponding result for a local minimum. If f assumes its local minimum at t_0, then there exist $\alpha, \beta \in \mathbb{R}^+$ such that $\Delta_\alpha [f] (t) \geqslant 0$ and $\nabla_\beta [f] (t) \leqslant 0$.

Theorem 2 (The α, β-symmetric Fermat theorem for stationary points). *Let $f :$ $[a,b] \to \mathbb{R}$ be a continuous function. If f assumes a local extremum at $t_0 \in \,]a,b[$, then there exist two positive real numbers α and β such that $D_{\alpha,\beta} [f] (t_0) = 0$.*

Proof. We prove the case where f assumes a local maximum at t_0. Then there exist $\alpha_1, \beta_1 \in \mathbb{R}^+$ such that $\Delta_{\alpha_1} [f] (t_0) \leqslant 0$ and $\nabla_{\beta_1} [f] (t_0) \geqslant 0$. If $f (t_0 + \alpha_1) = f (t_0 - \beta_1)$, then $D_{\alpha_1,\beta_1} [f] (t_0) = 0$. If $f (t_0 + \alpha_1) \neq f (t_0 - \beta_1)$, then let us choose $\gamma = \min \{\alpha_1, \beta_1\}$. Suppose (without loss of generality) that $f (t_0 - \gamma) > f (t_0 + \gamma)$. Then, $f (t_0) > f (t_0 - \gamma) > f (t_0 + \gamma)$, and since f is continuous, by the intermediate value theorem, there exists ρ such that $0 < \rho < \gamma$ and $f (t_0 + \rho) = f (t_0 - \gamma)$. Therefore, $D_{\rho,\gamma} [f] (t_0) = 0$. □

Theorem 3 (The α, β-symmetric Rolle mean value theorem). *Let $f : [a,b] \to \mathbb{R}$ be a continuous function with $f (a) = f (b)$. Then there exist $\alpha, \beta \in \mathbb{R}^+$ and $c \in \,]a,b[$ such that $D_{\alpha,\beta} [f] (c) = 0$.*

Proof. If $f = const$, then the result is obvious. If f is not a constant function, then there exists $t \in \,]a,b[$ such that $f (t) \neq f (a)$. Since f is continuous on the compact set $[a,b]$, f has an extremum $M = f (c)$ with $c \in \,]a,b[$. Since c is also a local extremizer, then, by Theorem 2, there exist $\alpha, \beta \in \mathbb{R}^+$ such that $D_{\alpha,\beta} [f] (c) = 0$. □

Theorem 4 (The α, β-symmetric Lagrange mean value theorem). *Let $f : [a,b]$ $\to \mathbb{R}$ be a continuous function. Then there exist $c \in \,]a,b[$ and $\alpha, \beta \in \mathbb{R}^+$ such that $D_{\alpha,\beta} [f] (c) = \frac{f(b)-f(a)}{b-a}$.*

Proof. Let function g be defined on $[a,b]$ by $g (t) = f (a) - f (t) + (t - a) \frac{f(b)-f(a)}{b-a}$. Clearly, g is continuous on $[a,b]$ and $g (a) = g (b) = 0$. Hence, by Theorem 3, there exist $\alpha, \beta \in \mathbb{R}^+$ and $c \in \,]a,b[$ such that $D_{\alpha,\beta} [g] (c) = 0$. Since

$$
\begin{aligned}
D_{\alpha,\beta} [g] (t) &= \frac{g (t + \alpha) - g (t - \beta)}{\alpha + \beta} \\
&= \frac{1}{\alpha + \beta} \left(f (a) - f (t + \alpha) + (t + \alpha - a) \frac{f (b) - f (a)}{b - a} \right) \\
&\quad - \frac{1}{\alpha + \beta} \left(f (a) - f (t - \beta) + (t - \beta - a) \frac{f (b) - f (a)}{b - a} \right) \\
&= \frac{1}{\alpha + \beta} \left(f (t - \beta) - f (t + \alpha) + (\alpha + \beta) \frac{f (b) - f (a)}{b - a} \right) \\
&= \frac{f (b) - f (a)}{b - a} - D_{\alpha,\beta} [f] (t),
\end{aligned}
$$

we conclude that $D_{\alpha,\beta} [f] (c) = \frac{f(b)-f(a)}{b-a}$. □

Theorem 5 (The α, β-symmetric Cauchy mean value theorem). *Let $f, g : [a, b]$ $\to \mathbb{R}$ be continuous functions. Suppose that $D_{\alpha, \beta}[g](t) \neq 0$ for all $t \in]a, b[$ and all $\alpha, \beta \in \mathbb{R}^+$. Then there exists $\bar{\alpha}, \bar{\beta} \in \mathbb{R}^+$ and $c \in]a, b[$ such that $\frac{f(b)-f(a)}{g(b)-g(a)} = \frac{D_{\bar{\alpha},\bar{\beta}}[f](c)}{D_{\bar{\alpha},\bar{\beta}}[g](c)}$.*

Proof. From condition $D_{\alpha, \beta}[g](t) \neq 0$ for all $t \in]a, b[$ and all $\alpha, \beta \in \mathbb{R}^+$ and the α, β-symmetric Rolle mean value theorem (Theorem 3), it follows that $g(b) \neq g(a)$. Let us consider function F defined on $[a, b]$ by $F(t) = f(t) - f(a) - \frac{f(b)-f(a)}{g(b)-g(a)}[g(t) - g(a)]$. Clearly, F is continuous on $[a, b]$ and $F(a) = F(b)$. Applying the α, β-symmetric Rolle mean value theorem to the function F, we conclude that there exist $\bar{\alpha}, \bar{\beta} \in \mathbb{R}^+$ and $c \in]a, b[$ such that

$$0 = D_{\bar{\alpha},\bar{\beta}}[F](c) = D_{\bar{\alpha},\bar{\beta}}[f](c) - \frac{f(b) - f(a)}{g(b) - g(a)} D_{\bar{\alpha},\bar{\beta}}[g](c),$$

proving the intended result. □

Acknowledgements This work was supported by *FEDER* funds through *COMPETE*— Operational Programme Factors of Competitiveness ("Programa Operacional Factores de Competitividade")—and by Portuguese funds through the *Center for Research and Development in Mathematics and Applications* (University of Aveiro) and the Portuguese Foundation for Science and Technology ("FCT—Fundação para a Ciência e a Tecnologia"), within project PEst-C/MAT/UI4106/2011 with COMPETE number FCOMP-01-0124-FEDER-022690. Brito da Cruz was also supported by FCT through the Ph.D. fellowship SFRH/BD/33634/2009.

References

1. Agarwal, R., Bohner, M., O'Regan, D., Peterson, A.: Dynamic equations on time scales: a survey. J. Comput. Appl. Math. **141**(1–2), 1–26 (2002)
2. Almeida, R., Torres, D.F.M.: Nondifferentiable variational principles in terms of a quantum operator. Math. Methods Appl. Sci. **34**(18), 2231–2241 (2011)
3. Bohner, M., Peterson, A.: Dynamic Equations on Time Scales. Birkhäuser, Boston (2001)
4. Brito da Cruz, A.M.C., Martins, N., Torres, D.F.M.: Higher-order Hahn's quantum variational calculus. Nonlinear Anal. **75**(3), 1147–1157 (2012)
5. Cresson, J., Frederico, G.S.F., Torres, D.F.M.: Constants of motion for non-differentiable quantum variational problems. Topol. Methods Nonlinear Anal. **33**(2), 217–231 (2009)
6. Ernst, T.: The different tongues of q-calculus. Proc. Est. Acad. Sci. **57**(2), 81–99 (2008)
7. Jackson, F.H.: q-difference equations. Am. J. Math. **32**(4), 305–314 (1910)
8. Kac, V., Cheung, P.: Quantum Calculus. Universitext. Springer, New York (2002)
9. Malinowska, A.B., Torres, D.F.M.: On the diamond-alpha Riemann integral and mean value theorems on time scales. Dyn. Syst. Appl. **18**(3–4), 469–481 (2009)
10. Malinowska, A.B., Torres, D.F.M.: The Hahn quantum variational calculus. J. Optim. Theor. Appl. **147**(3), 419–442 (2010)
11. Milne-Thomson, L.M.: The Calculus of Finite Differences. Macmillan and Co. Ltd., London (1951)
12. Thomson, B.S.: Symmetric Properties of Real Functions. Monographs and Textbooks in Pure and Applied Mathematics, vol. 183. Dekker, New York (1994)

Stability of Nonlinear Differential Systems with Right-Hand Side Depending on Markov's Process

Irada Dzhalladova

Abstract In this paper we investigated sufficient conditions for stability of solutions of systems of nonlinear differential equations with right-hand side depending on Markov's process. The basic role in proof has Lyapunov functions. Nontrivial illustrative example is given.

Keywords Sufficient condition • Asymptotic stability of solution • Systems of nonlinear differential equations • Markov's process

1 Introduction

We deduced equations for definition of Lyapunov functions for non-linear system of differential equations with right-hand side depending on stochastic Markov's process with finite value. Then we can find sufficient conditions of stability of solutions.

2 Statement of the Problem

Let $(\Omega \equiv \{\omega\}, \Im, \mathbb{P}, \mathbb{F} \equiv \{\mathbb{F}_t, t \geq 0\})$ is probability basis [1]. We will study the systems of differential equations for $X(t) \equiv X(t, \omega)$

$$\frac{dX(t)}{dt} = F(t, X(t), \varsigma(t)), \tag{1}$$

I. Dzhalladova (✉)
Department of Mathematics, V. Getman Kiev National Economic University,
prospect Peremogy 54/1, 01033 Kiev, Ukraine
e-mail: irada-05@mail.ru

S. Pinelas et al. (eds.), *Differential and Difference Equations with Applications*, Springer
Proceedings in Mathematics & Statistics 47, DOI 10.1007/978-1-4614-7333-6_30,
© Springer Science+Business Media New York 2013

where dim $X(t) = m$. Together with (1), we consider an initial condition

$$X(0) = x_0 \in E_m,$$

where E_m is m-dimensional Euclid space. We suppose that (1) has the zero solution, i.e., $F(t,0,\varsigma(t)) \equiv 0$. We suppose that $\varsigma(t)$ is Markov's stochastic process, where, takes values $\theta_1, \ldots, \theta_n$ with probability

$$p_k(t) = P\{\varsigma(t) = \theta_k\}, \quad k = 1,2,\ldots,n,$$

which satisfies the system of differential equation [5]

$$\frac{dp_k(t)}{dt} = \sum_{s=1}^{n} a_{ks}(t)p_s(t), \quad k = 1,2,\ldots,n,$$

where continuous coefficients $a_{ks}(t)$, $k,s = 1,\ldots,n$, satisfies known conditions (see [5]):

$$\sum_{k=1}^{n} a_{ks}(t) \equiv 0, \quad a_{ks}(t) \geq 0, \text{ for} k \neq s, \quad a_{kk} \leq 0.$$

We suppose that vector functions

$$F_s(t,x) \equiv F(t,x,\theta_s), \quad s = 1,2,\ldots,n, \quad x = (x_1,x_2,\ldots,x_n),$$

are continuous for $t \geq 0$, $\|x\| < \infty$ and exist as solutions of partial systems of differential equations

$$\frac{dx}{dt} = F_s(t,x), \quad s = 1,2,\ldots,n, \quad \|x\| < \infty,$$

and there are continuant for $t \geq 0$. As $\|x\|$, we denoted the Euclidean norm of vector x.

Let positive definite function $w(t,x,\varsigma)$ satisfies condition [4]

$$\lambda_1 \|x\|^2 \leq w_s(t,x,\varsigma) \leq \lambda_2 \|x\|^2, \tag{2}$$

where $0 < \lambda_1 \leq \lambda_s$, $\lambda_1, \lambda_2 \in R$, $\varsigma(t) = \theta_k$.

We introduce the description

$$w_s(t,x) = w(t,x,\theta_s), \quad s = 1,2,\ldots,n.$$

Then we rewriting condition (2) as

$$\lambda_1 \|x\|^2 \leq w_s(t,x) \leq \lambda_2 \|x\|^2 \quad s = 1,2,\ldots,n.$$

We suppose that zero solution of system of differential equation (1) is asymptotically stable in mean square for arbitrary realization of stochastic process $\varsigma(t)$ so that the infinite integral

$$I = \int_0^\infty \langle \|X(\tau)\|^2 \rangle \, d\tau$$

converge, where $\langle \cdots \rangle$ is the symbol of mean (average) value. We define Lyapunov function as (see [2,3])

$$v(t,x,\varsigma(t)) = \int_t^\infty \langle w(\tau,X(\tau),\varsigma(\tau)) | X(t) = x \rangle \, d\tau.$$

We define partial stochastic Lyapunov function as

$$v_s(t,x) = \int_t^\infty \langle w(\tau,X(\tau),\varsigma(\tau)) | X(t) = x, \varsigma(t) = \theta_s \rangle \, d\tau, \quad s = 1,2,\ldots,n. \quad (3)$$

Density of distribution $f(t,x,\varsigma)$ of continuous discrete stochastic process $(X(t),\varsigma(t))$ has the form

$$f(t,x,\varsigma) = \sum_{k=1}^n f_k(t,x)\delta(\varsigma - \theta_k),$$

where $f_k(t,x)$ are partial functions of distribution [3] and $\delta(\cdot)$ is a Dirac function. If functions $v_s(t,x)$ know from (3), and then a value of functional

$$v = \int_0^\infty \langle w(t,X(t),\varsigma(t)) \rangle \, dt$$

for $X(t) = x, \varsigma(t) = \theta_s$, then we can find using the formula [3]

$$v = \int_{E_m} \sum_{s=1}^n v_s(0,x) f_s(0,x) dx, \quad dx \equiv dx_1 \ldots dx_m,$$

where E_m is m-dimensional phase space of variables $x_1,\ldots,x_m$. Then true the more general expression

$$\int_t^\infty \langle w(\tau,X(\tau),\varsigma(t)) \rangle \, d\tau = \int_{E_m} \sum_{s=1}^n v_s(t,x) f_s(t,x) dx. \quad (4)$$

3   Main Result

We deduce equation which satisfies partial stochastic Lyapunov functions $v_s(t,x)$ from (3). Let argument t has infinitesimal increment $h > 0$. Then the Markov's stochastic process $\varsigma(t)$ during the time $\Delta t = h > 0$ with probability $1 + ha_{ss}(t)$ stay on state θ_s, and with probability $ha_{ks}(t)$, it goes over from state θ_s to state θ_k for $k \neq s$. In this case the solution of system (1) goes over from state x to state $x + hF_s(t,x)$.

So we have the equations

$$v_s(t,x) = hw_s(t,x)$$

$$+ v_s \left(t + h, x + hF_s(t,x) + h \sum_{k=1}^{n} a_{ks}(t) v_k(t + h, x + hF_s(t,x)) + O(h^2) \right),$$

$s = 1, \ldots, n$, where $O(\cdot)$ is Landau symbol. Resolve this equations according to power of h and compare coefficients by h. We get the system of partial differential equations

$$\frac{\partial v_s(t,x)}{\partial t} + \frac{D v_s(t,x)}{Dx} F_s(t,x) + w_s(t,x) + \sum_{k=1}^{n} a_{ks}(t) v_k(t,x) = 0, \qquad (5)$$

$s = 1, \ldots, n$.

From (4) follows the theorem.

Theorem. *Let's some functions* $w_s(t,x)$, $s = 1, \ldots, n$, *which satisfies condition* (2), *exist a solution* $v_s(t,x)$, $s = 1, \ldots, n$, *of system* (5), *where satisfies condition*

$$\lambda_3 \|x\|^2 \leq v_s(t,x) \leq \lambda_4 \|x\|^2,$$

$0 < \lambda_3 \leq \lambda_4$, $\lambda_3, \lambda_4 \in R$, $s = 1, \ldots, n$. *Then the zero solution of system* (1) *is uniformly asymptotically stable in mean square, in accordance with bounded infinite integral*

$$\int_t^{\infty} \langle w(\tau, X(\tau), \varsigma(\tau)) | X(t) = x \rangle \, d\tau \leq \lambda_4 \|x\|^2.$$

4   Particular Case

In autonomous case the system of (1) has the form

$$\frac{dX(t)}{dt} = F(X(t), \varsigma(t)) \qquad (6)$$

and

$$F_s(x) = F(x, \theta_s), \quad s = 1, 2, \ldots, n,$$

where $\varsigma(t)$ is homogenous Markov's process with finite numbers of options. This Markov's process is defined by system of differential equations

$$\frac{dp_k(t)}{dt} = \sum_{s=1}^{n} a_{ks} p_s(t), \quad k = 1, \ldots, n.$$

Partial stochastic Lyapunov functions are defined by forms

$$v_s(t) = \int_0^\infty \langle w(X(\tau), \varsigma(\tau)) | X(0) = x, \varsigma(t) = \theta_s \rangle \, d\tau, \ s = 1, \ldots, n. \tag{7}$$

Then system of (5) has the form

$$\frac{Dv_s(x)}{Dx} F_k(x) + w_s(x) + \sum_{k=1}^{n} a_{ks} v_k(x) = 0, \quad s = 1, \ldots, n. \tag{8}$$

Suppose that we have positive definite functions $w_s(x)$, $v_s(x)$, $s = 1, \ldots, n$ that satisfy conditions

$$u_1(x) \le v_s(x) \le u_2(x), \quad u_3(x) \le w_s(x) \le u_4(x),$$

where $u_i(x), i = 1, 2, 3, 4$ are continuous differential functions. From (7) follows inequalities

$$\int_0^\infty \langle u_4(X(t)) \rangle \, dt \ge \langle u_1(X(0)) \rangle. \tag{9}$$

$$\int_0^\infty \langle u_3(X(t)) \rangle \, dt \le \langle u_2(X(0)) \rangle. \tag{10}$$

From previous formulas follows the next theorem. If for some positive definite functions $w_s(x)$, $s = 1, \ldots, n$, exist positive functions $v_s(x)$, $s = 1, \ldots, n$, where, satisfies the system (8), then zero solution of system (6) asymptotically stable in mean square and inequalities (9) and (10) holds.

5   Example

We decided the stability of solutions of system of differential equations

$$\frac{dX(t)}{dt} = -\beta X(t) + F(X(t), \varsigma(t)), \beta > 0,$$

where $\varsigma(t)$ is stochastic Markov's process. This process takes two values θ_1, θ_2 with probability $p_k = P\{\varsigma(t) = \theta_k\}$, $k = 1, 2$, and satisfies the next system of equations for $\lambda > 0$

$$\frac{dp_1(t)}{dt} = -\lambda p_1(t) + \lambda p_2(t),$$

$$\frac{dp_2(t)}{dt} = \lambda p_1(t) - \lambda p_2(t).$$

In this case the stochastic right-hand side has the form

$$F_1(x) = F(x, \theta_1) = \begin{pmatrix} -w_1 x_2 - x_1^3 \\ w_1 x_1 - x_2^3 \end{pmatrix},$$

$$F_2(x) = F(x, \theta_2) = \begin{pmatrix} w_2 x_2 - x_1^3 \\ -w_2 x_1 - x_2^3 \end{pmatrix},$$

where $x = (x_1, x_2)$.
Solution: We define positive definite functions as

$$w_1(x) = w_2(x) = x_1^2 + x_2^2 + \beta^{-1}\left(x_1^4 + x_2^4\right).$$

Then it is easy to prove that the solution of (8) is positive definite functions:

$$v_1(x) = v_2(x) = \frac{1}{2\beta}\left(x_1^2 + x_2^2\right).$$

From convergence integral

$$v = \int_0^\infty \left\langle x_1^2(t) + x_2^2(t) + \beta^{-1}\left(x_1^4(t) + x_2^4(t)\right)\right\rangle dt$$

follows that converges to zero for $n \to +\infty$ not only the second moment but also the forth moment of stochastic solutions. Then follows that zero solution is asymptotically stable in mean square.

References

1. Capiński, M., Zastawniak, T.: Probability Through Problems. Springer, New York (2003). ISBN 0-387-950063-X
2. Dzhalladova, I.A.: Stochastic Lyapunov's functions functions for the non-linear differential equations. Ukr. Math. J. **54**(1), 13 (2002)

3. Dzhalladova, I.A.: Optimization of Stochastic Systems, pp. 284. KNEU Press, Kiev (2005) (in Ukraine)
4. Gusak, D., Kukush, A., Kulik, A., Mishura, Y., Pilipenko, A.: Theory of Stochastic Processes. Springer, New York (2010). ISBN 978-0-387-87861-4
5. Tikhonov, V.I., Mironov, M.A.: Markov's Processes, pp. 488. Sov. Radio, Moscow (1977) (in Russian)

Oscillation of Third-Order Differential Equations with Mixed Arguments

Jozef Džurina and Blanka Baculíková

Abstract In this paper we establish new comparison theorems for deducing property (A) and the oscillation of the third-order nonlinear functional differential equation with mixed arguments

$$\left[a(t)\left[x'(t)\right]^{\gamma}\right]'' + q(t)f(x[\tau(t)]) + p(t)h(x[\sigma(t)]) = 0$$

from the oscillation of a set of suitable first-order delay/advanced equations under condition $\int^{\infty} a^{-1/\gamma}(s)\,ds = \infty$.

Keywords Third-order differential equations • Comparison theorem • Oscillation • Nonoscillation

1 Introduction

We study property (A) and the oscillation of the third-order functional differential equations

$$\left[a(t)\left[x'(t)\right]^{\gamma}\right]'' + q(t)f(x[\tau(t)]) + p(t)h(x[\sigma(t)]) = 0, \tag{E}$$

where $a, q, \tau, p, \sigma \in C([t_0, \infty))$, $f, h \in C((-\infty, \infty))$, and we will always assume that

(H_1) γ is the ratio of two positive odd integers,
(H_2) $a(t), q(t), p(t)$ are positive,

J. Džurina (✉) • B. Baculíková
Department of Mathematics, Technical University of Košice, Letná 9,
04200 Košice, Slovakia, Europe
e-mail: jozef.dzurina@tuke.sk; blanka.baculikova@tuke.sk

S. Pinelas et al. (eds.), *Differential and Difference Equations with Applications*, Springer
Proceedings in Mathematics & Statistics 47, DOI 10.1007/978-1-4614-7333-6_31,
© Springer Science+Business Media New York 2013

(H_3) $\tau(t) \leq t$, $\sigma(t) \geq t$, $\tau(t)$, $\sigma(t)$ nondecreasing, $\lim\limits_{t\to\infty} \tau(t) = \infty$,

(H_4) $xf(x) > 0$, $xh(x) > 0$, $f'(x) \geq 0$, and $h'(x) \geq 0$ for $x \neq 0$.

We shall study canonical form of Eq. (E), i.e., it is assumed

$$\int_{t_0}^{\infty} a^{-1/\gamma}(s)\,ds = \infty.$$

Moreover, in some results, we shall require the following additional assumptions

(H_5) $-f(-xy) \geq f(xy) \geq f(x)f(y)$ for $xy > 0$,
(H_6) $-h(-xy) \geq h(xy) \geq h(x)h(y)$ for $xy > 0$.

By a solution of Eq. (E) we mean a function $x(t) \in C^1((T_x, \infty))$, $T_x \geq t_0$, which has the property $a(t)(x'(t))^\gamma \in C^2((T_x, \infty))$ and satisfies Eq. (E) on $[T_x, \infty)$. We consider only those solutions $x(t)$ of Eq. (E) which satisfy $\sup\{|x(t)| : t \geq T\} > 0$ for all $T \geq T_x$. We assume that Eq. (E) possesses such a solution. A solution of Eq. (E) is called oscillatory if it has arbitrarily large zeros on $[T_x, \infty)$; otherwise, it is called to be nonoscillatory. Equation (E) is said to be oscillatory if all its solutions are oscillatory.

Before we present a definition of property (A) of Eq. (E), we introduce classification of the nonoscillatory solutions of Eq. (E).

Lemma 1. *Let $x(t)$ be a nonoscillatory solution of Eq. (E). Then $x(t)$ satisfies one of the following conditions:*

(C_2) $x(t)x'(t) > 0$, $x(t)\left[a(t)\left[x'(t)\right]^\gamma\right]' > 0$, $x(t)\left[a(t)\left[x'(t)\right]^\gamma\right]'' < 0$.
(C_0) $x(t)x'(t) < 0$, $x(t)\left[a(t)\left[x'(t)\right]^\gamma\right]' > 0$, $x(t)\left[a(t)\left[x'(t)\right]^\gamma\right]'' < 0$,

eventually.

Proof. The result is a modification of the well-known lemma of Kiguradze (see, e.g., [3, 8, 9]) and so its proof is left to the reader.   □

It is known (see, e.g., [7]) that the ordinary differential equation

$$x'''(t) + p(t)x(t) = 0$$

always possesses a nonoscillatory solution satisfying the case (C_0) of Lemma 1. Consequently, property (A) of Eq. (E) is defined as follows: We say that Eq. (E) enjoys property (A) if its nonoscillatory solution $x(t)$ satisfies (C_0).

Our first aim is to exclude the case (C_2) to establish criteria for property (A). On the other hand, intended comparison theorems will permit us to eliminate also the case (C_0). Therefore, we will be able to present also new oscillation criteria for Eq. (E).

Our technique is based on the comparison theorems in which we compare studied equation with the first-order advanced/delayed equations, so that the oscillation of these first-order equations yields studied properties of Eq. (E). This method essentially simplifies the examination of the third-order differential equations.

Remark 1. All functional inequalities considered in this chapter are assumed to hold eventually, that is, they are satisfied for all t large enough.

2 Main Results

It is convenient to prove our main results by means of a series of lemmas, as follows: The first one is recalled from [3] and presents a useful relationship between an existence of positive solutions of the advanced differential inequality and the corresponding advanced differential equation.

Lemma 2. *Suppose p, σ, and h satisfy (H_2), (H_3), and (H_4), respectively. If the first-order advanced differential inequality*

$$z'(t) - p(t)h(z(\sigma(t))) \geq 0 \tag{1}$$

has an eventually positive solution, so does the advanced differential equation

$$z'(t) - p(t)h(z(\sigma(t))) = 0. \tag{2}$$

The second one is recalled from [10] and shows equivalence between the existence of a positive solution of the delay differential inequality and the corresponding delay differential equation.

Lemma 3. *Suppose q, τ, and f satisfy (H_2), (H_3), and (H_4), respectively. If the first-order delay differential inequality*

$$z'(t) + q(t)f(z(\tau(t))) \leq 0 \tag{3}$$

has an eventually positive solution, so does the delay differential equation

$$z'(t) + q(t)f(z(\tau(t))) = 0. \tag{4}$$

Now, we are prepared to offer three new criteria for property (A) and then we extend these results to cover also the oscillation of Eq. (E).

Theorem 1. *If the first-order advanced differential equation*

$$z'(t) - (t - t_1)^{1/\gamma} a^{-1/\gamma}(t) \left[\int_t^\infty p(s)\,\mathrm{d}s \right]^{1/\gamma} h^{1/\gamma}(z[\sigma(t)]) = 0 \tag{E_1}$$

is oscillatory, then Eq. (E) has property (A).

Proof. Assume the contrary, let $x(t)$ be a nonoscillatory solution of Eq. (E) satisfying (C_2), for $t \geq t_1$. We may assume that $x(t)$ is positive. It follows from Eq. (E) that

$$\left[a(t)\left[x'(t)\right]^{\gamma}\right]'' + p(t)h(x[\sigma(t)]) \leq 0. \tag{5}$$

Integrating Eq. (5) from t to ∞, one gets

$$\left(a(t)\left[x'(t)\right]^{\gamma}\right)' \geq \int_t^{\infty} p(s)h(x[\sigma(s)])\,ds \geq h(x[\sigma(t)]) \int_t^{\infty} p(s)\,ds. \tag{6}$$

On the other hand, since $a(t)\left[x'(t)\right]^{\gamma}$ is decreasing, it is easy to check that

$$a(t)\left[x'(t)\right]^{\gamma} \geq \int_{t_1}^t \left(a(s)\left[x'(s)\right]^{\gamma}\right)'\,ds \geq \left(a(t)\left[x'(t)\right]^{\gamma}\right)'(t-t_1). \tag{7}$$

Using the last inequalities in Eq. (6), we find that $x(t)$ is a positive solution of the first-order advanced differential inequality

$$x'(t) \geq (t-t_1)^{1/\gamma}a^{-1/\gamma}(t)\left[\int_t^{\infty} p(s)\,ds\right]^{1/\gamma} h^{1/\gamma}(z[\sigma(t)]).$$

By Lemma 2, we deduce that the corresponding differential equation (E_1) has also a positive solution, which is a contradiction. □

Applying any criterion for the oscillation of Eq. (E_1), we immediately get a sufficient condition for property (A).

Corollary 1. *Assume that*

$$\frac{h^{1/\gamma}(u)}{u} \geq 1, \quad |u| \geq 1 \tag{8}$$

and

$$\liminf_{t \to \infty} \int_t^{\sigma(t)} (u-t_1)^{1/\gamma}a^{-1/\gamma}(u)\left[\int_u^{\infty} p(s)\,ds\right]^{1/\gamma} du > \frac{1}{e}. \tag{9}$$

Then Eq. (E) has property (A).

Proof. It is easy to see Eq. (9) implies

$$\int_{t_0}^{\infty} (u-t_1)^{1/\gamma}a^{-1/\gamma}(u)\left[\int_u^{\infty} p(s)\,ds\right]^{1/\gamma} du = \infty. \tag{10}$$

Taking into account Theorem 1, we shall show that Eq. (E_1) is oscillatory. Assume the converse; let Eq. (E_1) have an eventually positive solution $z(t)$. Then $z'(t) > 0$ and so $z(\sigma(t)) > c > 0$. Integrating Eq. (E_1) from t_1 to t, we have

$$z(t) \geq \int_{t_1}^{t} (u - t_1)^{1/\gamma} a^{-1/\gamma}(u) \left[\int_{u}^{\infty} p(s) \, ds \right]^{1/\gamma} h^{1/\gamma}(z[\sigma(u)]) \, du$$

$$\geq h^{1/\gamma}(c) \int_{t_1}^{t} (u - t_1)^{1/\gamma} a^{-1/\gamma}(u) \left[\int_{u}^{\infty} p(s) \, ds \right]^{1/\gamma} du,$$

which together with Eq. (10) ensures $z(t) \to \infty$ as $t \to \infty$. Therefore, $z(t) \geq 1$, eventually. Using Eq. (8) in Eq. (E_1), we see that $z(t)$ is a positive solution of the differential inequality

$$z'(t) - (t - t_1)^{1/\gamma} a^{-1/\gamma}(t) \left[\int_{t}^{\infty} p(s) \, ds \right]^{1/\gamma} z[\sigma(t)] \geq 0. \tag{11}$$

On the other hand, by Theorem 2.4.1 in [9], condition (9) guarantees that Eq. (11) has no positive solutions. This is a contradiction and we conclude that Eq. (E) has property (A). $\qquad\Box$

Remark 2. Condition (8) imposed on the function h permits to set $h(u) = u^{\beta}$, $\beta \geq \gamma$. Note that the condition $h^{1/\gamma}(u)/u \geq 1$, $u \neq 0$ required in [6] does not allow it.

For our next result, we need an additional function $\alpha(t) \in C^1([t_0, \infty))$ such that

$$\alpha'(t) \geq 0, \quad \alpha(t) < t, \quad \text{and} \quad \alpha(\sigma(t)) > t. \tag{12}$$

Theorem 2. *Let (H_6) hold. Assume that there exists a function $\alpha(t) \in C^1([t_0, \infty))$ such that Eq. (12) holds. If the first-order advanced differential equation*

$$z'(t) - \left\{ h\left[\int_{\alpha(\sigma(t))}^{\sigma(t)} a^{-1/\gamma}(s) \, ds \right] \int_{t}^{\infty} p(s) \, ds \right\} h\left(z^{1/\gamma}(\alpha[\sigma(t)]) \right) = 0 \tag{E_2}$$

is oscillatory, then Eq. (E) has property (A).

Proof. Assuming the converse, we let $x(t)$ to be a positive solution of Eq. (E) satisfying (C_2), eventually. Then, proceeding exactly as in the proof of Theorem 1, we are led to Eq. (6). Furthermore, the monotonicity of $y(t) = a(t)[x'(t)]^{\gamma} > 0$ implies

$$x(t) \geq \int_{\alpha(t)}^{t} \left(a(s)(x'(s))^{\gamma} \right)^{1/\gamma} a^{-1/\gamma}(s) \, ds$$

$$\geq y^{1/\gamma}(\alpha(t)) \int_{\alpha(t)}^{t} a^{-1/\gamma}(s) \, ds, \tag{13}$$

eventually. Combining Eq. (13) together with Eq. (6), we find that

$$y'(t) \geq \left\{ h \left[\int_{\alpha(\sigma(t))}^{\sigma(t)} a^{-1/\gamma}(s)\,ds \right] \int_t^\infty p(s)\,ds \right\} h\left(y^{1/\gamma}(\alpha\,[\sigma(t)]) \right).$$

By Lemma 2, the corresponding differential equation (E_2) also has a positive solution. This contradicts our assumption and we conclude that Eq. (E) enjoys property (A). □

Using the similar arguments as in the proof of Corollary 1, we get the following criterion for property (A) of Eq. (E).

Corollary 2. *Assume that (H_6), Eqs. (8) and (12) hold. If*

$$\liminf_{t\to\infty} \int_t^{\alpha(\sigma(t))} h\left[\int_{\alpha(\sigma(u))}^{\sigma(u)} a^{-1/\gamma}(s)\,ds \right] \left(\int_u^\infty p(s)\,ds \right) du > \frac{1}{e}, \qquad (14)$$

then Eq. (E) has property (A).

We offer another criterion for property (A) of Eq. (E) in which we employ the *"delay part"* of Eq. (E).

Theorem 3. *Assume that (H_5) holds. If the first-order delay differential equation*

$$z'(t) + q(t)f\left[\int_{t_1}^{\tau(t)} a^{-1/\gamma}(s)(s-t_1)^{1/\gamma}\,ds \right] f\left(z^{1/\gamma}[\tau(t)] \right) = 0 \qquad (E_3)$$

is oscillatory, then Eq. (E) has property (A).

Proof. Assume that Eq. (E) has not property (A). Then there has to exist its nonoscillatory solution $x(t)$ satisfying (C_2). We may assume that $x(t) > 0$. Thus, Eq. (E) implies

$$\left[a(t)\,[x'(t)]^\gamma \right]'' + q(t)f(x[\tau(t)]) \leq 0. \qquad (15)$$

It follows from Eq. (7) that $y(t) = \left(a(t)\,[x'(t)]^\gamma \right)' > 0$ satisfies

$$x'(t) \geq \frac{y^{1/\gamma}(t)}{a^{1/\gamma}(t)}(t-t_1)^{1/\gamma}. \qquad (16)$$

Integrating the last inequality, from t_1 to $\tau(t)$, we get, in view of the monotonicity of $y(t)$, that

$$x(\tau(t)) \geq y^{1/\gamma}(\tau(t)) \int_{t_1}^{\tau(t)} \frac{(s-t_1)^{1/\gamma}}{a^{1/\gamma}(s)}\,ds. \qquad (17)$$

Setting Eq. (17) into Eq. (15), we see that $y(t)$ is a positive solution of the delay differential inequality

$$y(t) + q(t)f\left[\int_{t_1}^{\tau(t)} a^{-1/\gamma}(s)(s-t_1)^{1/\gamma}ds\right] f\left(y^{1/\gamma}[\tau(t)]\right) \leq 0.$$

It follows from Lemma 3 that the corresponding differential equation (E$_3$) also has a positive solution. This is a contradiction with our assumption for oscillation of equation (E$_3$). Therefore, we deduce that Eq. (E) has property (A). □

Corollary 3. *Assume that* (H_5) *holds. If*

$$\frac{f(u^{1/\gamma})}{u} \geq 1, \quad 0 < |u| \leq 1 \tag{18}$$

and

$$\liminf_{t \to \infty} \int_{\tau(t)}^{t} q(u)f\left[\int_{t_1}^{\tau(u)} a^{-1/\gamma}(s)(s-t_1)^{1/\gamma}ds\right]du > \frac{1}{e}. \tag{19}$$

Then Eq. (E) has property (A).

Now, we offer three criteria for the elimination of the nonoscillatory solutions of Eq. (E) satisfying the case (C_2) of Lemma 1. Combining our outcoming results together with our previous results, we get nine oscillatory criteria for Eq. (E).

For our next results, we employ an additional function $\xi(t) \in C^1([t_0, \infty))$ such that

$$\xi'(t) \geq 0, \quad \xi(t) > t, \quad \text{and} \quad \tau(\xi(\xi(t))) < t. \tag{20}$$

Theorem 4. *Assume that there exists a function* $\xi(t) \in C^1([t_0, \infty))$ *such that Eq. (20) holds. Denote* $\eta_1(t) = \tau(\xi(\xi(t)))$. *If the first-order delay equation*

$$z'(t) + \left\{a^{-1/\gamma}(t)\left[\int_t^{\xi(t)} \int_u^{\xi(u)} q(s)\,ds\,du\right]^{1/\gamma}\right\} f^{1/\gamma}(z[\eta_1(t)]) = 0 \tag{E$_4$}$$

is oscillatory, then Eq. (E) has no nonoscillatory solution satisfying the (C_0).

Proof. Assuming the converse, we let $x(t)$ to be a positive solution of Eq. (E), satisfying (C_0). Integration of Eq. (15) from t to $\xi(t)$ yields

$$\left(a(t)\left(x'(t)\right)^\gamma\right)' \geq \int_t^{\xi(t)} q(s_1)f(x(\tau(s_1)))\,ds_1$$

$$\geq f(x[\tau(\xi(t))])\int_t^{\xi(t)} q(s_1)\,ds_1. \tag{21}$$

Integrating from t to $\xi(t)$ once more, we get

$$-a(t)\left(x'(t)\right)^{\gamma} \geq \int_t^{\xi(t)} f(x[\tau(\xi(s_2))]) \int_{s_2}^{\xi(s_2)} q(s_1)\,ds_1\,ds_2$$

$$\geq f(x[\eta_1(t)]) \int_t^{\xi(t)} \int_{s_2}^{\xi(s_2)} q(s_1)\,ds_1\,ds_2.$$

This means that $x(t)$ is a positive solution of the delay differential inequality

$$x'(t) + a^{-1/\gamma}(t) \left[\int_t^{\xi(t)} \int_u^{\xi(u)} q(s)\,ds\,du \right]^{1/\gamma} f^{1/\gamma}(x[\eta_1(t)]) \leq 0.$$

Lemma 3 ensures that the corresponding differential equation (E$_4$) has also a positive solution satisfying the case (C_0). This contradiction finishes the proof. □

The following result is immediate.

Corollary 4. *Assume that Eq. (18) holds and there is a function* $\xi(t) \in C^1([t_0, \infty))$ *such that Eq. (20) is satisfied. If*

$$\liminf_{t \to \infty} \int_{\tau(t)}^t a^{-1/\gamma}(v) \left[\int_v^{\xi(v)} \int_u^{\xi(u)} q(s)\,ds\,dv \right]^{1/\gamma} du > \frac{1}{e}, \qquad (22)$$

then Eq. (E) has no nonoscillatory solution satisfying (C_0).

We are able to present another result for the elimination of (C_0).

Theorem 5. *Let* (H_5) *hold. Assume that there exists a function* $\xi(t) \in C^1([t_0, \infty))$ *such that Eq. (20) holds. Denote* $\eta_2(t) = \xi(\tau(\xi(t)))$. *If the first-order delay equation*

$$z'(t) + \left\{ f\left[\int_{\tau(\xi(t))}^{\eta_2(t)} a^{-1/\gamma}(s)\,ds \right] \int_t^{\xi(t)} q(s)\,ds \right\} f\left(z^{1/\gamma}[\eta_2(t)] \right) = 0 \qquad (E_5)$$

is oscillatory, then Eq. (E) has no nonoscillatory solution satisfying (C_0).

Proof. Assume that Eq. (E) possesses a positive solution $x(t)$ satisfying (C_0). It follows from the monotonicity of $y(t) = -a(t)\left[x'(t)\right]^{\gamma}$ that

$$x(t) \geq \int_t^{\xi(t)} \left[-a^{1/\gamma}(s)x'(s) \right] a^{-1/\gamma}(s)\,ds \geq y^{1/\gamma}(\xi(t)) \int_t^{\xi(t)} a^{-1/\gamma}(s)\,ds,$$

which together with Eq. (21) yields that $y(t)$ is a positive solution of the delay differential inequality

$$-y'(t) \geq \left\{ f\left[\int_{\tau(\xi(t))}^{\eta_2(t)} a^{-1/\gamma}(s)\,ds \right] \int_t^{\xi(t)} q(s)\,ds \right\} f\left(y^{1/\gamma}[\eta_2(t)] \right).$$

By Lemma 3, we conclude that the corresponding differential equation (E_5) has also a positive solution. This contradiction finishes the proof. □

Corollary 5. *Let* (H_5) *hold. Assume that Eq.* (18) *holds and there is a function* $\xi(t) \in C^1([t_0, \infty))$ *such that Eq.* (20) *is satisfied. If*

$$\liminf_{t \to \infty} \int_{\eta_2(t)}^t \left\{ f\left[\int_{\tau(\xi(u))}^{\eta_2(u)} a^{-1/\gamma}(s)\,ds \right] \int_u^{\xi(u)} q(s)\,ds \right\} du > \frac{1}{e}, \qquad (23)$$

then Eq. (E) *has no nonoscillatory solution satisfying* (C_0).

We provide the last one result for the exclusion of (C_0).

Theorem 6. *Let* (H_5) *hold. Assume that there exists a function* $\xi(t) \in C^1([t_0, \infty))$ *such that Eq.* (20) *holds. Denote* $\eta_3(t) = \xi(\xi(\tau(t)))$. *If the first-order delay equation*

$$z'(t) + \left\{ q(t) f\left[\int_{\tau(t)}^{\xi(\tau(t))} \frac{(\xi(s)-s)^{1/\gamma}}{a^{1/\gamma}(s)}\,ds \right] \right\} f\left(z^{1/\gamma}[\eta_3(t)] \right) = 0 \qquad (E_6)$$

is oscillatory, then Eq. (E) *has no nonoscillatory solution satisfying* (C_0).

Proof. Assume that Eq. (E) possesses a positive solution $x(t)$ satisfying (C_0). It is easy to verify that since $y(t) = \left(a(t)\, [x'(t)]^\gamma \right)' > 0$ is decreasing, it satisfies

$$-a(t)\, [x'(t)]^\gamma \geq \int_t^{\xi(t)} \left(a(s)\, [x'(s)]^\gamma \right)'\,ds \geq y(\xi(t))(\xi(t) - t)$$

or equivalently

$$-x'(t) \geq \frac{y^{1/\gamma}(\xi(t))}{a^{1/\gamma}(t)} (\xi(t) - t)^{1/\gamma}.$$

An integration from t to $\xi(t)$ yields

$$x(t) \geq y^{1/\gamma}(\xi(\xi(t))) \int_t^{\xi(t)} \frac{(\xi(s) - s)^{1/\gamma}}{a^{1/\gamma}(s)}\,ds. \qquad (24)$$

Setting Eq. (24) into Eq. (15), we see that

$$y'(t) + \left\{ q(t) f\left[\int_{\tau(t)}^{\xi(\tau(t))} \frac{(\xi(s)-s)^{1/\gamma}}{a^{1/\gamma}(s)}\,ds \right] \right\} f\left(y^{1/\gamma}[\eta_3(t)] \right) \leq 0.$$

By Lemma 3, the corresponding differential equation (E_6) has also a positive solution. This is a contradiction and the proof is complete. □

Corollary 6. *Let* (H_5) *hold. Assume that Eq. (18) holds and there is a function* $\xi(t) \in C^1([t_0, \infty))$, *such that Eq. (20) is satisfied. If*

$$\liminf_{t \to \infty} \int_{\eta_3(t)}^t q(u) f \left[\int_{\tau(u)}^{\xi(\tau(u))} \frac{(\xi(s) - s)^{1/\gamma}}{a^{1/\gamma}(s)} ds \right] du > \frac{1}{e}, \tag{25}$$

then Eq. (E) has no nonoscillatory solution satisfying (C_0).

Picking up our previous theorems, we immediately get the following oscillation criterion for Eq. (E).

Theorem 7. *Assume that at least one of the Eqs.* (E_1)–(E_3) *is oscillatory and at the same time at least one of the Eqs.* (E_4)–(E_6) *is oscillatory, then Eq. (E) is oscillatory.*

We illustrate all our results in the following example:

Example 1. Consider the third-order nonlinear differential equation with mixed arguments

$$\left(t^{-1} \left(x'(t) \right)^3 \right)'' + \frac{q}{t^6} x^3(\beta t) + \frac{p}{t^6} x^3(\delta t) = 0, \tag{26}$$

with $q > 0$, $p > 0$, $0 < \beta < 1$, $\delta > 1$. Setting $\alpha(t) = \omega t$, $\omega = (1 + \delta)/(2\delta)$, conditions (9), (14), (19) reduce to

$$p^{1/3} \ln \delta > \frac{5^{1/3}}{e}, \tag{27}$$

$$p \delta^4 \left(1 - \omega^{4/3} \right)^3 \ln(\beta \delta) > \left(\frac{4}{3} \right)^3 \frac{5}{e}, \tag{28}$$

$$q \beta^5 \ln \left(\frac{1}{\beta} \right) > \left(\frac{5}{3} \right)^3 \frac{1}{e}, \tag{29}$$

respectively. Then, by Corollaries 1–3, Eq. (26) enjoys property (A), provided that at least one of the conditions (27)–(29) holds.

On the other hand, we set $\xi(t) = \lambda t$, with $\lambda = (\sqrt{\beta} + 1)/(2\sqrt{\beta})$. Then conditions (22)–(25) convert to

$$\left[q \left(1 - \frac{1}{\lambda^4} \right) \left(1 - \frac{1}{\lambda^5} \right) \right]^3 \ln \left(\frac{1}{\beta \lambda^2} \right) > \frac{5^{1/3}}{e}, \tag{30}$$

$$\frac{q \beta^4}{\lambda} \left(\lambda^5 - 1 \right) \left(\lambda^{4/3} - 1 \right)^3 \ln \left(\frac{1}{\beta \lambda^2} \right) > \left(\frac{4}{3} \right)^3 \frac{5}{e}, \tag{31}$$

$$q \beta^5 (\lambda - 1) \left(\lambda^{5/3} - 1 \right)^3 \ln \left(\frac{1}{\beta \lambda^2} \right) > \left(\frac{5}{3} \right)^3 \frac{1}{e}, \tag{32}$$

respectively. Then, by Theorem 7 and Corollaries 1–6, Eq. (26) is oscillatory, provided that at least one of Eqs. (27)–(29) holds and at the same time at least one of Eqs. (30)–(32) is satisfied. Therefore, considering all possible combinations, we obtain nine independent oscillatory criteria for the studied equation.

3 Summary

In this paper, we have presented new comparison theorems for deducing property (A) and the oscillation of Eq. (E) from the oscillation of a set of the suitable first-order delay/advanced differential equation.

The presented method essentially simplifies the examination of the third-order equations, and what is more, it supports backward the research on the first-order delay/advanced differential equations. Our results here extend and complement latest ones of Grace et al. [6], Agarwal et al. [1,2], Cecchi et al. [5], and the presents authors [4].

References

1. Agarwal, R.P., Shien, S.L., Yeh, C.C.: Oscillation criteria for second order retarded differential equations. Math. Comput. Model. **26**, 1–11 (1997)
2. Agarwal, R.P., Grace, S.R., Smith, T.: Oscillation of certain third order functional differential equations. Adv. Math. Sci. Appl. **16**, 69–94 (2006)
3. Baculíková, B.: Properties of third order nonlinear functional differential equations with mixed arguments. Abstr. Appl. Anal. **2011**, 1–15 (2011)
4. Baculíková, B., Džurina, J.: Oscillation of third-order neutral differential equations. Math. Comput. Model. **52**, 215–226 (2010)
5. Cecchi, M., Došlá, Z., Marini, M.: On third order differential equations with property A and B. J. Math. Anal. Appl. **231**, 509–525 (1999)
6. Grace, S.R., Agarwal, R.P., Pavani, R., Thandapani, E.: On the oscillation of certain third order nonlinear functional differential equations. Appl. Math. Comput. **202**, 102–112 (2008)
7. Hartman, P., Wintner, A.: Linear differential and difference equations with monotone solutions. Am. J. Math. **75**, 731–743 (1953)
8. Kusano, T., Naito, M.: Comparison theorems for functional differential equations with deviating arguments. J. Math. Soc. Japan **3**, 509–533 (1981)
9. Ladde, G.S., Lakshmikantham, V., Zhang, B.G.: Oscillation Theory of Differential Equations with Deviating Arguments. Marcel Dekker, New York (1987)
10. Philos, Ch.G.: On the existence of nonoscillatory solutions tending to zero at ∞ for differential equations with positive delay. Arch. Math. **36**, 168–178 (1981)

On Estimates for the First Eigenvalue of the Sturm–Liouville Problem with Dirichlet Boundary Conditions and Integral Condition

Svetlana Ezhak

Abstract Estimates of the first eigenvalue λ_1 of the Sturm–Liouville problem with Dirichlet boundary conditions and integral condition to the potential are obtained.

1 Introduction

Consider the Sturm–Liouville problem:

$$y''(x) + \sigma Q(x)y(x) + \lambda y(x) = 0, \tag{1}$$

$$y(0) = y(1) = 0, \tag{2}$$

where $\sigma = \pm 1$, $Q(x)$ is a nonnegative bounded function on $[0,1]$ such that

$$\int_0^1 Q^\alpha(x)dx = 1, \quad \alpha \neq 0. \tag{3}$$

A function $y(x)$ is called a solution of problems (1) and (2) if it is defined on $[0,1]$, it satisfies condition (2), its derivative $y'(x)$ is absolutely continuous, and equation (1) holds almost everywhere on $(0,1)$.

We estimate the first eigenvalue λ_1 of this problem for different values of α.

Remark 1. The problem for the equation $y''(x) + \lambda Q(x)y(x) = 0$, $Q(x)$ being a nonnegative bounded summable function on $[0,1]$ satisfying (3), and condition (2) was considered in [1]. The authors obtained the first eigenvalue λ_1 of this problem for different values of α.

S. Ezhak (✉)
Moscow State University of Economics, Statistics and Informatics,
Nezhinskaya str. 7, Moscow, Russia
e-mail: SEzhak@mesi.ru

S. Pinelas et al. (eds.), *Differential and Difference Equations with Applications*, Springer
Proceedings in Mathematics & Statistics 47, DOI 10.1007/978-1-4614-7333-6_32,
© Springer Science+Business Media New York 2013

The problem for the equation $y''(x) - Q(x)y(x) + \lambda y(x) = 0$, $Q(x) \in A_\alpha$, and conditions (2) was considered in [2]. But λ_1 was investigated only for $\alpha \geq 1$.

Consider the functional

$$R[Q,y] = \frac{\int_0^1 y'^2(x)dx - \sigma \int_0^1 Q(x)y^2(x)dx}{\int_0^1 y^2(x)dx}. \tag{4}$$

According to the variation principle, we obtain

$$\lambda_1 = \inf_{y(x) \in H_0^1(0,1)} R[Q,y],$$

where $H_0^1(0,1)$ is a function space, defined on $(0,1)$, satisfying (2) and having the generalized derivative of the first order.

Put

$$m_\alpha = \inf_{Q(x) \in A_\alpha} \lambda_1, \quad M_\alpha = \sup_{Q(x) \in A_\alpha} \lambda_1,$$

where A_α is the set of the nonnegative bounded on $[0,1]$ functions such that $\int_0^1 Q^\alpha(x)dx = 1$.

2 Results

Theorem 1. *Case* $\sigma = -1$.

1. If $\alpha > 1$*, then* $m_\alpha = \pi^2$*,* $M_\alpha < \infty$*, and there exist functions* $u(x) \in H_0^1(0,1)$ *and* $Q(x) \in A_\alpha$ *such that*

$$\inf_{y(x) \in H_0^1(0,1)} R[Q,y] = R[Q,u] = M_\alpha.$$

2. If $\alpha = 1$*, then* $m_1 = \pi^2$*,* $M_1 = \frac{\pi^2}{2} + 1 + \frac{\pi}{2}\sqrt{\pi^2 + 4}$*, and there exist functions* $u(x) \in H_0^1(0,1)$ *and* $Q(x) \in A_\alpha$ *such that*

$$\inf_{y(x) \in H_0^1(0,1)} R[Q,y] = R[Q,u] = M_1.$$

3. If $0 < \alpha < 1$*, then* $m_\alpha = \pi^2$*,* $M_\alpha = \infty$*.*
4. If $\alpha < 0$*, then* $m_\alpha > \pi^2$*,* $M_\alpha = \infty$*, and there exist functions* $u(x) \in H_0^1(0,1)$ *and* $Q(x) \in A_\alpha$ *such that*

$$\inf_{y(x) \in H_0^1(0,1)} R[Q,y] = R[Q,u] = m_\alpha.$$

Theorem 2. *Case* $\sigma = +1$.

1. *If* $\alpha > 1$, *then* $m_\alpha \geq \frac{\pi^2}{2}$, $M_\alpha = \pi^2$, *and there exist functions* $u(x) \in H_0^1(0,1)$ *and* $Q(x) \in A_\alpha$ *such that*

$$\inf_{y(x)\in H_0^1(0,1)} R[Q,y] = R[Q,u] = m_\alpha.$$

2. *If* $\alpha = 1$, *then* $M_1 = \pi^2$, $m_1 = \lambda_*$, *where* $\lambda_* \in (0, \pi^2)$ *is the solution to the equation* $2\sqrt{\lambda} = tg\left(\frac{\sqrt{\lambda}}{2}\right)$. *Here* m_1 *is attained at* $Q(x) = \delta\left(x - \frac{1}{2}\right)$ *(* $Q(x)$ *doesn't belong to* $H_0^1(0,1)$ *).*
3. *If* $1/2 \leq \alpha < 1$, *then* $m_\alpha = -\infty$, $M_\alpha = \pi^2$.
4. *If* $1/3 \leq \alpha < 1/2$, *then* $m_\alpha = -\infty$, $M_\alpha \leq \pi^2$.
5. *If* $0 < \alpha < 1/3$, *then* $m_\alpha = -\infty$, $M_\alpha < \pi^2$.
6. *If* $\alpha < 0$, *then* $m_\alpha = -\infty$, $M_\alpha < \pi^2$, *and there exist functions* $u(x) \in H_0^1(0,1)$ *and* $Q(x) \in A_\alpha$, *such that*

$$\inf_{y(x)\in H_0^1(0,1)} R[Q,y] = R[Q,u] = M_\alpha.$$

3   Proofs of Some Results

Proof. Case $\sigma = -1$.

1. Note that $m_\alpha \geq \pi^2$ for any α, $\alpha \neq 0$.
2. Suppose $\alpha > 1$. Consider the functional

$$G[y] = \frac{\int_0^1 y'^2(x)dx + \left(\int_0^1 |y(x)|^p dx\right)^{2/p}}{\int_0^1 y^2(x)dx}, \; p = \frac{2\alpha}{\alpha - 1}. \tag{5}$$

Using Hölder inequality and (3), we obtain

$$\inf_{y(x)\in H_0^1(0,1)} R[Q,y] \leq \inf_{y(x)\in H_0^1(0,1)} G[y]. \tag{6}$$

Denote

$$m = m(\alpha) = \inf_{y(x)\in H_0^1(0,1)} G[y].$$

From (6), we have $M_\alpha \leq m$. To prove $M_\alpha = m$, we need the following lemma:

Lemma. *Suppose* $\alpha > 1$ $(p = \frac{2\alpha}{\alpha-1} > 2)$ *and* $m = \inf_{y(x) \in H_0^1(0,1)} G[y]$. *Then there exists function* $u(x) \in H_0^1(0,1)$ *which is positive on* $(0,1)$ *satisfies the equation*

$$u''(x) - u^{p-1}(x) + mu(x) = 0, \tag{7}$$

and the conditions

$$u(0) = u(1) = 0, \tag{8}$$

$$\int_0^1 u^p(x)dx = 1 \tag{9}$$

such that $m = G[u]$.

Main idea for lemma proof.

1. Note that $G[y] \geq 0$ for any $y(x) \in H_0^1(0,1)$. Put $m = \inf_{y(x) \in H_0^1(0,1)} G[y]$. Denote $\Gamma = \{y(x) : y(x) \in H_0^1(0,1), \int_0^1 |y(x)|^p dx = 1\}$.
   We prove that there exists $u(x) \in \Gamma$ such that $G[u] = m$.

   (a) Note that for any $y(x) \in \Gamma$ we have $G[y] \geq C\|y(x)\|_{H_0^1(0,1)}^2$, when C—a positive constant.

   (b) A minimizing sequence for $G[y]$ in some set of functions is called such sequence $\{y_k\}$ that $G[y_k] \to m$ as $k \to \infty$. We can show that the minimizing sequence $\{y_k\}$ for $G[y]$ exists in Γ.

   (c) Let us prove that there exists such a function $u(x) \in \Gamma$ that $G[u] = m$.
       As $\{y_k\}$ is a bounded sequence in the separable Hilbert space $H_0^1(0,1)$, it contains a subsequence $\{z_k\}$ converging weakly in $H_0^1(0,1)$ to some function $u(x)$ (it follows, e.g., from [1], Theorem 20, Chap. 1), and $\|u(x)\|_{H_0^1(0,1)}^2 \leq \frac{1}{C}(m+1)$. As the space $H_0^1(0,1)$ is compactly embedded in the space $C(0,1)$, and it, in turn, is embedded in $L_p(0,1)$, where $p \geq 1$, there exists a subsequence $\{u_k\}$ of the sequence $\{z_k\}$, converging strongly in $C(0,1)$. Then the subsequence $\{u_k\}$ converges strongly in $L_p(0,1)$ to the function $u(x)$. Hence,

$$G[u_k] = \frac{\int_0^1 (u_k'(x))^2 dx + \left(\int_0^1 |u_k(x)|^p dx\right)^{2/p}}{\int_0^1 u_k^2(x)dx}.$$

We have as $k \to \infty$

$$\left(\int_0^1 |u_k(x)|^p dx\right)^{2/p} \to \left(\int_0^1 |u(x)|^p dx\right)^{2/p}, \quad \int_0^1 u_k^2(x)dx \to \int_0^1 u^2(x)dx.$$

We can show that $\|u(x)\|_{L_2(0,1)} \neq 0$.

From the sequence $\{u_k(x)\}$, we choose a subsequence $\{w_k(x)\}$ such that $w'_k(x)$ converges weakly to $u'(x)$ in $L_2(0,1)$. Such sequence $\{w_k(x)\}$ exists, since $u'_k(x)$ is bounded in $L_2(0,1)$ (it follows from the bounded-ness of the sequence $\{u_k(x)\}$ in $H^1_0(0,1)$ and the definition of the norm $\|u_k(x)\|_{H^1_0(0,1)}$). Consider $\int_0^1 (w'_k(x))^2 dx$. This sequence has a finite inferior limit as $(w'_k(x))^2 \geq 0$. Hence there exists a subsequence $\{v_k(x)\}$ of the sequence $\{w_k(x)\}$, such that

$$\lim_{k\to\infty} \int_0^1 (v'_k(x))^2 dx = \underline{\lim}_{k\to\infty} \int_0^1 (w'_k(x))^2 dx.$$

We obtain the sequence $\{v_k(x)\}$ such that $\lim_{k\to\infty} \|v'_k(x)\|_{L_2(0,1)}$ exists.

As $v'_k(x)$ converges weak to $u'(x)$ in $L_2(0,1)$, then from [3], Theorem 7, Chap. IV, we have

$$\|u'(x)\|_{L_2(0,1)} \leq \underline{\lim}_{k\to\infty} \|v'_k(x)\|_{L_2(0,1)} = \lim_{k\to\infty} \|v'_k(x)\|_{L_2(0,1)}.$$

As $G[y]$ is a continuous functional in $H^1_0(0,1)$, we obtain

$$G[u] \leq \frac{\lim_{k\to\infty} \int_0^1 (v'_k(x))^2 dx + \lim_{k\to\infty} \left(\int_0^1 |v_k(x)|^p dx\right)^{2/p}}{\lim_{k\to\infty} \left(\int_0^1 v_k^2(x) dx\right)} = \lim_{k\to\infty} G[v_k] = m.$$

Therefore we have $G[u] \leq m$. But since $m = \inf_{y(x)\in H^1_0(0,1)} G[y]$, we get $G[u] = m$.

2. Put $u(x) \in \Gamma$ and $G[u] = m$. Let us prove that $u(x)$ is positive on $(0,1)$ and satisfies (7)–(9).

Put $z(x) \in H^1_0(0,1)$. Consider the function

$$g(t) = \frac{\int_0^1 (u'(x) + tz(x))^2 dx + \left(\int_0^1 |u(x) + tz(x)|^p dx\right)^{2/p}}{\int_0^1 (u(x) + tz(x))^2 dx},$$

where $t \in R$. Since $g(0) = G[u] = \inf_{y(x)\in H^1_0(0,1)} G[y]$, then $g'(0) = 0$.

We differentiate $g(t)$ and put $t = 0$. Since $u(x) \in \Gamma$ and $G[u] = m$, we have

$$\int_0^1 u'(x)z'(x)dx + \int_0^1 |u(x)|^{p-1}z(x)\operatorname{sign} u(x)dx = m\int_0^1 u(x)z(x)dx. \qquad (10)$$

This equation holds for any $z(x) \in H^1_0(0,1)$. If $z(x) \in C_0^\infty$, then (10) means that $u'(x)$ has the generalized derivative

$$u''(x) = |u(x)|^{p-1}\operatorname{sign} u(x) - mu(x) = |u(x)|^{p-2}u(x) - mu(x).$$

Since $u(x)$ is continuous on $(0, 1)$, $u''(x)$ is also continuous on $(0, 1)$. Hence,

$$\int_0^1 u'(x)z'(x)dx = u'(x)z(x)|_0^1 - \int_0^1 u''(x)z(x)dx,$$

$$u'(1)z(1) - u'(0)z(0) - \int_0^1 u''(x)z(x)dx$$

$$= \int_0^1 |u(x)|^{p-1}z(x)\operatorname{sign} u(x)dx - m\int_0^1 u(x)z(x)dx,$$

$$\int_0^1 (u''(x) - |u(x)|^{p-1}\operatorname{sign} u(x) + mu(x))z(x)dx = 0.$$

As this equation holds for any $z(x) \in H_0^1(0, 1)$, we have

$$u''(x) - |u(x)|^{p-1}\operatorname{sign} u(x) + mu(x) = 0. \qquad (11)$$

As $G[u] = G[|u|]$, we can consider the sequence $\{u_k(x)\}$ and $u(x)$ be nonnegative. Besides, since $u(x) \in \Gamma$, the existence of $u(x) \in H_0^1(0, 1)$ satisfying (7) and (9) is proved. We can show that $u(x)$ is positive on $(0, 1)$ and satisfies (8).

To prove the existence and uniqueness of m, we investigate problems (7)–(9) and obtain that m is the solution of the system of the equations

$$\begin{cases} \int_0^H \dfrac{du}{\sqrt{mH^2 - mu^2 - \frac{2}{p}H^p + \frac{2}{p}u^p}} = \dfrac{1}{2}, \\ \int_0^H \dfrac{u^p(x)du}{\sqrt{mH^2 - mu^2 - \frac{2}{p}H^p + \frac{2}{p}u^p}} = \dfrac{1}{2}, \end{cases}$$

where $H = \max_{x \in [0,1]} u(x)$.

Lemma is proved. We prove that $M_\alpha = m$. We have

$$m = G[u] = \frac{\int_0^1 u'^2(x)dx + (\int_0^1 |u(x)|^p dx)^{2/p}}{\int_0^1 u^2(x)dx},$$

where $u(x)$ satisfies (7)–(9).

On the other hand,

$$M_\alpha = \sup_{Q(x) \in A_\alpha} \lambda_1 = \sup_{Q(x) \in A_\alpha} \inf_{y(x) \in H_0^1(0,1)} R[Q, y] \le m.$$

Since $u(x) \in H_0^1(0, 1)$ and $u^{\frac{2}{\alpha-1}}(x) \in A_\alpha$, substituting these values for $y(x)$ and $Q(x)$ in $R[Q, y]$, we receive

$$R[u^{\frac{2}{\alpha-1}}, u] = \frac{\int_0^1 u'^2(x)dx + \int_0^1 u^{\frac{2}{\alpha-1}}(x)u^2(x)dx}{\int_0^1 u^2(x)dx}$$

$$= \frac{\int_0^1 u'^2(x)dx + (\int_0^1 u^p(x)dx)^{2/p}}{\int_0^1 u^2(x)dx} = G[u] = m.$$

Thus we have the pair of functions $Q(x)$ and $y(x)$, so that the functional $R[Q, y]$ is equal to m. Hence $M_\alpha = m$.

3. Suppose $\alpha = 1$. Consider

$$L[y] = \frac{\int_0^1 y'^2(x)dx + \max_{x \in [0,1]} y^2(x)}{\int_0^1 y^2(x)dx}. \tag{12}$$

We have

$$\inf_{y(x) \in H_0^1(0,1)} R[Q, y] \le \inf_{y(x) \in H_0^1(0,1)} L[y]. \tag{13}$$

Let us prove that $M_1 = \frac{\pi^2}{2} + 1 + \frac{\pi}{2}\sqrt{\pi^2 + 4}$.

Consider the function

$$Q^*(x) = \begin{cases} 0, & 0 < x < \tau, \\ \gamma, & \tau < x < 1 - \tau, \\ 0, & 1 - \tau < x < 1, \end{cases}$$

and the function

$$y^*(x) = \begin{cases} \sin\sqrt{\gamma}x, & 0 < x < \tau, \\ \sin\sqrt{\gamma}\tau, & \tau < x < 1 - \tau, \\ \sin\sqrt{\gamma}(1-x), & 1 - \tau < x < 1, \end{cases}$$

where $\tau = \frac{\pi}{2\sqrt{\gamma}}$, $\gamma = \frac{\pi^2}{2} + 1 + \frac{\pi}{2}\sqrt{\pi^2 + 4}$.

Note that $y^*(x)$, $(y^*(x))'$ are continuous on $[0, 1]$, $y^*(0) = y^*(1) = 0$. The function $Q^*(x)$ satisfies (3). Thus, $y^*(x)$ is the first eigenfunction for problems (1)–(3), with the potential $Q(x) = Q^*(x)$, and γ be the first eigenvalue. Then $\gamma \le M_1 = \sup_{Q(x) \in A_\alpha} \lambda_1$. Since $L[y^*] = \gamma$, we have $\inf_{y(x) \in H_0^1(0,1)} L[y] \le \gamma$.

Thus, we have a sequence of inequalities:

$$\gamma \le M_1 = \sup_{Q(x) \in A_\alpha} \lambda_1 = \sup_{Q(x) \in A_\alpha} \inf_{y(x) \in H_0^1(0,1)} R[Q, y] \le$$

$$\le \sup_{Q(x) \in A_\alpha} \inf_{y(x) \in H_0^1(0,1)} L[y] = \inf_{y(x) \in H_0^1(0,1)} L[y] \le \gamma.$$

Hence, $M_1 = \gamma$, and M_1 is attained on the function $Q^*(x)$.

□

Remark 2. Note that we proved that constant $M_1 = \frac{\pi^2}{2} + 1 + \frac{\pi}{2}\sqrt{\pi^2 + 4}$ is the precise estimation of λ_1 from above. In [2] for M_1 only, the result $M_1 \leq \frac{\pi^2}{2} + 1 + \frac{\pi}{2}\sqrt{\pi^2 + 4}$ was formulated. The result $M_\alpha < \infty$ for $\alpha > 1$ is also obtained in [2].

Remark 3. The others results of Theorem 1 were proved in [4]. Theorem 2 was proved in [5].

Remark 4. In [2], the lemma was formulated, according to which $\lambda_n = n^2\lambda_1\left(\frac{1}{n^2}\right)$. Thus, the received results for λ_1 are applicable and for estimations λ_n.

References

1. Egorov, Yu., Kondratiev, V.: On spectral theory of elliptic operators. Operator Theory: Advances and Applications (English), vol. 89, p. 328. Birkhauser, Basel (1996)
2. Vinokurov, V.A., Sadovnichii, V.A.: On the range of variation of an eigenvalue when the potential is varied (English, Russian original). Dokl. Math. **68**(2), 247–252 (2003); translation from Dokl. Ross. Akad. Nauk **392**(5), 592–597 (2003)
3. Mikhlin S.G.: Variational Methods in Mathematical Physics (Russian). Nauka, Moscow (1970)
4. Ezhak, S.S.: On the estimates for the minimum eigenvalue of the Sturm–Liouville problem with integral condition (English). J. Math. Sci. New York **145**(5), 5205–5218 (2007); translation from Sovrem. Mat. Prilozh. **36**, 56–69 (2005)
5. Ezhak, S.S.: Extremal estimations for the minimum eigenvalue of the Sturm–Liouville problem with limited on potential (Russian). In: Proceedings of the International Miniconference of Qualitative Theory of Differential Equations and Applications. MESI, Moscow, pp. 42–64 (2009). ISBN 978-5-7764-0563-1

Boundary Value Problems for Schrödinger Operators on a Path Associated to Orthogonal Polynomials

A. Carmona, A.M. Encinas, and S. Gago

Abstract In this work, we concentrate on determining explicit expressions, via suitable orthogonal polynomials on the line, for the Green function associated with any regular boundary value problem on a weighted path, whose weights are determined by the coefficients of the three-term recurrence relation.

1 Introduction

In this work we analyze linear boundary value problems on a finite weighted path associated with Schrödinger operators with nonconstant potential. In spite of its relevance the Green function on a path has been obtained only for some boundary conditions, mainly for Dirichlet conditions or more generally for the so-called *Sturm–Liouville boundary conditions*; see [1–3, 5, 8]. Recently, some of the authors have obtained the Green function on a path for general boundary value problems related to Schrödinger operator with constant conductances and potential [4].

We aim here at determining explicit expressions, via suitable orthogonal polynomials on the line, for the Green function associated with any regular boundary value problem on a weighted path, whose weights are determined by the coefficients of the three-term recurrence relation defining the polynomials. Our study is similar to what is known for boundary value problems associated with ordinary differential equations [6, Chaps. 7,11,12].

A. Carmona • A.M. Encinas
Departament de Matemàtica Aplicada III, Universitat Politècnica de Catalunya, Mod. C2, Campus Nord, c/ Jordi Girona Salgado 1–3, 08034 Barcelona, Spain
e-mail: angeles.carmona@upc.edu; andres.marcos.encinas@upc.edu

S. Gago (✉)
Departament de Matemàtica Aplicada III, Universitat Politècnica de Catalunya, EUETIB, Campus Urgell, c/ Comte Urgell 187, 08036 Barcelona, Spain
e-mail: silvia.gago@upc.edu

S. Pinelas et al. (eds.), *Differential and Difference Equations with Applications*, Springer Proceedings in Mathematics & Statistics 47, DOI 10.1007/978-1-4614-7333-6_33,
© Springer Science+Business Media New York 2013

The boundary value problems considered here are of two types. Corresponding to the cases in which the boundary has either one or two vertices. In each case, it is essential to describe the solutions of the Schröndinger equation on the interior nodes of the path. We show that it is possible to obtain explicitly such solutions in terms of the chosen orthogonal polynomials. As an immediate consequence of this property, we can easily characterize those boundary value problems that are regular and then we obtain their corresponding Green functions.

2 Schrödinger Operators and Orthogonal Polynomials

Let $\{A_n\}_{n=0}^{\infty}$ be a real positive sequence and $\{B_n\}_{n=0}^{\infty}$ a real sequence. Consider $\{\mathscr{R}_n\}_{n=0}^{\infty}$ be a sequence of real orthogonal polynomials satisfying the following recurrence relation:

$$\mathscr{R}_n(x) = (A_n x + B_n)\mathscr{R}_{n-1}(x) - C_n \mathscr{R}_{n-2}(x), \; n \geq 2, \tag{1}$$

where $C_n = \frac{A_n}{A_{n-1}}$. If k_n denotes the leading coefficient of $\mathscr{R}_n(x)$, then $A_n = \frac{k_n}{k_{n-1}}$, $C_n = \frac{k_n k_{n-2}}{k_{n-1}^2}$, for $n \geq 2$ and $\Pi_{i=1}^n C_i = \frac{A_n}{A_0}$.

Choosing a pair of initial polynomials $\mathscr{R}_0(x)$ and $\mathscr{R}_1(x)$, the recurrence relation in Eq. (1) leads to a family of orthogonal polynomials and allows us to extend the recurrence relation for $n = 1$. For instance, the families $\{P_n\}_{n=0}^{\infty}$ and $\{Q_n\}_{n=0}^{\infty}$, such that $Q_0(x) = P_0(x) = 1$, $Q_1(x) = A_1 x + B_1$, and $P_1(x) = \frac{A_0}{A_0 + A_1} Q_1(x)$, satisfy that $P_{-1}(x) = P_1(x)$ and $Q_{-1}(x) = 0$. From now on, the families $\{P_n\}_{n=0}^{\infty}$ and $\{Q_n\}_{n=0}^{\infty}$ will be called *first-kind and second-kind orthogonal polynomials*, respectively, as for $A_n = 2$, $B_n = 0$, $n \geq 0$, they are the Chebyshev polynomials of first and second kind. On the other hand, consider a path P_{n+2} with $n + 2$ vertices, with vertex set $V = \{0, \ldots, n+1\}$ and let $\mathscr{C}(V)$ be the vector space of real functions. Given $c \in \mathscr{C}(V)$ such that $c(k) > 0$ for any $k = 0, \ldots, n$, we define the conductance on P_{n+2} as $c : V \times V \to [0, +\infty)$ such that $c(k, k+1) = c(k+1, k) = c(k)$ and $c(k, m) = 0$ otherwise. The Schrödinger operator with potential $q \in \mathscr{C}(V)$ is the self-adjoint operator $\mathscr{L}_q : \mathscr{C}(V) \to \mathscr{C}(V)$ defined by

$$\mathscr{L}_q(u)(0) = c(0)[u(0) - u(1)] + q(0)u(0),$$

$$\mathscr{L}_q(u)(k) = c(k)[u(k) - u(k+1)] + c(k-1)[u(k) - u(k-1)] + q(k)u(k), \tag{2}$$

$$\mathscr{L}_q(u)(n+1) = c(n)[u(n+1) - u(n)] + q(n+1)u(n+1).$$

Thus, for a fixed $x \in \mathbb{R}$, if we consider the recurrence relation of Eq. (1) and we choose $c(k) = \frac{A_0}{A_{k+1}}$ for each $k = 0, \ldots, n$ and $q_x(k) = \frac{A_0(A_{k+1}x + B_{k+1} - 1)}{A_{k+1}} - \frac{A_0}{A_k}$ for each $k = 0, \ldots, n+1$, the corresponding Schrödinger operator is given by

$$\mathcal{L}_{q_x}(u)(0) = \left[\frac{A_0(A_1x+B_1)}{A_1} - 1\right]u(0) - \frac{A_0}{A_1}u(1),$$

$$\mathcal{L}_{q_x}(u)(k) = \frac{A_0(A_{k+1}x+B_{k+1})}{A_{k+1}}u(k) - \frac{A_0}{A_{k+1}}u(k+1) - \frac{A_0}{A_k}u(k-1), \quad k = 1,\ldots,n,$$

$$\mathcal{L}_{q_x}(u)(n+1) = \frac{A_0(A_{n+2}x+B_{n+2}-1)}{A_{n+2}}u(n+1) - \frac{A_0}{A_{n+1}}u(n).$$

If $F = \{1,\ldots,k\}$, given a $f \in \mathscr{C}(V)$, the *Schrödinger equation on F with data*
f is $\mathcal{L}_{q_x}(u) = f$ on F. Analogously, the equation $\mathcal{L}_{q_x}(u) = 0$ on F is called
homogeneous Schrödinger equation on F.
From now on, we consider only $x \neq -\frac{B_1}{A_1}$, i.e., such that $P_1(x) \neq 0$.

Lemma 1. *Consider the functions* $u(k) = P_k(x)$ *and* $v(k) = Q_k(x)$, $k \in V$. *Then, for*
any $k \in V$ *the Wronskian is* $w[u,v](k) = \frac{A_{k+1}}{A_0}P_1(x)$, *and hence,* $\{u,v\}$ *is a basis of*
the solution space of the homogeneous Schrödinger equation on F. Moreover, the
Green function of the homogenous Schrödinger equation is

$$g_x[k,s] = \frac{1}{P_1(x)}[P_k(x)Q_s(x) - P_s(x)Q_k(x)], \quad k,s \in V. \tag{3}$$

Therefore, the general solution of the Schrödinger equation on F with data $f \in$
$\mathscr{C}(V)$ is determined by

$$u(k) = \alpha P_k(x) + \beta Q_k(x) + \sum_{s=1}^{k} g_x[k,s]f(s), \quad k \in V,$$

where $\alpha, \beta \in \mathbb{R}$.

3 Two-Side Boundary Value Problems

Given $d_i \in \mathbb{R}$, $i \in \partial F = \{0,1,n,n+1\}$, a linear boundary condition on F with
coefficients d_i is a linear map $\mathscr{B} : \mathscr{C}(V) \to \mathbb{R}$ such that

$$\mathscr{B}(u) = d_0 u(0) + d_1 u(1) + d_n u(n) + d_{n+1} u(n+1), \text{ for any } u \in \mathscr{C}(V).$$

A *two-side boundary value problem on F* consists in finding $u \in \mathscr{C}(V)$ such that

$$\mathcal{L}_{q_x}(u) = f \text{ on } F, \quad \mathscr{B}_1(u) = g_1, \quad \mathscr{B}_2(u) = g_2, \tag{4}$$

for given $f \in \mathscr{C}(V)$ and $g_1, g_2 \in \mathbb{R}$, where the boundary conditions $\mathscr{B}_1$ and $\mathscr{B}_2$
are linearly independent. The problem is *semi-homogeneous* when $g_1 = g_2 = 0$
and *homogeneous* if besides $f = 0$. Problem (4) is *regular* if the corresponding
homogenous boundary value problem has the null function as its unique solution.

A function $y \in \mathscr{C}(V)$ is a solution of the homogeneous boundary value problem iff $y = \alpha u + \beta v$, where $\alpha, \beta \in \mathbb{R}$ and $\{u, v\}$ is a basis of solutions of the homogeneous equation on V, satisfies $\mathscr{B}_i(y) = \alpha \mathscr{B}_i(u) + \beta \mathscr{B}_i(v) = 0$, for $i = 1, 2$. Then, Problem (4) is regular iff $\mathscr{B}_1(u)\mathscr{B}_2(v) - \mathscr{B}_2(u)\mathscr{B}_1(v) \neq 0$, and hence, it also holds that the boundary value problem is regular iff for any data $f \in \mathscr{C}(V)$, $g_1, g_2 \in \mathbb{R}$ it has a unique solution. Moreover, for $u(k) = P_k(x)$ and $v(k) = Q_k(x)$, $k \in V$,

$$P_{\mathscr{B}}(x) = \mathscr{B}_1(u)\mathscr{B}_2(v) - \mathscr{B}_2(u)\mathscr{B}_1(v) = \sum_{i,j \in B} p_{ij} u(i) v(j) = P_1(x) \sum_{\substack{i < j \\ i,j \in B}} p_{i,j} g_x[i,j],$$

where $p_{ij} = d_{1i} d_{2j} - d_{2i} d_{1j}$ for all $i, j \in B = \{0, 1, n, n+1\}$ and g_x is the Green function defined in Eq. (3). The following lemma shows that two-side boundary problems can be restricted to the study of the semi-homogeneous ones.

Lemma 2. *Consider* $\alpha, \beta, \gamma, \delta \in \mathbb{R}$ *such that* $d_{j1}\alpha + d_{j2}\beta + d_{j3}\gamma + d_{j4}\delta = g_j$, *for* $j = 1, 2$, *then* $u \in \mathscr{C}(V)$ *verifies the Schrödinger equation* $\mathscr{L}_{q_x}(u) = f$ *on* F, *together with the boundary conditions* $\mathscr{B}_1(u) = g_1$ *and* $\mathscr{B}_2(u) = g_2$, *iff the function* $v = u - \alpha \varepsilon_0 - \beta \varepsilon_1 - \gamma \varepsilon_n - \delta \varepsilon_{n+1}$ *verifies that*

$$\mathscr{L}_{q_x}(v) = f + \left(\frac{A_0}{A_1}\alpha - \frac{A_0}{A_2}(A_2 x + B_2)\beta \right) \varepsilon_1 + \frac{A_0}{A_2}\beta \varepsilon_2 + \frac{A_0}{A_n}\gamma \varepsilon_{n-1}$$

$$+ \left(\frac{A_0}{A_{n+1}}\delta - \frac{A_0}{A_{n+1}}(A_{n+1}x + B_{n+1})\gamma \right) \varepsilon_n$$

on F *and* $\mathscr{B}_1(u) = \mathscr{B}_2(u) = 0$.

The solution of any regular semi-homogeneous boundary problem can be obtained by considering its resolvent kernel; i.e., the function $G_{q_x} \in \mathscr{C}(V \times F)$ s.t.

$$\mathscr{L}_{q_x}(G_{q_x}(\cdot, s)) = \varepsilon_s \text{ on } F, \quad \mathscr{B}_1(G_{q_x}(\cdot, s)) = \mathscr{B}_2(G_{q_x}(\cdot, s)) = 0, \quad s \in F.$$

This function is called the *Green function* for Problem (4). Thus, for any $f \in \mathscr{C}(V)$, the unique solution of the semi-homogeneous boundary problem with data f is

$$u(k) = \sum_{s=1}^{n} G_{q_x}(k, s) f(s).$$

Theorem 1. *The boundary problem (4) is regular iff* $P_{\mathscr{B}}(x) \neq 0$, *and then, its Green function is given for any* $1 \leq s \leq n$ *and* $0 \leq k \leq n+1$ *by*

$$G_{q_x}(k,s) = \frac{P_1(x)}{P_{\mathscr{B}}(x)} \left[p_{n,n+1} \frac{A_{n+1}}{A_0} g_x[s,k] + \sum_{i=0}^{1} \sum_{j=n}^{n+1} p_{i,j} g_x[k,i] g_x[j,s] \right] + \begin{cases} 0, & k \leq s \\ g_x[k,s], & k \geq s. \end{cases}$$

Proof. Observe from Lemma 1 that for a fixed $s \in F$, $k \in V$, the Green function G_{q_x} of the BVP Eq. (4) is given by

$$G_{q_x}(k,s) = a(s)P_k(x) + b(s)Q_k(x) + \begin{cases} 0 & \text{if } k < s, \\ g_x[k,s] & \text{if } k \geq s. \end{cases}$$

Therefore for a fixed $s \in F$ we just have to solve the system

$$\begin{pmatrix} \mathscr{B}_1(P_k(x)) & \mathscr{B}_1(Q_k(x)) \\ \mathscr{B}_2(P_k(x)) & \mathscr{B}_2(Q_k(x)) \end{pmatrix} \begin{pmatrix} a(s) \\ b(s) \end{pmatrix} = - \begin{pmatrix} \mathscr{B}_1(g_x[k,s]) \\ \mathscr{B}_2(g_x[k,s]) \end{pmatrix}.$$

For $i = 1,2$, we have $\mathscr{B}_i(g_x[k,s]) = d_{in}g_x[n,s] + d_{in+1}g_x[n+1,s]$, which implies

$$P_{\mathscr{B}}(x)a(s) = \sum_{i=0}^{1} \sum_{j=n}^{n+1} p_{ij}g_x[j,s]Q_i(x) - p_{nn+1}g_x[n,n+1]Q_s(x),$$

$$P_{\mathscr{B}}(x)b(s) = -\sum_{i=0}^{1} \sum_{j=n}^{n+1} p_{ij}g_x[j,s]P_i(x) + p_{nn+1}g_x[n,n+1]P_s(x).$$

Finally we obtain

$$\frac{P_{\mathscr{B}}(x)}{P_1(x)}[a(s)P_k(x) + b(s)Q_k(x)] = p_{nn+1}g_x[n,n+1]g_x[s,k]$$

$$+ \sum_{i=0}^{1} g_x[i,k] \left(\sum_{j=n}^{n+1} g_x[s,j] \right). \qquad \square$$

In what follows we study the more usual boundary value problems appearing in the literature with proper name; that is, unilateral, Dirichlet, and Neumann problems or more generally, Sturm–Liouville problems.

The pair of boundary conditions $(\mathscr{B}_1, \mathscr{B}_2)$ is called *unilateral* if either $d_{1,j} = d_{2,j} = 0$ for any $j \in \{n, n+1\}$ (initial value problem) or $d_{1,i} = d_{2,i} = 0$ for any $i \in \{0,1\}$ (final value problem). Any unilateral pair verifies that $p_{0j} = p_{1j} = 0$, for $j \in \{n, n+1\}$, and thus, $P_{\mathscr{B}}(x) = \frac{P_1(x)}{A_0}(A_1p_{0,1} + A_{n+1}p_{n,n+1})$. In addition, either $p_{n,n+1} = 0$ and $p_{0,1} \neq 0$ or $p_{n,n+1} \neq 0$ and $p_{0,1} = 0$, since the boundary conditions are linearly independent, which implies that unilateral boundary problems are regular. Therefore, any unilateral pair is equivalent to either $(u(0), u(1))$ for initial value problems or $(u(n), u(n+1))$ for final value problems.

Corollary 1. *The Green function for the initial boundary value problem is given by*

$$G_{q_x}(k,s) = \begin{cases} 0, & k \leq s, \\ g_x[k,s], & k \geq s, \end{cases}$$

and the Green function for the final boundary value problem is

$$G_{q_x}(k,s) = \begin{cases} g_x[s,k], & k \le s, \\ 0, & k \ge s, \end{cases}$$

where $1 \le s \le n$ *and* $0 \le k \le n+1$.

The boundary conditions are called *Sturm–Liouville conditions*, when $d_{1j} = d_{2i} = 0$, for $i \in \{0,1\}$, $j \in \{n, n+1\}$, that is, when

$$\mathscr{B}_1(u) = au(0) + bu(1) \quad \text{and} \quad \mathscr{B}_2(u) = cu(n) + du(n+1), \tag{5}$$

where $a,b,c,d \in \mathbb{R}$ are such that $(|a| + |b|)(|c| + |d|) > 0$. The most popular Sturm–Liouville conditions are the so-called *Dirichlet boundary conditions*, that correspond to take $b = c = 0$, and *Neumann boundary conditions*, that correspond to take $b = -a$ and $d = -c$.

Corollary 2. *Given* $a,b,c,d \in \mathbb{R}$ *such that* $(|a| + |b|)(|c| + |d|) > 0$ *and the Sturm–Liouville boundary conditions, then*

$$P_{\mathscr{B}}(x) = a\Big[d\big(Q_{n+1}(x) - P_{n+1}(x)\big) + c\big(Q_n(x) - P_n(x)\big)\Big]$$
$$\times b\Big[P_1(x)\big(dQ_{n+1}(x) + cQ_n(x)\big) - Q_1(x)\big(dP_{n+1}(x) + cP_n(x)\big)\Big],$$

and the Green function for the Sturm–Liouville boundary value problem is

$$G_{q_x}(k,s) = \frac{1}{P_1(x)P_{\mathscr{B}}(x)}\Big[b\big(Q_1(x)P_k(x) - Q_k(x)P_1(x)\big) + a\big(P_k(x) - Q_k(x)\big)\Big]$$
$$\times \Big[\big(dQ_{n+1}(x) + cQ_n(x)\big)P_s(x) - \big(dP_{n+1}(x) + cP_n(x)\big)Q_s(x)\Big],$$

for any $0 \le k \le s \le n$ *and* $1 \le s$, *whereas*

$$G_{q_x}(k,s) = \frac{1}{P_1(x)P_{\mathscr{B}}(x)}\Big[b\big(Q_1(x)P_s(x) - Q_s(x)P_1(x)\big) + a\big(P_s(x) - Q_s(x)\big)\Big]$$
$$\times \Big[\big(dQ_{n+1}(x) + cQ_n(x)\big)P_k(x) - \big(dP_{n+1}(x) + cP_n(x)\big)Q_k(x)\Big],$$

for any $n+1 \ge k \ge s \ge 1$ *and* $s \le n$.

As a consequence, the boundary polynomial for the Dirichlet problem is

$$P_{\mathscr{B}}(x) = ad\big(Q_{n+1}(x) - P_{n+1}(x)\big),$$

and hence, it is regular iff $Q_{n+1}(x) \ne P_{n+1}(x)$, *and the Green's function is given by*

$$G_{q_x}(k,s) = \begin{cases} \dfrac{\big(P_k(x) - Q_k(x)\big)\big(Q_{n+1}(x)P_s(x) - P_{n+1}(x)Q_s(x)\big)}{P_1(x)\big(Q_{n+1}(x) - P_{n+1}(x)\big)}, & k \leq s \\[3mm] \dfrac{\big(P_s(x) - Q_s(x)\big)\big(Q_{n+1}(x)P_k(x) - P_{n+1}(x)Q_k(x)\big)}{P_1(x)\big(Q_{n+1}(x) - P_{n+1}(x)\big)}, & k \geq s. \end{cases}$$

Finally, for Neumann boundary problem, the boundary polynomial is

$$P_{\mathscr{B}}(x) = ac\Big[\big(1 - Q_1(x)\big)\big(P_{n+1}(x) - P_n(x)\big) - \big(1 - P_1(x)\big)\big(Q_{n+1}(x) - Q_n(x)\big)\Big],$$

and the Green function, $G_{q_x}(k,s)$, for the Neumann problem is

$$\frac{\big[(1 - Q_1(x))P_k(x) - (1 - P_1(x))Q_k(x)\big]\big[Q_s(x)\big(P_{n+1}(x) - P_n(x)\big) - P_s(x)\big(Q_{n+1}(x) - Q_n(x)\big)\big]}{P_1(x)\Big[\big(1 - Q_1(x)\big)\big(P_{n+1}(x) - P_n(x)\big) - \big(1 - P_1(x)\big)\big(Q_{n+1}(x) - Q_n(x)\big)\Big]}$$

for any $0 \leq k \leq s \leq n$ and $1 \leq s$, whereas

$$\frac{\big[(1 - Q_1(x))P_s(x) - (1 - P_1(x))Q_s(x)\big]\big[Q_k(x)\big(P_{n+1}(x) - P_n(x)\big) - P_k(x)\big(Q_{n+1}(x) - Q_n(x)\big)\big]}{P_1(x)\Big[\big(1 - Q_1(x)\big)\big(P_{n+1}(x) - P_n(x)\big) - \big(1 - P_1(x)\big)\big(Q_{n+1}(x) - Q_n(x)\big)\Big]}$$

for any $n + 1 \geq k \geq s \geq 1$ and $s \leq n$.

4 One-Side Boundary Problems

In this section we analyze one-side boundary value problems; i.e., the boundary conditions are located at one side of the path P_{n+2}. So if we consider the vertex subset $\hat{F} = \{0, 1, \ldots, n\}$, the linear map $\mathscr{B} : \mathscr{C}(V) \to \mathbb{R}$ such that

$$\mathscr{B}(u) = au(n) + bu(n+1), \text{ for any } u \in \mathscr{C}(V)$$

is a *linear one-side boundary condition on* $\hat{F}$ with coefficients $a, b \in \mathbb{R}$, wherever $|a| + |b| > 0$. Moreover, a *one-side boundary value problem on* $\hat{F}$ consists in finding $u \in \mathscr{C}(V)$ such that

$$\mathscr{L}_{q_x}(u) = f \text{ on } \hat{F}, \quad \mathscr{B}(u) = g, \tag{6}$$

for a given $f \in \mathscr{C}(V)$ and $g \in \mathbb{R}$. The problem is semi-homogenous when $g = 0$ and homogeneous if, in addition, $f = 0$. Again, the one-side boundary value problem is regular if the corresponding homogeneous problem has the null function as its unique solution; equivalently, Eq. (6) is regular iff for any data $f \in \mathscr{C}(V)(V)$ and $g \in \mathscr{R}$ it has a unique solution. In this case, the *Green function for the one-side boundary value problem* (6) is the function $G_{q_x} \in \mathscr{C}(V \times \hat{F})$ characterized by

$$\mathscr{L}_{q_x}(G_{q_x}(\cdot,s)) = \varepsilon_s \quad \text{on } \hat{F}, \qquad \mathscr{B}(G_{q_x}(\cdot,s)) = 0, \qquad \text{for any } s \in \hat{F}. \tag{7}$$

The analysis of one-side boundary value problems can be easily derived from the study of two-side boundary value problems by observing that Eq. (6) can be rewritten as the following two-side Sturm–Liouville problem

$$\mathscr{L}_{q_x}(u) = f, \quad \text{on } F, \quad \left[A_0 Q_1(x) - A_1\right]u(0) - A_0 u(1) = A_1 f(0), \quad \mathscr{B}(u) = g. \tag{8}$$

Therefore, we can reduce the analysis of one-side boundary value problems to the analysis of semi-homogeneous Sturm–Liouville problems.

Lemma 3. *Given $g \in \mathscr{R}$, then for any $f \in \mathscr{C}(V)$ the function $u \in \mathscr{C}(V)$ satisfies that $\mathscr{L}_{q_x}(u) = f$ on $\hat{F}$ and $\mathscr{B}(u) = g$ iff the function $v = u + \frac{A_1}{A_0}f(0)\varepsilon_1 - \dfrac{g(a\varepsilon_n + b\varepsilon_{n+1})}{a^2 + b^2}$ satisfies $\left[A_0 Q_1(x) - A_1\right]v(0) - A_0 v(1) = \mathscr{B}(v) = 0$ and on F*

$$\mathscr{L}_{q_x}(v) = f + \frac{A_1 f(0)}{A_2}\left((A_2 x + B_2)\varepsilon_1 - \varepsilon_2\right)$$
$$+ \frac{A_0 g}{(a^2 + b^2)}\left(\left(a(A_{n+1}x + B_{n+1}) - b\right)\frac{\varepsilon_n}{A_{n+1}} - a\frac{\varepsilon_{n-1}}{A_n}\right).$$

Corollary 3. *Given the one-side boundary value problem (6), then*

$$P_{\mathscr{B}}(x) = A_1\left[\left(P_1(x) - 1\right)\left(bQ_{n+1}(x) + aQ_n(x)\right) + bP_{n+1}(x) + aP_n(x)\right]$$

and the Green function is

$$G_{q_x}(k,s) = \frac{\left[Q_k(x)\left(1 - P_1(x)\right) - P_k(x)\right]}{P_1(x)\left[\left(P_1(x) - 1\right)\left(bQ_{n+1}(x) + aQ_n(x)\right) + bP_{n+1}(x) + aP_n(x)\right]}$$
$$\times \left[\left(bQ_{n+1}(x) + aQ_n(x)\right)P_s(x) - \left(bP_{n+1}(x) + aP_n(x)\right)Q_s(x)\right]$$

for any $0 \le k \le s \le n$, whereas

$$G_{q_x}(k,s) = \frac{\left[Q_s(x)\left(1 - P_1(x)\right) - P_s(x)\right]}{P_1(x)\left[\left(P_1(x) - 1\right)\left(bQ_{n+1}(x) + aQ_n(x)\right) + bP_{n+1}(x) + aP_n(x)\right]}$$
$$\times \left[\left(bQ_{n+1}(x) + aQ_n(x)\right)P_k(x) - \left(bP_{n+1}(x) + aP_n(x)\right)Q_k(x)\right]$$

for any $n + 1 \ge k \ge s \ge 0$ and $s \le n$.

Acknowledgement This work has been partly supported by the Spanish Research Council (Comisión Interministerial de Ciencia y Tecnología,) under projects MTM2010-19660 and MTM 2008-06620-C03-01/MTM.

References

1. Agarwal, R.P.: Difference Equations and Inequalities. Marcel Dekker, Basel (2000)
2. Bârsan, V., Cojocaru, S.: Green functions for atomic wires. Rom. Rep. Phys. **58**, 123–127 (2006)
3. Bass, R.: The Green's function for a finite linear chain. J. Math. Phys. **26**, 3068–3069 (1985)
4. Bendito, E., Carmona, A., Encinas, A.M.: Eigenvalues, eigenfunctions and Green's functions on a path via Chebyshev polynomials. Appl. Anal. Discrete Math. **3**, 182–302 (2009)
5. Chung, F.R.K., Yau, S.T.: Discrete Green's functions. J. Combin. Theory A **91**, 191–214 (2000)
6. Coddington, E.A., Levinson, N.: Theory of Ordinary Differential Equations. McGraw-Hill, New York (1955)
7. García-Moliner, F.: Why Green function for matching? Microelectron. J. **36**, 876–881 (2005)
8. Jirari, A.: Second-order Sturm-Liouville difference equations and orthogonal polynomials. Mem. Am. Math. Soc. **542**, 138 (1995)

Centers in a Quadratic System Obtained from a Scalar Third Order Differential Equation

Adriana Buică, Isaac A. García, and Susanna Maza

Abstract In this paper it is shown that $(0,0,0)$ is a center for

$$\dot{x} = y, \quad \dot{y} = z, \quad \dot{z} = -\frac{1}{a}z - a\,(2x+1)y - x(x+1)$$

and that $(-1,0,0)$ is a center for

$$\dot{x} = y, \quad \dot{y} = z, \quad \dot{z} = -axz - \frac{1}{a}y - x(x+1),$$

(where $a > 0$) giving in this way a positive answer to questions raised in the paper *Analysis of a quadratic system obtained from a scalar third order differential equation*, Electron. J. Diff. Equat. **2010**(161) (2010).

Keywords Inverse Jacobi multiplier • Center manifold • Center problem

1 Introduction and Statement of the Results

The starting point is the scalar third-order differential equation

$$\dddot{x} + f(x)\ddot{x} + g(x)\dot{x} + h(x) = 0, \tag{1}$$

A. Buică
Department of Mathematics, Babeş-Bolyai University, Kogălniceanu 1,
400084 Cluj-Napoca, Romania
e-mail: abuica@math.ubbcluj.ro

I.A. García (✉) • S. Maza
Departament de Matemàtica, Universitat de Lleida, Avda. Jaume II 69, 25001 Lleida, Spain
e-mail: garcia@matematica.udl.cat; smaza@matematica.udl.cat

S. Pinelas et al. (eds.), *Differential and Difference Equations with Applications*, Springer
Proceedings in Mathematics & Statistics 47, DOI 10.1007/978-1-4614-7333-6_34,
© Springer Science+Business Media New York 2013

with f and g arbitrary polynomials of degree 1 and h a polynomial of degree 2. Without loss of generality we can take $h(x) = x(x+1)$ when h has two real zeros. We will associate to (1) the quadratic differential systems in $\mathbb{R}^3$

$$\dot{x} = y, \ \dot{y} = z, \ \dot{z} = -f(x)z - g(x)y - h(x). \tag{2}$$

A *Hopf point* of (2) is a singularity that possesses two complex eigenvalues $\pm i$ with zero real part and one nonzero real eigenvalue. System (2) having a singular point of Hopf type at the origin has a local two-dimensional *center manifold* $W^c(0)$. This manifold is invariant for (2) locally (only for sufficiently small $|x|$ and $|y|$), and for any $k \geq 1$ there exists $\tilde{h}$ of class C^k near the origin such that

$$\tilde{h}(0,0) = 0, \ D\tilde{h}(0,0) = 0,$$

$D\tilde{h}(x,y)$ being the Jacobian matrix of $\tilde{h}$, and

$$W^c(0) = \{(x,y,\tilde{h}(x,y)) \in \mathbb{R}^3 : (x,y) \text{ in a small neighborhood of } (0,0)\}.$$

The center problem for system (2) at the Hopf type singularity consists in detecting when the singular point becomes either a *center* or a *focus* for the flow of system (2) restricted to the center manifold. We say that the singular point is a *center* of (2) if all the orbits on $W^c(0)$ near the origin are periodic and a *focus* if they spiral around it. The classical procedure for the solution to the center problem can be found in [1,4], while the projection method for the calculation of the Lyapunov constants is given in [7].

Recall here that the Lyapunov constants are the coefficients of the Taylor series of the displacement map (Poincaré return map minus the identity), so that its vanishing is a necessary condition for having a center. But essentially, the main problem is the following: Let $\mathfrak{R} \subset \mathbb{R}[\lambda]$ be the ring of real polynomials whose variables are the coefficients $\lambda \in \mathbb{R}^p$ of some polynomial differential family (2). The Bautin ideal $\mathfrak{J}$ is the ideal of $\mathfrak{R}$ generated by all the Lyapunov constants. Using Hilbert's basis theorem, it follows that $\mathfrak{J}$ is finitely generated. Thus, there are $\{B_1, B_2, \ldots, B_r\} \subset \mathfrak{J}$ such that $\mathfrak{J} = (B_1, B_2, \ldots, B_r)$. Such a set of generators is called a basis of $\mathfrak{J}$ when r is the minimum number of the ideal generators. In this case, we say that $r = \dim \mathfrak{J}$. For the concrete family of polynomial systems (2), an open problem nowadays is the determination of $\mathfrak{J}$.

We recall that, by using the blow-up technique, the problem to characterize the local phase portrait near an isolated singular point of a planar vector field can be solved except when the singularity is monodromic; that is, it is either a focus or a center.

Some aspects of the dynamics of system (2) are studied in [5]. Under conditions on the set of parameters, the origin and the point $(-1,0,0)$ are Hopf points of system (2). The authors of [5] prove that the first three Lyapunov coefficients vanish at these Hopf points; hence, they conjecture that they are centers. In this work, analyzing the vector field (2) with the techniques developed in [3], we show two families of centers, solving in this way the two conjectures formulated in [5]. We have the following result:

Theorem 1. *Consider the following four-parameter family of quadratic differential systems in $\mathbb{R}^3$*

$$\dot{x} = y , \ \dot{y} = z , \ \dot{z} = -f(x)z - g(x)y - h(x) , \tag{3}$$

where $f(x) = a_1 x + a_0$, $g(x) = b_1 x + b_0$, and $h(x) = x(x+1)$ and the parameters $(a_0, a_1, b_0, b_1) \in \mathbb{R}^4$. We have the following center conditions:

(i) The point $(0,0,0)$ is a center of system (3) if $b_0 > 0$, $a_0 = 1/b_0$, $a_1 = 0$, and $b_1 = 2b_0$.

(ii) The point $(-1,0,0)$ is a center of system (3) if $a_0 = b_1 = 0$, $b_0 = 1/a_1$, and $a_1 > 0$.

The proof of Theorem 1 is based on the properties of an inverse Jacobi multiplier of system (3) as studied in [3]. Since this proof is based on the main results of the preprint [3], we write in the last section of this paper an appendix in order to give an alternative self-contained proof of Theorem 1.

We shortly present the properties of the inverse Jacobi multiplier function developed in [3] below.

Let $D \subseteq \mathbb{R}^n$ be an open subset and $\mathscr{Y} = \sum_{i=1}^{n} f_i(x)\partial_{x_i}$ be a $C^1(D)$ vector field with $x = (x_1, \ldots, x_n) \in D$. A C^1 function $V : D \to \mathbb{R}$ is said to be an *inverse Jacobi multiplier* of $\mathscr{Y}$ if it is not locally null and it satisfies the linear first-order partial differential equation

$$\mathscr{Y}V = V \operatorname{div}\mathscr{Y},$$

here $\operatorname{div}\mathscr{Y} = \sum_{i=1}^{n} \partial f_i(x)/\partial x_i$ is the divergence of the vector field $\mathscr{Y}$. A good reference to the theory of inverse Jacobi multipliers is [2]. See also [6] for a summary. The next result is a simple consequence of the main results proved in [3].

Corollary 1 ([3]). *Assume that the linear part of the analytic vector field $\mathscr{Y}$ in $\mathbb{R}^3$ has the block diagonal representation*

$$C = \begin{pmatrix} A & 0 \\ 0 & \lambda \end{pmatrix}, \ A = \begin{pmatrix} 0 & -1 \\ 1 & 0 \end{pmatrix},$$

where $\lambda \in \mathbb{R}\setminus\{0\}$. Then the origin is a center for $\mathscr{Y}$ if and only if there exists a local analytic inverse Jacobi multiplier $V(x,y,z)$ of $\mathscr{Y}$ near the origin having the following Taylor expansion

$$V(x,y,z) = z + \cdots , \quad \text{where the dots indicate terms of order two or higher.}$$

2 Proof of Theorem 1

The singularities of (3) are $p_0 = (0,0,0)$ and $p_1 = (-1,0,0)$. In addition, these singular points are Hopf points in the following cases:

- Taking $b_0 > 0$, $a_0 = 1/b_0$, the origin of system (3) has associated eigenvalues $-1/b_0$ and $\pm i\sqrt{b_0}$.
- Taking $a_0 - a_1 < 0$ and $(a_0 - a_1)(b_0 - b_1) = -1$, the singularity $(-1,0,0)$ of system (3) has associated eigenvalues $a_1 - a_0$ and $\pm i/\sqrt{a_1 - a_0}$.

Under the parameter restrictions $b_0 > 0$, $a_0 = 1/b_0$, $a_1 = 0$, and $b_1 = 2b_0$ of statement (i), system (3) possesses the inverse Jacobi multiplier:

$$V(x,y,z) = z + b_0 x(x+1).$$

This can be easily seen by verifying that this function is a solution of the linear partial differential equation:

$$y\frac{\partial V}{\partial x} + z\frac{\partial V}{\partial y} - (z/b_0 + b_0(2x+1)y + x(x+1))\frac{\partial V}{\partial z} = -V/b_0.$$

We do the linear change of variables

$$(x,y,z) \rightarrow \frac{1}{1+b_0^3}\begin{pmatrix} -b_0 & 0 & b_0^3 \\ -b_0^{5/2} & -(1+b_0^3)b_0^{1/2} & -b_0^{3/2} \\ b_0 & 0 & 1 \end{pmatrix}\begin{pmatrix} x \\ y \\ z \end{pmatrix}$$

and the rescaling of time $t \rightarrow -\sqrt{b_0}\, t$ bringing the linear part of (3) at the origin to its canonical form. In short, system (3) is written in the form

$$\dot{x} = -y + \frac{1}{1+b_0^3}\left(b_0^{1/2}x^2 + 2b_0^2 xy - 2b_0^5 yz - b_0^{13/2}z^2\right),$$

$$\dot{y} = x + \frac{1}{1+b_0^3}\left(-b_0^{-1}x^2 - 2b_0^{1/2}xy + 2b_0^{7/2}yz + b_0^5 z^2\right), \tag{4}$$

$$\dot{z} = b_0^{-3/2}z + \frac{1}{1+b_0^3}\left(b_0^{-5/2}x^2 + 2b_0^{-1}xy - 2b_0^2 yz - b_0^{7/2}z^2\right),$$

having the inverse Jacobi multiplier:

$$V(x,y,z) = z + (x - b_0^3 z)^2/(b_0(1+b_0^3)).$$

Applying Corollary 1, the origin is a center of system (4) and consequently of system (3) proving statement (i).

Under the parameter restrictions $a_0 = b_1 = 0$, $b_0 = 1/a_1$, and $a_1 > 0$ of statement (ii), system (3) possesses the inverse Jacobi multiplier:

$$V(x,y,z) = 1 + x + a_1 z.$$

This can be easily seen by verifying that this function is a solution of the linear partial differential equation:

$$y\frac{\partial V}{\partial x} + z\frac{\partial V}{\partial y} - (a_1 xz + y/a_1 + x(x+1))\frac{\partial V}{\partial z} = -a_1 xV.$$

Firstly, we translate the singular point $(-1,0,0)$ to the origin with the change $(x,y,z) \to (x+1,y,z)$. After, we do the linear change of variables

$$(x,y,z) \to \frac{1}{1+a_1^3}\begin{pmatrix} -a_1^2 & 0 & 1 \\ -a_1^{1/2} & a_1^{-1/2}+a_1^{5/2} & -a_1^{3/2} \\ a_1^2 & 0 & a_1^3 \end{pmatrix}\begin{pmatrix} x \\ y \\ z \end{pmatrix}$$

and the rescaling of time $t \to a_1^{-1/2} t$ bringing the linear part of the system to canonical form. In short we obtain that system (3) becomes

$$\dot{x} = -y + \frac{1}{a_1^{7/2}}z(a_1^3 x - z),$$

$$\dot{y} = x - a_1 xz + \frac{z^2}{a_1^2}, \tag{5}$$

$$\dot{z} = \frac{z(a_1^2 + a_1^3 x - z)}{\sqrt{a_1}}.$$

The origin of this system is trivially a center because $W^c(0) = \{z = 0\}$ and the system reduced to the center manifold is the linear center $\dot{x} = -y$, $\dot{y} = x$. Then statement (ii) is proved. Other proof of this fact follows applying Corollary 1, because an inverse Jacobi multiplier of this system is $V(x,y,z) = z$.

3 Appendix

In this section we present an alternative proof of Theorem 1. We start with the explicit knowledge of an inverse Jacobi multiplier $V(x,y,z)$ in all the cases, as stated before. Next the main idea is first to obtain from V the explicit expression of an analytic center manifold $W^c(0)$ and finally to check that the singularity of the reduced system to the center manifold is in fact a center. We only do this procedure in the proof of statement (i) of Theorem 1 because statement (ii) has been trivially proved without resorting to [3].

To prove statement (i) of Theorem 1 we must study the center problem at the origin for the equivalent system (4). Recall that this system admits the inverse Jacobi multiplier $V(x,y,z) = b_0(1 + b_0^3)z + (x - b_0^3 z)^2$. Therefore, $\{V(x,y,z) = 0\}$ defines an invariant algebraic surface of (4) which passes through the origin and is tangent to the plane $\{z = 0\}$ at this point. In particular, this means that this invariant surface is tangent to the center eigenspace, the (x,y)-plane, at the origin; hence, in a neighborhood of the origin forms a local center manifold. Indeed, solving $V = 0$ for z and inserting into the first two equations in (4), we obtain the following expression of the reduced system (4) to the center manifold in local coordinates:

$$\dot{x} = P(x,y) = -y + \frac{1}{4b_0^{7/2}(1 + b_0^3)} f_+(x) g(x,y),$$

$$\dot{y} = Q(x,y) = x + \frac{1}{4b_0^5(1 + b_0^3)} f_-(x) g(x,y),$$

where $f_\pm(x) = \pm 1 \pm b_0^3 \mp \sqrt{1 + b_0^3}\sqrt{1 + b_0^3 - 4b_0^2 x}$ and $g(x,y) = -1 - b_0^3 + 4b_0^2 x + \sqrt{1 + b0^3}\sqrt{1 + b_0^3 - 4b_0^2 x} + 4b_0^{7/2} y$. It is straightforward to check that the function $v(x,y) = \sqrt{1 + b_0^3 - 4b0^2 x}$ is an inverse integrating factor of this reduced system; that is, the rescaled system $\dot{x} = P(x,y)/v(x,y)$, $\dot{y} = Q(x,y)/v(x,y)$ is Hamiltonian. Since $v(x,y)$ is a nonvanishing analytic function near the origin, this implies that the reduced system to the center manifold possesses a local analytic first integral around the origin. Hence, the origin becomes a center.

References

1. Andronov, A.A. et al.: Theory of Bifurcations of Dynamic Systems on a Plane. Wiley, New York (1973)
2. Berrone, L.R., Giacomini, H.: Inverse Jacobi multipliers. Rend. Circ. Mat. Palermo (2) **52**, 77–130 (2003)
3. Buică, A., García, I.A., Maza, S.: Existence of inverse Jacobi multipliers around Hopf points in $\mathbb{R}^3$: emphasis on the center problem. To appear in J. Diff. Equat. **252**, 6324–6336 (2012)
4. Chicone, C.: Ordinary Differential Equations with Applications, 2nd edn. Springer, New York (2006)
5. Dias, F.S., Mello, L.F.: Analysis of a quadratic system obtained from a scalar third order differential equation. Electron. J. Diff. Equat. **2010**(161), 1–25 (2010)
6. García, I.A., Grau, M.: A survey on the inverse integrating factor. Qual. Theory Dyn. Syst. **9**, 115–166 (2010)
7. Kuznetsov, Y.A.: Elements of Applied Bifurcations Theory, 2nd edn. Springer, New York (1998)

Three Solutions for Systems of n Fourth-Order Partial Differential Equations

Shapour Heidarkhani

Abstract In this paper, we shall establish the existence of at least three weak solutions for a class of systems of n fourth-order partial differential equations coupled with Navier boundary conditions. The technical approach is fully based on a very recent three critical points theorem.

Keywords Three solutions • Critical point • Variational methods • $(p_1,\ldots,p_n)$-biharmonic • Navier boundary value problem

1 Introduction

In this work, based on a recent three critical points theorem due to Bonanno and Marano [5], we study the existence of at least three weak solutions for the nonlinear elliptic system of n fourth-order partial differential equations under Navier boundary conditions

$$\begin{cases} \Delta(|\Delta u_i|^{p_i-2}\Delta u_i) - \alpha_i\Delta_{p_i}u_i + \beta_i|u_i|^{p_i-2}u_i = \lambda F_{u_i}(x,u_1,\ldots,u_n) & \text{in } \Omega, \\ u_i = \Delta u_i = 0 & \text{on } \partial\Omega \end{cases} \tag{1}$$

for $1 \leq i \leq n$, where $\Delta_{p_i}u_i = \operatorname{div}(|\nabla u_i|^{p_i-2}\nabla u_i)$ is the p_i-Laplacian operator, α_i and β_i for $1 \leq i \leq n$ are positive constants, $\Omega \subset \mathbb{R}^N (N \geq 1)$ is a nonempty bounded open set with smooth boundary $\partial\Omega$, $p_i > \max\{1, \frac{N}{2}\}$ for $1 \leq i \leq n$, $\lambda > 0$, $F : \Omega \times \mathbb{R}^n \to \mathbb{R}$ is a function such that $F(.,t_1,\ldots,t_n)$ is continuous in Ω for all $(t_1,\ldots,t_n) \in \mathbb{R}^n$,

S. Heidarkhani (✉)
Department of Mathematics, Razi University, 67149 Kermanshah, Iran

School of Mathematics, Institute for Research in Fundamental Sciences (IPM),
P.O. Box: 19395-5746, Tehran, Iran
e-mail: s.heidarkhani@razi.ac.ir; sh.heidarkhani@yahoo.com

S. Pinelas et al. (eds.), *Differential and Difference Equations with Applications*, Springer   411
Proceedings in Mathematics & Statistics 47, DOI 10.1007/978-1-4614-7333-6_35,
© Springer Science+Business Media New York 2013

$F(x, ., \ldots, .)$ is C^1 in $\mathbb{R}^n$ for every $x \in \Omega$ and $F(x, 0, \ldots, 0) = 0$ for all $x \in \Omega$, and F_t denotes the partial derivative of F with respect to t. System (1) is called $(p_1, \ldots, p_n)$-biharmonic.

There seems to be increasing interest in studying fourth-order boundary value problems, because the static form change of beam or the sport of rigid body can be described by a fourth-order equation, and specially a model to study traveling waves in suspension bridges can be furnished by the fourth-order equation of nonlinearity, so it is important to Physics. In [11], Lazer and Mckenna have pointed out that this type of nonlinearity furnishes a model to study traveling waves in suspension bridges. More general nonlinear fourth-order elliptic boundary value problems have been studied [1–4, 6–10, 12–14] in recent years.

Here and in the next section, X will denote the Cartesian product of n Sobolev spaces $W^{2,p_i}(\Omega) \cap W_0^{1,p_i}(\Omega)$ for $i = 1, \ldots, n$, i.e., $X = (W^{2,p_1}(\Omega) \cap W_0^{1,p_1}(\Omega)) \times \ldots \times (W^{2,p_n}(\Omega) \cap W_0^{1,p_n}(\Omega))$ endowed with the norm $\|(u_1, \ldots, u_n)\| = \sum_{i=1}^{n} \|u_i\|_{p_i}$, where $\|u_i\|_{p_i} = (\int_\Omega |\Delta u_i(x)|^{p_i} dx + \alpha_i \int_\Omega |\nabla u_i(x)|^{p_i} dx + \beta_i \int_\Omega |u_i(x)|^{p_i} dx)^{1/p_i}$ for $1 \leq i \leq n$.

We say that $u = (u_1, \ldots, u_n)$ is a weak solution to system (1) if $u = (u_1, \ldots, u_n) \in X$ and

$$\int_\Omega \sum_{i=1}^{n} \Big(|\Delta u_i(x)|^{p_i-2} \Delta u_i(x) \Delta v_i(x) + \alpha_i |\nabla u_i(x)|^{p_i-2} \nabla u_i(x) \nabla v_i(x)$$

$$+ \beta_i |u_i(x)|^{p_i-2} u_i(x) v_i(x) \Big) dx$$

$$- \lambda \int_\Omega \sum_{i=1}^{n} F_{u_i}(x, u_1(x), \ldots, u_n(x)) v_i(x) dx = 0$$

for every $(v_1, \ldots, v_n) \in X$. For other basic notations and definitions, we refer the reader to [18].

2   Main Results

First we here recall for the reader's convenience Theorem 2.6 of [5] (see also [15–17] for related results) which is our main tool to transfer the existence of three solutions of system (1) into the existence of critical points of the Euler functional:

Theorem 1 (see [5, Theorem 2.6]). *Let X be a reflexive real Banach space, let $\Phi : X \longrightarrow \mathbb{R}$ be a sequentially weakly lower semicontinuous, coercive, and continuously Gâteaux differentiable whose Gâteaux derivative admits a continuous inverse on X^*, and let $\Psi : X \longrightarrow \mathbb{R}$ be a sequentially weakly upper semicontinuous and continuously Gâteaux differentiable functional whose Gâteaux derivative is compact. Assume that there exist $r \in \mathbb{R}$ and $u_1 \in X$ with $0 < r < \Phi(u_1)$, such that* *(i)* $\sup_{u \in \Phi^{-1}(]-\infty, r])} \Psi(u) < r \frac{\Psi(u_1)}{\Phi(u_1)}$.

(ii) For each $\lambda \in \Lambda_r :=] \frac{\Phi(u_1)}{\Psi(u_1)}, \frac{r}{\sup_{u \in \Phi^{-1}(]-\infty, r])} \Psi(u)} [$, the functional $\Phi - \lambda \Psi$ is coercive.

Then, for each $\lambda \in \Lambda_r$, the functional $\Phi - \lambda \Psi$ has at least three distinct critical points in X.

Let us recall that for $1 \le i \le n$, $W_0^{1,p_i}(\Omega)$ is compactly embedded in $C^0(\overline{\Omega})$ if $p_i > N$ and that for $1 \le i \le n$, $W^{2,p_i}(\Omega)$ is compactly embedded in $C^0(\overline{\Omega})$ if $p_i > \max\{1, \frac{N}{2}\}$. Put

$$k = \max \left\{ \sup_{u_i \in W^{2,p_i}(\Omega) \cap W_0^{1,p_i}(\Omega) \setminus \{0\}} \frac{\max_{x \in \overline{\Omega}} |u_i(x)|^{p_i}}{\|u_i\|_{p_i}^{p_i}}; \text{ for } 1 \le i \le n \right\}. \tag{2}$$

For $p_i > \max\{1, \frac{N}{2}\}$ for $1 \le i \le n$, since the embedding $W^{2,p_i}(\Omega) \cap W_0^{1,p_i}(\Omega) \hookrightarrow C^0(\overline{\Omega})$ for $1 \le i \le n$ is compact, one has $k < +\infty$. For all $\gamma > 0$ we denote by $K(\gamma)$ the set

$$\left\{ (t_1, \ldots, t_n) \in R^n : \sum_{i=1}^n \frac{|t_i|^{p_i}}{p_i} \le \gamma \right\}. \tag{3}$$

Now, we formulate our main result as follows:

Theorem 2. Assume that there exist a positive constant r and a function $w = (w_1, \ldots, w_n) \in X$ such that

(A1) $\sum_{i=1}^n \frac{\|w_i\|_{p_i}^{p_i}}{p_i} > r$.

(A2) $\int_\Omega \sup_{(t_1, \ldots, t_n) \in K(kr)} F(x, t_1, \ldots, t_n) dx < (r \prod_{i=1}^n p_i) \frac{\int_\Omega F(x, w_1(x), \ldots, w_n(x)) dx}{\sum_{i=1}^n \prod_{j=1, j \ne i}^n p_j \|w_i\|_{p_i}^{p_i}}$,

where $K(kr) = \{(t_1, \ldots, t_n) | \sum_{i=1}^n \frac{|t_i|^{p_i}}{p_i} \le kr\}$ [see (3)] and k is given by (2).

(A3) $km(\Omega) \limsup_{|t_1| \to +\infty, \ldots, |t_n| \to +\infty} \frac{F(x, t_1, \ldots, t_n)}{\sum_{i=1}^n \frac{|t_i|^{p_i}}{p_i}} < \frac{\int_\Omega \sup_{(t_1, \ldots, t_n) \in K(kr)} F(x, t_1, \ldots, t_n) dx}{r}$ uniformly with respect to $x \in \Omega$, where $m(\Omega)$ is the Lebesgue measure of the set Ω.

Then, for each

$$\lambda \in \Lambda_1 := \left] \frac{\frac{\sum_{i=1}^n \prod_{j=1, j \ne i}^n p_j \|w_i\|_{p_i}^{p_i}}{\prod_{i=1}^n p_i}}{\int_\Omega F(x, w_1(x), \ldots, w_n(x)) dx}, \frac{r}{\int_\Omega \sup_{(t_1, \ldots, t_n) \in K(kr)} F(x, t_1, \ldots, t_n) dx} \right[$$

system (1) admits at least three distinct weak solutions in X.

Proof. Put $\Phi(u) = \sum_{i=1}^n \frac{\|u_i\|_{p_i}^{p_i}}{p_i}$ and $\Psi(u) = \int_\Omega F(x, u_1(x), \ldots, u_n(x)) dx$ for each $u = (u_1, \ldots, u_n) \in X$. Of course, Φ is a continuously Gâteaux differentiable and sequentially weakly lower semi-continuous functional whose Gâteaux derivative admits a continuous inverse on X^*, and Ψ is a sequentially weakly upper semicontinuous and continuously Gâteaux differentiable functional whose Gâteaux derivative is compact (for details, see [10, Theorem 2.3]). In particular, the derivatives at the point $u = (u_1, \ldots, u_n) \in X$ are the functionals $\Phi'(u), \Psi'(u) \in X^*$, given by

$$\Phi'(u)(v) = \int_\Omega \sum_{i=1}^n |\Delta u_i(x)|^{p_i-2} \Delta u_i(x) \Delta v_i(x) dx$$

$$+ \alpha_i \int_\Omega \sum_{i=1}^n |\nabla u_i(x)|^{p_i-2} \nabla u_i(x) \nabla v_i(x) dx$$

$$+ \beta_i \int_\Omega \sum_{i=1}^n |u_i(x)|^{p_i-2} u_i(x) v_i(x) dx$$

and

$$\Psi'(u)(v) = \int_\Omega \sum_{i=1}^n F_{u_i}(x, u_1(x), \ldots, u_n(x)) v_i(x) dx$$

for every $v = (v_1, \ldots, v_n) \in X$, respectively. From (A1) we get $0 < r < \Phi(w)$. From (2) for each $(u_1, \ldots, u_n) \in X$, $\sup_{x \in \Omega} |u_i(x)|^{p_i} \le k \|u_i\|_{p_i}^{p_i}$ for $i = 1, \ldots, n$, then

$$\sup_{x \in \Omega} \sum_{i=1}^n \frac{|u_i(x)|^{p_i}}{p_i} \le k \sum_{i=1}^n \frac{\|u_i\|_{p_i}^{p_i}}{p_i} \tag{4}$$

for each $u = (u_1, \ldots, u_n) \in X$, and so using (4), we obtain

$$\Phi^{-1}(]-\infty, r]) = \{(u_1, \ldots, u_n) \in X; \Phi(u_1, \ldots, u_n) \le r\}$$

$$= \left\{ (u_1, \ldots, u_n) \in X; \sum_{i=1}^n \frac{\|u_i\|_{p_i}^{p_i}}{p_i} \le r \right\}$$

$$\subseteq \left\{ (u_1, \ldots, u_n) \in X; \sum_{i=1}^n \frac{|u_i(x)|^{p_i}}{p_i} \le kr \text{ for all } x \in \Omega \right\};$$

therefore, owing to the Assumption (A2), we have

$$\sup_{u \in \Phi^{-1}(]-\infty, r])} \Psi(u) = \sup_{(u_1, \ldots, u_n) \in \Phi^{-1}(]-\infty, r])} \int_\Omega F(x, u_1(x), \ldots, u_n(x)) dx$$

$$\le \int_\Omega \sup_{(t_1, \ldots, t_n) \in K(kr)} F(x, t_1, \ldots, t_n) dx$$

$$< r \frac{\int_\Omega F(x, w_1(x), \ldots, w_n(x)) dx}{\sum_{i=1}^n \frac{\|w_i\|_{p_i}^{p_i}}{p_i}}$$

$$= r \frac{\Psi(w)}{\Phi(w)}.$$

Furthermore, from (A3) there exist two constants τ, $\eta \in \mathbb{R}$ with $\tau < \dfrac{\int_\Omega \sup_{(t_1,\ldots,t_n)\in K(kr)} F(x,t_1,\ldots,t_n)dx}{r}$ such that

$$km(\Omega)F(x,t_1,\ldots,t_n) \leq \tau \sum_{i=1}^n \frac{|t_i|^{p_i}}{p_i} + \eta \quad \text{for all } x \in \Omega \text{ and for all } (t_1,\ldots,t_n) \in R^n.$$

Fix $(u_1,\ldots,u_n) \in X$. Then

$$F(x,u_1(x),\ldots,u_n(x)) \leq \frac{1}{km(\Omega)} \left(\tau \sum_{i=1}^n \frac{|u_i(x)|^{p_i}}{p_i} + \eta \right) \quad \text{for all } x \in \Omega. \quad (5)$$

Now, in order to prove the coercivity of the functional $\Phi - \lambda\Psi$, first we assume that $\tau > 0$. So, for any fixed $\lambda \in \Lambda_1$, from (4) and (5) we have

$$\Phi(u) - \lambda\Psi(u) = \sum_{i=1}^n \frac{\|u_i\|_{p_i}^{p_i}}{p_i} - \lambda \int_\Omega F(x,u_1(x),\ldots,u_n(x))dx$$

$$\geq \sum_{i=1}^n \frac{\|u_i\|_{p_i}^{p_i}}{p_i} - \frac{\lambda\tau}{km(\Omega)} \left(\sum_{i=1}^n \frac{1}{p_i} \int_\Omega |u_i(x)|^{p_i}dx \right) - \frac{\lambda\eta}{k}$$

$$\geq \sum_{i=1}^n \frac{\|u_i\|_{p_i}^{p_i}}{p_i} - \frac{\lambda\tau}{km(\Omega)} \left(km(\Omega) \sum_{i=1}^n \frac{\|u_i\|_{p_i}^{p_i}}{p_i} \right) - \frac{\lambda\eta}{k}$$

$$= \sum_{i=1}^n \frac{\|u_i\|_{p_i}^{p_i}}{p_i} - \lambda\tau \sum_{i=1}^n \frac{\|u_i\|_{p_i}^{p_i}}{p_i} - \frac{\lambda\eta}{k}$$

$$\geq \left(1 - \frac{\tau r}{\int_\Omega \sup_{(t_1,\ldots,t_n)\in K(kr)} F(x,t_1,\ldots,t_n)dx} \right) \sum_{i=1}^n \frac{\|u_i\|_{p_i}^{p_i}}{p_i} - \frac{\lambda\eta}{k},$$

and thus,

$$\lim_{\|u\|\to+\infty.} (\Phi(u) - \lambda\Psi(u)) = +\infty,$$

On the other end, if $\tau \leq 0$, clearly, we get $\lim_{\|u\|\to+\infty}(\Phi(u) - \lambda\Psi(u)) = +\infty$. Both cases lead to the coercivity of functional $\Phi - \lambda\Psi$. Therefore, we can apply Theorem 1. Taking into account that the weak solutions of system (1) are exactly the solutions of equation $\Phi'(u_1,\ldots,u_n) - \lambda\Psi'(u_1,\ldots,u_n) = 0$, it follows the conclusion. □

Now we want to present a verifiable consequence of the main result where the test function w is specified.

Following the construction given in [13], fix $x^0 \in \Omega$ and pick r_1, r_2 with $0 < r_1 < r_2$ such that $S(x^0, r_1) \subset S(x^0, r_2) \subseteq \Omega$, where $S(x^0, r_i)$ denotes the ball with center at x^0 and radius of r_i for $i = 1, 2$.

Put

$$\omega_i = \omega_i(N, p_i, r_1, r_2) = \frac{2\pi^{\frac{N}{2}}}{\Gamma(\frac{N}{2})} \int_{r_1}^{r_2} |(N+2)\xi^2 - (N+1)(r_1+r_2)\xi + Nr_1r_2|^{p_i}\xi^{N-1}d\xi \tag{6}$$

for $1 \le i \le n$,

$$\vartheta_i = \vartheta_i(N, p_i, r_1, r_2) = \int_{S(x^0,r_2)\setminus S(x^0,r_1)} |d(x,x^0)(l^2 - (r_1+r_2)l + r_1r_2)|^{p_i} dx \tag{7}$$

for $1 \le i \le n$, where $l = \sqrt{\sum_{i=1}^{N}(x_i - x_i^0)^2}$ and $d(x,x^0) = \sum_{i=1}^{N}(x_i - x_i^0)$, and

$$\kappa_i = \kappa_i(N, p_i, r_1, r_2) = \frac{2\pi^{\frac{N}{2}}}{(12)^{p_i}\Gamma(\frac{N}{2})}\Big(\frac{r_1^N}{N} + \int_{r_1}^{r_2} \Big|\Big(3(\xi^4 - r_2^4) - 4(r_1+r_2)(\xi^3 - r_2^3)$$
$$+ 6r_1r_2(\xi^2 - r_2^2)\Big)\Big|^{p_i}\xi^{N-1}d\xi\Big) \tag{8}$$

for $1 \le i \le n$.

Corollary 1. *Assume that there exist two positive constant v and τ with*

$$\sum_{i=1}^{n}\left(\frac{(12\tau)^{p_i}\prod_{j=1, j\neq i}^{n} p_j}{(r_2 - r_1)^{3p_i}(r_1 + r_2)^{p_i}}(\omega_i + \alpha_i\vartheta_i + \beta_i\kappa_i)\right) > \frac{v}{k}$$

such that

(B1) $F(x, t_1, \ldots, t_n) \ge 0$ for each $(x, t_1, \ldots, t_n) \in (\Omega \setminus S(x^0, r_1)) \times [0, \tau]^n$.

(B2) $\int_\Omega \sup_{(t_1,\ldots,t_n)\in K(\frac{v}{\prod_{i=1}^{n} p_i})} F(x, t_1, \ldots, t_n)dx < \frac{v}{k}\dfrac{\int_{S(x^0,r_1)} F(x,\tau,\ldots,\tau)dx}{\left(\sum_{i=1}^{n}\frac{(12\tau)^{p_i}\prod_{j=1, j\neq i}^{n} p_j}{(r_2-r_1)^{3p_i}(r_1+r_2)^{p_i}}(\omega_i+\alpha_i\vartheta_i+\beta_i\kappa_i)\right)}$,

where ω_i, ϑ_i, and κ_i are given by (6), (7), and (8), respectively, and

$$K\left(\tfrac{v}{\prod_{i=1}^{n} p_i}\right) = \{(t_1,\ldots,t_n)\mid \sum_{i=1}^{n}\tfrac{|t_i|^{p_i}}{p_i} \le \tfrac{v}{\prod_{i=1}^{n} p_i}\} \text{ [see (3)].}$$

(B3) $km(\Omega)\limsup_{|t_1|\to+\infty,\ldots,|t_n|\to+\infty} \dfrac{F(x,t_1,\ldots,t_n)}{\sum_{i=1}^{n}\frac{|t_i|^{p_i}}{p_i}} < \dfrac{\int_\Omega \sup_{(t_1,\ldots,t_n)\in K(\frac{v}{\prod_{i=1}^{n} p_i})} F(x,t_1,\ldots,t_n)dx}{\frac{v}{k\prod_{i=1}^{n} p_i}}$

uniformly with respect to $x \in \Omega$. Then, for each

$$\lambda \in \Lambda_2 := \left]\frac{\frac{\left(\sum_{i=1}^{n}\frac{(12\tau)^{p_i}\prod_{j=1,j\neq i}^{n} p_j}{(r_2-r_1)^{3p_i}(r_1+r_2)^{p_i}}(\omega_i+\alpha_i\vartheta_i+\beta_i\kappa_i)\right)}{\prod_{i=1}^{n} p_i}}{\int_{S(x^0,r_1)} F(x,\tau,\ldots,\tau)dx}, \frac{\frac{v}{k\prod_{i=1}^{n} p_i}}{\int_\Omega \sup_{(t_1,\ldots,t_n)\in K(\frac{v}{\prod_{i=1}^{n} p_i})} F(x,t_1,\ldots,t_n)dx}\right[,$$

system (1) *has at least three weak solutions in X.*

Proof. We put $w(x) = (w_1(x), \ldots, w_n(x))$ such that for $1 \leq i \leq n$,

$$
w_i(x) = \begin{cases} 0 & \text{if } x \in \Omega \setminus S(x^0, r_2) \\ \frac{\tau\left(3(l^4 - r_2^4) - 4(r_1 + r_2)(l^3 - r_2^3) + 6r_1 r_2(l^2 - r_2^2)\right)}{(r_2 - r_1)^3(r_1 + r_2)} & \text{if } x \in S(x^0, r_2) \setminus S(x^0, r_1) \\ \tau & \text{if } x \in S(x^0, r_1) \end{cases} \quad (9)
$$

and $r = \frac{v}{k \prod_{i=1}^n p_i}$. We have

$$
\frac{\partial w_i(x)}{\partial x_i} = \begin{cases} 0 & \text{if } x \in \Omega \setminus S(x^0, r_2) \cup S(x^0, r_1) \\ \frac{12\tau\left(l^2(x_i - x_i^0) - (r_1 + r_2)l(x_i - x_i^0) + r_1 r_2(x_i - x_i^0)\right)}{(r_2 - r_1)^3(r_1 + r_2)} & \text{if } x \in S(x^0, r_2) \setminus S(x^0, r_1) \end{cases}
$$

and

$$
\frac{\partial^2 w_i(x)}{\partial x_i^2} = \begin{cases} 0 & \text{if } x \in \Omega \setminus S(x^0, r_2) \cup S(x^0, r_1) \\ \frac{12\tau\left(r_1 r_2 + (2l - r_1 - r_2)(x_i - x_i^0)^2/l - (r_2 + r_1 - l)l\right)}{(r_2 - r_1)^3(r_1 + r_2)} & \text{if } x \in S(x^0, r_2) \setminus S(x^0, r_1) \end{cases}
$$

and so

$$
\sum_{i=1}^N \frac{\partial w_i(x)}{\partial x_i} = \begin{cases} 0 & \text{if } x \in \Omega \setminus S(x^0, r_2) \cup S(x^0, r_1) \\ \frac{12\tau\left(l^2 d(x, x^0) - (r_1 + r_2)ld(x, x^0) + r_1 r_2 d(x, x^0)\right)}{(r_2 - r_1)^3(r_1 + r_2)} & \text{if } x \in S(x^0, r_2) \setminus S(x^0, r_1) \end{cases}
$$

and

$$
\sum_{i=1}^N \frac{\partial^2 w_i(x)}{\partial x_i^2} = \begin{cases} 0 & \text{if } x \in \Omega \setminus S(x^0, r_2) \cup S(x^0, r_1) \\ \frac{12\tau\left((N+2)l^2 - (N+1)(r_1 + r_2)l + Nr_1 r_2\right)}{(r_2 - r_1)^3(r_1 + r_2)} & \text{if } x \in S(x^0, r_2) \setminus S(x^0, r_1). \end{cases}
$$

It is easy to see that $w = (w_1, \ldots, w_n) \in X$ and, in particular, one has

$$
\|w_i\|_{p_i}^{p_i} = \frac{(12\tau)^{p_i}}{(r_2 - r_1)^{3p_i}(r_1 + r_2)^{p_i}}(\omega_i + \alpha_i \vartheta_i + \beta_i \kappa_i) \quad (10)
$$

for $1 \leq i \leq n$. However, taking into account that $\sum_{i=1}^n \prod_{j=1, j\neq i}^n p_j(\sigma_i + \alpha_i \theta_i + \beta_i \rho_i)\tau^{p_i} > \frac{v}{k}$, we see that $\sum_{i=1}^n \frac{\|w_i\|_{p_i}^{p_i}}{p_i} > r$, which is Assumption (A1).
Since $0 \leq w_i(x) \leq \tau$ for each $x \in \Omega$ for $1 \leq i \leq n$, the condition (B1) ensures that

$$
\int_{\Omega \setminus S(x^0, r_2)} F(x, w_1(x), \ldots, w_n(x))dx + \int_{S(x^0, r_2) \setminus S(x^0, r_1)} F(x, w_1(x), \ldots, w_n(x))dx \geq 0.
$$

$$(11)$$

Moreover, from (B2), (10), and (11), we have

$$\int_{\Omega} \sup_{(t_1,\ldots,t_n)\in K(\frac{\nu}{\prod_{i=1}^n p_i})} F(x,t_1,\ldots,t_n)dx < \frac{\nu \int_{S(x^0,r_1)} F(x,\tau,\ldots,\tau)dx}{k \sum_{i=1}^n \left(\frac{(12\tau)^{p_i}\prod_{j=1,j\neq i}^n p_j}{(r_2-r_1)^{3p_i}(r_1+r_2)^{p_i}}(\omega_i+\alpha_i\vartheta_i+\beta_i\kappa_i) \right)}$$

$$\leq \frac{\nu}{k} \frac{\int_{\Omega} F(x,w_1(x),\ldots,w_n(x))dx}{\sum_{i=1}^n \prod_{j=1,j\neq i}^n p_j||w_i||_{p_i}^{p_i}}$$

$$= (r\prod_{i=1}^n p_i)\frac{\int_{\Omega} F(x,w_1(x),\ldots,w_n(x))dx}{\sum_{i=1}^n \prod_{j=1,j\neq i}^n p_j||w_i||_{p_i}^{p_i}},$$

namely, the Assumption (A2) is satisfied. Also, from (B3) we arrive at (A3). Hence, using Theorem 2, we have the desired conclusion. □

Finally, we point out the following remarkable consequence of Corollary 1.

Corollary 2. *Let $F : R^n \to R$ be a C^1-function such that $F(0,\ldots,0) = 0$. Assume that there exist two positive constant ν and τ with*

$$\sum_{i=1}^n \left(\frac{(12\tau)^{p_i}\prod_{j=1,j\neq i}^n p_j}{(r_2-r_1)^{3p_i}(r_1+r_2)^{p_i}}(\omega_i+\alpha_i\vartheta_i+\beta_i\kappa_i) \right) > \frac{\nu}{k}$$

such that

(C1) $F(t_1,\ldots,t_n) \geq 0$ *for each* $(t_1,\ldots,t_n) \in [0,\tau]^n$.

(C2) $m(\Omega)\max_{(t_1,\ldots,t_n)\in K(\frac{\nu}{\prod_{i=1}^n p_i})} F(t_1,\ldots,t_n) < \frac{\nu}{k} \frac{r_1^N \frac{\pi^{N/2}}{\Gamma(1+N/2)} F(\tau,\ldots,\tau)}{\left(\sum_{i=1}^n \frac{(12\tau)^{p_i}\prod_{j=1,j\neq i}^n p_j}{(r_2-r_1)^{3p_i}(r_1+r_2)^{p_i}}(\omega_i+\alpha_i\vartheta_i+\beta_i\kappa_i) \right)}.$

(C3) $km(\Omega)\limsup_{|t_1|\to+\infty,\ldots,|t_n|\to+\infty} \frac{F(t_1,\ldots,t_n)}{\sum_{i=1}^n \frac{|t_i|^{p_i}}{p_i}} < \frac{m(\Omega)\max_{(t_1,\ldots,t_n)\in K(\frac{\nu}{\prod_{i=1}^n p_i})} F(t_1,\ldots,t_n)}{\frac{\nu}{k\prod_{i=1}^n p_i}}$

uniformly with respect to $x \in \Omega$. Then, for every

$$\lambda \in \Lambda_3 := \left] \frac{\left(\sum_{i=1}^n \frac{(12\tau)^{p_i}\prod_{j=1,j\neq i}^n p_j}{(r_2-r_1)^{3p_i}(r_1+r_2)^{p_i}}(\omega_i+\alpha_i\vartheta_i+\beta_i\kappa_i) \right)}{\prod_{i=1}^n p_i} \middle/ r_1^N \frac{\pi^{N/2}}{\Gamma(1+N/2)}F(\tau,\ldots,\tau), \frac{\frac{\nu}{k\prod_{i=1}^n p_i}}{m(\Omega)\max_{(t_1,\ldots,t_n)\in K(\frac{\nu}{\prod_{i=1}^n p_i})} F(t_1,\ldots,t_n)} \right[$$

systems

$$\begin{cases} \Delta(|\Delta u_i|^{p_i-2}\Delta u_i) - \alpha_i\Delta_{p_i}u_i + \beta_i|u_i|^{p_i-2}u_i = \lambda F_{u_i}(u_1,\ldots,u_n) & \text{in } \Omega, \\ u_i = \Delta u_i = 0 & \text{on } \partial\Omega \end{cases}$$

for $1 \leq i \leq n$ admit at least three weak solutions in X.

Proof. Set $F(x,t_1,\ldots,t_n) = F(t_1,\ldots,t_n)$ for all $x \in \Omega$ and $t_i \in R$ for $1 \le i \le n$. Clearly, from (C1) and (C3), we arrive at (B1) and (B3), respectively. In particular, since $m(S(x^0,r_1)) = r_1^N \frac{\pi^{N/2}}{\Gamma(1+N/2)}$, Assumption (C2) follows that Assumption (B2) is fulfilled. So, we have the conclusion by using Corollary 2. $\square$

Acknowledgement This research was in part supported by a grant from IPM (No. 90470020).

References

1. Afrouzi, G.A., Heidarkhani, S., O'Regan, D.: Existence of three solutions for a doubly eigenvalue fourth-order boundary value problem Taiwanese. J. Math. **15**(1), 201–210 (2011)
2. Bonanno, G., Di Bella, B.: A boundary value problem for fourth-order elastic beam equations. J. Math. Anal. Appl. **343**, 1166–1176 (2010)
3. Bonanno, G., Di Bella, B.: A fourth-order boundary value problem for a Sturm-Liouville type equation. Appl. Math. Comput. **217**, 3635–3640 (2010)
4. Bonanno, G., Di Bella, B.: Infinitely many solutions for a fourth-order elastic beam equation. Nonlinear Diff. Equat. Appl. NoDEA **18**(3), 357–368 (2011). doi:10.1007/s00030-011-0099-0
5. Bonanno, G., Marano, S.A.: On the structure of the critical set of non-differentiable functions with a weak compactness condition. Appl. Anal. **89**, 1–10 (2010)
6. Bonanno, G., Di Bella, B., O'Regan, D.: Non-trivial solutions for nonlinear fourth-order elastic beam equations. Comput. Math. Appl. **62**(4), 1862–1869 (2011). doi:10.1016/j.camwa.2011.06.029
7. Candito, P., Livrea, R.: Infinitely many solutions for a nonlinear Navier boundary value problem involving the p-biharmonic. Studia Univ. Babeş-Bolyai Math. **LV**(4), 41–51 (2010)
8. Candito, P., Molica Bisci, G.: Multiple solutions for a Navier boundary value problem involving the p-biharmonic. Discrete Contin. Dyn. Syst. Ser. S **5**(4) (2012). doi:10.3934/dsdss.2012.5.741
9. Graef, J.R., Heidarkhani, S., Kong, L.: Multiple solutions for a class of $(p_1,\ldots,p_n)$-biharmonic systems (preprint)
10. Heidarkhani, S., Tian, Y., Tang, C.-L.: Existence of three solutions for a class of $(p_1,\ldots,p_n)$-biharmonic systems with Navier boundary conditions. Ann. Polon. Math. (to appear)
11. Lazer, A.C., McKenna, P.J.: Large amplitude periodic oscillations in suspension bridges: Some new connections with nonlinear analysis. SIAM Rev. **32**, 537–578 (1990)
12. Li, L., Tang, C.-L.: Existence of three solutions for (p,q)-biharmonic systems. Nonlinear Anal. **73**, 796–805 (2010)
13. Li, C., Tang, C.-L.: Three solutions for a Navier boundary value problem involving the p-biharmonic. Nonlinear Anal. **72**,1339–1347 (2010)
14. Liu, H., Su, N.: Existence of three solutions for a p-biharmonic problem. Dyn. Contin. Discrete Impuls. Syst. Ser. A Math. Anal. **15**(3), 445–452 (2008)
15. Ricceri, B.: Existence of three solutions for a class of elliptic eigenvalue problem. Math. Comput. Model. **32**, 1485–1494 (2000)
16. Ricceri, B.: On a three critical points theorem. Arch. Math. (Basel) **75**, 220–226 (2000)
17. Ricceri, B.: A three critical points theorem revisited. Nonlinear Anal. **70**, 3084–3089 (2009)
18. Zeidler, E.: Nonlinear functional analysis and its applications, vol. II. Springer, Berlin (1985)

The Fučík Spectrum: Exploring the Bridge Between Discrete and Continuous World

Gabriela Holubová and Petr Nečesal

Abstract In this paper, we would like to point out some similarities of interesting structures of Fučík spectra for continuous and discrete operators. We propose a simple algorithm that allows us to complete the reconstruction of the Fučík spectrum in the case of small order matrices. Finally, we point out some properties of the Fučík spectrum in general unifying terms and concepts.

Keywords Fucik spectrum • Scalar differential operators • Matrices • Hilbert lattice • Convex cone • Inadmissible areas

1 Introduction

In this paper, we review the Fučík spectrum of several linear operators L, i.e., the set

$$\Sigma(L) := \left\{ (\alpha, \beta) \in \mathbb{R}^2 : Lu = \alpha u^+ - \beta u^- \text{ has a nontrivial solution} \right\}, \quad (1)$$

where u^+ and u^- denote the positive and the negative parts of u. Our goal is to figure out some interesting common phenomena concerning the structure of the Fučík spectrum for operators L defined on both finite and infinite dimensional spaces. Let us note that for $\alpha = \beta$, the problem $Lu = \alpha u^+ - \beta u^-$ reads as the linear eigenvalue problem $Lu = \lambda u$ with $\lambda = \alpha = \beta$, thus $\left\{ (\lambda, \lambda) \in \mathbb{R}^2 : \lambda \in \sigma(L) \right\} \subset \Sigma(L)$, where $\sigma(L)$ denotes the point spectrum of L. The Fučík spectrum was introduced by Fučík [2] and Dancer [1] in the study of the solvability of sublinear boundary value problems. According to our best knowledge, R. Švarc was the first one who investigated the Fučík spectrum for matrices, recall here at least [10, 11]. Among other results, R. Švarc shows in details how to reconstruct the nontrivial Fučík spectrum for a 4×4 matrix.

G. Holubová • P. Nečesal
University of West Bohemia, Univerzitní 22, 306 14 Plzeň, Czech Republic
e-mail: hasgabriela@kma.zcu.cz; haspnecesal@kma.zcu.cz

S. Pinelas et al. (eds.), *Differential and Difference Equations with Applications*, Springer Proceedings in Mathematics & Statistics 47, DOI 10.1007/978-1-4614-7333-6_36,
© Springer Science+Business Media New York 2013

421

2 The Fučík Spectrum for Scalar Differential Operators

In this section, we recall the Fučík spectra of three particular differential operators
on $L^2(0,\pi)$. For any $u = u(x) \in L^2(0,\pi)$, we define $u^+(x) = \max\{u(x),0\}$ and
$u^-(x) = \max\{-u(x),0\}$ for a.e. $x \in (0,\pi)$.

Example 1. Let us consider the following differential operator with nonlocal
boundary condition

$$L^\eta u(x) := -u''(x),$$

$$\mathrm{dom}(L^\eta) := \left\{ u \in H^2(0,\pi) : u(0) = 0,\ (1-\eta)u(\pi) + \eta \int_0^\pi u(x)\,\mathrm{d}x = 0 \right\}.$$

For $\eta = 0$, the operator L^η stands for the self-adjoint differential operator with
Dirichlet boundary conditions, for which the corresponding Fučík spectrum $\Sigma(L^\eta)$
is well known and is given as the union of countably many curves given by
$\frac{n}{\sqrt{\alpha}} + \frac{n}{\sqrt{\beta}} = 1$, $\frac{n-1}{\sqrt{\alpha}} + \frac{n}{\sqrt{\beta}} = 1$, $\frac{n}{\sqrt{\alpha}} + \frac{n-1}{\sqrt{\beta}} = 1$, $n \in \mathbb{N}$ (see Fig. 1, left). For $\eta > 0$, the
operator L^η is non-self-adjoint, the Fučík spectrum consists of two smooth curves
$\mathcal{C}^+$ and $\mathcal{C}^-$ such that $\mathcal{C}^+ \cap \mathcal{C}^- = \left\{(\lambda,\lambda) : \mathbb{R}^2 : \lambda \in \sigma(L^\eta)\right\}$ (see [8] for detailed
description of these curves).

Example 2. Let us denote by $\mathcal{P} := \{0,\eta,\xi,\pi\}$ the partition of the interval $[0,\pi]$
and let us define the following four-point differential operator

$$L_4 u(x) := -u''(x), \quad \mathrm{dom}(L_4) := \left\{u \in H^2(\mathcal{P}) : B_i(u) = 0,\ i = 1,\dots,6\right\},$$

where (note that $u = (u_1, u_2, u_3)$)

$$B_1(u) := u_2(\xi) - u_3(\xi),\ B_3(u) := u_2'(\xi) - u_3'(\xi),\ B_5(u) := u_1'(0) - u_3'(\xi),$$
$$B_2(u) := u_1(\eta) - u_2(\eta),\ B_4(u) := u_1'(\eta) - u_2'(\eta),\ B_6(u) := u_1(\eta) - u_3(\pi).$$

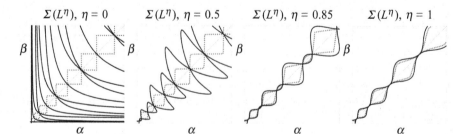

$\Sigma(L^\eta)$, $\eta = 0$ $\qquad$ $\Sigma(L^\eta)$, $\eta = 0.5$ $\qquad$ $\Sigma(L^\eta)$, $\eta = 0.85$ $\qquad$ $\Sigma(L^\eta)$, $\eta = 1$

Fig. 1 The Fučík spectrum of L^η for various settings of η

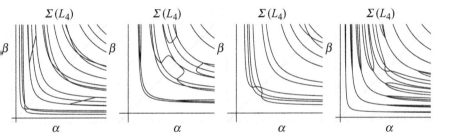

Fig. 2 The Fučík spectrum of L_4 for various settings of ξ and η

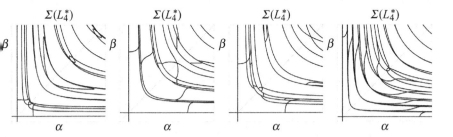

Fig. 3 The Fučík spectrum of L_4^* for various settings of ξ and η

The complete explicit analytic description of the Fučík spectrum of L_4 is provided in [3,4] (see Fig. 2). The adjoint operator L_4^* of L_4 is a four-point differential operator

$$L_4^* u(x) = -u''(x), \quad \mathrm{dom}(L_4^*) = \left\{ u \in H^2(\mathcal{P}) : B_i^*(u) = 0,\ i = 1,\ldots,6 \right\},$$

where the adjoint boundary terms are given by

$$
\begin{aligned}
B_1^*(u) &= u_2(\xi) - u_3(\xi) + u_1(0), & B_4^*(u) &= u_1'(\eta) - u_2'(\eta) + u_3'(\pi), \\
B_2^*(u) &= u_1(\eta) - u_2(\eta), & B_5^*(u) &= u_1'(0), \\
B_3^*(u) &= u_2'(\xi) - u_3'(\xi), & B_6^*(u) &= u_3(\pi).
\end{aligned}
$$

Hence, the Fučík spectrum problems for both operators L_4 and L_4^* read as

$$
\begin{cases}
u'' + \alpha u^+ - \beta u^- = 0, \\
u'(0) = u'(\xi), \\
u(\eta) = u(\pi),
\end{cases}
\qquad
\begin{cases}
u'' + \alpha u^+ - \beta u^- = 0, \\
u'(0) = u(\pi) = 0, \quad u(\eta) = u(\pi), \\
u(\eta-) = u(\eta+), \quad u'(\eta-) = u'(\eta+) - u(\pi), \\
u(\xi-) = u(\xi+) + u(0), \quad u'(\xi-) = u'(\xi+).
\end{cases}
$$

Figure 3 contains examples of the Fučík spectrum of L_4^* for the same settings of inner points ξ and η as for L_4 in Fig. 2. Notice that the structures of $\Sigma(L_4)$ and $\Sigma(L_4^*)$ differ essentially in spite of the fact that $\sigma(L_4) = \sigma(L_4^*)$. For more details see [5].

3 The Fučík Spectrum for Matrices

In this section, we consider L as a linear operator on the finite dimensional space $\mathbb{R}^n$, $n \in \mathbb{N}$, i.e., L is represented by a real $n \times n$ matrix A. For any $u \in \mathbb{R}^n$, we define $\mathbf{u}^{\pm} := [u_1^{\pm}, \ldots, u_n^{\pm}]^t$ with $u_i^{+} := \max\{u_i, 0\}$ and $u_i^{-} := \max\{-u_i, 0\}$ for $i = 1, \ldots, n$. Let S be the set of all n-dimensional vectors s such that $s_i = \pm 1$, $i = 1, \ldots, n$. Then we have

$$\Sigma(A) = \bigcup_{s \in S} \Sigma_s(A), \qquad \Sigma(A) \subset \Theta(A) := \bigcup_{s \in S} \Theta_s(A),$$

$$\Sigma_s(A) := \big\{(\alpha, \beta) \in \mathbb{R}^2 : [A - \alpha \chi(s^{+}) - \beta \chi(s^{-})]u = 0 \text{ has a nontrivial} \\ \text{solution } u \text{ such that } u_i s_i \geq 0 \text{ for } i = 1, \ldots, n\},$$

$$\Theta_s(A) := \big\{(\alpha, \beta) \in \mathbb{R}^2 : [A - \alpha \chi(s^{+}) - \beta \chi(s^{-})]u = 0 \text{ has a nontrivial} \\ \text{solution } u\},$$

where $\chi(s^{+})$ and $\chi(s^{-})$ are characteristic diagonal matrices, which have only nonzero entries equal to 1 if and only if the corresponding component of s is $+1$ or -1, respectively.

In the case of 2×2 matrices, we have $\Theta(A) = \Theta_{++} \cup \Theta_{+-} \cup \Theta_{-+} \cup \Theta_{--}$, where

$$\begin{aligned}
\Theta_{++} &= \big\{(\alpha, \beta) \in \mathbb{R}^2 : (a - \alpha)(d - \alpha) = bc\}, \\
\Theta_{+-} &= \big\{(\alpha, \beta) \in \mathbb{R}^2 : (a - \alpha)(d - \beta) = bc\}, \\
\Theta_{-+} &= \big\{(\alpha, \beta) \in \mathbb{R}^2 : (a - \beta)(d - \alpha) = bc\}, \\
\Theta_{--} &= \big\{(\alpha, \beta) \in \mathbb{R}^2 : (a - \beta)(d - \beta) = bc\},
\end{aligned} \qquad A = \begin{bmatrix} a & b \\ c & d \end{bmatrix},$$

and there are ten qualitatively different types of the Fučík spectrum $\Sigma(A)$ (see Fig. 4). In the case of a general $n \times n$ matrix A, we use the following procedure. For fixed $s \in S$ and $\alpha \in \mathbb{R}$, we look for all $\beta \in \mathbb{R}$ such that the generalized eigenvalue

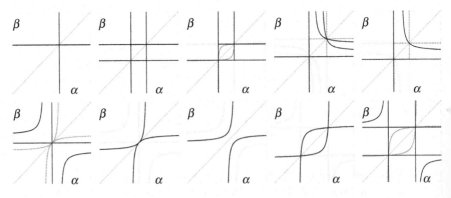

Fig. 4 The Fučík spectrum $\Sigma(A)$ (*black*) and the set $\Theta(A)$ (*gray*) for 2×2 matrices

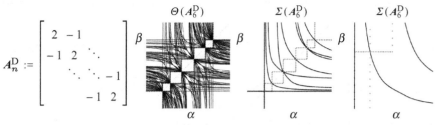

$$A_n^D := \begin{bmatrix} 2 & -1 & & \\ -1 & 2 & \ddots & \\ & \ddots & \ddots & -1 \\ & & -1 & 2 \end{bmatrix}$$

Fig. 5 The set $\Theta(A_6^D)$ and the Fučík spectrum $\Sigma(A_6^D) \subset \Theta(A_6^D)$

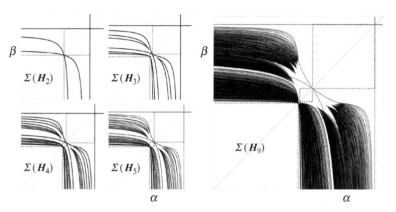

Fig. 6 The Fučík spectrum of the $n \times n$ Hilbert matrices H_n for $n = 2,3,4,5,9$

problem $[A - \alpha\chi(s^+)]u = \beta\chi(s^-)u$ has a nontrivial solution u with $u_i s_i \geq 0$ for $i = 1,\dots,n$. The computational complexity of this procedure is exponential, we have to solve 2^n generalized eigenvalue problems.

Example 3. Let us consider the three-diagonal matrix A_n^D, which is the discrete equivalent of the continuous Dirichlet differential operator. Its set $\Theta(A_n^D)$ and the Fučík spectrum $\Sigma(A_n^D)$ are illustrated in Fig. 5 for $n = 6$. Let us note that the Fučík curves lose some properties typical for the continuous case, e.g., convexity (for details, see [9]).

Example 4. The Hilbert matrix H_n with entries $H_{ij} = \frac{1}{i+j-1}$ is an example of the operator, for which the Fučík spectrum $\Sigma(H_n) \subset \Theta(H_n)$ has almost the same rich structure as the set $\Theta(H_n)$ (see Fig. 6).

Let us close this section by simple examples which demonstrate that it is quite complicated to obtain the global description of the Fučík spectrum structure even in the case of small matrices.[1]

---

[1] There are more than 300 qualitatively different patterns of the Fučík spectrum for 3×3 matrices (cf. the simple situation of 2×2 matrices in Fig. 4).

$$A_5 = \begin{bmatrix} 1\ 1\ \dots\ 1 \\ 2\ 2\ \dots\ 2 \\ \vdots\ \vdots\ \dots\ \vdots \\ 5\ 5\ \dots\ 5 \end{bmatrix}$$

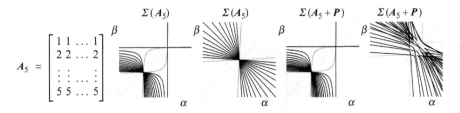

Fig. 7 The Fučík spectrum of A_5 and of the perturbed matrix $A_5 + P$

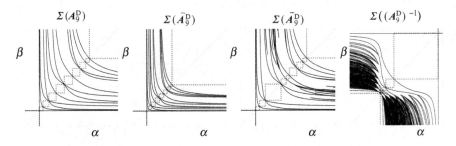

Fig. 8 The change of the Fučík spectrum of A_9^D if we modify the eigenvalues of A_9^D and leave their eigenvectors

Example 5. Let us consider the 5×5 matrix A_5 with entries $A_{ij} = i$, which has two eigenvalues 0 and 15 and the eigenvalue 0 has the geometric multiplicity 4. The Fučík spectrum consists of 16 curves, two of them are straight lines going through the point (15, 15) and the remaining 14 curves emanate from the origin $(0,0)$. If we perturb A_5 by a diagonal matrix $P = \mathrm{diag}(0.01, 0.02, 0.03, 0.04, 0.05)$, then the structure of the Fučík spectrum changes qualitatively. The matrix $A_5 + P$ has five distinct simple eigenvalues which give arise to 10 curves going through the diagonal $\alpha = \beta$. Moreover, there are other 20 curves of the Fučík spectrum $\Sigma(A_5 + P)$ which do not intersect the diagonal. Both Fučík spectra $\Sigma(A_5)$ and $\Sigma(A_5 + P)$ are depicted in Fig. 7.

Example 6. Let us consider the Jordan decomposition of the matrix $A_n^D = \mathbf{V}J\mathbf{V}^{-1}$, where J is the Jordan matrix (see Example 3 for the definition of A_n^D). If we multiply the Jordan matrix J by $\varepsilon > 0$, it is straightforward to predict how the Fučík spectrum $\Sigma(A_n^D)$ changes to $\Sigma(\tilde{A}_n^D)$, where $\tilde{A}_n^D = V(\varepsilon J)V^{-1}$. Such a modification of J changes only the scale of the axes (see Fig. 8 for $n = 9$). Similarly, the change of J to $J + \varepsilon I$, $\varepsilon \neq 0$, causes only the shift of the origin of the axes. Qualitatively more complicated and unpredictable situation appears if we shift only one eigenvalue of J or if we take the inverse matrix J^{-1}. See Fig. 8 for the Fučík spectrum of $\bar{A}_9^D = V(J + \mathrm{diag}(0,0,0,\varepsilon,0,\dots,0))V^{-1}$, $\varepsilon \neq 0$, and the Fučík spectrum of the inverse matrix $(A_9^D)^{-1} = VJ^{-1}V^{-1}$. In both cases, the Fučík spectra $\Sigma(\bar{A}_9^D)$ and $\Sigma((A_9^D)^{-1})$ significantly differ from $\Sigma(A_9^D)$ in spite of the fact that there is an obvious transformation between the point spectra $\sigma(A_9^D)$, $\sigma(\bar{A}_9^D)$, and $\sigma((A_9^D)^{-1})$.

4 Properties of the Fučík Spectrum in General

We can link the previous continuous and discrete examples and treat the Fučík spectrum in a general setting. For this purpose, let H be an arbitrary Hilbert space over the field of real numbers $\mathbb{R}$ *ordered* by a *closed convex cone* K such that $(H, \langle \cdot, \cdot \rangle, K)$ is a *Hilbert lattice* (see [7]). Thus, for any element $u \in H$, we can define its positive and negative parts by $u^+ := P_K(u)$ and $u^- := -P_{-K}(u)$, where P_K denotes the orthogonal projection onto K. Further, we consider an arbitrary *linear operator* $L : \text{dom}(L) \subset H \to H$ with $\text{dom}(L)$ *dense in* H. In this setting, we can again study the structure of the Fučík spectrum $\Sigma(L)$ defined by (1).[2] In particular, for any general operator L, it is possible to detect areas, where the Fučík spectrum $\Sigma(L)$ cannot be located (for more details and other properties of $\Sigma(L)$, see [6]):

1. Let $\lambda \in \mathbb{R}$ be the *principal* eigenvalue of the adjoint operator L^*, i.e., $L^*v = \lambda v$ with $v \in \text{Int}(K)$. Then

$$\{(\alpha, \beta) \in \mathbb{R}^2 : (\alpha - \lambda)(\beta - \lambda) < 0\} \cap \Sigma(L) = \emptyset.$$

2. Let $\varepsilon \notin \sigma(L)$ and let us denote $d(\varepsilon) := \left\| (L - \varepsilon I)^{-1} \right\|^{-1}$. Then

$$\{(\varepsilon - td(\varepsilon), \varepsilon + td(\varepsilon)) \in \mathbb{R}^2 : t \in (-1, 1)\} \cap \Sigma(L) = \emptyset.$$

Illustration of the *inadmissible areas* given by the above properties is provided in Figs. 1 and 4–9 (see the regions bounded by orange dashed curves).

Acknowledgement The authors were supported by the Ministry of Education, Youth and Sports of the Czech Republic, Research Plan MSM4977751301, by grant ME09109 (program KONTAKT) and partially by the European Regional Development Fund (ERDF), project "NTIS—New Technologies for Information Society", European Centre of Excellence, CZ.1.05/1.1.00/02.0090.

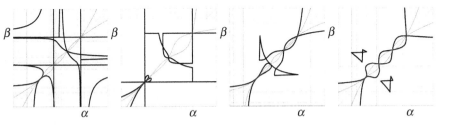

Fig. 9 Examples of Fučík spectra with strange structures for 5×5 matrices: bifurcations off the diagonal, isola formations, etc

---

[2]Let us note that examples in Sect. 2 are treated in $H = L^2(0, \pi)$ ordered by $K = \{u \in H : u = u(x) \geq 0 \text{ a.e. } x \in (0, \pi)\}$, and examples in Sect. 3 are treated in $H = \mathbb{R}^n$ ordered by $K = (\mathbb{R}_+)^n$.

References

1. Dancer, E.N.: On the Dirichlet problem for weakly non-linear elliptic partial differential equations. Proc. Roy. Soc. Edinb. Sect. A **76**(4), 283–300 (1976/77)
2. Fučík, S.: Solvability of nonlinear equations and boundary value problems. Mathematics and its Applications, vol. 4. D. Reidel Publishing Co., Dordrecht (1980)
3. Holubová, G., Nečesal, P.: Nontrivial Fučík spectrum of one non-selfadjoint operator. Nonlinear Anal. **69**(9), 2930–2941 (2008)
4. Holubová, G., Nečesal, P.: Nonlinear four-point problem: non-resonance with respect to the Fučík spectrum. Nonlinear Anal. **71**(10), 4559–4567 (2009)
5. Holubová, G., Nečesal, P.: The Fučík spectra for multi-point boundary-value problems. Electron. J. Differ. Equat. Conf. **18**, 33–44 (2010)
6. Holubová, G., Nečesal, P.: Fučík spectrum in general. To appear in Topol. Methods Nonlinear Anal. (2011)
7. Hyers, D.H., Isac, G., Rassias, T.M.: Topics in Nonlinear Analysis and Applications. World Scientific Publishing Co. Inc., River Edge (1997)
8. Sergejeva, N.: On nonlinear spectra for some nonlocal boundary value problems. Math. Model. Anal. **13**(1), 87–97 (2008)
9. Stehlík, P.: Discrete Fučík spectrum - anchoring rather than pasting. Boundary Value Problems, 2013:67 (2013)
10. Švarc, R.: The solution of a Fučík's conjecture. Commentat. Math. Univ. Carol. **25**, 483–517 (1984)
11. Švarc, R.: Two examples of the operators with jumping nonlinearities. Commentat. Math. Univ. Carol. **30**(3), 587–620 (1989)

The Displacement of a Sliding Bar Subject to Nonlinear Controllers

Gennaro Infante and Paolamaria Pietramala

It is our pleasure to dedicate this paper to Professor Ravi Agarwal.

Abstract We discuss the existence of positive solutions for a fourth-order differential equation subject to nonlinear and nonlocal boundary conditions, which models a sliding bar. Our approach allows the involved nonlinearity to be singular. Our main ingredient is the theory of fixed-point index.

1 Introduction

In this note we discuss the existence of positive solutions for the ODE

$$u^{(4)}(t) = g(t)f(t, u(t)), \ t \in (0, 1), \tag{1}$$

subject to the nonlocal, nonlinear boundary conditions (BCs)

$$u'(0) + H_1(\alpha_1[u]) = 0, \ \theta_1 u'(1) + u(\xi_1) = H_2(\alpha_2[u]), \tag{2}$$

$$u'''(0) = 0, \ \theta_2 u'''(1) + u''(\xi_2) = 0, \tag{3}$$

where, for $i = 1, 2, \xi_i \in [0, 1]$, H_i is a continuous function and α_i is a linear functional on the space $C[0, 1]$ given by a Stieltjes integral, namely, $\alpha_i[u] = \int_0^1 u(s) \, dA_i(s)$.

G. Infante (✉) • P. Pietramala
Dipartimento di Matematica, Università della Calabria, 87036
Arcavacata di Rende, Cosenza, Italy
e-mail: gennaro.infante@unical.it; pietramala@unical.it

S. Pinelas et al. (eds.), *Differential and Difference Equations with Applications*, Springer
Proceedings in Mathematics & Statistics 47, DOI 10.1007/978-1-4614-7333-6_37,
© Springer Science+Business Media New York 2013

This type of BCs is fairly general and includes, when $H_i(w) = (w)$, m-point and integral BCs, namely, $\alpha_i[u] = \sum_{j=1}^{m} \alpha_j u(\eta_j)$ and $\alpha_i[u] = \int_0^1 \alpha_i(s)u(s)\,ds$; in the case of fourth-order equations, nonlocal and nonlinear conditions are widely studied objects; see, for example, [6, 9, 11, 15, 19, 20, 24, 27–29] for nonlocal BCs and [1–5, 7, 15, 25] for nonlinear BCs and references therein.

One motivation for studying the boundary value problem (BVP) (1)–(3) is that it occurs in the deformations of an elastic beam of length 1, when we suppose that, along its length, a load is added to cause deformations and that the beam has some feedback controllers. For example, the six-point BCs

$$u'(0) + H_1(u(\xi_3)) = \theta_1 u'(1) + u(\xi_1) - H_2(u(\xi_4)) = u'''(0) = \theta_2 u'''(1) + u''(\xi_2) = 0$$

mean that the shear force vanishes at $t = 0$ and at $t = 1$ is related to the bending moment in another point of the beam; moreover, some other sensors provide feedback control to the angular attitude at $t = 0$ and at $t = 1$.

Here we prove the existence of multiple positive solutions of (1)–(3) under suitable oscillatory behavior of f. Our approach is to rewrite the BVP (1)–(3) as a perturbed Hammerstein integral equation of the form

$$u(t) = \gamma_1(t)H_1(\alpha_1[u]) + \gamma_2(t)H_2(\alpha_2[u]) + \int_0^1 k(t,s)g(t)f(s,u(s))ds.$$

This type of perturbed integral equation, with f nonsingular, has been studied in [12]. Lan [23] studied the Hammerstein case with a singular f, and his results were exploited in [14], where the Hammerstein integral equation was perturbed by two *linear* perturbations.

We prove our results by means of the classical fixed-point index theory and make use of results and ideas from the papers [12–14, 23, 26].

2 The Boundary Value Problem

We firstly consider the BVP, studied by GI and Webb in [17],

$$-u''(t) = g(t)f(t,u(t)), \quad u'(0) = 0, \quad \theta u'(1) + u(\xi) = 0, \tag{4}$$

where $\xi \in [0,1]$. The solution of the BVP (4) can be written as

$$u(t) = \int_0^1 k_{\theta,\xi}(t,s)g(s)f(s,u(s))ds, \tag{5}$$

where

$$k_{\theta,\xi}(t,s) = \theta + \begin{cases} \xi - s, & s \le \xi \\ 0, & s > \xi \end{cases} - \begin{cases} t - s, & s \le t \\ 0, & s > t. \end{cases}$$

It is known (see [17]) that the case $\theta + \xi > 1$ leads to the existence of *positive solutions*.

We now utilize the integral formulation (5) to build a new Green's function that will be useful in the sequel.

Lemma 1. *Let* $\theta_1 + \xi_1 > 1$ *and* $\theta_2 + \xi_2 > 1$. *The Green's function* $k(t,s)$ *for the linear fourth-order BVP*

$$u^{(4)}(t) = y(t), \ t \in (0,1),$$
$$u'(0) = 0, \ u'''(0) = 0, \ \theta_1 u'(1) + u(\xi_1) = 0, \ \theta_2 u'''(1) + u''(\xi_2) = 0 \tag{6}$$

is given by

$$k(t,s) = \theta_2\left(\theta_1 + \frac{\xi_1^2}{2} - \frac{t^2}{2}\right) - \frac{\theta_1}{2}(1-s)^2 + \begin{cases} (\theta_1 + \frac{\xi_1^2}{2} - \frac{t^2}{2})(\xi_2 - s), & s \le \xi_2 \\ 0, & s > \xi_2 \end{cases}$$

$$- \begin{cases} \frac{1}{6}(\xi_1 - s)^3, & s \le \xi_1 \\ 0, & s > \xi_1 \end{cases} + \begin{cases} \frac{1}{6}(t - s)^3, & s \le t \\ 0, & s > t. \end{cases} \tag{7}$$

Moreover, for $(t,s) \in [0,1] \times [0,1]$, *we have*

$$c_1 k(0,s) \le k(t,s) \le k(0,s) := \Phi(s),$$

where

$$c_1 := 1 - \frac{1}{2\theta_1 + \xi_1^2}.$$

Proof. The Green's function $k(t,s)$ for the fourth-order problem (6), in a similar way as in [8], is given by

$$k(t,s) = \int_0^1 k_{\theta_1,\xi_1}(t,v) k_{\theta_2,\xi_2}(v,s) dv.$$

Since

$$k_{\theta_1,\xi_1}(t,v) k_{\theta_2,\xi_2}(v,s) = \theta_1\theta_2 + \theta_1(\xi_2 - s)\chi_{[0,\xi_2]}(s) - \theta_1(v - s)\chi_{[0,v]}(s)$$

$$+ \theta_2(\xi_1 - v)\chi_{[0,\xi_1]}(v) + (\xi_2 - s)\chi_{[0,\xi_2]}(s)\chi_{[0,\xi_1]}(v)$$

$$- (\xi_1 - v)(v - s)\chi_{[0,v]}(s)\chi_{[0,\xi_1]}(v) - \theta_2(t - v)\chi_{[0,t]}(v)$$

$$- (t - v)(\xi_2 - s)\chi_{[0,\xi_2]}(s)\chi_{[0,t]}(v) + (t - v)(v - s)\chi_{[0,v]}(s)\chi_{[0,t]}(v),$$

we obtain (7). A suitable upper bound for $k(t,s)$ is given by $\max_{t \in [0,1]} k(t,s)$ for each fixed s. Since

$$(\partial/\partial t)k(t,s) = -\int_0^1 k_{\theta_2,\xi_2}(v,s)dv$$

is nonpositive in $[0,1] \times [0,1]$, the maximum for each fixed s occurs when $t = 0$. By evaluating

$$\min_{t,s \in [0,1]} \frac{k(t,s)}{k(0,s)} = \frac{k(1,1)}{k(0,1)} = 1 - \frac{1}{2\theta_1 + \xi_1^2},$$

we obtain the sharp value of c_1.   □

We can now introduce two auxiliary functions γ_1 and γ_2, so that we may seek solutions of the BVP (1)–(3) as solutions of the integral equation

$$u(t) = \gamma_1(t)H_1(\alpha_1[u]) + \gamma_2(t)H_2(\alpha_2[u]) + \int_0^1 k(t,s)g(s)f(s,u(s))ds := T(u), \quad (8)$$

where $k(t,s)$ is as in (7).

Let $\gamma_1(t) := \theta_1 + \xi_1 - t$ and $\gamma_2(t) := 1$ be the unique solutions of

$$\gamma_1^{(4)}(t) = 0, \; \gamma_1'(0) + 1 = \gamma_1'''(0) = \theta_1\gamma_1'(1) + \gamma_1(\xi_1) = \theta_2\gamma_1''(1) + \gamma_1'(\xi_2) = 0,$$

$$\gamma_2^{(4)}(t) = 0, \; \gamma_2(0) = \gamma_2''(0) = \theta_1\gamma_2'(1) + \gamma_2(\xi_1) - 1 = \theta_2\gamma_2''(1) + \gamma_2'(\xi_2) = 0.$$

Note that $\|\gamma_1\| = \theta_1 + \xi_1$ and $\gamma_1(t) \geq c_2\|\gamma_1\|$ and $\gamma_2(t) \geq c_3\|\gamma_2\|$ for $t \in [0,1]$, where

$$c_2 := (1 - \frac{1}{\theta_1 + \xi_1}), \; c_3 := 1.$$

In the sequel of the paper we assume the following assumptions:

- There exist constants $0 \leq r_1 < r_2$ such that $f(t,u) : [0,1] \times [r_1,r_2] \to [0,\infty)$ is continuous.
- $g\Phi \in L^1[0,1]$, $g \geq 0$ almost everywhere, and $\int_0^1 \Phi(s)g(s)\,ds > 0$.
- H_1, H_2 are positive continuous functions such that there exist $h_{11}, h_{12}, h_{21}, h_{22} \in [0,\infty)$ with

$$h_{11}v \leq H_1(v) \leq h_{12}v \quad \text{and} \quad h_{21}v \leq H_2(v) \leq h_{22}v,$$

for every $v \geq 0$.

- $\alpha_1[\cdot]$, $\alpha_2[\cdot]$ are positive bounded linear functionals on $C[0,1]$ given by

$$\alpha_i[u] = \int_0^1 u(s)\,dA_i(s), \ \ i = 1,2,$$

involving Stieltjes integrals with *positive* measures dA_i.
- $h_{12}\alpha_1[\gamma_1] < 1, \ h_{22}\alpha_2[\gamma_2] < 1.$
- $D_2 := (1 - h_{12}\alpha_1[\gamma_1])(1 - h_{22}\alpha_2[\gamma_2]) - h_{12}h_{22}\alpha_1[\gamma_2]\alpha_2[\gamma_1] > 0.$ This implies $D_1 := (1 - h_{11}\alpha_1[\gamma_1])(1 - h_{21}\alpha_2[\gamma_2]) - h_{11}h_{21}\alpha_1[\gamma_2]\alpha_2[\gamma_1] > 0.$

The assumptions above enable us to use the cone

$$K = \left\{ u \in C[0,1] : \min_{t \in [0,1]} u(t) \ge c\|u\| \right\},$$

where

$$c = \min\{c_1, c_2, c_3\} = 1 - \frac{1}{\theta_1 + \xi_1},$$

a type of cone firstly used by Krasnosel'skiĭ(see, e.g., [21]) and Guo (see, e.g.,). [10].

Since the operator T is not well defined on $C[0,1]$, we extend f in a similar way to that of Lan [23] and define $\tilde{f}(t,u) : [0,1] \times [0,\infty) \to [0,\infty)$ as

$$\tilde{f}(t,u) := \begin{cases} f(t,r_1), & \text{if } 0 \le u \le r_1, \\ f(t,u), & \text{if } r_1 \le u \le r_2, \\ f(t,r_2), & \text{if } r_2 \le u < \infty \end{cases}$$

and consider the operator

$$\tilde{T}u(t) := \gamma_1(t)H_1(\alpha_1[u]) + \gamma_2(t)H_2(\alpha_2[u]) + \int_0^1 k(t,s)g(s)\tilde{f}(s,u(s))\,ds.$$

First of all, note that $\tilde{T}$ maps K into K and is compact and, by construction,

$$\tilde{T}u = Tu \text{ for } u \in C(r_1,r_2),$$

where

$$C(r_1,r_2) = \{u \in K : r_1 \le u(t) \le r_2 \text{ for } t \in [0,1]\}.$$

Clearly, fixed points of $\tilde{T}$ in $C(r_1,r_2)$ provide solutions to the original problem. For our index calculations we shall use the following open bounded sets (relative to K):

$$K_\rho = \{u \in K : \|u\| < \rho\}, \ V_\rho = \{u \in K : \min_{t \in [0,1]} u(t) < \rho\}.$$

Note that $K_\rho \subset V_\rho \subset K_{\rho/c}$. The set V_ρ was introduced in [18] and is equal to the set called $\Omega_{\rho/c}$ in [22]. We make use of the following numbers:

$$\bar{f}^{0,\rho} := \sup_{0 \le u \le \rho, 0 \le t \le 1} \frac{\tilde{f}(t,u)}{\rho}, \quad \tilde{f}_{\rho,\rho/c} := \inf_{\rho \le u \le \rho/c, 0 \le t \le 1} \frac{\tilde{f}(t,u)}{\rho},$$

$$\frac{1}{m} := \sup_{t \in [0,1]} \int_0^1 k(t,s)g(s)\,ds, \quad \frac{1}{M} := \inf_{t \in [0,1]} \int_0^1 k(t,s)g(s)\,ds,$$

and use the notation

$$\mathcal{K}_i(s) := \int_0^1 k(t,s)\,dA_i(t).$$

We utilize the following Lemma from [12], which provides conditions on the value of the fixed-point index on some sets.

Lemma 2 ([12]).

1. *Assume that there exists $\rho > 0$ such that*

$$\tilde{f}_{\rho,\rho/c} \left(\left(\frac{c_2\|\gamma_1\|}{D_1}(1 - h_{21}\alpha_2[\gamma_2]) + \frac{c_3\|\gamma_2\|}{D_1} h_{11}\alpha_2[\gamma_1] \right) \int_0^1 \mathcal{K}_1(s)g(s)\,ds \right.$$
$$\left. + \left(\frac{c_2\|\gamma_1\|}{D_1} h_{21}\alpha_1[\gamma_2] + \frac{c_3\|\gamma_2\|}{D_1}(1 - h_{11}\alpha_1[\gamma_1]) \right) \int_0^1 \mathcal{K}_2(s)g(s)\,ds + \frac{1}{M} \right) > 1. \tag{9}$$

   Then the fixed-point index, $i_K(\tilde{T}, V_\rho)$, is 0.
2. *Assume that there exists $\rho > 0$ such that*

$$\bar{f}^{0,\rho} \left(\left(\frac{\|\gamma_1\|}{D_2}(1 - h_{22}\alpha_2[\gamma_2]) + \frac{\|\gamma_2\|}{D_2} h_{12}\alpha_2[\gamma_1] \right) \int_0^1 \mathcal{K}_1(s)g(s)\,ds \right.$$
$$\left. + \left(\frac{\|\gamma_1\|}{D_2} h_{22}\alpha_1[\gamma_2] + \frac{\|\gamma_2\|}{D_2}(1 - h_{12}\alpha_1[\gamma_1]) \right) \int_0^1 \mathcal{K}_2(s)g(s)\,ds + \frac{1}{m} \right) < 1. \tag{10}$$

   Then $i_K(\tilde{T}, K_\rho) = 1$.

The above Lemma leads to the following new result on existence of multiple positive solutions for (8).

Theorem 1. *Equation (8) has one positive solution in $C(r_1, r_2)$ if either of the following conditions hold:*

(S_1) *There exist ρ_1, ρ_2 with $r_1 \le c\rho_1 < \rho_1 < \rho_2 \le cr_2$ such that (10) is satisfied for ρ_1 and (9) is satisfied for ρ_2.*

(S_2) *There exist ρ_1, ρ_2 with $r_1 \leq \rho_1 < c\rho_2 < \rho_2 \leq r_2$ such that (9) is satisfied for ρ_1 and (10) is satisfied for ρ_2.*

Equation (8) has two positive solutions in $C(r_1, r_2)$ if one of the following conditions hold:

(O_1) *There exist ρ_1, ρ_2, ρ_3 with $r_1 \leq c\rho_1 < \rho_1 < \rho_2 < c\rho_3 < \rho_3 \leq r_2$ such that (10) is satisfied for ρ_1, (9) is satisfied for ρ_2, and (10) is satisfied for ρ_3.*

(O_2) *There exist ρ_1, ρ_2, ρ_3 with $r_1 \leq \rho_1 < c\rho_2 < c\rho_3 < \rho_3 \leq cr_2$ such that (9) is satisfied for ρ_1, (10) is satisfied for ρ_2, and (9) is satisfied for ρ_3.*

Remark 1. By similar arguments it is possible to state results valid for three or more positive solutions (see, for example, [22]) and for nonlinearities with more than one singularity (see [13]).

2.1 Example

We consider the BVP

$$u^{(4)}(t) = g(t)f(t, u(t)), \; u'(0) + H_1(u(\xi_3)) = \theta_1 u'(1) + u(\xi_1) - H_2(u(\xi_4)) = 0,$$

$$u'''(0) = 0, \;\; \theta_2 u'''(1) + u''(\xi_2) = 0,$$

where $\xi_3, \xi_4 \in [0,1]$ and the functions H_i are defined as in [16], namely,

$$H_i(w) = \begin{cases} \frac{1}{4i}w, & 0 \leq w \leq 1, \\ \frac{1}{8i}w + \frac{1}{8i}, & w \geq 1. \end{cases}$$

In this case we have $h_{11} = 1/8$, $h_{12} = 1/4$, $h_{21} = 1/16$, and $h_{22} = 1/8$. By fixing $\xi_1 = 1/5$, $\xi_2 = 4/5, \xi_3 = 2/5, \xi_4 = 3/5, \theta_1 = 5/6, \theta_2 = 2/7$, we obtain by direct calculation

$$1/m = \max_{t \in [0,1]} \int_0^1 k(t,s)\,ds = \max_{t \in [0,1]} \left\{ \frac{23809}{63000} - \frac{53t^2}{175} + \frac{t^4}{24} \right\} = \frac{23809}{63000},$$

$$1/M = \min_{t \in [0,1]} \int_0^1 k(t,s)\,ds = \frac{3677}{31500},$$

$$\alpha_1[\gamma_1] = \frac{19}{30}, \quad \alpha_1[\gamma_2] = 1, \quad \alpha_2[\gamma_1] = \frac{13}{30}, \quad \alpha_2[\gamma_2] = 1,$$

$$\int_0^1 \mathcal{K}_1(s)\,ds = \int_0^1 k(\xi_3, s)\,ds = \frac{104117}{315000}, \quad \int_0^1 \mathcal{K}_2(s)\,ds = \int_0^1 k(\xi_4, s)\,ds = \frac{43201}{157500}.$$

The conditions (10) and (9) read $\bar{f}^{0,\rho_1} < 0.8269$ and $\tilde{f}_{\rho_2,\rho_2/c} > 2.2525$. Since (S_1) holds, from Theorem 1 it follows that this BVP has a nontrivial solution in K. A nonlinearity that verifies (S_1), for example, is the function

$$f(t,u) = \begin{cases} 1/(256u), & u \leq 1/4, \\ u^3, & u > 1/4, \end{cases}$$

with the choice $r_1 = 11/465$, $\rho_1 = 11/15$, $\rho_2 = 8/5$, and $r_2 \geq 248/5$.

References

1. Agarwal, R.P.: On fourth order boundary value problems arising in beam analysis. Diff. Integr. Equat. **2**, 91–110 (1989)
2. Alves, E., Ma, T.F., Pelicer, M.L.: Monotone positive solutions for a fourth order equation with nonlinear boundary conditions. Nonlinear Anal. **71**, 3834–3841 (2009)
3. Amster, P., Cárdenas Alzate, P.P.: A shooting method for a nonlinear beam equation. Nonlinear Anal. **68**, 2072–2078 (2008)
4. Cabada, A., Minhós, F.: Fully nonlinear fourth-order equations with functional boundary conditions. J. Math. Anal. Appl. **340**, 239–251 (2008)
5. Cabada, A., Pouso, R.L., Minhós, F.: Extremal solutions to fourth-order functional boundary value problems including multipoint conditions. Nonlinear Anal. Real World Appl. **10**, 2157–2170 (2009)
6. Eggensperger, M., Kosmatov, N.: Positive solutions of a fourth-order multi-point boundary value problem. Commun. Math. Anal. **6**, 22–30 (2009)
7. Franco, D., O'Regan, D., Perán, J.: Fourth-order problems with nonlinear boundary conditions. J. Comput. Appl. Math. **174**, 315–327 (2005)
8. Graef, J.R., Yang, B.: Existence and nonexistence of positive solutions of fourth order nonlinear boundary value problems. Appl. Anal. **74**, 201–214 (2000)
9. Graef, J.R., Qian, C., Yang, B.: A three point boundary value problem for nonlinear fourth order differential equations. J. Math. Anal. Appl. **287**, 217–233 (2003)
10. Guo, D., Lakshmikantham, V.: Nonlinear Problems in Abstract Cones. Academic, New York (1988)
11. Henderson, J., Ma, D.: Uniqueness of solutions for fourth-order nonlocal boundary value problems. Bound. Value Probl. **2006**, 12 (2006)
12. Infante, G.: Nonlocal boundary value problems with two nonlinear boundary conditions. Commun. Appl. Anal. **12**, 279–288 (2008)
13. Infante, G.: Positive solutions of some nonlinear BVPs involving singularities and integral BCs. Discrete Contin. Dyn. Syst. Ser. S **1**, 99–106 (2008)
14. Infante, G.: Positive solutions of nonlocal boundary value problems with singularities. Discrete Contin. Dyn. Syst. **suppl.**, 377–384 (2009)
15. Infante, G., Pietramala, P.: A cantilever equation with nonlinear boundary conditions. Electron. J. Qual. Theory Diff. Equat. **Spec. Ed.** I(15), 1–14 (2009)
16. Infante, G., Pietramala, P.: Existence and multiplicity of non-negative solutions for systems of perturbed Hammerstein integral equations. Nonlinear Anal. **71**, 1301–1310 (2009)
17. Infante, G., Webb, J.R.L.: Loss of positivity in a nonlinear scalar heat equation. NoDEA Nonlinear Diff. Equat. Appl. **13**, 249–261 (2006)
18. Infante, G., Webb, J.R.L.: Nonlinear nonlocal boundary value problems and perturbed Hammerstein integral equations. Proc. Edinb. Math. Soc. **49**, 637–656 (2006)

19. Karna, B.K., Kaufmann, E.R., Nobles, J.: Comparison of eigenvalues for a fourth-order four-point boundary value problem. Electron. J. Qual. Theory Differ. Equat. **2005**, 9 (2005)
20. Kelevedjiev, P.S., Palamides, P.K., Popivanov, N.I.: Another understanding of fourth-order four-point boundary-value problems, Electron. J. Diff. Equat. **2008**(47), 15 (2008)
21. Krasnosel'skiĭ, M.A., Zabreĭko, P.P.: Geometrical Methods of Nonlinear Analysis. Springer, Berlin (1984)
22. Lan, K.Q.: Multiple positive solutions of semilinear differential equations with singularities. J. London Math. Soc **63**, 690–704 (2001)
23. Lan, K.Q.: Multiple positive solutions of Hammerstein integral equations and applications to periodic boundary value problems. Appl. Math. Comput. **154**, 531–542 (2004)
24. Ma, R., Chen, T.: Existence of positive solutions of fourth-order problems with integral boundary conditions. Bound. Value Probl. **2011**, 17 (2011). Art. ID 297578
25. Pietramala, P.: A note on a beam equation with nonlinear boundary conditions. Bound. Value Probl. **2011**, 14 (2011). Art. ID 376782
26. Webb, J.R.L., Infante, G.: Positive solutions of nonlocal boundary value problems: a unified approach. J. London Math. Soc. **74**, 673–693 (2006)
27. Webb, J.R.L., Infante, G.: Nonlocal boundary value problems of arbitrary order, J. London Math. Soc. **79**, 238–258 (2009)
28. Webb, J.R.L., Infante, G., Franco, D.: Positive solutions of nonlinear fourth-order boundary-value problems with local and non-local boundary conditions. Proc. Roy. Soc. Edinb. Sect. A **138**, 427–446 (2008)
29. Webb, J.R.L., Zima, M.: Multiple positive solutions of resonant and non-resonant nonlocal fourth order boundary value problems. Glasg. Math. J. **54**, 225–240 (2012)

Abstract Bifurcation Theorems and Applications to Dynamical Systems with Resonant Eigenvalues

Vladimir Jaćimović

Abstract We reconsider the abstract bifurcation theorem stated in Andronov et al. (Teoriya Kolebaniy, 1959) and investigate some known and unknown corollaries. We focus on the case of Fredholm operators with zero index and find out that the main result is meaningful only for certain dimensions of the critical subspace (namely, 1, 2, 4, and 8). This particularity is due to certain algebraic and topological aspects of the problem. Finally, we provide some interesting applications to the system of ODE's and abstract integral equation.

1 Introduction

Interest in the concept of bifurcation in science is steadily growing, mainly due to the rising number of applications. In particular, it has been observed that numerous important phenomena and mechanisms in science can be explained in terms of bifurcation theory.

Hopf bifurcation is probably one of the most studied cases of bifurcation phenomena. One reason is that Hopf bifurcation is very often observed in various dynamical systems, with very different applications. Hopf bifurcation theorem for ODEs is first stated in classical papers by Hopf [10] and Andronov et al. [3]. In the consequent decades Hopf bifurcation for ODE's has been studied in many details and this mathematical theory can be considered as completed by the 1970s.

However, Hopf bifurcation appears not only in the systems described by ODE's. On the contrary, Hopf bifurcation in infinite-dimensional dynamical systems has attracted attention of mathematicians until present time. Early theorems on Hopf bifurcation for semigroups are stated in [7, 13]. More details and applications to

V. Jaćimović (✉)
University of Montenegro, Cetinjski put bb., 81000 Podgorica, Montenegro
e-mail: vladimir@jacimovic.me

S. Pinelas et al. (eds.), *Differential and Difference Equations with Applications*, Springer Proceedings in Mathematics & Statistics 47, DOI 10.1007/978-1-4614-7333-6_38,
© Springer Science+Business Media New York 2013

functional differential equations can be found in [8]. At the same time, other authors have studied existence of Hopf bifurcation in some Navier–Stokes equations and reaction-diffusion systems (see, for instance, [11, 14]).

Taking into account such different contexts where bifurcations (including Hopf bifurcation) can appear, one fairly general method of dealing with such problems is by proving abstract bifurcation theorems and applying it to various dynamical systems in appropriate functional setting. For infinite-dimensional dynamical systems, one typically needs additional technical tools, such as center manifold reduction (see [5, 8, 16]).

That is the approach we will adopt here. Namely, in the next section we will state bifurcation theorem for the abstract problem with Fredholm zero index operator. Further investigation demonstrates that this theorem is meaningful only for certain dimensions of critical subspace (1, 2, 4, or 8). In Sect. 3, we will employ this theorem to obtain abstract Hopf bifurcation theorem [2] and "bifurcation from simple eigenvalue" [6]. Besides, our abstract theorem covers two more cases that are not treated in the literature (at least, to our knowledge). In the last section, we briefly consider applications of abstract theorem to the system of ODE's and AIE (abstract integral equations), the last being example of infinite-dimensional dynamical system.

2   Abstract Bifurcation Theorem

Let X and Y be the Banach spaces and $F : R^s \times X \to Y$ a smooth map. Consider the equation

$$F(\mu, x) = 0, \tag{1}$$

and suppose that $x_0 \in X$ is the solution of Eq. (1), that is, $F(\mu, x_0) = 0, \forall \mu \in R^s$.

The point (μ_0, x_0) is called a *bifurcation point* of Eq. (1) if there exists a sequence $\{(\mu_k, x_k)\} \subset R^s \times X$ converging to (μ_0, x_0) and such that $F(\mu_k, x_k) = 0, \forall k = 1, 2, \ldots$.

From the Implicit Function Theorem it follows that (μ_0, x_0) can be bifurcation point only if $\frac{\partial F}{\partial x}(\mu_0, x_0)$ is not invertible. Therefore, violation of Implicit Function Theorem conditions is a simple necessary condition for existence of bifurcation.

We start by standard assumptions:

(A1) The map F is C^2 in the neighborhood of the point (μ_0, x_0).
(A2) $F(\mu, x_0) = 0, \forall \mu \in R^s$.
       Set: $V = Ker\frac{\partial F}{\partial x}(\mu_0, x_0)$ and $R = Im\frac{\partial F}{\partial x}(\mu_0, x_0)$.
       Denote by W a complementary subspace of V in X and by Z a complementary subspace of R in Y.
       Then one has

$$X = V \oplus W, Y = R \oplus Z.$$

Introduce two more assumptions:

(A3) $dimV = dimZ = m \neq 0$, with some finite integer m.

(A4) In addition, assume that $s = m$ (i.e., the number of parameters equals the dimension of kernel and the codimension of image).

Note that assumption (A3) restricts our consideration to the case when $\frac{\partial F}{\partial x}(\mu_0, x_0)$ is Fredholm operator with the zero index.

Throughout the paper we will assume that each finite-dimensional vector space is equipped with an inner product denoted by $\langle \cdot, \cdot \rangle$.

For each $y \in Z$, we define the linear operator $\mathscr{A}_y : Ker\frac{\partial F}{\partial x}(\mu_0, x_0) \to R^s$ by the following relation:

$$\langle \eta, \mathscr{A}_y h \rangle = \langle y, P\frac{\partial^2 F}{\partial \mu \partial x}(\mu_0, x_0)[\eta, h]\rangle, \ \forall \eta \in R^s.$$

The following two theorems are stated in [12]:

Theorem 1. *Under assumptions (A1)–(A4), if*

$$rank\mathscr{A}_y = m, \forall y \in Z, y \neq 0,$$

then (μ_0, x_0) is a bifurcation point for the Eq. (1).

Theorem 2. *Under the assumptions (A1)–(A4), the case $rank\mathscr{A}_y = m, \forall y \in Z \setminus \{0\}$ is possible only if $m = 1, 2, 4, 8$.*

Note that the first theorem is corollary of more general abstract bifurcation theorem stated in [4].

3 Corollaries: Bifurcation from Simple Eigenvalue and Abstract Hopf Bifurcation Theorem

We start this section by reformulating Theorem 1 in the way that will be more convenient for further discussion.

Denote by P projection onto subspace Z parallel to R.

Theorem 3. *Suppose F satisfies (A1)–(A4) and set vector parameter $\mu = (\mu_1, \ldots, \mu_m)^T$. Moreover, assume that for any nonzero vector $v \in V$, vectors*

$$P\frac{\partial^2 F}{\partial x \partial \mu_1}(\mu_0, x_0)v, P\frac{\partial^2 F}{\partial x \partial \mu_2}(\mu_0, x_0)v, \ldots, P\frac{\partial^2 F}{\partial x \partial \mu_m}(\mu_0, x_0)v \qquad (2)$$

are linearly independent.

Then (μ_0, x_0) is bifurcation point for F.

In the setting, adopted by introducing assumptions (A3) and (A4), Theorem 3 is just a modification of Theorem 1. This formulation also provides better insight into Theorem 2. Indeed, notice that Eq. (2) can be viewed as linear vector fields on Z. Recalling celebrated Adams theorem [1] about maximal number of linearly independent vector fields on spheres, we conclude that the assumption about linear independence of Eq. (2) can hold only if $m = \rho(m)$, i.e., when $m = 1, 2, 4$, or 8. (Here $\rho(m)$ is Adams number.)

Hence, there are essentially four cases covered by the Theorem 3. In the rest of this section we demonstrate that the first two cases ($m = 1$ and $m = 2$) yield classical results from bifurcation theory.

Case $m = 1$: Bifurcation from Simple Eigenvalue

Consider Theorem 3 for the case $dimV = m = 1$. In that case μ is a scalar parameter, and we require the projection $P\frac{\partial^2 F}{\partial x \partial \mu}(\mu_0, x_0)v$ to be linearly independent (i.e., nonzero) vector for $v \in V$. Therefore,

$$P\frac{\partial^2 F}{\partial x \partial \mu}(\mu_0, x_0)v \neq 0, \, v \in Ker\frac{\partial F}{\partial x}(\mu_0, x_0),$$

where P stands for the projection onto complementary subspace of $Im\frac{\partial F}{\partial x}(\mu_0, x_0)$. This further implies that

$$\frac{\partial^2 F}{\partial x \partial \mu}(\mu_0, x_0)v \notin Im\frac{\partial F}{\partial x}(\mu_0, x_0). \tag{3}$$

Thus, when $dimV = dimZ = 1$, Theorem 3 transfers into classical theorem on "bifurcation from simple eigenvalue" (see [7]), stating that Eq. (3) is a sufficient condition for bifurcation.

Case $m = 2$: Abstract Hopf Bifurcation Theorem

Consider the case when $dimV = m = 2$. Then $\mu = (\mu_1, \mu_2)^T$ is two-dimensional vector parameter, and Theorem 3 requires two vector fields

$$P\frac{\partial^2 F}{\partial x \partial \mu_1}(\mu_0, x_0)v \text{ and } P\frac{\partial^2 F}{\partial x \partial \mu_2}(\mu_0, x_0)v$$

to be linearly independent.

This yields interesting bifurcation result, named an *abstract Hopf bifurcation theorem* in the book [2]. In the same book, the classical Hopf bifurcation theorem for the system of ODE's is derived from abstract Hopf bifurcation theorem. Application of the abstract bifurcation theorem to the bifurcation of small periodic orbits in the system of ODE's requires reformulating the original problem in terms of appropriate function spaces. Also, in order to prove classical Hopf bifurcation by applying Theorem 3, one needs two parameters. This is usually achieved by explicit introduction of the internal parameter ω corresponding to the frequency of the periodic orbits (along with external parameter μ).

Cases $m = 4$ and $m = 8$: Resonant Bifurcations

The cases when $m = 4$ and $m = 8$ are more involved and (unlike the first two cases) do not transfer into classical facts from the bifurcation theory. In particular, these cases arise when the linearization of dynamical system has purely imaginary eigenvalues of geometric multiplicities 2 and 4 (i.e., center manifold is of real dimension 4 and 8), respectively. For these cases, Theorem 3 requires higher number of (external or internal) parameters and satisfaction of some special conditions on parameters that can be seen as an analogue of transversality condition in classical Hopf bifurcation theorem. As these cases require cumbersome calculations we limit ourselves to some examples of finite- and infinite-dimensional dynamical systems exhibiting bifurcations at resonant pairs of purely imaginary eigenvalues. This is exposed in the next section.

Remark: Relation to the Hopf Invariant

We just briefly mention that Theorem 3 obviously relates bifurcations under consideration to the famous Hopf invariant (named after Hopf [9]). Indeed, it is well known that linearly independent vector fields on spheres are closely related to important topological concepts such as Hopf map and Hopf fibrations of spheres. In particular, the case $m = 2$ would correspond to the famous Hopf fibration:

$$S^3 \xrightarrow{S^1} S^2.$$

This unexpected relation can be relevant for the possible physical applications. Hopf fibration is known to be an adequate model for description of many important phenomena observed in physics, including classical mechanics and quantum information theory. For some interesting examples, see [15]. However, we will study this relation and some physical applications in detail elsewhere.

4   Application to ODE's and AIE

In this section we briefly demonstrate some applications of abstract bifurcation theorems to the systems of ODE's and AIE (abstract integral equations). We omit almost all calculations and techniques and focus on the result only.

We restrict our consideration to the case $m = 4$ only. Examples with $m = 8$ can be studied in analogous way, but calculations are cumbersome in that case (involving matrices 8×8).

We will apply Theorem 1 to the system of 4 ODE's in the simple form that would be sufficient for further application to AIE. Employing Theorem 1 to formulate the bifurcation result for the system of ODE's in more general form would lead to the certain technical difficulties. The most important is that (since eigenvalue $\pm i\omega_0$ is not simple) one cannot guarantee differentiability of eigenvalues and eigenvectors w.r. to parameters. Nevertheless, Theorem 1 can be applied in straightforward way for any given system of ODE's.

Let us start with the system of 4 ODE's in the form

$$\dot{x} = M(\sigma, \eta, v)x + G(x). \tag{4}$$

Here, $x = (x_1, x_2, x_3, x_4)^T$ is four-dimensional vector, $G : R^4 \to R^4$ is C^2 mapping, and M is matrix depending on three scalar parameters:

$$M(\sigma, \eta, v) = \begin{pmatrix} a_{11}(\sigma) & \omega_0 & a_{13}(\eta) & a_{14}(v) \\ -\omega_0 & a_{22}(\sigma) & a_{23}(v) & a_{24}(\eta) \\ a_{31}(\eta) & a_{32}(v) & a_{33}(\sigma) & \omega_0 \\ a_{41}(v) & a_{42}(\eta) & -\omega_0 & a_{44}(\sigma) \end{pmatrix}.$$

The $\omega_0 > 0$ is a constant. Introducing assumptions on G and M:

(B1) $G(0) = 0$; $dG(0) = 0$.

(B2) For the critical values of parameters σ_0, η_0 and v_0, the matrix $M(\sigma_0, \eta_0, v_0)$ has two pairs of purely imaginary eigenvalues, namely,

$$M(\sigma_0, \omega_0, \eta_0, v_0) = \begin{pmatrix} 0 & \omega_0 & 0 & 0 \\ -\omega_0 & 0 & 0 & 0 \\ 0 & 0 & 0 & \omega_0 \\ 0 & 0 & -\omega_0 & 0 \end{pmatrix}.$$

Notice that the matrix $M(\sigma_0, \eta_0, v_0)$ has two pairs of purely imaginary eigenvalues $\pm i\omega_0$, hence the non-resonance condition (crucial for classical Hopf bifurcation) is violated.

(B3) Finally, we introduce the last assumption, which is an analogue of the transversality condition in classical Hopf bifurcation:

$$c_1(a'_{11}(\sigma_0) + a'_{22}(\sigma_0)) = a'_{33}(\sigma_0) + a'_{44}(\sigma_0) \neq 0;$$

$$c_2(a'_{13}(\eta_0) + a'_{24}(\eta_0)) = a'_{31}(\eta_0) + a'_{42}(\eta_0) \neq 0;$$

$$c_3(a'_{32}(v_0) - a'_{41}(v_0)) = a'_{14}(v_0) - a'_{23}(v_0) \neq 0,$$

where c_1, c_2, and c_3 are arbitrary real constants satisfying $c_1 < 0, c_2 = -c_3 > 0$ or $c_1 = -1, c_2 > 0, c_3 < 0$.

Lemma 1. *Suppose (B1)–(B3) are satisfied. Then for ε sufficiently small there exist C^1-functions $\sigma^*(\varepsilon), \omega^*(\varepsilon), \eta^*(\varepsilon)$, and $v^*(\varepsilon)$, taking values in R and C^1-function $x^*(\varepsilon)$, taking values in $C(R, R^n)$, such that:*

(a) At parameter values $\sigma = \sigma^*(\varepsilon), \eta = \eta^*(\varepsilon)$, and $v = v^*(\varepsilon)$, $x^*(\varepsilon)(t)$ is a nontrivial periodic solutions of Eq. (4) with periods $\frac{2\pi}{\omega^*(\varepsilon)}$.

(b) $\sigma^*(0) = \sigma_0, \omega^*(0) = \omega_0, \eta^*(0) = \eta_0, v^*(0) = v_0$.

(c) The amplitude of the orbit $x^*(\varepsilon)(t)$ tends to 0 as $\varepsilon \to 0$.

The proof of Lemma is application of Theorem 1 in appropriate functional setting. First of all, define nonlinear operator:

$$F(x, \mu, \omega, \eta, v) = \omega \dot{x} - M(\mu, \eta, v) - G(x)$$

on the space of $C_0^1[0, \frac{2\pi}{\omega_0}]$ of continuously differentiable, four-dimensional vector functions, that are $\frac{2\pi}{\omega_0}$-periodic and vanish at 0 and $\frac{2\pi}{\omega_0}$. Hence, the problem is stated in the form (1), where $\mu = (\sigma, \omega, \eta, v)$ and $F : R^4 \times C_0^1[0, \frac{2\pi}{\omega_0}] \to C_0[0, \frac{2\pi}{\omega_0}]$.

The remaining part of the proof consists in differentiating operator F at critical point in order to construct 4×4 matrix A_y of an operator $\mathscr{A}_y$ and verifying that A_y is regular for all $y = (y_1, y_2, y_3, y_4)^T \neq 0$. However, we omit these calculations here in order to proceed with the application of abstract theorems to the infinite-dimensional dynamical system.

We consider the system described by AIE. We will rely on technique and notations exposed in [8]. In particular, we employ "sun-star" machinery and corresponding notations to describe adjoint semigroup. Hence, we write AIE in the following form:

$$u(t) = T(t-s)u(s) + \int_s^t T^{\odot*}(t-\tau)R(u(\tau), \sigma, \eta, v)d\tau. \tag{5}$$

Here, $u(\cdot) \in X$, X is a real Banach space, T is C_0-semigroup on X, while σ, η, v are three scalar parameters. Note that we consider the case of finite delay s. Nonlinearity in Eq. (5) is given by the mapping $R(\cdot, \cdot) : X \times R^3 \to X^{\odot*}$. We demand R to be C^2-smooth in both variables.

Assume that $u(\cdot) \equiv 0$ is an equilibrium solution of Eq. (5) for all values of parameters, namely:

C1) $R(0, \sigma, \eta, v) = 0$, for all $\sigma, \eta, v \in R$.

Furthermore, consider the critical values of parameters:

C2) $D_u R(0, \sigma_0, \eta_0, \mu_0) = 0$.

Denote by A generator of the semigroup T and assume:

C3) A has exactly two pairs of purely imaginary eigenvalues $\pm i\omega_0$, and the eigenspace corresponding to these two pairs of eigenvalues has real dimension 4.

Denote by ϕ_1, ϕ_2 linearly independent eigenvectors of A at the eigenvalue $\pm i\omega_0$. Also, let $\phi_1^\odot, \phi_2^\odot \in X^\odot$ be linearly independent eigenvectors of A^* at $i\omega_0$. We can choose eigenvectors in such a way to have

$$\langle \phi_1^\odot, \phi_1 \rangle = \langle \phi_2^\odot, \phi_2 \rangle = 1;$$
$$\langle \phi_2^\odot, \phi_1 \rangle = \langle \phi_1^\odot, \phi_2 \rangle = 0.$$

C4) Finally, assume that

$$c_1 Re\langle \phi_1^\odot, D_\sigma D_u R(0,\sigma_0,\eta_0,v_0)\phi_1\rangle = Re\langle \phi_2^\odot, D_\sigma D_u R(0,\sigma_0,\eta_0,v_0)\phi_2\rangle \neq 0;$$

$$c_2 Re\langle \phi_1^\odot, D_\eta D_u R(0,\sigma_0,\eta_0,v_0)\phi_2\rangle = Re\langle \phi_2^\odot, D_\eta D_u R(0,\sigma_0,\eta_0,v_0)\phi_1\rangle \neq 0;$$

$$c_3 Im\langle \phi_2^\odot, D_v D_u R(0,\sigma_0,\eta_0,v_0)\phi_1\rangle = Im\langle \phi_1^\odot, D_v D_u R(0,\sigma_0,\eta_0,v_0)\phi_2\rangle \neq 0;$$

where c_1, c_2, and c_3 are real constants satisfying $c_1 < 0, c_2 = -c_3 > 0$ or $c_1 = -1, c_2 > 0, c_3 < 0$.

Lemma 2. *Suppose (C1)–(C4) are satisfied. Then, for ε sufficiently small, there exist C^1-functions $\sigma^*(\varepsilon), \omega^*(\varepsilon), \eta^*(\varepsilon)$, and $v^*(\varepsilon)$ taking values in R and C^1-function $x^*(\varepsilon)$, taking values in $C(R,R^n)$, such that:*

(a) At parameter values $\sigma = \sigma^*(\varepsilon), \eta = \eta^*(\varepsilon)$, and $v = v^*(\varepsilon)$, $x^*(\varepsilon)(t)$ is a nontrivial periodic solutions of Eq. (4) with periods $\frac{2\pi}{\omega^*(\varepsilon)}$.
(b) $\sigma^*(0) = \sigma_0, \omega^*(0) = \omega_0, \eta^*(0) = \eta_0, v^*(0) = v_0$.
(c) The amplitude of the orbit $x^*(\varepsilon)(t)$ tends to 0 as $\varepsilon \to 0$.

As we pointed out in the Introduction, one needs some additional technique to deal with infinite-dimensional dynamical systems. Fortunately, methods of reduction of this system to center manifold in order to obtain finite-dimensional system are ready, with all tricky aspects more or less successfully treated. We present the lines of the proof here, omitting almost all technicalities. Necessary details are explained in [8].

The idea of the proof is to decompose the space X according to the spectrum of A and to use center manifold reduction to pass to finite-dimensional system. In our case it will be the system described by four real ODE's that can be written as two complex ODE's. After that we can refer to previous Lemma.

In fact, subspace X_0, that is, tangent to center manifold, corresponds to eigenvalues $\pm i\omega_0$. We will represent this four-dimensional real subspace by two complex basis vectors in the following way:

$$X_0 = \{z_1\phi_1 + \bar{z}_1\bar{\phi}_1 + z_2\phi_2 + \bar{z}_2\bar{\phi}_2 | z_1, z_2 \in C\}$$

in order to get system of two complex ODE's for complex-valued functions z_1 and z_2:

$$\dot{z}_1 = (i\omega_0 + \langle \phi_1^\odot, D_1 R(0,\mu_0,\eta_0,v_0)\phi_1\rangle)z_1 + \langle \phi_1^\odot, D_1 R(0,\mu_0,\eta_0,v_0)\bar{\phi}_1\rangle\bar{z}_1$$
$$+ \langle \phi_1^\odot, D_1 R(0,\mu_0,\eta_0,v_0)\phi_2\rangle z_2 + \langle \phi_1^\odot, D_1 R(0,\mu_0,\eta_0,v_0)\bar{\phi}_2\rangle\bar{z}_2;$$

$$\dot{z}_2 = \langle \phi_2^\odot, D_1 R(0,\mu_0,\eta_0,v_0)\phi_1\rangle z_1 + \langle \phi_2^\odot, D_1 R(0,\mu_0,\eta_0,v_0)\bar{\phi}_1\rangle\bar{z}_1$$
$$+ (i\omega_0 + \langle \phi_2^\odot, D_1 R(0,\mu_0,\eta_0,v_0)\phi_2\rangle)z_2 + \langle \phi_2^\odot, D_1 R(0,\mu_0,\eta_0,v_0)\bar{\phi}_2\rangle\bar{z}_2.$$

Decomposing this system as system of 4 real-valued ODE's, we can apply previous Lemma and obtain desired result. In particular, one can find out that assumption (A3) transfers into (B4).

Acknowledgements The author would like to thank the anonymous referee for numerous valuable comments and suggestions.

References

1. Adams, J.F.: Vector fields on spheres. Ann. of Math. **75**, 603–622 (1962)
2. Ambrosetti, A., Prodi, G.: A Primer of Nonlinear Analysis. Cambridge University Press, Cambridge (1993)
3. Andronov, A.A., Vit, A.A., Khaikin, S.E.: Teoriya Kolebaniy. Fizmatgiz, Moscow (1959)
4. Arutyunov, A.V., Izmailov, A.F., Jaćimović, V.: New bifurcation theorems via second order optimality conditions. J. Math. Anal. Appl. **359**, 752–764 (2009)
5. Chow, S.N., Lu, K.: Invariant manifolds for flows in Banach spaces. J. Diff. Equat. **74**, 285–317 (1988)
6. Crandall, M.C., Rabinowitz, P.H.: Bifurcation from simple eigenvalue. J. Funct. Anal. **48**, 321–340 (1971)
7. Crandall, M.C., Rabinowitz, P.H.: The Hopf bifurcation theorem in infinite dimensions. Arch. Rat. Mech. Anal. **67**, 53–72 (1978)
8. Diekmann, O., van Gils, S.A., Verduyun Lunel, S.M., Walther, H.-O.: Delay Equations: Functional, Complex and Nonlinear Analysis. Springer, New York (1995)
9. Hopf, H.: Uber die Abbildungen der dreidimensionalen Sphäre auf die Kugelflche. Math. Ann. **104**, 631–665 (1931)
10. Hopf, E.: Abzweigung einer periodischen Lösung von einer stationaeren Lösung eines differential systems. Ber. Math. Phys. Sachs. Akad. **94**, 3–22 (1942)
11. Iudovich, V.I.: Investigation of auto-oscillations of a continuous medium occurring at loss of stability of a stationary mode. Prikl. Mat. Mekh. **36**, 450–459 (1972)
12. Jaćimović, V.: Abstract Hopf bifurcation theorem and further extensions via second variation. Nonlinear Anal. TMA **73**(8), 2426–2432 (2010)
13. Marsden, J.: The Hopf bifurcation for nonlinear semigroups. Bull. Am. Math. Soc. **79**, 537–541 (1973)
14. Sattinger, D.H.: Bifurcation of periodic orbits of the Navier–Stokes system. Arch. Rat. Mech. Anal. **41**, 66–80 (1971)
15. Urbantke, H.K.: Hopf fibration seven times in physics. J. Geom. Phys. **46**, 125–150 (2003)
16. Vanderbauwhede, A., Ioss, G.: Center manifold theory in infinite dimensions. In: Jones, C., Kirchgraber, U., Walther, H.O. (eds.) Dynamics Reported: New Series, vol. 1, pp. 125–163. Springer, New York (1992)

Oscillation of Difference Equations with Impulses

Fatma Karakoç

Abstract This paper is concerned with a second-order linear impulsive difference equation with continuous variable. Sufficient conditions for the oscillation of impulsive difference equation are obtained.

Keywords Oscillation • Difference equation • Impulse • Continuous variable

AMS Subject Classification: 34K11, 34K45

1 Introduction

In this paper we consider second-order linear impulsive difference equations of the form

$$\Delta_\tau^2 x(t) + \Delta_\tau x(t) + x(t) + p(t)x(t-\sigma) = 0, \ t \neq t_n, \tag{1}$$

$$x(t_n^+) - x(t_n^-) = L_n x(t_n^-), \ n \in \mathbb{N} = \{1,2,\dots\}, \tag{2}$$

F. Karakoç (✉)
Department of Mathematics, Ankara University, 06100 Ankara, Turkey
e-mail: fkarakoc@ankara.edu.tr

S. Pinelas et al. (eds.), *Differential and Difference Equations with Applications*, Springer
Proceedings in Mathematics & Statistics 47, DOI 10.1007/978-1-4614-7333-6_39,
© Springer Science+Business Media New York 2013

where $\Delta_\tau x(t) = x(t+\tau) - x(t)$; τ, σ are positive constants; $x(t_n^+) = \lim\limits_{t \to t_n^+} x(t)$, and
$x(t_n^-) = \lim\limits_{t \to t_n^-} x(t)$, $p \in C(\mathbb{R}^+, \mathbb{R}^+)$, $\mathbb{R}^+ = (0, \infty)$, $0 < t_1 < t_2 < \dots < t_n < t_{n+1} <$
...are fixed points with $\lim\limits_{n \to \infty} t_n = +\infty$, $\{L_n\}$ is a sequence of positive real numbers.
It is well known that impulsive equations appear as a natural description of
the observed evolution phenomena of several real-world problems [6, 7]. There
has been rich literature on the oscillation of impulsive differential equations [1–
4]. On the other hand, in recent years oscillation of difference equations with
continuous variables has been investigated intensively [8, 10–14]. But to the best
of our knowledge, there has been only a few works on the oscillation of impulsive
difference equations with continuous variables [5, 9].

In this paper, our aim is to establish sufficient conditions for the oscillation of
second-order impulsive difference equation with continuous variable. We shall construct a nonimpulsive inequality, and using it we shall obtain sufficient conditions
for the oscillation.

Definition 1. A function $x : [-\sigma, \infty) \to \mathbb{R}$ is called a solution of Eqs. (1) and (2) if

(a) For $t \neq t_n$, $n \in \mathbb{N}$, x is continuous and satisfies Eq. (1).
(b) For $t = t_n$, $x(t_n^+)$ and $x(t_n^-)$ exist and satisfy Eq. (2) with $x(t_n^-) = x(t_n)$.

Definition 2. If a function $x(t)$ is positive (negative) for all large values of t, then it
is said that $x(t)$ is eventually positive (negative). A solution $x(t)$ of Eqs. (1) and (2)
is called oscillatory if it is neither eventually positive nor eventually negative.

2 Main Results

Let $x(t)$ be a solution of Eqs. (1) and (2). Define

$$z(t) = x(t) \prod_{0 \leq t_m < t} (1 + L_m)^{-1}, \ t \geq 0.$$

As usual, the symbol $\prod\limits_{a \leq t_m < b} a_m$ denotes the product of members of the sequence
$\{a_m\}$ over m such that $t_m \in [a, b) \cap \{t_n : n \in \mathbb{N}\}$. If $[a, b) \cap \{t_n : n \in \mathbb{N}\} = \emptyset$ or $a > b$,
then we use the convention that $\prod\limits_{a \leq t_m < b} a_m = 1$.
It can be seen that the function $z(t)$ is continuous at t_k, $k = 1, 2, \dots$ Indeed,

$$z(t_k^-) = x(t_k^-) \prod_{0 \leq t_m < t_k^-} (1 + L_m)^{-1}$$

$$= z(t_k),$$

and

$$z(t_k^+) = x(t_k^+) \prod_{0 \le t_m < t_k^+} (1 + L_m)^{-1}$$

$$= x(t_k^+) \prod_{0 \le t_m < t_k} (1 + L_m)^{-1} (1 + L_k)^{-1}$$

$$= z(t_k),$$

where we have used the impulse condition (2).

Define

$$v(t) = \frac{1}{\tau} \int_{t+\tau}^{t+2\tau} z(u)\,du, \quad t \ge 0. \tag{3}$$

Lemma 1. *If $x(t)$ is an eventually positive solution of Eqs. (1) and (2), then $v(t) > 0$, and $v'(t) \le 0$ eventually.*

Proof. Let $x(t) > 0$ for $t \ge T$, here T is a sufficiently large number. Then it is clear that $v(t) > 0$ for $t \ge T$. From Eq. (3) we obtain

$$v'(t) = \frac{1}{\tau}[z(t+2\tau) - z(t+\tau)]$$

$$= \frac{1}{\tau} \prod_{T \le t_m < t+\tau} (1 + L_m)^{-1} \left[x(t+2\tau) \prod_{t+\tau \le t_m < t+2\tau} (1 + L_m)^{-1} - x(t+\tau) \right]. \tag{4}$$

Now from Eq. (1), we have

$$x(t+2\tau) - x(t+\tau) < 0.$$

Since $0 < \prod_{t+\tau \le t_m < t+2\tau} (1 + L_m)^{-1} \le 1$, we also have

$$x(t+2\tau) \prod_{t+\tau \le t_m < t+2\tau} (1 + L_m)^{-1} < x(t+\tau). \tag{5}$$

Using Eqs. (4) and (5), we obtain $v'(t) < 0$ for $t \ge T$, $t \ne t_m$. Since $v(t)$ is continuous, it follows that $v'(t) \le 0$ for $t \ge T$. $\quad\square$

Remark 1. If $x(t)$ is eventually negative solution of Eqs. (1) and (2), then $v(t) < 0$, and $v'(t) \ge 0$ eventually.

Let $\sigma = k\tau + \theta$, $k \in \mathbb{N}$, $\theta \in [0, \tau)$, and $q(t) = \min_{t+\tau \le s \le t+2\tau} p(s)$.

Lemma 2. *If $x(t)$ is an eventually positive solution of Eqs. (1) and (2), then $v(t)$ defined by Eq. (3) eventually satisfies the inequality*

$$v(t+2\tau) - v(t+\tau) \prod_{T \le t_m < t+3\tau} (1+L_m) + v(t) + q(t)v(t-k\tau) \le 0. \qquad (6)$$

Proof. Let $x(t) > 0$, $t \ge T$. Then from Eq. (3) we get

$$v(t+2\tau) - v(t+\tau) \prod_{T \le t_m < t+3\tau} (1+L_m) + v(t) + q(t)v(t-\sigma)$$

$$\le \frac{1}{\tau} \left\{ \int_{t+\tau}^{t+2\tau} x(u+2\tau)du - \int_{t+\tau}^{t+2\tau} x(u+\tau)du + \int_{t+\tau}^{t+2\tau} x(u)du + \int_{t+\tau}^{t+2\tau} p(u)x(u-\sigma)du \right\} = 0. \qquad (7)$$

On the other hand, in view of Lemma 1, we have

$$v(t-\sigma) \ge v(t-k\tau).$$

Using the above inequality, we easily obtain Eq. (6) from Eq. (7). The proof is complete. □

Remark 2. Let $x(t)$ be an eventually negative solution of Eqs. (1) and (2). Then $v(t)$ defined by Eq. (3) eventually satisfies the inequality

$$v(t+2\tau) - v(t+\tau) \prod_{T \le t_m < t+3\tau} (1+L_m) + v(t) + q(t)v(t-k\tau) \ge 0.$$

Theorem 1. *Assume that the following conditions are satisfied:*

$(H1)$ $\limsup\limits_{t \to \infty} \prod\limits_{T \le t_m < t+3\tau} (1+L_m) = L < \infty.$

$(H2)$ $\liminf\limits_{t \to \infty} q(t) = K > L^{k+2} \frac{(k+1)^{k+1}}{(k+2)^{k+2}}.$

Then every solution of Eqs. (1) and (2) is oscillatory.

Proof. Suppose to the contrary that $x(t)$ is a nonoscillatory solution of Eqs. (1) and (2). We may assume without any loss of generality that $x(t)$ is eventually positive. From Eq. (6), we have

$$\frac{v(t+2\tau)}{v(t+\tau)} - \prod_{T \le t_m < t+3\tau} (1+L_m) \le -q(t)\frac{v(t-k\tau)}{v(t+\tau)}$$

$$= -q(t)\prod_{j=0}^{k} \frac{v(t-j\tau)}{v(t-(j-1)\tau)}. \qquad (8)$$

Define

$$\alpha(t) = \frac{v(t)}{v(t+\tau)}, \quad t \ge T.$$

Since $v'(t) \leq 0$, it is clear that $\alpha(t) \geq 1$. From Eq. (8), we have

$$\frac{1}{\alpha(t+\tau)} + q(t) \prod_{j=0}^{k} \alpha(t-j\tau) \leq \prod_{T \leq t_m < t+3\tau} (1+L_m). \tag{9}$$

In view of (H1) and (H2), inequality (9) implies that $\alpha(t)$ is bounded. Let $\beta = \liminf_{t \to \infty} \alpha(t)$. Taking the inferior limit on both sides of Eq. (9), we obtain

$$1 + K\beta^{k+2} \leq \beta L.$$

This inequality implies that

$$\beta > \frac{1}{L} \text{ and } \frac{K\beta^{k+2}}{\beta L - 1} \leq 1. \tag{10}$$

Using the fact that

$$\min_{\beta > \frac{1}{L}} \frac{\beta^{k+2}}{\beta L - 1} = \frac{1}{L^{k+2}} \frac{(k+2)^{k+2}}{(k+1)^{k+1}},$$

we obtain from Eq. (10) that

$$\frac{1}{L^{k+2}} \frac{(k+2)^{k+2}}{(k+1)^{k+1}} \leq \frac{1}{K},$$

which however contradicts (H2). If $x(t)$ is an eventually negative solution of Eqs. (1) and (2), we can lead to a contradiction by similar method. The proof is complete.

$\square$

Theorem 2. *In addition to* (H1), *assume that the following condition is satisfied:*

$$\limsup_{t \to \infty} q(t) > L^{k+2}. \tag{11}$$

Then every solution of Eqs. (1) *and* (2) *is oscillatory.*

Proof. Suppose to the contrary that $x(t)$ is a nonoscillatory solution of Eqs. (1) and (2). We may assume without any loss of generality that $x(t)$ is eventually positive. From Eq. (6), we have

$$v(t+2\tau) \leq v(t+\tau) \prod_{T \leq t_m < t+3\tau} (1+L_m). \tag{12}$$

Using Eq. (12), we obtain

$$v(t+\tau) \le v(t-k\tau) \prod_{i=1}^{k+1} \prod_{T \le t_m < t-(i-3)\tau} (1+L_m).$$

Now using the above inequality from Eq. (6), we get

$$q(t)v(t-k\tau) \le v(t+\tau) \prod_{T \le t_m < t+3\tau} (1+L_m)$$

$$\le v(t-k\tau) \prod_{T \le t_m < t+3\tau} (1+L_m)^{k+2}.$$

From the last inequality, we have

$$q(t) \le \prod_{T \le t_m < t+3\tau} (1+L_m)^{k+2}. \tag{13}$$

Taking the superior limit on both sides of Eq. (13), we obtain

$$\limsup_{t \to \infty} q(t) \le L^{k+2},$$

which however contradicts Eq. (11). If $x(t)$ is an eventually negative solution of Eqs. (1) and (2), we can lead to a contradiction by the similar method. The proof is complete. □

Remark 3. If $x(t_n^+) = x(t_n^-)$ for all $n \in \mathbb{N}$, then $L = 1$ and the assertions of Theorems 1 and 2 are valid for nonimpulsive equation.

Corollary 1. *Assume that* $\sum\limits_{m=1}^{\infty} L_m < \infty$, *and* $K > \dfrac{(k+1)^{k+1}}{(k+2)^{k+2}}$, *then every solution of Eqs.* (1) *and* (2) *is oscillatory.*

Corollary 2. *Assume that* $\sum\limits_{m=1}^{\infty} L_m < \infty$. *If* $\limsup\limits_{t \to \infty} q(t) > 1$, *then every solution of Eqs.* (1) *and* (2) *is oscillatory.*

Example 1. Consider the linear impulsive difference equation with continuous variable

$$\begin{cases} \Delta_{1/2}^2 x(t) + \Delta_{1/2} x(t) + x(t) + (e^{-t}+2)x(t-\tfrac{1}{3}) = 0, \ t \ne t_n, \\ x(t_n^+) - x(t_n^-) = \dfrac{1}{n(n+1)} x(t_n), \ t_n = n, \ n \in \mathbb{N}, \end{cases} \tag{14}$$

where $\tau = 1/2$, $\sigma = 1/3$, $p(t) = e^{-t}+2$, $L_n = \dfrac{1}{n(n+1)}$, $n \in \mathbb{N}$. By Corollary 1, every solution of Eq. (14) is oscillatory.

References

1. Agarwal, R.P., Karakoç, F.: A survey on oscillation of impulsive delay differential equations. Comput. Math. Appl. **60**, 1648–1685 (2010)
2. Agarwal, R.P., Grace, S.R., O'Regan, D.: Oscillation Theory for Second Order Dynamic Equations. Taylor Francis, New York (2003)
3. Agarwal, R.P., Karakoç, F., Zafer, A.: A survey on oscillation of impulsive ordinary differential equations. Adv. Diff. Equat. **2010**, 52 (2010). Article ID 354841
4. Bainov, D.D., Simeonov, P.S.: Oscillation Theory of Impulsive Differential Equations. International Publications, Orlando (1998)
5. Jiang, Z., Xu, Y., Lin, L.: Oscillation of solution of impulsive difference equation with continuous variable. Dyn. Contin. Discrete Impuls. Syst. Ser. A Math. Anal.**13**A(Part 2 suppl.), 587–592 (2006)
6. Lakshmikantham, V., Bainov, D.D., Simeonov, P.S.: Theory of Impulsive Differential Equations. World Scientific, Singapore (1998)
7. Samoilenko, A.M., Perestyuk, N.A.: Impulsive Differential Equations. World Scientific, Singapore (1995)
8. Wang, P., Wu, M.: Oscillation of certain second order nonlinear damped difference equations with continuous variable. Appl. Math. Lett. **20**(6), 637–644 (2007)
9. Wei, G., Shen, J.: Oscillation of solutions of impulsive neutral difference equations with continuous variable. Dyn. Contin. Discrete Impuls. Syst. Ser. A Math. Anal. **13**(1), 147–152 (2006)
10. Wu, S., Hou, Z.: Oscillation criteria for a class of neutral difference equations with continuous variable. J. Math. Anal. Appl. **290**(1), 316–323 (2004)
11. Zhang, B.G., Yan, J., Choi, S.K.: Oscillation for difference equations with continuous variable. Comput. Math. Appl. **36**(9), 11–18 (1998)
12. Zhang, B.G.: Oscillation of a class of difference equations with continuous arguments. Appl. Math. Lett. **14** 557–561 (2001)
13. Zhang, B.G., Lian, F.Y.: Oscillation criteria for certain difference equations with continuous variables. Indian J. Pure Appl. Math. **37**(6), 325–341 (2006)
14. Zhang, Z., Bi, P., Chen, J.: Oscillation of a second order nonlinear difference equation with continuous variable. J. Math. Anal. Appl. **255**(1), 349–357 (2001)

On Estimates of the First Eigenvalue for the Sturm–Liouville Problem with Symmetric Boundary Conditions and Integral Condition

Elena Karulina

Abstract We consider the Sturm–Liouville problem with symmetric boundary conditions and an integral condition. We estimate the first eigenvalue λ_1 of this problem for different values of the parameters.

1 Introduction

Consider the Sturm–Liouville problem:

$$y'' - q(x)y + \lambda y = 0, \tag{1}$$

$$\begin{cases} y'(0) - k^2 y(0) = 0, \\ y'(1) + k^2 y(1) = 0, \end{cases} \tag{2}$$

where $q(x)$ belongs to the set A_γ ($\gamma \neq 0$) of nonnegative bounded summable functions on $[0,1]$ such that

$$\int_0^1 q^\gamma(x)dx = 1.$$

We estimate the first eigenvalue $\lambda_1(q)$ of this problem for different values of γ and k.

E. Karulina (✉)
Moscow State University of Economics, Statistics and Informatics, Nezhinskaya str. 7,
Moscow, Russia
e-mail: karulinaes@yandex.ru

S. Pinelas et al. (eds.), *Differential and Difference Equations with Applications*, Springer   457
Proceedings in Mathematics & Statistics 47, DOI 10.1007/978-1-4614-7333-6_40,
© Springer Science+Business Media New York 2013

According to the variation principle $\lambda_1(q) = \inf\limits_{y \in H_1(0,1) \setminus \{0\}} R(q,y)$, where

$$R(q,y) = \frac{\int\limits_0^1 y'^2(x)dx + \int\limits_0^1 q(x)y^2(x)dx + k^2\left(y^2(0) + y^2(1)\right)}{\int\limits_0^1 y^2(x)dx}. \qquad (3)$$

Put $m_\gamma = \inf\limits_{q \in A_\gamma} \lambda_1(q)$, $M_\gamma = \sup\limits_{q \in A_\gamma} \lambda_1(q)$.

Remark 1. Dirichlet problem for the Eq. (1), $q(x) \in A_\gamma$, was considered in [3, 7]. Different problems for the equation $y'' + \lambda q(x)y = 0$, $q(x) \in A_\gamma$ was considered in [2, 6].

2   Results

Theorem 1. *The following assertions are valid:*

1. *If $\gamma \in (-\infty, 0) \cup (0,1)$, then $M_\gamma = +\infty$.*
2. *If $\gamma = 1$, then $M_1 = \xi_*$, where ξ_* is the solution to the equation* $\arctan\dfrac{k^2}{\sqrt{\xi}} = \dfrac{\xi - 1}{2\sqrt{\xi}}$.
3. *Suppose $\gamma > 1$, then*

   (a) *For $k = 0$, we have $M_\gamma = 1$, and this estimate is attained at $q(x) \equiv 1$.*
   (b) *For $k \neq 0$, we have the following inequalities:*

   $$M_\gamma \leq 1 + 2k^2, \quad M_\gamma \leq \pi^2 + 2,$$

   and there exist such functions $u(x) \in H_1(0,1)$ and $q_*(x) \in A_\gamma$ that $R(q_*, u) = M_\gamma$.

Theorem 2. *The following assertions are valid:*

1. *If $\gamma > 1$, then $m_\gamma = \lambda_1^0$, where λ_1^0 is the first eigenvalue of the problem for the equation $y'' + \lambda y = 0$ with conditions (2) and this estimate is attained by $y(x) = C_1 \cos\sqrt{\lambda_1^0}x + C_2 \sin\sqrt{\lambda_1^0}x$, where $C_2 = C_1 k^2 / \sqrt{\lambda_1^0}$.*
2. *Suppose $\gamma > 0$, then*

   (a) *For all k, we have $m_\gamma \leq \pi^2$.*
   (b) *For $k \to \infty$, we have $m_\gamma \to \pi^2$, also $m_\gamma \geq \pi^2 - \frac{4\pi^2}{k^2} + O\left(\frac{1}{k^4}\right)$.*

3. *If $\gamma \leq 1$, then $m_\gamma \geq 1/4$.*

3 Proofs of Some Results

Theorem 1 and proposition 3 of the Theorem 2 were proved in [4,5]. Here we prove propositions 1 and 2 of the Theorem 2.

Suppose $\gamma > 1$ and $k \neq 0$. Let us prove that $m_\gamma = \lambda_1^0$, where λ_1^0 is the first eigenvalue of the problem for the equation $y'' + \lambda y = 0$ with conditions (2).

Proof. Consider Sturm–Liouville problem

$$y''(x) + \lambda y(x) = 0, \tag{4}$$

with conditions (2). Let λ_1^0 be the first eigenvalue of this problem.

According to the variation principle, $\lambda_1^0 = \inf\limits_{y \in H_1(0,1) \setminus \{0\}} R(0, y)$, where

$$R(0, y) = \frac{\int\limits_0^1 y'^2(x)dx + k^2 \left(y^2(0) + y^2(1) \right)}{\int\limits_0^1 y^2(x)dx}.$$

For problem (1), (2), we estimate the first eigenvalue λ_1 using λ_1^0:

$$\lambda_1(q) = \inf\limits_{y \in H_1(0,1) \setminus \{0\}} R(q, y) \geq \inf\limits_{y \in H_1(0,1) \setminus \{0\}} R(0, y) = \lambda_1^0.$$

Therefore $m_\gamma = \inf\limits_{q(x) \in A_\gamma} \lambda_1(q) \geq \lambda_1^0$.

On the other hand, we have

$$m_\gamma = \inf\limits_{q \in A_\gamma}\left(\inf\limits_{y \in H_1(0,1) \setminus \{0\}} R(q, y) \right) \leq \inf\limits_{q \in A_\gamma}\left(\frac{\int\limits_0^1 y_1'^2 dx + k^2 \left(y_1^2(0) + y_1^2(1) \right)}{\int\limits_0^1 y_1^2 dx} + \right.$$

$$\left. + \frac{\int\limits_0^1 q(x)y_1^2 dx}{\int\limits_0^1 y_1^2 dx} \right) = \inf\limits_{q \in A_\gamma} \left(\lambda_1^0 + \frac{\int\limits_0^1 q(x)y_1^2 dx}{\int\limits_0^1 y_1^2 dx} \right) \leq \lambda_1^0 + \frac{\int\limits_0^1 q_\varepsilon(x)y_1^2 dx}{\int\limits_0^1 y_1^2 dx},$$

where $y_1(x) = C_1 \cos\sqrt{\lambda_1^0}x + C_2 \sin\sqrt{\lambda_1^0}x$ is the first eigenfunction of problem (4), (2),

$$q_\varepsilon(x) = \begin{cases} \varepsilon^{-1/\gamma}, & 0 < x < \varepsilon, \\ 0, & \varepsilon < x < 1. \end{cases}$$

Note that constants N_1, N_2 exist such that $N_1 \geq (y_1(x))^2 \geq N_2 > 0$ at $x \in [0,1]$. Hence

$$\lambda_1^0 + \frac{\int_0^1 q_\varepsilon(x)y_1^2 dx}{\int_0^1 y_1^2 dx} \leq \lambda_1^0 + \frac{N_1}{N_2} \cdot \varepsilon^{1-1/\gamma} \to \lambda_1^0 \quad \text{as } \varepsilon \to 0.$$

Therefore, $m_\gamma = \lambda_1^0$. $\square$

Suppose $\gamma > 0$. Let us prove that $m_\gamma \to \pi^2$ as $k \to \infty$.

(a) First let us prove that if $\gamma > 0$, then $m_\gamma \leq \pi^2$.

Proof. Put

$$y_\delta(x) = \begin{cases} \sin \frac{\pi x}{\delta}, & 0 < x < \delta, \\ 0, & \delta < x < 1 \end{cases} \quad \text{and} \quad q_\delta(x) = \begin{cases} 0, & 0 < x < \delta, \\ (1-\delta)^{-\frac{1}{\gamma}}, & \delta < x < 1, \end{cases}$$

where $\delta \to 1 - 0$.

Then we have $R(q_\delta, y_\delta) = \dfrac{\frac{\pi^2}{2\delta} + 0 + k^2 \sin^2 \frac{\pi}{\delta}}{\delta/2}$.

Therefore we obtain

$$m_\gamma = \inf_{q \in A_\gamma} \left[\inf_{y \in H_1(0,1)\setminus\{0\}} R(q,y) \right] \leq R(q_\delta, y_\delta) \to \pi^2 \quad \text{as } \delta \to 1-0.$$

$\square$

(b) Now let us prove that if $\gamma > 0$, then $m_\gamma \to \pi^2$ as $k \to \infty$.

Proof. Consider λ_1^0—the first eigenvalue of problem (4), (2), which at $k^2 > \pi/2$ is the minimal positive solution for the equation

$$\tan \sqrt{\lambda} = \frac{2\sqrt{\lambda}k^2}{\lambda - k^4}. \tag{5}$$

This solution tends to $\pi^2 - 0$ as $k \to \infty$. Since $m_\gamma \geq \lambda_1^0$, we get that $m_\gamma \to \pi^2$ as $k \to \infty$.

Note that $\lambda_1^0 \to 0$ as $k^2 \to 0$. $\square$

Suppose $\gamma > 0$. Let us prove that $m_\gamma \geq \pi^2 - \frac{4\pi^2}{k^2} + O\left(\frac{1}{k^4}\right)$ as $k \to \infty$.

Proof. Let us make in Eq. (5) the change of variables $z = 1/k^2$, $t = \sqrt{\lambda}$, then this equation has the form

$$\tan t = -\frac{2tz}{1 - t^2 z^2}. \tag{6}$$

The function $F(z,t) = \tan t + \dfrac{2tz}{1 - t^2 z^2}$ is continuous with all its derivatives at the point $(z_0, t_0) = (0, \pi)$, is equal to zero at this point, and its derivative $F_t'(z_0, t_0) \neq 0$. Therefore in some neighbourhood of the point (z_0, t_0), there exists a unique solution $t(z)$ of the Eq. (6), and function $t(z)$ is continuous with all its derivatives of any order. Hence in a neighbourhood of the point (z_0, t_0) we have Taylor formula:

$$t(z) = t(z_0) + \frac{t'(z_0)}{1!}(z - z_0) + \cdots + \frac{t^{(n-1)}(z_0)}{(n-1)!}(z - z_0)^{n-1} + O((z - z_0)^n)$$

for any $n \in \mathbb{N}$. We find the coefficients for the first two terms:

$$t(z_0) = t_0 = \pi, \qquad t'(z_0) = -\frac{F_z'(z_0, t_0)}{F_t'(z_0, t_0)} = -2\pi.$$

So we obtain $t(z) = \pi - 2\pi z + O(z^2)$.

Therefore we get $m_\gamma \geq t^2 = \pi^2 - \dfrac{4\pi^2}{k^2} + O\left(\dfrac{1}{k^4}\right)$ as $k \to \infty$. $\square$

4 Appendix

It's possible to construct the graphs representing the dependence of m_γ and M_γ from k^2 for different values of γ (Figs. 1–4). Moreover, in the cases where we can't obtain the accurate estimates, we can define the regions in the $k^2 O\lambda$-plane, which m_γ and M_γ belong to. For constructing of these graphs we need the propositions of the Theorems 1 and 2 and some additional estimates.

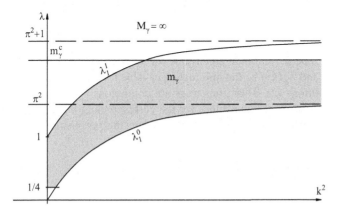

Fig. 1 $\gamma < 0$

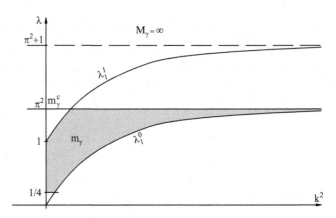

Fig. 2 $\gamma \in (0;1)$

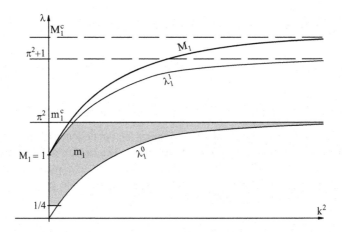

Fig. 3 $\gamma = 1$

1. Let $\lambda_1^0(k^2)$ be the minimal eigenvalue for the problem $y'' + \lambda y = 0$ with conditions (2), and $\lambda_1^1(k^2)$ be the minimal eigenvalue for the problem $y'' - y + \lambda y = 0$ with conditions (2). These functions are continuous by k^2, increase by k^2 and tend to the minimal eigenvalues for the corresponding Dirichlet problems (see [1]). The function $\lambda_1^0(k^2)$ was considered in the propositions 1 and 2 of the Theorem 2, and note that $\lambda_1^1 = \lambda_1^0 + 1$.

So for all γ we have

$$M_\gamma \geq \lambda_1^1 \geq m_\gamma \geq \lambda_1^0.$$

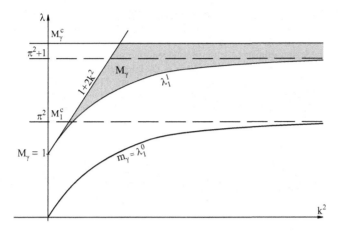

Fig. 4 $\gamma > 1$

2. We have

$$
\lambda_1 = \inf_{y \in H_1(0,1) \setminus \{0\}} \frac{\int_0^1 y'^2(x)dx + \int_0^1 q(x)y^2(x)dx + k^2 \left(y^2(0) + y^2(1) \right)}{\int_0^1 y^2(x)dx}
$$

$$
\leq \inf_{y \in H_1^0(0,1) \setminus \{0\}} \frac{\int_0^1 y'^2(x)dx + \int_0^1 q(x)y^2(x)dx + k^2 \left(y^2(0) + y^2(1) \right)}{\int_0^1 y^2(x)dx}
$$

$$
= \inf_{y \in H_1^0(0,1) \setminus \{0\}} \frac{\int_0^1 y'^2(x)dx + \int_0^1 q(x)y^2(x)dx}{\int_0^1 y^2(x)dx} = \lambda_1^c,
$$

where λ_1^c is the minimal eigenvalue for the Dirichlet problem for Eq. (1), where $q(x) \in A_\gamma$. Therefore $m_\gamma \leq m_\gamma^c$ and $M_\gamma \leq M_\gamma^c$, where m_γ^c and M_γ^c are the corresponding estimates for λ_1^c, obtained for all γ in [3].

References

1. Courant, R., Hilbert, D.: Methods of Mathematical Physics, vol. 1 and 2, p. 560. Wiley, New York (1989)
2. Egorov, Yu., Kondratiev, V.: On spectral theory of elliptic operators (English). Operator Theory: Advances and Applications, vol. 89, p. 328. Birkhauser, Basel (1996)

3. Ezhak, S.S.: On the estimates for the minimum eigenvalue of the Sturm–Liouville problem with integral condition (English). J. Math. Sci. New York **145**(5), 5205–5218 (2007); translation from Sovrem. Mat. Prilozh. **36**, 56–69 (2005)
4. Karulina, E.: Some estimates for the first eigenvalue of the Sturm–Liouville problem with symmetric boundary conditions. Proceedings of the International miniconference of Qualitative Theory of Differential Equations and Applications, pp. 116–123. MESI, Moscow (2010). ISBN:978-5-7764-0607-2
5. Karulina, E.: Some estimates for the minimal eigenvalue of the Sturm–Liouville problem with third-type boundary conditions. J. Math. Bohemica, Praha, Czech Republic **136**(4), 377–384 (2011)
6. Muryshkina, O.V.: On estimates for the first eigenvalue of the Sturm–Liouville problem with symmetric boundary conditions (Russian). Vestnik Molodyh Uchenyh. – 3'2005. Series: Applied mathematics and Mechanics. – 1'2005, p. 36–52
7. Vinokurov, V.A., Sadovnichii, V.A.: On the range of variation of an eigenvalue when the potential is varied (English. Russian original). Dokl. Math. **68**(2), 247–252 (2003); translation from Dokl. Ross. Akad. Nauk **392**(5), 592–597 (2003)

On Polyhedral Estimates for Trajectory Tubes of Differential Systems with a Bilinear Uncertainty

Elena K. Kostousova

Abstract The paper deals with the state estimation problem in control theory under set-membership uncertainty. We consider linear systems of ordinary differential equations (ODE) with parallelepiped-valued uncertainties in initial states and interval uncertainties in coefficients of the system. As a result we have the uncertainty of the bilinear type and essentially nonlinear problem. We construct internal and external estimates for trajectory tubes of such systems. Using discrete-time approximations and techniques of the "polyhedral calculus" and passing to the limit in the discrete-time estimates, we obtain nonlinear ODE systems which describe the evolution of the parallelotope-valued estimates for reachable sets (time cross-sections of the trajectory tubes). The main results are obtained for internal estimates. The properties of the obtained ODE systems are investigated; existence and uniqueness of solutions and also nondegeneracy of estimates are established. Results of numerical simulations are presented.

1 Introduction

The problem of constructing trajectory tubes is an essential theme in control theory under set-membership uncertainty [16]. Since practical construction of these tubes may be cumbersome, different numerical methods are devised for this cause, in particular methods based on approximations of sets either by arbitrary polytopes with a large number of vertices or by unions of points. Such methods may require much calculations, especially for large dimension systems.

Another techniques use ellipsoidal calculus, interval analysis or "polyhedral calculus", the latter of which operates with parallelepipeds and parallelotopes as

---

E.K. Kostousova (✉)

Institute of Mathematics and Mechanics, Ural Branch of the Russian Academy of Sciences, 16 S. Kovalevskaja street, Ekaterinburg 620990, Russia

e-mail: kek@imm.uran.ru

S. Pinelas et al. (eds.), *Differential and Difference Equations with Applications*, Springer Proceedings in Mathematics & Statistics 47, DOI 10.1007/978-1-4614-7333-6_41,
© Springer Science+Business Media New York 2013

basic sets and extends, in this sense, the interval analysis (see, e.g. [3, 6, 8–10, 14–17, 19] and references therein). Fair results in this area were obtained for linear systems with set-valued additive uncertain inputs.

It is also important to study linear systems when system matrices are uncertain too. This leads to the bilinear uncertainty and additional difficulties due to nonlinearity of the problem (in particular, reachable sets, i.e. cross-sections of trajectory tubes, can be non-convex). There are some results for such systems with different types of bounds on uncertainties (see, e.g. [2, 4, 7]), including constructing external ellipsoidal estimates (in particular, [3, 19]) and external interval (or box-valued) estimates (e.g. [9, 15, 17]).

We construct polyhedral (parallelepiped-valued and parallelotope-valued) estimates for reachable sets and trajectory tubes of differential systems with parallelepiped-valued uncertainties in initial states and interval uncertainties in coefficients of the system. The work continues the researches [11–13]. In contrast to classical interval analysis [1], faces of our estimates may be not parallel to the coordinate planes. The main results are obtained for internal estimates. Using discrete-time approximations, techniques of the "polyhedral calculus" and passage to the limit, we obtain nonlinear systems of ordinary differential equations (ODE) which describe the evolution of the internal parallelotope-valued estimates for reachable sets. The properties of the obtained ODE systems are investigated; existence and uniqueness of solutions and also nondegeneracy of estimates are established. ODE for external estimates were obtained earlier [11]. Here they are rewritten, for unification, in the form for parallelotopes. Results of numerical simulations are presented.

The following notation is used below: $\mathbb{R}^n$ is n-dimensional vector space; $\top$ is the transposition symbol; $\|x\|_2 = (x^\top x)^{1/2}$, $\|x\|_\infty = \max_{1 \le i \le n} |x_i|$ are vector norms for $x = (x_1, x_2, \ldots, x_n)^\top \in \mathbb{R}^n$; $e^i = (0, \ldots, 0, 1, 0, \ldots, 0)^\top$ is the unit vector oriented along the axis $0x_i$ (the unit stands at i-position); $e = (1, 1, \ldots, 1)^\top$; $\mathbb{R}^{n \times m}$ is the space of real $n \times m$-matrices $A = \{a_i^j\} = \{a^j\}$ (with columns a^j); I is the unit matrix; 0 is the zero matrix (vector); $\text{Abs}\, A = \{|a_i^j|\}$ for $A = \{a_i^j\}$; $\text{diag}\, \pi$, $\text{diag}\, \{\pi_i\}$ is the diagonal matrix A with $a_i^i = \pi_i$ (π_i are the components of the vector π); $\det A$ is the determinant of $A \in \mathbb{R}^{n \times n}$.

2 Problem Formulation

Consider the system ($x \in \mathbb{R}^n$ is the state)

$$\dot{x} = A(t)x + w(t), \quad t \in T = [0, \theta], \tag{1}$$

where the input (control) $w(t) \in \mathbb{R}^n$ is a given Lebesgue measurable function; the initial state $x(0) = x_0 \in \mathbb{R}^n$ and the measurable matrix function $A(t) \in \mathbb{R}^{n \times n}$ are unknown but subjected to given set-valued constraints

$$x_0 \in \mathscr{X}_0, \tag{2}$$

$$A(t) \in \mathscr{A}(t) = \{A \in \mathbb{R}^{n \times n} | \underline{A}(t) \le A \le \overline{A}(t)\}, \quad \text{a.e. } t \in T, \tag{3}$$

the matrix functions $\underline{A}(t)$, $\overline{A}(t)$ are assumed to be continuous. Matrix and vector inequalities ($\le, <, \ge, >$) here and below are understood componentwise. The interval constraints (3) can be rewritten in the form

$$A(t) \in \mathscr{A}(t) = \{A \,|\, \text{Abs}\,(A - \tilde{A}(t)) \le \hat{A}(t)\}, \quad \tilde{A} = (\underline{A} + \overline{A})/2, \quad \hat{A} = (\overline{A} - \underline{A})/2. \tag{4}$$

Let $\mathscr{X}(t) = \mathscr{X}(t, 0, \mathscr{X}_0)$ be a *reachable set* of system (1)–(3) at time $t > 0$ that is the set of those points $x \in \mathbb{R}^n$, for each of which there exists a pair $\{x_0, A(\cdot)\}$ that satisfies (2), (3) and generates a solution $x(\cdot)$ of (1) that satisfies $x(t) = x$. The multivalued function $\mathscr{X}(t)$, $t \in T$, is known as the *trajectory tube* $\mathscr{X}(\cdot)$.

We presume the given set $\mathscr{X}_0$ to be a parallelepiped (then the sets $\mathscr{X}(t)$ are not obliged to be parallelepipeds) and look for external and internal parallelepiped-valued or parallelotope-valued (shorter, *polyhedral*) estimates $\mathscr{P}^{\pm}(t)$ for $\mathscr{X}(t)$.

By a *parallelepiped* $\mathscr{P}(p, P, \pi) \subset \mathbb{R}^n$ we mean a set such that $\mathscr{P} = \mathscr{P}(p, P, \pi) = \{x \in \mathbb{R}^n | x = p + \sum_{i=1}^n p^i \pi_i \xi_i, \|\xi\|_\infty \le 1\}$, where $p \in \mathbb{R}^n$; $P = \{p^i\} \in \mathbb{R}^{n \times n}$ is such that $\det P \ne 0$, $\|p^i\|_2 = 1\}$[1]; $\pi \in \mathbb{R}^n$, $\pi \ge 0$. It may be said that p is the centre of the parallelepiped; P is the orientation matrix; p^i are the "directions"; and π_i are the values of its "semi-axes". We call a parallelepiped *nondegenerate* if all $\pi_i > 0$.

By a *parallelotope* $\mathscr{P}[p, \bar{P}] \subset \mathbb{R}^n$ we mean a set $\mathscr{P} = \mathscr{P}[p, \bar{P}] = \{x \in \mathbb{R}^n | x = p + \bar{P}\xi, \|\xi\|_\infty \le 1\}$, where $p \in \mathbb{R}^n$ and the matrix $\bar{P} = \{\bar{p}^i\} \in \mathbb{R}^{n \times m}$, $m \le n$, may be singular. We call a parallelotope $\mathscr{P}$ *nondegenerate*, if $m = n$ and $\det \bar{P} \ne 0$.

Each parallelepiped $\mathscr{P}(p, P, \pi)$ is a parallelotope $\mathscr{P}[p, \bar{P}]$ with $\bar{P} = P \text{diag} \pi$, and each nondegenerate parallelotope is a parallelepiped with $P = \bar{P} \text{diag} \{\|\bar{p}^i\|_2^{-1}\}$, $\pi_i = \|\bar{p}^i\|_2$ or, in a different way, with $P = \bar{P}$, $\pi = e$, where $e = (1, 1, \ldots, 1)^\top$.

We call $\mathscr{P}$ an *external (internal) estimate* for $\mathscr{Q} \subset \mathbb{R}^n$ if $\mathscr{P} \supseteq \mathscr{Q}$ ($\mathscr{P} \subseteq \mathscr{Q}$).

Assumption 1. The set $\mathscr{X}_0 = \mathscr{P}_0 = \mathscr{P}[p_0, \bar{P}_0] = \mathscr{P}(p_0, P_0, \pi_0)$ is a parallelepiped.

Problem 1. Find some external $\mathscr{P}^+(t)$ and internal $\mathscr{P}^-(t)$ polyhedral (parallelotope-valued) estimates[2] for reachable sets $\mathscr{X}(t)$: $\mathscr{P}^-(t) \subseteq \mathscr{X}(t) \subseteq \mathscr{P}^+(t)$, $t \in T$.

3   Auxiliary Discrete-Time Systems: Primary Estimates

We will obtain differential equations for the estimates. We follow arguments similar to [16]. The first step in this way is to construct estimates for reachable sets $\mathscr{X}[k]$

---

[1] The normality condition $\|p^i\|_2 = 1$ may be omitted to simplify formulas (particulary, it ensures the uniqueness of the representation of a parallelepiped with nonzero values of semi-axes).

[2] Our estimates will satisfy the generalised semigroup property [16] which is analogues to the well-known semigroup property for $\mathscr{X}(t)$.

of auxiliary discrete-time systems (the Euler approximations of the initial system):

$$x[k] = A[k]x[k-1] + w[k], \quad k = 1,2,\ldots,N; \quad x[0] \in \mathscr{P}_0;$$
$$w[k] = h_N \int_{t_{k-1}}^{t_k} w(\tau)d\tau; \tag{5}$$
$$A[k] \in \mathscr{A}[k] = \{I + h_N A \,|\, A \in \mathscr{A}(t_{k-1})\},$$

where $t_k = kh_N$, $h_N = \theta N^{-1}$. It is known that $\mathscr{X}[k]$ satisfy the recurrence relations

$$\mathscr{X}[k] = \mathscr{A}[k] \circ \mathscr{X}[k-1] + w[k], \quad k = 1,2,\ldots,N, \quad \mathscr{X}[0] = \mathscr{P}_0, \tag{6}$$

which involve some operation with sets (*multiplying an interval matrix* $\mathscr{A} = \{A \in \mathbb{R}^{n \times n} | \underline{A} \leq A \leq \overline{A}\}$ *on a set* $\mathscr{X} \subset \mathbb{R}^n$): $\mathscr{A} \circ \mathscr{X} = \{y \in \mathbb{R}^n | y = Ax, A \in \mathscr{A}, x \in \mathscr{X}\}$.

Therefore we can calculate polyhedral estimates for $\mathscr{X}[k]$ if we are able to construct primary external and internal polyhedral estimates for $\mathscr{A} \circ \mathscr{P}$, where $\mathscr{P}$ is a parallelotope. The ways of constructing such estimates are described in [11–13]. Then we have recurrence relations [11–13] for polyhedral estimates $\mathscr{P}^{\pm}[k]$ for $\mathscr{X}[k]$. Passing to the limit as $N \to \infty$ ($h_N \to 0$) we obtain the corresponding nonlinear ODE systems for parallelotopes $\mathscr{P}^{\pm}(t)$. These equations are considered below.

4 Internal Estimates

The formal passage to the limit for $\mathscr{P}^-[k]$ gives the following ODE system which describes the dynamics of parallelotopes $\mathscr{P}^-(t) = \mathscr{P}[p^-(t), \bar{P}^-(t)]$:

$$\frac{dp^-}{dt} = \tilde{A}(t)p^- + w(t), \quad p^-(0) = p_0; \tag{7}$$

$$\frac{d\bar{P}^-}{dt} = \left(\tilde{A}(t) + \operatorname{diag}\alpha(t, \bar{P}^-; J(t))\right)\bar{P}^-, \quad \bar{P}^-(0) = \bar{P}_0,$$
$$\alpha_i(t, \bar{P}^-; J(t)) = \hat{a}_i^{j_i}(t)\,\eta_{j_i}(t, \bar{P}^-)/(e^{i^\top}(\operatorname{Abs}\bar{P}^-)e), \quad i = 1,2,\ldots,n, \tag{8}$$
$$\eta(t, \bar{P}^-) = \max\{0, \operatorname{Abs} p^-(t) - (\operatorname{Abs}\bar{P}^-)e\}$$

(the operation of maximum is understood componentwise). Here $\{j_1, j_2, \ldots, j_n\} = J$ is an arbitrary permutation of numbers $\{1,2,\ldots,n\}$. Let $\mathbb{J}$ be the set of all Lebesgue measurable vector functions $J(\cdot)$ with values $J(t)$ being arbitrary permutations of numbers $\{1,2,\ldots,n\}$.

Theorem 1. *Let all the assumptions about system* (1), (2), *and* (4), *mentioned above, be satisfied and $\mathscr{P}_0$ be a nondegenerate parallellotope ($\det \bar{P}_0 \neq 0$). Then system* (7) *and* (8) *has a unique solution on $T = [0, \theta]$ whatever is $J(\cdot) \in \mathbb{J}$, and* *parallelotopes $\mathscr{P}^-(t) = \mathscr{P}[p^-(t), \bar{P}^-(t)]$ are internal nondegenerate estimates for the reachable sets $\mathscr{X}(t)$ of system (1), (2), and (4): $\mathscr{P}^-(t) \subseteq \mathscr{X}(t)$, $t \in T$.*

Sketch of the Proof. Since there is an operation of division in (8), we obtain estimates which show that solutions of (8) cannot leave some domain where the Caratheodory conditions are satisfied and the right-hand side of (8) is Lipschitz in state variables. Then the existence, uniqueness and extendability of a solution follow from the known results [5, pp. 7, 8, 10]. The mentioned estimates also guarantee that $\det \bar{P}(t) \neq 0$. The proof of the inclusion $\mathscr{P}^-(t) \subseteq \mathscr{X}(t)$ is similar to [10, Theorem 4.1] with applying constructions from the proof of [13, Theorem 3.3]. $\quad\square$

Remark 1. Obviously, we have $\mathscr{X}(t) \supseteq \mathscr{X}^0(t) \equiv \mathscr{P}^{0-}(t)$, $t \in T$, where $\mathscr{X}^0(t)$ are reachable sets of system (1) under assumptions $x_0 \in \mathscr{P}_0$ and $A(\cdot) \equiv \tilde{A}(\cdot)$, and parallelotopes $\mathscr{P}^{0-}(t)$ are determined by (7) and (8) when $\alpha \equiv 0$. We call these parallelotopes $\mathscr{P}^{0-}(t)$ *trivial internal estimates for $\mathscr{X}(t)$.*

Compare estimates $\mathscr{P}^-(t)$ satisfying (7) and (8) with $\mathscr{P}^{0-}(t)$ in the sense of volume. We would remind that volume of a nondegenerate parallelotope $\mathscr{P} = \mathscr{P}[p, \bar{P}] \subset \mathbb{R}^n$ is equal to $\operatorname{vol} \mathscr{P} = 2^n |\det \bar{P}|$.

Corollary 1. *Under conditions of Theorem 1 we have*

(i) $\operatorname{vol} \mathscr{P}^-(t) = \operatorname{vol} \mathscr{P}^{0-}(t) \exp \psi(t)$, $t \in T$, where $\psi(t) = \int_0^t e^\top \alpha(\tau, \bar{P}^-(\tau); J(\tau)) d\tau$.
Therefore $\operatorname{vol} \mathscr{P}^-(t) \geq \operatorname{vol} \mathscr{P}^{0-}(t)$; *and* $\operatorname{vol} \mathscr{P}^-(t) > \operatorname{vol} \mathscr{P}^{0-}(t)$ *iff* $\psi(t) > 0$.

(ii) *If it is turned out that* $\mathscr{P}^-(t) \ni 0$ *for all* $t \in T$, *then* $\mathscr{P}^-(t) \equiv \mathscr{P}^{0-}(t)$, $t \in T$.

Remark 2. We can choose $J(\cdot)$ in different ways. A simple way is to apply a "local" optimisation. Fix a natural number N and introduce a grid T_N of times $\tau_k = k h_N$, $k = 0, 1, \ldots, N$, $h_N = \theta N^{-1}$. Let us, for each $\tau_k \in T_N$, solve the optimisation problem $e^\top \alpha(\tau_k, \bar{P}^-(\tau_k); J) \to \max_J$ over all possible permutations $J = \{j_1, j_2, \ldots, j_n\}$ assuming that $\bar{P}^-(\tau_k)$ has already been found. Then we can sequentially construct the piecewise constant function $J(t) \equiv J(\tau_k) \in \operatorname{Argmax}_J e^\top \alpha(\tau_k, \bar{P}^-(\tau_k); J)$, $t \in (\tau_k, \tau_{k+1})$, $k = 0, 1, \ldots, N-1$, and find $\bar{P}^-(\cdot)$. Note that the described procedure is not obliged to give the estimates $\mathscr{P}^-(t)$ with maximal volume even if $N \to \infty$.

5 External Estimates

In [11], the ODE systems of two types were obtained for external estimates for $\mathscr{X}(t)$ in the form of parallelepipeds $\mathscr{P}^+(t) = \mathscr{P}(p^+(t), P(t), \pi^+(t))$, where $P(t) \in \mathbb{R}^{n \times n}$, $t \in T$, is an arbitrary continuously differentiable matrix function such that

$$\det P(t) \neq 0, \quad t \in T. \tag{9}$$

For the unification with the description of the internal estimates, let us rewrite the mentioned ODE systems in the form for parallelotopes and make it only for more accurate estimates of the II type. Consider the following ODE system:

$$\frac{dp^+}{dt} = \dot{P}P^{-1}p^+ + P(\Phi^{(+)} - \Phi^{(-)})/2 + w, \quad p^+(0) = p_0;$$

$$\frac{d\bar{P}^+}{dt} = \dot{P}P^{-1}\bar{P}^+ + P\,\mathrm{diag}\left((\Phi^{(+)} + \Phi^{(-)})/2\right), \quad \bar{P}^+(0) = P(0)\,\mathrm{diag}\left(\mathrm{Abs}\,(P(0)^{-1}\bar{P}_0)\,e\right),$$

$$\text{where } \Phi_i^{(\pm)} = \max_{\xi \in \Xi_i^\pm} \left(\pm P^{-1}(\tilde{A} - \dot{P}P^{-1})x + \mathrm{Abs}\,(P^{-1})\hat{A}\,\mathrm{Abs}\,x\right)_i,$$

$$x = p^+ + \bar{P}^+\xi; \quad \Xi_i^\pm = \{\xi \mid \xi \in \mathbf{E}(\mathscr{P}[0,I]), \ \xi_i = \pm 1\}, \quad i = 1,2,\dots,n,$$

(10)

symbol $\mathbf{E}(\mathscr{P})$ denotes the set of all vertices of a parallelotope, namely, the set of points of the form $x = p + \sum_{i=1}^m \bar{p}^i\xi_i$, $\xi_i \in \{-1,1\}$.

Theorem 2. *Let all the above assumptions be satisfied and $P(t) \in \mathbb{R}^{n\times n}$ be an arbitrary continuously differentiable function satisfying (9). Then the system (10) has a unique solution on T, and the sets $\mathscr{P}^+(t) = \mathscr{P}[p^+(t), \bar{P}^+(t)]$ are the external estimates for the reachable sets $\mathscr{X}(t)$ of system (1), (2), and (4): $\mathscr{X}(t) \subseteq \mathscr{P}^+(t)$, $t \in T$.*

Remark 3. In fact, Theorem 2 describes the whole family of estimates where the function $P(\cdot)$ is a parameter. Some heuristic ways of choosing $P(\cdot)$ were indicated in [11] (in particular, to find $P(\cdot)$ from relations $\dot{P} = \tilde{A}(t)P$, $P(0) = P_0$, or put $P(t) \equiv I$).

6 Examples

Consider some examples of constructing the estimates. The estimates were calculated using the Euler approximations (5) with $N = 100$ (in fact, the estimates for $\mathscr{X}[k]$ are presented in figures below). But it would be emphasised that different schemes of approximation can be used for solving the obtained differential systems and finding the estimates. Some more examples can also be found in [11–13].

Example 1. Let $\tilde{A}(t) \equiv \begin{bmatrix} -0.5 & 2 \\ 1 & -0.5 \end{bmatrix}$; $\hat{A}(t) \equiv \begin{bmatrix} 0 & 1.5 \\ 0.5 & 0 \end{bmatrix}$; $w(t) \equiv 0$; $\mathscr{P}_0 = \mathscr{P}((1,1.5)^\top, I,$

$(0.05, 0.05)^\top)$. The system of such type may be interpreted as the Richardson arms race model [18] known in political science. Figure 1a presents tubes formed by external and internal estimates for $\mathscr{X}[k]$. Figure 1b shows the initial set $\mathscr{P}_0$ (*dashed line*), three external estimates for $\mathscr{X}[N]$ (*thin lines*) and the internal one. For comparing, the trivial internal estimate $\mathscr{P}^{0-}[N]$ is shown too (*dashed line*). The reachable set belongs to the intersection of external estimates and contains the internal ones.

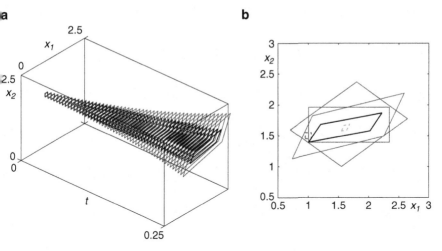

Fig. 1 External and internal estimates for $\mathscr{X}[\cdot]$ (**a**) and $\mathscr{X}[N]$ (**b**) in Example 1

Example 2. Let $\tilde{A} \equiv \begin{bmatrix} -1 & 0 & 5 \\ 1 & -1 & 0 \\ 0 & 1 & -1 \end{bmatrix}$; $\hat{A} \equiv \begin{bmatrix} 0 & 0 & 3 \\ 0 & 0 & 0 \\ 0 & 0 & 0 \end{bmatrix}$; $w \equiv (-0.6, -0.4, -0.2)^\top$; $\mathscr{P}_0 = \mathscr{P}((1,1,1)^\top, I, (0.2, 0.2, 0.2)^\top)$. The system of such type may be interpreted as a simple ecological model of dynamics of a number of microorganisms which have 3 stages of development, provide division at the last stage and produce from 2 to 8 descendants [20, p. 112]. Estimates for $\mathscr{X}[\cdot]$ and $\mathscr{X}[N]$ are shown in Fig. 2 which is similar to Fig. 1. Since parallelotopes in this problem are three-dimensional, we present their two-dimensional projections on coordinate plains.

It must be admitted that the proposed estimates may turn out to be rather conservative. But we can calculate them easily via integration of the ODE, and they can give useful information, while it is hard to calculate exact reachable sets. Improved external (possibly nonconvex) estimates in the form of the union of parallelotopes can be constructed for the case of systems with constant coefficients [11].

Acknowledgements The work was supported by the Program of the Presidium of the Russian Academy of Sciences "Mathematical Theory of Control" under the support of the Ural Branch of RAS (project 09-P-1-1014) and by the Russian Foundation for Basic Research (grant 09-01-00223).

References

1. Alefeld, G., Herzberger, J.: Introduction to Interval Computations. Academic, New York (1983)
2. Barmish, B.R., Sankaran, J.: The propagation of parametric uncertainty via polytopes. IEEE Trans. Automat. Control. AC-**24**, 346–349 (1979)

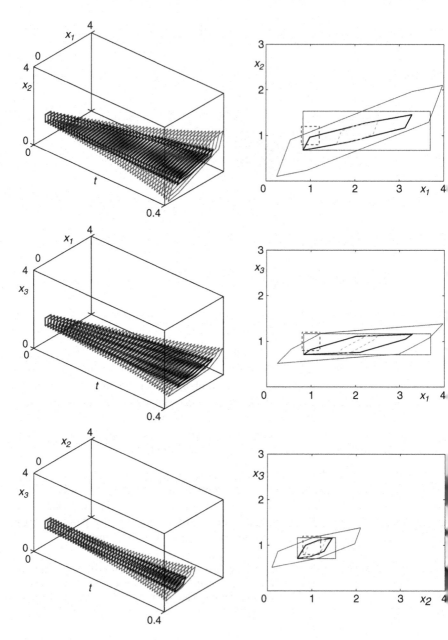

Fig. 2 Projections of external and internal estimates for $\mathscr{X}[\cdot]$ and $\mathscr{X}[N]$ in Example 2

3. Chernousko, F.L., Rokityanskii, D.Ya.: Ellipsoidal bounds on reachable sets of dynamical systems with matrices subjected to uncertain perturbations. J. Optim. Theory Appl. **104**, 1–19 (2000)

4. Digailova, I.A., Kurzhanski, A.B.: On the joint estimation of the model and state of an under-determined system from the results of observations (Russian). Dokl. Akad. Nauk. **384**, 22–27 (2002); Transl. as Dokl. Math. **65**, 459–464 (2002)
5. Filippov, A.F.: Differential Equations with Discontinuous Right-hand Sides (Russian). Nauka, Moscow (1985)
6. Filippova, T.F.: Trajectory tubes of nonlinear differential inclusions and state estimation problems. J. Concr. Appl. Math. **8**, 454–469 (2010)
7. Filippova, T.F., Lisin, D.V.: On the estimation of trajectory tubes of differential inclusions. Proc. Steklov Inst. Math. Suppl. **2**, S28–S37 (2000)
8. Gusev, M.I.: Estimates of reachable sets of multidimensional control systems with nonlinear interconnections. Proc. Steklov Inst. Math. Suppl. **2**, S134–S146 (2010)
9. Kornoushenko, E.K.: Interval coordinatewise estimates for the set of accessible states of a linear stationary system I–IV (Russian). Avtom. Telemekh. (5), 12–22 (1980); (12), 10–17 (1980); (10), 47–52 (1982); (2), 81–87 (1983). Transl. as Autom. Remote Control
10. Kostousova, E.K.: External and internal estimation of attainability domains by means of parallelotopes (Russian). Vychisl. Tekhnol. **3**(2), 11–20 (1998)
11. Kostousova, E.K.: Outer polyhedral estimates for attainability sets of systems with bilinear uncertainty (Russian). Prikl. Mat. Mekh. **66**, 559–571 (2002). Transl. as J. Appl. Math. Mech. **66**, 547–558 (2002)
12. Kostousova, E.K.: On polyhedral estimates for reachable sets of discrete-time systems with bilinear uncertainty (Russian). Avtom. Telemekh. (9), 49–60 (2011). Transl. as Autom. Remote Control. **72**(9), 1841–1851 (2011)
13. Kostousova, E.K.: On polyhedral estimates for trajectory tubes of dynamical discrete-time systems with multiplicative uncertainty. In: Proceedings of 8th AIMS Conference, Discrete Contin. Dyn. Syst. Dynamical Systems, Differential Equations and Applications, vol. Suppl, pp. 864–873, 2011
14. Kostousova, E.K., Kurzhanski, A.B.: Guaranteed estimates of accuracy of computations in problems of control and estimation (Russian). Vychisl. Tekhnol. **2**(1), 19–27 (1997)
15. Kuntsevich, V.M., Kurzhanski, A.B.: Calculation and control of attainability sets for linear and certain classes of nonlinear discrete systems (Russian). Problemy Upravlen. Inform. (1), 5–21 (2010). Transl. as J. Automation and Inform. Sci. **42**, 1–18 (2010)
16. Kurzhanski, A.B., Vályi, I.: Ellipsoidal Calculus for Estimation and Control. Birkhäuser, Boston (1997)
17. Nazin, S.A., Polyak, B.T.: Interval parameter estimation under model uncertainty. Math. Comput. Model. Dyn. Syst. **11**, 225–237 (2005)
18. Nikolskii, M.S.: On controllable variants of the Richardson model in political science (Russian). Trudy Instituta Matematiki i Mekhaniki UrO RAN. **17**(1), 121–128 (2011)
19. Polyak, B.T., Nazin, S.A., Durieu, C., Walter, E.: Ellipsoidal parameter or state estimation under model uncertainty. Automatica J. IFAC. **40**, 1171–1179 (2004)
20. Zaslavskii, B.G., Poluektov, R.A.: Control of Ecological Systems (Russian). Nauka, Moscow (1988)

Analysis and Computational Approximation of a Forward–Backward Equation Arising in Nerve Conduction

P.M. Lima, M.F. Teodoro, N.J. Ford, and P.M. Lumb

Abstract This paper is concerned with the approximate solution of a nonlinear mixed-type functional differential equation (MTFDE) arising from nerve conduction theory. The equation considered describes conduction in a myelinated nerve axon. We search for a solution defined on the whole real axis, which tends to given values at $\pm\infty$. The numerical algorithms, developed previously by the authors for linear problems, were upgraded to deal with the case of nonlinear problems on unbounded domains. Numerical results are presented and discussed.

Keywords Mixed-type functional differential equation • Asymptotic analysis • Newton method • Nerve conduction

1 Introduction

This paper is concerned with a nonlinear MTFDE of the form

$$RCv'(t) = f(v(t)) + v(t - \tau) + v(t + \tau) - 2v(t), \tag{1}$$

P.M. Lima (✉)
Department of Matemática/CEMAT, Instituto Superior Técnico, Universidade
Técnica de Lisboa, Av. Rovisco Pais, 1049-001 Lisboa, Portugal
e-mail: plima@math.ist.utl.pt

M.F. Teodoro
Department of Matemática, Instituto Politécnico de Setúbal, Estefanilha,
2910-761 Setúbal, Portugal
e-mail: mteodoro@est.ips.pt

N.J. Ford • P.M. Lumb
Department of Mathematics, University of Chester, Parkgate Road, Chester CH4BJ, UK
e-mail: njford@chester.ac.uk; p.lumb@chester.ac.uk

S. Pinelas et al. (eds.), *Differential and Difference Equations with Applications*, Springer
Proceedings in Mathematics & Statistics 47, DOI 10.1007/978-1-4614-7333-6_42,
© Springer Science+Business Media New York 2013

where R and C are constants and f is a given function, as described below. We are interested in a solution of (1), increasing on $]-\infty,\infty[$, which satisfies the conditions

$$\lim_{t\to-\infty} v(t) = 0, \qquad \lim_{t\to\infty} v(t) = 1, \qquad v(0) = 0.5. \qquad (2)$$

The problem (1) and (2) was analysed in [1–3], where its physical meaning is explained in detail. The unknown v represents the transmembrane potential at a node in a myelinated axon. The function f reflects the current-voltage model, which is given by

$$f(v) = bv(v-a)(1-v), \qquad (3)$$

where $b > 0$ and $0 < a < 1/2$. R and C are respectively the nodal resistivity and the nodal capacity. The mathematical formulation can be derived from an electric circuit model which assumes the so-called pure saltatory conduction (PSC). This means that the myelin has such high resistance and low capacitance that it completely insulates the membrane; therefore, if a node is sufficiently stimulated and its transmembrane potential reaches a certain threshold level, ionic currents are generated which excite the neighbouring node. As a consequence, this node also attains the mentioned threshold potential. In this way, the process propagates across the nerve axon, giving the impression of an excitation jumping node to node.

In the cited work of Chi, Bell and Hassard the problem (1) and (2) was thoroughly investigated, both from the analytical and numerical point of view. A computational algorithm was proposed and the first numerical results (as far as we are aware) have been obtained for a nonlinear MTFDE.

In this paper, we continue the analytical and numerical investigation of (1), called the discrete Fitzhugh-Nagumo equation. In Sect. 2, we analyse the asymptotic behaviour of its solutions at infinity. In Sect. 3, we discuss the computational methods for its numerical approximation. In Sect. 4, we present some numerical results and we finish with some conclusions in Sect. 5.

2   Asymptotic Approximation at Infinity

Since in the next section we describe computational methods for the numerical solution of the given problem, it is necessary to study the asymptotic behaviour of the required solution at infinity, so that we can define the domain where this solution needs to be computed.

An extensive analysis of this behaviour has been provided in [3], so here we will just recall the main results from that paper.

Let us first consider the case where $t \to -\infty$. According to the conditions (2), $v(-\infty) = 0$, so that in order to linearise (1) about this point, we first use the Taylor expansion for function f:

$$f(v) = f(0) + vf'(0) + \frac{v^2}{2}f''(0) + O(v^3) = vf'(0) + \frac{v^2}{2}f''(0) + O(v^3),$$

where f is given by (3).

As usual, in order to obtain a characteristic equation for (1) at $-\infty$, we must replace f by the main term of its Taylor expansion and assume that v is replaced by

$$w_1(t) = \varepsilon_1 e^{\lambda(t+L)}, \tag{4}$$

where L is a sufficiently large parameter and ε_1 is an estimate for $v_1(-L)$. In this way we obtain the equation

$$\lambda + 2 - f'(0) - 2cosh(\lambda\tau) = 0. \tag{5}$$

This equation has two real roots; since we are interested in a function w_1 that tends to 0 at $-\infty$, we choose the positive one, which we denote by λ_1.

The case where $t \to \infty$ can be handled in an analogous way. In this case, we have the following Taylor expansion for f:

$$f(v) = (v-1)f'(1) + \frac{(v-1)^2}{2}f''(1) + O((v-1)^3).$$

Moreover, as $t \to +\infty$, we assume that v is replaced by

$$w_2(t) = 1 - \varepsilon_2 e^{\lambda(t-L)}, \tag{6}$$

where ε_2 is an estimate of $1 - v_2(L)$. In this way we obtain the characteristic equation

$$\lambda + 2 - f'(1) - 2cosh(\lambda\tau) = 0. \tag{7}$$

In this case we choose the negative root of the characteristic equation λ_2, in order to have $w_2(t) \to 1$, as $t \to +\infty$.

Now we have obtained two representations of the solution of our problem (4) and (6), which can be used to approximate the solution, for $t < -L$ and $t > L$, respectively, where L is a sufficiently large number. According to the form of (1), L must be a multiple of the delay τ; in our computations we have used $L = 2\tau$ or $L = 3\tau$, which is large enough to obtain a reasonable accuracy.

These representations of the solution are used in the computational methods to replace the boundary conditions (2). In the next section we will show how this can be achieved.

3 Computational Methods

3.1 Numerical Methods for Linear Boundary Value Problems

Boundary value problems (BVP) for linear mixed-type functional differential equations (MTFDE) have been developed in [5–7, 9]. In these papers, we have considered equations of the form

$$x'(t) = \alpha(t)x(t) + \beta(t)x(t-1) + \gamma(t)x(t+1), \tag{8}$$

where x is the unknown function and α, β, γ are known functions. MTFDEs of the considered form contain deviating (advanced and delayed) arguments and for this reason are known also as forward–backward equations.

The authors of [4] have developed a new approach to the analysis of the autonomous case. They have analysed MTFDEs as BVP, that is, they have considered the problem of finding a differentiable solution on a certain real interval $[0, k-1]$, given its values on the intervals $[-1, 0]$ and $(k-1, k]$. Assuming that such a solution exists, they have introduced a numerical algorithm to compute it. In [9], a numerical algorithm based on the collocation method was proposed for the solution of such BVPs. In [6, 7] these methods were extended to the nonautonomous case, and a new algorithm, based on the least squares method, was introduced. In [5], a new numerical algorithm was proposed, based on the decomposition of the solution into a growing and a decaying component. This approach, which is based on the analytical results of [8], provides a way of reducing the ill conditioning of the BVP.

The algorithm developed in [6, 9], which will be applied to the solution of the present problem, is based on the so-called ODE approach: the solution is sought as the sum of two terms, one of which is defined from the initial data (boundary conditions and equation) and the other one must be computed as a linear combination of known basis functions (usually splines). Using the method of steps, the problem of computing this term can be reduced to the solution of a BVP for a kth order ODE (k is the length of the interval where the solution must be computed). This last problem can be solved by standard methods of numerical analysis, such as the collocation or the finite elements method. As shown in [6], the error of these methods, when applied to linear equations such as (8) on a bounded interval, is of order h^2.

The numerical method described in this paper, although based on this approach, has two new features: (1) it allows us to deal with problems on the whole real axis (where the boundary conditions are given at infinity) and (2) nonlinear equations are considered. In the next sections we will explain how to handle this case.

3.2   Numerical Solution by the Newton Method

Once we know the approximate solution of the equation for $t \geq L$ and $t \leq -L$ (as described in Sec. 2), the problem is reduced to a BVP on $[-L, L]$, where L is a multiple of τ.

The nonlinear problem can be reduced to a sequence of linear problems by means of the Newton method.

In the ith iteration of the Newton method, we have to solve a linear equation of the form:

$$RCv'_{i+1}(t) - f'(v_i)(v_{i+1}(t) - v_i(t)) - L(v_{i+1}(t)) = f(v_i(t)), \qquad t \in [-L, L], \quad (9)$$

where

$$L(v(t)) = v(t + \tau) + v(t - \tau) - 2v(t).$$

We search for a monotone solution v_{i+1} which satisfies the boundary conditions

$$\begin{aligned} v_{i+1}(t) &= w_1(t), & t \in [-L - \tau, -L], \\ v_{i+1}(t) &= w_2(t), & t \in [L, L + \tau], \end{aligned} \qquad (10)$$

where w_1 and w_2 are given by (4) and (6), respectively.

In order to compute an initial approximation v_0, which enables the convergence of the Newton iteration process, we need the values of $\lambda_1, \lambda_2, \tau, \varepsilon_1$ and ε_2.

These values can be obtained by solving a system of five nonlinear equations:

$$\begin{aligned} &\lambda_1 + 2 - F'(0) - 2\cosh(\lambda_1 \tau) = 0 \\ &\lambda_2 + 2 - F'(1) - 2\cosh(\lambda_2 \tau) = 0 \\ &\lim_{t \to 0-} v(t) = 1/2 \\ &\lim_{t \to 0+} v(t) = 1/2 \\ &\lim_{t \to 0-} v'(t) = \lim_{t \to 0+} v'(t). \end{aligned} \qquad (11)$$

The values of v in this system are computed, using the method of steps and assuming that v satisfies the obtained asymptotic expansions, when $t < -L$ and $t > L$. More precisely, if $v(t)$ is defined at $[-L - \tau, -L]$ by (10), then we can define it on the interval $[-L, -L + \tau]$ and on the following intervals using the recurrence formula:

$$v(t + \tau) = 2v(t) + RCv'(t) - v(t - \tau) + g(v(t)), \qquad (12)$$

where $g(v) = -f(v)$, if v is the solution of the nonlinear equation (1), and $g(v) = -f(v_i) - f'(v_i)(v - v_i)$, if v is the solution of the Newton iterates (9) (here v_i is the preceding iterate).

In the same way, starting from the definition of v at $[L, L+\tau]$ by (10), this function can be defined on $[L-\tau, L]$ by the "backwards" formula:

$$v(t-\tau) = 2v(t) + RCv'(t) - v(t+\tau) + g(v(t)). \tag{13}$$

The system (11) is solved again at each iterate of the Newton method, in order to update the parameter values.

4 Numerical Results

The algorithm was implemented in the form of a MATLAB code. In this section we present and discuss some numerical results.

The test case considered in our computations is one of the cases discussed in [3]. We consider (1), with $R = C = 1$, where f is defined by (3), with $a = 0.05$ and $b = 15$. In this case, by solving system (11), with $L = 2\tau$, we obtain $\lambda_1 = 5.98$, $\lambda_2- = -6.35$, $\varepsilon_1 = 0.00794$, $\varepsilon_2 = 0.00635$ and $\tau = 0.360$.

The graphs of some Newton iterates of the solution are displayed in Fig. 1.

For comparison, we have solved the problem numerically using two finite intervals: with $L = 2\tau$ and $L = 3\tau$.

In Table 1, we display the results with $L = 2\tau$. We denote by N the number of grid points at each subinterval of length τ in the collocation method and by h the corresponding stepsize, so that $h = \tau/N$. Let v_h be the approximate solution.

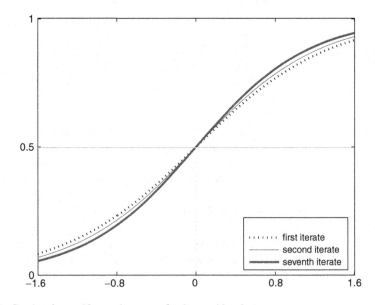

Fig. 1 Graphs of some Newton iterates v_i for the considered test case

Table 1 Estimates of errors and convergence orders, for the solution (first two columns) and for its derivative (last two columns), in the case $L = 2\tau$

N	ε	p	ε	p
20	8.510 E-5	1.89	1.999 E-3	1.99
40	2.178 E-5	1.97	5.046 E -4	1.98
80	5.482 E-6	1.99	1.265 E-4	1.99
160	1.373 E-6	2.00	3.164 E-5	1.99

Table 2 Estimates of errors and convergence orders, for the solution (first two columns) and for its derivative (last two columns), in the case $L = 3\tau$

N	ε	p	ε	p
20	8.440 E-6	1.93	1.821 E-4	1.45
40	2.150 E-6	1.97	5.068 E -5	1.85
80	5.422 E-7	1.99	1.312 E-5	1.95
160	1.360 E-7	1.99	3.324 E-6	1.98

obtained with stepsize h. Then, as a measure of the error we use the Euclidean 2-norm $\varepsilon_h = \|v_h - v_{2h}\|_2$. As an estimate of the convergence order, as usual, we take

$$p = \frac{\log_2(\varepsilon_h)}{\log_2(\varepsilon_{2h})}.$$

In Table 2, we display analogous results, obtained with $L = 3\tau$.

In the overall scheme, the error depends upon the choices of h and L. When we solve the problem numerically on $[-L, L]$, we assume that $v(-L) = \varepsilon_1$, according to (4), and $v(L) = 1 - \varepsilon_2$, according to (6). Taking into account the linearisation process the error of this approximation is of order ε_1^2, in the first case, and ε_2^2, in the second. Therefore the choice of L imposes a limitation on the overall accuracy of the scheme and limits the benefit of reducing the error of the collocation method, which is $O(h^2)$, beyond the point at which the error in the numerical scheme is smaller than the error introduced by the choice of L. Since the values of ε_1 and ε_2 are not known a priori, the value of L must be adjusted experimentally, and it depends on the values of a and b. For the case illustrated by Tables 1 and 2, a choice of $L = 3\tau$ yields a sufficiently small error that the convergence rate of the collocation method is recovered in the estimates.

In both cases, the numerical results suggest that the convergence order of the collocation method is 2, which is in agreement with the theoretical results obtained in [6,7].

An important issue that can be analysed from the numerical results for (1) is the dependence of the propagation speed of the nerve impulses on the parameters a, b of the function f. As remarked in [3], the propagation speed is inversely proportional to the delay τ, which is computed by our algorithm. In Table 3, we observe how this parameter varies with a and b.

From Table 3 we conclude that the propagation speed (inverse of τ) increases with b and when a tends to 0, which is in agreement with the previously obtained results (see [3]).

Table 3 The dependence of τ on a and b

a	$b = 10$	$b = 12$	$b = 14$	$b = 16$
0	0.459649	0.435100	0.403125	0.377948
0.02	0.483611	0.454817	0.421840	0.398989
0.04	0.512678	0.464581	0.445605	0.410900
0.06	0.547146	0.499670	0.462720	0.439170
0.08	0.567360	0.527444	0.490468	0.470200
0.10	0.610765	0.565340	0.530310	0.499321

5 Conclusions

The numerical experiments have shown that the Newton method provides a fast iterative scheme and with convergence from a good initial approximation.

The numerical results also suggest that the collocation method provides second-order convergence, as was observed earlier for linear problems.

Comparing with numerical algorithms used by other authors to solve similar problems (in particular, [3]), the present algorithm has the advantage that the convergence of the iteration process does not require the use of the continuation method (where the algorithm is first applied to a test problem with known exact solution and then this problem is transformed into the target one, by smoothly changing a certain parameter).

In the near future we intend to improve the algorithm and to carry out new numerical investigations of the problem.

Acknowledgements M. F. Teodoro acknowledges financial support from FCT, through grant SFRH/BD/37528/2007. The research of N. Ford and P. Lumb has been supported by a University of Chester International Research Excellence Award, funded under the Santander Universities scheme.

References

1. Bell, J.: Behaviour of some models of myelinated axons. IMA J. Math. Appl. Med. Biol. **1** 149–167 (1984)
2. Bell, J., Cosner, C.: Threshold conditions for a diffusive model of a myelinated axon. J. Math Biol. **18**, 39–52 (1983)
3. Chi, H., Bell, J., Hassard, B.: Numerical solution of a nonlinear advance-delay-differential equation from nerve conduction theory. J. Math. Biol. **24**, 583–601 (1986)
4. Ford, N., Lumb, P.: Mixed-type functional differential equations: a numerical approach. J Comput. Appl. Math. **229**, 471–479 (2009)
5. Ford, N., Lumb, P., Lima, P., Teodoro, F.: The numerical solution of forward–backward differential equations: Decomposition and related issues. J. Comput. Appl. Math. **234**, 2745–2756 (2010)
6. Lima, P., Teodoro, F., Ford, N., Lumb, P.: Analytical and numerical investigation of mixed-type functional differential equations. J. Comput. Appl. Math. **234**, 2826–2837 (2010)

7. Lima, P., Teodoro, F., Ford, N., Lumb, P.: Finite element solution of a linear mixed-type functional differential equation. Numer. Algorithms **55**, 301–320 (2010)

8. Mallet-Paret, J., Verduyn Lunel, S.: Mixed-type functional differential equations, holomorphic factorization and applications. Proceedings of Equadiff 2003, International Conference on Differential Equations, HASSELT 2003, pp. 73–89. World Scientific, Singapore (2005)

9. Teodoro, F., Lima, P., Ford, N., Lumb, P.: New approach to the numerical solution of forward–backward equations. Front. Math. China **4**(1), 155–168 (2009)

Multiple Solutions for a Class of Degenerate Quasilinear Elliptic Systems

G.A. Afrouzi and S. Mahdavi

Abstract Some existence and multiplicity results of weak solutions are established for a class of degenerate quasilinear elliptic system by using Ekeland's variational principle, the Mountain Pass Theorem and the Critical Point Theory.

1 Introduction

Let us consider the problem

$$-\operatorname{div}(h_1(x)|\nabla u|^{p-2}\nabla u) = \lambda\, a(x)|u|^{p-2}u + \gamma r(x)\alpha|u|^{\alpha-2}u|v|^{\beta} \quad \text{in } \Omega$$

$$-\operatorname{div}(h_2(x)|\nabla v|^{q-2}\nabla v) = \mu\, b(x)|v|^{q-2}v + \gamma r(x)\beta|u|^{\alpha}|v|^{\beta-2}v \quad \text{in } \Omega$$

$$u = v = 0 \qquad \text{on } \partial\Omega \qquad (1)$$

where Ω is a bounded smooth domain in $\mathbf{R}^N$; λ, μ and γ are nonnegative parameters; $0 \le a, b \in L^{\infty}(\Omega)$ are weight functions; $1 < p, q < \infty$ if $N = 1, 2$ or $1 < p, q < N$ if $N \ge 3$ and α, β are positive constants satisfying $\frac{\alpha}{p} + \frac{\beta}{q} = 1$.

We observe that there exists a vast literature on nonuniformly nonlinear elliptic problems in bounded or unbounded domains. Many authors studied the existence of solutions for such problems (equations or systems); for instance, see [5–7]. We also observe that nonlinear boundary conditions have only been considered in recent years (see [2–4]).

G.A. Afrouzi (✉) • S. Mahdavi
Department of Mathematics, Faculty of Mathematical Sciences,
University of Mazandaran, Babolsar, Iran
e-mail: afrouzi@umz.ac.ir; smahdavi@umz.ac.ir

S. Pinelas et al. (eds.), *Differential and Difference Equations with Applications*, Springer   485
Proceedings in Mathematics & Statistics 47, DOI 10.1007/978-1-4614-7333-6_43,
© Springer Science+Business Media New York 2013

In [1] Afrouzi et al. motivated by the paper of Ou and Tang [10] in which Laplacian system was discussed, obtained three solutions for

$$-\Delta_p u = \lambda a(x)|u|^{p-2}u + F_u(x,u,v) \ \ in \ \ \Omega$$

$$-\Delta_q v = \mu b(x)|v|^{q-2}v + F_v(x,u,v) \ \ in \ \ \Omega \tag{2}$$

as the parameter λ and μ approach λ_1 and μ_1 from the left, respectively. Inspired by [1] and [11], we have the goal in this paper of extending these results to some degenerate elliptic systems.

Let h_1, h_2 be positive weight functions a.e. in Ω such that

$$h_1 \in L^1_{loc}(\Omega), \ \ h_1^{-s} \in L^1(\Omega), \ \ s \in (\frac{N}{p}, \infty) \cap [\frac{1}{p-1}, \infty) , \tag{3}$$

$$h_2 \in L^1_{loc}(\Omega), \ \ h_2^{-s'} \in L^1(\Omega), \ \ s' \in (\frac{N}{q}, \infty) \cap [\frac{1}{q-1}, \infty) . \tag{4}$$

We define $W_0^{1,p}(\Omega, h_1) \left(W_0^{1,q}(\Omega, h_2) \right)$ as being the completion of $C_0^\infty(\Omega)$ with respect to the norm defined by

$$\|u\|_{h_1,p} = (\int_\Omega h_1(x)|\nabla u|^p dx)^{\frac{1}{p}} \ \left(\|v\|_{h_2,q} = (\int_\Omega h_2(x)|\nabla v|^q dx)^{\frac{1}{q}} \right) \ \ \forall u \in C_0^\infty(\Omega)$$

and set $H = W_0^{1,p}(\Omega, h_1) \times W_0^{1,q}(\Omega, h_2)$. It is clear that H is a reflexive Banach space under the norm $\|w\|_H = \|u\|_{h_1,p} + \|v\|_{h_2,q}$ for all $w = (u,v) \in H$. We shall assume that, unless otherwise stated, integrals are over Ω. We recall some facts about the homogeneous degenerate eigenvalue problem

$$-\operatorname{div}(h_1(x)|\nabla u|^{p-2}\nabla u) = \lambda \ a(x)|u|^{p-2}u \ \ in \ \Omega$$

$$u = 0 \qquad\qquad on \ \partial\Omega \tag{5}$$

where Ω is a bounded domain in $\mathbf{R}^N$, $1 < p \le N$ and h_1 satisfy (3). With the number s given in (3) we define

$$p_s = \frac{ps}{s+1}, \ p_s^* = \frac{Np_s}{N - p_s} = \frac{Nps}{N(s+1) - ps} > p.$$

We assume that the coefficient function a satisfy

$$\text{meas}\{x \in \Omega : a(x) > 0\} > 0, \ a \in L^{\frac{r_1}{r_1 - p}}, \ \text{for some} \ p \le r_1 < p_s^*.$$

The author in [9] established the existence of sequence of positive eigenvalues $\{\lambda_k\}_{k\in N}$ where λ_k is determined by the following way. Let

$$M_1 = \left\{ u \in W_0^{1,p}(\Omega,h_1) \,\middle|\, \int a(x)|u|^p dx = 1 \right\},$$

$$I_1(u) = \int h_1(x)|\nabla u|^p dx \qquad u \in W_0^{1,p}(\Omega,h_1)$$

then

$$\lambda_k = \inf_{A_1 \in \Sigma_k} \sup_{u \in A_1} I_1(u) , \tag{6}$$

where $\Sigma_k = \{A_1 \subset M_1 :$ there exists a continuous, odd and surjective $h : S^{k-1} \to A_1\}$
and S^{k-1} denotes the unit sphere in $\mathbf{R}^k$. It has been proved in Chap. 3 of [8] that the principal eigenvalue λ_1 is simple and isolated and all eigenfunctions corresponding to λ_1 do not change sign in Ω. It is obvious that

$$\int h_1(x)|\nabla u|^p dx \geq \lambda_1 \int a(x)|u|^p dx \quad \forall u \in W_0^{1,p}(\Omega,h_1) . \tag{7}$$

Similarly, we consider the eigenvalue problem

$$-\operatorname{div}(h_2(x)|\nabla v|^{q-2}\nabla v) = \mu b(x)|v|^{q-2}v \quad \text{in } \Omega$$

$$v = 0 \qquad\qquad \text{on } \partial\Omega \tag{8}$$

where h_2 satisfy condition (4), $\operatorname{meas}\{x \in \Omega : b(x) > 0\} > 0$, $b \in L^{\frac{r_2}{r_2-q}}$, for some $q \leq r_2 < q_{s'}^*$ where $q_{s'}^* = \frac{Nqs'}{N(s'+1)-qs'} > q$. Let

$$M_2 = \left\{ v \in W_0^{1,q}(\Omega,h_2) \,\middle|\, \int b(x)|v|^q dx = 1 \right\}, \quad I_2(u) = \int h_2(x)|\nabla v|^q dx \quad v \in W_0^{1,q}(\Omega,h_2).$$

By a standard argument, problem (8) has a sequence of eigenvalues with the variational characterization

$$\mu_k = \inf_{A_2 \in \Sigma_k'} \sup_{v \in A_2} I_2(v) , \tag{9}$$

where $\Sigma_k' = \{A_2 \subset M_2 :$ there exists a continuous, odd and surjective $h : S^{k-1} \to A_2\}$. We also have

$$\int h_2(x)|\nabla v|^q dx \geq \mu_1 \int b(x)|v|^q dx \quad \forall v \in W_0^{1,q}(\Omega,h_2) . \tag{10}$$

Besides, the corresponding normalised eigenfunction φ_1 belongs to $W_0^{1,p}(\Omega, h_1)$ and ψ_1 belongs to $W_0^{1,q}(\Omega, h_2)$. Let

$$W' = \{(u,v) \in H \quad \int_\Omega a|\varphi_1|^{p-2}\varphi_1 u = 0, \quad \int_\Omega b|\psi_1|^{q-2}\psi_1 v = 0\}.$$

We can easily prove that W' is complementary subspace of $W = \langle \varphi_1 \rangle \times \langle \psi_1 \rangle$. Therefore we have the direct sum $H = W \oplus W'$. The main results in this paper are the following theorems.

Theorem 1. *For $\lambda < \lambda_1$ and $\mu < \mu_1$ sufficiently close to λ_1 and μ_1, problem (1) has at least three solutions.*

Theorem 2. *For $\lambda_k < \lambda < \lambda_{k+1}$ and $\mu_k < \mu < \mu_{k+1}$, problem (1) has at least one solution.*

2   Preliminaries

Let $I : H \to \mathbf{R}$ be the functional defined by

$$I(u,v) = \frac{1}{p}\int h_1(x)|\nabla u|^p dx + \frac{1}{q}\int h_2(x)|\nabla v|^p dx - \frac{\lambda}{p}\int a(x)|u|^p dx$$

$$- \frac{\mu}{q}\int b(x)|v|^q dx - \gamma \int r(x)|u|^\alpha |v|^\beta dx . \tag{11}$$

We see that $I \in C^1(H, \mathbf{R})$ and $(u,v) \in H$ is a weak solution of problem (1) if and only if (u,v) is a critical point of I. It is well known that the following lemma holds.

Lemma 1 ([12, Lemma 2.1]). *Assume that Ω is a bounded domain in $\mathbf{R}^N$ and the weight h_1 satisfy (3). Then the following embeddings hold:*

(i) $W_0^{1,p}(\Omega, h_1(x)) \hookrightarrow L^{p_s^*}(\Omega)$ *continuously for $1 < p_s^* < N$.*
(ii) $W_0^{1,p}(\Omega, h_1(x)) \hookrightarrow L^r(\Omega)$ *compactly for $r \in [1, p_s^*)$.*

Putting

$U := \{u \in W_0^{1,p}(\Omega, h_1); \int_\Omega a|\varphi_1|^{p-2}\varphi_1 u dx = 0\}$,
$V := \{v \in W_0^{1,q}(\Omega, h_2); \int_\Omega b|\psi_1|^{q-2}\psi_1 v dx = 0\}$.

U and V are closed subspace, and $W_0^{1,p}(\Omega, h_1) = U \oplus \langle \phi_1 \rangle$ and $W_0^{1,q}(\Omega, h_2) = V \oplus \langle \psi_1 \rangle$ hold. Moreover, we set

$$\bar{\lambda} = \inf_{u \in U \backslash \{0\}} \frac{\|u\|_{h_{1,p}}}{\|u\|_{L^p}^p}, \quad \bar{\mu} = \inf_{v \in V \backslash \{0\}} \frac{\|v\|_{h_{2,q}}}{\|v\|_{L^q}^q} . \tag{12}$$

We can show that $\lambda_1 < \overline{\lambda}$ and $\mu_1 < \overline{\mu}$ by contradiction. Since $\lambda_1 \leq \dfrac{\|u\|_{h_1,p}}{\|u\|_{L^p}^p}$ holds for each $u \neq 0$, we assume that $\lambda_1 = \overline{\lambda}$, i.e. $\overline{\lambda} = \inf_{u \in W_0^{1,p}(\Omega,h_1)\setminus\{0\}} \dfrac{\|u\|_{h_1,p}}{\|u\|_{L^p}^p}$.

Then, we may suppose that there exist $\{u_n\}_n \subset U$ and $u \in W_0^{1,p}(\Omega,h_1)$ such that $\|u_n\|_{L^p} = 1$, $\lim_{n\to\infty} \|u_n\|_{h_1,p} = \overline{\lambda}$ and

$$u_n \rightharpoonup u \in W_0^{1,p}(\Omega,h_1); \quad \text{hence,} \quad u_n \to u \text{ in } L^p. \tag{13}$$

Since U is weakly closed and u_n strongly converges to u in L^p, $u \in U$ and $\|u\|_{L^p} = 1$ hold. Using weak lower semicontinuity of the norm and the variational characterization of λ_1, we get

$$\lambda_1 \leq \|u\|_{h_1,p} \leq \liminf_{n\to\infty} \|u_n\|_{h_1,p} = \overline{\lambda} = \lambda_1$$

which implies that $u = \pm\varphi_1$. This contradicts $\pm\varphi_1 \notin U$. Analogously, we can prove $\mu_1 < \overline{\mu}$.

3 Proof of Theorems

We will prove Theorem 1 by using Ekeland's variational principal and the Mountain Pass Theorem.

Proof. The proof will be divided into four steps.

Step 1. The functional I is coercive in H, I is bounded from below on W' and there is a constant m, independent of λ, μ, such that $\inf_{W'} I \geq m$.
For $\lambda < \lambda_1$ and $\mu < \mu_1$, from the definition of λ_1, μ_1, (18) and Young's inequality, we get

$$I(u,v) \geq \frac{1}{p}(1 - \frac{\lambda}{\lambda_1} - \gamma\|r(x)\|_\infty \alpha S_1^p) \int_\Omega h_1(x)|\nabla u|^p dx$$

$$+ \frac{1}{q}(1 - \frac{\mu}{\mu_1} - \gamma\|r(x)\|_\infty \beta S_2^q) \int_\Omega h_2(x)|\nabla v|^q dx.$$

where S_1, S_2 are the embedding constants of $W_0^{1,p}(\Omega,h_1) \hookrightarrow L^p(\Omega)$, $W_0^{1,q}(\Omega,h_2) \hookrightarrow L^q(\Omega)$, respectively. Letting $\gamma = \frac{1}{2}\min\{\frac{\lambda_1-\lambda}{\lambda_1\|r(x)\|_\infty \alpha S_1^p}, \frac{\mu_1-\mu}{\mu_1\|r(x)\|_\infty \beta S_2^q}\}$, it follows that I is coercive in H. Similarly, from (16), we obtain

$$I(u,v) \geq \frac{1}{p}(1 - \frac{\lambda_1}{\overline{\lambda}} - \gamma\|r(x)\|_\infty \alpha S_1^p) \int_\Omega h_1(x)|\nabla u|^p dx$$

$$+ \frac{1}{q}(1 - \frac{\mu_1}{\overline{\mu}} - \gamma\|r(x)\|_\infty \beta S_2^q)) \int_\Omega h_2(x)|\nabla v|^q dx.$$

Let $\gamma = \frac{1}{2}\min\{\frac{\overline{\lambda}-\lambda_1}{\overline{\lambda}\|r(x)\|_\infty \alpha S_1^p}, \frac{\overline{\mu}-\mu_1}{\overline{\mu}\|r(x)\|_\infty \beta S_2^q}\}$; hence, I is coercive in W and I is bounded from below on W', and, moreover, there is a constant m, independent of λ, μ, such that $\inf_{W'} I \geq m$.

Step 2. If $\lambda < \lambda_1$ and $\mu < \mu_1$ are sufficiently close to λ_1, μ_1, we have $t_1^- < 0 < t_1^+$, $t_2^- < 0 < t_2^+$ such that $I(t_1^\pm \varphi_1, t_2^\pm \psi_1) < m$.

Step 3. If $\lambda < \lambda_1$, the functional I satisfies the (PS) condition.

If $\{z_n\} = \{(u_n, v_n)\}$ is a (PS) sequence of I, $\{(u_n, v_n)\}$ must be bounded. Then passing to a subsequence if necessary, there exists $z = (u, v) \in H$ such that

$$(u_n, v_n) \rightharpoonup (u, v) \text{ weakly in } H, \ (u_n, v_n) \to (u, v) \text{ strongly in } L^p(\Omega) \times L^q(\Omega).$$

So there exists a strictly decreasing subsequence ε_n, $\lim_{n\to\infty} \varepsilon_n = 0$ such that

$$|I'(u_n, v_n)(u_n - u, 0)| = \Big| \int h_1(x)|\nabla u_n|^{p-2}\nabla u_n \nabla(u_n - u)dx$$

$$-\lambda \int a(x)|u_n|^{p-2}u_n(u_n - u)dx - \alpha\gamma \int r(x)|u|^{\alpha-1}|v|^\beta(u_n - u)dx\Big|$$

$$\leq \varepsilon_n \|(u_n - u, 0)\|_H . \tag{14}$$

Since $u_n \to u$ in $L^p(\Omega)$, $v_n \to v$ in $L^q(\Omega)$, we have

$$\lim_{n\to\infty} \int a(x)|u_n|^{p-2}u_n(u_n-u)dx \leq \lim_{n\to\infty} \|a\|_{\frac{r_1}{r_1-p}} \left(\int |u_n|^{r_1}dx\right)^{\frac{p-1}{r_1}} \left(\int |u_n-u|^{r_1}dx\right)^{\frac{1}{r_1}} = 0 . \tag{15}$$

$$\int r(x)|u|^{\alpha-1}|v|^\beta(u_n-u)dx \leq \left(\int |u_n|^p dx\right)^{\frac{\alpha-1}{p}} \left(\int |v|^q dx\right)^{\frac{\beta}{q}} \left(\int |u_n-u|^p dx\right)^{\frac{1}{p}} \to 0 \tag{16}$$

Combining (14) with (15) and (16), we get

$$\lim_{n\to\infty} \int h_1(x)|\nabla u_n|^{p-2}\nabla u_n(\nabla u_n - \nabla u)dx = 0.$$

Subtracting

$$\int h_1(x)|\nabla u|^{p-2}\nabla u(\nabla u_n - \nabla u)dx$$

(which converges to zero as n tends to infinity), we conclude that

$$\lim_{n\to\infty} \int h_1(x)(|\nabla u_n|^{p-2}\nabla u_n - |\nabla u|^{p-2}\nabla u)(\nabla u_n - \nabla u)dx = 0 , \tag{17}$$

Hölder's inequality, and substituting $z_n = h_1^{\frac{1}{p}}u_n, z = h_1^{\frac{1}{p}}u$ in Lemma in [11], we obtain $\|u_n - u\|_{h_1, p} \to 0$, for $p > 1$, as $n \to \infty$, that is, $u_n \to u$ in $W_0^{1,p}(\Omega, h_1)$ as $n \to \infty$. Similarly , we obtain $v_n \to v$ in $W_0^{1,q}(\Omega, h_2)$ as $n \to \infty$.

Consequently, the functional I satisfies the (PS) condition $\lambda < \lambda_1$, $\mu < \mu_1$. In addition, let

$$\sum_{\pm} = \{z \in H : z = \pm(t_1\varphi_1, t_2\psi_1) + w \quad \text{with} \quad t_1, t_2 > 0 \text{ and } w \in W'\}.$$

I satisfies $(PS)_{c,\Sigma_+}$ and $(PS)_{c,\Sigma_-}$ for all $c < m$.

Let $\{z_n\} \subset \Sigma_+$ satisfy $I(z_n) \to c < m$ and $I'(z_n) \to 0$ an $n \to \infty$. Since I is coercive, there is $z \in H$ such that $\|z_n\|_H \to \|z\|_H$ strongly in H. If $z \in \partial\Sigma_+ = W'$, from $\inf_{W'} I \geq m$, we get $I_\lambda(z_n) \to c \geq m$, which is impossible. Hence $z \in \Sigma_+$ and I satisfies the $(PS)_{c,\Sigma_+}$ condition. Similarly we have that $(PS)_{c,\Sigma_-}$ holds for all $c < m$. If $\lambda < \lambda_1, \mu < \mu_1$ is sufficiently close to λ_1, we get $-\infty < \inf_{\Sigma_+} I < m$, which implies that I is bounded below in Σ_+. Consequently, from Ekeland's variational principle, there exists $\{z_n\} \subset \Sigma_+$ such that $I(z_n) \to \inf_{\Sigma_+} I$ and $I'(z_n) \to 0$ as $n \to \infty$. Since I satisfies $(PS)_{c,\Sigma_+}$ for all $c < m$, there is $z^+ \in \Sigma_+$ such that $I(z^+) = \inf_{\Sigma_+} I$, that is, the infimum is attained in Σ_+. A similar conclusion holds in Σ_-. So I has two distinct critical points, denoted by z^+, z^-.

As in [10], we can obtain the third critical point z of I by applying Mountain Pass Theorem such that $I(z) = c \geq m$. $\qquad\square$

Proof. To prove Theorem 2, we will verify the functional I satisfying the condition of Theorem 3 in [11]. Following similar argument as in step 3, we may prove that the functional I satisfies the Cerami condition.

It follows from definition of λ_k and μ_k, there exist $A_1 \in \Sigma_k$ and $A_2 \in \Sigma'_k$ such that $\sup_{u \in A_1} I_1(u) = m_1 \in (\lambda_k, \lambda)$ and $\sup_{v \in A_2} I_2(v) = m_2 \in (\mu_k, \mu)$, respectively.

From (11), we obtain

$$I(t^{\frac{1}{p}}u, t^{\frac{1}{q}}v) \leq \frac{t}{p} \int h_1(x)|\nabla u|^p dx + \frac{t}{q} \int h_2(x)|\nabla v|^q dx$$

$$-\frac{t\lambda}{p} \int a(x)|u|^p dx - \frac{t\mu}{q} \int b(x)|v|^q dx$$

$$+|\gamma \int r(x)|t^{\frac{1}{p}}u|^\alpha|t^{\frac{1}{q}}v)|^\beta dx| \leq \frac{t}{p}(m_1 - \lambda + \gamma\|r(x)\|_\infty \alpha S_1^p m_1)$$

$$+\frac{t}{q}(m_2 - \mu + \gamma\|r(x)\|_\infty \beta S_2^q m_2) \tag{18}$$

for $(u,v) \in A_1 \times A_2, t > 0$. Set

$$F_{K+1} = \{u \in W_0^{1,p}(\Omega, h_1) : \int h_1(x)|\nabla u|^p dx \geq \lambda_{k+1} \int a(x)|u|^p dx\}$$

$$F'_{K+1} = \{v \in W_0^{1,q}(\Omega, h_2) : \int h_2(x)|\nabla v|^q dx \geq \mu_{k+1} \int b(x)|v|^q dx\}.$$

For $(u, v) \in F_{K+1} \times F'_{K+1}$ we have

$$I(u, v) \geq \left(\frac{1}{p} - \frac{\lambda}{p\lambda_{k+1}} - \frac{\gamma \|r(x)\|_\infty \alpha S_1^p}{p}\right) \int h_1(x) |\nabla u|^p dx +$$

$$\left(\frac{1}{q} - \frac{\mu}{q\mu_{k+1}} - \frac{\gamma \|r(x)\|_\infty \beta S_2^q}{q}\right) \int h_2(x) |\nabla v|^q dx \qquad (19)$$

Let $\gamma = \frac{1}{2} \min\{\frac{\lambda - m_1}{m_1 \|r(x)\|_\infty \alpha S_1^p}, \frac{\mu - m_2}{m_2 \|r(x)\|_\infty \beta S_2^q}, \frac{\lambda_{k+1} - \lambda}{\lambda_{k+1} \|r(x)\|_\infty \alpha S_1^p}, \frac{\mu_{k+1} - \mu}{\mu_{k+1} \|r(x)\|_\infty \beta S_2^q}\}$. Then we get

$$\beta := \inf_{(u,v) \in F_{K+1} \times F'_{K+1}} I(u, v) \qquad (20)$$

and set $T > 0$ such that

$$\alpha := \max_{(u,v) \in A_1 \times A_2, t \geq T} I(t^{\frac{1}{p}} u, t^{\frac{1}{q}} v) < \beta \qquad (21)$$

by (18), (19). Now let $TA := \{(t^{\frac{1}{p}} u, t^{\frac{1}{q}} v) : (u, v) \in A_1 \times A_2, t \geq T\}$. Set $Q = B_k(B_k$ represents the closed unit ball in R^k), $\partial Q = S^{k-1}$, $\Gamma = \{h \in C^0(S^{k-1}, H) : h$ is odd and $h(S^{k-1}) \subset TA\}$. For any $h \in \Gamma$, (20), (21) we obtain $h(S^{k-1}) \cap (F_{K+1} \times F'_{K+1})$ $= \emptyset$ which shows that Γ is a subset of $C(S^{k-1}, H \backslash (F_{K+1} \times F'_{K+1}))$. Let $\Gamma^* = \{h \in C^0(B_k, H) : h|_{S^{k-1}} \in \Gamma\}$. Then Γ^* is nonempty, and $h(B_k) \cap (F_{K+1} \times F'_{K+1}) \neq \emptyset$ for all $h \in \Gamma^*$. In fact by the definition of $\sum_k, \sum'_k$, there exist continuous odd surjection $h_1 : S^{k-1} \to A_1, h_2 : S^{k-1} \to A_2$. So we can define $h : S^{k-1} \to A_1 \times A_2$ by $h = (h_1, h_2)$ Define $\overline{h} : B_k \to H$ by $\overline{h}(ts) = tTh(s)$ for any $s \in S^{k-1}$ and any $t \in [0, 1]$. Thus $\overline{h} \in \Gamma^*$ If there exists $(u, v) \in h(B_k)$ such that $\int a(x) |u|^p dx = 0, \int b(x) |v|^q dx = 0$, we get $h(B_k) \cap (F_{K+1} \times F'_{K+1}) \neq \emptyset$. Otherwise we consider the map $\hat{h} : S^k \to E$ by

$$\hat{h}(x_1, \ldots, x_{k+1}) = \begin{cases} \pi \circ h(x_1, \ldots, x_k) & x_{k+1} \geq 0 \\ -\pi \circ h(-x_1, \ldots, -x_k) & x_{k+1} < 0 \end{cases} \qquad (22)$$

where $\pi(u, v) = \left(\frac{u}{\int a(x) |u|^p dx}, \frac{v}{\int b(x) |v|^q dx}\right)$. It is not difficult to verify that $\hat{h}(S^k) \in F_{K+1} \times F'_{K+1}$. Therefore,

$$\int h_1(x) |\nabla u_0|^p dx \geq \lambda_{k+1} \int a(x) |u_0|^p dx, \int h_2(x) |\nabla v_0|^q dx \geq \mu_{k+1} \int b(x) |v_0|^q dx$$

for some $(u_0, v_0) \in \hat{h}(S^k)$, i.e. $(u_0, v_0) \in F_{K+1} \times F'_{K+1}$. Notice that $\pi \circ h(x) \in F_{K+1} \times F'_{K+1}$ implies that $h(x) \in F_{K+1} \times F'_{K+1}$. Then $h(B_k) \cap \{F_{K+1} \times F'_{K+1}\} \neq \emptyset$. Hence S^k and $F_{K+1} \times F'_{K+1}$ are Γ-linking. The condition of Theorem (2.5) is satisfied. So Theorem 2 holds for $\lambda_k < \lambda < \lambda_{k+1}, \mu_k < \mu < \mu_{k+1}$ with the critical value

$$c := \inf_{h \in \Gamma^*} \sup_{x \in B_k} I(h(x)).$$

References

1. Afrouzi, G.A., Mahdavi, S.: Naghizadeh, Z., Existence of multiple solutions for a class of (p,q)-Laplacian systems. Nonlinear Anal. **72**, 2243–2250 (2010)
2. Bonanno, G., Molica Bisci, G., Rădulescu, V.: Existence of three solutions for a non-homogeneous Neumann problem through Orlicz-Sobolev spaces. Nonlinear Anal. **74**(14), 4785– 4795 (2011)
3. Bonanno, G., Molica Bisci, G., Rădulescu, V.: Infinitely many solutions for a class of nonlinear eigenvalue problems in Orlicz-Sobolev spaces. C. R. Acad. Sci. Paris, Ser. I **349**, 263–268 (2011)
4. Bonanno, G., Molica Bisci, G., Rădulescu, V: Arbitrarily small weak solutions for a nonlinear eigenvalue problem in Orlicz-Sobolev spaces. Monatsh Math. 1–14 (2011) DOI 10.1007/s00605-010-0280-2.
5. Caldiroli, P., Musina, R.: On a variational degenerate elliptic problem. Nonlinear Diff. Equ. Appl. **7**, 189–199 (2000)
6. Chung, N.T.: Existence of weak solution for a nonuniformly elliptic nonlinear system in R^N. Electronic J. Differ. Equat. **119**, 1–10 (2008)
7. Chung, N.T.: On the existence of weak solutions for a degenerate and singular elliptic systems in R^N. Acta. Appl. Math. **7**, 47–56 (2010)
8. Drábek, P., Kufner, A., Niclosi F.: Quasilinear Elliptic Equations with Degenerate and Singularities. de Gruyter Series in Nonlinear Analysis and Applications. vol. 5, Walter and Gruyter and Company, Berlin (1997)
9. Lê, A., Schmitt, K.: Variational eigenvalues of degenerate eigenvalue problem for weighted p-laplacian. Adv. Nonlinear Stud. **5**, 553–565 (2005)
10. Ou, Z.Q., Tang, C.L.: Existence and multiplicity results for some elliptic system at resonance. Nonlinear Anal. **71**, 2660–2666 (2009)
11. Zhao, X.X., Tang, C.L.: Resonance problems for (p,q)-Laplacian systems. Nonlinear Anal. **72**, 1019–1030 (2010)
12. Zographopoulos N.B.: On the principal eigenvalue of degenerate quasilinear elliptic systems. Math. Nachr. **281**(9), 1351–1365 (2008)

A Symmetric Nörlund Sum with Application to Inequalities

Artur M.C. Brito da Cruz, Natália Martins and Delfim F.M. Torres

Abstract Properties of an α, β-symmetric Nörlund sum are studied. Inspired in the work by Agarwal et al., α, β-symmetric quantum versions of Hölder's, Cauchy–Schwarz's and Minkowski's inequalities are obtained.

1 Introduction

The symmetric derivative of function f at point x is defined as $\lim_{h \to 0}(f(x+h) - f(x-h))/(2h)$. The notion of symmetrically differentiable is interesting because if a function is differentiable at a point, then it is also symmetrically differentiable, but the converse is not true. The best-known example of this fact is the absolute value function: $f(x) = |x|$ is not differentiable at $x = 0$ but is symmetrically differentiable at $x = 0$ with symmetric derivative zero [6].

Quantum calculus is, roughly speaking, the equivalent to traditional infinitesimal calculus but without limits [4]. Therefore, one can introduce the symmetric quantum derivative of f at x by $(f(x+h) - f(x-h))/(2h)$. As in any calculus, it is then natural to develop a corresponding integration theory.

A.M.C. Brito da Cruz (✉)
Escola Superior de Tecnologia de Setúbal, Instituto Politécnico de Setúbal, Estefanilha,
2910-761 Setúbal, Portugal

Department of Mathematics, University of Aveiro,
Center for Research and Development in Mathematics and Applications,
3810-193 Aveiro, Portugal
e-mail: artur.cruz@estsetubal.ips.pt

N. Martins • D.F.M. Torres
Department of Mathematics, University of Aveiro,
Center for Research and Development in Mathematics and Applications,
3810-193 Aveiro, Portugal
e-mail: natalia@ua.pt; delfim@ua.pt

S. Pinelas et al. (eds.), *Differential and Difference Equations with Applications*, Springer   495
Proceedings in Mathematics & Statistics 47, DOI 10.1007/978-1-4614-7333-6_44,
© Springer Science+Business Media New York 2013

The main goal of this paper is to study the properties of a general symmetric quantum integral that we call, due to the so-called Nörlund sum [4], the α, β-symmetric Nörlund sum.

The paper is organised as follows. In Section 2 we define the forward and backward Nörlund sums. Then, in Section 3, we introduce the α, β-symmetric Nörlund sum and give some of its properties. We end with Section 4, proving α, β-symmetric versions of Hölder's, Cauchy–Schwarz's and Minkowski's inequalities.

2 Forward and Backward Nörlund Sums

This section is dedicated to the inverse operators of the α-forward and β-backward differences, $\alpha > 0$, $\beta > 0$, defined, respectively, by

$$\Delta_\alpha [f] (t) := \frac{f(t+\alpha) - f(t)}{\alpha}, \quad \nabla_\beta [f] (t) := \frac{f(t) - f(t-\beta)}{\beta}.$$

Definition 1. Let $I \subseteq \mathbb{R}$ be an interval such that $a, b \in I$ with $a < b$ and $\sup I = +\infty$. For $f : I \to \mathbb{R}$ and $\alpha > 0$ we define the Nörlund sum (the α-forward integral) of f from a to b by

$$\int_a^b f(t) \Delta_\alpha t = \int_a^{+\infty} f(t) \Delta_\alpha t - \int_b^{+\infty} f(t) \Delta_\alpha t,$$

where $\displaystyle\int_x^{+\infty} f(t) \Delta_\alpha t = \alpha \sum_{k=0}^{+\infty} f(x+k\alpha)$, provided the series converges at $x = a$ and $x = b$. In that case, f is said to be α-forward integrable on $[a,b]$. We say that f is α-forward integrable over I if it is α-forward integrable for all $a, b \in I$.

Until Definition 2 (the backward/nabla case), we assume that I is an interval of $\mathbb{R}$ such that $\sup I = +\infty$. Note that if $f : I \to \mathbb{R}$ is a function such that $\sup I < +\infty$, then we can extend function f to $\tilde{f} : \tilde{I} \to \mathbb{R}$, where $\tilde{I}$ is an interval with $\sup \tilde{I} = +\infty$, in the following way: $\tilde{f}|_I = f$ and $\tilde{f}|_{\tilde{I} \setminus I} = 0$.

Using the techniques of Aldwoah in his Ph.D. thesis [2], it can be proved that the α-forward integral has the following properties:

Theorem 1. *If $f, g : I \to \mathbb{R}$ are α-forward integrable on $[a,b]$, $c \in [a,b]$, $k \in \mathbb{R}$, then:*

1. $\displaystyle\int_a^a f(t) \Delta_\alpha t = 0.$

2. $\displaystyle\int_a^b f(t) \Delta_\alpha t = \int_a^c f(t) \Delta_\alpha t + \int_c^b f(t) \Delta_\alpha t$, *when the integrals exist.*

3. $\displaystyle\int_a^b f(t) \Delta_\alpha t = -\int_b^a f(t) \Delta_\alpha t.$

4. kf *is α-forward integrable on $[a,b]$ and* $\displaystyle\int_a^b kf(t) \Delta_\alpha t = k \int_a^b f(t) \Delta_\alpha t.$

5. $f + g$ is α-forward integrable on $[a,b]$ and

$$\int_a^b (f+g)(t)\Delta_\alpha t = \int_a^b f(t)\Delta_\alpha t + \int_a^b g(t)\Delta_\alpha t.$$

6. If $f \equiv 0$, then $\int_a^b f(t)\Delta_\alpha t = 0$.

Theorem 2. *Let $f : I \to \mathbb{R}$ be α-forward integrable on $[a,b]$. If $g : I \to \mathbb{R}$ is a nonnegative α-forward integrable function on $[a,b]$, then fg is α-forward integrable on $[a,b]$.*

Proof. Since g is α-forward integrable, then both series $\alpha \sum_{k=0}^{+\infty} g(a+k\alpha)$ and $\alpha \sum_{k=0}^{+\infty} g(b+k\alpha)$ converge. We want to study the nature of series $\alpha \sum_{k=0}^{+\infty} fg(a+k\alpha)$ and $\alpha \sum_{k=0}^{+\infty} fg(b+k\alpha)$. Since there exists an order $N \in \mathbb{N}$ such that $|fg(b+k\alpha)| \leqslant g(b+k\alpha)$ and $|fg(a+k\alpha)| \leqslant g(a+k\alpha)$ for all $k > N$, then both $\alpha \sum_{k=0}^{+\infty} fg(a+k\alpha)$ and $\alpha \sum_{k=0}^{+\infty} fg(b+k\alpha)$ converge absolutely. The intended conclusion follows. $\square$

Theorem 3. *Let $f : I \to \mathbb{R}$ and $p > 1$. If $|f|$ is α-forward integrable on $[a,b]$, then $|f|^p$ is also α-forward integrable on $[a,b]$.*

Proof. There exists $N \in \mathbb{N}$ such that $|f(b+k\alpha)|^p \leqslant |f(b+k\alpha)|$ and $|f(a+k\alpha)|^p \leqslant |f(a+k\alpha)|$ for all $k > N$. Therefore, $|f|^p$ is α-forward integrable on $[a,b]$. $\square$

Theorem 4. *Let $f,g : I \to \mathbb{R}$ be α-forward integrable on $[a,b]$. If $|f(t)| \leqslant g(t)$ for all $t \in \{a+k\alpha : k \in \mathbb{N}_0\}$, then for $b \in \{a+k\alpha : k \in \mathbb{N}_0\}$ one has*

$$\left| \int_a^b f(t)\Delta_\alpha t \right| \leqslant \int_a^b g(t)\Delta_\alpha t.$$

Proof. Since $b \in \{a+k\alpha : k \in \mathbb{N}_0\}$, there exists k_1 such that $b = a + k_1\alpha$. Thus,

$$\left| \int_a^b f(t)\Delta_\alpha t \right| = \left| \alpha \sum_{k=0}^{+\infty} f(a+k\alpha) - \alpha \sum_{k=0}^{+\infty} f(a+(k_1+k)\alpha) \right|$$

$$= \left| \alpha \sum_{k=0}^{+\infty} f(a+k\alpha) - \alpha \sum_{k=k_1}^{+\infty} f(a+k\alpha) \right| = \left| \alpha \sum_{k=0}^{k_1-1} f(a+k\alpha) \right|$$

$$\leqslant \alpha \sum_{k=0}^{k_1-1} |f(a+k\alpha)| \leqslant \alpha \sum_{k=0}^{k_1-1} g(a+k\alpha)$$

$$= \alpha \sum_{k=0}^{+\infty} g(a+k\alpha) - \alpha \sum_{k=k_1}^{+\infty} g(a+k\alpha) = \int_a^b g(t)\Delta_\alpha t. \qquad \square$$

Corollary 1. *Let $f,g : I \to \mathbb{R}$ be α-forward integrable on $[a,b]$ with $b = a+k\alpha$ for some $k \in \mathbb{N}_0$.*

1. *If $f(t) \geqslant 0$ for all $t \in \{a+k\alpha : k \in \mathbb{N}_0\}$, then $\int_a^b f(t)\Delta_\alpha t \geqslant 0$.*
2. *If $g(t) \geqslant f(t)$ for all $t \in \{a+k\alpha : k \in \mathbb{N}_0\}$, then $\int_a^b g(t)\Delta_\alpha t \geqslant \int_a^b f(t)\Delta_\alpha t$.*

We can now prove the following fundamental theorem of the α-forward calculus.

Theorem 5 (Fundamental Theorem of Nörlund Calculus). *Let* $f : I \to \mathbb{R}$ *be* α-*forward integrable over* I. *Let* $x \in I$ *and define* $F(x) := \int_a^x f(t) \Delta_\alpha t$. *Then,* $\Delta_\alpha [F](x) = f(x)$. *Conversely,* $\int_a^b \Delta_\alpha [f](t) \Delta_\alpha t = f(b) - f(a)$.

Proof. If $G(x) = -\int_x^{+\infty} f(t) \Delta_\alpha t$, then

$$\Delta_\alpha [G](x) = \frac{G(x+\alpha) - G(x)}{\alpha} = \frac{-\alpha \sum_{k=0}^{+\infty} f(x+\alpha+k\alpha) + \alpha \sum_{k=0}^{+\infty} f(x+k\alpha)}{\alpha}$$

$$= \sum_{k=0}^{+\infty} f(x+k\alpha) - \sum_{k=0}^{+\infty} f(x+(k+1)\alpha) = f(x).$$

Therefore, $\Delta_\alpha [F](x) = \Delta_\alpha \left(\int_a^{+\infty} f(t) \Delta_\alpha t - \int_x^{+\infty} f(t) \Delta_\alpha t \right) = f(x)$. Using the definition of α-forward difference operator, the second part of the theorem is also a consequence of the properties of Mengoli's series. Since

$$\int_a^{+\infty} \Delta_\alpha [f](t) \Delta_\alpha t = \alpha \sum_{k=0}^{+\infty} \Delta_\alpha [f](a+k\alpha) = \alpha \sum_{k=0}^{+\infty} \frac{f(a+k\alpha+\alpha) - f(a+k\alpha)}{\alpha}$$

$$= \sum_{k=0}^{+\infty} \left(f(a+(k+1)\alpha) - f(a+k\alpha) \right) = -f(a)$$

and $\int_b^{+\infty} \Delta_\alpha [f](t) \Delta_\alpha t = -f(b)$, it follows that

$$\int_a^b \Delta_\alpha [f](t) \Delta_\alpha t = \int_a^{+\infty} f(t) \Delta_\alpha t - \int_b^{+\infty} f(t) \Delta_\alpha t = f(b) - f(a).$$

$\square$

Corollary 2 (α-Forward Integration by Parts). *Let* $f, g : I \to \mathbb{R}$. *If* fg *and* $f\Delta_\alpha [g]$ *are* α-*forward integrable on* $[a,b]$, *then*

$$\int_a^b f(t) \Delta_\alpha [g](t) \Delta_\alpha t = f(t) g(t) \Big|_a^b - \int_a^b \Delta_\alpha [f](t) g(t+\alpha) \Delta_\alpha t.$$

Proof. Since $\Delta_\alpha [fg](t) = \Delta_\alpha [f](t) g(t+\alpha) + f(t) \Delta_\alpha [g](t)$, then

$$\int_a^b f(t) \Delta_\alpha [g](t) \Delta_\alpha t = \int_a^b \left(\Delta_\alpha [fg](t) - \Delta_\alpha [f](t) g(t+\alpha) \right) \Delta_\alpha t$$

$$= \int_a^b \Delta_\alpha [fg](t) \Delta_\alpha t - \int_a^b \Delta_\alpha [f](t) g(t+\alpha) \Delta_\alpha t$$

$$= f(t) g(t) \Big|_a^b - \int_a^b \Delta_\alpha [f](t) g(t+\alpha) \Delta_\alpha t.$$

$\square$

Remark 1. Our study of the Nörlund sum is in agreement with the Hahn quantum calculus [2, 3, 5]. In [4] $\int_a^b f(t)\Delta_\alpha t = \alpha [f(a)+f(a+\alpha)+\cdots+f(b-\alpha)]$ for $a < b$ such that $b-a \in \alpha\mathbb{Z}$, $\alpha \in \mathbb{R}^+$. In contrast with [4], our definition is valid for any two real points a,b and not only for those points belonging to the time scale $\alpha\mathbb{Z}$. The definitions (only) coincide if the function f is α-forward integrable on $[a,b]$.

Similarly, we introduce the β-backward integral.

Definition 2. Let I be an interval of $\mathbb{R}$ such that $a,b \in I$ with $a < b$ and $\inf I = -\infty$. For $f : I \to \mathbb{R}$ and $\beta > 0$ we define the β-backward integral of f from a to b by

$$\int_a^b f(t)\nabla_\beta t = \int_{-\infty}^b f(t)\nabla_\beta t - \int_{-\infty}^a f(t)\nabla_\beta t,$$

where $\int_{-\infty}^x f(t)\nabla_\beta t = \beta \sum_{k=0}^{+\infty} f(x-k\beta)$, provided the series converges at $x = a$ and $x = b$. In that case, f is called β-backward integrable on $[a,b]$. We say that f is β-backward integrable over I if it is β-backward integrable for all $a,b \in I$.

The β-backward Nörlund sum has similar results and properties as the α-forward Nörlund sum. In particular, the β-backward integral is the inverse operator of ∇_β.

3 The α,β-Symmetric Nörlund Sum

We define the α,β-symmetric integral as a linear combination of the α-forward and the β-backward integrals.

Definition 3. Let $f : \mathbb{R} \to \mathbb{R}$ and $a,b \in \mathbb{R}$, $a < b$. If f is α-forward and β-backward integrable on $[a,b]$, $\alpha,\beta \geq 0$ with $\alpha + \beta > 0$, then we define the α,β-symmetric integral of f from a to b by

$$\int_a^b f(t)\,d_{\alpha,\beta}t = \frac{\alpha}{\alpha+\beta} \int_a^b f(t)\Delta_\alpha t + \frac{\beta}{\alpha+\beta} \int_a^b f(t)\nabla_\beta t.$$

Function f is α,β-symmetric integrable if it is α,β-symmetric integrable for all $a,b \in \mathbb{R}$.

Remark 2. Note that if $\alpha \in \mathbb{R}^+$ and $\beta = 0$, then $\int_a^b f(t)\,d_{\alpha,\beta}t = \int_a^b f(t)\Delta_\alpha t$ and we do not need to assume in Definition 3 that f is β-backward integrable; if $\alpha = 0$ and $\beta \in \mathbb{R}^+$, then $\int_a^b f(t)\,d_{\alpha,\beta}t = \int_a^b f(t)\nabla_\beta t$ and we do not need to assume that f is α-forward integrable.

Example 1. Let $f(t) = 1/t^2$. Then $\int_1^3 \frac{1}{t^2} d_{2,2}t = \frac{10}{9}$.

The α, β-symmetric integral has the following properties:

Theorem 6. *Let $f, g : \mathbb{R} \to \mathbb{R}$ be α, β-symmetric integrable on $[a,b]$. Let $c \in [a,b]$ and $k \in \mathbb{R}$. Then:*

1. $\int_a^a f(t) d_{\alpha,\beta} t = 0$.

2. $\int_a^b f(t) d_{\alpha,\beta} t = \int_a^c f(t) d_{\alpha,\beta} t + \int_c^b f(t) d_{\alpha,\beta} t$, *when the integrals exist.*

3. $\int_a^b f(t) d_{\alpha,\beta} t = -\int_b^a f(t) d_{\alpha,\beta} t$.

4. kf *is α, β-symmetric integrable on $[a,b]$ and* $\int_a^b kf(t) d_{\alpha,\beta} t = k \int_a^b f(t) d_{\alpha,\beta} t$.

5. $f + g$ *is α, β-symmetric integrable on $[a,b]$ and*

$$\int_a^b (f+g)(t) d_{\alpha,\beta} t = \int_a^b f(t) d_{\alpha,\beta} t + \int_a^b g(t) d_{\alpha,\beta} t.$$

6. fg *is α, β-symmetric integrable on $[a,b]$ provided g is a nonnegative function.*

Proof. These results are easy consequences of the α-forward and β-backward integral properties. □

The next result follows immediately from Theorem 3 and the corresponding β-backward version.

Theorem 7. *Let $f : \mathbb{R} \to \mathbb{R}$ and $p > 1$. If $|f|$ is α, β-symmetric integrable on $[a,b]$, then $|f|^p$ is also α, β-symmetric integrable on $[a,b]$.*

Theorem 8. *Let $f, g : \mathbb{R} \to \mathbb{R}$ be α, β-symmetric integrable functions on $[a,b]$, $\mathscr{A} := \{a + k\alpha : k \in \mathbb{N}_0\}$ and $\mathscr{B} := \{b - k\beta : k \in \mathbb{N}_0\}$. For $b \in \mathscr{A}$ and $a \in \mathscr{B}$ one has:*

1. *If $|f(t)| \leqslant g(t)$ for all $t \in \mathscr{A} \cup \mathscr{B}$, then $\left| \int_a^b f(t) d_{\alpha,\beta} t \right| \leqslant \int_a^b g(t) d_{\alpha,\beta} t$.*

2. *If $f(t) \geqslant 0$ for all $t \in \mathscr{A} \cup \mathscr{B}$, then $\int_a^b f(t) d_{\alpha,\beta} t \geqslant 0$.*

3. *If $g(t) \geqslant f(t)$ for all $t \in \mathscr{A} \cup \mathscr{B}$, then $\int_a^b g(t) d_{\alpha,\beta} t \geqslant \int_a^b f(t) d_{\alpha,\beta} t$.*

Proof. It follows from Theorem 4 and Corollary 1 and the corresponding β-backward versions. □

In Theorem 9 we assume that $a, b \in \mathbb{R}$ with $b \in \mathscr{A} := \{a + k\alpha : k \in \mathbb{N}_0\}$ and $a \in \mathscr{B} := \{b - k\beta : k \in \mathbb{N}_0\}$, where $\alpha, \beta \in \mathbb{R}_0^+$, $\alpha + \beta \neq 0$.

Theorem 9 (Mean Value Theorem). *Let $f, g : \mathbb{R} \to \mathbb{R}$ be bounded and α, β-symmetric integrable on $[a,b]$ with g nonnegative. Let m and M be the infimum and*

he supremum, respectively, of function f. Then, there exists a real number K satis-
ying the inequalities $m \leqslant K \leqslant M$ such that $\int_a^b f(t)g(t)d_{\alpha,\beta}t = K \int_a^b g(t)d_{\alpha,\beta}t$.

Proof. Since $m \leqslant f(t) \leqslant M$ for all $t \in \mathbb{R}$ and $g(t) \geqslant 0$, then $mg(t) \leqslant f(t)g(t) \leqslant Mg(t)$ for all $t \in \mathscr{A} \cup \mathscr{B}$. All functions mg, fg and Mg are α,β-symmetric integrable on $[a,b]$. By Theorems 6 and 8, $m\int_a^b g(t)d_{\alpha,\beta}t \leqslant \int_a^b f(t)g(t)d_{\alpha,\beta}t \leqslant M\int_a^b g(t)d_{\alpha,\beta}t$. If $\int_a^b g(t)d_{\alpha,\beta}t = 0$, then $\int_a^b f(t)g(t)d_{\alpha,\beta}t = 0$; if $\int_a^b g(t)d_{\alpha,\beta}t > 0$, hen $m \leqslant \frac{\int_a^b f(t)g(t)d_{\alpha,\beta}t}{\int_a^b g(t)d_{\alpha,\beta}t} \leqslant M$. Therefore, the middle term of these inequalities is equal o a number K, which yields the intended result. $\qquad\square$

4 α,β-Symmetric Integral Inequalities

nspired in the work by Agarwal et al. [1], we now present α,β-symmetric versions of Hölder, Cauchy–Schwarz and Minkowski inequalities. As before, we assume that $a,b \in \mathbb{R}$ with $b \in \mathscr{A} := \{a+k\alpha : k \in \mathbb{N}_0\}$ and $a \in \mathscr{B} := \{b-k\beta : k \in \mathbb{N}_0\}$, where $\alpha,\beta \in \mathbb{R}_0^+$, $\alpha+\beta \neq 0$.

Theorem 10 (Hölder's Inequality). *Let* $f,g : \mathbb{R} \to \mathbb{R}$ *and* $a,b \in \mathbb{R}$ *with* $a < b$. *If* $|f|$ *and* $|g|$ *are* α,β-*symmetric integrable on* $[a,b]$, *then*

$$\int_a^b |f(t)g(t)|d_{\alpha,\beta}t \leqslant \left(\int_a^b |f(t)|^p d_{\alpha,\beta}t\right)^{\frac{1}{p}} \left(\int_a^b |g(t)|^q d_{\alpha,\beta}t\right)^{\frac{1}{q}}, \qquad (1)$$

where $p > 1$ *and* $q = p/(p-1)$.

Proof. For $\alpha,\beta \in \mathbb{R}_0^+$, $\alpha+\beta \neq 0$, the following inequality holds: $\alpha^{\frac{1}{p}}\beta^{\frac{1}{q}} \leqslant \frac{\alpha}{p} + \frac{\beta}{q}$. Without loss of generality, suppose that $\left(\int_a^b |f(t)|^p d_{\alpha,\beta}t\right)\left(\int_a^b |g(t)|^q d_{\alpha,\beta}t\right) \neq 0$ note that both integrals exist by Theorem 7). Set $\xi(t) = |f(t)|^p / \int_a^b |f(\tau)|^p d_{\alpha,\beta}\tau$ and $\gamma(t) = |g(t)|^q / \int_a^b |g(\tau)|^q d_{\alpha,\beta}\tau$. Since both functions α and β are α,β-symmetric integrable on $[a,b]$, then

$$\int_a^b \frac{|f(t)|}{\left(\int_a^b |f(\tau)|^p d_{\alpha,\beta}\tau\right)^{\frac{1}{p}}} \frac{|g(t)|}{\left(\int_a^b |g(\tau)|^q d_{\alpha,\beta}\tau\right)^{\frac{1}{q}}} d_{\alpha,\beta}t = \int_a^b \xi(t)^{\frac{1}{p}}\gamma(t)^{\frac{1}{q}}d_{\alpha,\beta}t$$

$$\leqslant \int_a^b \left(\frac{\xi(t)}{p} + \frac{\gamma(t)}{q}\right) d_{\alpha,\beta}t$$

$$= \frac{1}{p}\int_a^b \left(\frac{|f(t)|^p}{\int_a^b |f(\tau)|^p d_{\alpha,\beta}\tau}\right) d_{\alpha,\beta}t + \frac{1}{q}\int_a^b \left(\frac{|g(t)|^q}{\int_a^b |g(\tau)|^q d_{\alpha,\beta}\tau}\right) d_{\alpha,\beta}t = 1. \quad \square$$

The particular case $p = q = 2$ of (1) gives the Cauchy–Schwarz inequality.

Corollary 3 (Cauchy–Schwarz's Inequality). *Let* $f, g : \mathbb{R} \to \mathbb{R}$ *and* $a, b \in \mathbb{R}$ *with* $a < b$. *If* f *and* g *are* α, β-*symmetric integrable on* $[a, b]$, *then*

$$\int_a^b |f(t) g(t)| \, d_{\alpha,\beta} t \leqslant \sqrt{\left(\int_a^b |f(t)|^2 \, d_{\alpha,\beta} t \right) \left(\int_a^b |g(t)|^2 \, d_{\alpha,\beta} t \right)}.$$

We prove the Minkowski inequality using Hölder's inequality.

Theorem 11 (Minkowski's Inequality). *Let* $f, g : \mathbb{R} \to \mathbb{R}$ *and* $a, b, p \in \mathbb{R}$ *with* $a < b$ *and* $p > 1$. *If* f *and* g *are* α, β-*symmetric integrable on* $[a, b]$, *then*

$$\left(\int_a^b |f(t) + g(t)|^p \, d_{\alpha,\beta} t \right)^{\frac{1}{p}} \leqslant \left(\int_a^b |f(t)|^p \, d_{\alpha,\beta} t \right)^{\frac{1}{p}} + \left(\int_a^b |g(t)|^p \, d_{\alpha,\beta} t \right)^{\frac{1}{p}}.$$

Proof. If $\int_a^b |f(t) + g(t)|^p \, d_{\alpha,\beta} t = 0$, then the result is trivial. Suppose that

$$\int_a^b |f(t) + g(t)|^p \, d_{\alpha,\beta} t \neq 0.$$

One has

$$\int_a^b |f(t) + g(t)|^p \, d_{\alpha,\beta} t = \int_a^b |f(t) + g(t)|^{p-1} |f(t) + g(t)| \, d_{\alpha,\beta} t$$

$$\leqslant \int_a^b |f(t)| |f(t) + g(t)|^{p-1} \, d_{\alpha,\beta} t + \int_a^b |g(t)| |f(t) + g(t)|^{p-1} \, d_{\alpha,\beta} t.$$

Applying Hölder's inequality (Theorem 10) with $q = p/(p-1)$, we obtain

$$\int_a^b |f(t) + g(t)|^p \, d_{\alpha,\beta} t \leqslant \left(\int_a^b |f(t)|^p \, d_{\alpha,\beta} t \right)^{\frac{1}{p}} \left(\int_a^b |f(t) + g(t)|^{(p-1)q} \, d_{\alpha,\beta} t \right)^{\frac{1}{q}}$$

$$+ \left(\int_a^b |g(t)|^p \, d_{\alpha,\beta} t \right)^{\frac{1}{p}} \left(\int_a^b |f(t) + g(t)|^{(p-1)q} \, d_{\alpha,\beta} t \right)^{\frac{1}{q}}$$

$$= \left[\left(\int_a^b |f(t)|^p \, d_{\alpha,\beta} t \right)^{\frac{1}{p}} + \left(\int_a^b |g(t)|^p \, d_{\alpha,\beta} t \right)^{\frac{1}{p}} \right] \left(\int_a^b |f(t) + g(t)|^{(p-1)q} \, d_{\alpha,\beta} t \right)^{\frac{1}{q}}.$$

Therefore,

$$\frac{\int_a^b |f(t) + g(t)|^p \, d_{\alpha,\beta} t}{\left(\int_a^b |f(t) + g(t)|^{(p-1)q} \, d_{\alpha,\beta} t \right)^{\frac{1}{q}}} \leqslant \left(\int_a^b |f(t)|^p \, d_{\alpha,\beta} t \right)^{\frac{1}{p}} + \left(\int_a^b |g(t)|^p \, d_{\alpha,\beta} t \right)^{\frac{1}{p}}.$$

Our α, β-symmetric calculus is more general than the standard h-calculus. In particular, all our results give, as corollaries, results in the classical quantum h-calculus by choosing $\alpha = h > 0$ and $\beta = 0$.

Acknowledgements This work is supported by FEDER and Portuguese funds, COMPETE reference FCOMP-01-0124-FEDER-022690, and CIDMA and FCT, project PEst-C/MAT/UI4106/2011. Brito da Cruz is also supported by FCT through the Ph.D. fellowship SFRH/BD/33634/2009.

References

1. Agarwal, R., Bohner, M., Peterson, A.: Inequalities on time scales: a survey. Math. Inequal. Appl. **4**(4), 535–557 (2001)
2. Aldwoah, K.A.: Generalized time scales and associated difference equations, Ph.D. thesis, Cairo University (2009)
3. Brito da Cruz, A.M.C., Martins, N., Torres, D.F.M.: Higher-order Hahn's quantum variational calculus, Nonlinear Anal. **75**(3), 1147–1157 (2012)
4. Kac, V., Cheung, P.: Quantum Calculus. Springer, New York (2002)
5. Malinowska, A.B., Torres, D.F.M.: The Hahn quantum variational calculus. J. Optim. Theory Appl. **147**(3), 419–442 (2010)
6. Thomson, B.S.: Symmetric Properties of Real Functions, Dekker, New York (1994)

Numerical Range and Bifurcation Points of a Family of Rational Function

Helena Melo and João Cabral

Abstract Using the Numerical Range Theory we make some interesting observations about the behavior of the dynamics of the family of rational function $f_\lambda(x)$ given by

$$f_\lambda(x) = 1 - \frac{2\lambda}{x^2 + \lambda - 4}$$

in the neighborhood of the bifurcation point.

Mathematics Subject Classification 2010: 37M20

1 Introduction

The goal of this work is to propose an alternative new study for the family of real rational maps

$$f_\lambda(x) = 1 - \frac{2\lambda}{x^2 + \lambda - 4} \tag{1}$$

applying the Numerical Range Theory to the dynamics of the map f_λ through iteration.

International Conference on Differential and Difference Equations and Applications Iteration Theory
July 4–8, 2011, Ponta Delgada, Portugal

H. Melo (✉) • J. Cabral
Universidade dos Açores, Ponta Delgada, Portugal
e-mail: hmelo@uac.pt; jcabral@uac.pt

S. Pinelas et al. (eds.), *Differential and Difference Equations with Applications*, Springer
Proceedings in Mathematics & Statistics 47, DOI 10.1007/978-1-4614-7333-6_45,
© Springer Science+Business Media New York 2013

If we replace $4 + \lambda$ by a, and replace $4 - \lambda$ by b, then we obtain

$$f(x) = \frac{x^2 - a}{x^2 - b}.$$ (2)

Let

$$F = \begin{bmatrix} a_0 & a_1 & a_2 & 0 \\ 0 & a_0 & a_1 & a_2 \\ b_0 & b_1 & b_2 & 0 \\ 0 & b_0 & b_1 & b_2 \end{bmatrix},$$ (3)

as described in Milnor [2]. Each map f, in the space Rat_2, can be expressed as

$$f(z) = \frac{p(z)}{q(z)} = \frac{a_0 x^2 + a_1 x + a_2}{b_0 x^2 + b_1 x + b_2}$$ (4)

where a_0 and b_0 are not simultaneously zero and $p(z)$, $q(z)$ have no common roots. Milnor [2] states that we can obtain a rough description of the topology of this space Rat_2 that can be identified with the Zariski open subset of *complex projective 5-space* consisting of all points

$$(a_0 : a_1 : a_2 : b_0 : b_1 : b_2) \in CP^5$$ (5)

for which the resultant $res(p,q) = det(F)$ is nonzero.

By Melo and Cabral [1] we can build the matrix

$$B = \begin{bmatrix} A & O_2 \\ O_2 & A \end{bmatrix}$$ (6)

such that $A = \begin{bmatrix} 1 & -a \\ 1 & -b \end{bmatrix}$ and O_2 is the null matrix of order 2.

Using the Numerical Range Theory, $W(A)$ is the numerical range of the matrix A that can be defined as

$$W(A) = \frac{w^* . A . w}{w^* . w}, w \in \mathbb{C}^n \underbrace{\backslash (0, \dots, 0)}_{n}$$ (7)

where w^* is the transpose conjugate of w.

We have $W(A) = W(B)$ as an important result (see Melo and Cabral [1] for proper demonstration), and the boundary curve of A, $\partial W(A)$, is a curve of class two, in this case, an ellipse, whose two foci coincide with the eigenvalues of A.

Replacing again a by $4 + \lambda$ and b by $4 - \lambda$, we obtain the matrix A_λ where

$$A_\lambda = \begin{bmatrix} 1 & -\lambda - 4 \\ 1 & \lambda - 4 \end{bmatrix}.$$ (8)

2 Merging the Map f_λ in the Numerical Range of Matrix A_λ

In the real rational maps (1) we take a particular case making $0 < \lambda < 4$, with $\lambda \in \mathbb{R}$. This map f_λ has domain $D = \mathbb{R} \backslash_{\{\pm\sqrt{4-\lambda}\}}$ and has images on the set $\hat{D} = (-\infty, 1) \cup \left[\frac{4+\lambda}{4-\lambda}, +\infty\right)$.

Let

$$\Lambda = D \times \hat{D} \tag{9}$$

and

$$graph(f_\lambda) = \{(x,y) \in \mathbb{R}^2 : x \in D, y = f_\lambda(x)\}. \tag{10}$$

So, the set $graph(f_\lambda)$ is the typical plot of the map $f_\lambda(x)$ in $\mathbb{R}^2$.

Let

$$C = \{v \in \mathbb{C}^2 : v = (x, if_\lambda(x))\}, \tag{11}$$

$$\Psi = \left\{z \in \mathbb{C} : z = \frac{v^*.A_\lambda.v}{v^*.v}, v \neq 0, v \in C\right\}, \tag{12}$$

and the functions

$$\begin{aligned} \theta : D &\longrightarrow \Lambda \\ x &\longmapsto (x, f_\lambda(x)) \end{aligned} \tag{13}$$

$$\begin{aligned} V : \quad \Lambda &\longrightarrow C \\ (x, f_\lambda(x)) &\longmapsto (x, if_\lambda(x)) \end{aligned} \tag{14}$$

$$\begin{aligned} \Xi : \quad C &\longrightarrow \Psi \\ (x, if_\lambda(x)) &\longmapsto z_\lambda(x). \end{aligned} \tag{15}$$

So, we can define the function

$$\varepsilon_\lambda(x) = \begin{cases} \Xi \circ V \circ \theta(x) &, \text{if } x \in D \\ 1 &, \text{if } x \in \{-\infty, \infty\}. \\ -4 + \lambda &, \text{if } x \in \{-\sqrt{4-\lambda}, +\sqrt{4-\lambda}\}. \end{cases} \tag{16}$$

By *Rayleigh quotient* we obtain the elements of set Ψ, the complex numbers $z_\lambda(x)$, as in Fig. 1,

$$z_\lambda(x) = \frac{\left[x - if_\lambda(x)\right].A_\lambda.\begin{bmatrix} x \\ if_\lambda(x) \end{bmatrix}}{\left[x - if_\lambda(x)\right].\begin{bmatrix} x \\ if_\lambda(x) \end{bmatrix}}. \tag{17}$$

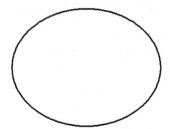

Fig. 1 The set $\Psi \subset \mathbb{C}$

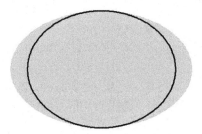

Fig. 2 The set $W(A) \subset \mathbb{C}$

Fig. 3 $\Psi \subset W(A)$

Figure 2 is the representation of the numerical range of the matrix A_λ, and Fig. 3 shows clearly that the set Ψ is a subset of the numerical range of the matrix A_λ.

The function $z_\lambda(x)$ is a complex function such that $g_\lambda(x) = Re(z_\lambda(x))$, the real part of $z_\lambda(x)$, and $h_\lambda(x) = Im(z_\lambda(x))$, the imaginary part of $z_\lambda(x)$, so $z_\lambda(x) = g_\lambda(x) + ih_\lambda(x)$.

The functions

$$g_\lambda(x) = \frac{-(4-\lambda)(\lambda+4)^2 + (4-\lambda)(12+\lambda)x^2 - 3(4-\lambda)x^4 + x^6}{(x^2 - \lambda - 4)^2 + x^2(x^2 + \lambda - 4)^2} \tag{18}$$

and

$$h_\lambda(x) = \frac{(5+\lambda)((\lambda^2 - 16)x + 8x^3 - x^5)}{(x^2 - \lambda - 4)^2 + x^2(x^2 + \lambda - 4)^2} \tag{19}$$

where $0 < \lambda < 4$, with $\lambda \in \mathbb{R}$, are continuous functions, so $z_\lambda(x)$ is continuous in $\mathbb{C} \cup \{\infty\}$.

Let

$$\Omega = \left\{ \frac{\left(x - \frac{3-\lambda}{2}\right)^2}{\left(\frac{5-\lambda}{2}\right)^2} + \frac{y^2}{\left(\frac{5+\lambda}{2}\right)^2} = 1, (x,y) \in \mathbb{R}^2, 0 < \lambda < 4 \right\} \tag{20}$$

that can be represented also in $\mathbb{C}$ as

$$\left\{ z \in \mathbb{C} : \left| z - \frac{-3 + \lambda + i2\sqrt{5\lambda}}{2} \right| + \left| z - \frac{-3 + \lambda - i2\sqrt{5\lambda}}{2} \right| = 5 + \lambda, 0 < \lambda < 4 \right\}. \tag{21}$$

So $z_\lambda(x) = g_\lambda(x) + ih_\lambda(x)$ is one element of Ω.

Since each element $x \in \mathbb{R} \cup \{\infty\}$ is transformed by $\varepsilon_\lambda(x)$ through $z_\lambda(x)$ continuously in Ω, which is continuous also, then all values of the real line have a match in Ω covering all Ω. So Ω is dense.

The $f_\lambda(x)$ elements are related also with Ω through the construction of $z_\lambda(x)$ giving us a way to analyze the dynamics of $f_\lambda(x)$, not in the real axis but in Ω.

Analyzing the function $g_\lambda(x)$ we have maximum values at

$$x_2 = -x_8 = -\sqrt{4 + \lambda}$$
$$x_4 = -x_6 = -\sqrt{4 + 2\lambda - \sqrt{16\lambda + 5\lambda^2}}$$

and minimum values at

$$x_1 = -x_9 = -\sqrt{4 + 2\lambda + \sqrt{16\lambda + 5\lambda^2}}$$
$$x_3 = -x_7 = -\sqrt{4 - \lambda}$$
$$x_5 = 0$$

see Fig. 4.

As demonstrated in Melo and Cabral [1], the zeros of $f_\lambda(x)$ are the relative maximum of $g_\lambda(x)$, and the relative minimum of $f_\lambda(x)$ or discontinuities of $f_\lambda(x)$ are relative minimums of $g_\lambda(x)$. This way we can associate the monotonicity of $f_\lambda(x)$ and the monotonicity of $g_\lambda(x)$.

On the real axis, we can organize these values as such:

$$-\infty < x_1 < x_2 < x_3 < x_4 < x_5 < x_6 < x_7 < x_8 < 1 < x_9 < +\infty.$$

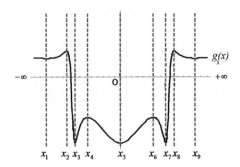

Fig. 4 Abscissas of maximum and minimum of $g_\lambda(x)$

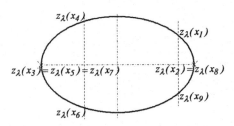

Fig. 5 Graph($\varepsilon_\lambda(x)$)

With some calculus we obtain

$$z_\lambda(x_2) = z_\lambda(x_8) = 1 + 0i$$
$$z_\lambda(x_3) = z_\lambda(x_5) = z_\lambda(x_7) = -4 + \lambda + 0i$$
$$z_\lambda(x_1) = g_{\lambda 1} + ih_{\lambda 1},$$

where $g_{\lambda 1} = Re(z_\lambda(x_1)) \neq 0$, and $h_{\lambda 1} = Im(z_\lambda(x_1)) \neq 0$,

$$z_\lambda(x_9) = \overline{z_\lambda(x_1)}$$
$$z_\lambda(x_4) = g_{\lambda 2} + ih_{\lambda 2},$$

where $g_{\lambda 2} = Re(z_\lambda(x_4)) \neq 0$, and $h_{\lambda 2} = Im(z_\lambda(x_4)) \neq 0$,

$$z_\lambda(x_6) = \overline{z_\lambda(x_4)}.$$

We can observe this complex values represented in Ω as showed in Fig. 5.

Through the transformation $\Xi \circ V$, we have a relation between Λ and Ψ, and if a generic element of Ψ is also reciprocally an element of Ω, then $\Psi = \Omega = graph(\varepsilon_\lambda(x))$.

3 Association Between $(x, f_\lambda(x))$ and $(g_\lambda(x), h_\lambda(x))$

Let $\hat{E}$ be the ordered set of the nonzero critical points of $g_\lambda(x)$ that are not zeros or discontinuous of $f_\lambda(x)$; thus, $\hat{E} = \{x_1, x_4, x_6, x_9\}$.

Each value $x \in \mathbb{R} \cup \{\infty\}$ is transformed by $\varepsilon(x)$ in one point of Ω (see Fig. 6).

The image of $-\infty$ is clearly $\varepsilon_\lambda(-\infty) = 1 + 0i$, and *walking* through the real line, following the natural order of the real values, $\varepsilon_\lambda(x)$ will *walk* in the arch of the ellipse Ω following the counterclockwise orientation until x pass through the first element of $\hat{E}$. Then the walk in Ω will be clockwise, restoring the counterclockwise order after x pass through the second element of $\hat{E}$ and so on, until $+\infty$ where $\varepsilon_\lambda(+\infty) = 1 + 0i$.

The orientation of the rotation, *the walk*, of $\varepsilon_\lambda(x)$, in Ω is related by $(-1)^\phi$, where ϕ is the sum of elements of $\hat{E}$ that x cross in the real line.

For example, the orientation of *the walk* of $\hat{E}$ in Ω is $(-1)^0 = +1$, positive, when the walk of x in the real line doesn't cross any element of $\hat{E}$. And the orientation of the walk of $\varepsilon_\lambda(x)$ in Ω is $(-1)^3 = -1$, negative, when the walk of x in the real line cross three elements of $\hat{E}$.

Let $\hat{I}_1 = [-\infty, x_1]$, $\hat{I}_2 = (x_1, x_4]$, $\hat{I}_3 = (x_4, x_6]$, $\hat{I}_4 = (x_6, x_9]$, and $\hat{I}_5 = (x_9, \infty]$. So Fig. 7 shows the representation on Ω of $\varepsilon_\lambda(\hat{I}_1)$, Fig. 8 shows $\varepsilon_\lambda(\hat{I}_2)$, and so on (Fig. 9).

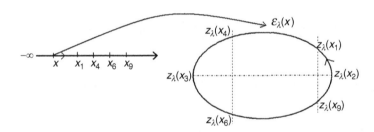

Fig. 6 Walking on the $\varepsilon_\lambda(x)$

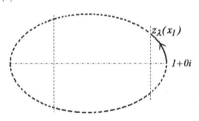

Fig. 7 $\varepsilon_\lambda(\hat{I}_1)$

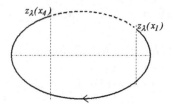

Fig. 8 $\varepsilon_\lambda(\hat{I}_2)$

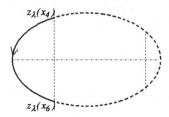

Fig. 9 $\varepsilon_\lambda(\hat{I}_3)$

4   The Border Collision Bifurcation

The fixed point of the real rational function $f_\lambda(x)$ in Ψ is the complex number

$$\frac{\left[x - ix \right].A_\lambda.\begin{bmatrix} x \\ ix \end{bmatrix}}{\left[x - ix \right].\begin{bmatrix} x \\ ix \end{bmatrix}} = \frac{-3+\lambda}{2} + i\frac{-5-\lambda}{2}. \tag{22}$$

Let $g_\lambda(x) = \frac{-3+\lambda}{2}$.

We have the solutions x', x'' and x''', respectively: (Fig. 10)

$$x' = \frac{13 - 3\lambda + \sqrt[3]{k} + \sqrt[3]{k^2}}{3\sqrt[3]{k}}$$

$$x'' = \frac{13 - 13i\sqrt{3} - (3 - 3i\sqrt{3})\lambda + 2\sqrt{k} + \sqrt[3]{k^2} + i\sqrt{3}\sqrt[3]{k^2}}{6\sqrt[3]{k}} \tag{23}$$

$$x''' = \frac{13 + 13i\sqrt{3} - (3 + 3i\sqrt{3})\lambda + 2\sqrt{k} + \sqrt[3]{k^2} - i\sqrt{3}\sqrt[3]{k^2}}{6\sqrt[3]{k}}$$

where $k = 35 + 18\lambda + 3\sqrt{3}\sqrt{-36 + 103\lambda - \lambda^2 + \lambda^3}$.

The value of x is a real number if and only if $k \neq 0$ and $-36 + 103\lambda - \lambda^2 + \lambda^3 \geq 0$.

With these conditions we have (Fig. 11)

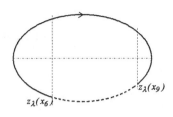

Fig. 10 $\varepsilon_\lambda(\hat{I}_4)$

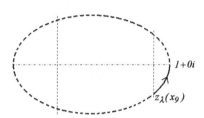

Fig. 11 $\varepsilon_\lambda(\hat{I}_5)$

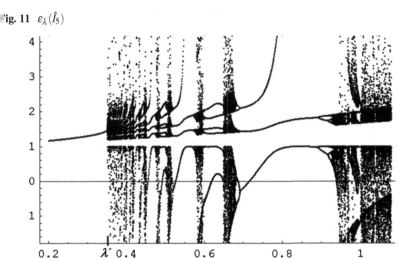

Fig. 12 Border collision bifurcation in λ'

$$\lambda' = \frac{1}{3}\left(1 - 308\sqrt[3]{\frac{2}{47 + 1017\sqrt{113}}} + \frac{1}{2}\sqrt[3]{47 + 1017\sqrt{113}}\right),$$

or $\lambda' \approx 0.35028855281187055$.

With this value of λ', we have a border collision bifurcation as showed in Fig. 12.

If $0 < \lambda < \lambda'$, the real rational function has three fixed points. If $\lambda = \lambda'$, then exist two fixed points, but if $\lambda' < \lambda < 4$, there is only one fixed point. Figures 13, 14, and 15 showed that.

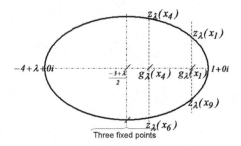

Fig. 13 $0 < \lambda < \lambda'$

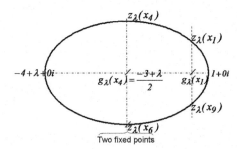

Fig. 14 $\lambda = \lambda'$

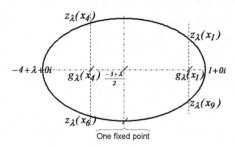

Fig. 15 $\lambda' < \lambda < 4$

In resume the value of λ' where there is a border collision bifurcation indicates how many fixed point has the real rational function.

Figure 16 shows the position of the complex number $z_\lambda(x)$ by the variation of λ.

We observed that the curve obtained by $z_\lambda(x_6)$ intersects the curve obtained by the value that represents the fixed point:

$$z_\lambda(x') = \frac{-3+\lambda}{2}.$$

We build a 3D model where we can see the border bifurcation points as geometric places in $\mathbb{R}^3$ that correspond to the intersection of the lines. In Fig. 16 we see

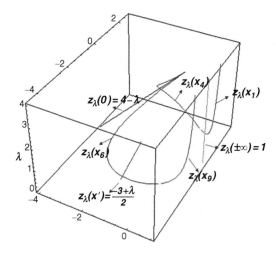

Fig. 16 $0 < \lambda < 4, \lambda \in \mathbb{R}$

an example of the presence of one of these bifurcation points. We intend to build a more generic model that will describe, as accurate as possible, the dynamics of this family of maps.

References

1. Melo, H., Cabral, J.: Numerical Range, Numerical Radii and the Dynamics of a Rational Function, Discrete Dynamics and Difference Equations. Proceedings of the twelfth International Conference on Difference Equations and Applications, pp 336–344 (2010)
2. Milnor, J.: Geometry and Dynamics of Quadratic Rational Maps, with an Appendix. In: Milnor, J., Lei, T (eds.) A.K. Peters, Ltd, Wellesley (1993)

Existence of Solutions for a Class of Semilinear Elliptic Systems via Variational Methods

G.A. Afrouzi and M. Mirzapour

Abstract This is concerned with the existence of solutions to a class of semilinear elliptic systems of the form

$$\begin{cases} -div(a(x)\nabla u) = \lambda F_u(x,u,v) \text{ in } \Omega, \\ -div(b(x)\nabla v) = \lambda F_v(x,u,v) \text{ in } \Omega, \\ u = v = 0 \qquad\qquad\qquad \text{ on } \partial\Omega, \end{cases}$$

where the domain Ω is a bounded domain in $R^N (N > 2)$. Using variational argument, we prove some existence results based on the Minimum principle and Mountain pass theorem of A. Ambrosetti and P. Rabinowitz.

Keywords Semilinear elliptic systems • Variational methods • Minimum principle • Mountain pass theorem

1 Preliminaries

In this paper, we deal with the existence of solutions to a class of semilinear elliptic systems of the form

G.A. Afrouzi (✉) • M. Mirzapour
Department of Mathematics, Faculty of Mathematical Sciences,
University of Mazandaran, Babolsar, Iran
e-mail: afrouzi@umz.ac.ir; mirzapour@stu.umz.ac.ir

S. Pinelas et al. (eds.), *Differential and Difference Equations with Applications*, Springer
Proceedings in Mathematics & Statistics 47, DOI 10.1007/978-1-4614-7333-6_46,
© Springer Science+Business Media New York 2013

$$\begin{cases} -div(a(x)\nabla u) = \lambda F_u(x,u,v) & \text{in } \Omega, \\ -div(b(x)\nabla v) = \lambda F_v(x,u,v) & \text{in } \Omega, \\ u = v = 0 & \text{on } \partial\Omega, \end{cases} \tag{1}$$

where the domain Ω is a bounded domain in $R^N (N > 2)$, the weights $a(x)$, $b(x)$ are measurable nonnegative weights on Ω, $(F_u, F_v) = \nabla F$ stands for the gradients of F in the variables $(u,v) \in R^2$, and λ is a parameter.

Recently, many authors have studied the existence of nontrivial solutions for such problems (see [4–8, 11–13] and their references) because several physical phenomena related equilibrium of continuous media are modeled by these elliptic problems (see [3]).

In [14], N.B. Zographopoulos studied a class of degenerate potential semilinear elliptic systems of the form

$$\begin{cases} -div(a(x)\nabla u) = \lambda\mu(x)|u|^{\gamma-1}|v|^{\delta+1}u & \text{in } \Omega, \\ -div(b(x)\nabla v) = \lambda\mu(x)|u|^{\gamma+1}|v|^{\delta-1}v & \text{in } \Omega, \\ u = v = 0 & \text{on } \partial\Omega, \end{cases} \tag{2}$$

where $\lambda \in R$, the exponents γ, δ are positive, and $\mu(x)$ may change sign. He proved the existence of at least one solution for the system (2) under suitable assumption on the data. Let λ_1 be the first eigenvalue of the system (2) (see [14] for $\mu(x) \equiv 1$). Then $\lambda_1 > 0$ and is given by

$$\lambda_1 = \inf_{(u,v)\in W\backslash\{(0,0)\}} \frac{\int_\Omega (\frac{\gamma+1}{p}a(x)|\nabla u|^2 + \frac{\delta+1}{q}b(x)|\nabla v|^2)dx}{\int_\Omega |u|^{\gamma+1}|v|^{\delta+1}dx}, \tag{3}$$

where γ, δ satisfying the condition $(\mathbf{F_1})$ below. The associated eigenfunction (u_0, v_0) is componentwise nonnegative and is unique (up to multiplication by a nonzero scalar) (see [6]).

In recent paper [2], G.A. Afrouzi et al. have studied the existence of solutions for semilinear problem (1) under conditions:

$(\mathbf{F_1})$ There exist positive constant c_1, c_2 such that

$$|F_t(x,t,s)| \leq c_1 t^\gamma s^{\delta+1}, \qquad |F_s(x,t,s)| \leq c_2 t^{\gamma+1}s^\delta$$

for all $(t,s) \in R^2$, a.e. $x \in \Omega$ and some $\gamma, \delta > 1$ with $\frac{\gamma+1}{p} + \frac{\delta+1}{q} = 1$ and $\gamma + 1 < p < 2_s^*$, $\delta + 1 < q < 2_s^*$.

$(\mathbf{F_2})$ There exist positive constant c and $2 < \alpha, \beta < 2_s^*$ such that

$$|F(x,t,s)| \leq c(1 + |t|^\alpha + |s|^\beta).$$

$(\mathbf{F_3})$ There exist $R > 0$, θ and θ' with $\frac{1}{2_s^*} < \theta, \theta' < \frac{1}{2}$ such that

$$0 < F(x,t,s) \leq \theta t F_t(x,t,s) + \theta' s F_s(x,t,s)$$

for all $x \in \overline{\Omega}$ and $|t|, |s| \geq R$.

($\mathbf{F_4}$) There exist $\overline{\alpha} > 2$, $\overline{\beta} > 2$ and $\varepsilon > 0$ such that

$$|F(x,t,s)| \leq c(|t|^{\overline{\alpha}} + |s|^{\overline{\beta}})$$

for all $x \in \overline{\Omega}$ and $|t|, |s| \leq \varepsilon$.

The condition ($\mathbf{F_1}$) plays an important role in proving the existence of solution by using the Minimum principle (see [10, p. 4, theorem 1.2]) and ($\mathbf{F_2}$) $-$ ($\mathbf{F_4}$) in proving that the functional satisfies the geometry of the Mountain pass theorem of A. Ambrosetti and P. Rabinowitz [3].

Motivated by the results in [9], our main goal in this paper is to illustrate how the ideas in [7, 13] can be applied to handle the problem of existence of nontrivial solution for system (1) without the conditions ($\mathbf{F_2}$) $-$ ($\mathbf{F_4}$).

Throughout this work, we assume the weights a, $b \in L^1_{loc}(\Omega)$, a^{-s}, $b^{-s} \in L^1(\Omega)$, $s \in (\frac{N}{2}, \infty) \cap [1, \infty)$. With the number s we define

$$2_s = \frac{2s}{s+1}, \quad 2_s^* = \frac{N2_s}{N - 2_s} = \frac{N2_s}{N(s+1) - 2s} > 2.$$

We define the Hilbert spaces $W_0^{1,2}(\Omega, a)$ and $W_0^{1,2}(\Omega, b)$ as the closures of $C_0^\infty(\Omega)$ with respect to the norms

$$\|u\|_a^2 = \int_\Omega a(x)|\nabla u|^2 dx \quad \text{for all} \quad u \in C_0^\infty(\Omega),$$

$$\|v\|_b^2 = \int_\Omega b(x)|\nabla v|^2 dx \quad \text{for all} \quad v \in C_0^\infty(\Omega).$$

Set $W = W_0^{1,2}(\Omega, a) \times W_0^{1,2}(\Omega, b)$. It is clear that W is a Hilbert space under the norm

$$\|(u,v)\|_W = \|u\|_a + \|v\|_b \quad \text{for all} \quad (u,v) \in W,$$

and with respect to the scalar product

$$\langle \varphi, \psi \rangle_W = \int_\Omega (a(x)\nabla\varphi_1\nabla\psi_1 + b(x)\nabla\varphi_2\nabla\psi_2)dx$$

for all $\phi = (\varphi_1, \varphi_2)$, $\psi = (\psi_1, \psi_2) \in W$ (see [1]).

Then W is a uniformly convex space. Moreover, the continuous embedding

$$W \hookrightarrow (W^{1,2_s})^2$$

holds with $2_s = \frac{2s}{s+1}$ (cf. Example 1.3, [9]), and we have the Sobolev's embedding $W \hookrightarrow (L^{2_s^*}(\Omega))^2$. We notice that the compact embedding

$$W \hookrightarrow L^r(\Omega) \times L^t(\Omega)$$

holds provided that $1 \leq r,t < 2_s^*$.

Next, we assume that $F(x,t,s)$ is a C^1—functional on $\Omega \times [0,\infty) \times [0,\infty) \rightarrow R$ satisfying the hypothesis:

(**F$_5$**)

$$\limsup_{|(t,s)|\to 0} \frac{2\lambda F(x,t,s)}{|t|^{\gamma+1}|s|^{\delta+1}} < \lambda_1 < \liminf_{|(t,s)|\to\infty} \frac{2\lambda F(x,t,s)}{|t|^{\gamma+1}|s|^{\delta+1}}$$

where λ_1 is defined by (3).

Definition 1. We say that $(u,v) \in W$ is a weak solution of system (1) if and only i

$$\int_\Omega (a(x)\nabla u \nabla \varphi + b(x)\nabla v \nabla \psi)dx = \lambda \int_\Omega (F_u(x,u,v)\varphi + F_v(x,u,v)\psi)dx$$

for all $(\varphi, \psi) \in W$.

The functional corresponding to problem (1) is

$$I_\lambda(u,v) = \frac{1}{2}\int_\Omega (a(x)|\nabla u|^2 + b(x)|\nabla v|^2)dx - \lambda \int_\Omega F(x,u,v)dx. \qquad (4$$

It is easy to see that the functional $I(u,v)$ is well defined and is of class C^1 in W Thus, weak solutions of (1) are exactly the critical points of the functional I_λ.

Now, we can describe our main results as follows:

Theorem 1. *Suppose that the condition* (**F$_1$**) *is satisfied. Then there exists* $\bullet$ *constant* $\underline{\lambda} > 0$ *such that for all* $\lambda < \underline{\lambda}$, *system (1) has a weak solution.*

Theorem 2. *In addition suppose that the condition* (**F$_1$**), (**F$_5$**) *are satisfied. The* *for* $\lambda \leq \lambda_1$ *problem (1) has a nontrivial solution.*

2 Proof of Theorem 1

We need to prove the following Lemmas and then apply the Minimum principle.

Lemma 1 (see [2]). *The functional* I_λ *given by (4) is weakly lower semicontinuou* *in* W.

To prove this Lemma, we suppose that a sequence $\{(u_m,v_m)\}$ converges weakly t (u,v) in W. By the weak lower semicontinuity of the norms in the spaces $W_0^{1,2}(\Omega,a$ and $W_0^{1,2}(\Omega,b)$, we deduce that

$$\liminf_{m\to\infty}\int_\Omega \left[a(x)|\nabla u_m|^2 + b(x)|\nabla v_m|^2\right]dx \geq \int_\Omega \left[a(x)|\nabla u|^2 + b(x)|\nabla v|^2\right]dx. \qquad (5$$

Next using the Mean value theorem and Holder's inequality, we prove that

$$\lim_{m \to \infty} \int_\Omega F(x, u_m, v_m)dx = \int_\Omega F(x, u, v)dx. \tag{6}$$

Lemma 2 (see [2]). *The functional I_λ given by (4) is coercive and bounded below in W.*

Proof. By $(\mathbf{F_1})$, there exists $c_3 > 0$ such that for all $(t, s) \in R^2$ and a.e. $x \in \Omega$, we have

$$|F(x, t, s)| \leq c_3 |t|^{\gamma+1} |s|^{\delta+1}.$$

Using the Young's inequality and embedding theorems on the term $\int_\Omega F(x, u, v)dx$, we can write

$$I_\lambda(u, v) \geq \left(\frac{1}{2} - \lambda c \frac{\gamma+1}{p} \right) \|u\|_a^2 + \left(\frac{1}{2} - \lambda c \frac{\delta+1}{q} \right) \|v\|_b^2,$$

where c is a positive constant. Let $\underline{\lambda} = \min \left\{ \frac{p}{2c(\gamma+1)}, \frac{q}{2c(\delta+1)} \right\} > 0$; then for all $\lambda < \underline{\lambda}$, we conclude that $I_\lambda(u, v) \to \infty$, provided that $\|(u, v)\| \to \infty$. □

By Lemmas (1) and (2), applying the Minimum principle, the functional I_λ attains its minimum, and thus system (1) admits at least one weak solution.

Proof of Theorem 2

In the following we prove two Lemmas to construct the geometry of the Mountain Pass theorem due to Ambrosetti and Rabinowitz [3].

Lemma 3. *The functional I_λ given by (4) satisfies the Palais-Smale condition in W.*

Proof. Let $\{(u_m, v_m)\}$ be a Palais-Smale sequence for the functional I_λ; thus there exists $c_4 > 0$ such that

$$|I_\lambda(u_m, v_m)| \leq c_4 \qquad \text{for any} \quad m \in N, \tag{7}$$

and there exists a strictly decreasing sequence $\{\varepsilon_m\}_{m=1}^\infty$, $\lim_{m \to \infty} \varepsilon_m = 0$, such that

$$|\langle I'_\lambda(u_m, v_m), (\xi, \eta) \rangle| \leq \varepsilon_m \|(\xi, \eta)\| \qquad \text{for any} \quad m \in N \quad \text{and for any} \quad (\xi, \eta) \in W. \tag{8}$$

By Lemma (2), we deduce that I_λ is coercive; relation (7) implies that the sequence $\{(u_m, v_m)\}$ is bounded in W. Since W is a Hilbert space, there exists $(u, v) \in W$ such that, passing to subsequence, still denote by $\{(u_m, v_m)\}$, it converges weakly to (u, v) in W and strongly in $L^p(\Omega) \times L^q(\Omega)$.

Choosing $(\xi, \eta) = (u_m - u, 0)$ in (8), we have

$$\left| \int_{\Omega} a(x) |\nabla u_m| \nabla(u_m - u) - \lambda \int_{\Omega} F_u(x, u_m, v_m)(u_m - u) \right| \leq \varepsilon_m \|u_m - u\|. \quad (9)$$

Using the condition $(\mathbf{F_1})$ combined with Holder's inequality, we conclude that

$$\int_{\Omega} F_u(x, u_m, v_m)|u_m - u|dx \leq c_1 \int_{\Omega} |u_m|^{\gamma} |v_m|^{\delta+1} |u_m - u|dx$$

$$\leq c_1 \|u_m\|_{L^p}^{\gamma} \|v_m\|_{L^q}^{\delta+1} \|u_m - u\|_{L^p}. \quad (10)$$

It follows from relations (9) and (10) that

$$\lim_{m \to \infty} \int_{\Omega} a(x) |\nabla u_m| \nabla(u_m - u)dx = 0$$

subtracting

$$\int_{\Omega} a(x) |\nabla u| (\nabla u_m - \nabla u)dx,$$

we obtain

$$0 = \lim_{m \to \infty} \int_{\Omega} a(x)(|\nabla u_m| - |\nabla u|)(\nabla u_m - \nabla u)dx \geq \lim_{m \to \infty} (\|u_m\|_a - \|u\|_a)^2 \geq 0,$$

which implies that $\|u_m\|_a \to \|u\|_a$. The uniform convexity of $W_0^{1,2}(\Omega, a)$ yields that u_m converges strongly to u in $W_0^{1,2}(\Omega, a)$. Similarly, we obtain $v_m \to v$ in $W_0^{1,2}(\Omega, b)$ as $n \to \infty$.   □

By Lemma (3), we obtain that the functional I_λ satisfies (PS)-condition (compactness condition). Now we verify that the functional I_λ has the geometry of the Mountain pass theorem.

Lemma 4. *Under assumptions* $(\mathbf{F_1})$ *and* $(\mathbf{F_5})$ *the functional* I_λ *satisfies:*

(i) *There exists* ρ, $\sigma > 0$ *such that* $\|(u, v)\|_H = \rho$ *implies* $I(u, v) \geq \sigma > 0$.
(ii) *There exists* $(z_1, z_2) \in W$ *such that* $\|(z_1, z_2)\|_H > \rho$ *and* $I(z_1, z_2) \leq 0$.

Proof. (i) From the left-hand side of $(\mathbf{F_5})$, there exists $\rho > 0$ such that

$$\lambda F(x, u, v) < \frac{1}{2} \lambda_1 |u|^{\gamma+1} |v|^{\delta+1}$$

provided that $\|u\|_a + \|v\|_b = \rho$ which will be chosen later.

By (3) and the variational characterization of the principle eigenvalue λ_1, we have

$$\lambda \int_{\Omega} F(x, u, v)dx \leq \frac{1}{2} \left[\frac{\gamma+1}{p} \int_{\Omega} a(x) |\nabla u|^2 dx + \frac{\delta+1}{q} \int_{\Omega} b(x) |\nabla v|^2 dx \right]$$

$$\leq \frac{1}{2}\max\{\frac{\gamma+1}{p}, \frac{\delta+1}{q}\}\left(||u||_a^2 + ||v||_b^2\right)$$

$$\leq \frac{1}{2}\left(||u||_a^2 + ||v||_b^2\right).$$

Then, there exist σ, $\rho > 0$ such that $I(u,v) \geq \sigma > 0$ if $||u||_a + ||v||_b = \rho$.

ii) From the right-hand side of (F_5), we get for $\varepsilon > 0$ and t sufficiently large that

$$\lambda F(x, tu_0, tv_0) \geq (\lambda_1 + \varepsilon)t^{2+\gamma+\delta}|u_0|^{\gamma+1}|v_0|^{\delta+1}$$

$$\geq (\lambda_1 + \varepsilon)t^2|u_0|^{\gamma+1}|v_0|^{\delta+1},$$

where (u_0, v_0) is the eigenfunction pair corresponding to the principle eigenvalue λ_1. Then

$$I(tu_0, tv_0) = \frac{t^2}{2}(||u_0||_a^2 + ||v_0||_b^2) - \lambda \int_\Omega F(x, tu_0, tv_0)dx$$

$$\leq \frac{t^2}{2}(||u_0||_a^2 + ||v_0||_b^2) - (\lambda_1 + \varepsilon)t^2 \int_\Omega |u_0|^{\gamma+1}|v_0|^{\delta+1}dx$$

$$\leq -t^2\varepsilon \int_\Omega |u_0|^{\gamma+1}|v_0|^{\delta+1}dx.$$

We conclude that

$$I(tu_0, tv_0) \to -\infty \quad as \quad t \to +\infty,$$

and thus there exists a constant t_0 large enough such that $I(t_0u_0, t_0v_0) < 0$. □

Consequently, the functional I_λ has a nonzero critical point, and the nonzero critical point of I_λ is precisely the nontrivial solution of problem (1).

References

1. Adams, R.A.: Sobolev Spaces. Academic Press, New York (1975)
2. Afrouzi, G.A., Mirzapour, M., Zographopoulos, N.B.: Existence results for a class of semilinear elliptic systems. Theor. Math. Appl. **2**, 77–86 (2012)
3. Ambrosetti, A., Rabinowitz, P.H.: Dual variational methods in critical point theory and applications. J. Funct. Anal. **14**, 349–381 (1973)
4. Boccardo, L., De Figueiredo, D.G.: Some remarks on a system of quasilinear elliptic equations. Nonlinear Differ. Equat. Appl. **9**, 309–323 (2002)
5. P. Caldiroli, R. Musina, On a variational degenerate elliptic problem. Nonlinear Differ. Equat. Appl. **7**, 187–199 (2000)
6. Chung, N.T., Toan, H.Q.: On a class of degenerate and singular elliptic systems in bounded domain. J. Math. Anal. Appl. **360**, 422–431 (2009)
7. Costa, D.G.: On a class of elliptic systems in R^N. Electron. J. Differ. Equat. **07**, 1–4 (1994)

8. Djellit, A. , Tas, S.: Existence of solutions for a class of elliptic systems in R^N involving th p-Laplacian. Electron. J. Differ. Equat. **56**, 1–8 (2003)
9. Drabek, P., Kufner, A., Nicolosi, F.: Quasilinear elliptic equations with degenarate an singularities. de Gruyter Series in Nonlinear Analysis and Applications, vol. 5, Walter d Gruyter and Company, Berlin (1997)
10. Struwe, M.: Variational Methods. Applications to Nonlinear Partial Differential Equations an Hamiltonian Systems. Fourth Edition, Springer, Berlin (2008)
11. Zhang, G., Wang, Y.: Some existence results for a class of degenerate semilinear ellipti systems. J. Math. Anal. Appl. **333**, 904–918 (2007)
12. Willem, M.: Minimax Theorems. Birkhauser, Boston (1996)
13. Zographopoulos, N.B.: On a class of degenerate potential elliptic system. Nonlinear Diff. Equ Appl. **11**, 191–199 (2004)
14. Zographopoulos, N.B.: p-Laplacian systems on resonance. Appl. Anal. **83**, 509–519 (2004)

Periodic Boundary Value Problems For Systems of First-Order Differential Equations with Impulses

M. Mohamed, H.S. Ahmad, and M.S.M. Noorani

Abstract In this paper, we prove the existence and uniqueness for systems of first-order impulsive differential equations with periodic boundary conditions. To establish such results, sufficient conditions of limit forms are given.

Keywords Periodic boundary value problems • Impulsive equations • Fixed-point theory

1 Introduction

In recent years, the solvability of the periodic boundary value problems (PBVPs for short) of first-order impulsive differential equations was studied by many authors; see the pioneer solutions on the theory of impulsive differential equations [1, 2, 6, 15], the papers, and the references there in [5, 7–9, 13]. Motivated by the studies in [3, 16], we study the existence and uniqueness of solutions to the following first-order differential system with periodic boundary conditions

$$x'(t) + a(t)x(t) = F(t, x(t)), \ a.e. \ t \in [0, N], \ t \neq t_1, \tag{1}$$

$$x(0) = x(N), 0 < N \in R, \tag{2}$$

M. Mohamed (✉)
Fakulti Sains Komputer dan Matematik, Universiti Teknologi MARA Pahang,
Bandar Jengka, Pahang, Malaysia
e-mail: mesliza@pahang.uitm.edu.my

H.S. Ahmad,
Universiti Teknologi MARA (Perlis), 02600 Arau, Malaysia
e-mail: huda847@perlis.uitm.edu.my

M.S.M. Noorani
Pusat Pengajian Sains Matematik, 43600 UKM Bangi, Selangor, Malaysia
e-mail: msn@pkris.cc.ukm.my

S. Pinelas et al. (eds.), *Differential and Difference Equations with Applications*, Springer
Proceedings in Mathematics & Statistics 47, DOI 10.1007/978-1-4614-7333-6_47,
© Springer Science+Business Media New York 2013

where

$$F : [0,N] \times R^n \to R^n \text{ iscontinuouson } (t,p) \in [0,N] \backslash \{t_1\} \times R^n \text{andpossibly}$$

$$\text{nonlinear}, a(t) \in C([0,N],R) \text{ with } \alpha(t) = \int_0^t a(t)dt \neq 0. \tag{3}$$

For the sake of simplicity (as in [3]), we consider only one impulse at $t = t_1 \in (0,N)$
An arbitrary finite number of impulses can be addressed similarly. The impulse a
$t = t_1$ is given by a continuous function $I_1 : R^n \to R^n$ with

$$x(t_1^+) = x(t_1^-) + I_1(x(t_1)), \ t_1 \in (0,N), \ t_1 \text{ fixed}, \tag{4}$$

using the notation $x(t_1^-) := lim_{t \to t_1^-} x(t)$ and $x(t_1^+) := lim_{t \to t_1^+} x(t)$. We assume that

$$F(t_1^+,x) := lim_{t \to t_1^+} F(t,x) \text{ and } F(t_1^-,x) := lim_{t \to t_1^-} F(t,x)$$

both exist with $F(t_1^-,x) = F(t_1,x)$.

This article is organized as follows. Section 2 presents some preliminary idea
associated with the impulsive BVP (1)–(4). Sections 3 and 4 contain the mai
results of the paper and are devoted to the existence and uniqueness of solution
to (1)–(4). There, sufficient conditions of limit forms are developed and applied
in conjunction with Schaeffer's theorem (see[10], Theorem 4.4.12) and contractio
mapping theorem (see[10], Theorem 3.4.10) to prove the existence and uniquenes
of solutions to (1)–(4). The sufficient conditions of limit forms to establis
such results introduced by Zhang and Yan [16] are generalized to the impulsiv
problems (1)–(4). The new results complement and extend those of [3, 4, 11, 12] i
the sense that sufficient conditions of limit forms are given to establish the result
of existence and uniqueness of solutions, whereas the theorem in [14] permitte
superlinear growth of $\| F(t,p) \|$ in $\| p \|$ in (1); our results apply to systems o
impulsive BVPs as in [3], unlike the papers [4, 11, 12]. The main idea rely on nove
differential inequalities and a priori bound on solutions to a certain family of integra
operator equations, with the operator being compact.

2   Preliminary Results

We introduce and denote the Banach space $PC([0,N];R^n)$ by

$$PC([0,N];R^n) := \{u : [0,N] \to R^n, u \in C([0,N] \backslash t_1;R^n), u \text{isleftcontinuousat} t = t_1,$$

$$\text{the right} - \text{hand limit } u(t_1^+) \text{ exists}\}$$

with norm

$$\| x \| = \left(\sum_{i=1}^n | x_i |^2 \right)^{\frac{1}{2}}, \ | x_i | = \max_{0 \le t \le N} | x_i(t) |, \ i = 1, \cdots, n.$$

If $x,y \in R^n$, then $< x,y >$ denotes the usual inner product.

Lemma 1 (Schaeffer's theorem [6,Theorem 4.4.12). *] Let X be a normed space and T:X → X be a completely continuous map. If the set*

$$S := \{x \in X : x = \lambda Tx, \lambda \in [0,1]\} \tag{5}$$

s bounded, then T has at least one fixed point.

In this section the impulsive BVP (1)–(4) is reformulated as an appropriate integral equation so that potential solutions to the integral equation will be solutions to the impulsive BVP (1)–(4). The motivation of this approach is to define a suitable operator, with fixed-point of the operator corresponding to the solution of the BVP (1)–(4).

The following results are included to keep the paper self-contained for the benefit of the reader.

Lemma 2. *Consider the impulsive BVP (1)–(4) with $\alpha(N) \neq 0, \alpha(t) = \int_0^t a(s)ds \neq 0$ or $t \in [0,N]$. Let $F : [0,N] \times R^n \to R^n$ and $I_1 : R^n \to R^n$ both be continuous.*

(i) If $x \in PC^1([0,N];R^n)$ is a solution of (1)–(4), then

$$x(t) = \int_0^N g(t,s)F(s,x(s))ds + g(t,t_1)I_1(x(t_1)), \ t \in [0,N], \tag{6}$$

where

$$g(t,s) = \begin{cases} \frac{e^{-[\alpha(t)-\alpha(s)]}}{1-e^{-\alpha(N)}}, & 0 \leq s \leq t \leq N \\ \frac{e^{-[\alpha(N)+\alpha(t)-\alpha(s)]}}{1-e^{-\alpha(N)}}, & 0 \leq t < s \leq N \end{cases} . \tag{7}$$

(ii) If $x \in PC([0,N];R^n)$ satisfies (6), then $x \in PC^1([0,N];R^n)$ and x is a solution of (1)–(4).

Proof. The proof follows similar lines to that of [11], Lemma 2.1. (i) Let $x \in PC^1([0,N];R^n)$ and from (1) consider

$$\frac{d}{dt}(xe^{\alpha(t)}) = e^{\alpha(t)}x' + e^{\alpha(t)}a(t)x = e^{\alpha(t)}F(t,x),$$

where the multiplication of the (possibly) vector values x,x' and F by the scalar-valued $e^{\alpha(t)}$ is done in a component-wise fashion. Integrating the above expression from t_1 to t with $t \in (t_1,N)$, we have

$$x(t)e^{\alpha(t)} = x(t_1^+)e^{\alpha(t_1)} + \int_{t_1}^t e^{\alpha(s)}F(s,x(s))ds.$$

A similar integration from 0 to t_1 shows that

$$x(t_1^-)e^{\alpha(t_1)} = x(0) + \int_0^{t_1} e^{\alpha(s)}F(s,x(s))ds.$$

Hence, adding the two previous expressions, we then have

$$x(t)e^{\alpha(t)} = x(0) + x(t_1^+)e^{\alpha(t_1)} - x(t_1^-)e^{\alpha(t_1)} + \int_0^t e^{\alpha(s)}F(s,x(s))ds$$

$$= x(0) + e^{\alpha(t_1)}I_1(x(t_1)) + \int_0^t e^{\alpha(s)}F(s,x(s))ds. \qquad (8$$

Letting $t = N$ in the previous and using the boundary conditions, we obtain

$$x(N)e^{\alpha(N)} = x(0) + e^{\alpha(t_1)}I_1(x(t_1)) + \int_0^N e^{\alpha(s)}F(s,x(s))ds$$

$$= e^{\alpha(N)}(x(0)).$$

A rearrangement then gives

$$x(0) = \frac{e^{\alpha(t_1)}}{e^{\alpha(N)} - 1}I_1(x(t_1)) + \frac{1}{e^{\alpha(N)} - 1}\int_0^N e^{\alpha(s)}F(s,x(s))ds,$$

which is substituted in (8) and a rearrangement leads to (6). [

(ii) Let $x \in PC([0,N];R^n)$ be a solution to (6). Since F is continuous, it is easy to se
that $x \in PC^1([0,N];R^n)$. To verify that x also satisfies the impulsive BVP (1)–(4
just differentiate (6).

Lemma 3. *Consider the impulsive BVP (1)–(4) with $\alpha(t) = \int_0^t a(s)ds \neq 0$ fo
$t \in [0,N]$. Let $F : [0,N] \times R^n \to R^n$ and $I_1 : R^n \to R^n$ both be continuous. Let g b
defined as in Lemma 2 and consider the mapping $T : PC([0,N];R^n) \to PC([0,N];R^n$
defined by*

$$Tx(t) := \int_0^N g(t,s)F(s,x(s))ds + g(t,t_1)I_1(x(t_1)), \ t \in [0,N]; \qquad (9$$

*If T has a fixed-point p, that is, $Tp = p$ for some $p \in PC([0,N];R^n)$, then this fixed
point p is also a solution to the impulsive BVP (1)–(4).*

Proof. The result immediately follows from Lemma 2. Obviously, from (7), ther
is a constant G such that

$$\max_{0 \leq s,t \leq N, t \neq t_1} | g(t,s), g(t,t_1) | = G. \qquad (10$$

Our topologically inspired fixed-point theorem that will be used to guarantee th
existence of at least one fixed-point of T requires that T be a compact map [10
pp. 54–55. We now illustrate that this is true for the above T in (9). [

Lemma 4. *Consider the impulsive BVP (1)–(4) with $\alpha(N) \neq 0, \alpha(t) = \int_0^t a(s)ds \neq$
for $t \in [0,N]$. Let $F : [0,N] \times R^n \to R^n$ and $I_1 : R^n \to R^n$ both be continuous. The
$T : PC([0,N];R^n) \to PC([0,N];R^n)$ is completely continuous, where T is defin
in (9).*

Proof. (i) Let $x_n, x \in A$ and $lim_{n \to \infty} x_n = x$. Then for $t \in [0,N]$,

$$\| Tx_n(t) - Tx(t) \| \leq \int_0^N g(t,s) \| F(s,x_n(s)) - F(s,x(s)) \| ds$$

$$+ g(t,t_1)[I_1(x_n(t)) - I_1(x(t))]$$

$$\leq G \int_0^N \| F(s,x_n(s)) - F(s,x(s)) \| ds + [I_1(x_n(t)) - I_1(x(t))].$$

Since $F : [0,N] \times R^n \to R^n$ and $I_1 : R^n \to R^n$ both be continuous, we have

$$\| Tx_n - Tx \| \to 0, \quad as\ n \to \infty.$$

That is, T is continuous. (ii) Let $A \subset PC([0,N];R^n)$ be a bounded set, that is, there is an $L > 0$ such that for any $\{x_n\} \in A$, $\| x_n(t) \| \leq L$. Since $F : [0,N] \times R^n \to R^n$ and $I_1 : R^n \to R^n$ both be continuous, there exists constants K_1, K such that $\| F(t,x_n(t)) \| \leq K$, $| I_1(x_n(t_1)) | \leq K_1$ for all n where $t \in [0,N]$.

$$\| Tx_n(t) \| \leq \| \int_0^N g(t,s)F(s,x_n(s))ds \| + \| g(t,t_1)I_1(x_n(t_1)) \|$$

$$\leq K \int_0^N \| g(t,s) \| ds + K_1 \| g(t,t_1) \|$$

$$\leq KNG + K_1 G.$$

This shows that $T(A)$ is a bounded set of $PC([0,N],R^n)$. Since A is a bounded subset of $PC([0,N],R^n)$, the set $T(A)$ is relatively compact. Hence, T is completely continuous operator. □

3 Existence

In this section we obtain some new existence results for (1)–(4). The ideas use:

Theorem 1. *Assume that (3) holds and one of the following conditions holds:*

(i) $\| F(t,x) \|$ *is bounded on* $[0,N] \times R^n$.

(ii) *There exist function* $V \in C^1(R^n, [0,\infty))$ *and bounded function* $h(t,x) \in C([0,N] \times R^n; R^n)$ *such that for* $t \in [0,N]$ *and* $\lambda \in (0,1]$ *uniformly,*

$$lim_{\|x\| \to \infty} inf \frac{< \nabla V(x(t)), \lambda F(t,x(t)) - a(t)x(t) > + \| h(t,x(t)) \|}{\lambda \| F(t,x(t)) \|} > 0 \quad (11)$$

and

$$lim_{x \to \infty} \frac{I_1(x)}{x} = W. \quad (12)$$

Then boundary value problem (1)–(4) has at least one solution.

Proof. From Lemma 3, we see that BVP (1)–(4) is equivalent to the integral (6). Let us define $T : PC([0,N];R^n) \to PC([0,N];R^n)$ by

$$Tx(t) = \int_0^N g(t,s)F(s,x(s))ds + g(t,t_1)I_1(x(t_1)), \ t \in [0,N]. \tag{13}$$

Now we apply Schaeffer's theorem to prove that BVP (1)–(4) has at least one solution. Hence, we need to prove that the set

$$S_\lambda := \{x \in PC([0,N];R^n) : x = \lambda Tx, \lambda \in [0,1]\} \tag{14}$$

is a bounded set with the bound being independent of $\lambda \in [0,1]$. Then we can conclude existence of at least one fixed point $x \in PC([0,N];R^n)$ of T. In consequently from Lemma 3 , BVP (1)–(4) has at least one solution $x \in PC([0,N];R^n)$. From Lemma 3 and (13), it is easy to see that $x = \lambda Tx$ is equivalent to BVP

$$x'(t) + a(t)x(t) = \lambda F(t,x(t)), \ a.e. \ t \in [0,N], \ t \neq t_1, \tag{15}$$

$$x(0) = x(N) \tag{16}$$

$$x(t_1^+) = x(t_1^-) + \lambda I_1(x(t_1)), \ t \in (0,N), \ t_1 \text{ fixed.} \tag{17}$$

If (11) holds, suppose that $\lambda = 0$, then $x = 0$; if $\lambda \in (0,1]$, assume, for the sake of contradiction, that S_λ is unbounded. Thus, there exists $\{x_n(t)\} \in S_\lambda$ for $t \in [0,N]$ such that $lim_{n\to\infty} \| x_n(t) \| = \infty$ and $lim_{n\to\infty} \| F(t,x_n(t)) \| = \infty$. Hence, from (11) there exists constant $N > 0$ satisfying that for $n > N$ and $t \in [0,N]$ and $\lambda \in (0,1]$ uniformly,

$$\frac{< \nabla V_n(t), \lambda F(t,x_n(t)) - a(t)x_n(t) > + h(t,x_n(t))}{\lambda F(t,x_n(t))} \geq \delta > 0, \tag{18}$$

where δ is independent of $\lambda \in (0,1]$. In view of (12), there is an $n \geq N$, such that

$$I_1(x_n) \leq (W + \varepsilon)x_n.$$

Taking into account (10), (17), and (18), we have that for each $n \geq N, t \in [0,N] \ \lambda \in (0,1]$,

$$\| x_n(t) \| = \lambda \| Tx_n \| \leq \lambda G \| F(s,x_n)ds \| + \lambda \| I_1(x_n(t_1)) \|$$

$$\leq \frac{G}{\delta} \int_0^N \Big[< \nabla V(x_n(s)), \lambda F(s,x_n(s)) - a(s)x_n(s) > ds$$

$$+ \int_0^N \| h(s,x_n(s)) \| \Big] + \lambda \| I_1(x_n(t_1)) \|$$

$$\leq \frac{G}{\delta} \Big[\int_0^N \frac{d}{ds} V(x_n(s))ds + \int_0^N \| h(s,x_n(s)) \| \Big] + \lambda \| I_1(x_n(t_1)) \|$$

$$= \frac{G}{\delta}[V(x_n(N))-V(x_n(0))]+\frac{G}{\delta}\int_0^N \| h(s,x_n(s)) \| \, ds+ \| I_1(x_n(t_1)) \|$$

$$\leq \frac{G}{\delta}\int_0^N \| h(s,x_n(s)) \| \, ds+(W+\varepsilon)x_n$$

which contradict that $h(t,x(t))$ is bounded on $[0,N] \times R^n$. Therefore, S_λ is bounded and P is independent of $\lambda \in [0,1]$. This proves that the BVP (1)–(4) has at least one solution. The proof is complete. □

In particular, choose, respectively, $V(x) = e^{\|x\|}$ and $V(x) =\| x \|^\alpha, \alpha \geq 2$; from Theorem 1, the following result can be obtained. ([16]) In Theorem 1, replace (11) by the condition below; then Theorem 1 is still valid, where

$$lim_{\|x\|\to\infty} \, inf \frac{e^{\|x\|}/ \| x \|< x(t),\lambda F(t,x(t))-a(t)x(t)>+ \| h(t,x(t)) \|}{\lambda \| F(t,x(t)) \|} > 0 \quad (19)$$

$$lim_{\|x\|\to\infty} \, inf \frac{\alpha\| x \|^{\alpha-2}<x(t),\lambda F(t,x(t))-a(t)x(t)>+ \| h(t,x(t)) \|}{\lambda \| F(t,x(t)) \|} > 0. \quad (20)$$

Proof. We only proof the corollary for (19) only. The proof for (20) is given in [16]. Let $V(x) = e^{\|x\|}$. Since

$$\frac{d}{dt} \| x \|= \frac{\sum_i^n x_i x'}{\| x \|},$$

it follows that

$$\frac{d}{dt}V(x(t)) = \frac{e^{\|x\|}}{\| x \|} < x(t),x'(t) > = \frac{e^{\|x\|}}{\| x \|} < x(t),\lambda F(t,x(t)) - a(t)x(t) > .$$

The proof is complete. □

Uniqueness

In this section, we will establish uniqueness results of the solutions of the BVP (1)–(4). Consider the Banach space X defined in Sect. 1. Assume that $F(t,x)$ satisfies Lipschitz condition with respect to x; that is, there exists constant L such that

$$\| F(t,x) - F(t,y) \| \leq L_1 \| x-y \|, \, t \in [0,N]$$
$$| I_1(x) - I_1(y) | \leq L_2 | x-y | \quad (21)$$

hold for any $(t,x),(t,y) \in [0,N] \times R^n$.

Theorem 2. *Assume that (1) and (21) hold and $G(L_1N+L_2) < 1$ where G is defined by (10). Then the BVP (1)–(4) has exactly one solution.*

Proof. Let us define $T : PC([0,N];R^n) \to PC([0,N];R^n)$ by (13). Consider the map

$$Tx(t) := \int_0^N g(t,s)F(s,x(s))ds + g(t,t_1)I_1(x(t_1)), \quad t \in [0,N]; \qquad (22$$

$$Ty(t) := \int_0^N g(t,s)F(s,y(s))ds + g(t,t_1)I_1(y(t_1)), \quad t \in [0,N]. \qquad (23$$

Then

$$\begin{aligned}
\| Tx(t) - Ty(t)) \| &\leq \int_0^N |g(t,s)| \, \| [F(s,x(s)) - F(s,y(s))] \| \, ds \\
&\quad + |g(t,t_1)[I_1(x(t_1)) - I_1(y(t_1))]| \\
&\leq \int_0^N GL_1 \, \| x(s) - y(s) \| \, ds + GL_2 |x(t_1) - y(t_1)| \\
&= GL_1 \, \| x - y \| \, N + GL_2 |x(t_1) - y(t_1)| \\
&\leq G(L_1 N + L_2) \, \| x - y \|
\end{aligned}$$

Hence, $\| Tx(t) - Ty(t) \| \leq \| x - y \|$ where $G(L_1 N + L_2) < 1$. This implies that T i contractive mapping. By the fixed-point theorem of Banach, the map T has uniqu fixed point.   [

5 Example

In this section we consider the scalar-valued differential equation as an example.

Example 1. Consider the impulsive BVP given by

$$x' - tx = x^3 + t^2 \qquad (24$$

$$x(0) = x(1) \qquad (25$$

$$x(t_1^+) = x(t_1^-) + 3x(t_1), \qquad (26$$

where x is scalar-valued and two given functions, $V(t,x) = x + 5$ and $h(t,x) = 1 - x$ The above BVP has at least one solution.

Proof.

$$\begin{aligned}
\frac{<\nabla V(x(t)), \lambda F(t,x(t)) - a(t)x(t)> + \| h(t,x(t)) \|}{\lambda \| F(t,x(t)) \|} &= \frac{<1, \lambda(x^3 + t^2) + tx> + |1 - x|}{\lambda |x^3 + t^2|} \\
&= \frac{\lambda(x^3 + t^2) + tx + |1 - x|}{\lambda |x^3 + t^2|} \\
&= \frac{\lambda(x^3 + t^2) + tx}{\lambda |x^3 + t^2|} + \frac{|1 - x|}{\lambda |x^3 + t^2|}
\end{aligned}$$

Taking the limit as $x \to \infty$,

$$lim_{x \to \infty} \left[\frac{\lambda(x^3 + t^2) + tx}{\lambda \mid x^3 + t^2 \mid} + \frac{\mid 1 - x \mid}{\lambda \mid x^3 + t^2 \mid} \right] = lim_{x \to \infty} \frac{\lambda(x^3 + t^2) + tx}{\lambda \mid x^3 + t^2 \mid} + lim_{x \to \infty} \frac{\mid 1 - x \mid}{\lambda \mid x^3 + t^2 \mid}$$

$$= 1 + 0 = 1 > 0$$

Also,

$$lim_{x \to \infty} \frac{I_1(x)}{x} = lim_{x \to \infty} \frac{3x}{x} = 3 > 0.$$

Thus, all of the conditions of Theorem 1 hold and the solvability follows. □

Acknowledgements This research is supported by Kementerian Pengajian Tinggi and Fundamental Research Grant Scheme ($600 - RMI/ST/FRGS5/3/Fst(22/2010)$) of Malaysia.

References

1. Bainov, D.D., Simeonov, P.S.: Systems with Impulsive Effect: Stability, Theory and Applications. Ellis, Horwood, Chichester, (1989)
2. Borysenko, S., Kololapov, V., Obolenskii, Y.: Stability of Processes Under Continuous and Discrete Disturbances, vol. 198, Naukova Dumka Kiev (1998) (in Russia)
3. Chen, J., Tisdell, C.C., Yuan, R.: On the solvability of periodic boundary value problems with impulse. J. Math. Anal. Appl. **331**, 902–912 (2007)
4. Franco, D., Nieto, J.J.: Maximum principles for periodic impulsive first order problems. J. Comput. Appl. Math. **88**, 144–159 (1998)
5. He, Z., Yu, J.: Periodic boundary value problems for first order impulsive ordinary differential equations, J. Math. Anal. Appl. **272**, 67–78 (2002)
6. Laksmikantham, V., Bainov, D.D., Simenov, P.S.: Theory of Impulsive Differential Equations. World Scientific, Singapore (1989)
7. Li, J., Nieto, J.J., Shen, J.: Impulsive periodic boundary value problems of first-order differential equations. J. Math. Anal. Appl. **325**, 226–236 (2007)
8. Liu, Y.: Further results on periodic boundary value problems for nonlinear first order impulsive functional differential equations. J. Math. Anal. Appl. **327**, 435–452 (2007)
9. Liu, Y.: Positive solutions of periodic boundary value problems for nonlinear first-order impulsive differential equations. Nonlinear Anal. **70**, 2106–2122 (2009)
10. Lloyd, N.G.: Degree theory. Cambridge Tracts in Mathematics. Vol. 73, Cambridge University Press, Cambridge, New York, Melbourne (1978)
11. Nieto, J.J.: Basic theory for nonresonance impulsive periodic problems of first order. J. Math. Anal. Appl. **205**, 423–433 (1997)
12. Nieto, J.J.: Periodic boundary value problems for first order impulsive ordinary differential equations. Nonlinear Appl. **51**, 1223–1239 (2002)
13. Nieto, J.J.: Impulsive resonance periodic problems of first order. Appl. Math. Lett. **15**(4), 489–493 (2002)
14. Nieto, J.J., Tisdell, C.: Existence and uniqueness of solutions to first-order systems of nonlinear impulsive boundary value problems with sub–, superlinear or linear growth, Electronic J. Differ. Equation **2007**(105), 1–14 (2007)

15. Samoilenko, A.M., Perestyuk, N.A.: Differential Equations with Impulsive Excitation, vol.286 Vyshcha Shkola, Kiev (1987) (in Russia)
16. Zhang, F., Yan, J.: Resonance and nonresonance periodic value problems of first order differential systems. Discrete Dynam. Nat. Soc. **2010**, 11, Article ID 863193 doi:10.1155/2010/863193.

Numerical Methods for Multi-term Fractional Boundary Value Problems

N.J. Ford and M.L. Morgado

Abstract This paper discusses the issues of existence and uniqueness of solution and the structural stability of boundary value problems for multi-term fractional differential equations. For the numerical solution of such problems we propose a shooting algorithm.

1 Introduction

Recently, in [7], we have investigated boundary value problems for single-term fractional differential equations. We have established sufficient conditions for the existence and uniqueness of the solution of problems of the form

$$D_*^{\alpha} y(t) = f(t, y(t)), \quad t \in [0, T] \tag{1}$$

$$y(a) = y_a, \tag{2}$$

where we have considered $D^{\alpha} y(t)$ as the derivative of order α, $0 < \alpha < 1$, of $y(t)$ in the Caputo sense, f is a continuous function on a suitable domain satisfying a Lipschitz condition with respect to its second argument, and $a > 0$. The case where $a = 0$ had been studied previously in [3]. In that case, problem (1) and (2) corresponds to an initial value problem, and according to the results obtained

N.J. Ford (✉)
University of Chester, Chester, CH1 4BJ, UK
e-mail: n.ford@chester.ac.uk

M.L. Morgado
CEMAT/IST and UTAD, Quinta de Prados 5001-801, Vila Real, Portugal
e-mail: luisam@utad.pt

S. Pinelas et al. (eds.), *Differential and Difference Equations with Applications*, Springer Proceedings in Mathematics & Statistics 47, DOI 10.1007/978-1-4614-7333-6_48,
© Springer Science+Business Media New York 2013

in both papers, we see that there are substantial differences in the analysis and in the numerical treatment of the two cases: $a = 0$ or $a \neq 0$. When $a \neq 0$ the approximate solution was obtained by using a shooting algorithm. To be more precise, we have considered the initial value problem

$$D_*^\alpha y(t) = f(t, y(t)), \quad t \in [0, T]$$

$$y(0) = y_0,$$

and for a certain value of y_0, we have determined its numerical solution using standard initial value problem solvers. Then we used an iterative scheme to find the appropriate y_0, for which the solution of the initial value problem passes through the point (a, y_a).

When instead of (1), we have a multi-term differential equation, that is, we have a differential equation involving different orders of derivatives of the unknown function y, the approach used to provide an analytical and numerical analysis is to reduce it to a system of low-order single-term equations. That was the technique used in [6], where the authors considered initial value problems of the form

$$D^\alpha y(t) = f(t, y(t), D^{\beta_1} y(t), D^{\beta_2} y(t), \ldots, D^{\beta_n} y(t)), \quad t \in [0, T] \tag{3}$$

$$y^{(k)}(0) = y_0^{(k)}, \quad k = 0, \ldots, \lceil \alpha \rceil - 1, \tag{4}$$

where $\alpha > \beta_1 > \beta_2 > \ldots > \beta_n$, $\alpha - \beta_1 \leq 1$, $\beta_j - \beta_{j-1} \leq 1$, $0 < \beta_n \leq 1$ and D^α denotes the Caputo differential operator of order $\alpha \notin \mathbb{N}$ ([1]), which is defined by $D^\alpha y(t) := {}^{RL}D^\alpha (y - T[y])(t)$ where $T[y]$ is the Taylor polynomial of degree $\lfloor \alpha \rfloor$ for y, centered at 0, and ${}^{RL}D^\alpha$ is the Riemann-Liouville derivative of order α [8]. The latter is defined by ${}^{RL}D^\alpha := D^{\lceil \alpha \rceil} J^{\lceil \alpha \rceil - \alpha}$, with J^β being the Riemann-Liouville integral operator,

$$J^\beta y(t) := \frac{1}{\Gamma(\beta)} \int_0^t (t - s)^{\beta - 1} y(s) ds,$$

and $D^{\lceil \alpha \rceil}$ is the classical integer order derivative. Here, $\lfloor \alpha \rfloor$ denotes the biggest integer smaller than α, and $\lceil \alpha \rceil$ represents the smallest integer greater than or equal to α.

In that paper, the authors proved that if all the orders of the derivatives appearing in (3) are rational numbers, and defining $N = \alpha M$, M the least common multiple of the denominators of $\alpha, \beta_1, \ldots, \beta_n$ and $\gamma = 1/M$, problem (3) and (4) is equivalent to the following system of equations:

$$D^\gamma y_1(t) = y_2(t), \ldots, D^\gamma y_{N-1}(t) = y_N(t)$$

$$D^\gamma y_N(t) = f\left(t, y_{\frac{\beta_1}{\gamma}+1}(t), \ldots, y_{\frac{\beta_n}{\gamma}+1}(t)\right), \tag{5}$$

together with conditions

$$y_j(0) = \begin{cases} y_0^{(k)} & \text{, if } j = kM+1 \text{ for some } k \in \mathbb{N} \\ 0 & \text{,else,} \end{cases} \quad (6)$$

า the following sense:

Whenever $Y = (y_1,\ldots,y_N)^T$ with $y_1 \in C^{\lceil\alpha\rceil}[0,T]$ is the solution of (5)–(6), the function $y = y_1$ solves the multi-term equation (3) and satisfies the conditions (4). Whenever $y \in C^{\lceil\alpha\rceil}[0,T]$ is a solution of (3)–(4), the vector-valued function $Y = (y_1,\ldots,y_N)^T$ satisfies the system (5) and the conditions (6).

Jote that, as a consequence of the definition of the Caputo differential operator, $_j(0)$ must vanish whenever, for any $k \in \mathbb{N}$, $j \neq kM+1$. The following lemma was roved in [4]:

Lemma 1. *Let $y \in C^k[0,T]$ for some $T > 0$ and some $k \in \mathbb{N}$, and let $0 < q < k$, $\notin \mathbb{N}$. Then $D^q y(0) = 0$.*

Here, we intend to extend the results obtained in all of these papers, when onsidering multi-term boundary value problems of the form

$$D^\alpha y(t) = f(t, y(t), D^{\beta_1} y(t), D^{\beta_2} y(t), \ldots, D^{\beta_n} y(t)), \quad t \in [0,T] \quad (7)$$

$$y^{(k)}(a) = y_a^{(k)} \quad k = 0, \ldots, \lceil\alpha\rceil - 1, \quad (8)$$

where $a > 0$ and $\alpha > \beta_1 > \beta_2 > \ldots > \beta_n$, $\alpha - \beta_1 \leq 1$, $\beta_j - \beta_{j-1} \leq 1$, $0 < \beta_n \leq 1$.

The paper is organized in the following way: in Sect. 2 we reduce our problem า to a system of low-order equations, establishing sufficient conditions for the xistence and uniqueness of the solution. In Sect. 3 we propose a numerical dgorithm to approximate the solution of problems (7) and (8). We end with some umerical results and some conclusions.

Existence, Uniqueness, and Structural Stability

f all the orders of the derivatives appearing in (7) are rational numbers, following าe approach in [6], we can easily transform problem (7) and (8) into an equivalent ystem of equations with lower order. Let M be the least common multiple of the enominators of the derivatives appearing in the equation; define $\gamma = 1/M$ and $N = kM$. We can then state the following result:

Theorem 1. *Equation (7) equipped with conditions (8) is equivalent to the system f N equations (5) together with conditions*

$$y_j(a) = \begin{cases} y_a^{(k)} & \text{, if } j = kM+1 \text{ for some } k \in \mathbb{N} \\ y_a^{(j)} & \text{,else,} \end{cases} \quad (9)$$

where $y_a^{(j)}$ are suitable constants, in the following sense:

- *Whenever $Y = (y_1, \ldots, y_N)^T$ with $y_1 \in C^{\lceil \alpha \rceil}[0, a]$ is the solution of (5), (9), the function $y = y_1$ solves the multi-term equation (7) and satisfies the conditions (8)*
- *Whenever $y \in C^{\lceil \alpha \rceil}[0, a]$ is a solution of (7)–(8), the vector-valued function $Y = (y_1, \ldots, y_N)^T$ satisfies the system (5) and the conditions (9).*

Proof. This theorem is a simple generalization of Theorem 2.1 in [6] to the case where $a > 0$. ∎

Note that in system (5) all the differential equations have derivatives with order between zero and one; therefore, it is useful to recall the results obtained in [7]. In that paper we proved that if the function f in (1) is continuous and satisfies a Lipschitz condition with Lipschitz constant $L > 0$ with respect to its second argument, and if $\frac{2La^\alpha}{\Gamma(\alpha+1)} < 1$, then the boundary value problem (1) and (2) has unique solution on $[0, a]$. We have also proved that, in that case, problem (1) and (2) is equivalent to the following integral equation:

$$y(t) = y(a) - \frac{1}{\Gamma(\alpha)} \int_0^a (a-s)^{\alpha-1} f(s, y(s)) ds + \frac{1}{\Gamma(\alpha)} \int_0^t (t-s)^{\alpha-1} f(s, y(s)) ds.$$
$$\tag{10}$$

As pointed out in [7], after proving the existence of $y(0)$, existence and uniqueness results for $t > a$ could be inherited from the corresponding initial value problem theory.

In what follows, we extend these results to multi-order fractional differential equations. The existence and uniqueness results are immediate, taking into account Theorem 1.

Theorem 2 (Existence and uniqueness (muti-term, commensurate orders)).
Let the continuous function f in (7) satisfy a uniform Lipschitz condition, with Lipschitz constant L, in all its arguments except for the first on a suitable domain D. Assume that $\alpha, \beta_1, \ldots, \beta_n \in \mathbb{Q}$ and $\frac{2La^\gamma}{\Gamma(\gamma+1)} < 1$. Then, problem (7) and (8) has unique continuous solution on an interval $[0, T]$ of the real line.

When we have non-commensurate orders, as pointed out in [6], there is no system of fractional equations that exactly corresponds to the original problem, and to overcome this difficulty, we use the well-known fact that any real number can be approximated arbitrarily closely by a rational number. Hence, we can approximate a fractional differential equation with nonrational orders in its derivatives by a fractional differential equation whose orders, being rational, are as close as we choose to the original orders.

In order to do that, we need to be sure that under small variations in the orders of the derivatives α and β_j in (7), a uniform bound on the change in the solution can be provided in any compact interval.

For the sake of simplicity, we shall now consider the case where we have only two orders of derivatives in (7), and $\alpha \leq 2$, that is, we consider the following problem:

$$D^{\alpha}y(t) = f(t, y(t), D^{\beta}y(t)), \quad 0 < \beta < \alpha \leq 2 \tag{11}$$

$$y(a) = y_a, \quad y'(a) = y_a^{(1)}. \tag{12}$$

'he generalization to multi-order fractional differential equations is straightforward.
imilarly to our approach in [7], we begin by recalling a well-known result in
'ractional calculus.

_emma 2. *If the function f is continuous, then the initial value problem*

$$D^{\alpha}y(t) = f(t, y(t), D^{\beta}y(t)), \quad 0 < \beta \leq 1 < \alpha \leq 2$$

$$y(0) = y_0, \quad y'(0) = y_0^{(1)}$$

; equivalent to the following Volterra integral equation

$$y(t) = y(0) + y'(0)t + \frac{1}{\Gamma(\alpha)} \int_0^t (t-s)^{\alpha-1} f(s, y(s), D^{\beta}y(s))ds. \tag{13}$$

If the conditions of Theorem 2 are fulfilled, the solution of problem (11) and (12)
xists and is unique and continuous on $[0, a]$. In particular, $y(0)$ and $y'(0)$ exist and
re unique, and taking (13) into account, they are given by

$$y(0) = y(a) - ay'(a) + \frac{a}{\Gamma(\alpha-1)} \int_0^a (a-s)^{\alpha-2} f(s, y(s), D^{\beta}y(s))ds -$$

$$- \frac{1}{\Gamma(\alpha)} \int_0^a (a-s)^{\alpha-1} f(s, y(s), D^{\beta}y(s))ds$$

$$y'(0) = y'(a) - \frac{1}{\Gamma(\alpha-1)} \int_0^a (a-s)^{\alpha-2} f(s, y(s), D^{\beta}y(s))ds,$$

nd therefore, one can conclude that the boundary value problem (11) and (12) is
quivalent to the following integral equation:

$$y(t) = y(a) + y'(a)(t-a) - \frac{(t-a)}{\Gamma(\alpha-1)} \int_0^a (a-s)^{\alpha-2} f(s, y(s), D^{\beta}y(s))ds -$$

$$- \frac{1}{\Gamma(\alpha)} \int_0^a (a-s)^{\alpha-1} f(s, y(s), D^{\beta}y(s))ds + \tag{14}$$

$$+ \frac{1}{\Gamma(\alpha)} \int_0^t (t-s)^{\alpha-1} f(s, y(s), D^{\beta}y(s))ds.$$

Ve have just proved the following theorem:

'heorem 3. *Assume that all the conditions of Theorem 2 are satisfied. Then, the
oundary value problem (11) and (12) is equivalent to the integral equation (14).*

Therefore, as happens in the single-term case (see [7]), here we also have an exact correspondence between boundary and initial value problems, and it follows that properties such as existence and uniqueness of the solution in the non-commensurate case and structural stability can be inherited from the initial value theory (see [6]) allowing us to conclude the following result.

Theorem 4 (Structural stability). *Let y and z be the unique solutions of the following BVPs:*

$$D^{\alpha}y(t) = f(t, y(t), D^{\beta_1}y(t), D^{\beta_2}y(t), \ldots, D^{\beta_n}y(t)), \quad t \in [0, T]$$

$$y^{(k)}(a) = y_a^{(k)}, \quad k = 0, \ldots, \lceil \alpha \rceil - 1,$$

and

$$D^{\tilde{\alpha}}z(t) = f(t, z(t), D^{\tilde{\beta}_1}z(t), D^{\tilde{\beta}_2}z(t), \ldots, D^{\tilde{\beta}_n}z(t)), \quad t \in [0, T]$$

$$z^{(k)}(a) = y_a^{(k)}, \quad k = 0, \ldots, \lceil \alpha \rceil - 1,$$

respectively, where $|\alpha - \tilde{\alpha}| < \varepsilon$ *and* $\left|\beta_j - \tilde{\beta}_j\right| < \varepsilon$, $j = 1, \ldots, n$. *Then*

$$\|y - z\|_{L_\infty[0,T]} = O(\varepsilon), \quad \varepsilon \to 0.$$

3 Numerical Methods and Results

Now, we present a numerical algorithm for the solution of multi-term BVPs (7) and (8). Our approach is based on the equivalence of such problems with a system of equations of lower order, as explained in Sect. 2.

Let us explain our approach through an example. Consider the linear test problem

$$D^2y(t) + D^{0.5}y(t) + y(t) = t^3 + 6t + \frac{3.2t^{2.5}}{\Gamma(0.5)}, \quad 0 \le t \le 1 \tag{15}$$

$$y(0.1) = 0.001, y'(0.1) = 0.03$$

whose analytical solution is known and given by $y(t) = t^3$.

First, we convert this problem into the equivalent linear system of equations

$$D^{0.5}y_1(t) = y_2(t), \quad D^{0.5}y_2(t) = y_3(t), \quad D^{0.5}y_3(t) = y_4(t)$$

$$D^{0.5}y_4(t) = -y_1(t) - y_2(t) + t^3 + 6t + \frac{3.2t^{2.5}}{\Gamma(0.5)} \tag{16}$$

together with the conditions

$$y_1(0.1) = 0.001, \quad y_2(0.1) = y_{2a}, \quad y_3(0.1) = 0.03, \quad y_4(0.1) = y_{4a},$$

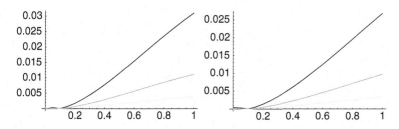

Fig. 1 Absolute errors using method 1 (*left*) and method 2 (*right*) with $h = 1/20$ (*black*), $h = 1/40$ (*gray*) and $h = 1/80$ (*light gray*)

Table 1 Values of the absolute error at $t = 1$ and estimates of the convergence order (EOC)					
h	Method 1	EOC	Method 2	EOC
1/10	0.0871717		0.0720353	
1/20	0.0309596	1.49	0.0267395	1.43
1/40	0.0110865	1.48	0.00970896	1.46
1/80	0.00396478	1.48	0.00348684	1.48

where y_{2a} and y_{4a} are unknown constants.

Since the solution of (15) is unique, $y_1(0)$, $y_2(0)$, $y_3(0)$ and $y_4(0)$ exist and are unique, and moreover, taking Lemma 1 into account, we must have $y_2(0) = y_4(0) = 0$.

Therefore, given the initial value problem (16) equipped with the conditions

$$y_1(0) = y_{10}, \quad y_2(0) = 0, \quad y_3(0) = y_{30}, \quad y_4(0) = 0,$$

we can determine its approximate solution, say $(\tilde{y}_1, \tilde{y}_2, \tilde{y}_3, \tilde{y}_4)$, using any standard numerical initial solver. Finally, we adjust the unknowns y_{10}, y_{30}, y_{2a} and y_{4a} in order to be satisfied the following system of equations:

$$e_1 := \tilde{y}_1(0.1) - 0.001 = 0, \ e_2 := \tilde{y}_2(0.1) - y_{2a} = 0$$
$$e_3 := \tilde{y}_3(0.1) - 0.03 = 0, \ e_4 := \tilde{y}_4(0.1) - y_{4a} = 0.$$

In our numerical experiments we have used the Adams method ([5]) and the fractional backward difference method ([2]) to solve the initial value problems. From now on, we will denote these methods by method 1 and method 2, respectively. In Fig. 1 we present the absolute error at the discretization points. In Table 1 we can observe the expected convergence order of the two methods.

4 Conclusions

In this paper we propose an algorithm for the numerical solution of multi-term fractional boundary value problems. The idea is to rewrite the problem as an equivalent system of differential equations of lower order. This is also useful to provide results on the existence and uniqueness of solutions and on the important issue of structural stability. For lack of space, here we do not illustrate numerically the effect on the solution resulting from a perturbation in the order of the derivative however, it may be remarked that a small change on the order of the derivative can have several consequences such as increasing of the size of the system and reducing of the order of convergence of the numerical method (since the order of the derivatives of the system may decrease significantly). Thus, it should be noticed that when solving a given problem numerically, it will be necessary to balance several factors, such as the approximation of the order of the derivative and the choice of the step-size h in the numerical method.

Acknowledgements M. L. Morgado acknowledges financial support from FCT, Fundação para Ciência e Tecnologia, through grant SFRH/BPD/46530/2008.

References

1. Caputo, M.: Elasticity e Dissipazione, Zanichelli, Bologna (1969)
2. Diethelm, K.: An algorithm for the numerical solution of differential equations of fractional order. Electr. Trans. Numer. Anal. **5**, 1–6 (1997)
3. Diethelm, K., Ford, N.J.: Analysis of fractional differential equations. J. Math. Anal. Appl. **265** 229–248 (2002)
4. Diethelm, K., Ford, N.J.: Numerical solution of the Bagley-Torvik equation. BIT **42**, 490–507 (2002)
5. Diethelm, K., Ford, N.J., Freed, A.D.: predictor-corrector approach for the numerical solution of fractional differential equations. Nonlinear Dynam. **29**, 3–22 (2002)
6. Diethelm, K., Ford, N.J.: Multi-order fractional differential equations and their numerical solution. Appl. Math. Comp. **154**, 621–640 (2004)
7. Ford, N.J., Morgado, M.L.: Fractional boundary value problems: analysis and numerical methods. Fract. Calc. Appl. Anal. **14**(4), 554–567 (2011)
8. Samko, S.G., Kilbas, A.A., Marichev, O.I.: Fractional Integrals and Derivatives: Theory and Applications. Gordon and Breach, Yverdon (1993)

Topological Structures for Studying Dynamic Equations on Time Scales

Bonita A. Lawrence and Ralph W. Oberste-Vorth

Abstract We construct the topological framework within which we can study the solution space for a given dynamic equation on time scales. We call these the Hausdorff-Fell topologies. The space of finite time scales is dense in the space of all time scales under the Hausdorff-Fell topology. The natural projection from solutions to their domains is a homeomorphism when all solutions are unique.

Keywords Time scales • Dynamic equations • Hyperspaces • Hausdorff-Fell topology

1 Introduction

A *time scale* is a nonempty closed subset of the real numbers. Stefan Hilger developed the calculus on times scales in 1988 in [5]. For real-valued functions on a time scale, Hilger defined the Δ-derivative. Let $\mathbb{T}$ be a time scale and let $t \in \mathbb{T}$ with $t < \sup \mathbb{T}$. Suppose that $f : \mathbb{T} \to \mathbb{R}$ is a function. The Δ-*derivative of f at t*, if it is defined, is denoted

$$f^\Delta(t).$$

We do not give details here; [1] is a thorough introduction to the calculus on time scales. The Δ-derivative does not always exist. However, if a function, f, is Δ-differentiable at x, then the Δ-derivative is well known in two cases: when the time

B.A. Lawrence
Department of Mathematics, Marshall University, Huntington, WV 25755, USA
e-mail: lawrence@marshall.edu

R.W. Oberste-Vorth
Department of Mathematics and Computer Science, Indiana State University,
Terre Haute, IN 47809, USA
e-mail: ralph.oberste-vorth@indstate.edu

S. Pinelas et al. (eds.), *Differential and Difference Equations with Applications*, Springer Proceedings in Mathematics & Statistics 47, DOI 10.1007/978-1-4614-7333-6_49,
© Springer Science+Business Media New York 2013

scale is either $\mathbb{R}$ or $\mathbb{Z}$. If $\mathbb{T} = \mathbb{R}$, then

$$f^{\Delta}(t) = f'(t);$$

the Δ-derivative is the usual derivative. If $\mathbb{T} = \mathbb{Z}$, then

$$f^{\Delta}(t) = \Delta f(t);$$

the Δ-derivative is the difference operator.

Generalizing differential and difference equations, we are interested in equation with Δ-derivatives in place of derivatives and differences. These are called *dynami equations*.

Suppose we wish to study a given dynamic equation such as the following initi value problem:

$$x^{\Delta} = f(t, x), \quad x(t_0) = x_0, \tag{1}$$

where $f : A \times B \to \mathbb{R}$ is continuous. The solution function, $x(t)$, depends on the tim scale, $\mathbb{T}$: $\mathbb{T}$ is a subset of A, and is the domain of x. We would like to examine how the solution of (1) depends on the time scale.

How are the solutions of the initial value problem (1) related and changing as th time scale changes? Consider the following example.

Example 1. In [7], we considered

$$x^{\Delta} = 4x\left(\tfrac{3}{4} - x\right), \quad x(0) = x_0. \tag{2}$$

On the Eulerian time scales, $\mu\mathbb{Z}_+$, for $0 < \mu \leq 1$, solving Eq. (2) is equivalent t iterating

$$L_{\mu}(x) = 4\mu x \left(\tfrac{3\mu+1}{4\mu} - x\right).$$

For $\mu = 1$, we have

$$L_1(x) = 4x(1 - x)$$

defined on $\mathbb{Z}$. On the other hand, as μ tends to 0, the solutions tend toward the solution of the logistic differential equation on $\mathbb{R}_+$. For $0 < \mu \leq 1$, L_{μ} is topologically conjugate to

$$Q_c(x) = x^2 + c,$$

where $c = \tfrac{1}{4}(1 - 9\mu^2)$. Every $\mu \in (0, 1]$ corresponds to exactly one $c \in [-2, 1/4)$:

$$c = -2 \iff \mu = 1 \quad \text{and} \quad c \to 1/4 \text{ as } \mu \to 0.$$

This example shows that the solutions of differential and difference equations ca be related by passing through the time scales. In this particular example, all of th dynamics of real quadratic polynomials are displayed!

In Example 1, the domain of the solutions on Eulerian time scales is treated as a parameter of a family of dynamical systems. We do not know what happens when non-Eulerian time scales (*i.e.*, not $\mu\mathbb{Z}$) are allowed. Nonunique solutions can make things even more complicated.

This suggests the following approach. For any given initial value problem, treat the time scales as a parameter. Let $CL(\mathbb{R})$ denote the set of all time scales and let $\mathscr{S}$ denote the set of all solutions of the initial value problem on all possible time scales. Consider the canonical projection:

$$
\begin{array}{c}
\mathscr{S} \\
\Big\downarrow \pi \\
CL(\mathbb{R}).
\end{array}
\qquad (3)
$$

That is, an element of $\mathscr{S}$, a solution $x : \mathbb{T} \to \mathbb{R}$, projects to its domain, $\mathbb{T}$. What can we say about this projection, especially when there are nonunique solutions? Under what conditions is there unique lifting? Can we follow two different paths from the same starting point (a solution on the initial time scale) to different solutions following the same path of time scales? Can a loop in $CL(\mathbb{R})$ lift to a path that is not a loop? How can this approach help us to understand the changes in dynamics of solutions caused by changes in their time scales? In order to make sense of these questions, we must first discuss the topologies on these sets.

From the point of view of applications, the topological properties of the space of time scales should be important.

2   Hyperspace Topologies

Given a topological space X, researchers in hyperspace theory use the following notation:

$$
CL(X) = \{ A \subset X \mid A \neq \emptyset \text{ and } A \text{ is closed in } X \}.
$$

We are especially interested in the set of time scales, $CL(\mathbb{R})$, and the set of real-valued functions on time scales.

There are several well-known topologies in use in hyperspace theory. Among these are the Hausdorff metric topology (for a metric space) introduced in [4] and the Vietoris topology introduced in [9]. See [6] for a good introduction to these hyperspace topologies.

We need a topology on $CL(\mathbb{R})$ that makes sense from a dynamical systems point of view. It has been noted that neither the topology induced by the Hausdorff metric nor the hit-and-miss topology described by Vietoris satisfies this condition. Consider the following sequences as (see [7]).

1) $[-n,n]$ does not converge to $\mathbb{R}$ with respect to the Hausdorff metric.
2) $\mathbb{Z} + \frac{1}{n}$ does not converge to $\mathbb{Z}$ with respect to the Vietoris topology.

We need to specify the topologies on the space of time scales and its space of real-valued functions, where we find the space of solutions of a dynamic equation parameterized by the domain.

Let us briefly move to a more general setting.

Definition 1. A *hyperspace* is a set of nonempty closed subsets of a topological space X. The set of all nonempty closed subsets of X is denoted $CL(X)$. We also use the notation $\overline{CL(X)} = CL(X) \cup \{\emptyset\}$.

For example, $CL(\mathbb{R})$ is the set of all time scales and $\overline{CL(\mathbb{R})} = CL(\mathbb{R}) \cup \{\emptyset\}$ is the set of all closed subsets of $\mathbb{R}$, including all time scales and the empty set.

From here on, let X be a metric space. We give two equivalent ways of describing the dynamically appropriate topology on $\overline{CL(X)}$.

Firstly, a dynamical point of view suggests that the limit, $\mathbb{T}$, of a sequence $\{\mathbb{T}_n\}$ in $\overline{CL(X)}$ should be defined by the following:

(1) If $t \in \mathbb{T}$, then every open set U such that $t \in U$ intersects all but finitely many $\mathbb{T}_n$'s.
(2) If every open set U containing t intersects infinitely many $\mathbb{T}_n$'s, then $t \in \mathbb{T}$.

This is sometimes called *L-convergence* (see [6]). The topology generated by L-convergence (in the sense that closed sets contain their sequential limit points) is sometimes called the *Hausdorff topology* since these properties were studied in [4].

Secondly, the hit-and-miss nature of the Vietoris topology leads to the Fell topology (see [2]). The Fell topology on $CL(X)$ is generated by the "hit sets" U^- for all open subsets U of X and the "miss sets" V^+ for all cocompact subsets V of X. That is, every closed set $\mathbb{A}$ in U^- intersects the open set U and every closed set $\mathbb{A}$ in V^+ misses the compact set $X - V$:

$$U^- = \{\mathbb{A} \in CL(X) | \mathbb{A} \cap U \neq \emptyset\}$$

and

$$V^+ = \{\mathbb{A} \in CL(X) | \mathbb{A} \subset V\}$$
$$= \{\mathbb{A} \in CL(X) | \mathbb{A} \cap (X - V) = \emptyset\}.$$

It was shown in [8] that the Hausdorff and Fell topologies agree.

Theorem 1. *Let X be metrizable. Let $\{\mathbb{T}_n\}$ be a sequence in $\overline{CL(X)}$. $\{\mathbb{T}_n\}$ converges to $\mathbb{T}$ in the Fell topology if and only if $\{\mathbb{T}_n\}$ L converges to $\mathbb{T}$.*

Hence, we call this topology the *Hausdorff-Fell topology* on $CL(X)$. By convergence in $CL(X)$, we will mean convergence with respect to the Hausdorff-Fell topology on $CL(X)$.

8 The Role of Finite Time Scales

A time scale is *totally discrete* if all of its points are isolated (it has no accumulation points). These correspond with variable step-size Euler's methods of first-order differential equations. We denote the subspace of totally discrete time scales in $CL(\mathbb{R})$ by $CL_D(\mathbb{R})$.

It was shown in [3] that with respect to the Hausdorff metric, $CL_D(\mathbb{R})$ is dense in $CL(\mathbb{R})$.

We can improve on this in the Hausdorff-Fell topology. We denote the subspace of finite time scales in $CL(\mathbb{R})$ by $CL_F(\mathbb{R})$. So,

$$CL_F(\mathbb{R}) \subset CL_D(\mathbb{R}) \subset CL(\mathbb{R}).$$

Theorem 2. $CL_F(\mathbb{R})$ *is dense in* $CL(\mathbb{R})$.

Proof. Choose a time scale $\mathbb{T} \in CL(\mathbb{R})$. We will construct a sequence $\{\mathbb{T}_n\}$ in $CL_F(\mathbb{R})$ that L converges to $\mathbb{T}$.

Fix n. Set $\mathbb{S}_n = \mathbb{T} \cap [-n, n]$. Consider the collection of open intervals

$$N\left(t, \frac{1}{n}\right) = \left(t - \frac{1}{n}, t + \frac{1}{n}\right)$$

for all $t \in \mathbb{S}_n$. Since $\mathbb{S}_n$ is compact, there exist $t_{1_n}, t_{2_n}, \dots, t_{k_n}$ such that

$$N\left(t_{1_n}, \frac{1}{n}\right), N\left(t_{2_n}, \frac{1}{n}\right), \dots, N\left(t_{k_n}, \frac{1}{n}\right)$$

covers $\mathbb{S}_n$. Set $\mathbb{T}_n = \{t_{1_n}, t_{2_n}, \dots, t_{k_n}\}$.

It is easy to verify that $\{\mathbb{T}_n\}$ L converges to $\mathbb{T}$. $\square$

There are many interesting facts about $CL(X)$ and $\overline{CL(X)}$ with the Hausdorff-Fell topology. For example:

If X is Hausdorff, then $\overline{CL(X)}$ is compact and $\overline{CL(X)}$ is the one-point compactification of $CL(X)$.

If X is compact metric space, then the Hausdorff metric topology, Vietoris topology, and Hausdorff-Fell topology are equivalent.

This generalizes to metric spaces. From the point of view of applications, Theorem 2 may not be surprising. It is impractical to do more than finitely many computations even when examining asymptotic behavior. In practice, using the Hausdorff metric or the Vietoris topology on a large but bounded subset of $\mathbb{R}$ suffices and is much the same.

4   The Space of Continuous Functions on Time Scales

Recall that for topological spaces X and Y, a subbasis for the *compact-open topolog*
on the set, $C(X,Y)$, of continuous functions from X to Y is

$$S(K,U) = \{ f \in C(X,Y) \mid f(K) \subset U \}$$

for all compact subset K of X and open subsets U of Y.

Since we are interested in function spaces over variable domains, we must unit
the standard function spaces. For a closed subset K of X, a function $f : K \to Y$ ca
be thought of as a *partial function* from X to Y—the domain of definition is K rathe
than X. By a *partial mapping*, we will mean a continuous partial function. The so
of all partial mappings from X to Y is

$$C_p(X,Y) = \cup \{ C(K,Y) \mid K \in \mathrm{CL}(X) \}.$$

So, $C_p(\mathbb{R}, \mathbb{R})$ is the space of all continuous real-valued functions on time scales.

Suppose that X and Y are metric spaces. So $X \times Y$ is metrizable. We wish t
give a topology on $C_p(X,Y)$ that is consistent with the compact-open topology o
$C(X,Y)$.

Consider the function $\mathrm{Gr} : C_p(X,Y) \to \mathrm{CL}(X \times Y)$ that sends each partia
mapping to its graph. Note that Gr is injective. We can pull back the Hausdorf
Fell topology on $\mathrm{CL}(X \times Y)$ to give a topology on $C_p(X,Y)$.

In this sense, we can consider the topology of the solution space of an initi
value problem on dynamic equations on time scales. For example, the following i
true by design:

Theorem 3. *Let $\mathscr{B}$ be a subset of $\mathrm{CL}(\mathbb{R})$. If the initial value problem (1) has uniqu
solutions for all $\mathbb{T} \in \mathscr{B}$ and $\mathscr{S}$ is the set of all such solutions, then the projection
in (3) maps $\mathscr{S}$ homeomorphically onto $\mathscr{B}$.*

No examples that are more complicated are understood.

References

1. Bohner, M., Peterson, A.: Dynamic Equations on Time Scales: An Introduction with Applica
   tions, Birkhäuser, Boston (2001)
2. Fell, J.: A Hausdorff topology for the closed subsets of a locally compact non-Hausdorff spac
   Proc. Amer. Math. Soc. **13**, 472–476 (1962)
3. Hall, K., Oberste-Vorth, R.: Totally discrete and Eulerian time scales, in Difference Equation
   Special Functions and Orthogonal Polynomials. In: Elaydi, S., Cushing, J., Lasser, R., Ruffin
   A., Papageorgiou, V., Van Assche, W. (eds.) 462–470. World Scientific, New Jersey (2007)
4. Hausdorff, F.: Grundzüge der Mengenlehre, Verlag von Veit, Leipzig (1914)
5. Hilger, S.: Ein Masskettenkalkül mit Anwendung auf Zentrumsmannigfaltigkeiten, Ph.l
   Thesis, Universität Würzburg (1988)
6. Illanes, A., Nadler, S. Jr.: Hyperspaces: Fundamentals and Recent Advances, Marcel Dekke
   New York (1999)

. Oberste-Vorth, R.: The Fell topology on the space of time scales for dynamic equations. Adv. Dyn. Syst. Appl. **3**, 177–184 (2008)
. Oberste-Vorth, R.: The Fell topology for dynamic equations on time scales. Nonlinear Dyn. Syst. Theory **9**, 399–406 (2009)
. Vietoris, L.: Bereiche zweiter Ordnung. Monatsh. Math. **32**, 258–280 (1922)

Existence of Nonoscillatory Solutions of the Discrete FitzHugh-Nagumo System

Ana Pedro and Pedro Lima

Abstract In this work, we are concerned with a system of two functional differential equations of mixed type (with delays and advances), known as the discrete FitzHugh-Nagumo equations, which arises in the modeling of impulse propagation in a myelinated axon:

$$C\frac{dv}{dt}(t) = \frac{1}{R}(v(t+\tau) - 2v(t) + v(t-\tau)) + f(v(t)) - w(t)$$
$$\frac{dw}{dt} = \sigma v(t) - \gamma w(t). \tag{1}$$

In the case $\gamma = \sigma = 0$, this system reduces to a single equation, which is well studied in the literature. In this case it is known that for each set of the equation parameters (within certain constraints), there exists a value of τ (delay) for which the considered equation has a monotone solution v satisfying certain conditions at infinity. The main goal of the present work is to show that for sufficiently small values of the coefficients in the second equation of system (1), this system has a solution (v, w) whose first component satisfies certain boundary conditions and has similar properties to the ones of v, in the case of a single equation. With this purpose we linearize the original system as $t \to -\infty$ and $t \to \infty$ and analyze the corresponding characteristic equations. We study the existence of nonoscillatory solutions based on the number and nature of the roots of these equations.

A. Pedro (✉)
Departamento de Matemática, Faculdade de Ciências e Tecnologia da Universidade Nova de Lisboa, Quinta da Torre, 2825-114 Monte da Caparica
e-mail: anap@fct.unl.pt

P. Lima
Departamento de Matemática, Instituto Superior Técnico, Universidade Técnica de Lisboa, Av. Rovisco Pais, 1049-001 Lisboa,
e-mail: plima@math.ist.utl.pt

S. Pinelas et al. (eds.), *Differential and Difference Equations with Applications*, Springer Proceedings in Mathematics & Statistics 47, DOI 10.1007/978-1-4614-7333-6_50,
© Springer Science+Business Media New York 2013

Keywords Discrete Fitzhugh-Nagumo equations • Non-oscillatory solution • Mixed-type functional-differential equations

1 Introduction

The modeling of propagation of impulses in nerve axons was started in the work of FitzHugh [5, 6] and J.Nagumo and his co-authors [10]. In the particular case of myelinated axons, studied by J. Bell, for example, in [1] and [2], the nerve membrane is insulated by a substance called myelin; therefore only a small part of it is exposed to the extracellular medium at the nodes of Ranvier. In this case the propagation of nerve impulses may be modeled by the following system of differential-difference equations:

$$C\frac{dv_k}{dt} = \frac{1}{R}(v_{k+1} - 2v_k + v_{k-1}) + f(v_k) - w_k$$
$$\frac{dw_k}{dt} = \sigma v_k - \gamma w_k, \tag{2}$$

where v_k and w_k are the potential and the recovery variables, respectively, at the k-th node of Ranvier (k is an integer). System (2) was derived from the so-called HH model, introduced by Hodgkin and Huxley [7], which describes the excitation and flow of electric current through a nerve fiber. The HH-model consists of a system of four ODEs with four unknowns, which of them has a certain physical meaning. In order to make this system more tractable analytically, FitzHugh [5] has reduced it to a system of two equations, which in the case of space discretization yield system (2). Therefore, this last system is not derived directly from physiological principles and not all its variables can be identified physically. However it contains much of the relevant behavior expected in physiological models. As said above v describes the membrane potential, while w_k is responsible for "accommodation and refractoriness" [5]. Moreover, R and C in the first equation can be identified with axoplasmical resistance and nodal membrane capacitance, respectively, while γ and σ are positive constants used to describe the dynamics of the recovery processes. The function f represents a current-voltage relation and is given by

$$f(v) = bv(1 - v)(v - a), \tag{3}$$

where b is positive, $0 < a < 1/2$. Assuming that the Ranvier nodes are identical and uniformly spaced (with internodal length L) and that the impulse propagates at certain constant speed θ, we must have

$$v_k(t) = v_{k+1}(t + \tau),$$

where $\tau = L/\theta$. Then, omitting the index k, system (2) may be written as a system of mixed-type functional differential equations

$$C\frac{dv}{dt}(t) = \frac{1}{R}(v(t+\tau) - 2v(t) + v(t-\tau)) + f(v(t)) - w(t)$$
$$\frac{dw}{dt} = \sigma v(t) - \gamma w(t). \tag{4}$$

The case $\gamma = \sigma = 0$ in (2) and (4) has been subject of detailed analysis. Physically, this corresponds to the case where the variation of the recovery variable w is negligible and the system reduces to a single mixed-type functional differential equation. In [2] sufficient conditions on the parameters of the equation are given so that it has a unique monotone solution v, which satisfies the following conditions:

$$\lim_{t \to -\infty} v(t) = 0, \qquad \lim_{t \to \infty} v(t) = 1, \qquad v(0) = \frac{1}{2}. \tag{5}$$

In [3], a numerical method was proposed and numerical results were presented for this equation.

Concerning system (4), as far as we know, there are few available results about the existence of nonoscillatory solutions. The main purpose of the present paper is to analyze this problem, starting with the case where γ and σ are close to zero (i.e. considering that the recovery variable w changes slowly). With this purpose, we linearize the system at its stationary points and study the corresponding characteristic equations.

The outline of this paper is as follows: in Sect. 2, we present some results of oscillation theory for linear systems of mixed-type functional differential equations. In Sect. 3 we apply the obtained results to system (4) after linearizing it at its stationary points. Finally, in Sect. 4 we present some conclusions.

Oscillation Theory for Systems of Linear Mixed-Type Functional Differential Equations

An important part of the analysis of linear functional differential equations is devoted to the problem of existence of oscillatory and nonoscillatory solutions of such equations. Recently, some works have been devoted to the oscillatory behaviour of linear systems of mixed-type functional differential equations [4, 11]. For our purposes we will need only one result on a particular case, which is concerned with systems of two equations, with one delay and one advance. The possibility of extending this result to a more general context will be considered in a separate paper.

The result that we will apply in Sect. 3 of this paper may be formulated as the following theorem.

Theorem 1. *Consider a system of two mixed-type functional differential equations of the form:*

$$\frac{dx}{dt}(t) = Ax(t-r) + Bx(t+\tau) + Dx(t), \qquad t \in \mathbb{R} \tag{6}$$

where A,B,D are 2×2 real-valued matrices and r and τ are positive constants. Suppose the following conditions are satisfied:

1. $Det\,A = Det\,B = 0$.
2. $Tr(A)$ and $Tr(B)$ are different from 0 and have the same sign.

Then system (6) has at least one nonoscillatory solution.

Proof. We begin by noting that the characteristic equation of the considered system has the form

$$F(\lambda,\tau,r) = \lambda^2 - Q_2(\lambda)\exp(-2\lambda r) - Q_1(\lambda)\exp(-\lambda r) -$$

$$\tag{7}$$

$$R_2(\lambda)\exp(2\lambda\tau) - R_1(\lambda)\exp(\lambda\tau) - S\exp(\lambda(-r+\tau)) - Q_0(\lambda) = 0,$$

where

$$Q_2(\lambda) = -Det\,A, \qquad R_2(\lambda) = -Det\,B,$$

$$Q_1(\lambda) = \lambda(a_{11}+a_{22}) + a_{21}d_{12} + a_{12}d_{21} - a_{11}d_{22} - d_{11}a_{22},$$

$$R_1(\lambda) = \lambda(b_{11}+b_{22}) + b_{21}d_{12} + b_{12}d_{21} - b_{11}d_{22} - d_{11}b_{22},$$

$$Q_0(\lambda) = \lambda(d_{11}+d_{22}) - Det\,D,$$

$$S = -a_{11}b_{22} - b_{11}a_{22} + a_{21}b_{12} + a_{12}b_{21}.$$

Our argument is based on the fact that, under the conditions of the theorem, for sufficiently large λ we have $sign(F(\lambda,\tau,r)) = -sign(F(-\lambda,\tau,r))$. Actually, from condition 1 it follows that $Q_2(\lambda) = R_2(\lambda) = 0$. In this case, as $|\lambda| \to \infty$, the function $F(\lambda,\tau,r)$ is dominated by either $R_1(\lambda)\exp(\lambda\tau)$, $Q_1(\lambda)\exp(-\lambda r)$ or $S\exp(\lambda(-r+\tau))$. Concerning the term with S, as $\lambda \to \pm\infty$, it may decay or grow exponentially, but in this case, it grows slower than the terms with $Q_1(\lambda)$ or $R_1(\lambda)$. Suppose now that $Tr(A) > 0$; then, by condition 2, $Tr(B) > 0$; hence, as $\lambda \to -\infty$, $Q_1(\lambda)\exp(-\lambda r)$ tends to $-\infty$ and $R_1(\lambda)\exp(\lambda\tau)$ tends to 0; therefore $F(\lambda,\tau,r) \to +\infty$. On the other hand, as $\lambda \to +\infty$, $R_1(\lambda)\exp(\lambda\tau)$ tends to $+\infty$ and $Q_1(\lambda)\exp(-\lambda r)$ tends to 0; therefore $F(\lambda,\tau,r) \to -\infty$. In the same way, one shows that if $Tr(A) < 0$, $Tr(B) < 0$ we have $F(\lambda,\tau,r) \to -\infty$, as $\lambda \to -\infty$, and $F(\lambda,\tau,r) \to +\infty$, as $\lambda \to +\infty$.

In any case, this means that $F(\lambda,\tau,r)$ has at least one real root and therefore system (6) has at least one nonoscillatory solution.

3 Application to the Discrete FitzHugh-Nagumo System

The stationary points of the discrete Fitzhugh-Nagumo system (4) are the roots of the system

$$\begin{cases} f(v) - w = 0, \\ \sigma v - \gamma w = 0. \end{cases} \tag{8}$$

This system may be also written in the form

$$\begin{cases} w = f(v), \\ w = \frac{\sigma v}{\gamma}. \end{cases} \tag{9}$$

Let us assume that $\gamma > 0$ and $0 \leq \frac{\sigma}{\gamma} < \delta$, where δ is a sufficiently small positive constant. From Fig. 1 we easily conclude that, besides the trivial solution $v = 0, w = 0$, system (8) has also the non-zero solution (v^*, w^*), such that $a < v^* < 1$, $v^* = f(v^*)$. As $\sigma \to 0$, we have $v^* \to 1$, $w^* \to 0$, so that in the limit case we obtain $v = 1, w = 0$. Assuming that the stationary point (v^*, w^*) exists, we search for a monotone solution of the FitzHugh-Nagumo equations which satisfies the conditions:

$$\begin{aligned} \lim_{t \to -\infty} v(t) = \lim_{t \to -\infty} w(t) = 0, \\ \lim_{t \to +\infty} v(t) = v^*, \lim_{t \to +\infty} w(t) = w^*. \end{aligned} \tag{10}$$

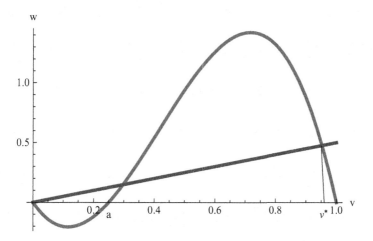

Fig. 1 Finding the roots of system (8) in the case $a = 0.25$, $b = 15$, $\gamma = 1$, $\sigma = 0.5$. The *curve* is the graphic of $f(v)$ and the *straight line* represents $w = \frac{\sigma}{\gamma}v$. The point v^* is the closest to $(1,0)$ root of the system

We will now write the linear systems resulting from the linearization of (4) a $(0,0)$ and (v^*, w^*). In matrix form, this system can be written as

$$\begin{bmatrix} \frac{dv}{dt} \\ \frac{dw}{dt} \end{bmatrix} = \begin{bmatrix} \frac{1}{R} & 0 \\ 0 & 0 \end{bmatrix} \begin{bmatrix} v(t-\tau) \\ w(t-\tau) \end{bmatrix} + \begin{bmatrix} \frac{1}{R} & 0 \\ 0 & 0 \end{bmatrix} \begin{bmatrix} v(t+\tau) \\ w(t+\tau) \end{bmatrix} + \begin{bmatrix} \frac{-2}{R}+c & -1 \\ \sigma & -\gamma \end{bmatrix} \begin{bmatrix} v(t) \\ w(t) \end{bmatrix}, \quad (11$$

where $c = f'(0)$, in the case of the stationary point $(0,0)$, and $c = f'(v^*)$, in th case of the stationary point (v^*, w^*). For system (11) we obtain the characteristi equation:

$$\lambda^2 - \frac{\lambda+\gamma}{R}\exp(-\lambda\tau) - \frac{\lambda+\gamma}{R}\exp(\lambda\tau) - \lambda\left[(\frac{-2}{R}+c-\gamma)\right] + \gamma\left[(\frac{2}{R}-c)\right] + \sigma = 0 \quad (12$$

We are interested in a solution of the original system which satisfies the boundar conditions (10). If such a monotone solution exists, (12) must have at least on positive root, in the case of $c = f'(0)$, and one negative root, in the case of $c = f'(v^*)$

We begin with the case $\gamma = \sigma = 0$. In this case, as we have remarked above, th system reduces to a single equation. The asymptotic behavior of the solutions of thi equation was analysed, for example, in [3].

The characteristic equation may then be reduced to the form:

$$\lambda^2 - \frac{\lambda}{R}(\exp(-\lambda\tau) + \exp(\lambda\tau)) - \lambda\left[(-\frac{2}{R}+c)\right] = 0, \quad (13$$

where $c = f'(0)$, in the case of the stationary point $v = 0$, and $c = f'(1)$, in the cas of the stationary point $v = 1$. As remarked in the cited paper, in both cases Eq. (13 has a positive and a negative root, which makes possible to obtain the asymptoti behaviour of a monotone solution. Let us denote by λ_+ the positive root of Eq. (13 (in the case of $c = f'(0)$) and by λ_- the negative root of Eq. (13) (in the case c $c = f'(1)$). Then a monotone solution of the considered equation, if exists, mus have the asymptotic behavior:

$$v(t) = C_1 \exp(\lambda_+ t)(1 + o(1)), \qquad \text{as } t \to -\infty; \quad (14$$

$$v(t) = C_2 \exp(\lambda_- t)(1 + o(1)), \qquad \text{as } t \to \infty, \quad (15$$

where C_1 and C_2 are constants.

Analogously in the case of system (4) we are looking for a nonoscillator solution, which must have the asymptotic behavior

$$\begin{bmatrix} v(t) \\ w(t) \end{bmatrix} = \begin{bmatrix} a_1 \exp(\lambda_+ t)(1 + o(1)) \\ a_2 \exp(\lambda_+ t)(1 + o(1)) \end{bmatrix}, \qquad \text{as } t \to -\infty; \quad (16$$

and

$$\begin{bmatrix} v(t) \\ w(t) \end{bmatrix} = \begin{bmatrix} v^* - b_1 \exp(\lambda_- t)(1 + o(1)) \\ w^* - b_2 \exp(\lambda_- t)(1 + o(1)) \end{bmatrix}, \quad \text{as} \quad t \to \infty. \tag{17}$$

In (16), λ_+ is a positive root of Eq. (12) with $c = f'(0)$, while in (17) λ_- is a negative root of the same equation with $c = f'(w^*)$. Therefore, the existence of a nonoscillatory solution of system (4), having the asymptotic behaviour defined by (16) and (17), depends on the existence of the characteristic roots λ_- and λ_+. It is much more difficult to establish the existence of such roots in the case of a system than in the case of a single equation (when $\gamma = \sigma = 0$).

Let us now verify that the linear system (11) satisfies the conditions of Theorem of Sect. 2. Actually, we have $Det(A) = Det(B) = 0$ and $Tr(A) = Tr(B) = 1/R$. Therefore, each of the mentioned systems has at least a nonoscillatory solution or, by other words, Eq. (12) has at least one real root.

Moreover, in the case $\gamma = \sigma = 0$ we know that Eq. (13) has exactly three real roots: $\lambda_1 < 0$, $\lambda_2 = 0$ and $\lambda_3 > 0$.

Let us now consider σ and γ such that $\sqrt{\sigma^2 + \gamma^2} < M$ (where M is sufficiently small); then, in the case $c = f'(0)$ Eq. (12) still has at least one real root λ_+, as needed for the construction of the solution to the nonlinear system. Consider now the case where $c = f'(w^*)$, where w^* is a root of system (8). In order to guarantee that this system has a root of the needed form we must have $0 < \sigma/\gamma < \delta$, where δ is a certain constant. Therefore we must consider only pairs (γ, σ) which satisfy this condition. Moreover, there exists a constant $M' > 0$ such that if $\sqrt{\sigma^2 + \gamma^2} < M'$, Eq. (12) still has, at least, one negative root λ_-. In conclusion, combining the inequalities $\sqrt{\sigma^2 + \gamma^2} < \min(M, M')$ and $0 < \sigma/\gamma < \delta$, we obtain the conditions on γ and σ under which the roots λ_- and λ_+ exist.

As known from the case $\gamma = \sigma = 0$ (without recovery process) the existence of nonoscillatory solutions to Eq. (4) assures the existence of a traveling wave to the system of Eq. (2), propagating with a certain constant speed. The above analysis shows that such a traveling wave may exist also in the case where the second equation is considered, provided that the intensity of the recovery processes does not become too large. In particular, σ (which corresponds to the growing rate of the recovery variable w_k) should be sufficiently small compared with γ (the decay rate for the same variable).

4   Examples and Conclusions

A computational algorithm was implemented in Mathematica to obtain numerical approximations of system (4). This algorithm is based in the same ideas as the numerical method described in [9], for the case of a single equation ($\gamma = \sigma = 0$). This numerical method, on its turn, results from the application to nonlinear problems of an approach previously developed for linear forward-backward equations [8].

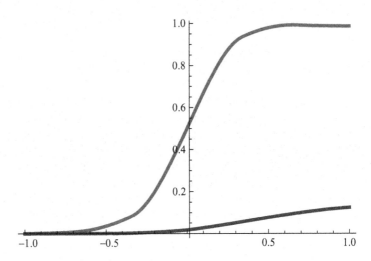

Fig. 2 Graphics of the functions v (*upper line*) and w (*lower line*) for the given example

First we fix an initial value v_0 for v, such that v_0 is a monotone function, $v_0 \to 0$ as $t \to -\infty$ and $v_0 \to v^*$, as $t \to +\infty$. With this purpose, we have set $v_0(t) = \frac{v^*}{2}\left(tanh(\frac{\lambda_0}{2}t) + 1\right)$, where λ_0 is the positive characteristic root of the problem with $\gamma = \sigma = 0$. Then, assuming that $v = v_0$ and solving the second equation of system (4), we obtain an initial approximation for w: $w_0(t) = \sigma e^{-\gamma t} \int_{-\infty}^{t} v_0(s)e^{\gamma s}ds$. Finally, assuming that $w = w_0$, the first equation of system (4) is solved numerically, using the algorithm described in [9], thus obtaining approximations of v, λ_-,λ_+ and τ. This computation is then iterated until the norm of the difference between two successive iterates is sufficiently small.

Next we present a numerical example which was solved using this algorithm. The algorithm is still being improved with respect to accuracy, so that the numerical results we present are still rough approximations. However, they give at least a correct idea of the qualitative behavior of the solution.

Example 1. Let us consider the case where $a = 0.05$, $b = 15$, $\sigma = 0.2$, $\gamma = 1$,$R = C = 1$. We need to analyze the asymptotic behavior of the solutions of system (4) in this case. In order to obtain the stationary point (v^*,w^*), we must solve system (8). The needed root of the system in this case is $v^* = 0.985751$, $w^* = 0.19715$.

For the given example, after two iterations we obtain the approximate following values:

$$\lambda_+ = 6.498, \quad \lambda_- = -6.403, \quad \tau = 0.34096.$$

The graphics of the corresponding numerical approximations of v and w are displayed in Fig. 2.

The results of our analysis, confirmed by this numerical example, lead us to the following conclusions.

In the case of small values of σ, γ, $\gamma \neq 0$, the Fitzhugh-Nagumo discrete equations ave a stationary point (v^*, w^*) such that $v^* \approx 1$, $w^* = \frac{\sigma}{\gamma} v^*$. At this stationary oint, the characteristic equation has a negative root, indicating the existence of nonoscillatory solution $(v(t), w(t))$, which tends to (v^*, w^*) as $t \to \infty$.

For the same values of σ, γ the characteristic equation at the stationary point $0, 0)$ has a positive root, indicating the existence of a nonoscillatory solution $v(t), w(t))$ of the discrete FitzHugh-Nagumo equations, which tends to $(0, 0)$ as $\to -\infty$.

An interesting feature of the numerical results is that in the presence of recovery rocesses, v may not be monotone (although tending monotonically to w^* at ifinity). While in the case of $\gamma = \sigma = 0$, this function is increasing on the whole real xis, and when the recovery processes are taken into consideration, the numerical pproximation of v gives a function that attains a maximum at a certain point and ien decreases monotonically to w^*.

In the near future, we intend to continue the investigation of oscillatory and onoscillatory behavior of mixed-type functional differential systems. On the other and, based on this analysis, we plan to improve the described computational iethod for the approximation of the discrete Fitzhugh-Nagumo system.

References

1. Bell, J.: Some threshold results for models of myelinated nerves. Math. Biosci. **54**, 181–90 (1981)
2. Bell, J.: Behaviour of some models of myelinated axons. IMA J. Math. Appl. Med. Biol. **1**, 149–167 (1984)
3. Chi, H., Bell, J., Hassard, B.: Numerical solution of a nonlinear advance-delay-differential equation from nerve conduction theory. J. Math. Biol. **24**, 583–601 (1986)
4. Ferreira, J.M., Pinelas, S.: Oscillatory mixed differential systems. Funkc. Ekvac. **53**, 1–20 (2010)
5. FitzHugh, R.: Impulses and physiological states in theoretical models of nerve membrane. Biophys. J. **1**, 445–466 (1961)
6. FitzHugh, R.: Computation of impulse initiation and saltatory condition in a myelinated nerve fiber. Biophys. J. **2**, 11–21 (1962)
7. Hodgkin, A., Huxley, A.: A qualitative description of nerve current and its application to conduction and excitation in nerve. J. Physiol. **117**, 500–544 (1952)
8. Lima, P., Teodoro, F., Ford, N., Lumb, P.: Analytical and numerical investigation of mixed-type functional differential equations. J. Comput. Appl. Math. **234**, 2826–2837 (2010)
9. Lima, P., Teodoro, F., Ford, N., Lumb, P.: Analysis and computational approximation of a forward-backward equation arising in nerve conduction, in S. Pinela s et al., editors, Differential and Difference Equations with Applications. Springer Proceedings in Mathematics and Statistics, **47** (2013)
10. Nagumo, J., Arimoto, S., Yoshizawa, S.: An active pulse transmission line simulating nerve axon. Proc. IRE **50**, 2061–2070 (1962)
11. Pinelas, S.: Oscillations and nonoscillations in mixed differential equations with monotonic delays and advances. Adv. Dyn. Syst. Appl. **4**, 107–121 (2009)

On Extinction Phenomena for Parabolic Systems

T. Gramchev, M. Marras, and S. Vernier–Piro

Abstract We investigate the extinction phenomena for some linear combinations of components of the vector-valued solutions to classes of semilinear parabolic systems. The crucial assumption on simultaneous splitting of the matrix-valued elliptic operators and the nonlinear source term allow us to uncouple the systems into a linear part and a scalar nonlinear equation depending on the solutions of the linear part. We propose necessary conditions and sufficient conditions on the existence of the extinction time for the solutions. We recapture as particular case previous results and apply our abstract theorem to a class of 3×3 systems appearing as models in chemical engineering.

Keywords Parabolic systems • Uncoupled systems • Extinction time

MSC classification: 35K50 (primary), 35K57 (secondary)

Introduction and Statement of the New Results

Broadly speaking, the aim of this paper is to study the (non)existence of extinction solutions of parabolic initial boundary value problems to generalise the aforementioned results for general classes of parabolic systems of the following type:

T. Gramchev • S. Vernier–Piro (✉)
Dipartimento di Matematica e Informatica, Università di Cagliari, via Ospedale 72,
09124 Cagliari, Italy
e-mail: todor@unica.it; svernier@unica.it

M. Marras
Dipartimento di Matematica e Informatica, Università di Cagliari, Viale Merello 92,
09123 Cagliari, Italy
e-mail: mmarras@unica.it

S. Pinelas et al. (eds.), *Differential and Difference Equations with Applications*, Springer
Proceedings in Mathematics & Statistics 47, DOI 10.1007/978-1-4614-7333-6_51,
© Springer Science+Business Media New York 2013

$$\partial_t \vec{u} + A\vec{u} = \vec{f}(t,x,\vec{u}), \qquad t > 0, x \in \Omega \tag{1}$$

$$\vec{u}(0,x) = \vec{u}^0(x), \qquad x \in \Omega \tag{2}$$

and boundary conditions either of Dirichlet-type

$$\vec{u}(t,x) = \vec{g}(t,x), \qquad x \in \partial\Omega \tag{3}$$

or the Robin-type

$$\partial_\nu \vec{u} + B\vec{u}(t,x) = \vec{g}(t,x), \qquad x \in \partial\Omega, t > 0, \tag{4}$$

where $\Omega \subset \mathbb{R}^n$ is an open bounded domain with smooth boundary $\partial\Omega$, $\vec{u} = (u_1,\ldots,u_d)$, $\vec{f} = (f_1,\ldots,f_d)$ smooth and are real-valued, and $\nu = \nu_x, x \in \partial\Omega$, stand for the unit outer normal vector to $\partial\Omega$. Let M_d(respectively $M_{k\times l}$) be the space $d \times d$ ($k \times l$) real matrices; A stands for a second-order $d \times d$ matrix-valued operator

$$A = A(x,\partial_x) = \{A^{\mu\nu}(x,\partial_x)\}^d_{\mu,\nu=1}, \quad A^{\mu\nu}(x,\partial_x)v = \sum_{j,k=1}^n \partial_{x_j}(a^{\mu\nu}_{jk}(x)\partial_{x_k}v) \tag{5}$$

with $a^{\mu\nu}_{jk} \in C^\infty(\bar{\Omega} : M_d(\mathbb{R}))$, $a^{\mu\nu}(x) := \{a^{\mu\nu}_{jk}\}_{j,k=1,\ldots,n}$, $\mu,\nu = 1,\ldots,d$, are symmetric, and, in the case of Robin-type boundary conditions,

$$B = B(x) = \{b_{\mu\nu}(x)\}^d_{\mu,\nu=1}. \tag{6}$$

We define the principal symbol $\{A^{\mu\nu;0}(x,\xi)\}^d_{j,k=1}$ of $A^{\mu\nu}(x,\partial_x)$ by

$$A^{\mu\nu;0}(x,\xi) = \sum_{j,k=1}^n a^{\mu\nu}_{jk}(x)(-i\xi_j)(-i\xi_k) = -\sum_{j,k=1}^n a^{\mu\nu}_{jk}(x)\xi_j\xi_k \tag{7}$$

and the principal symbol $A^0(x,\xi) = \{A^{\mu\nu;0}(x,\xi)\}^d_{\mu,\nu=1}$. We assume that the system is uniformly parabolic ([7, 8]), namely, the eigenvalues $\lambda_\ell(x,\xi)$, $\ell = 1,\ldots,d$ of $A^0(x,\xi)$ are simple and positive and belong to $C^\infty(\bar{\Omega} \times \mathbb{R}^n \setminus 0)$, are positively homogeneous of order 2 and for some $C > 0$ satisfy the uniform ellipticity condition

$$C^{-1}|\xi|^2 \le \lambda_\ell(x,\xi) \le C|\xi|^2, \quad x \in \bar{\Omega}, \xi \in \mathbb{R}^n \setminus 0. \tag{8}$$

Remark 1.1. Let $A = -A_0\Delta$, $A_0 \in M_d(\mathbb{R})$. Then A is elliptic if the eigenvalues of A_0 are positive and distinct. We also recapture as particular cases the elliptic $2 \times$ systems considered in [17, 18].

Parabolic systems of reaction-diffusion type model a great number of physical and chemical phenomena. Among properties of the vector solution of the system blow-up phenomena has been widely investigated ([9, 10, 12, 16]).

Another important property is the phenomena of the "extinction in finite time" for ιe solution, i. e. there exists a time $t^* > 0$ (extinction time) such that for every $t \geq t^*$

$$u(x,t) \equiv 0$$

'14,15]). As a model case we consider a system appearing in chemical engineering:

$$\partial_t u_1 - \mu \Delta u_1 = u_1^m e^{-1/u_2} g(u_3) \tag{9}$$

$$\partial_t u_2 - \nu \Delta u_2 = q u_1^m e^{-1/u_2} g(u_3) \tag{10}$$

$$\partial_t u_3 - \rho \Delta u_3 = p u_1^m e^{-1/u_2} g(u_3) \tag{11}$$

'here m, μ, ν, ρ, p, q are positive constants, g is smooth negative function and u_j, $= 1, 2, 3$ are required to be positive. It is a kinetic model of an irreversible reaction ιvolving two reactant species with concentration u_1, u_2 and the temperature u_3. ystems (9)–(11) are an extension of the case 2×2 considered in [17, 18]. For the ιse of one equation, we refer to the paper [1], where the extinction is investigated ρr the porous medium equation with absorption ($m < 1$)(see also [2, 3, 5]). Other ɛsults for the extinction time t^* are obtained in [13] in a very general contest. There ιe authors Ragnedda, Vernier-Piro and Vespri consider a class of singular parabolic ɾoblems with Dirichlet boundary conditions and derive estimate from above and ˈom below for the L^2-norm of the solution in terms of the extinction time t^* and the ˌsymptotic behaviour when the solution approaches t^*.

This paper is organised as follows: In Sect. 2 we propose sufficient conditions ρr reducing initial boundary value problem (1),(2) with (3) or (4) to uncoupling ɔrmal forms. Section 3 deals with necessary condition for extinction of some ɪnear combinations of the components $u_1(t,x), \ldots, u_d(t,x)$, while Sect. 4 proposes ɪfficient conditions.

Reduction to Uncoupled Normal Form

ˈe assume the following condition on the degeneracy of the right-hand side: there ⅹists a unitary vector $\vec{\varkappa} \in \mathbb{R}^d$ such that

$$\vec{f}(t,x,z) = f(t,x,z)\vec{\varkappa}, \quad t \geq 0, x \in \bar{\Omega}, z \in \mathbb{R}^d \tag{12}$$

ˈhere

$$f(t,x,z) = \|\vec{f}(t,x,z)\|, \quad t \geq 0, x \in \bar{\Omega}, z \in R^d. \tag{13}$$

Clearly (12) is equivalent to say that

$$\vec{f}(t,x,z) \perp \Gamma, \quad t \geq 0, x \in \Omega, z \in \mathbb{R}^d \tag{14}$$

where the hyperplane Γ is orthogonal to $\vec{\varkappa}$, namely,

$$\Gamma := \{\vec{r} \in \mathbb{R}^d; \, \vec{\varkappa} \cdot \vec{r} = 0\}. \tag{15}$$

We denote by π_Γ the orthogonal projection on Γ

$$\pi_\Gamma : \mathbb{R}^d \longrightarrow \Gamma, \tag{16}$$

clearly, for every $\vec{r}$ we have the following uniquely determined decomposition:

$$\vec{r} = \pi_\Gamma(\vec{r}) + (\vec{r} \cdot \vec{\varkappa})\vec{\varkappa}, \quad \vec{r} \in \mathbb{R}^d. \tag{17}$$

Next, we introduce the second condition for the uncoupling of the system into linear $(d-1) \times (d-1)$ system and a scalar nonlinear equation.

We suppose that there exists a nonsingular matrix $Q = \{Q_{\mu\nu}\}_{\mu,\nu=1}^d$ such that

$$QA(x,\partial_x)Q^{-1} = \begin{pmatrix} \tilde{A}(x,\partial_x) & 0_{d-1\times 1} \\ 0_{1\times d-1} & L(x,\partial_x) \end{pmatrix}, \tag{18}$$

$$Q\vec{\varkappa} = \vec{e}_d = \begin{pmatrix} 0 \\ \vdots \\ 0 \\ 1 \end{pmatrix} \tag{19}$$

and in the case of Robin-type boundary conditions,

$$QB(t,x)Q^{-1} = \begin{pmatrix} \tilde{B}(t,x) & 0_{d-1\times 1} \\ 0_{1\times d-1} & b(t,x) \end{pmatrix}, \quad t \geq 0, x \in \partial\Omega, \tag{20}$$

where $\tilde{A}(x,\partial_x)$ is $(d-1) \times (d-1)$ matrix-valued elliptic operator, L is a second order scalar elliptic operator and $\tilde{B}(t,x) \in C([0,+\infty[\times\partial\Omega : M_{d-1}(\mathbb{R})), b \in \tilde{B}(t,x) \in C([0,+\infty[\times\partial\Omega : \mathbb{R}).$

We observe that, by standard linear algebra arguments, (18), (19) lead to

$$Q = \begin{pmatrix} Q' \\ \vec{\varkappa}^{tr} \end{pmatrix}, \tag{21}$$

where $Q' \in M_{(d-1)\times d}(\mathbb{R})$ while the last row $(q_{d1},\ldots,q_{dd})$ coincides with $\vec{\varkappa}^{tr}$.

We will use the change of the variables $\vec{v} = Q\vec{u}$.

Now we state the first main result of our paper.

Theorem 2.1. *Assume that* (12), (18), (19), (20) *hold. Then* $\vec{u}(t,x)$ *satisfies th system* (1) *iff* $\vec{v}(t,x) = (\vec{v}'(t,x), v_d(t,x))$ *satisfies the following uncoupled system linear (respectively, semilinear) for* $\vec{v}'(t,x)$ *(respectively,* v_d*):*

$$\partial_t \vec{v'} + \tilde{A}(x, \partial_x)\vec{v'} = 0, \qquad t \geq 0, x \in \Omega \tag{22}$$

ith initial data

$$\vec{v'}(0,x)) = \vec{v'}^0(x) := Q'\vec{u}^0(x), \qquad x \in \Omega \tag{23}$$

nd Dirichlet boundary condition

$$\vec{v'}(t,x) = \vec{h'}(t,x) := Q'\vec{g}(t,x), \quad t \geq 0, x \in \Omega \tag{24}$$

r the corresponding Robin-type boundary condition

$$\partial_\nu \vec{v'}(t,x) + \tilde{B}\vec{v'}(t,x) = \vec{h'} := Q'\vec{g}(t,x), \quad t \geq 0, x \in \Omega \tag{25}$$

espectively:

$$\partial_t v_d + L(x, \partial_x)v_d = g_{\vec{v}}(t,x,v_d) := \left(Q\vec{f}(t,x,Q^{-1}\vec{v}(t,x)) \right)_d \quad t \geq 0, x \in \Omega \tag{26}$$

$$v_d(0,x)) = v_d^0(x) := \vec{\varkappa} \cdot \vec{u}^0(x), \qquad x \in \Omega \tag{27}$$

ith the Dirichlet boundary condition

$$v_d(t,x) = h_d(t,x) := \vec{\varkappa} \cdot \vec{g}(t,x), \quad t \geq 0, x \in \Omega \tag{28}$$

r the corresponding Robin-type boundary condition

$$\partial_\nu v_d(t,x) + b(t,x)v_d(t,x) = h_d(t,x) := \vec{\varkappa} \cdot \vec{g}(t,x), \quad t \geq 0, x \in \Omega.) \tag{29}$$

roof. Set

$$\vec{v}(t,x) = Q\vec{u}(t,x), \quad \vec{u}(t,x) = Q^{-1}\vec{v}(t,x). \tag{30}$$

In particular, (19), (30) and the orthogonality condition (15) yield that

$$\vec{v}(t,x) := Q\vec{u}(t,x) = \begin{pmatrix} \vec{v'}(t,x) \\ v_d(t,x) \end{pmatrix}, \qquad t \geq 0, x \in \Omega, \tag{31}$$

$$\vec{v'} := Q' \cdot \vec{u}(t,x) \in \mathbb{R}^{d-1}, \qquad t \geq 0, x \in \Omega, \tag{32}$$

$$v_d(t,x) = \vec{\varkappa} \cdot \vec{u}(t,x). \tag{33}$$

n the other hand, since Q is orthogonal, we have

$$Q' \cdot \vec{\varkappa} = 0 \in \mathbb{R}^{d-1}. \tag{34}$$

Plugging (30) into the system (1) and applying Q^{-1} to (1), we get, using th
hypotheses (18), (19), that

$$\partial_t \vec{v} + QAQ^{-1}\vec{v} = \partial_t \vec{v} + \begin{pmatrix} \tilde{A}(x,\partial_x) & 0_{d-1 \times 1} \\ 0_{1 \times d-1} & L(x,\partial_x) \end{pmatrix} \vec{v}$$

$$= Q\vec{f}(t,x,Q^{-1}\vec{v}(t,x)) = f(t,x,Q^{-1}\vec{v}(t,x))Q\vec{\varkappa} = f(t,x,Q^{-1}\vec{v}(t,x))\vec{e}_d \quad (3\text{!})$$

which proves the uncoupling of the system into a linear part and a scalar nonline
one. Similarly, in the case of Robin-type boundary condition, using (20), we obtai
(25), (29). The proof is complete.   ∎

Remark 2.2. As a particular case, we capture the systems of two equations consi
ered by Van Der Mee and Vernier-Piro [17]. Next, we point out that the system of
equations (9), (10), (11) satisfies (12) while (18), (19) hold, and as a consequenc
the system can be uncoupled, iff $\mu = v = \rho$.

3 Necessary Condition for the Existence of Extinction

The uncoupling of the system allows us to obtain immediately a necessary conditio
for the existence of extinction solution of the nonlinear component $v_d(t,x)$.

Proposition 3.1. *Suppose that* $v_d(t,x) = 0$ *for some* $t \geq t^* > 0$. *Then*

$$\left(Q\vec{f}(t,x,Q^{-1}(\vec{v}'(t,x),0)^{tr}) \right)_d \equiv 0$$

for some solutions of the linear part.

Proof. We transform the problem for v_d and observe that in the case clearly th
necessary condition is given by

$$F_d(t,x,(v'(t,x),0)) \equiv 0,$$

where $\vec{F}(t,x,\vec{v}(t,x)) = (F_1(t,x,\vec{v}(t,x)),\ldots,F_d(t,x,\vec{v}(t,x)))^{tr} = Q\vec{f}(t,x,Q^{-1}\vec{v}(t,x))$
                                                                                      ∎

4 Existence of Extinction Time

In this section we prove that the extinction phenomena may occur only for
plus linear combination of $v_1,\ldots,v_{d-1}$, solutions of the linear system. Moreove
in the framework of the L^2 spaces, we introduce conditions on data sufficient for th
solution to vanish in finite time, deriving an upper bound for the extinction time.

We have the following result.

Theorem 4.1. *Consider the system* (1), *with fixed initial data* (2) *and zero Dirichlet boundary condition. Suppose that*

$$\vec{u}'(t,x) = Q^{-1}\begin{pmatrix} \vec{v}'(t,x) \\ 0 \end{pmatrix} \tag{36}$$

e fixed solution of the linear system. Set

$$\vec{u}^{\varkappa}(t,x) = v_d(t,x)\vec{\varkappa} = (\vec{\varkappa} \cdot \vec{u}(t,x))\vec{\varkappa}. \tag{37}$$

uppose that $w \mapsto g_{\vec{v}}(t,x,w)$, defined in (26), *is monotone decreasing function atisfying for some $C = C_{\vec{v}} > 0$, $\delta \in]0,1[$ the following estimate:*

$$g_{\vec{v}}(t,x,w) \le -C|w|^{1-\delta}. \tag{38}$$

Then $|\vec{u}^{\varkappa}(t,x)| = |v_d||\vec{\varkappa}|$ vanishes at some finite time t^* with t^* bounded by

$$t^* \le C_1 |\vec{u}^{\varkappa,0}|_{C(\Omega)}^{\delta}, \tag{39}$$

here $C_1 > 0$ is independent of the initial data.

roof. We pass to the scalar equation for v_d. We observe that the condition (39) mplies that $g_{\vec{v}}$ is bounded by a monotone decreasing function. Arguing as in [1], in ect. 3 and in [17], if $\bar{U}(t)$ is the solution of the problem

$$\begin{cases} \bar{U}_t \le -C|\bar{U}|^{1-\delta} \\ \bar{U}(0) = |\vec{u}^{\varkappa,0}|_{C(\Omega)}, \end{cases}$$

en $\bar{U}(t)$ is a super solution of the analogous problem for v_d, with no diffusion erm. If $\bar{U}$ vanishes at time T^*, after an integration, we easily see that

$$T^* \le \frac{1}{C\delta}|\vec{u}^{\varkappa,0}|_{C(\Omega)}^{\delta}. \tag{40}$$

y comparison principle also v_d (and $|\vec{u}^{\varkappa}(t,x)|$) vanishes at time $t^* \le T^*$ and (39) proved.   □

xample 4.2. Suppose that the nonlinear term is given by

$$\vec{f}(t,x,\vec{u}) = -(\vec{\varkappa}\vec{u})|\vec{\varkappa}\vec{u}|^{m-1},$$

here $m > 0$. Then the extinction occurs if $m < 1$, while for $m \ge 1$ the extinction in nite time does not occur.

We can provide also an abstract extinction in the framework of the L^2 spaces nder additional conditions on A and the nonlinear term.

Proposition 4.3. *Suppose that the elliptic matrix-valued operator is symmetric namely,*

$$(A\vec{w}, \vec{w})_{L^2(\Omega)} \geq \beta \|\vec{w}(t,.)\|^2_{L^2(\Omega)}, \quad w \in H^1_0(\Omega), \ \beta > 0, \tag{41}$$

and the nonlinear term satisfies the following dissipation-type estimate:

$$\vec{f}(t,x,z) \cdot z \leq -C|z|^{2-\delta}, \quad t \geq 0, x \in \bar{\Omega}, z \in \mathbb{R}^d \tag{42}$$

for some $C > 0$, $\delta \in]0,1[$. Then, if $\vec{u}(t,x)$ is a solution as in the previous theorem which belongs to $C([0,+\infty[: (H^1_0(\Omega))^d)$, then $\vec{u}(t,x)$ vanishes for $t \geq t^*$ with t bounded by

$$t^* \leq \frac{1}{\gamma |\Omega|^{\frac{\delta}{2}}} \|\vec{u}^0\|^\delta_{L^2(\Omega)} \tag{43}$$

where $\gamma > 0$ is independent of the initial data.

Proof. The argument is standard. We have, after multiplying the system by $\vec{u}$ and then integrating over Ω,

$$\frac{1}{2}\frac{d}{dt}(\|\vec{u}(t,.)\|^2_{L^2(\Omega)}) + (A\vec{u}, \vec{u})_{L^2(\Omega)} = (\vec{f}(t,.,\vec{u}(t,.)), \vec{u}(t,.))_{L^2(\Omega)} \tag{44}$$

and obtain, in view of (41), (42), the estimate

$$\frac{d}{dt}(\|\vec{u}(t,.)\|^2_{L^2(\Omega)}) \leq -C_1 \|\vec{u}(t,.)\|^{2-\delta}_{L^{2-\delta}(\Omega)}. \tag{45}$$

Clearly $\|\vec{u}(t,.)\|_{L^2(\Omega)}$ is decreasing and

$$\lim_{t \to t^*} \|\vec{u}(t,.)\|_{L^2(\Omega)} \geq 0, \quad t^* =: \sup\{t > 0 : u(t,.) \not\equiv 0\}. \tag{46}$$

Integration from 0 to t leads to

$$\|\vec{u}(t,.)\|^2_{L^2(\Omega)} \leq \|\vec{u}^0\|^2_{L^2(\Omega)} - C_1 \int_0^t \|\vec{u}(\tau,.)\|^{2-\delta}_{L^{2-\delta}(\Omega)} d\tau. \tag{47}$$

The Hölder inequality implies

$$\|\vec{u}(t,.)\|^2_{L^{2-\delta}(\Omega)} \leq |\Omega|^{\frac{\delta}{2-\delta}} \|\vec{u}(t,.)\|^2_{L^2(\Omega)}. \tag{48}$$

Thus, combining (48) and (47), we obtain

$$\|\vec{u}(t,.)\|^2_{L^{2-\delta}(\Omega)} \leq |\Omega|^{\frac{\delta}{2-\delta}} \left(\|\vec{u}^0\|^2_{L^2(\Omega)} - C_1 \int_0^t \|\vec{u}(\tau,.)\|^{2-\delta}_{L^{2-\delta}(\Omega)} d\tau \right). \tag{49}$$

Setting $\omega(t) = \int_0^t \|\vec{u}(\tau,.)\|_{L^{2-\delta}(\Omega)}^{2-\delta} d\tau$, with $q = \dfrac{2}{2-\delta} > 1$, (49) becomes

$$\omega'(t) \leq |\Omega|^{\frac{\delta}{2}} \left(\|\vec{u}^0\|_{L^2(\Omega)}^2 - C_1 \omega(t) \right)^{1/q}. \tag{50}$$

ince $\omega(0) = 0$ and $\omega(t) < \|\vec{u}^0\|_{L^2(\Omega)}^2 / C_1$, $t < t^*$, by setting $C_0 = \|\vec{u}^0\|_{L^2(\Omega)}^2$, then
ategration of the differential inequality and the identity

$$\int_0^{C_0/C_1} \left(\|\vec{u}^0\|_{L^2(\Omega)}^2 - C_1 \eta \right)^{-1/q} d\eta = \frac{2}{\delta C_1} \|\vec{u}^0\|_{L^2(\Omega)}^\delta \tag{51}$$

nplies the desired estimate (43). □

emark 4.4. We note that nonlinearity with sublinear growth near 0 might lead to roblems with the uniqueness of the solution (e.g. cf. [4] for semilinear parabolic quations with nonlinear term with sublinear growth; see also [6, 11] and the ·ferences therein for the role of the sublinear growth for the existence of compactly upported solitary waves). Finally, we mention that more general results will be iven in a future work.

References

1. Bandle, C., Nanbu, T., Stakgold, I.: Porous medium equation with absorption. SIAM J. Math. Anal. **29**, 1268–1278 (1998)
2. Bandle, C., Sperb, R.P., Stakgold, I.: Diffusion and reaction with monotone kinetics. Nonlinear Anal. **8**, 321–333 (1984)
3. Bandle, C., Stakgold, I.: The formation of the dead core in parabolic reaction-diffusion problems. Trans. Amer. Math. Soc. **286**, 275–293 (1984)
4. Biagioni, H.A., Cadeddu, L., Gramchev, T.: Semilinear parabolic equations with singular initial data in anisotropic weighted spaces. Differ. Integ. Equat. **12**, 613–636 (1999)
5. Bobisud, L.E., Stakgold, I.: Dead cores in nonlinear reaction-diffusion systems. Nonlinear Anal. **11**, 1219–1227 (1987)
6. Gaeta, G., Gramchev, T., Walcher, S.: Compact solitary waves in linearly elastic chains with non-smooth on-site potential. J. Phys. A **40**, 4493–4509 (2007)
7. Ladyzhenskaja, O.A., Solonnikov, V.A., Uralceva, N.N.: Linear and Quasilinear Equations of Parabolic Type, Transl. Math. Monographs vol.23, American Mathematical Society, Providence, RI, (1968)
8. Pao, C.V.: On nonlinear reaction-diffusion systems. J. Math. Anal. Appl. **87**, 165–198 (1982)
9. Payne, L.E., Philippin, G.A., Vernier Piro, S.: Blow-up phenomena for a semilinear heat equation with nonlinear bouandary condition, I. Z. Angew. Math. Phys. **61**, 999–1007 (2010)
10. Payne, L.E., Philippin, G.A., Vernier Piro, S.: Blow-up phenomena for a semilinear heat equation with nonlinear bouandary condition, II. Nonlinear Anal. **73**, 971–978 (2010)
11. Popivanov, P., Slavova, A.: Nonlinear waves. An introduction. Series on Analysis, Applications and Computation, 4. World Scientific Publishing Company Private Limited, Hackensack, NJ (2011)

12. Quittner, P., Souplet, P.: Superlinear parabolic problems. Blow-up, global existence and stead states, Birkhäuser Advanced Texts, Basel (2007)

13. Ragnedda, F., Vernier Piro, S., Vespri, V.: The asymptotic profile of solutions of a class o singular parabolic equations. Prog. Nonlinear Diff. Equat. Appl. **60**, 577–589 (2011) Spring Basel AG

14. Stakgold, I.: Functions and Boundary Value Problems, 2nd Edn. Wiley-Interscienc New York (1998)

15. Stakgold, I.: Reaction-diffusion problems in chemical engineering. In: Fasano, A., Primiceri M. (eds.) Nonlinear Diffusion Problems, Lecture Notes in Mathematics vol. 1224, Springe Berlin (1985)

16. Vazquez, J.L.: The problems of blow-up for nonlinear heat equations. Complete blow-up ar avalanche formation. Rend. Mat. Acc. Lincei **15**, 281–300 (2004)

17. van der Mee, C., Vernier-Piro, S.: Dead cores for time-dependent reactiondiffusion equation Nonlinear Anal. **47**, 2113–2123 (2001)

18. van der Mee, C., Vernier-Piro, S.: Dead cores for parabolic reaction diffusion problems. Dy Contin. Discrete I. serie A: Math. Anal. **10**, 139–147 (2003)

Conjugacy of Discrete Semidynamical Systems in the Neighbourhood of Invariant Manifold

Andrejs Reinfelds

Abstract The conjugacy of discrete semidynamical system and its partially decoupled discrete semidynamical system in Banach space is proved in the neighbourhood of the trivial invariant manifold.

The conjugacy for noninvertible mappings in Banach space was considered by B. Aulbach and B. M. Garay [1–3]. For noninvertible mappings in a complete metric space, it was extended and generalized by A. Reinfelds [4–8]. In this paper we consider the case when the linear part of the noninvertible mapping depends on the behaviour of variables in the neighbourhood of the invariant manifold.

Definition 1. Two discrete semidynamical systems $S^n, T^n \colon \mathbf{X} \to \mathbf{X}$ $(n \in \mathbb{N})$ are *conjugate*, if there exists a homeomorphism $H \colon \mathbf{X} \to \mathbf{X}$ such that

$$S^n \circ H(x) = H \circ T^n(x).$$

Definition 2. Two mappings $S, T \colon \mathbf{X} \to \mathbf{X}$ are *conjugate*, if there exists a homeomorphism $H \colon \mathbf{X} \to \mathbf{X}$ such that

$$S \circ H(x) = H \circ T(x).$$

It is easily verified that two discrete semidynamical systems S^n and T^n, generated by mappings S and T, are conjugate if and only if the mappings S and T are conjugate.

Let $\mathbf{E}$ and $\mathbf{F}$ be Banach spaces, $\mathbf{B}(a) = \{r \in \mathbf{F} \mid |r| \le a\}$ and $a > 0$. Consider the following mapping $S \colon \mathbf{E} \times \mathbf{B}(a) \to \mathbf{E} \times \mathbf{B}(a)$ defined by

$$x_1 = g(x) + \Psi(x, r) = X(x, r)$$

A. Reinfelds (✉)
Institute of Mathematics and Computer Science, Raiņa bulvāris 29, Rīga, Latvia

University of Latvia, Faculty of Physics and Mathematics, Zeļļu iela 8, Rīga, Latvia
e-mail: reinf@latnet.lv

S. Pinelas et al. (eds.), *Differential and Difference Equations with Applications*, Springer Proceedings in Mathematics & Statistics 47, DOI 10.1007/978-1-4614-7333-6_52, © Springer Science+Business Media New York 2013

$$r_1 = A(x)r + \Phi(x,r) = R(x,r), \tag{1}$$

where derivative of diffeomorphism $g \colon \mathbf{E} \to \mathbf{E}$ is uniformly continuous $\|Dg(x) - Dg(x')\| \le \omega(|x-x'|)$, $\sup_x \|A(x)\| + \varepsilon < 1$, mappings A, Ψ, and Φ are Lipschitzian

$$\|A(x) - A(x')\| \le \gamma|x-x'|,$$
$$|\Psi(x,r) - \Psi(x',r')| \le \varepsilon(|x-x'| + |r-r'|),$$
$$|\Phi(x,r) - \Phi(x',r')| \le \varepsilon(|x-x'| + |r-r'|)$$

and $\Phi(x,0) = 0$.

Theorem 1. *If $\sup_x(\|(Dg(x))^{-1}\|\|A(x))\|) + 4\varepsilon \sup_x \|(Dg(x))^{-1}\| < 1$, then there exists a continuous map $v \colon \mathbf{E} \times \mathbf{B}(\delta) \to \mathbf{E}$ that is Lipschitzian with respect to the second variable and such that the mappings (1) and*

$$x_1 = X(x,0)$$
$$r_1 = R(x + v(x,r), r) \tag{2}$$

are conjugated in a small neighbourhood of the invariant manifold $r = 0$.

We will seek the mapping establishing the conjugacy of (1) and (2) in the form

$$H(x,r) = (x + v(x,r), r).$$

We get the following functional equation:

$$X(x + v(x,r), r) = X(x,0) + v(X(x,0), R(x + v(x,r), r)) \tag{3}$$

or equivalently

$$\begin{aligned} v(x,r) = (Dg(x))^{-1} \big(& Dg(x)v(x,r) - X(x + v(x,r), r) \\ & + X(x,0) + v(X(x,0), R(x + v(x,r), r)) \big). \end{aligned}$$

The proof of the theorem consists of four lemmas.

Lemma 1. *The functional equation (3) has a unique solution in $\mathcal{M}_1$.*

Proof. The set of continuous maps $v \colon \mathbf{E} \times \mathbf{B}(\delta) \to \mathbf{E}$

$$\mathcal{M} = \left\{ v \in \mathbf{C}(\mathbf{E} \times \mathbf{B}(\delta), \mathbf{E}) \; \middle| \; \sup_{x,r} \frac{|v(x,r)|}{|r|} < +\infty \right\}$$

becomes a Banach space if we use the norm $\|v\| = \sup_{x,r} \frac{|v(x,r)|}{|r|}$. The set

$$\mathcal{M}_1 = \left\{ v \in \mathcal{M} \mid \|v\| \le 1 \text{ and } |v(x,r) - v(x,r')| \le |r-r'| \right\}$$

is a closed subset of the Banach space $\mathcal{M}$.

Let us consider the mapping $v \mapsto \mathcal{L}v$, $v \in \mathcal{M}_1$ defined by the equality

$$\mathscr{L}v(x,r) = (Dg(x))^{-1}v(X(x,0),R(x+v(x,r),r))$$
$$+(Dg(x))^{-1}\left(Dg(x)v(x,r) - g(x+v(x,r)) - \Psi(x+v(x,r),r) + g(x) + \Psi(x,0)\right).$$

irst we obtain

$$|\mathscr{L}v(x,r)| \le \|(Dg(x))^{-1}\|\left(|R(x+v(x,r),r)| + (\omega(|r|)+2\varepsilon)|r|\right)$$
$$\le \|(Dg(x))^{-1}\|\left(\|A(x)\| + 3\varepsilon + \omega(|r|) + \gamma|r|\right)|r|.$$

ere we used Hadamard's lemma

$$g(x') - g(x) = \int_0^1 Dg(x+\theta(x'-x))\,d\theta(x'-x).$$

ext we get

$$|\mathscr{L}v(x,r) - \mathscr{L}v(x,r')| \le \|(Dg(x))^{-1}\|\,|R(x+v(x,r),r) - R(x+v(x,r'),r')|$$
$$+\|(Dg(x))^{-1}\|\,|Dg(x)(v(x,r) - v(x,r')) - g(x+v(x,r)) + g(x+v(x,r'))|$$
$$+\|(Dg(x))^{-1}\|\,|\Psi(x+v(x,r),r) - \Psi(x+v(x,r'),r')|$$
$$\le \|(Dg(x))^{-1}\|\left(\|A(x)\| + 4\varepsilon + \omega(\max\{|r|,|r'|\}) + 2\gamma\max\{|r|,|r'|\}\right)|r-r'|.$$

n addition,

$$|\mathscr{L}v(x,r) - \mathscr{L}v'(x,r)| \le \|(Dg(x))^{-1}\|\,|R(x+v(x,r),r) - R(x+v'(x,r),r)|$$
$$+\|(Dg(x))^{-1}\|\,|v(X(x,0),R(x+v(x,r),r)) - v'(X(x,0),R(x+v(x,r),r))|$$
$$+\|(Dg(x))^{-1}\|\,|Dg(x)(v(x,r) - v'(x,r)) - g(x+v(x,r)) + g(x+v'(x,r))|$$
$$+\|(Dg(x))^{-1}\|\,|\Psi(x+v(x,r),r) - \Psi(x+v'(x,r),r)|$$
$$\le \|(Dg(x))^{-1}\|\left(\|A(x)\| + 3\varepsilon + \omega(|r|) + 2\gamma|r|\right)\|v-v'\|\,|r|.$$

e choose $\delta > 0$, where $\max\{|r|,|r'|\} = \delta \le a$, such that

$$\sup_x(\|(Dg(x))^{-1}\|\|A(x)\|) + (4\varepsilon + \omega(2\delta) + 4\gamma\delta)\sup_x\|(Dg(x))^{-1}\| < 1.$$

hen $\|\mathscr{L}v\| \le 1$, $|\mathscr{L}v(x,r) - \mathscr{L}v(x,r')| \le |r-r'|$, the mapping $\mathscr{L}$ is a contraction
nd consequently the functional equation (3) has unique solution in $\mathscr{M}_1$. $\qquad\square$

Next we will prove that the mapping H is a homeomorphism in the small
eighbourhood of the invariant manifold $r = 0$. Let us consider the functional
quation

$$X(x+v_1(x,r),0) = X(x,r) + v_1(X(x,r),R(x,r)) \tag{4}$$

or equivalently

$$v_1(x,r) = (Dg(x))^{-1}\left(Dg(x)v_1(x,r) - X(x+v_1(x,r),0)\right.$$
$$\left. + X(x,r) + v_1(X(x,r),R(x,r))\right).$$

Lemma 2. *The functional equation (4) has a unique solution in $\mathcal{M}_2$.*

Proof. The set

$$\mathcal{M}_2 = \{v \in \mathcal{M} \mid \|v\| \leq 1\}$$

is a closed subset of the Banach space $\mathcal{M}$.

Let us consider the mapping $v_1 \mapsto \mathcal{L}v_1$, $v_1 \in \mathcal{M}_2$ defined by the equality

$$\mathcal{L}v_1(x,r) = (Dg(x))^{-1}v_1(X(x,r),R(x,r))$$
$$+ (Dg(x))^{-1}(Dg(x)v_1(x,r) - g(x+v_1(x,r)) - \Psi(x+v_1(x,r),0) + g(x) + \Psi(x,r))$$

We have

$$|\mathcal{L}v_1(x,r)| \leq \|(Dg(x))^{-1}\| \left(|R(x,r)| + \omega(|r|) + 2\varepsilon\right)|r|$$
$$= \|(Dg(x))^{-1}\| \left(\|A(x)\| + 3\varepsilon + \omega(|r|)\right)|r|.$$

We obtain

$$|\mathcal{L}v_1(x,r) - \mathcal{L}v_1'(x,r)|$$
$$\leq \|(Dg(x))^{-1}\| |Dg(x)(v_1(x,r) - v_1'(x,r)) - g(x+v_1(x,r)) + g(x+v_1'(x,r))$$
$$-\Psi(x+v_1(x,r),0) - \Psi(x+v_1'(x,r),0)|$$
$$+ |(Dg(x))^{-1}(v_1(X(x,r),R(x,r)) - v_1'(X(x,r),R(x,r)))|$$
$$\leq \|(Dg(x))^{-1}\| (\|A(x)\| + \omega(|r|) + 2\varepsilon) \|v_1 - v_1'\| |r|.$$

We get $\|\mathcal{L}v_1\| \leq 1$, $\mathcal{L}$ is a contraction and consequently the functional equation (4) has a unique solution in $\mathcal{M}_2$.

Consider the mapping G defined by equality $G(x,r) = (x+v_1(x,r),r)$.

Lemma 3. $G \circ H = id$.

Proof. Let us consider the functional equation

$$X(x+v_2(x,r),0) = X(x,0) + v_2(X(x,0),R(x+v(x,r),r)) \tag{5}$$

or equivalently

$$v_2(x,r) = (Dg(x))^{-1}\left(Dg(x)v_2(x,r) - X(x+v_2(x,r),0)\right.$$
$$\left. + X(x,0) + v_2(X(x,0),R(x+v(x,r),r))\right).$$

is easily verified that the functional equation (5) has a trivial solution. Let us prove ᵉe uniqueness of the solution in $\mathscr{M}_3$, where

$$\mathscr{M}_3 = \{v_2 \in \mathscr{M} \mid \|v_2\| \leq 2\}$$

a closed subset of the Banach space $\mathscr{M}$. We get

$$|v_2(x,r)| \leq \|(Dg(x))^{-1}\| \, |Dg(x)v_2(x,r) - g(x + v_2(x,r)) + g(x)|$$
$$+ \|(Dg(x))^{-1}\| \, |\Psi(x + v_2(x,r),0) - \Psi(x,0)|$$
$$+ \|(Dg(x))^{-1}\| \, |v_2(X(x,0),R(x + v(x,r),r))|$$
$$\leq \|(Dg(x))^{-1}\| (\|A(x)\| + 2\varepsilon + \omega(\|v_2\| |r|) + \gamma \|v_2\| |r|)) \|v_2\| |r|.$$

follows that $v_2(x,r) \equiv 0$. The mapping $w_1 \in \mathscr{M}_3$, where

$$w_1(x,r) = v(x,r) + v_1(x + v(x,r),r)$$

ᵃlso satisfies the functional equation (5). Using the change of variables $x \mapsto x + (x,r) in (4), we get

$$X(x + w_1(x,r),0) = X(x + v(x,r),r) + v_1(X(x + v(x,r),r),R(x + v(x,r),r)).$$

ᶠsing (3), we obtain

$$X(x + w_1(x,r),0) = X(x,0) + v(X(x,0),R(x + v(x,r),r))$$
$$+ v_1(X(x,0) + v(X(x,0),R(x + v(x,r),r)),R(x + v(x,r),r))$$
$$= X(x,0) + w_1(X(x,0),R(x + v(x,r),r)).$$

ᶜonsequently, we have

$$v(x,r) + v_1(x + v(x,r),r) = 0.$$

We obtain that $G \circ H = id$.   □

ᴸemma 4. $H \circ G = id$.

roof. The set of continuous maps $v_3 \colon \mathbf{E} \times \mathbf{B}(\delta) \times \mathbf{B}(\delta) \to \mathbf{E}$

$$\mathscr{N} = \left\{ v_3 \in \mathbf{C}(\mathbf{E} \times \mathbf{B}(\delta) \times \mathbf{B}(\delta),\mathbf{E}) \,\middle|\, \sup_{x,r,z} \frac{|v_3(x,r,z)|}{\max(|r|,|z-r|)} < \infty \right\}$$

ᵉcomes a Banach space if we use the norm $\|v_3\| = \sup_{x,r,z} \frac{|v_3(x,r,z)|}{\max(|r|,|z-r|)}$. The set

$$\mathscr{N}_1 = \{v_3 \in \mathscr{N} \mid \|v_3\| \leq 1 \text{ and } |v_3(x,r,z) - v_3(x,r,z')| \leq |z - z'|\}$$

a closed subset of the Banach space $\mathscr{N}$.

Let us consider the functional equation

$$X(x,r) + v_3(X(x,r),R(x,r),R(x+v_3(x,r,z),z)) = X(x+v_3(x,r,z),z) \quad ($$

or equivalently

$$v_3(x,r,z) = (Dg(x))^{-1} (Dg(x)v_3(x,r,z) - g(x+v_3(x,r,z)) + g(x)$$
$$+\Psi(x,r) - \Psi(x+v_3(x,r,z),z) + v_3(X(x,r),R(x,r),R(x+v_3(x,r,z),z))).$$

Let us consider the mapping $v_3 \mapsto \mathscr{L}v_3$, $v_3 \in \mathscr{N}_1$ defined by the equality

$$\mathscr{L}v_3(x,r,z) = (Dg(x))^{-1} (Dg(x)v_3(x,r,z) - g(x+v_3(x,r,z)) + g(x)$$
$$-\Psi(x+v_3(x,r,z),z) + \Psi(x,r) + v_3(X(x,r),R(x,r),R(x+v_3(x,r,z),z))).$$

We obtain

$$|\mathscr{L}v_3(x,r,z)| \le \|(Dg(x))^{-1}\|(\omega(\max\{|r|,|z-r|\}) + 2\varepsilon)\max\{|r|,|z-r|\}$$
$$+\|(Dg(x))^{-1}\|\max\{|R(x,r)|,|R(x+v_3(x,r,z),z) - R(x,r)|\}$$
$$\le \|(Dg(x))^{-1}\|(\|A(x)\| + 4\varepsilon)\max\{|r|,|z-r|\}$$
$$+\|(Dg(x))^{-1}\|(\omega(\max\{|r|,|z-r|\}) + \gamma|z|)\max\{|r|,|z-r|\}.$$

In addition,

$$|\mathscr{L}v_3(x,r,z) - \mathscr{L}v_3(x,r,z')|$$
$$\le \|(Dg(x))^{-1}\|\|Dg(x)(v_3(x,r,z) - v_3(x,r,z')) - g(x+v_3(x,r,z)) + g(x+v_3(x,r,z'))|$$
$$+\|(Dg(x))^{-1}\|\|\Psi(x+v_3(x,r,z),z) - \Psi(x+v_3(x,r,z'),z')|$$
$$+\|(Dg(x))^{-1}\|\|R(x+v_3(x,r,z),z) - R(x+v_3(x,r,z'),z')|$$
$$\le \|(Dg(x))^{-1}\|(\omega(\max\{|r|,|z-r|,|z'-r|\}) + 2\varepsilon)|z-z'|$$
$$+\|(Dg(x))^{-1}\|(\|A(x)\| + 2\varepsilon + 2\gamma\max\{|r|,|z|,|z'-r|\})|z-z'|$$
$$= \|(Dg(x))^{-1}\|(\|A(x)\| + 4\varepsilon)|z-z'|$$
$$+\|(Dg(x))^{-1}\|(\omega(\max\{|r|,|z-r|,|z'-r|\}) + 2\gamma\max\{|r|,|z|,|z'-r|\})|z-z'$$

Let $v_3 \in \mathscr{N}_1$ and $v_3' \in \mathscr{N}_1 \cup \mathscr{N}_2$ where

$$\mathscr{N}_2 = \left\{ v_3' \in \mathscr{N} \;\middle|\; \sup_{x,|r|\le\delta,|z|\le\delta} |v_3'(x,r,z)| \le 2\delta \text{ and } |v_3(x,r,z) - v_3(x,r,z')| \le |z-z'| \right\}$$

We have

$$|\mathscr{L}v_3(x,r,z) - \mathscr{L}v_3'(x,r,z)|$$

$$\leq \|(Dg(x))^{-1}\|\|Dg(x)(v_3(x,r,z)-v_3'(x,r,z))-g(x+v_3(x,r,z))+g(x+v_3'(x,r,z))|$$

$$+\|(Dg(x))^{-1}\|\|\Psi(x+v_3(x,r,z),z)-\Psi(x+v_3'(x,r,z),z)|$$

$$+\|(Dg(x))^{-1}\|\|v_3(X(x,r),R(x,r),R(x+v_3(x,r,z),z))$$

$$-v_3'(X(x,r),R(x,r),R(x+v_3(x,r,z),z))|$$

$$+\|(Dg(x))^{-1}\|\|v_3'(X(x,r),R(x,r),R(x+v_3(x,r,z),z))$$

$$-v_3'(X(x,r),R(x,r),R(x+v_3'(x,r,z),z))|$$

$$\leq \|(Dg(x))^{-1}\|(\omega(\max\{|r|,|r-z|,2\delta\})+\varepsilon)\|v_3-v_3'\|\max\{|r|,|z-r|\}$$

$$+\|(Dg(x))^{-1}\|\max\{|R(x,r)|,|R(x+v_3(x,r,z),z)-R(x,r)|\}\|v_3-v_3'\|$$

$$+\|(Dg(x))^{-1}\|\|R(x+v_3(x,r,z),z)-R(x+v_3'(x,r,z),z)|$$

$$\leq \|(Dg(x))^{-1}\|(\omega(2\delta)+\varepsilon)\|v_3-v_3'\|\max\{|r|,|z-r|\}$$

$$+\|(Dg(x))^{-1}\|(\|A(x)\|+2\varepsilon+\gamma|z|)\|v_3-v_3'\|\max\{|r|,|z-r|\}$$

$$+\|(Dg(x))^{-1}\|(\varepsilon+\gamma|z|)\|v_3-v_3'\|\max\{|r|,|z-r|\}$$

$$= \|(Dg(x))^{-1}\|(\|A(x)\|+4\varepsilon+\omega(2\delta)+2\gamma|z|)\|v_3-v_3'\|\max\{|r|,|z-r|\}.$$

Then $\|\mathscr{L}v_3\| \leq 1$, $|\mathscr{L}v_3(x,r,z) - \mathscr{L}v_3(x,r,z')| \leq |z-z'|$, the mapping $\mathscr{L}$ is a ontraction and consequently the functional equation (6) has a unique solution in $\mathcal{V}_1$. Moreover, this solution is also unique in the closed subset $\mathcal{N}_2$. Let us note that

$$v_3(x,r,r) = 0.$$

The mapping $w_2 \in \mathcal{N}_2$, where

$$w_2(x,r,z) = v_1(x,r) + v(x+v_1(x,r),z)$$

atisfies (6). Using the change of variables $(x,r) \mapsto (x+v_1(x,r),z)$ in (3) we get

$$\mathcal{X}(x+w_2(x,r,z),z) = X(x+v_1(x,r),0) + v(X(x+v_1(x,r),0),R(x+w_2(x,r,z),z)).$$

sing (4) we obtain

$$X(x+w_2(x,r,z),z) = X(x,r)+v_1(X(x,r),R(x,r))$$

$$+v(X(x,r)+v_1(X(x,r),R(x,r)),R(x+w_2(x,r,z),z))$$

$$= X(x,r)+w_2(X(x,r),R(x,r),R(x+w_2(x,r,z),z)).$$

onsequently, we have

$$v_1(x,r)+v(x+v_1(x,r),r) = 0.$$

follows that $H \circ G = id$.

Finally we conclude that the mapping H is a homeomorphism establishing a onjugacy of the noninvertible mappings (1) and (2). □

Acknowledgements This work has been supported by the grant 09.1220 of the Latvian Counc of Science and by the ESF Project 2009/0223/1DP/1.1.1.2.0/09/APIA/VIAA/008.

References

1. Aulbach, B., Garay, B.M.: Linearization and decoupling of dynamical and semidynamic systems. In: Bainov, D., Covachov, V. (eds.) The Second Colloquium on Differential Equation pp. 15–27. World Scientific, Singapore (1992)
2. Aulbach, B., Garay B.M.: Linearizing the expanded part of noninvertible mappings. Z. Ange\ Math. Phys. **44**(3), 469–494 (1993)
3. Aulbach, B., Garay B.M.: Partial decoupling for noninvertible mappings. Z. Angew. Mat Phys. **45**(4), 505–542 (1994)
4. Reinfelds, A.: Partial decoupling of noninvertible mappings. Differ. Equ. Dyn. Syst. **2**(3) 205–215 (1994)
5. Reinfelds, A.: The reduction principle for discrete dynamical and semidynamical systems metric spaces. Z. Angew. Math. Phys. **45**(6), 933–955 (1994)
6. Reinfelds, A.: Partial decoupling of semidynamical systems. Latv. Univ. Zināt. Raksti **59** 54–61 (1994)
7. Reinfelds, A.: The reduction of discrete dynamical and semidynamical systems in metr spaces. In: Aulbach, B., Colonius F. (eds.) Six Lectures on Dynamical Systems, pp. 267–31 World Scientific, Singapore (1996)
8. Reinfelds, A.: Partial decoupling of semidynamical system in metric space. J. Tech. Un\ Plovdiv, Fundam. Sci. Appl., Ser. A, Pure Appl. Math. **5**, 33–40 (1997)

Γ-Convergence of Multiscale Periodic Energies Depending on the Curl of Divergence-Free Fields

Hélia Serrano

Abstract We study the Γ-convergence of a sequence of multiscale periodic quadratic energies, depending on the curl of solenoidal fields, whose associated Euler–Lagrange equations are the vector potential formulation of the stationary Maxwell equations, which may describe the magnetic properties of a multiscale periodic composite material.

Introduction

In this contribution, we are interested in the explicit characterization of the Γ-limit density of sequences of quadratic energies with linear perturbations of the type

$$E_\varepsilon(u) = \int_\Omega \left(\frac{A_\varepsilon(x)}{2} \operatorname{curl} u(x) \cdot \operatorname{curl} u(x) - b_\varepsilon(x) \cdot \operatorname{curl} u(x) \right) dx, \tag{1}$$

where Ω is an open bounded set in $\mathbb{R}^3$; $A_\varepsilon : \Omega \to \mathbb{R}^{3\times 3}$ is of the form

$$A_\varepsilon(x) = A\left(x, \frac{x}{\varepsilon}\right),$$

for some 3×3-matrix-valued function $A(x,y)$ Q-periodic in the second variable, with $Q = (0,1)^3$, such that there exist constants $\beta > \alpha > 0$ for which $\alpha|\rho|^2 \leq A(x,y)\rho \cdot \rho \leq \beta|\rho|^2$ for every $\rho \in \mathbb{R}^3$; and $b_\varepsilon : \Omega \to \mathbb{R}^3$ is given by

H. Serrano (✉)
Departamento de Matemáticas, E.T.S.I. Industriales, Universidad de Castilla-La Mancha,
Av. Camilo José Cela s/n, 13071 Ciudad Real, Spain
email: heliac.pereira@uclm.es

Pinelas et al. (eds.), *Differential and Difference Equations with Applications*, Springer   579
Proceedings in Mathematics & Statistics 47, DOI 10.1007/978-1-4614-7333-6_53,
© Springer Science+Business Media New York 2013

$$b_\varepsilon(x) = b\left(x, \frac{x}{\varepsilon}, \frac{x}{\varepsilon^2}\right),$$

for some vector-valued function $b \in L^2(\Omega \times \mathbb{R}^3 \times \mathbb{R}^3; \mathbb{R}^3)$ Q-periodic in the secon and third variables. The energies E_ε are well defined in the Hilbert space $X(\Omega$ given by

$$X(\Omega) = \left\{ w \in L^2(\Omega; \mathbb{R}^3) : \operatorname{curl} w \in L^2(\Omega; \mathbb{R}^3), \ \operatorname{div} w = 0 \text{ in } \Omega, \ w \cdot \mathbf{n} = 0 \text{ on } \partial\Omega \right\}$$

where $\mathbf{n}$ stands for the outward normal vector to $\partial\Omega$, with the norm $\|w\|^2_{X(\Omega)} =$ $\|w\|^2_{L^2} + \|\operatorname{curl} w\|^2_{L^2}$, see [9].

Our aim is to study the asymptotic behaviour, as the parameter ε goes to 0, of th family of quadratic functionals E_ε whose quadratic and linear coefficients, A_ε an b_ε, respectively, oscillate in different length scales. For such purpose we study th Γ-convergence of the sequence $\{E_\varepsilon\}$ and focus on the explicit characterization c the coefficients of the Γ-limit density. The Γ-convergence of sequences of periodi functionals defined in spaces of divergence-free fields, and not depending on th curl operator, was firstly addressed in [3] (see also [8]) through the usual machiner of Γ-convergence. The non-periodic case was studied in [17] through the study c the div-Young measures (see [15]) associated with non-periodic sequences of field (see [13, 16] in the case of curl-free fields).

Notice that, for each $\varepsilon > 0$, if u_ε is a minimizer of E_ε in $X(\Omega)$, then it is th solution of the second-order boundary value problem

$$\begin{cases} \operatorname{curl}(A_\varepsilon(x)\operatorname{curl} u_\varepsilon(x)) = \operatorname{curl} b_\varepsilon(x) & \text{in } \Omega \\ \operatorname{div} u_\varepsilon = 0 & \text{in } \Omega \\ u_\varepsilon \cdot \mathbf{n} = 0 & \text{on } \partial\Omega \\ (A_\varepsilon \operatorname{curl} u_\varepsilon) \times \mathbf{n} = b_\varepsilon \times \mathbf{n} & \text{on } \partial\Omega; \end{cases} \tag{2}$$

see [4]. Thus, if we are able to represent explicitly the Γ-limit of the sequence of er ergies E_ε, we may represent the homogenized problem and characterize the effectiv coefficients associated with the previous family of boundary value problems. In th way, the main motivation to study the Γ-convergence of quadratic functionals of th form (1) comes from the homogenization of second-order boundary value problem of type (2). On the other hand, the boundary value problem (2) may be considere as the vector potential formulation of a magnetostatic problem describing th magnetic properties of a composite anisotropic material in a perfect conductin media. Precisely, for each $\varepsilon > 0$, the field u_ε may be considered as the magnetic fiel of a composite material occupying a region Ω and with a periodic microstructu of relative size ε, subject to a current density of type $\operatorname{curl} b_\varepsilon$ and with magnet permeability $\mu_\varepsilon = A_\varepsilon^{-1}$; see [10].

Preliminaries

ι this section, we will present some basic concepts as well as some results
sed throughout this contribution, namely, the notion of Γ-convergence of integral
ınctionals (see [5–7]), the definition of multiscale Young measures (see [2,14]) and
ιe concept of multiscale convergence (see [1,11]).

** efinition 1.** The sequence of functionals E_ε defined in $X(\Omega)$ is said to Γ-converge
with respect to the weak topology in $X(\Omega)$) to the functional E if, for any u in
(Ω), it holds:

. For every sequence $\{u_\varepsilon\} \subset X(\Omega)$ such that $u_\varepsilon \rightharpoonup u$ in $X(\Omega)$ we have

$$\liminf_{\varepsilon \searrow 0} E_\varepsilon(u_\varepsilon) \geq E(u).$$

. There exists a sequence $\{u_\varepsilon\} \subset X(\Omega)$ such that $u_\varepsilon \rightharpoonup u$ in $X(\Omega)$ and

$$\lim_{\varepsilon \searrow 0} E_\varepsilon(u_\varepsilon) = E(u).$$

Following the ideas introduced in [13] to study the Γ-convergence through the
oung measures associated with relevant sequences, here we will study the Γ-
onvergence of the sequence $\{E_\varepsilon\}$ through a special type of Young measures: the
ıultiscale Young measures associated with divergence-free fields.

** efinition 2.** A family of probability measures $\{\mu_{x,y,z}\}_{x\in\Omega,(y,z)\in Q^2}$ supported on $\mathbb{R}^3$
. said to be the multiscale Young measure associated with the sequence of functions
$\varepsilon : \Omega \to \mathbb{R}^3$ if the joint Young measure $\theta = \{\theta_x\}_{x\in\Omega}$ associated with the sequence

$$\left\{ \left(u_\varepsilon(\cdot), \left\langle \frac{\cdot}{\varepsilon} \right\rangle, \left\langle \frac{\cdot}{\varepsilon^2} \right\rangle \right) \right\},$$

ιay be decomposed, for a.e. $x \in \Omega$ and $(y,z) \in Q^2$, as $\theta_x = \mu_{x,y,z} \otimes dz \otimes dy$, where
$v\rangle \in Q$ stands for the fractional part of $w \in \mathbb{R}^3$.

The following proposition will be the starting point to prove the lower limit
ιequality in the definition of Γ-convergence.

** roposition 1 (see [12]).** *If* $\{\mu_{x,y,z}\}_{x\in\Omega,(y,z)\in Q^2}$ *is the multiscale Young measure*
ssociated with the sequence $\{u_\varepsilon\}$, *then*

$$\text{m} \inf_{\varepsilon \searrow 0} \int_\Omega \psi\left(x, \left\langle \frac{x}{\varepsilon} \right\rangle, \left\langle \frac{x}{\varepsilon^2} \right\rangle, u_\varepsilon(x)\right) dx \geq \int_\Omega \int_{Q^2} \int_{\mathbb{R}^3} \psi(x,y,z,\rho)\, d\mu_{x,y,z}(\rho)\, dz\, dy\, dx,$$

›r every Carathéodory function $\psi : \Omega \times \mathbb{R}^3 \times \mathbb{R}^3 \times \mathbb{R}^3 \to \mathbb{R}$ *bounded from below.*

The notion of multiscale Young measure is intimately related with the notion of
ıultiscale convergence.

Definition 3. The sequence $\{u_\varepsilon\} \subset L^2(\Omega; \mathbb{R}^3)$ is said to be multiscale converge_
to a function $u_0 \in L^2(\Omega \times Q^2; \mathbb{R}^3)$ if, for any function $\varphi \in L^2\left[\Omega; C_{per}(Q^2; \mathbb{R}^3)\right]$,
holds

$$\lim_{\varepsilon \searrow 0} \int_\Omega u_\varepsilon(x) \cdot \varphi\left(x, \frac{x}{\varepsilon}, \frac{x}{\varepsilon^2}\right) dx = \int_\Omega \int_{Q^2} u_0(x, y, z) \cdot \varphi(x, y, z) \, dz \, dy \, dx.$$

The multiscale limit u_0 of a sequence $\{u_\varepsilon\}$ may be defined as the first mome_
of the multiscale Young measure associated with it, as follows.

Proposition 2 (see [14]). *Let $\{u_\varepsilon\}$ be a multiscale convergent sequence and u_0 $_$
its multiscale limit. If $\{\mu_{x,y,z}\}_{x\in\Omega,(y,z)\in Q^2}$ is the multiscale Young measure associate_
with $\{u_\varepsilon\}$, then $u_0 : \Omega \times Q^2 \to \mathbb{R}^3$ is the first moment of $\{\mu_{x,y,z}\}_{x\in\Omega,(y,z)\in Q^2}$ given _*

$$u_0(x, y, z) = \int_{\mathbb{R}^3} \rho \, d\mu_{x,y,z}(\rho).$$

Here, we are particularly interested in the multiscale convergence of sequence_
of divergence-free fields.

Proposition 3 (see [1]). *A function $u_0 \in L^2\left[\Omega; L^2_{per}(Q^2; \mathbb{R}^3)\right]$ is the multiscale lim_
of a sequence of divergence-free functions $\{u_\varepsilon\} \subset L^2(\Omega; \mathbb{R}^3)$ if and only if*

$$\operatorname{div}_z u_0(x, y, z) = 0, \quad \int_Q \operatorname{div}_y u_0(x, y, z) \, dz = 0, \quad \int_{Q^2} \operatorname{div}_x u_0(x, y, z) \, dz \, dy = 0.$$

We will focus on sequences of curls of divergence-free functions, that is, f_
any sequence of divergence-free functions $v_\varepsilon = \left(v_\varepsilon^1, v_\varepsilon^2, v_\varepsilon^3\right)$, we will consider th_
sequence $\{\operatorname{curl} v_\varepsilon\}$ where

$$\operatorname{curl} v_\varepsilon = \left(\frac{\partial v_\varepsilon^3}{\partial x_2} - \frac{\partial v_\varepsilon^2}{\partial x_3}, \frac{\partial v_\varepsilon^1}{\partial x_3} - \frac{\partial v_\varepsilon^3}{\partial x_1}, \frac{\partial v_\varepsilon^2}{\partial x_1} - \frac{\partial v_\varepsilon^1}{\partial x_2}\right).$$

Proposition 4. *Let $\{v_\varepsilon\}$ be a bounded sequence in $H^1(\Omega; \mathbb{R}^3)$ such that $\operatorname{div} v_\varepsilon =$
in Ω and $v_\varepsilon \cdot \mathbf{n} = 0$ on $\partial\Omega$, for every $\varepsilon > 0$. If $\{v_\varepsilon\}$ converges weakly to v _
$H^1(\Omega; \mathbb{R}^3)$, then:*

(i) $\{v_\varepsilon\}$ multiscale converges to v.
*(ii) There exist functions $v_1 : \Omega \times Q \to \mathbb{R}^3$ and $v_2 : \Omega \times Q^2 \to \mathbb{R}^3$, wi_*
$\int_Q \operatorname{curl}_y v_1(x, y) \, dy = 0$ *and* $\int_Q \operatorname{curl}_z v_2(x, y, z) \, dz = 0$, *such that $\{\operatorname{curl} v_\varepsilon\}$ mu_*
tiscale converges to $\operatorname{curl} v + \operatorname{curl}_y v_1 + \operatorname{curl}_z v_2$.

Main Result

In this section, we present and prove our main result on the characterization of the Γ-limit density of sequences of periodic quadratic integral functionals depending on the curl of divergence-free fields.

Theorem 1. *Let $A(x,y) = (a_{ij}(x,y))$ be a symmetric 3×3-matrix-valued function such that $a_{ij} \in L^{\infty}(\Omega \times Q)$ are Q-periodic in the second variable, and there exist constants $\beta > \alpha > 0$ for which $\alpha|\rho|^2 \leq A(x,y)\rho \cdot \rho \leq \beta|\rho|^2$, for every $\rho \in \mathbb{R}^3$. Let $b \in L^2(\Omega \times Q^2; \mathbb{R}^3)$ be Q-periodic in the second and third variables. Then, the sequence of energies E_{ε} defined in $X(\Omega)$ by*

$$E_{\varepsilon(u)} = \int_{\Omega} \left(\frac{A\left(x, \frac{x}{\varepsilon}\right)}{2} \operatorname{curl} u(x) \cdot \operatorname{curl} u(x) - b\left(x, \frac{x}{\varepsilon}, \frac{x}{\varepsilon^2}\right) \cdot \operatorname{curl} u(x) \right) dx$$

Γ-converges to the functional E defined in $X(\Omega)$ by

$$E(u) = \int_{\Omega} \left(\frac{A_0(x)}{2} \operatorname{curl} u(x) \cdot \operatorname{curl} u(x) - b_0(x) \cdot \operatorname{curl} u(x) + c(x) \right) dx.$$

The homogenized coefficient $A_0 : \Omega \to \mathbb{R}^{3 \times 3}$ is given by

$$A_0(x) = \int_Q A(x,y)\left(I_3 + \operatorname{curl}_y V(x,y)\right) dy,$$

here I_3 stands for the 3×3-identity matrix, and $\operatorname{curl}_y V(x,y) = (\operatorname{curl}_y v_i(x,y))_{1 \leq i \leq 3}$ is also a 3×3-matrix with v_i solution of

$$\begin{cases} \operatorname{curl}_y \left(A(x,y)\left(e_i + \operatorname{curl}_y v_i(x,y)\right) \right) = 0 & \text{in } Q \\ v_i(x, \cdot) \in H_0^1(Q; \mathbb{R}^3). \end{cases}$$

The effective linear coefficient $b_0 : \Omega \to \mathbb{R}^3$ is defined by

$$b_0(x) = \int_{Q^2} (A(x,y)\operatorname{curl}_y w(x,y) - b(x,y,z)) \cdot (I_3 + \operatorname{curl}_y V(x,y)) \, dz \, dy,$$

here w is the solution of the unit cell problem

$$\begin{cases} \operatorname{curl}_y (A(x,y)\operatorname{curl}_y w(x,y)) = \operatorname{curl}_y \left(\int_Q b(x,y,z)dz \right) & \text{in } Q \\ w(x, \cdot) \in H_0^1(Q; \mathbb{R}^3). \end{cases}$$

The independent term $c : \Omega \to \mathbb{R}$ is

$$c(x) = \int_{Q^2} \left(\frac{A(x,y)}{2} \left(\operatorname{curl}_y w(x,y) + \operatorname{curl}_z \tilde{w}(x,y,z) \right) - b(x,y,z) \right) \cdot$$

$$\left(\operatorname{curl}_y w(x,y) + \operatorname{curl}_z \tilde{w}(x,y,z) \right) \, dz \, dy,$$

where $\tilde{w}$ is the solution of the unit cell problem

$$\begin{cases} \operatorname{curl}_z \left(A(x,y) \operatorname{curl}_z \tilde{w}(x,y,z) \right) = \operatorname{curl}_z b(x,y,z) & \text{in } Q \\ \tilde{w}(x,y,\cdot) \in H_0^1(Q; \mathbb{R}^3). \end{cases}$$

Proof. In order to prove the Γ-convergence result, we will proceed in two step according to the definition of Γ-convergence.

First step: Let u be in $X(\Omega)$, and let $\{u_\varepsilon\}$ be any weak convergent sequence to in $X(\Omega)$. Since the sequence $\{\operatorname{curl} u_\varepsilon\}$ converges weakly to $\operatorname{curl} u$ in $L^2(\Omega; \mathbb{R}^3)$, w know there exists a multiscale Young measure $\mu = \{\mu_{x,y,z}\}_{x \in \Omega, (y,z) \in Q^2}$ associate with it. Thus, if we apply Proposition 1, it follows that

$$\liminf_{\varepsilon \searrow 0} \int_\Omega \left(\frac{A\left(x, \frac{x}{\varepsilon}\right)}{2} \operatorname{curl} u_\varepsilon(x) \cdot \operatorname{curl} u_\varepsilon(x) - b\left(x, \frac{x}{\varepsilon}, \frac{x}{\varepsilon^2}\right) \cdot \operatorname{curl} u_\varepsilon(x) \right) dx \geq$$

$$\geq \int_\Omega \int_Q \int_Q \int_{\mathbb{R}^3} \left(\frac{A(x,y)}{2} \rho \cdot \rho - b(x,y,z) \cdot \rho \right) d\mu_{x,y,z}(\rho) \, dz \, dy \, dx.$$

Now, we may apply the Jensen inequality so that

$$\int_{\mathbb{R}^3} \left(\frac{A(x,y)}{2} \rho \cdot \rho - b(x,y,z) \cdot \rho \right) d\mu_{x,y,z}(\rho) \geq$$

$$\geq \frac{A(x,y)}{2} \varphi(x,y,z) \cdot \varphi(x,y,z) - b(x,y,z) \cdot \varphi(x,y,z),$$

where we have defined the function $\varphi : \Omega \times Q^2 \to \mathbb{R}^3$ as the barycenter of μ, i.e.,

$$\varphi(x,y,z) = \int_{\mathbb{R}^3} \rho \, d\mu_{x,y,z}(\rho).$$

Since the sequence $\{\operatorname{curl} u_\varepsilon\}$ multiscale converges to φ in $L^2(\Omega \times Q^2; \mathbb{R}^3)$, it turn out that the multiscale limit φ may be written as

$$\varphi(x,y,z) = \operatorname{curl} u(x) + \operatorname{curl}_y u_1(x,y) + \operatorname{curl}_z u_2(x,y,z)$$

for some functions $u_1 : \Omega \times Q \to \mathbb{R}^3$ and $u_2 : \Omega \times Q^2 \to \mathbb{R}^3$. In this way, we have

$$\liminf_{\varepsilon\searrow 0}\int_\Omega\left(\frac{A\left(x,\frac{x}{\varepsilon}\right)}{2}\operatorname{curl}u_\varepsilon(x)\cdot\operatorname{curl}u_\varepsilon(x)-b\left(x,\frac{x}{\varepsilon},\frac{x}{\varepsilon^2}\right)\cdot\operatorname{curl}u_\varepsilon(x)\right)dx\geq$$

$$\geq\int_\Omega\int_{Q^2}\Bigg(\frac{A(x,y)}{2}\left(\operatorname{curl}u(x)+\operatorname{curl}_y u_1(x,y)+\operatorname{curl}_z u_2(x,y,z)\right)\cdot\left(\operatorname{curl}u(x)+\operatorname{curl}_y u_1(x,y)\right.$$

$$\left.+\operatorname{curl}_z u_2(x,y,z)\right)-b(x,y,z)\cdot\left(\operatorname{curl}u(x)+\operatorname{curl}_y u_1(x,y)+\operatorname{curl}_z u_2(x,y,z)\right)\Bigg)dz\,dy\,dx.$$

we take the infimum overall functions u_1 and u_2 in the last inequality, then

$$\liminf_{\varepsilon\searrow 0}\int_\Omega\left(\frac{A\left(x,\frac{x}{\varepsilon}\right)}{2}\operatorname{curl}u_\varepsilon(x)\cdot\operatorname{curl}u_\varepsilon(x)-b\left(x,\frac{x}{\varepsilon},\frac{x}{\varepsilon^2}\right)\cdot\operatorname{curl}u_\varepsilon(x)\right)dx\geq$$

$$\geq\int_\Omega\phi(x,\operatorname{curl}u(x))\,dx,$$

here the function $\phi:\Omega\times\mathbb{R}^3\to\mathbb{R}$ is given by

$$\phi(x,\rho)=$$

$\inf\Bigg\{\int_{Q^2}\Bigg(\frac{A(x,y)}{2}\big(\rho+\operatorname{curl}_y u_1(x,y)+\operatorname{curl}_z u_2(x,y,z)\big)\cdot\big(\rho+\operatorname{curl}_y u_1(x,y)+\operatorname{curl}_z u_2(x,y,z)\big)$

$$-b(x,y,z)\cdot\big(\rho+\operatorname{curl}_y u_1(x,y)+\operatorname{curl}_z u_2(x,y,z)\big)\Bigg)dz\,dy\ :$$

$$u_1\in L^2\left[\Omega;H_0^1(Q;\mathbb{R}^3)\right],\,u_2\in L^2\left[\Omega\times Q;H_0^1(Q;\mathbb{R}^3)\right]\Bigg\}.$$

Notice that, for fixed $x\in\Omega$ and $\rho\in\mathbb{R}^3$ and fixed u_2, the infimum u_1 is a solution of the unit cell problem

$$\begin{cases}\operatorname{curl}_y\left(A(x,y)\left(\rho+\operatorname{curl}_y u_1(x,y)\right)\right)=\operatorname{curl}_y\left(\int_Q b(x,y,z)dz\right) & \text{in } Q\\ u_1(x,\cdot)\in H_0^1(Q;\mathbb{R}^3).\end{cases}$$

we rewrite the infimum as $u_1(x,y)=\sum_{i=1}^3 v_i(x,y)\rho_i+w(x,y)$, where v_i is the solution of

$$\begin{cases}\operatorname{curl}_y\left(A(x,y)\left(e_i+\operatorname{curl}_y v_i(x,y)\right)\right)=0 & \text{in } Q\\ v_i(x,\cdot)\in H_0^1(Q;\mathbb{R}^3),\end{cases}$$

or every $i=1,2,3$, and w is the solution of

$$\begin{cases}\operatorname{curl}_y\left(A(x,y)\operatorname{curl}_y w(x,y)\right)=\operatorname{curl}_y\left(\int_Q b(x,y,z)dz\right) & \text{in } Q\\ w(x,\cdot)\in H_0^1(Q;\mathbb{R}^3),\end{cases}$$

then we may rewrite the density ϕ as

$$\phi(x,\rho) = \int_{Q^2} \left(\frac{A(x,y)}{2} \left(\rho + \sum_{i=1}^{3} \mathrm{curl}_y\, v_i(x,y)\rho_i + \mathrm{curl}_y\, w(x,y) + \mathrm{curl}_z\, u_2(x,y,z) \right) \right.$$

$$\left(\rho + \sum_{i=1}^{3} \mathrm{curl}_y\, v_i(x,y)\rho_i + \mathrm{curl}_y\, w(x,y) + \mathrm{curl}_z\, u_2(x,y,z) \right)$$

$$\left. - b(x,y,z) \cdot \left(\rho + \sum_{i=1}^{3} \mathrm{curl}_y\, v_i(x,y)\rho_i + \mathrm{curl}_y\, w(x,y) + \mathrm{curl}_z\, u_2(x,y,z) \right) \right) dz\, dy$$

where $u_2(x,y,\cdot)$ is the solution of problem

$$\begin{cases} \mathrm{curl}_z\, (A(x,y)\mathrm{curl}_z\, u_2(x,y,z)) = \mathrm{curl}_z\, b(x,y,z) & \text{in } Q \\ u_2(x,y,\cdot) \in H_0^1(Q;\mathbb{R}^3). \end{cases}$$

After some calculations, we conclude that the function ϕ may be written as th quadratic form

$$\phi(x,\rho) = \frac{A_0(x)}{2}\rho \cdot \rho - b_0(x) \cdot \rho + c(x)$$

where A_0, b_0 and c were defined previously. Therefore, we have achieved the lowe bound estimate: $\liminf_{\varepsilon \searrow 0} E_\varepsilon(u_\varepsilon) \geq E(u)$.

Second step: For each $u \in X(\Omega)$, let us take the minimizers u_1 and u_2 so that, fo a.e. $x \in \Omega$,

$$\phi(x,\mathrm{curl}\,u(x)) = \int_{Q^2} \left(\frac{A(x,y)}{2}\, (\mathrm{curl}\,u(x) + \mathrm{curl}_y\, u_1(x,y) + \mathrm{curl}_z\, u_2(x,y,z)) \cdot \right.$$

$$(\mathrm{curl}\,u(x) + \mathrm{curl}_y\, u_1(x,y) + \mathrm{curl}_z\, u_2(x,y,z))$$

$$\left. - b(x,y,z) \cdot (\mathrm{curl}\,u(x) + \mathrm{curl}_y\, u_1(x,y) + \mathrm{curl}_z\, u_2(x,y,z)) \right)\, dz\, dy.$$

We may define the sequence of functions $u_\varepsilon : \Omega \to \mathbb{R}^3$ by putting $u_\varepsilon(x) = u(x)$ $\varepsilon u_1\left(x,\frac{x}{\varepsilon}\right) + \varepsilon^2 u_2\left(x,\frac{x}{\varepsilon},\frac{x}{\varepsilon^2}\right)$ so that the sequence $\{\mathrm{curl}\,u_\varepsilon\}$ converges weakly to curl in $L^2(\Omega;\mathbb{R}^3)$, as well as $\{\mathrm{div}\,u_\varepsilon\}$ converges weakly to $\mathrm{div}\,u$ in $L^2(\Omega)$. Moreove $\{\mathrm{curl}\,u_\varepsilon\}$ multiscale converges to $\mathrm{curl}\,u + \mathrm{curl}_y\,u_1 + \mathrm{curl}_z\,u_2$ in $L^2(\Omega \times Q^2;\mathbb{R}^3)$. I this way, we know there exists a sequence $\{v_\varepsilon\}$ in $H^1(\Omega;\mathbb{R}^3)$ such that $\mathrm{div}\,v_\varepsilon =$ in Ω, and the norm $\|v_\varepsilon - u_\varepsilon\|_{H^1(\Omega;\mathbb{R}^3)}$ goes to 0 as ε vanishes. Taking into accou this sequence, we get the equality

$$\lim_{\varepsilon \to 0} E_\varepsilon(v_\varepsilon) = \int_\Omega \phi(x,\mathrm{curl}\,u(x))\, dx. \qquad \square$$

Acknowledgements This work was supported by Project MTM201019739 (Ministerio de Cienc e Innovación, Spain), and Grant TC20101856 (Universidad de Castilla-La Mancha, Spain).

References

1. Allaire, G., Briane, M.: Multiscale convergence and reiterated homogenization. Proc. Royal Soc. Edinb. **126A**, 297–342 (1996)
2. Ambrosio, L., Frid, H.: Multiscale Young measures in almost periodic homogenization and applications. Arch. Ration. Mech. Anal. **192**, 37–85 (2009)
3. Ansini, N., Garroni, A.: Γ-convergence of functionals on divergence-free fields. ESAIM Control Optim. Calc. Var. **13**, 809–828 (2007)
4. Auchmuty, G., Alexander, J.C.: L^2-well-posedness of 3d div-cul boundary value problems. Quart. Appl. Math **63**, 479–508 (2005)
5. Braides, A.: Γ-Convergence for Beginners. Oxford University Press, Oxford (2002)
6. Dal Maso, G.: An introduction to Γ-convergence, Birkhäuser (1993)
7. De Giorgi, E., Franzoni, T.: Su un tipo di convergenza degli integrali dell'energia per operatori ellittici del secondo ordine. Boll. Un. Mat. Ital. **58**, 842–850 (1975)
8. Fonseca, I., Krömer, S.: Multiple integrals under differential constraints: Two-scale convergence and homogenization. Indiana Univ. Math. J. **59**, 427–458 (2010)
9. Girault, V., Raviart, P.-A.: Finite Element Methods for the Navier–Stokes Equations: Theory and Algorithms. Springer, Berlin (1986)
10. Kong, J.A.: Electromagnetic Wave Theory. Wiley, New York (1986)
11. Nguetseng, G.: A general convergence result for a functional related to the theory of homogenization. SIAM J. Math. Anal. **20**, 608–623 (1989)
12. Pedregal, P.: Parametrized Measures and Variational Principles. Birkhäuser, Boston (1997)
13. Pedregal, P.: Γ-convergence through Young measures. SIAM J. Math. Anal. **36**, 423–440 (2004)
14. Pedregal, P.: Multi-scale Young measures. Trans. Am. Math. Soc. **358**, 591–602 (2005)
15. Pedregal, P.: Div-curl Young measures and optimal design in any dimension. Rev. Mat. Complut. **20**, 239–255 (2007)
16. Pedregal, P., Serrano, H.: Γ-convergence of quadratic functionals with oscillating linear perturbations. Nonlinear Anal. **70**, 4178–4189 (2009)
17. Serrano, H.: On Γ-convergence in divergence-free fields through Young measures. J. Math. Anal. Appl. **359**, 311–321 (2009)

A Geometric Proof of the Existence of Two Solutions of the Bahri–Coron Problem

Norimichi Hirano and Naoki Shioji

Abstract We give a proof of Clapp and Weth's existence theorem of two pairs of nontrivial solutions of the Bahri–Coron problem by using the degree theory.

1 Introduction

We consider the Bahri-Coron problem

$$\begin{cases} -\Delta u = |u|^{2^*-2}u & \text{in } \Omega, \\ u = 0 & \text{in } \partial\Omega, \end{cases} \tag{1}$$

where Ω is a bounded domain in $\mathbb{R}^N$ ($N \geq 3$) whose boundary is smooth and $2^* = 2N/(N-2)$. By Pohožaev's identity [15], we know that if Ω is star shaped, then problem (1) does not have a nontrivial solution. In the case that Ω has a nontrivial topology, Coron [4] and Bahri and Coron [1,2] showed the existence of a positive solution. Recently, Clapp and Weth [3] showed that under the same assumptions of the theorem in [4], problem (1) has at least two pairs of nontrivial solutions.

Theorem 1 (Clapp–Weth). *Let $N \in \mathbb{N}$ with $N \geq 3$, let $R_2 > R_1 > 0$, and let $\Omega \subset \mathbb{R}^N$ be a bounded domain such that $\partial\Omega$ is smooth,*

$$\Omega \supset \{x \in \mathbb{R}^N : R_1 < |x| < R_2\} \quad \text{and} \quad \{x \in \mathbb{R}^N : |x| < R_1\} \setminus \overline{\Omega} \neq \emptyset.$$

N. Hirano
Department of Mathematics, Graduate School of Environment and information Sciences, Yokohama National University, 79-7, Tokiwadai, Hodogaya-ku, Yokohama 240-8501, Japan
e-mail: hira0918@ynu.ac.jp

N. Shioji (✉)
Department of Mathematics, Faculty of Engineering, Yokohama National University, 79-5, Tokiwadai, Hodogaya-ku, Yokohama 240-8501, Japan
e-mail: shioji@ynu.ac.jp

Pinelas et al. (eds.), *Differential and Difference Equations with Applications*, Springer Proceedings in Mathematics & Statistics 47, DOI 10.1007/978-1-4614-7333-6_54, © Springer Science+Business Media New York 2013

If R_2/R_1 is large enough, then problem (1) *has at least two pairs of nontrivi solutions.*

In order to prove their result, they used a topological tool "fixed point transfer given in [7]. We note that Ge, Musso, and Pistoia [8, Theorem 1.2] showed th. problem (1) has some pairs of sign-changing solutions if Ω has two sufficient small holes; see also [8, Theorem 1.1]. Quite recently, the authors [9] studied th existence of two pairs of nontrivial solutions in the case that Ω is a contractib domain. In such a case, the existence of a positive solution was studied b Dancer [5], Ding [6], and Passaseo [14].

Theorem 2 (Hirano–Shioji). *Let $N \in \mathbb{N}$ with $N \geq 3$, let $R_2 > R_1 > 0$, and let $\tilde{\Omega}$ $\mathbb{R}^N$ be a bounded domain such that $\partial\tilde{\Omega}$ is smooth,*

$$\tilde{\Omega} \supset \{x \in \mathbb{R}^N : R_1 < |x| < R_2\} \quad \text{and} \quad 0 \notin \overline{\tilde{\Omega}}.$$

If $\eta_0 > 0$ is small enough and Ω is a domain such that $\partial\Omega$ is smooth and

$$\tilde{\Omega}_{\eta_0} \subset \Omega \quad \text{and} \quad \overline{\Omega} \subset \tilde{\Omega}_0,$$

then problem (1) *has at least two pairs of nontrivial solutions, where*

$$\tilde{\Omega}_\eta = \tilde{\Omega} \setminus \{x = (x', x_N) \in \mathbb{R}^N : x_N \geq 0, |x'| \leq \eta\} \quad \text{for } \eta \in [0, \infty).$$

They used the degree theory [10–13] instead of the fixed point transfer. In th article, we give a proof of Theorem 1 by the method employed in [9].

2 Proof of Theorem 1

For each nonempty open subset $G \subset \mathbb{R}^N$, we consider that $\mathcal{D}_0^{1,2}(G)$ is the completic of $C_0^\infty(G)$ with respect to the inner product $\int_G \nabla u(x) \nabla v(x)\,dx$ for $u, v \in C_0^\infty(G)$. W denote by (u, v) and $\|u\|$ the inner product of $u, v \in \mathcal{D}_0^{1,2}(\mathbb{R}^N)$ and the norm of u $\mathcal{D}_0^{1,2}(\mathbb{R}^N)$. We can consider that $\mathcal{D}_0^{1,2}(G) \subset \mathcal{D}_0^{1,2}(\mathbb{R}^N)$ by the zero extension. W define a functional $I_{\mathbb{R}^N} : \mathcal{D}_0^{1,2}(\mathbb{R}^N) \to \mathbb{R}$ by

$$I_{\mathbb{R}^N}(u) = \int_{\mathbb{R}^N} \left(\frac{1}{2}|\nabla u(x)|^2 - \frac{1}{2^*}|u(x)|^{2^*}\right) dx, \quad u \in \mathcal{D}_0^{1,2}(\mathbb{R}^N),$$

and we set its Nehari manifold as follows:

$$\mathcal{N}_{\mathbb{R}^N} = \left\{u \in \mathcal{D}_0^{1,2}(\mathbb{R}^N) \setminus \{0\} : \int_{\mathbb{R}^N} |\nabla u(x)|^2 dx = \int_{\mathbb{R}^N} |u(x)|^{2^*} dx\right\}.$$

We set $c_\infty = \inf\{I_{\mathbb{R}^N}(u) : u \in \mathcal{N}_{\mathbb{R}^N}\}$. For each $(\varepsilon, z) \in (0, \infty) \times \mathbb{R}^N$, we set

$$U_{\varepsilon,z}(x) = (N(N-2))^{\frac{N-2}{4}} \left(\frac{\varepsilon}{\varepsilon^2 + |x - z|^2} \right)^{\frac{N-2}{2}} \quad \text{for } x \in \mathbb{R}^N.$$

The following is obtained by Weth [16].

roposition 1 (Weth). *There exists $\varepsilon_1 \in (0, c_\infty)$ such that $I_{\mathbb{R}^N}$ does not have a ritical point $u \in \mathcal{D}_0^{1,2}(\mathbb{R}^N)$ such that $I_{\mathbb{R}^N}(u) \in (c_\infty, 2c_\infty + \varepsilon_1]$.*

Throughout this article, we consider $u^-(x) = \min\{u(x), 0\}$. We define

$$\hat{\mathcal{N}}_{\mathbb{R}^N} = \left\{ u \in \mathcal{N}_{\mathbb{R}^N} : u^+ \neq 0,\ u^- \neq 0 \right\},$$
$$\mathcal{N}_{\mathbb{R}^N,*} = \left\{ u \in \mathcal{N}_{\mathbb{R}^N} : u^+, u^- \in \mathcal{N}_{\mathbb{R}^N} \right\}.$$

We set

$$\tau(u) = \left(\frac{\int_{\tilde{\Omega}} |\nabla u(x)|^2 dx}{\int_{\tilde{\Omega}} |u(x)|^{2^*} dx} \right)^{\frac{1}{2^*-2}} \quad \text{for each } u \in \mathcal{D}_0^{1,2}(\mathbb{R}^N) \setminus \{0\},$$

$$\mathcal{T}(u) = \tau(u)u \,(\in \mathcal{N}_{\mathbb{R}^N}) \quad \text{for each } u \in \mathcal{D}_0^{1,2}(\mathbb{R}^N) \setminus \{0\},$$

$$\mathcal{T}_*(u) = \mathcal{T}(u^+) + \mathcal{T}(u^-) \,(\in \mathcal{N}_{\mathbb{R}^N,*}) \quad \text{for each } u \in \hat{\mathcal{N}}_{\mathbb{R}^N},$$

$$\alpha(t,v) = \mathcal{T}((1-t)v^+ + tv^-) \quad \text{for each } (t,v) \in [0,1] \times \mathcal{N}_{\mathbb{R}^N,*}.$$

We define $\mu \in C(\hat{\mathcal{N}}_{\mathbb{R}^N}, (0,1))$ by $\alpha(\mu(u), \mathcal{T}_*(u)) = u$ for each $u \in \hat{\mathcal{N}}_{\mathbb{R}^N}$. We note

$$\mu(\alpha(t,v)) = t \quad \text{for each } (t,v) \in (0,1) \times \mathcal{N}_{\mathbb{R}^N,*}. \tag{2}$$

For the proofs of Lemmas 1–3, see [9, Lemmas 2, 4, and 5]. Although the orresponding lemmas in [9] were studied on a bounded set $\tilde{\Omega}(\subset \mathbb{R}^N)$, the proofs ork similarly.

emma 1. *For each $\varepsilon > 0$, there exists $\delta > 0$ such that $\|u - \mathcal{T}_*(u)\| < \varepsilon$ for each $\in \hat{\mathcal{N}}_{\mathbb{R}^N}$ satisfying $I_{\mathbb{R}^N}(u) \leq 3c_\infty$ and $|\mu(u) - 1/2| < \delta$.*

emma 2. *There exist $\varepsilon_2 \in (0, \varepsilon_1)$ and $C_1 > 0$ such that*

$$\inf_{w \in \mathcal{N}_{\mathbb{R}^N,*}} \|u - w\| \geq 2C_1$$

r each $u \in \hat{\mathcal{N}}_{\mathbb{R}^N}$ satisfying $|I_{\mathbb{R}^N}(u) - 2c_\infty| \leq \varepsilon_2$ and $|\mu(u) - 1/2| = 1/4$.

Since they considered $I_{\mathbb{R}^N}$ instead of $I_{\hat{\Omega}}$ in [9, Lemma 4], for the lemma abov
the concentration compactness argument in [9, Lemma 4] should be modified :
follows: there exist $\{\lambda_n^1\}, \{\lambda_n^2\} \subset (0, \infty), \{z_n^1\}, \{z_n^2\} \subset \mathbb{R}^N$ such that $\{\lambda_n^i\}$ converge
to an element of $[0, \infty]$ for $i = 1, 2$, $\{z_n^i\}$ converges to an element of $\mathbb{R}^N$ or $|z_n^i| \to$
for $i = 1, 2$,

$$\max\left\{\frac{\lambda_n^1}{\lambda_n^2}, \frac{\lambda_n^2}{\lambda_n^1}, \frac{|z_n^1 - z_n^2|}{\sqrt{\lambda_n^1 \lambda_n^2}}\right\} \to \infty,$$

$$\left\|v_n^+ - U_{\lambda_n^1, z_n^1}\right\| \to 0 \quad \text{and} \quad \left\|v_n^- + U_{\lambda_n^2, z_n^2}\right\| \to 0.$$

With this modification, the proof of [9, Lemma 4] works similarly.

By Lemma 1 and the previous lemma, we can choose $\delta_1 \in (0, 1/4)$ such th
$\|u - \mathcal{T}_*(u)\| \leq C_1$ for each $u \in \hat{\mathcal{N}}_{\mathbb{R}^N}$ with $I_{\mathbb{R}^N}(u) \leq 3c_\infty$ and $|\mu(u) - 1/2| \leq \delta_1$.

Lemma 3. *There exist $\varepsilon_0 \in (0, \varepsilon_2)$ and $\delta_0 \in (0, \delta_1)$ such that*

$$I_{\mathbb{R}^N}(\alpha(t, \mathcal{T}_*(u))) \leq 2c_\infty$$

for each $t \in [0, 1]$ with $|t - 1/2| \geq \delta_1$ and $u \in \hat{\mathcal{N}}_{\mathbb{R}^N}$ satisfying $I_{\mathbb{R}^N}(u) \leq 2c_\infty + \varepsilon_0$ a
$|\mu(u) - 1/2| \leq \delta_0$.

We set $\delta_2 = 1/4$, $\underline{\mu}_i = 1/2 - \delta_i$ and $\overline{\mu}_i = 1/2 + \delta_i$ for $i = 0, 1, 2$, and

$$C_2 = \sup\left\{2\left(\|u\| + (Nc_\infty)^{-\frac{2}{N-2}}\|u\|^{2^*-1}\right) : u \in \hat{\mathcal{N}}_{\mathbb{R}^N},\right.$$

$$\left. I_{\mathbb{R}^N}(u) \leq 3c_\infty, \left|\mu(u) - \frac{1}{2}\right| \leq \delta_2\right\},$$

$$C_3 = \inf\left\{\frac{(2^* - 2)^2}{\left(2 + 2^*(Nc_\infty)^{-\frac{2}{N-2}}\|u\|^{2^*-2}\right)^2} \frac{\left|\frac{\partial}{\partial t}I_{\mathbb{R}^N}(\alpha(t, \mathcal{T}_*(u)))\big|_{t=\mu(u)}\right|^2}{\left\|\frac{\partial}{\partial t}\alpha(t, \mathcal{T}_*(u))\big|_{t=\mu(u)}\right\|^2} : \right.$$

$$\left. u \in \hat{\mathcal{N}}_{\mathbb{R}^N}, I_{\mathbb{R}^N}(u) \leq 3c_\infty, \delta_0 \leq \left|\mu(u) - \frac{1}{2}\right| \leq \delta_2\right\}.$$

By preparing similar lemmas as [9, Lemmas 6–8], we can find that C_2 and C_3 a
positive real numbers. Without loss of generality, we may assume

$$2\varepsilon_0 \leq \frac{C_1 C_3}{C_2}.$$

Let R_2, R_1, and Ω satisfy the assumptions of Theorem 1 and let R_2/R_1 be ufficiently large. We fix $\mathfrak{n} \in \mathbb{R}^N$ such that $|\mathfrak{n}| < R_1$ and $\mathfrak{n} \notin \overline{\Omega}$. By [3, Lemma 5] nd its proof, we can find two disjoint $\varepsilon_0/3$-punctured subdomains A_{R_1,R_0} and A_{R_0,R_2} f Ω, where $R_0 = \sqrt{R_1 R_2}$ and $A_{r,R} = \{x \in \mathbb{R}^N : r < |x| < R\}$. For the definition of ie ε-punctured subdomain of Ω, see [3, Definition 1]. From the definition, there re closed balls $B_1, B_2 \subset \mathbb{R}^N$ associated with A_{R_1,R_0}, A_{R_0,R_2}, respectively, and by the roof of [3, Lemma 3], we may assume that $\{x \in \mathbb{R}^N : |x| \leq R_1\} \subset B_1 \subset B_2$.

We set $I(u) = I_{\mathbb{R}^N}(u)$ in $\mathcal{D}_0^{1,2}(\Omega)$, $\mathcal{N} = \mathcal{N}_{\mathbb{R}^N} \cap \mathcal{D}_0^{1,2}(\Omega)$, $\hat{\mathcal{N}} = \hat{\mathcal{N}}_{\mathbb{R}^N} \cap \mathcal{D}_0^{1,2}(\Omega)$, nd $\mathcal{N}_* = \mathcal{N}_{\mathbb{R}^N,*} \cap \mathcal{D}_0^{1,2}(\Omega)$. We define

$$\beta(u) = \frac{\int_\Omega x |u(x)|^{2^*} dx}{\int_\Omega |u(x)|^{2^*} dx} \quad \text{for } u \in \mathcal{D}_0^{1,2}(\Omega) \setminus \{0\}.$$

Je set

$$\Gamma_1 = \{\gamma \in C(B_1, \mathcal{N}) : \beta(\gamma(x)) \neq \mathfrak{n} \ \forall x \in \partial B_1, \ \deg(\beta \circ \gamma; \operatorname{Int} B_1, \mathfrak{n}) \neq 0\},$$

$$c_1 = \inf_{\gamma \in \Gamma_1} \max_{x \in B_1} I(\gamma(x)) \quad \text{and} \quad \overline{c} = \inf\{I(u) : u \in \mathcal{N}, \ \beta(u) = \mathfrak{n}\},$$

'here $\deg(\beta \circ \gamma; \operatorname{Int} B_1, \mathfrak{n})$ stands for the Brouwer degree of $\beta \circ \gamma : B_1 \to \mathbb{R}^N$. For ie definition of this Brouwer degree, see [11, 12]. We note that the method in [11]) define the Brouwer degree with the Tietze extension theorem will be used later. Je can infer that

$$c_\infty < \overline{c} \leq c_1 \leq c_\infty + \frac{\varepsilon_0}{3}.$$

If problem (1) has two pairs of nontrivial solutions, there is nothing to prove. So 'e assume that it has at most one pair of nontrivial solutions. We can show that has a positive solution u_0 of (1) with $I(u_0) = c_1$; see [3, 9]. It yields that $\pm u_0$ re the only pair of nontrivial solutions of (1). We give the proof of Theorem 1 by ontradiction.

We set

$$\beta_+(u) = \beta(u^+) \quad \text{and} \quad \beta_-(u) = \beta(u^-) \quad \text{for } u \in \hat{\mathcal{N}}.$$

'e choose c_0 such that $c_\infty < c_0 < \overline{c}$. By [3, Lemma 4], we can find $g \in C(B_1 \times$ $_2; \mathcal{N}_*)$ which satisfies

·1) $I(g(x,y)) \leq 2c_\infty + 2\varepsilon_0/3$ for each $(x,y) \in B_1 \times B_2$.

·2) $I(g(x,y)) \leq c_1 + c_0$ for each $(x,y) \in \partial(B_1 \times B_2)$.

·3) $I(g(x,y)^+) \leq c_0$ and $\beta_+(g(x,y)) = x$ for each $(x,y) \in \partial B_1 \times B_2$.

·4) $I(g(x,y)^-) \leq c_0$ and $\beta_-(g(x,y)) = y$ for each $(x,y) \in B_1 \times \partial B_2$.

'e set $\varphi \in C\left(B_1 \times B_2 \times \left[\underline{\mu_2}, \overline{\mu_2}\right], \hat{\mathcal{N}}\right)$ by

$$\varphi(x,y,t) = \alpha(t, g(x,y)) \quad \text{for } (x,y,t) \in B_1 \times B_2 \times \left[\underline{\mu_2}, \overline{\mu_2}\right].$$

We define $\tilde{\sigma} \in C([0,\infty) \times \varphi(B_1 \times B_2 \times \left[\underline{\mu_2},\overline{\mu_2}\right]),\mathcal{N})$ by

$$\tilde{\sigma}(0,u) = u, \quad \frac{\partial}{\partial s}\tilde{\sigma}(s,u) = -\nabla_{\mathcal{N}} I(\tilde{\sigma}(s,u))$$

for $(s,u) \in [0,\infty) \times \varphi(B_1 \times B_2 \times \left[\underline{\mu_2},\overline{\mu_2}\right])$. We also define $\sigma : [0,\infty) \times \varphi(B_1 \times B_2 \times \left[\underline{\mu_2},\overline{\mu_2}\right]) \to \mathcal{N}$ by

$$\sigma(s,u) = \begin{cases} \tilde{\sigma}(s,u) & \text{if } I(\tilde{\sigma}(s,u)) > c_1 + c_0, \\ \tilde{\sigma}(\min\{t \geq 0 : I(\tilde{\sigma}(t,u)) \leq c_1 + c_0\},u) & \text{if } I(\tilde{\sigma}(s,u)) \leq c_1 + c_0 \end{cases}$$

for $(s,u) \in [0,\infty) \times \varphi(B_1 \times B_2 \times \left[\underline{\mu_2},\overline{\mu_2}\right])$.

For the proof of the following lemma, see [9, Lemma 12].

Lemma 4. *There hold the following:*

(σ1) $\sigma \in C([0,\infty) \times \varphi(B_1 \times B_2 \times \left[\underline{\mu_2},\overline{\mu_2}\right]),\mathcal{N})$.

(σ2) $I(\sigma(s,u)) \leq I(\sigma(s',u))$ *for all* $s,s' \in [0,\infty)$ *and* $u \in \varphi(B_1 \times B_2 \times \left[\underline{\mu_2},\overline{\mu_2}\right])$ *with* $s \geq s' \geq 0$.

(σ3) *There exists* $s_0 > 0$ *such that* $I(\sigma(s_0,u)) \leq c_1 + c_0$ *for all* $u \in \varphi(B_1 \times B_2 \times \left[\underline{\mu_2},\overline{\mu_2}\right])$.

(σ4) $\sigma(s,u) = u$ *for all* $s \in [0,\infty)$ *and* $u \in \varphi(B_1 \times B_2 \times \left[\underline{\mu_2},\overline{\mu_2}\right])$ *with* $I(u) \leq c_1 + c_0$.

(σ5) $\sigma(s,u) \in \hat{\mathcal{N}}$ *and* $\mu(\sigma(s,u)) \in \left[\underline{\mu_2},\overline{\mu_2}\right]$ *for all* $s \in [0,\infty)$ *and* $u \in \varphi(B_1 \times B_2 \times \left[\underline{\mu_2},\overline{\mu_2}\right])$.

We set

$$\varphi_s(x,y,t) = \sigma(s,\varphi(x,y,t)) \quad \text{for } (s,x,y,t) \in [0,s_0] \times B_1 \times B_2 \times \left[\underline{\mu_2},\overline{\mu_2}\right].$$

From (2), (g3), and (g4), we can see that

$$r(\beta_+,\beta_-,\mu) \circ \varphi_0(x,y,t) + (1-r)(x,y,t) \neq (\mathfrak{n},\mathfrak{n},1/2)$$

for each $(x,y,t) \in \partial(B_1 \times B_2 \times \left[\underline{\mu_2},\overline{\mu_2}\right])$, which yields $\deg((\beta_+,\beta_-,\mu) \circ \varphi_0; \text{Int}B_1 \times \text{Int}B_2 \times \left(\underline{\mu_2},\overline{\mu_2}\right),(\mathfrak{n},\mathfrak{n},1/2)) = 1$. Since from ($g$2) and ($\sigma$3), we have (β_+,β_-,μ) $\varphi_s(x,y,t) \neq (\mathfrak{n},\mathfrak{n},1/2)$ for each $(s,x,y,t) \in [0,s_0] \times \partial(B_1 \times B_2 \times \left[\underline{\mu_2},\overline{\mu_2}\right])$, by the homotopy invariance of the Brouwer degree, we have

$$\deg((\beta_+,\beta_-,\mu)\circ\varphi_{s_0};\operatorname{Int}B_1\times\operatorname{Int}B_2\times\left(\underline{\mu_2},\overline{\mu_2}\right),(\mathfrak{n},\mathfrak{n},1/2))=1.$$

Now, we give our proof of Theorem 1.

Proof of Theorem 1. Let $\varepsilon>0$. By Lemma 1, we can choose $\delta\in(0,\delta_1)$ such that

$$I((\varphi_{s_0}(x,y,t))^+)\geq I(\mathcal{T}((\varphi_{s_0}(x,y,t))^+))-\varepsilon,$$

for each $(x,y,t)\in B_1\times B_2\times\left(\underline{\mu_2},\overline{\mu_2}\right)$ with $|\mu((\varphi_{s_0}(x,y,t))^+)-1/2|\leq\delta$, and

$$I((\varphi_{s_0}(x,y,t))^-)\geq\overline{c}-\varepsilon$$

for each $(x,y,t)\in B_1\times B_2\times\left[\underline{\mu_2},\overline{\mu_2}\right]$ with $|\beta_-(\varphi_{s_0}(x,y,t))-\mathfrak{n}|\leq\delta$. We choose $\xi\in(0,\min\{\delta,\min_{(x,y,t)\in\partial G}|(\beta_+,\beta_-,\mu)\circ\varphi_{s_0}(x,y,t)-(\mathfrak{n},\mathfrak{n},1/2)|\}/2)$ and $\psi\in C^\infty(B_1\times B_2\times\left[\underline{\mu_2},\overline{\mu_2}\right],\mathbb{R}^{2N+1})$ such that

$$\max_{(x,y,t)\in B_1\times B_2\times\left[\underline{\mu_2},\overline{\mu_2}\right]}\left|\psi(x,y,t)-(\beta_+,\beta_-,\mu)\circ\varphi_{s_0}(x,y,t)\right|\leq\xi.$$

By the homotopy invariance of the Brouwer degree, we have

$$\deg(\psi;\operatorname{Int}B_1\times\operatorname{Int}B_2\times\left(\underline{\mu_2},\overline{\mu_2}\right),(\mathfrak{n},\mathfrak{n},1/2))=1.$$

We write $\psi=(\psi_+,\psi_-,\psi_\mu)$. Since $(\psi_-,\psi_\mu)\in C^\infty(B_1\times B_2\times\left[\underline{\mu_2},\overline{\mu_2}\right],\mathbb{R}^{2N+1})$, we can choose $(\mathfrak{n}',t')\in\operatorname{Int}B_2\times\left(\underline{\mu_2},\overline{\mu_2}\right)$ such that $|\mathfrak{n}'-\mathfrak{n}|\leq\delta/2$, $|t'-1/2|\leq\delta/2$ and $(\mathfrak{n}',t')$ is a regular value of (ψ_-,ψ_μ) and $(\psi_-,\psi_\mu)|_{\partial(B_1\times B_2\times\left[\underline{\mu_2},\overline{\mu_2}\right])}$. We set

$$G=(\psi_-,\psi_\mu)^{-1}(\mathfrak{n}',t').$$

Then G is an N-dimensional, smooth submanifold of $B_1\times B_2\times\left[\underline{\mu_2},\overline{\mu_2}\right]$ whose boundary is $\partial B_1\times\{(\mathfrak{n}',t')\}$. Since for each $(x,y,t)\in G$,

$$|\mu(\varphi_{s_0}(x,y,t))-1/2|\leq|\mu(\varphi_{s_0}(x,y,t))-\psi_\mu(x,y,t)|+|t'-1/2|\leq\delta,$$
$$|\beta_-(\varphi_{s_0}(x,y,t))-\mathfrak{n}'|\leq|\beta_-(\varphi_{s_0}(x,y,t))-\psi_-(x,y,t)|+|\mathfrak{n}'-\mathfrak{n}|\leq\delta,$$

we have

$$I((\varphi_{s_0}(x,y,t))^+)\geq I(\mathcal{T}((\varphi_{s_0}(x,y,t))^+))-\varepsilon\quad\text{for each }(x,y,t)\in G,$$

and

$$I((\varphi_{s_0}(x,y,t))^-) \geq \bar{c} - \varepsilon \quad \text{for each } (x,y,t) \in G.$$

Since $G \subset B_1 \times B_2 \times \left[\underline{\mu_2}, \overline{\mu_2}\right]$, $\partial G = \partial B_1 \times \{(\mathfrak{n}', t')\}$ and $(\mathfrak{n}', t')$ is a regular value of $(\psi_-, \psi_\mu)|_{\partial(B_1 \times B_2 \times [\underline{\mu_2}, \overline{\mu_2}])}$, we can find an N-dimensional, orientable, compact C^∞-manifold $\tilde{G} \subset \mathbb{R}^{2N+1}$ such that $G \subset \tilde{G}$, $\partial\tilde{G} = \emptyset$ and $\partial(\tilde{G} \setminus G) = \partial G$. For each $f \in C(G, \mathbb{R}^N)$ with $\mathfrak{n} \notin f(\partial G)$, we define $E(f) \in C(\tilde{G}, S^N)$ as follows. For such f, take a small open ball $V \subset B_1$ such that its center is $\mathfrak{n}$ and $\overline{V} \cap f(\partial G) = \emptyset$. Since we can naturally consider $\mathbb{R}^N \subset S^N$ and $S^N \setminus V$ is homeomorphic to $[0,1]^N$, by the Tietze extension theorem, we can extend the mapping $f|_{\partial G}$ to $\tilde{f} \in C(\tilde{G} \setminus \text{Int} G, S^N \setminus V)$. We define $E(f) \in C(\tilde{G}, S^N)$ by

$$E(f)(x) = \begin{cases} f(x,y,t) & \text{if } (x,y,t) \in G, \\ \tilde{f}(x,y,t) & \text{if } (x,y,t) \in \tilde{G} \setminus G. \end{cases}$$

Then by similar lines as those in [9], we can show $\deg(E(\psi_+|_G); \tilde{G}, S^N) \neq 0$ and $\deg(E(\beta \circ (\mathcal{T}(\varphi_{s_0}^+)|_G)); \tilde{G}, S^N) \neq 0$. For the definition of these Brouwer degrees, see [10,13]. We note that these values of the degrees do not depend on the extensions in $E(\cdot)$ by the homotopy invariance. We set

$$\Gamma_2 = \{\gamma \in C(G, \mathcal{N}) : \gamma = \mathcal{T}(\varphi_{s_0}^+) \text{ on } \partial G, \deg(E(\beta \circ \gamma); \tilde{G}, S^N) \neq 0\},$$

$$c_2 = \inf_{\gamma \in \Gamma_2} \max_{(x,y,t) \in G} I(\gamma(x,y,t)).$$

Since $\mathcal{T}(\varphi_{s_0}^+)|_G \in \Gamma_2$, $I(\mathcal{T}(\varphi_{s_0}^+)|_G(x, \mathfrak{n}', t')) = I(g(x, \mathfrak{n}')^+) \leq c_0$ for each $x \in \partial B_1$ from $(g3)$ and we assumed that there is only one positive solution, we can infer $c_2 = c_1$, which yields $\max_{(x,y,t) \in G} I(\mathcal{T}(\varphi_{s_0}^+)|_G(x,y,t)) \geq c_1$. Hence, we have

$$\max_{(x,y,t) \in B_1 \times B_2 \times [\underline{\mu_2}, \overline{\mu_2}]} I(\varphi_{s_0}(x,y,t)) \geq \max_{(x,y,t) \in G} I(\varphi_{s_0}(x,y,t))$$

$$\geq \max_{(x,y,t) \in G} I((\varphi_{s_0}(x,y,t))^+) + \min_{(x,y,t) \in G} I((\varphi_{s_0}(x,y,t))^-)$$

$$\geq \max_{(x,y,t) \in G} I(\mathcal{T}(\varphi_{s_0}^+)|_G(x,y,t)) - \varepsilon + \bar{c} - \varepsilon \geq c_1 + \bar{c} - 2\varepsilon.$$

From the arbitrariness of $\varepsilon > 0$, we have

$$\max_{(x,y,t) \in B_1 \times B_2 \times [\underline{\mu_2}, \overline{\mu_2}]} I(\varphi_{s_0}(x,y,t)) \geq c_1 + \bar{c},$$

which contradicts $(\sigma 3)$. Thus, we have shown that problem (1) has at least two pairs of solutions.

Acknowledgements The second author is partially supported by the Grant-in-Aid for Scientific Research (C) (No. 21540214) from Japan Society for the Promotion of Science.

References

1. Bahri, A., Coron, J.M.: Sur une équation elliptique non linéaire avec l'exposant critique de Sobolev. C. R. Acad. Sci. Paris Sér. I Math. **301**(7), 345–348 (1985)
2. Bahri, A., Coron, J.M.: On a nonlinear elliptic equation involving the critical Sobolev exponent: The effect of the topology of the domain. Commun. Pure Appl. Math. **41**(3), 253–294 (1988)
3. Clapp, M., Weth, T.: Two solutions of the Bahri-Coron problem in punctured domains via the fixed point transfer. Commun. Contemp. Math. **10**(1), 81–101 (2008)
4. Coron, J.M.: Topologie et cas limite des injections de Sobolev. C. R. Acad. Sci. Paris Sér. I Math. **299**(7), 209–212 (1984)
5. Dancer, E.N.: A note on an equation with critical exponent. Bull. London Math. Soc. **20**(6), 600–602 (1988)
6. Ding, W.Y.: Positive solutions of $\Delta u + u^{(n+2)/(n-2)} = 0$ on contractible domains. J. Partial Differ. Equ. **2**(4), 83–88 (1989)
7. Dold, A.: The fixed point transfer of fibre-preserving maps. Math. Z. **148**(3), 215–244 (1976)
8. Ge, Y., Musso, M., Pistoia, A.: Sign changing tower of bubbles for an elliptic problem at the critical exponent in pierced non-symmetric domains. Commun. Partial Differ. Equ. **35**(8), 1419–1457 (2010)
9. Hirano, N., Shioji, N.: Existence of two solutions for the Bahri–Coron problem in an annular domain with a thin hole. J. Funct. Anal. **261**(12), 3612–3632 (2011)
10. Hirsch, M.W.: Differential topology, Graduate Texts in Mathematics, vol. 33. Springer, New York (1994). Corrected reprint of the 1976 original
11. Hocking, J.G., Young, G.S.: Topology, 2nd edn. Dover Publications Inc., New York (1988)
12. Lloyd, N.G.: Degree theory. Cambridge University Press, Cambridge (1978). Cambridge Tracts in Mathematics, No. 73
13. Milnor, J.W.: Topology from the differentiable viewpoint. Princeton Landmarks in Mathematics. Princeton University Press, Princeton (1997). Based on notes by David W. Weaver, Revised reprint of the 1965 original
14. Passaseo, D.: Multiplicity of positive solutions of nonlinear elliptic equations with critical Sobolev exponent in some contractible domains. Manuscripta Math. **65**(2), 147–165 (1989)
15. Pohožaev, S.I.: On the eigenfunctions of the equation $\Delta u + \lambda f(u) = 0$. Dokl. Akad. Nauk SSSR **165**, 36–39 (1965)
16. Weth, T.: Energy bounds for entire nodal solutions of autonomous superlinear equations. Calc. Var. Partial Differ. Equ. **27**(4), 421–437 (2006)

Period-Two Solutions to Some Systems of Rational Difference Equations

Walter S. Sizer

Abstract We consider period-two solutions to various cases of the difference equations $x_{n+1} = \frac{A+Bx_n+Cy_n}{D+Ex_n+Fy_n}$, $y_{n+1} = \frac{J+Kx_n+Ly_n}{P+Qx_n+Ry_n}$. We give several examples of period-two solutions and in some cases categorize prime period-two solutions. We also apply our techniques to a single second-order rational difference equation to get circumstances leading to a prime period-two solution.

Keywords Difference equations • Periodic solutions

1 Introduction

Camouzis et al. in [1] ask whether there are prime period-two solutions to 55 special cases of the pair of rational difference equations $x_{n+1} = \frac{A+Bx_n+Cy_n}{D+Ex_n+Fy_n}$, $y_{n+1} = \frac{J+Kx_n+Ly_n}{P+Qx_n+Ry_n}$, where $x_0, y_0 > 0$, all constants are greater than or equal to 0, constants in each numerator and in each denominator are not all 0, and no numerator is a scalar multiple of its denominator. In Sect. 2 we give examples of several period-two solutions to these equations, noting cases of prime period two, and in Sect. 3 we give results which show negative answers for all the 55 special cases mentioned in [1] and show how the examples with prime period-two solutions in Sect. 2 were obtained. Section 4 looks at the case where we allow initial conditions to be 0. In Sect. 5 we apply the approach of Sect. 3 to a single second-order rational difference equation in x.

W.S. Sizer
Minnesota State University Moorhead, Moorhead, MN, USA
email: sizer@mnstate.edu

Pinelas et al. (eds.), *Differential and Difference Equations with Applications*, Springer Proceedings in Mathematics & Statistics 47, DOI 10.1007/978-1-4614-7333-6_55, © Springer Science+Business Media New York 2013

2 Examples of Period-Two Solutions

We give several examples of period-2 solutions.

Example 1. Any equilibrium solutions will be period-two solutions. It should be noted that some solutions with x_n, y_n in equilibrium involve complex equations, as with $x_{n+1} = \frac{5 + 6x_n + 11y_n}{4 + 3x_n + y_n}$, $y_{n+1} = \frac{2 + 4x_n + 3y_n}{3 + x_n + 2y_n}$, $x_0 = 3$, $y_0 = 2$.

Example 2. For the equation $x_{n+1} = \frac{K}{x_n}$, every solution is periodic of period 2. If $x_0 \neq \sqrt{K}$, the solution has prime period 2. A similar result holds for y.

Example 3. If x_n has period 2, for example, $x_{n+1} = \frac{K}{x_n}$, and $y_{n+1} = x_n$, (and with $y_0 = x_1$) then the system has period 2. The numbers in the sequence of y_n's are the same as in the sequence of x_n's but with a shift of subscript.

Example 4. The equations $x_{n+1} = ky_n$, $y_{n+1} = \frac{1}{k}x_n$ give period-2 solutions for all values of x_0, y_0. If $x_0 \neq \frac{1}{k}y_0$, the solution has prime period 2. If $k = 1$, the two sequences are the same but with a shift of subscript.

Example 5. The equations $x_{n+1} = \frac{K}{y_n}$, $y_{n+1} = \frac{K}{x_n}$ are periodic of period 2 for all choices of x_0 and y_0. The numbers in the two sequences will be different if $x_0 \neq y_0$, and the sequences will be of prime period two if $x_0 \neq \frac{K}{y_0}$.

Example 6. The equations $x_{n+1} = \frac{\sqrt{K}x_n}{y_n}$, $y_{n+1} = \frac{K}{y_n}$ have every solution periodic of period 2 and have prime period 2 unless $y_0 = \sqrt{K}$.

Example 7. The equations $x_{n+1} = \frac{10}{2x_n + y_n}$, $y_{n+1} = \frac{8}{x_n + \frac{4}{3}y_n}$, with $x_0 = 3$ and $y_0 = 4$ (so $x_1 = 1$, $y_1 = \frac{4}{3}$), give a prime period-2 solution. Some other choices of x_0, y_0 will not give period-2 solutions to these equations.

Example 8. The equations $x_{n+1} = \frac{6}{x_n}$, $y_{n+1} = \frac{\sqrt{12}y_n}{1 + x_n}$, with $x_0 = 2$, have period 2 for any value of y_0. In general the period-2 sequence $x_0, x_1, x_0, x_1, \ldots$ and the equation $y_{n+1} = \frac{Ly_n}{P + Qx_n}$, give a period-2 solution for any y_0 provided $(P + Qx_0)(P + Qx_1) = L^2$. Example 6 is also of this type, but with the roles of the variables reversed.

Example 9. Suppose $x_{n+1} = \frac{2}{x_n}$ with $x_0 = 1$ (so $x_1 = 2$ and the x's are periodic of prime period 2). Suppose $y_{n+1} = \frac{2 + x_n + y_n}{3 + 2x_n}$ with $y_0 = \frac{23}{34}$. Then the solution in y is periodic of prime period 2 (as is readily checked – and $y_1 = \frac{25}{34}$). For other values of y_0 solutions are not periodic of period 2.

Example 10. Suppose $x_{n+1} = \frac{6}{x_n}$ with $x_0 = 2$ (so $x_1 = 3$ and the x's are periodic of prime period 2). Let $y_{n+1} = \frac{5y_n}{2 + x_n + 3y_n}$ with $y_0 = \frac{1}{6}$. Then the solution in y is periodic of prime period 2 (again, readily checked, with $y_1 = \frac{5}{27}$), and solutions are not periodic of period 2 for other values of y_0.

Example 11. Suppose $x_{n+1} = \frac{4}{x_n}$ with $x_0 = 4$ (so $x_1 = 1$ and the x's are periodic of prime period 2). Suppose $y_{n+1} = \frac{2+x_n+y_n}{3+2x_n+4y_n}$. There is only one value of $y_0 > 0$ giving a period 2 solution, and this is $y_0 = \frac{1}{2}$. Setting $y_0 = \frac{1}{2}$, however, gives an equilibrium for y.

Example 12. Suppose $x_{n+1} = \frac{2+3x_n}{2+x_n}$, $y_{n+1} = \frac{4+x_n}{5y_n}$, $x_0 = 2$. We get a period-2 solution for all values of y_0; the x's are constant and the second equation is equivalent to $y_{n+1} = \frac{\frac{6}{5}}{y_n}$.

Example 13. Let $x_{n+1} = \frac{x_n+6y_n}{2+\frac{5}{2}x_n+y_n}$, $y_{n+1} = \frac{8x_n+3y_n}{2+x_n+\frac{11}{2}y_n}$, with $x_0 = 1$ and $y_0 = 2$. This is easily seen to generate sequences of prime period 2 in both variables. These initial values are the only ones with that property.

3 General Results

We derive results about sequences giving period-two solutions by actually starting by assuming we have prime period-two solutions and asking what the equations can look like that give these sequences. Thus, let $a, b, a, b, \ldots$ and $c, d, c, d, \ldots$ be two prime period-two sequences of positive real numbers. We ask what conditions on $A, B, C, D, E, F, J, K, L, P, Q, R$ (all assumed greater than or equal to 0) allow these sequences to appear as recursive sequences defined by the equations $x_{n+1} = \frac{A+Bx_n+Cy_n}{D+Ex_n+Fy_n}$, $y_{n+1} = \frac{J+Kx_n+Ly_n}{P+Qx_n+Ry_n}$.

Theorem 1. *If $a - b$ and $c - d$ have opposite signs, there are numerous choices for parameters which give these sequences as solutions to the recursive equations.*

Proof. Assume $a, b, a, b, \ldots$ and $c, d, c, d, \ldots$ arise from the given equations. Then $a = \frac{A+Bb+Cd}{D+Eb+Fd}$, or

$$Da + Eab + Fad = A + Bb + Cd. \tag{1}$$

Also, $b = \frac{A+Ba+Cc}{D+Ea+Fc}$, or

$$Db + Eab + Fcb = A + Ba + Cc. \tag{2}$$

Subtracting (2) from (1) gives

$$D(a-b) + F(ad-bc) = B(b-a) + C(d-c), \text{ or}$$
$$F(ad-bc) = (B+D)(b-a) + C(d-c). \tag{3}$$

Any values of $A, B, C, D, E,$ and F which satisfy (1) and (3), in conjunction with an equation giving the sequence $c, d, c, d, \ldots$, will give the first sequence as solution. If $a - b$, $c - d$ have opposite signs one can choose $B, D,$ and C to be any positive numbers satisfying the property that $(B+D)(b-a) + C(d-c)$ has the same sign as $ad - bc$. One can then solve to get a positive value of F, which leads to Eq. (3) being true. Substituting these into Eq. (1) and choosing A positive and large enough to guarantee that E will be positive then gives a solution to (1) and (2).

A similar analysis with the equation for $y_{n+1} = \frac{J+Kx_n+Ly_n}{P+Qx_n+Ry_n}$ leads to equations

$$Q(cb - ad) = K(b - a) + (L + P)(d - c), \tag{4}$$

and

$$Pc + Qcb + Rcd = J + Kb + Ld. \tag{5}$$

Again, if $a - b$, $c - d$ have opposite signs there are many solutions. $\square$

Theorem 2. *If $a - b$, $c - d$ have the same sign, one of the following is true:*

(a) The first equation reduces to $x_{n+1} = \frac{T}{x_n}$.

(b) The second equation reduces to $y_{n+1} = \frac{S}{y_n}$.

(c) The equations are $x_{n+1} = \frac{A}{Ex_n+Fy_n}$, $y_{n+1} = \frac{J}{Qx_n+Ry_n}$, and these are equivalent to the equations in (a) and (b).

Proof. Assume $a - b$, $c - d$ have the same sign. Assume both are positive. We consider three cases.

Case (i): $bc - ad > 0$.
In Eq. (4) the left-hand side is greater than or equal to 0, and the right-hand side is less than or equal to 0. The only possibility for the equation to be true is to have K, L, P, and Q all equal to 0. This gives result (b).

Case (ii): $bc - ad < 0$.
In this case Eq. (3) forces B, C, D, and F to all be 0, and we get (a).

Case (iii): $bc = ad$.
In this case Eq. (4) forces K, L, and P to all be 0, and Eq. (3) forces B, C, and D to all be 0. Thus we get the two equations in (c). Note also that $\frac{a}{c} = \frac{b}{d}$; call this value W. Then $a = Wc$ and $b = Wd$ (and so $x_n = Wy_n$ for all n, $y_n = \frac{1}{W}x_n$ for all n)

Thus $x_{n+1} = \frac{A}{Ex_n+Fy_n} = \frac{A}{Ex_n+\frac{F}{W}x_n}$, and the equation is equivalent to $x_{n+1} = \frac{T}{x_n}$ Likewise, the equation for y_{n+1} is equivalent to $y_{n+1} = \frac{S}{y_n}$.

This proof is easily adapted to the case where both $a - b$ and $c - d$ are negative.
                                                                                                             $\square$

Note that in the situations leading to cases (a) and (b) above – where $bc - ad \neq$ – there are again numerous choices of the coefficients which lead to prime period two sequences for the other variables.

In [1] the question is asked whether some systems covered by our equations hav prime period-two solutions. The authors specify in these systems which coefficien are 0 and which are nonzero. In all the cases asked about, either C or K is zero. Th solution of these problems is derived from the following result.

Theorem 3. *If C (or K) is 0, one of the following is true:*

(a) *The first equation reduces to* $x_{n+1} = \frac{T}{x_n}$.

(b) *The second equation reduces to* $y_{n+1} = \frac{S}{y_n}$.

(c) *The equations are* $x_{n+1} = \frac{A}{Ex_n+Fy_n}$, $y_{n+1} = \frac{J}{Qx_n+Ry_n}$, *and these are equivalent to the equations in (a) and (b).*

Proof. Assume $C = 0$ and $a > b$. Then the right side of Eq. (3) is less than or equal to 0, so $ad - bc \leq 0$ or $B = D = F = C = 0$. The second possibility gives conclusion (a); in the first situation, by Eq. (4), $d - c \geq 0$ (actually $d > c$ as the sequence is assumed to have prime period 2) or $L = P = K = 0$ and $cd - ab = 0$. The latter situation gives (b) or (c) as in the proof of Theorem 2, and the former gives a contradiction: $d > c$ and $a > b$ but $ad - bc \leq 0$.

The proof if $K = 0$ is similar.   □

None of the systems asked about in [1] satisfy the conclusion of Theorem 3, so none of the systems have prime period-two solutions.

4   With an Initial Condition Equal to 0

Interesting results also appear if we relax the restrictions on initial conditions, just requiring that $x_{-1}, x_0, y_{-1}, y_0 \geq 0$. If we require prime period-two solutions in only one variable, we get as one possibility the equations $x_{n+1} = \frac{Bx_n}{D+Ex_n+Fy_n}$, $y_{n+1} = \frac{J+Kx_n}{Qx_n+Ry_n}$, with $x_0 = 0$, $y_0 > 0$, $y_0 \neq \sqrt{\frac{J}{R}}$. This gives the sequence of x_i's equal to all 0's and the sequence of y_i's equal to $y_0, \frac{J}{Ry_0}, y_0, \frac{J}{Ry_0}, \ldots$.

The more interesting case is the one requiring both sequences to have prime period 2. We may assume then that the sequence of x_i's is $0, a, 0, a, \ldots$, and the sequence of y_i's is $c, d, c, d, \ldots$, where the values of a, c, and d are as yet undetermined. The equation for x_2 gives $0 = \frac{A+Ba+Cd}{D+Ea+Fd}$, and since the parameters are all nonnegative and $a > 0$, $d \geq 0$, we see that A and B are both 0. Since the numerator does not have all parameters equal to 0, this also means $C \neq 0$, and thus $d = 0$.

Since $d = 0$ and we assume both sequences have prime period 2, this means $c \neq 0$. An argument like the one in the last paragraph but working with the equation for y_1 then gives us $J = L = 0$ and $K \neq 0$.

The equations for x_1 and y_2 are then $a = \frac{Cc}{D+Fc}$ and $c = \frac{Ka}{P+Qa}$.. Solving for a and c gives $c = \frac{KC-PD}{PF+QC}$ and then $a = \frac{Cc}{D+Fc}$. Thus, if $A = B = J = L = 0$ and $KC - PD > 0$, we get the desired nonnegative solutions as shown. We have already seen that if we get solutions $0, a, 0, a, \ldots$ in x_i and $c, 0, c, 0, \ldots$ in y_i, then $A = B = J = L = 0$ and $KC - PD > 0$.

The condition $KC - PD > 0$ is unchanged if we interchange the roles of x and y, so nothing is gained by considering the sequence of x_i's to be $0, a, 0, a, \ldots$ and the sequence of y_i's to be $c, 0, c, 0, \ldots$.

Also note that, given fixed values of a and c greater than 0, we can choose C, I and F in numerous ways so that $a = \frac{Cc}{D+Fc}$, and then choosing P and $Q > 0$ as w like finds a value K so that $c = \frac{KC-PD}{PF+QC}$. Thus, we get numerous pairs of equation giving these prime period-2 solutions.

We have established the following result:

Theorem 4. *The equations* $x_{n+1} = \frac{A+Bx_n+Cy_n}{D+Ex_n+Fy_n}$, $y_{n+1} = \frac{J+Kx_n+Ly_n}{P+Qx_n+Ry_n}$ *have prime pe riod 2 solutions* $0, a, 0, a, \ldots$ *and* $c, 0, c, 0, \ldots$ *with* $a, c > 0$ *if and only if* $A = B = J$ $L = 0$ *and* $KC - PD > 0$. *Solutions of this sort are the only ones where one initi condition is 0 and both sequences have prime period two.*

Proof. See above.

Example 14. The equations $x_{n+1} = \frac{4y_n}{2+7x_n+6y_n}$, $y_{n+1} = \frac{5x_n}{3+8x_n+5y_n}$, with initial cond tions $x_0 = 0$, $y_0 = \frac{7}{25}$, give prime period-2 solutions with $x_1 = \frac{7}{23}$, $y_1 = 0$.

5 A Single Second-Order Rational Equation

In light of the interest of Radin [3] in prime period-two solutions of a single secon order rational difference equation, we show how our approach can be adapted to th situation. Consider the equation

$$x_{n+1} = \frac{A+Bx_n+Cx_{n-1}}{D+Ex_n+Fx_{n-1}}, \qquad (\cdot$$

with initial conditions x_{-1} and x_0 positive, and all coefficients nonnegativ neither numerator nor denominator identically zero, and the numerator not a scal multiple of the denominator. Again, starting from a prime period-two sequence s, $s, t, \ldots$, we ask what equations of the form (5) give rise to the sequence. We get

Theorem 5. *Given any prime period-two sequence* $s, t, s, t, \ldots$, *there are man varied choices of A, B, C, D, E, F which give rise to that sequence using Eq. (5).*

Proof. Suppose the given sequence satisfies Eq. (5). Then
$t = \frac{A+Bs+Ct}{D+Es+Ft}$, or

$$Dt + Est + Ft^2 = A + Bs + Ct, \qquad ($$

and likewise,

$$Ds + Est + Fs^2 = A + Bt + Cs. \qquad ($$

Subtracting and simplifying we get

$$(B+D)(t-s) + F(t-s)(t+s) = C(t-s). \qquad ($$

Thinking of s and t as fixed and the capital letters as variables, any solution to (7) and (9) gives a pair of equations leading to the given sequence. Since $t \neq s$, we can divide Eq. (9) by $t - s$ and rearrange it to get

$$F(t+s) = C - B - D. \tag{10}$$

Choosing B, C, and D positive with $C > B + D$ makes the right side of (10) positive, so we can solve for F and get a positive value. Given these values of D, B, C, and F, we can choose E so that the left side of (7) is greater than $Bt + Cs$, thus assuring that A will also be positive.   □

Example 15. Consider the equation $x_{n+1} = \frac{2+2x_n+8x_{n-1}}{3+5x_n+x_{n-1}}$. Then initial values $x_{-1} = 1$, $x_0 = 2$ give a prime period-two solution.

For further treatment of prime period-two solutions of this equation, see Kulenovic and Ladas [2].

We would like to thank the referee for suggesting the problem described in Sect. 4.

References

1. Camouzis, E., Kulenovic, M.R.S., Ladas, G., Merino, O.: Open problems and conjectures: Rational systems in the plane. J. Differ. Equ. Appl. **15**(3), 303–323 2009
2. Kulenovic, M.R.S., Ladas, G.: Dynamics of Second Order Rational Difference Equations. Chapman & Hall/CRC Press, Boca Raton (2002)
3. Michael Radin: On the Global Character of Solutions of the System $x_{n+1} = \frac{\alpha_1}{x_n+y_n}$, $y_{n+1} = \frac{\alpha_2}{Bx_n+y_n}$, $n = 0, 1, \ldots$. In: Contributed Paper, International Conference on Differential and Difference Equations and Applications, University of the Azores, Ponta Delgada, Portugal, 4–8 July 2011

Effect of Roughness on the Flow of a Fluid Against Periodic and Nonperiodic Rough Boundaries

J. Casado-Díaz, M. Luna-Laynez, and F.J. Suárez-Grau

Abstract In this paper we study the asymptotic behavior of a viscous fluid in a domain with a slightly and periodic rough boundary. Assuming the Navier condition on the rough boundary, we prove that if the roughness is not strong enough, then it makes appear a new term in the limit equation. Finally, we study this phenomenon in the case of a rough domain not necessarily periodic assuming a very general condition on the rough boundary.

Keywords Stokes equation • Rough boundary • Navier condition

1 Introduction

In [8] it was considered a viscous fluid in a rough domain Ω_ε with an impermeable rough boundary Γ_ε described by

$$ x_3 = -\varepsilon \Psi\left(\frac{x'}{\varepsilon}\right), \quad \forall x' \in \omega, $$

along this paper a point $x \in \mathbb{R}^3$ is decomposed as $x = (x', x_3)$ with $x' \in \mathbb{R}^2$ and $x_3 \in \mathbb{R}$) with ω a bounded open set of $\mathbb{R}^2$ and Ψ a smooth periodic function such that

$$ \text{Span}\left(\left\{\nabla\Psi(z') : z' \in \mathbb{R}^2\right\}\right) = \mathbb{R}^2. \tag{1} $$

Casado-Díaz • M. Luna-Laynez • F.J. Suárez-Grau (✉)
Dpto. Ecuaciones Diferenciales y Análisis Numérico, Universidad de Sevilla,
Tarfia s/n, 41012 Sevilla, Spain
e-mail: jcasadod@us.es; mllaynez@us.es; fjsgrau@us.es

Pinelas et al. (eds.), *Differential and Difference Equations with Applications*, Springer Proceedings in Mathematics & Statistics 47, DOI 10.1007/978-1-4614-7333-6_56, © Springer Science+Business Media New York 2013

Assuming that the velocity u_ε of the fluid satisfies the Navier condition,

$$u_\varepsilon(x) \in T_\varepsilon(x), \quad \text{on } \Gamma_\varepsilon, \quad \frac{\partial u_\varepsilon}{\partial \nu}(x) + \gamma u_\varepsilon(x) \in T_\varepsilon(x)^\perp, \quad \text{on } \Gamma_\varepsilon, \qquad (2)$$

where ν denotes the unitary outside normal vector to Ω_ε on Γ_ε, $T_\varepsilon(x)$ the tangent space in the point $x \in \Gamma_\varepsilon$, and γ a nonnegative friction constant, it was proved in the limit that the velocity u of the fluid vanishes on the boundary, i.e., it satisfies the adherence condition $u = 0$ on $\omega \times \{0\}$.

This result has been generalized in [2] for a nonperiodic boundary described by

$$x_3 = \Phi_\varepsilon(x'), \quad \forall x' \in \omega,$$

where Φ_ε converges weakly-$*$ to zero in $W^{1,\infty}(\omega)$ and it is such that the support of the Young's measure associated to $\nabla \Phi_\varepsilon$ contains two linearly independent vectors. Remark that this last condition implies that $\nabla \Phi_\varepsilon$ does not converge to zero in $L^1(\omega)^2$.

Our goal in Sect. 2 is to generalize the result in [8] to the case of a viscous fluid confined in a domain Ω_ε defined by

$$\Omega_\varepsilon = \left\{ x = (x', x_3) \in \omega \times \mathbb{R} : -\delta_\varepsilon \Psi\left(\frac{x'}{\varepsilon}\right) < x_3 < 1 \right\}, \qquad (3)$$

with a slightly rough boundary Γ_ε defined by

$$\Gamma_\varepsilon = \left\{ x = (x', x_3) \in \omega \times \mathbb{R} : x_3 = -\delta_\varepsilon \Psi\left(\frac{x'}{\varepsilon}\right) \right\}, \qquad (4)$$

where $\omega \subset \mathbb{R}^2$ is a Lipschitz bounded open set, $\Psi \in W^{2,\infty}_{loc}(\mathbb{R}^2)$ is periodic of period $Z' = (0,1)^2$, and $\delta_\varepsilon > 0$ satisfies $\lim_{\varepsilon \to 0} \dfrac{\delta_\varepsilon}{\varepsilon} = 0$.

Remark that in our case $\Phi_\varepsilon = \delta_\varepsilon \Psi(\frac{x'}{\varepsilon})$ converges strongly to zero in $W^{1,\infty}(\omega)$ and therefore the results in [2] do not apply.

Assuming the Navier boundary condition (2) on the oscillating boundary Γ_ε of period ε and amplitude δ_ε (with $\delta_\varepsilon \ll \varepsilon$), and denoting by

$$\lambda = \lim_{\varepsilon \to 0} \frac{\delta_\varepsilon}{\varepsilon^{\frac{3}{2}}} \in [0, +\infty], \qquad (5$$

(the limit exists at least for a subsequence) we show

- If $\lambda = +\infty$ and (1) holds, then the Navier and adherence boundary conditions are asymptotically equivalent. This extends the result obtained in [8] for $\delta_\varepsilon = \varepsilon$ the case when $\delta_\varepsilon / \varepsilon$ tends to zero and $\delta_\varepsilon / \varepsilon^{\frac{3}{2}}$ tends to infinity.
- If $\lambda = 0$, the roughness is so small that it has no effect on the limit problem.
- If $\lambda \in (0, +\infty)$, the roughness is not strong enough to obtain the adherence condition, but it is large enough to make appear a new friction term. Namely

denoting by $T(x)$ the tangent space in the point $x \in \omega \times \{0\}$, we obtain the following Navier boundary condition in the limit,

$$u_3(x) \in T(x), \quad \text{on } \omega \times \{0\}, \quad -\partial_3 u'(x) + \gamma u'(x) + \lambda^2 R u'(x) \in T^\perp(x), \quad \text{on } \omega \times \{0\},$$

where $R \in \mathbb{R}^{2 \times 2}$ is a symmetric and nonnegative matrix. The new term $\lambda^2 R$ is similar to the *strange term* obtained by D. Cioranescu and F. Murat in [13] for the homogenization of Dirichlet problems in perforated domains.

In relation with this last case, it has been studied in [5] the asymptotic behavior f viscous fluids confined in general rough domains, not necessarily periodic. In the articular case of a domain with a rough bottom described by

$$x_3 = \Psi_\varepsilon(x'), \ \forall x' \in \omega,$$

ith Ψ_ε converging weakly-$*$ to zero in $W^{1,\infty}(\omega)$, the results in [5] imply that there xist μ a nonnegative Borel measure, which vanishes in the sets of capacity zero nd can be infinity in compact sets of ω, and H a μ-measurable matrix evaluated anction such that the limit boundary condition is given by

$$_3(x) \in T(x), \quad \text{on } \omega \times \{0\}, \quad -\partial_3 u'(x) + \gamma u'(x) + H u' \mu(x) \in T^\perp(x), \quad \text{on } \omega \times \{0\}.$$

ur results provide an example where the extra term $H u' \mu$ is not zero. We refer [3, 4] for other examples of different nature for a ribbed boundary described by $_3 = \varepsilon \Psi(\frac{x_1}{\varepsilon})$.

The results obtained in Sect. 2 show that the Navier boundary condition provides new term in the limit equation. In Sect. 3 we study this phenomenon for domains ot necessarily periodic. At the place of the Stokes system for a sequence of near elliptic systems of M equations posed in varying open sets $\Omega_\varepsilon \subset \mathbb{R}^N$ with boundary condition similar to (2), where $T_\varepsilon(x)$ is replaced by an arbitrary linear nace $V_\varepsilon(x) \subset \mathbb{R}^M$. This abstract formulation has the advantage that it contains lot of classical boundary conditions. For instance, this permits us to study the symptotic behavior of linear elliptic systems in rough domains Ω_ε where we npose Dirichlet and Neumann boundary conditions on varying subsets of $\partial\Omega_\varepsilon$. his problem has been studied in [6, 7] for $\Omega_\varepsilon = \Omega$ fixed.

The resultsof Sect. 3 can be extended to viscous fluids. For the particular choice $_\varepsilon(x) = T_\varepsilon(x)$, it would recover the results in [5].

Asymptotic Behavior of a Viscous Fluid in the Rough Domain Ω_ε Defined by (3)

iven $f \in L^2(\omega \times \mathbb{R})^3$, we consider the following Stokes system posed in Ω_ε defined (3) satisfying the Navier condition on Γ_ε defined by (4), and adherence conditions the rest of the boundary $\partial\Omega_\varepsilon \setminus \Gamma_\varepsilon$,

$$\begin{cases} -\Delta u_\varepsilon + \nabla p_\varepsilon = f \text{ in } \Omega_\varepsilon, \quad \text{div } u_\varepsilon = 0 \text{ in } \Omega_\varepsilon \\ u_\varepsilon \in T_\varepsilon, \text{ on } \Gamma_\varepsilon, \quad \dfrac{\partial u_\varepsilon}{\partial v} + \gamma u_\varepsilon \in T_\varepsilon^\perp, \text{ on } \Gamma_\varepsilon, \quad u_\varepsilon = 0 \quad \text{on } \partial\Omega_\varepsilon \setminus \Gamma_\varepsilon. \end{cases} \quad (6$$

We prove the following existence and uniqueness result.

Theorem 1. *The system* (6) *has a unique solution* $(u_\varepsilon, p_\varepsilon) \in H^1(\Omega_\varepsilon)^3 \times L_0^2(\Omega_\varepsilon$ $(L_0^2(\Omega_\varepsilon)$ *denotes the space of functions in* $L^2(\Omega_\varepsilon)$ *whose integral in* Ω_ε *is zero Moreover, we prove that there exists* $C > 0$ *such that*

$$\|u_\varepsilon\|_{H^1(\Omega_\varepsilon)^3} + \|p_\varepsilon\|_{L^2(\Omega_\varepsilon)} \leq C, \quad \forall \varepsilon > 0.$$

Proof. See [15].

Our aim is to investigate the asymptotic behavior of the solution $(u_\varepsilon, p_\varepsilon)$. As consequence of Theorem 1, letting $\varepsilon \to 0$, we may infer that there exist $u \in H^1(\Omega)$ and $p \in L^2(\Omega)$ such that

$$u_\varepsilon \rightharpoonup u \text{ in } H^1(\Omega)^3, \qquad p_\varepsilon \rightharpoonup p \text{ in } L^2(\Omega), \quad \text{with } \Omega = \omega \times (0,1).$$

Then, our goal becomes identifying the limit problem satisfied by (u, p), whic is given in the following theorem.

Theorem 2. *The couple* (u, p) *is the unique solution of*

$$\begin{cases} -\Delta u + \nabla p = f \text{ in } \Omega, \quad \text{div } u = 0 \text{ in } \Omega \\ u = 0 \text{ on } \partial\Omega \setminus \Gamma, \end{cases} \quad ($$

plus a boundary condition for u *on* $\Gamma = \omega \times \{0\}$ *which depends on the paramet* λ *defined by* (5). *More precisely we have*

(i) If $\lambda = 0$, *then*

$$u_3 \in T \text{ on } \Gamma, \quad -\partial_3 u' + \gamma u' \in T^\perp \text{ on } \Gamma. \quad ($$

(ii) If $\lambda \in (0, +\infty)$, *then defining* $(\hat{\phi}^i, \hat{q}^i)$, $i = 1, 2$ *as a solution of*

$$\begin{cases} -\Delta_z \hat{\phi}^i + \nabla_z \hat{q}^i = 0, \quad \text{div}_z \hat{\phi}^i = 0 \text{ in } \mathbb{R}^2 \times (0, +\infty) \\ \hat{\phi}_3^i(z', 0) + \partial_{z_i} \Psi(z') = 0, \quad \partial_{z_3}(\hat{\phi}^i)'(z', 0) = 0, \quad a.e. \ z' \in \mathbb{R}^2 \\ \hat{\phi}^i(., z_3), \ \hat{q}^i(., z_3) \text{ periodic of period } Z', \quad a.e. \ z_3 \in (0, +\infty) \\ \hat{\phi}^i \in H^1(Z' \times (0, +\infty))^3, \ \hat{q}^i \in L^2(Z' \times (0, +\infty)), \end{cases} \quad ($$

and $R \in \mathbb{R}^{2 \times 2}$ *by*

$$R_{ij} = \int_{Z' \times (0, +\infty)} D_z \hat{\phi}^i : D_z \hat{\phi}^j \, dz, \quad \forall i, j \in \{1, 2\}, \quad (1$$

we have

$$u_3 \in T \text{ on } \Gamma, \quad -\partial_3 u' + \gamma u' + \lambda^2 R u' \in T^\perp \text{ on } \Gamma. \tag{11}$$

(iii)   If $\lambda = +\infty$, then defining

$$W = Span(\{\nabla \Psi(z') : z' \in Z'\})^\perp, \tag{12}$$

we have

$$u_3 \in T, \text{ on } \Gamma, \quad u' \in W, \text{ on } \Gamma, \quad -\partial_3 u' + \gamma u' \in W^\perp, \text{ on } \Gamma \tag{13}$$

Proof. See [15]. It is based on an original adaptation of the unfolding method, [1, 9, 12], which is very related to the two-scale convergence method. □

Remark 1. For $\lambda = 0$, Theorem 2 shows that the roughness of Γ_ε is very slight and so the solution $(u_\varepsilon, p_\varepsilon)$ of (6) behaves as if Γ_ε coincides with the plane boundary Γ. For $\lambda \in (0, +\infty)$ (critical size), the boundary condition satisfied by the limit u of u_ε on the tangent space to Γ contains the new term $\lambda^2 R u'$. In this case, the effect of the roughness of Γ_ε is not worthless and it makes to appear this new term in the limit. Finally, for $\lambda = +\infty$ the roughness of Γ_ε is so strong that the limit u of u_ε does not only satisfies the condition $u_3 \in T$, i.e., $u_3 = 0$, on Γ, but also its tangent velocity on Γ, u', is orthogonal to the vectors $\nabla \Psi(z')$, with $z' \in Z'$. In particular, if the space W defined by (12) has dimension 0, then u satisfies the adherence condition $u = 0$ on Γ.

Remark 2. The case $\lambda \in (0, +\infty)$ can be considered as the general one. In fact, if λ tends to zero or infinity in (11), we get (8) or (13), respectively.

Remark 3. Theorem 2 provides an approximation of $(u_\varepsilon, p_\varepsilon)$ in the weak topology of $H^1(\Omega)^3 \times L^2(\Omega)$. We refer to [9] for corrector results, i.e., strong approximation of the solution of the Stokes system (6), and to [11] for error estimates between the solution and its corrector.

Remark 4. The asymptotic behavior of viscous fluids confined in a thin domain of height h_ε with a rough boundary described by (4), such that $\delta_\varepsilon \ll \varepsilon \ll h_\varepsilon$, has been considered in [10], where we obtain a Reynolds system in the limit which shows that near the rough bottom Γ_ε the behavior of the fluid is similar to the one obtained in Theorem 2 but with λ replaced by $\lambda_{thin} = \lim_{\varepsilon \to 0} \dfrac{\delta_\varepsilon h_\varepsilon^{1/2}}{\varepsilon^{3/2}}$. Remark that $\lambda = \lambda_{thin}$ if $\varepsilon = 1$.

3 Asymptotic Behavior of Elliptic Partial System in General Rough Domains

In the previous section we have shown that the Navier boundary condition for the Stokes system provides a new term in the limit problem. In this section we study this phenomenon for linear elliptic systems in rough domains $\Omega_\varepsilon \subset \mathbb{R}^N$, where Ω_ε is not necessarily a periodic structure.

We consider a sequence of Lipschitz open sets $\Omega_\varepsilon \subset \mathbb{R}^N$ which converges to a Lipschitz open set $\Omega \subset \mathbb{R}^N$ in the following sense: For every $\rho > 0$, there exists $\varepsilon_0 > 0$ such that for every $\varepsilon \in (0, \varepsilon_0)$, we have

$$\Omega^{\rho^-} = \{x \in \Omega : d(x, \partial\Omega) > \rho\} \subset \Omega_\varepsilon \subset \{x \in \mathbb{R}^N : d(x, \overline{\Omega}) < \rho\} = \Omega^{\rho^+}. \quad (14)$$

We denote by $\tilde{\Omega}$ an open set containing strictly Ω.

In Ω_ε, we consider the following homogenization problem:

$$\begin{cases} -\mathrm{div}ADu_\varepsilon = f & \text{in } \Omega_\varepsilon \\ u_\varepsilon \in V_\varepsilon, \quad \forall x \in \partial\Omega_\varepsilon, \quad ADu_\varepsilon \cdot v \in V_\varepsilon^\perp, \quad \forall x \in \partial\Omega_\varepsilon, \end{cases} \quad (15)$$

where A belongs to $L^\infty(\tilde{\Omega}; \mathscr{T}_{M \times N})$ ($\mathscr{T}_{M \times N}$ is the space of linear applications from the space of matrices $\mathscr{M}_{M \times N}$ into itself), V_ε is an arbitrary sequence of functions from $\partial\Omega_\varepsilon$ into the set of linear subspaces of $\mathbb{R}^M$, v denotes the unitary outside normal vector to Ω_ε on $\partial\Omega_\varepsilon$, and the second member f is a function in $L^2(\tilde{\Omega})^M$.

We also assume the following ellipticity condition: there exists $\alpha > 0$ such that

$$\begin{cases} \alpha \|v\|_{H^1(\Omega_\varepsilon)^M}^2 \leq \displaystyle\int_{\Omega_\varepsilon} ADv : Dv \, dx, \\ \forall v \in H^1(\Omega_\varepsilon)^M, \ v \in V_\varepsilon, \ \text{a.e. } x \in \partial\Omega_\varepsilon. \end{cases}$$

Observe that this ellipticity condition is written in an integral form instead of a pointwise one. This is more convenient for systems where the pointwise and integral ellipticity conditions are not equivalent. In particular it permits to deal with the linear elasticity system, where the tensor only depends on the symmetric part of the derivative.

Assuming that $V_\varepsilon(x) = T_\varepsilon(x)$, with $T_\varepsilon(x)$ the tangent space in the point $x \in \partial\Omega_\varepsilon$ the oscillating boundary condition in (15) is equivalent to the Navier boundary condition [see (2)] considered in Sect. 2. Some other choices of V_ε are also interesting, see [15]. For example, taking S_ε an arbitrary subset of $\partial\Omega_\varepsilon$, and defining V_ε as $V_\varepsilon(x) = \{0\}$ for $x \in S_\varepsilon$, and $V_\varepsilon(x) = \mathbb{R}^N$ for $x \in \partial\Omega_\varepsilon \setminus S_\varepsilon$, the problem represents the homogenization of elliptic partial systems with Dirichlet and Neumann conditions on varying subsets of the boundary.

Our main result in this section is the following theorem.

Theorem 3. *There exist a subsequence of ε, still denoted by ε, a Borel measure μ in $\partial\Omega$ which vanishes on the sets of null capacity, a μ-measurable function R : $\partial\Omega \to \mathscr{M}_{M\times M}$, with*

$$R\xi\cdot\xi \geq 0, \ |R\xi\cdot\eta| \leq \beta(R\xi\cdot\xi)^{\frac{1}{2}}(R\eta\cdot\eta)^{\frac{1}{2}}, \quad \forall\xi,\eta\in\mathbb{R}^N, \ \mu\text{-a.e. in } \partial\Omega,$$

for some $\beta > 0$, and an application V from $\partial\Omega$ into the set of linear subspaces of $\mathbb{R}^M$, satisfying

$$\alpha\|v\|^2_{H^1(\Omega)^M} \leq \int_\Omega ADv:Dv\,dx + \int_{\partial\Omega} Rv\cdot v\,d\mu, \quad \forall v\in H^1(\Omega)^M, \ v\in V \text{ q.e. on } \partial\Omega,$$

with the following property: For every $f \in L^2(\Omega)^M$, the unique solution of (15) converges weakly in $H^1(\Omega^{\rho^-})$, for every $\rho > 0$, to the unique solution $u \in H^1(\Omega)^M$ of problem

$$\begin{cases} -div\,ADu = f & in \ \Omega \\ u\in V, \quad q.e.\,x\in\partial\Omega, \quad \int_{\partial\Omega} Ru\cdot u\,d\mu < +\infty \\ ADu\cdot v + Ru\mu \in V^\perp, & q.e. \ in \ \partial\Omega. \end{cases} \tag{16}$$

Proof. The proof is given in [15]. It is based on a representation theorem strongly related with the one given in [14]. □

Using the ideas considered to prove Theorem 3, and assuming that Ω_ε satisfies the uniform cone condition (see [5]), we can prove the following result relative to the asymptotic behavior of viscous fluids in rough domains not necessarily periodic.

Theorem 4. *There exist a measure μ, a μ-measure function R, and an application V in the statements of Theorem 3 such that for every $f \in L^2(\tilde\Omega)^M$, the unique solution of*

$$\begin{cases} -\Delta u_\varepsilon + u_\varepsilon + \nabla p_\varepsilon = f \ in \ \Omega_\varepsilon, \quad div\,u_\varepsilon = 0 \ in \ \Omega_\varepsilon \\ u_\varepsilon\in V_\varepsilon \ on \ \partial\Omega_\varepsilon, \quad \dfrac{\partial u_\varepsilon}{\partial v} + \gamma u_\varepsilon \in V_\varepsilon^\perp \ on \ \partial\Omega_\varepsilon \end{cases} \tag{17}$$

converges weakly in $H^1(\Omega^{\rho^-})^3 \times L^2(\Omega^{\rho^-})$, for every $\rho > 0$, to the unique solution $(u,p) \in H^1(\Omega)^3 \times L^2(\Omega)$ of problem

$$\begin{cases} -\Delta u + u + \nabla p = f & in \ \Omega \\ u\in V, \quad q.e. \ in \ x\in\partial\Omega, \quad \int_{\partial\Omega} Ru\cdot u\,d\mu < +\infty \\ \dfrac{\partial u}{\partial v} + \gamma u + Ru\mu \in V^\perp, & q.e. \ in \ \partial\Omega. \end{cases}$$

Remark 5. For the particular choice $V_\varepsilon(x) = T_\varepsilon(x)$ for every $x \in \partial\Omega_\varepsilon$, we would cover the results given in [5].

Acknowledgements This work has been partially supported by the project MTM2011-24457 of the "Ministerio de Ciencia e Innovación" and the research group of the "Junta de Andalucía" FQM309.

References

1. Arbogast, T., Douglas, J., Hornung U.: Derivation of the double porosity model of single phase flow via homogenization theory. SIAM J. Math. Anal. **21**, 823–836 (1990)
2. Bucur, D., Feireisl, E., Nečasová, S., Wolf, J.: On the asymptotic limit of the Navier–Stokes system on domains with rough boundaries. J. Differ. Equ. **244**, 2890–2908 (2008)
3. Bucur, D., Feireisl, E., Nečasová, S.: Influence of wall roughness on the slip behavior of viscous fluids. Proc. Royal Soc. Edinburgh A **138**(5), 957–973 (2008)
4. Bucur, D., Feireisl, E., Nečasová, S.: On the asymptotic limit of flows past a ribbed boundary. J. Math. Fluid Mech. **10**(4), 554–568 (2008)
5. Bucur, D., Feireisl, E., Nečasová, S.: Boundary behavior of viscous fluids: Influence of wall roughness and friction-driven boundary conditions. Arch. Rational Mech. Anal. **197**, 117–138 (2010)
6. Calvo-Jurado, C., Casado-Díaz, J., Luna-Laynez, M.: Homogenization of elliptic problems with Dirichlet and Neumann conditions imposed on varying subsets. Math. Methods Appl. Sci. **30**(14), 1611–1625 (2007)
7. Calvo-Jurado, C., Casado-Díaz, J., Luna-Laynez, M.: Asymptotic behavior of nonlinear systems in varying domains with boundary conditions on varying sets. ESAIM Control Optim. Calc. Var. **15**, 49–67 (2009)
8. Casado-Díaz, J., Fernández-Cara, E., Simon, J.: Why viscous fluids adhere to rugose walls: A mathematical explanation. J. Differ. Equ. **189**(2), 526–537 (2003)
9. Casado-Díaz, J., Luna-Laynez, M., Suárez-Grau, F.J.: Asymptotic behavior of a viscous fluid with slip boundary conditions on a slightly rough wall. Math. Mod. Meth. Appl. Sci. **20**, 121–156 (2010)
10. Casado-Díaz, J., Luna-Laynez, M., Suárez-Grau, F.J.: A viscous fluid in a thin domain satisfying the slip condition on a slightly rough boundary. C. R. Acad. Sci. Paris Ser. I **348**, 967–971 (2010)
11. Casado-Díaz, J., Luna-Laynez, M., Suárez-Grau, F.J.: Estimates for the asymptotic expansion of a viscous fluid satisfying Navier's law on a rugous boundary. Math. Meth. Appl. Sci. **34**, 1553–1561 (2011)
12. Cioranescu, D., Damlamian, A., Griso, G.: Periodic unfolding and homogenization. C. R. Acad. Sci. Paris Sér. I **335**, 99–104 (2002)
13. Cioranescu, D., Murat, F.: Un terme trange venu d'ailleurs. Nonlinear partial differentia equations and their applications. In: Brzis H., Lions J.L. (eds.) Collge de France seminar, vols II and III. Research Notes in Mathematics 60 and 70, pp. 98–138 and 154–78. Pitman, Londo (1982)
14. Dal Maso, G., Defranceschi, A., Vitali, E.: A characterization of C^1-convex sets in Sobole spaces. Manuscripta Math. **75**, 247–272 (1992)
15. Suárez-Grau, F.J.: Comportamiento asintótico de fluidos viscosos con condiciones de desliz miento sobre fronteras rugosas. Ph.D. Dissertation, Universidad de Sevilla (2011)

A Double Complex Construction and Discrete Bogomolny Equations

Volodymyr Sushch

Abstract We study discrete models which are generated by the self-dual Yang–Mills equations. Using a double complex construction, we construct a new discrete analog of the Bogomolny equations. Discrete Bogomolny equations, a system of matrix-valued difference equations, are obtained from discrete self-dual equations. The gauge invariance of the discrete model is established.

Keywords Discrete model • Difference equations • Bogomolny equations • Yang-Mills equations

Introduction

This work is concerned with discrete model of the $SU(2)$ self-dual Yang–Mills equations described in [11]. It is well known that the self-dual Yang–Mills equations admit reduction to the Bogomolny equations [1]. Let A be an $SU(2)$-connection in $\mathbb{R}^3$. This means that A is an $su(2)$-valued 1-form and we can write

$$A = \sum_{i=1}^{3} A_i(x)\mathrm{d}x^i, \tag{1}$$

where $A_i : \mathbb{R}^3 \to su(2)$. Here $su(2)$ is the Lie algebra of $SU(2)$. The connection A is also called a gauge potential with the gauge group $SU(2)$ (see [8] for more details). Given the connection A, we define the curvature 2-form F by

$$F = \mathrm{d}A + A \wedge A, \tag{2}$$

Sushch (✉)
Koszalin University of Technology, Sniadeckich 2, 75-453 Koszalin, Poland
e-mail: volodymyr.sushch@tu.koszalin.pl

Pinelas et al. (eds.), *Differential and Difference Equations with Applications*, Springer Proceedings in Mathematics & Statistics 47, DOI 10.1007/978-1-4614-7333-6_57,
© Springer Science+Business Media New York 2013

where $\wedge$ denotes the exterior multiplication of differential forms. Let $\Phi : \mathbb{R}^3 \to su(2$ be a scalar field (a Higgs field). The Bogomolny equations are a set of nonline partial differential equations, where unknown is a pair (A, Φ). These equations ca be written as

$$F = *d_A \Phi, \tag{3}$$

where $*$ is the Hodge star operator on $\mathbb{R}^3$ and d_A is the covariant exterior differenti operator. This operator is defined by the formula

$$d_A \Omega = d\Omega + A \wedge \Omega + (-1)^{r+1} \Omega \wedge A,$$

where Ω is an arbitrary $su(2)$-valued r-form.

Let us now consider the connection A on $\mathbb{R}^4$. We define A to be

$$A = \sum_{i=1}^{3} A_i(x) dx^i + \Phi(x) dx^4, \tag{4}$$

where A_i and Φ are independent of x^4. In other words, the scalar field Φ is identifie with a fourth component A_4 of the connection A. It is easy to check that if the pa (A, Φ) satisfies Eq. (3), then the connection (4) is a solution of the self-dual equatic

$$F = *F. \tag{5}$$

In fact, the Bogomolny equations can be obtained from the self-dual equations t using dimensional reduction from $\mathbb{R}^4$ to $\mathbb{R}^3$ [1].

The aim of this paper is to construct a discrete model of Eq. (3) that preserves th geometric structure of the original continual object. This means that speaking of discrete model, we mean not only the direct replacement of differential operato by difference ones but also a discrete analog of the Riemannian structure ov a properly introduced combinatorial object. The idea presented here is strongl influenced by the book by Dezin [3]. Using a double complex construction, w construct a new discrete analog of the Bogomolny equations. In much the same wa as in the continual case, these discrete equations are obtained from discrete se dual equations. The gauge invariance of the discrete model is proved. We contin the investigations [10, 11], where discrete analogs of the self-dual and anti-self-du equations on a double complex are studied. It should be noted that there are mal other approaches to discretization of Yang–Mills theories. As the list of pape on the subject is very large, we content ourselves by referencing the works [2, 4 7, 9]. In these papers some other discrete versions of the Bogomolny equations a studied.

2 Double Complex Construction

The double complex construction is described in [10]. For the convenience of the reader we briefly repeat the relevant material from [10] without proofs. Let the tensor product $C(n) = C \otimes \ldots \otimes C$ of a 1-dimensional complex C be a combinatorial model of the Euclidean space $\mathbb{R}^n$. The 1-dimensional complex C is defined in the following way. Let C^0 denote the real linear space of 0-dimensional chains generated by basis elements x_i (points), $i \in \mathbb{Z}$. It is convenient to introduce the shift operator τ in the set of indices by

$$\tau i = i + 1.$$

We denote the open interval $(x_i, x_{\tau i})$ by e_i. We regard the set $\{e_i\}$ as a set of basis elements of the real linear space C^1 of 1-dimensional chains. Then the 1-dimensional complex (combinatorial real line) is the direct sum of the spaces introduced above: $C = C^0 \oplus C^1$. The boundary operator ∂ on the basis elements of C is given by

$$\partial x_i = 0, \qquad \partial e_i = x_{\tau i} - x_i. \tag{6}$$

The definition is extended to arbitrary chains by linearity.

Multiplying the basis elements x_i and e_i of C in various ways, we obtain the basis elements of $C(n)$. Let $s_k^{(r)} = s_{k_1} \otimes \ldots \otimes s_{k_n}$, where $k = (k_1, \ldots, k_n)$ and $k_i \in \mathbb{Z}$, be an arbitrary r-dimensional basis element of $C(n)$. The product contains exactly r of 1-dimensional elements e_{k_i} and $n - r$ of 0-dimensional elements x_{k_i}. The superscript (r) also uniquely determines an r-dimensional basis element of $C(n)$. For example, the 1-dimensional e_k^i and 2-dimensional ε_k^{ij} basis elements of $C(3)$ can be written as

$$e_k^1 = e_{k_1} \otimes x_{k_2} \otimes x_{k_3}, \quad e_k^2 = x_{k_1} \otimes e_{k_2} \otimes x_{k_3}, \quad e_k^3 = x_{k_1} \otimes x_{k_2} \otimes e_{k_3},$$

$$\varepsilon_k^{12} = e_{k_1} \otimes e_{k_2} \otimes x_{k_3}, \quad \varepsilon_k^{13} = e_{k_1} \otimes x_{k_2} \otimes e_{k_3}, \quad \varepsilon_k^{23} = x_{k_1} \otimes e_{k_2} \otimes e_{k_3},$$

where $k = (k_1, k_2, k_3)$ and $k_i \in \mathbb{Z}$.

Now we consider a dual object of the complex $C(n)$. Let $K(n)$ be a cochain complex with $gl(2, \mathbb{C})$-valued coefficients, where $gl(2, \mathbb{C})$ is the Lie algebra of the group $GL(2, \mathbb{C})$. We suppose that the complex $K(n)$, which is a conjugate of $C(n)$, has a similar structure: $K(n) = K \otimes \ldots \otimes K$, where K is a dual of the 1-dimensional complex C. We will write the basis elements of K as x^i, e^i. Then an arbitrary basis element of $K(n)$ is given by $s^k = s^{k_1} \otimes \ldots \otimes s^{k_n}$, where s^{k_i} is either x^{k_i} or e^{k_i}. For an r-dimensional cochain $\varphi \in K(n)$, we have

$$\varphi = \sum_k \sum_r \varphi_k^{(r)} s_{(r)}^k, \tag{7}$$

here $\varphi_k^{(r)} \in gl(2, \mathbb{C})$. We will call cochains forms, emphasizing their relationship with the corresponding continual objects, differential forms.

We define the pairing operation for arbitrary basis elements $\varepsilon_k \in C(n)$, $s^k \in K(n)$ by the rule

$$< \varepsilon_k, \, as^k >= \begin{cases} 0, & \varepsilon_k \neq s_k \\ a, & \varepsilon_k = s_k, \quad a \in gl(2, \mathbb{C}). \end{cases} \tag{8}$$

Here for simplicity the superscript (r) is omitted. The operation (8) is linearly extended to cochains.

The operation ∂ induces the dual operation d^c on $K(n)$ in the following way:

$$< \partial \varepsilon_k, \, as^k >=< \varepsilon_k, \, \mathrm{ad}^c s^k > . \tag{9}$$

For example, if φ is a 0-form, i.e., $\varphi = \sum_k \varphi_k x^k$, where $x^k = x^{k_1} \otimes \ldots \otimes x^{k_n}$, then

$$\mathrm{d}^c \varphi = \sum_k \sum_{i=1}^{n} (\Delta_i \varphi_k) e_i^k, \tag{10}$$

where e_i^k is the 1-dimensional basis elements of $K(n)$ and

$$\Delta_i \varphi_k = \varphi_{\tau_i k} - \varphi_k. \tag{11}$$

Here the shift operator τ_i acts as

$$\tau_i k = (k_1, \ldots, \tau k_i, \ldots, k_n).$$

The coboundary operator d^c is an analog of the exterior differentiation operator d.

Introduce a cochain product on $K(n)$. We denote this product by $\cup$. In terms of the homology theory this is the so-called Whitney product. For the basis elements of 1-dimensional complex K, the $\cup$-product is defined as follows:

$$x^i \cup x^i = x^i, \quad e^i \cup x^{\tau i} = e^i, \quad x^i \cup e^i = e^i, \quad i \in \mathbb{Z},$$

supposing the product to be zero in all other cases. By induction we extend this definition to basis elements of $K(n)$ (see [10] for details). For example, for the 1-dimensional basis elements $e_i^k \in K(3)$ we have

$$e_1^k \cup e_2^{\tau_1 k} = \varepsilon_{12}^k, \quad e_1^k \cup e_3^{\tau_1 k} = \varepsilon_{13}^k, \quad e_2^k \cup e_3^{\tau_2 k} = \varepsilon_{23}^k,$$

$$e_2^k \cup e_1^{\tau_2 k} = -\varepsilon_{12}^k, \quad e_3^k \cup e_1^{\tau_3 k} = -\varepsilon_{13}^k, \quad e_3^k \cup e_2^{\tau_3 k} = -\varepsilon_{23}^k. \tag{12}$$

To arbitrary forms the $\cup$-product be extended linearly. Note that the components of forms multiply as matrices. It is worth pointing out that for any forms $\varphi, \psi \in K(n)$ the following relation holds:

$$\mathrm{d}^c(\varphi \cup \psi) = \mathrm{d}^c \varphi \cup \psi + (-1)^r \varphi \cup \mathrm{d}^c \psi, \tag{13}$$

where r is the dimension of a form φ. For the proof we refer the reader to [3]. Relation (13) is a discrete analog of the Leibniz rule for differential forms.

Let us now together with the complex $C(n)$ consider its "double," namely, the complex $\tilde{C}(n)$ of exactly the same structure. Define the one-to-one correspondence

$$* : C(n) \to \tilde{C}(n), \qquad * : \tilde{C}(n) \to C(n) \tag{14}$$

the following way:

$$* : s_k^{(r)} \to \pm \tilde{s}_k^{(n-r)}, \qquad * : \tilde{s}_k^{(r)} \to \pm s_k^{(n-r)}, \tag{15}$$

here $\tilde{s}_k^{(n-r)} = *s_{k_1} \otimes \ldots \otimes *s_{k_n}$ and $*s_{k_i} = \tilde{e}_{k_i}$ if $s_{k_i} = x_{k_i}$ and $*s_{k_i} = \tilde{x}_{k_i}$ if $s_{k_i} = $
$_i$. We let the plus sign in (15) if a permutation of $(1,\ldots,n)$ with $(1,\ldots,n) \to$
$r),\ldots,(n-r))$ is representable as the product of an even number of transpositions
and the minus sign otherwise.

The complex of the cochains $\tilde{K}(n)$ over the double complex $\tilde{C}(n)$ has the same
structure as $K(n)$. Note that forms $\varphi \in K(n)$ and $\tilde{\varphi} \in \tilde{K}(n)$ have both the same
components. The operation (14) induces the respective mapping

$$* : K(n) \to \tilde{K}(n), \qquad * : \tilde{K}(n) \to K(n) \tag{16}$$

the rule: $<\tilde{c}, *\varphi> = <\tilde{c}, \varphi>$, $<c, *\tilde{\psi}> = <*c, \tilde{\psi}>$, where $c \in C(n)$, $\tilde{c} \in$
(n), $\varphi \in K(n)$, $\tilde{\psi} \in \tilde{K}(n)$. For example, for the 2-dimensional basis elements $\varepsilon_{ij}^k \in$
(3) we have

$$*\varepsilon_{12}^k = \tilde{e}_3^k, \quad *\varepsilon_{13}^k = -\tilde{e}_2^k, \quad *\varepsilon_{23}^k = \tilde{e}_1^k. \tag{17}$$

This operation is a discrete analog of the Hodge star operation. Similarly to the
continual case, we have $** \varphi = (-1)^{r(n-r)} \varphi$ for any discrete r-form $\varphi \in K(n)$.

Finally, for convenience we introduce the operation

$$\tilde{\iota} : K(n) \to \tilde{K}(n), \qquad \tilde{\iota} : \tilde{K}(n) \to K(n) \tag{18}$$

setting $\tilde{\iota} s_{(r)}^k = \tilde{s}_{(r)}^k$, $\tilde{\iota} \tilde{s}_{(r)}^k = s_{(r)}^k$. It is easy to check that the following hold:

$$\tilde{\iota}* = *\tilde{\iota}, \quad \tilde{\iota} d^c = d^c \tilde{\iota}, \quad \tilde{\iota} \varphi = \tilde{\varphi}, \quad \tilde{\iota}\tilde{\iota}\varphi = \varphi, \quad \tilde{\iota}(\varphi \cup \psi) = \tilde{\iota}\varphi \cup \tilde{\iota}\psi,$$

here $\varphi, \psi \in K(n)$.

Discrete Bogomolny Equations

et us consider a discrete $su(2)$-valued 0-form $\Phi \in K(3)$. We put

$$\Phi = \sum_k \Phi_k x^k, \tag{19}$$

where $\Phi_k \in su(2)$ and $x^k = x^{k_1} \otimes x^{k_2} \otimes x^{k_3}$ is the 0-dimensional basis element c $K(3)$, $k = (k_1, k_2, k_3)$, $k_i \in \mathbb{Z}$. We define a discrete $SU(2)$-connection A to be

$$A = \sum_k \sum_{i=1}^3 A_k^i e_i^k, \tag{20}$$

where $A_k^i \in su(2)$ and e_i^k is the 1-dimensional basis element of $K(3)$.

On account of (7), an arbitrary discrete 2-form $F \in K(3)$ can be written a follows:

$$F = \sum_k \sum_{i<j} F_k^{ij} \varepsilon_{ij}^k = \sum_k \left(F_k^{12} \varepsilon_{12}^k + F_k^{13} \varepsilon_{13}^k + F_k^{23} \varepsilon_{23}^k \right), \tag{2}$$

where $F_k^{ij} \in gl(2, \mathbb{C})$ and ε_{ij}^k is the 2-dimensional basis element of $K(3)$. Define discrete analog of the curvature form (2) by

$$F = d^c A + A \cup A. \tag{2}$$

By the definition of d^c (9) and using (12) we have

$$d^c A = \sum_k \sum_{i<j} (\Delta_i A_k^j - \Delta_j A_k^i) \varepsilon_{ij}^k, \tag{2}$$

$$A \cup A = \sum_k \sum_{i<j} (A_k^i A_{\tau_i k}^j - A_k^j A_{\tau_j k}^i) \varepsilon_{ij}^k. \tag{2}$$

Recall that Δ_i is the difference operator (11). Combining (23) and (24) with (21 we obtain

$$F_k^{ij} = \Delta_i A_k^j - \Delta_j A_k^i + A_k^i A_{\tau_i k}^j - A_k^j A_{\tau_j k}^i. \tag{2}$$

It should be noted that in the continual case the curvature form F takes valu in the algebra $su(2)$ for any $su(2)$-valued connection form A. Unfortunately, this not true in the discrete case because, generally speaking, the components $A_k^i A_{\tau_i k}^j$ $A_k^j A_{\tau_j k}^i$ of the form $A \cup A$ in (22) do not belong to $su(2)$. For a definition of th $su(2)$-valued discrete curvature form, we refer the reader to [11].

Define a discrete analog of the exterior covariant differential operator d_A as

$$d_A^c \varphi = d^c \varphi + A \cup \varphi + (-1)^{r+1} \varphi \cup A,$$

where φ is an arbitrary r-form (7) and A is given by (20). Then for the 0-form (19 we obtain

$$d_A^c \Phi = d^c \Phi + A \cup \Phi - \Phi \cup A. \tag{2}$$

Using (10) and the definition of $\cup$, we can rewritten (26) as follows:

$$d_A^c \Phi = \sum_k \sum_{i=1}^3 (\Delta_i \Phi_k + A_k^i \Phi_{\tau_i k} - \Phi_k A_k^i) e_i^k. \tag{27}$$

Applying the operation $*$ (16) to this expression and by (17) we find

$$* d_A^c \Phi = \sum_k (\Delta_1 \Phi_k + A_k^1 \Phi_{\tau_1 k} - \Phi_k A_k^1) \tilde{\varepsilon}_{23}^k$$

$$- \sum_k (\Delta_2 \Phi_k + A_k^2 \Phi_{\tau_2 k} - \Phi_k A_k^2) \tilde{\varepsilon}_{13}^k$$

$$+ \sum_k (\Delta_3 \Phi_k + A_k^3 \Phi_{\tau_3 k} - \Phi_k A_k^3) \tilde{\varepsilon}_{12}^k. \tag{28}$$

Now suppose that Φ in the form (19) is a discrete analog of the Higgs field. Then the discrete analog of the Bogomolny equation (3) is given by the formula

$$F = \tilde{\iota} * d_A^c \Phi, \tag{29}$$

where $\tilde{\iota}$ is the operation (17). From (21) and (28) it follows immediately that Eq. (29) is equivalent to the following difference equations:

$$\begin{aligned} F_k^{12} &= \Delta_3 \Phi_k + A_k^3 \Phi_{\tau_3 k} - \Phi_k A_k^3, \\ F_k^{13} &= -\Delta_2 \Phi_k - A_k^2 \Phi_{\tau_2 k} + \Phi_k A_k^2, \\ F_k^{23} &= \Delta_1 \Phi_k + A_k^1 \Phi_{\tau_1 k} - \Phi_k A_k^1. \end{aligned} \tag{30}$$

Consider now the discrete curvature form (22) in the 4-dimensional case, i. e., $F \in K(4)$. The discrete analog of the self-dual Eq. (5) can be written as follows:

$$F = \tilde{\iota} * F. \tag{31}$$

By the definition of $*$ for the 2-dimensional basis elements $\varepsilon_{ij}^k \in K(4)$, we have

$$*\varepsilon_{12}^k = \tilde{\varepsilon}_{34}^k, \quad *\varepsilon_{13}^k = -\tilde{\varepsilon}_{24}^k, \quad *\varepsilon_{14}^k = \tilde{\varepsilon}_{23}^k,$$

$$*\varepsilon_{23}^k = \tilde{\varepsilon}_{14}^k, \quad *\varepsilon_{24}^k = -\tilde{\varepsilon}_{13}^k, \quad *\varepsilon_{34}^k = \tilde{\varepsilon}_{12}^k.$$

Using this we may compute $*F$:

$$*F = \sum_k \left(F_k^{12} \tilde{\varepsilon}_{34}^k - F_k^{13} \tilde{\varepsilon}_{24}^k + F_k^{14} \tilde{\varepsilon}_{23}^k + F_k^{23} \tilde{\varepsilon}_{14}^k - F_k^{24} \tilde{\varepsilon}_{13}^k + F_k^{34} \tilde{\varepsilon}_{12}^k \right).$$

Then Eq. (31) becomes

$$F_k^{12} = F_k^{34}, \qquad F_k^{13} = -F_k^{24}, \qquad F_k^{14} = F_k^{23}. \tag{32}$$

Let the discrete connection 1-form $A \in K(4)$ be given by

$$A = \sum_k \sum_{i=1}^{3} A_k^i e_i^k + \sum_k \Phi_k e_4^k, \tag{33}$$

where $A_k^i \in su(2)$, $\Phi_k \in su(2)$ and $k = (k_1, k_2, k_3, k_4)$, $k_i \in \mathbb{Z}$. Note that here we put $A_k^4 = \Phi_k$ and Φ_k are the components of the discrete Higgs field. Suppose that the connection form (33) is independent of k_4, i.e.,

$$\Delta_4 A_k^i = 0, \qquad \Delta_4 \Phi_k = 0 \tag{34}$$

for any $i = 1, 2, 3$ and $k = (k_1, k_2, k_3, k_4)$. Substituting (34) into (25) yields

$$F_k^{i4} = \Delta_i \Phi_k + A_k^i \Phi_{\tau_i k} - \Phi_k A_k^i, \qquad i = 1, 2, 3.$$

Putting these expressions in Eq. (32) we obtain Eq. (30).

Thus, we have the following:

Theorem 1. *The discrete Bogomolny equation (29) and the discrete self-dual Eq. (31) are equivalent.*

Let us consider the $SU(2)$-valued 0-form

$$h = \sum_k h_k x^k, \tag{35}$$

where $h_k \in SU(2)$ and $x^k = x^{k_1} \otimes x^{k_2} \otimes x^{k_3}$ is the 0-dimensional basis element of $K(3)$. By analogy with classical Yang–Mills theories, we define a gauge transformation for the discrete potential $A \in K(3)$ and discrete field $\Phi \in K(3)$ as

$$A' = h \cup d^c h^{-1} + h \cup A \cup h^{-1}, \tag{36}$$

$$\Phi' = h \cup \Phi \cup h^{-1}, \tag{37}$$

where h^{-1} is the 0-form with inverse components (inverse matrices) of h. Suppose that the components $h_k \in SU(2)$ of (35) satisfy the following conditions:

$$h_{\tau_1 \tau_2 k} = h_{\tau_3 k}, \qquad h_{\tau_1 \tau_3 k} = h_{\tau_2 k}, \qquad h_{\tau_2 \tau_3 k} = h_{\tau_1 k} \tag{38}$$

for all $k = (k_1, k_2, k_3)$, $k_i \in \mathbb{Z}$. It is easy to check that the set of forms (35) satisfying conditions (38) is a group under $\cup$-product.

Theorem 2. *The discrete Bogomolny equation (29) is invariant under the gauge transformation (36) and (37), where h satisfies condition (38).*

Proof. Rewrite Eq. (29) in the form

$$\tilde{\imath} * F - d_A^c \Phi = 0. \tag{39}$$

The proof is based on Theorem 4.3 and Lemma 4.6 in [11]. Under the transformation (36) the curvature form (22) changes as

$$F' = h \cup F \cup h^{-1}.$$

Using conditions (38) and Lemma 4.6 of [11] we have

$$\tilde{\imath} * F' = \tilde{\imath} * (h \cup F \cup h^{-1}) = h \cup \tilde{\imath} * F \cup h^{-1}. \tag{40}$$

Since $d^c h \cup h^{-1} = -h \cup d^c h^{-1}$ by (13), (26), (36), and (37), we compute

$$d_{A'}^c \Phi' = d_{A'}^c (h \cup \Phi \cup h^{-1}) = h \cup d^c \Phi \cup h^{-1}$$
$$+ h \cup A \cup \Phi \cup h^{-1} - h \cup \Phi \cup A \cup h^{-1} = h \cup d_A^c \Phi \cup h^{-1}. \tag{41}$$

Comparing (40) and (41) we obtain

$$\tilde{\imath} * F' - d_{A'}^c \Phi' = h \cup (\tilde{\imath} * F - d_A^c \Phi) \cup h^{-1}.$$

Thus, if the pair (A, Φ) is a solution of Eq. (29), then (A', Φ') is also a solution of (29). □

References

1. Atiyah, M., Hitchin, N.: Geometry and Dynamics of Magnetic Monopoles, Princeton University Press, Princeton (1988)
2. Cherrington, J.W., Christensen, J.D.: A dual non-abelian Yang–Mills amplitude in four dimensions. Nucl. Phys. B **813**(FS), 370–382 (2009)
3. Dezin, A.A.: Multidimensional Analysis and Discrete Models, CRC Press, Boca Raton (1995)
4. Gross, D.J., Nekrasov, N.A.: Monopoles and strings in noncommutative gauge theory. J. High Energy Phys. **4**(7B), 034-0–034-33 (2000)
5. Kampmeijer, L., Slingerland, J.K., Schroers, B.J., Bais, F.A.: Magnetic charge lattices, moduli spaces and fusion rules. Nucl. Phys. B **806**, 386–435 (2009)
6. Koikawa, T.: Discrete and continuous Bogomolny equations through the deformed algebra. Phys. Lett. A **256**(4), 284–290 (1999)
7. Murray, M.K., Singer, M.A.: On the complete integrability of the discrete Nahm equations. Commun. Math. Phys. **210**, 497–519 (2000)
8. Nash, C., Sen, S.: Topology and Geometry for Physicists, Academic, London (1989)

9. Oeckl, R.: Discrete Gauge Theory: From Lattices to TQFT, Imperial College Press, London (2005)
10. Sushch V.: A gauge-invariant discrete analog of the Yang–Mills equations on a double complex. Cubo A Math. J. **8**(3), 61–78 (2006)
11. Sushch V.: Self-dual and anti-self-dual solutions of discrete Yang–Mills equations on a double complex. Cubo A Math. J. **12**(3), 99–120 (2010)

Pseudo-Differential Equations on Manifolds with Non-smooth Boundaries

Vladimir B. Vasilyev

Abstract We discuss the different variants of multidimensional Riemann boundary problem and suggest to use wave factorization to obtain solvability conditions for pseudo-differential equations in model non-smooth domains.

Keywords Pseudo-Differential equation • Riemann boundary problem • Wave factorization

1 Classical Case

The classical Riemann boundary problem in its simplest form [4], is to find piecewise analytic function, more precisely the function which is analytic in upper and lower complex plane and which satisfies the linear relation on a straight real line

$$\Phi^+(t) = G(t)\Phi^-(t) + g(t), \quad t \in \mathbf{R}. \tag{1}$$

If $\Phi(z)$, $z \in \mathbf{C} \setminus \mathbf{R}$ is analytic function, then $\Phi^\pm(t)$ denote its boundary values on $\mathbf{R}$ ($z = x + iy$, $y \to 0\pm$), $G(t)$ is called coefficient of the Riemann problem, and $g(t)$ and $G(t)$ are given functions on $\mathbf{R}$.

The solution of the problem (1) is constructed with the help of factorization of function G and Cauchy-type integral (one-dimensional singular integral):

$$\Phi(z) = \frac{1}{2\pi i} \int\limits_{-\infty}^{+\infty} \frac{\varphi(t)}{z-t} dt, \quad z \in \mathbf{C} \setminus \mathbf{R}. \tag{2}$$

B. Vasilyev (✉)
Chair of Pure Mathematics, Lipetsk State Technical University,
Moskovskaya 30, Lipetsk 398600, Russia
e-mail: vladimir.b.vasilyev@gmail.com

Pinelas et al. (eds.), *Differential and Difference Equations with Applications*, Springer Proceedings in Mathematics & Statistics 47, DOI 10.1007/978-1-4614-7333-6_58,
© Springer Science+Business Media New York 2013

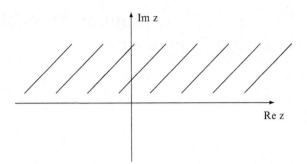

Fig. 1 One-dimensional case

Definition 1. Factorization of function $G(t)$ is called its representation in the form

$$G(t) = G_+(t)G_-(t), \tag{3}$$

where the functions $G_\pm(t)$ admit analytic continuation in upper and lower complex half-plane.

Key point for solving the problem (1) takes the formulas for limit boundary values for integral (2) and Sokhotski formulas

$$\Phi_+(t) - \Phi_-(t) = \varphi(t),$$

$$\Phi_+(t) + \Phi_-(t) = \frac{1}{\pi i} \text{ v.p.} \int_{-\infty}^{+\infty} \frac{\varphi(t)}{t - \tau} d\tau. \tag{4}$$

The last integral is treated in principal value sense and is called Hilbert transform of function φ :

$$(H\varphi)(t) = \frac{1}{\pi i} \text{ v.p.} \int_{-\infty}^{+\infty} \frac{\varphi(t)}{t - \tau} d\tau. \tag{5}$$

This problem permits a series of multidimensional generalizations. I will talk on these variants step by step.

2 Generalizations

For one-dimensional case the upper and lower half-plane is a set of complex numbers of the type

$$\mathbf{R} \pm i\mathbf{R}_+,$$

Im $z \subset \mathbf{R}_+$, and $\mathbf{R}_\pm$ is unique simplest one-dimensional cone.

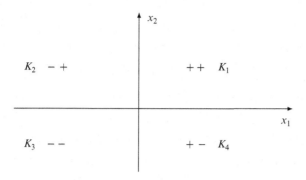

Fig. 2 Two-dimensional case

For two-dimensional case the situation is more complicated. The first generalizations were related to so-called bicylindrical domains [8, 11].

In this picture the imaginary part of two-dimensional bicylindrical domains is shown, and the two-dimensional Riemann problem is stated by the following way. We seek a function $\Phi(z_1, z_2)$ which is analytic in the four domains of complex space $\mathbf{C}^2$ of type $\mathbf{R}^2 + i\mathbf{K_m}$, $m = 1, 2, 3, 4$ (these domains are called radial tube domains over the cones K_m [7, 19]), and for which their boundary values (there are four boundary values in this case) satisfy the linear relation

$$A(x_1, x_2)\Phi^{++}(x_1, x_2) + B(x_1, x_2)\Phi^{-+}(x_1, x_2) +$$

$$+C(x_1, x_2)\Phi^{--}(x_1, x_2) + D(x_1, x_2)\Phi^{+-}(x_1, x_2) = f(x_1, x_2), \qquad (6)$$

$$(x_1, x_2) \in \mathbf{R}^2.$$

Although for one-dimensional case the problem (1) is completely solvable by the Cauchy-type integral (2), the two-dimensional analogue of Cauchy-type integral

$$\Phi(z_1, z_2) = \frac{1}{4\pi^2} \int\limits_{\mathbf{R}^2} \frac{\varphi(t_1, t_2)dt_1 dt_2}{(t_1 - z_1)(t_2 - z_2)} \qquad (7)$$

oesn't help for solving the problem (6).

For some special cases only it is possible constructing the solution with the help f the integral (7).

Another variant of multidimensional generalization of the Riemann problem was uggested by V.S. Vladimirov [18] (I will use the picture 2), which coincides with ae problem (1) in one-dimensional case and is formulated by the following way. inding the function $\Phi(z_1, z_2)$ which is analytic in radial tube domains [7,19] $T(K_1)$, (K_2) over the cones K_1, K_2, respectively, and for which boundary values satisfy the lear relation

$$\Phi_{++}(x_1, x_2) = G(x_1, x_2)\Phi_{--}(x_1, x_2) + g(x_1, x_2), \quad (x_1, x_2) \in \mathbf{R}^2, \qquad (8)$$

at this statement for such problem doesn't take into account the domains $T(K_2)$, (K_4).

All these problems mentioned above were solved by factorization method, namely, by function decomposition into the product of two factors admitting an analytic continuation into appropriate domain. In this way the different functional classes for solution were described, this sufficient factorization conditions were obtained, but in my point of view, no one from this multidimensional generalizations had future development and serious application.

I will describe now one variant of multidimensional Riemann problem and will show what consequences we can have stating from this statement and existence of a special factorization.

So, for simplicity, we consider the space $L_2(\mathbf{R}^2)$ and the space $A(\mathbf{R}^2)$ which is consisting of analytic functions in radial tube domain $T(K_1)$ and satisfying the condition [7, 19]

$$\sup_{y \in K_1} \int_{\mathbf{R}^2} |f(x+iy)|^2 dx < +\infty.$$

$B(\mathbf{R}^2)$ is an orthogonal complement of $A(\mathbf{R}^2)$ in the space $L_2(\mathbf{R}^2)$, so that

$$A(\mathbf{R}^2) \oplus B(\mathbf{R}^2) = \mathbf{L_2}(\mathbf{R}^2).$$

Further, the statement of multidimensional Riemann problem is a precise copy of (1):

$$\Phi^+(t) = G(t)\Phi^-(t) + g(t), \quad t \in \mathbf{R}^2 \tag{9}$$

with the one difference, the function Φ^+ is sought in $A(\mathbf{R}^2)$, and the function Φ^- is sought in the space $B(\mathbf{R}^2)$.

3 Equations

It was shown [21, 22] such statement for multidimensional boundary Riemann problem must give certain meaning to studying solvability of pseudo-differential equations in model non-smooth domains. If we consider the pseudo-differential equation

$$(Au)(x) = f(x), \quad x \in D, \tag{10}$$

in multidimensional domain $D \subset \mathbf{R}^m$, $m \geq 2$, for which its boundary is a smooth surface, then a model problem is the equation in the half-space:

$$(Au)(x) = f(x), \quad x \in \mathbf{R}^m_+. \tag{11}$$

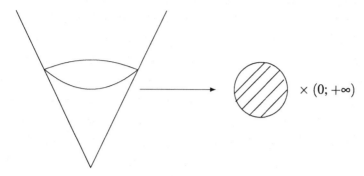

Fig. 3 Transformation of cone

Here A stands for a pseudo-differential operator with symbol non-depending on the pole x (the equation with "frozen coefficients"). Equation (2) in Fourier images is reduced to one-dimensional singular integral equation with the parameter $\xi' = (\xi_1, \ldots, \xi_{m-1})$, i.e., to the Riemann problem (1) which is solved by factorization method. This situation is studied in detail in papers' series of M.I. Vishik and G.I. Eskin [3, 16, 17], is fixed algebraically by L. Boutet de Monvel [1], and moved up to index theorem (see also S.Rempel, B.-W. Schulze [13]). But existence of one conical point only on a boundary forbids to use this theory.

I wrote many times on another approach to studying solvability for pseudo-differential equations in domains with conical points and wedges, but now I would like to speak on principal difference of my papers from other authors (V.G. Maz'ya [5, 6], B.A. Plamenevski [9, 10], B.-W. Schulze [14], and many others).

In all papers the conical domain (see Fig. 3) is treated as the direct product of a circle and a half-axis (but in my point of view, it is a cylinder, see for example [12]), then they apply the Mellin transform on half-axis, and the initial problem is reduced to a problem in a domain with a smooth boundary with operator-valued symbol. That follows further it is like the generalization of well-known results on operator symbol case. Of course, my approach is generalization also, but it is a generalization on dimension space, and the principal difference is that I don't divide the cone, and it is treated as an emergent thing.

In this way we meet the multidimensional Riemann boundary problem mentioned above; it permits to construct very interesting theory of pseudo-differential equations and boundary value problems in domains, for which their boundaries have singularities of "cone" and "wedge" type.

Let $C_+^a = \{x \in \mathbf{R}^{\mathbf{m}} : \mathbf{x_m} > \mathbf{a}|\mathbf{x'}|, \mathbf{x'} = (\mathbf{x_1}, \ldots, \mathbf{x_{m-1}}), \mathbf{a} > \mathbf{0}\}$ be a cone in m-dimensional space, $\overset{*}{C_+^a}$ be a conjugate cone, and $C_-^a = -C_+^a$, $T(\overset{*}{C_+^a})$ be a radical tube domain over the cone C_+^a [7, 19]. The model pseudo-differential equation in the cone C_+^a is the equation of type

$$(Au)(x) = f(x), \quad x \in C_+^a, \tag{12}$$

where A is pseudo-differential operator with the symbol $A(\xi)$, $\xi \in \mathbf{R^m}$, satisfying the condition

$$c_1 \leq |A(\xi)(1 + |\xi|)^{-\alpha}| \leq c_2, \tag{13}$$

c_1, c_2 are positive constants, and $\alpha \in \mathbf{R}$ is roughly speaking the order of a pseudo-differential operator.

Definition 2. Wave factorization of a symbol $A(\xi)$ with respect to C_+^a is called its representation in the form

$$A(\xi) = A_{\neq}(\xi)A_{=}(\xi),$$

and the factors $A_{\neq}(\xi), A_{=}(\xi)$ have satisfy the following conditions:

(1) $A_{\neq}(\xi), A_{=}(\xi)$ are defined on R^m without maybe the points $\{\xi \in \mathbf{R^m} : a\xi_m^2 = |\xi'|^2\}$.

(2) $A_{\neq}(\xi), A_{=}(\xi)$ admit an analytical condition into radial tube domains $T(\overset{*}{C}_{\pm}^a)$ over the cones $\overset{*}{C}_{\pm}^a$ respectively which satisfy the estimates

$$\left|A_{\neq}^{\pm 1}(\xi + i\tau)\right| \leq c(1 + |\xi| + |\tau|)^{\pm æ},$$

$$\left|A_{\neq}^{\pm 1}(\xi - i\tau)\right| \leq c(1 + |\xi| + |\tau|)^{\pm(\alpha - æ)}, \quad \tau \in \overset{*}{C}_+^a.$$

The number $æ \in \mathbf{R}$ is called index of wave factorization.

Existence of wave factorization permits to obtain the solution of multidimensional Riemann problem (9) by the special integral

$$(G_m u)(x) = \lim_{\tau \to 0+} \int_{\mathbf{R^m}} \frac{u(y', y_m)dy'dy_m}{(|x' - y'|^2 - a^2(x_m - y_m + i\tau)^2)^{m/2}}. \tag{14}$$

The integral G_m is a multidimensional analogue of the Cauchy-type integra[l] (more precisely, its limit case corresponding to boundary values). It looks as [a] convolution which kernel is Fourier image of C_+^a-indicator. But this multiplier i[s] not integrable function, and we need to go out into complex plane to destroy th[e] divergence (see [15]). Definition (14) is one of the possible definitions for suc[h] singular integral. Of course, it is very desirable to give this definition for re[al] variables (as principal value type of Cauchy integral like one-dimensional case[)] but I would like to note such definition was used in classical papers [2].

So, what can we obtain for solvability of Eq. (12), if we have wave factorizatio[n] for the symbol $A(\xi)$?

I will enumerate main conclusions which we can obtain (see [21, 22] for detail[s]) starting from existence of wave factorization for the symbol $A(\xi)$. We consid[er]

Sobolev–Slobodetski space $H^s(C_+^a)$ (there are functions from $H^s(\mathbf{R^m})$ with support in C_+^a). We study Eq. (3) in the space $H^s(C_+^a)$, and the right-hand side is fixed in the space $H_0^{s-\alpha}(C_+^0)$ [21, 22].

(1) The index of wave factorization determines fully the solvability cases for Eq. (12). If the solution is unique ($\text{æ} - s = \delta$, $|\delta| < 1/2$), then it can be written by integral (14). For the case $\text{æ} - s = n + \delta$, $n \in \mathbf{Z}$, $n > 0$, $|\delta| < 1/2$, there are many solutions, but we have the formula for a general solution which includes $2n$ arbitrary functions from corresponding Sobolev–Slobodetski spaces. Last, $\text{æ} - s = n + \delta$, $n \in \mathbf{Z}$, $n > 0$, $|\delta| < 1/2$, Eq. (3) is overdetermined, and the solvability conditions are given.

(2) There are many interesting applied problems, particularly, the diffraction problem of a spatial wave on a plane screen, and the problem of indentation of a wedge-shaped punch into elastic half-space. These problems are reduced to two-dimensional Eq. (12) for which the wave factorization for its symbol is constructed explicitly. Earlier these problems solved approximately, or solution's construction was very hard.

(3) It is very problematic that wave factorization exists for every symbol $A(\xi)$ satisfying (13). The author proved the class of such symbols is very wide. But we can't give the constructive algorithm for wave factorization in this time although in one-dimensional case (3), such factorization can be constructed by Cauchy-type integral (2).

(4) Although we have pessimistic point (3), the optimistic point (2) permits to construct wave factorization needed for two-dimensional case and the Laplacian, and taking into account point (1) to consider the classical Dirichlet and Neumann problems in a plane case. By transformations series (first Fourier, then Mellin transforms), the boundary value problems were reduced to equivalent system of linear algebraic equations. The unique solvability was verified by direct calculation for determinant needed. For a general case the unique solvability condition for such system of linear algebraic equation was called angle (conical) Shapiro–Lopatinski condition.

These results from (1) to (4) show the direction to more complicated singularities, so-called "thin" singularities.

These cases include such situations like plane cut in a space. The first preliminary sketches are presented in [23], and the author hopes to obtain something like (1)–(4) in this case also.

From my point of view any singularity corresponds to a certain distribution. It will be the distribution

$$\frac{a\Gamma(m/2)}{2\pi^{\frac{m+2}{2}}} \frac{1}{\left(|\xi'|^2 - a^2(\xi_m + i0)^2\right)^{m/2}}, \tag{15}$$

or the cone C_+^a, Γ is Euler function [21, 22].

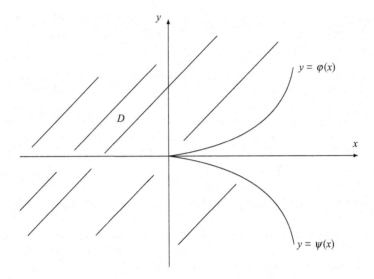

Fig. 4 Outer cusp point

If we try to find a limit under $a \to +\infty$, then we must obtain the distribution corresponding to singularity of one-dimensional cut (as a ray) in a plane. I calculated these limits for some cases both two-dimensional and multidimensional and obtained some interesting formulas. I give some of these results (see [23] for details).

4 Thin Singularities

Let us consider a two-dimensional domain of out cusp point (see Fig. 4) with vanishing angle, so that the functions $\varphi(x), \psi(x)$ are continuously differentiable on $[0, +\infty)$ and $\varphi'(0) = \psi'(0) = 0$. Obviously such domain will be diffeomorphic to $\mathbf{R}^2 \setminus [0, +\infty)$. Indeed the diffeomorphism for origin's neighborhood can be defined by the formulas

$$\begin{cases} \xi = x \\ \eta = y - \varphi(x) \end{cases}$$

for the points from the first quadrant and

$$\begin{cases} \xi = x \\ \eta = y - \psi(x) \end{cases}$$

for the points from the fourth quadrant, but the points from second and third quadrants must be in their own places. The Jacobian for such transformation will be the following:

$$\frac{D(\xi,\eta)}{D(x,y)} = \begin{vmatrix} 1 & 0 \\ -\varphi'(x) & 1 \end{vmatrix}, \qquad \frac{D(\xi,\eta)}{D(x,y)} = \begin{vmatrix} 1 & 0 \\ -\psi'(x) & 1 \end{vmatrix}$$

for the second and fourth quadrants and equals to 1 for the second and third quadrants. The Jacobian is continuous in origin's neighborhood and equals to 1 at the origin.

If in the origin's neighborhood we transfer to coordinates (ξ,η), then the singular integral operator with Calderon–Zygmund kernel

$$u(x) \longmapsto \int_D K(x, x-y)u(y)dy$$

is quasi-equivalent [21] to the operator

$$u(x) \longmapsto \int_{\mathbf{R}^2} K(0, \xi-\eta)u(\eta)d\eta.$$

Because for invertibility of the last operator we need nothing excluding ellipticity, if we construct for the Lebesgue integrable functions, for example, $L^2(\mathbf{R}^2)$), then we conclude the question on Noether property for the operator considered is solved. According to [21] it can be shown that ellipticity condition implies the index of such operator is vanishing.

If we consider the domain with singularity of inner cusp point type (see Fig. 5), then obviously the previous arguments don't work, and we suggest to use some "asymptotical" ideas. Such singularity can be treated as a limit case of cone when its size tends to zero. Here we give initial conclusions and results related to this approach.

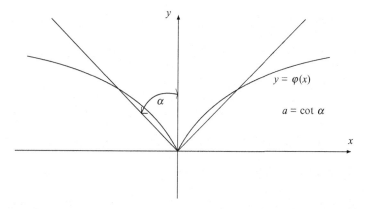

Fig. 5 Inner cusp point and approximating cone

We begin from two-dimensional case. The problem is to know what is Fourier image of multiplication operator on characteristic function of positive half-axis y. Analytically this multiplier is

$$m(x,y) = \begin{cases} 1, & x = 0, \ y > 0, \\ 0, & \text{in other cases.} \end{cases}$$

It is obviously a priori the Fourier image for such multiplier is a convolution operator for some distribution, and the distribution must be homogeneous of order -2.

The angle of size α is the set $\{(x,y) \in \mathbf{R}^2 : y > a|\mathbf{x}|\}$, $\mathbf{a} = \cot \alpha$, and then we need the asymptotic $(\alpha \to 0)$, i.e., $a \to \infty$. The distribution corresponding to such multiplier is [21, 22]

$$\begin{aligned} &\tfrac{1}{2}\delta(\xi) + K_a(\xi_1, \xi_2), \\ &K_a(\xi_1, \xi_2) = \tfrac{a}{2\pi^2} \tfrac{1}{\xi_1^2 - a^2(\xi_2 + i0)^2}, \end{aligned} \tag{16}$$

where $\xi = (\xi_1, \xi_2)$, $\delta(\xi)$ is Dirac mass function.

We need to find

$$\lim_{a \to \infty} \frac{a}{2\pi^2} \frac{1}{\xi_1^2 - a^2 \xi_2^2}$$

in distribution sense. Let $\varphi(\xi) \in S(\mathbf{R}^2)$ (Schwartz class of infinitely differentiable rapidly decreasing at infinity functions), and then we have [23]

Theorem 1. *The following formula holds:*

$$\lim_{a \to \infty} \frac{a}{2\pi^2} \frac{1}{\xi_1^2 - a^2 \xi_2^2} = \frac{i}{2\pi} \mathrm{P} \frac{1}{\xi_1} \otimes \delta(\xi_2), \tag{17}$$

where the notation for distribution P is taken from V.S. Vladimirov's book [20], and $\otimes$ denotes the direct product of distributions.

So distribution (17) is that corresponds to half-infinite crack (of course with mass supplement).

If we find another asymptotic for distribution (16) $a \to 0$, then we have

$$\lim_{a \to \infty} \frac{a}{2\pi^2} \frac{1}{\xi_1^2 - a^2 \xi_2^2} = \frac{1}{2\pi i} \delta(\xi_1) \otimes \mathrm{P} \frac{1}{\xi_2}, \tag{18}$$

and it corresponds to half-plane case (see [3]).

Now we will speak on another asymptotic related to multi-wedge angle. The simplest variant of such angle is the following: $\{x \in \mathbf{R}^3 : x_3 > a|x_1| + b|x_2|\}$, and we have two parameters a, b. If the parameters tend to 0 or ∞, we obtain new type of thin singularities.

The distribution corresponding to such angle is [21, 22]

$$K_{a,b}(\xi_1, \xi_2, \xi_3) = \frac{4iab}{(2\pi)^3} \frac{\xi_3}{\left(\xi_1^2 - a^2\xi_3^2\right)\left(\xi_2^2 - b^2\xi_3^2\right)}.$$

We consider the different relations between a and b.

Theorem 2. *The following formula holds:*

$$\lim_{b \to \infty} \frac{4iab\,\xi_3}{(2\pi)^3 \left(\xi_1^2 - a^2\xi_3^2\right)\left(\xi_2^2 - b^2\xi_3^2\right)} = \frac{i}{2\pi}\delta(\xi_1) \otimes P\frac{1}{\xi_2} \otimes \delta(\xi_3).$$

Analogously one can obtain

Theorem 3. *The equality*

$$\lim_{a \to \infty} \frac{4iab\,\xi_3}{(2\pi)^3 \left(\xi_1^2 - a^2\xi_3^2\right)\left(\xi_2^2 - b^2\xi_3^2\right)} = \frac{i}{2\pi}P\frac{1}{\xi_1} \otimes \delta(\xi_2) \otimes \delta(\xi_3)$$

is valid.

Theorem 4. *The equality*

$$\lim_{b \to 0} \frac{4iab}{(2\pi)^3} \frac{\xi_3}{\left(\xi_1^2 - a^2\xi_3^2\right)\left(\xi_2^2 - b^2\xi_3^2\right)} = \delta(\xi_2) \otimes K_a(\xi_1, \xi_3)$$

is valid

[see formula (16)].

Theorem 5. *The equality*

$$\lim_{a \to 0} \frac{4iab}{(2\pi)^3} \frac{\xi_3}{\left(\xi_1^2 - a^2\xi_3^2\right)\left(\xi_2^2 - b^2\xi_3^2\right)} = \delta(\xi_1) \otimes K_b(\xi_2, \xi_3)$$

holds.

Theorem 6. *The equality*

$$\lim_{\substack{a \to 0 \\ b \to 0}} \frac{4iab}{(2\pi)^3} \frac{\xi_3}{\left(\xi_1^2 - a^2\xi_3^2\right)\left(\xi_2^2 - b^2\xi_3^2\right)} == \frac{1}{2\pi i}\delta(\xi') \otimes P\frac{1}{\xi_3}, \qquad \xi' = (\xi_1, \xi_2),$$

holds.

The last result corresponds to half-space case $x_3 > 0$ [3].

5   Conclusion

The author hopes such experiments will help to explain how to formulate the Noether property condition for multidimensional singular integral and pseudo-differential equations in domains with singularities mentioned. As we see the limit operator is more simple than initial ones. It may be this point will permit to find convenient form for these conditions.

References

1. Boutet de Monvel, L.: Boundary problems for pseudodifferential operators. Acta Math. **126**, 11–51 (1971)
2. Dyn'kin, E.M.: Methods of theory of singular integrals (Hilbert transform and Calderon–Zygmund theory). In: Results in science and technics, Modern Probl. Math. Fundamental branches, vol. 15, pp. 197–292. VINITI, Moscow (1986, in Russian)
3. Eskin, G.: Boundary value problems for elliptic pseudodifferential equations. AMS Providence, RI, (1981)
4. Gakhov, F.D.: Boundary Problems. Moscow, Nauka (1977 in Russian)
5. Kozlov, V.A., Maz'ya, V.G., Rossmann, J.: Elliptic boundary value problems in domains with point singularities. Am. Math. Soc. Math. Surv. Monogr. **52**, 1–414 (1997)
6. Maz'ya, V.G., Plamenevsky, B.A.: Elliptic boundary value problems on manifolds with singularities. Probl. Math. Anal. **6**, (1977). Leningrad St. Univ. Press, 85–145 (Russian)
7. Bochner, S., Martin, U.T.: Functions of several complex variables. Moscow, Publ. Foreign Lit. (1951 in Russian)
8. Kakichev, V.A.: Boundary linear conjugation problems for holomorphic functions in bicylindrical domains. In: Function theory, functional analysis and their applications, vol. 5, pp. 37–58. Kharkov University Press, Kharkov (1967 in Russian)
9. Nazarov, S.A., Plamenevsky, B.A.: Elliptic problems in domains with piecewise smooth boundaries. Walter de Gruyter, Berlin (1994)
10. Plamenevsky, B.A.: Algebras of pseudodifferential operators. Nauka, Moscow (1986 in Russian).
11. Rabinovich, V.S.: Multidimensional Wiener-Hopf equation for cones. In: Theory of functions Functional analysis and their Applications, vol. 5, pp. 37–58. Kharkov University Press Kharkov (1967 in Russian).
12. Rabinovich, V.S.: Pseudodifferential equations in unbounded domains with conic structure a infinity. Matem. Sbornik **89**, 77–96 (1969 in Russian)
13. Rempel, S., Schulze, B.-W.: Index theory of elliptic boundary problems. Akademie, Berli (1982)
14. Schulze, B.-W.: Pseudo-differential boundary value problems, conical singularities, an asymptotics. Akademie, Berlin (1994)
15. Stein, E.M., Weiss, G.: Introduction to Fourier Analysis on Euclidean Spaces. University Pres Princeton (1971)
16. Vishik, M.I.: Elliptic convolution equations in a bounded domain and their applications. Pro Int. Congr. Math., Moscow, 1966. Moscow, 1968, 409–420 (Russian).
17. Vishik, M.I., Eskin, G.I.: Convolution equations in a bounded domain. Russian Math. Surve **20**, 89–152 (1965 in Russian)
18. Vladimirov, V.S.: Linear conjugation problem for holomorphic functions. Izvestiya Acad. S USSR, Ser. math. **29**, 807–834 (1965 in Russian)

19. Vladimirov, V.S.: Methods of Functions Theory of Several Complex Variables. Nauka, Moscow (1964 in Russian)
20. Vladimirov, V.S.: Distributions in mathematical physics. Nauka, Moscow (1979 in Russian)
21. Vasil'ev, V.B.: Wave factorization of elliptic symbols: theory and applications. In: Introduction to the Boundary Value Problems in Non-smooth Domains. Kluwer Academic Publishers, Boston (2000)**
22. Vasilyev, V.B.: Fourier multipliers, pseudo differential equations, wave factorization, boundary value problems. 2nd Edition, Moscow, 2010. (Russian)
23. Vasilyev, V.B.: Asymptotical analysis of singularities for pseudo differential equations in canonical non-smooth domains. In: Constanda, C., Harris P.J. (eds.) Integral Methods in Science and Engineering. Computational and Analytic Aspects, pp. 379–390. Birkhäuser, Boston (2011)

Positive Solutions for a Kind of Singular Nonlinear Fractional Differential Equations with Integral Boundary Conditions

Junfang Zhao and Hairong Lian

Abstract In this paper, we study the following fractional differential equation:

$$D_{0+}^{\alpha}u(t) + f(t, u(t), (\phi u)(t), (\psi u)(t)) = 0, \quad 0 < t < 1,$$

with integral boundary condition:

$$u(0) = 0, \quad u(1) = \int_0^1 g(s)u(s)\,\mathrm{d}s,$$

where $1 < \alpha < 2$; $(\phi u)(t) = \int_0^t \gamma(t,s)u(s)\,\mathrm{d}s$; $(\psi u)(t) = \int_0^t \delta(t,s)u(s)\,\mathrm{d}s$; $0 < \int_0^1 g(s)$ $u(s)\,\mathrm{d}s < 1$; f satisfies the Carathéodory conditions on $[0,1] \times \mathscr{B}$, $\mathscr{B} = (0, +\infty) \times [0, +\infty)^2$; f is positive; $f(t,x,y,z)$ is singular at $x = 0$; and D^{α} is the standard Riemann–Liouville fractional derivative. By using the fixed point theorems on cones, we get the existence of positive solution.

Keywords Positive solution • Fractional differential equation • Singular • Boundary value problem

Introduction

In this paper, we are concerned with the following fractional differential equation with integral boundary conditions:

$$D^{\alpha}u(t) + f(t, u(t), (\phi u)(t), (\psi u)(t)) = 0, \quad 0 < t < 1, \tag{1}$$

$$u(0) = 0, \quad u(1) = \int_0^1 g(s)u(s)\,\mathrm{d}s, \tag{2}$$

Zhao (✉) • H. Lian
School of Science, China University of Geoscience, Beijing 100083, PR China
e-mail: zhao_junfang@163.com; lianhr@126.com

Pinelas et al. (eds.), *Differential and Difference Equations with Applications*, Springer
Proceedings in Mathematics & Statistics 47, DOI 10.1007/978-1-4614-7333-6_59,
© Springer Science+Business Media New York 2013

where $1 < \alpha < 2; (\phi u)(t) = \int_0^t \gamma(t,s)u(s)\mathrm{d}s; (\psi u)(t) = \int_0^t \delta(t,s)u(s)\mathrm{d}s;$
$0 < \int_0^1 g(s)s^{\alpha-1}\mathrm{d}s < 1;$ f satisfies the Carathéodory conditions on $[0,1] \times \mathscr{B}$,
$\mathscr{B} = (0,+\infty) \times [0,+\infty) \times [0,+\infty);$ f is positive; $f(t,x,y,z)$ is singular at $x = 0$; and
D^α is the standard Riemann–Liouville fractional derivative.

Fractional differential equations have been of great interest recently. It is caused both by the intensive development of the theory of fractional calculus and by the applications of such constructions in various sciences such as physics, mechanics, chemistry, and engineering. For details, see the references therein.

In [5], Bai and Lü investigate the existence and multiplicity of positive solutions for nonlinear fractional differential equation boundary value problem:

$$D_{0+}^\alpha u(t) + f(t,u(t)) = 0, \quad 0 < t < 1,$$

$$u(0) = u(1) = 0,$$

where $1 < \alpha \leq 2$ is a real number and D_{0+}^α is the standard Riemann–Liouville differentiation. By means of some fixed point theorems on cone, they get the existence and multiplicity results of positive solutions.

In [4], Bashir and Sivasundaram study the existence and uniqueness of solutions for a four-point nonlocal boundary value problem of nonlinear integro-differential equations of fractional order $q \in (1,2]$:

$$\begin{cases} {}^cD^q x(t) + f(t,x(t),(\phi x)(t),(\psi x)(t)) = 0, & 0 < t < 1, \\ x'(0) + ax(\eta_1) = 0, & bx'(1) + x(\eta_2) = 0, 0 < \eta_1 \leq \eta_2 < 1, \end{cases}$$

where ${}^cD$ is Caputo's fractional derivative. Their results are based on some standard fixed point theorems.

In [1], Agarwal, O'regan, and Stanek investigate the existence of positive solutions for the singular fractional boundary value problem:

$$\begin{cases} D^\alpha u(t) + f(t,u(t),D^\mu u(t)) = 0, & 0 < t < 1, \\ u(0) = u(1) = 0, \end{cases}$$

where $1 < \alpha < 2, 0 < \mu \leq \alpha - 1$, D^α is the standard Riemann–Liouville fractional derivative. By means of a fixed point theorem on a cone, the existence of positive solutions is obtained.

For more references on fractional differential equations, we refer the readers to [2,3,6–11].

Motivated by the papers mentioned above, in this paper, we considered

$$D_{0+}^\alpha u(t) + f(t,u(t),(\phi u)(t),(\psi u)(t)) = 0, \quad 0 < t < 1,$$

with integral boundary condition:

$$u(0) = 0, \quad u(1) = \int_0^1 g(s)u(s)\mathrm{d}s,$$

where $1 < \alpha < 2$, $(\phi u)(t) = \int_0^t \gamma(t,s)u(s)\mathrm{d}s$, $(\psi u)(t) = \int_0^t \delta(t,s)u(s)\mathrm{d}s$, $0 < \int_0^1 g(s)u(s)\mathrm{d}s < 1$.

We will assume throughout:

(H_1) $f \in \mathrm{Car}([0,1] \times \mathscr{B})$, $\mathscr{B} = (0,\infty) \times [0,+\infty) \times [0,+\infty)$, $\lim_{x \to 0^+} f(t,x,y,z) = \infty$ for a.e. $t \in [0,1]$ and all $y \in \mathbf{R}$, and there exists a positive constant m such that $f(t,x,y,z) \geq m(1-t)^{2-\alpha}$ for a.e. $t \in [0,1]$ and all $(x,y,z) \in \mathscr{B}$.

(H_2) f fulfills the estimate $f(t,x,y,z) \leq \gamma(t)(q(x) + p(x) + p(1) + w(y) + v(z))$ for a.e. $t \in [0,1]$ and all $(x,y,z) \in \mathscr{B}$, where $\gamma \in L^1[0,1]$; $q,w,v \in C[0,\infty]$ are positive; q is nonincreasing and p,w,v are nondecreasing. Further,

$$\int_0^1 \gamma(t)q(K(1-t^{\alpha-1}))\mathrm{d}t < \infty, \quad K = \frac{m}{2\Gamma(\alpha-1)}, \quad \lim_{x \to \infty} \frac{p(x)+w(x)+v(x)}{x} = 0.$$

2 Background Materials and Some Lemmas

In this section, for convenience, we present the main definitions for fractional calculus theory. These definitions can be found in the recent literature.

Definition 2.1. The fractional integral of order $\alpha > 0$ of a function $y : (0,+\infty) \to R$ is given by

$$I_{0+}^{\alpha} y(t) = \frac{1}{\Gamma(\alpha)} \int_0^t (t-s)^{\alpha-1} y(s)\mathrm{d}s$$

provided the right side is pointwise defined on $(0,+\infty)$.

Definition 2.2. The fractional derivative of order $\alpha > 0$ of a continuous function $\cdot : (0,+\infty) \to R$ is given by

$$D_{0+}^{\alpha} y(t) = \frac{1}{\Gamma(n-\alpha)} \left(\frac{\mathrm{d}}{\mathrm{d}t}\right)^n \int_0^t \frac{y(s)}{(t-s)^{\alpha-n+1}}\mathrm{d}s,$$

where $n = [\alpha] + 1$, provided that the right side is pointwise defined on $(0,+\infty)$.

Lemma 2.1. *Let* $\alpha > 0$. *If we assume* $u \in C(0,1) \cap L(0,1)$, *then the fractional differential equation*

$$D_{0+}^{\alpha} u(t) = 0$$

as $u(t) = C_1 t^{\alpha-1} + C_2 t^{\alpha-2} + \cdots + C_N t^{\alpha-N}$, $C_i \in R$, $i = 1,2,\cdots,N$ *as unique solutions where N is the smallest integer equal to or greater than α.*

As $D_{0+}^{\alpha} I_{0+}^{\alpha} u = u$ for all $u \in C(0,1) \cap L(0,1)$. From Lemma 2.1 we deduce the following law of composition.

Lemma 2.2. *Assume that* $u \in C(0,1) \cap L(0,1)$ *with a fractional derivative of order* $\alpha > 0$ *that belongs to* $C(0,1) \cap L(0,1)$. *Then*

$$I_{0+}^{\alpha} D_{0+}^{\alpha} u(t) = u(t) + C_1 t^{\alpha-1} + C_2 t^{\alpha-2} + \cdots + C_N t^{\alpha-N},$$

for some $C_i \in R$, $i = 1, 2, \cdots, N$, *has* $u(t) = C_1 t^{\alpha-1} + C_2 t^{\alpha-2} + \cdots + C_N t^{\alpha-N}$, $C_i \in R$, $i = 1, 2, \cdots, N$, *as unique solutions, where* N *is the smallest integer equal to or greater than* α.

In the following, we present Green's function of fractional differential equation boundary value problem.

Lemma 2.3. *Given* $y \in L^1[0,1]$ *and* $1 < \alpha \leq 2$, *the unique solution for*

$$D_{0+}^{\alpha} u(t) + y(t) = 0, \ 0 < t < 1, \tag{3}$$

$$u(0) = 0, \ u(1) = \int_0^1 g(s)u(s)ds, \tag{4}$$

is

$$u(t) = \int_0^1 G(t,s)y(s)ds, \tag{5}$$

where

$$G(t,s) = \frac{1}{\rho} \begin{cases} (1-s)^{\alpha-1} - \int_s^1 g(\tau)(s-\tau)^{\alpha-1}d\tau, \ 0 \leq t \leq s \leq 1, \\[2mm] (1-s)^{\alpha-1} - \int_s^1 g(\tau)(s-\tau)^{\alpha-1}d\tau \\[2mm] \quad - (1 - \int_0^1 g(s)s^{\alpha-1}ds)(t-s)^{\alpha-1}, \ 0 \leq s \leq t \leq 1, \end{cases} \tag{6}$$

and $\rho = \Gamma(\alpha)(1 - \int_0^1 g(s)s^{\alpha-1}ds)$.

Proof We may apply Lemma 2.1 to reduce $D_{0+}^{\alpha} u(t) + y(t) = 0$ to an equivalent integral equation

$$u(t) = -I_{0+}^{\alpha} y(t) + c_1 t^{\alpha-1} + c_2 t^{\alpha-2},$$

for some $c_1, c_2 \in R$. Consequently, the general solution of Eq. (3) is

$$u(t) = -\int_0^t \frac{(t-s)^{\alpha-1}}{\Gamma(\alpha)} y(s)ds + c_1 t^{\alpha-1} + c_2 t^{\alpha-2}. \tag{}$$

By (4), there is $c_2 = 0$, then we have

$$u(t) = -\int_0^t \frac{(t-s)^{\alpha-1}}{\Gamma(\alpha)} y(s)\mathrm{d}s + c_1 t^{\alpha-1}.$$

And so we have

$$\int_0^1 g(t)u(t)\mathrm{d}t = \int_0^1 g(t)\left[-\left(\int_0^t \frac{(t-s)^{\alpha-1}}{\Gamma(\alpha)} y(s)\mathrm{d}s + c_1 s^{\alpha-1}\right)\right]\mathrm{d}t. \tag{8}$$

Letting $t = 1$, we find that

$$u(1) = -\int_0^1 \frac{(1-s)^{\alpha-1}}{\Gamma(\alpha)} y(s)\mathrm{d}s + c_1. \tag{9}$$

By the boundary condition we have

$$c_1 = \frac{1}{\Gamma(\alpha)(1-\int_0^1 g(s)s^{\alpha-1}\mathrm{d}s)}\left(\int_0^1 [(1-s)^{\alpha-1} - \int_s^1 g(\tau)(\tau-s)^{\alpha-1}\mathrm{d}\tau] y(s)\mathrm{d}s\right). \tag{10}$$

Substituting (10) into (7), we get

$$u(t) = \frac{1}{\Gamma(\alpha)}\left[-\int_0^t (t-s)^{\alpha-1} y(s)\mathrm{d}s + \frac{1}{1-\int_0^1 g(s)s^{\alpha-1}\mathrm{d}s}\right.$$

$$\left.\left(\int_0^1 [(1-s)^{\alpha-1} - \int_s^1 g(\tau)(\tau-s)^{\alpha-1}\mathrm{d}\tau] y(s)\mathrm{d}s\right)\right] \tag{11}$$

$$= \frac{1}{\rho}\int_0^1 G(t,s)y(s)\mathrm{d}s,$$

where $G(t,s)$ is as defined by (6). The proof is complete. $\qquad\square$

Lemma 2.4. *Suppose* $0 < \int_0^1 g(s)s^{\alpha-1}\mathrm{d}s < 1$, *then:*

(i) $G(t,s) > 0$, $t,s \in (0,1)$.

(ii) $G(t,s) < G(s,s)$.

(iii) $\max\limits_{0 \le t,s \le 1} G(t,s) \le 1, t,s \in [0,1]$.

Proof (i) If $0 \le t \le s \le 1$,

$$(1-s)^{\alpha-1} - \int_s^1 g(\tau)(\tau-s)^{\alpha-1}\mathrm{d}\tau = (1-s)^{\alpha-1} - \int_s^1 g(\tau)\tau^{\alpha-1}(1-\frac{s}{\tau})^{\alpha-1}\mathrm{d}\tau$$

$$\ge (1-s)^{\alpha-1} - (1-s)^{\alpha-1}\int_0^1 g(\tau)\tau^{\alpha-1}\mathrm{d}\tau$$

$$= (1-s)^{\alpha-1}(1-\int_0^1 g(\tau)\tau^{\alpha-1}\mathrm{d}\tau) > 0.$$

It is clear here that $G(t,s) \leq G(s,s)$.
If $0 \leq s \leq t \leq 1$,

$$(1-s)^{\alpha-1} - \int_s^1 g(\tau)(\tau-s)^{\alpha-1}d\tau - (1 - \int_0^1 g(s)s^{\alpha-1}ds)(t-s)^{\alpha-1}$$

$$> (1-s)^{\alpha-1}(1 - \int_0^1 g(\tau)\tau^{\alpha-1}d\tau) - (1 - \int_0^1 g(s)s^{\alpha-1}ds)t^{\alpha-1}(1-\frac{s}{t})^{\alpha-1}$$

$$= (1-s)^{\alpha-1}(1 - \int_0^1 g(s)s^{\alpha-1}ds)(1-t^{\alpha-1}) > 0.$$

And it is also clear here that $G(t,s) \leq G(s,s)$. □

Lemma 2.5. *If $u(t)$ is the solution to BVP* (3) *and* (4), *then*

(i) $u(t) \geq 0$.
(ii) $u(t) \geq (1-t^{\alpha-1})(1 - \int_0^1 g(t)t^{\alpha-1}ds)\|u\|$.

Proof

(i) Since $u(t)$ is the solution to BVP (3) and (4), therefore,

$$u'(t) = -\frac{1}{\Gamma(\alpha-1)} \int_0^t (t-s)^{\alpha-2}y(s)ds < 0;$$

thus, $u(t)$ is strictly decreasing on $(0,1]$, and so $u(t) \geq u(1), t \in [0,1]$. By the boundary condition, we have

$$u(1) = \int_0^1 g(s)u(s)ds \geq u(1) \int_0^1 g(s)ds \geq 0;$$

therefore, $u(t) \geq 0, t \in [0,1]$.

(ii) Since $u(t)$ is the solution to BVP (3) and (4), then by Lemma 2.3 and by the proof of Lemma 2.4 , we have

$$u(t) = \int_0^1 G(t,s)y(s)ds$$

$$\geq \frac{1-t^{\alpha-1}}{\Gamma(\alpha)} \int_0^1 (1-s)^{\alpha-1}y(s)ds$$

$$\geq \frac{1-t^{\alpha-1}}{\Gamma(\alpha)} \int_0^1 \frac{(1-s)^{\alpha-1}}{G(s,s)}G(s,s)y(s)ds \qquad (12$$

$$\geq \frac{1-t^{\alpha-1}}{\Gamma(\alpha)} \int_0^1 \frac{(1-s)^{\alpha-1}}{G(s,s)}G(s,s)y(s)ds$$

$$\geq (1-t^{\alpha-1})(1 - \int_0^1 g(t)t^{\alpha-1}dt)\|u\|.$$

The proof is complete.

Lemma 2.6. *Let positive constants m and K be as in* (H_1) *and* (H_2) *and let* $r \in L^1[0,1]$*and* $r(t) \geq m(1-t)^{2-\alpha}$ *for a.e.* $t \in [0,1]$. *Then*

$$\int_0^1 G(t,s)r(s)ds \geq K(1-t^{\alpha-1}) \quad for \ t \in [0,1]. \tag{13}$$

Proof From the proof of Lemma 2.4, we can see that

$$G(t,s) \geq \frac{1}{\Gamma(\alpha)}(1-s)^{\alpha-1}(1-t^{\alpha-1}), \quad for \ t,s \in [0,1]. \tag{14}$$

Thus, we have

$$\int_0^1 G(t,s)r(s)ds \geq m \int_0^1 G(t,s)(1-s)^{2-\alpha}ds$$

$$\geq \frac{m}{\Gamma(\alpha)}(1-t^{\alpha-1}) \int_0^1 (1-s)^{\alpha-1}(1-s)^{2-\alpha}ds \tag{15}$$

$$= \frac{m}{2\Gamma(\alpha)}(1-t^{\alpha-1}),$$

denote $K = \frac{m}{2\Gamma(\alpha)}$, then this completes the proof. □

3 Auxiliary Regular Problem

In this section, we consider the following auxiliary boundary value problem.

Since (1) is a singular equation, we use regularization and sequential techniques for the existence of a positive solution of problem (1) and (2). For this end, for each $n \in \mathbf{N}$, define f_n by the formula

$$f_n(t,x,y,z) = \begin{cases} f(t,x,y,z) & if \ x \geq \frac{1}{n}, \\ f(t,\frac{1}{n},y,z) & if \ 0 < x < \frac{1}{n}. \end{cases}$$

Then $f_n \in Car([0,1] \times \mathscr{B}_*)$, $\mathscr{B}_* = [0,\infty)^3 = [0,\infty) \times [0,\infty) \times [0,\infty)$, and conditions (H_1) and (H_2) give

$$\left. \begin{array}{l} f_n(t,x,y) \leq \eta(t)(q(\frac{1}{n}) + p(x) + p(1) + w(y) + v(z)) \\ for \ a.e.t \in [0,1] \ and \ all \ (x,y,z) \in \mathscr{B}_*, \end{array} \right\} \tag{16}$$

$$
\left.
\begin{aligned}
f_n(t,x,y) &\leq \eta(t)(q(x)+p(x)+p(1)+w(y)+v(z)) \\
&\text{for } a.e.t \in [0,1] \text{ and all } (x,y,z) \in \mathscr{B}_*.
\end{aligned}
\right\}
\tag{17}
$$

We discuss the regular fractional differential equation:

$$
D_0^\alpha u(t) + f_n(t,u(t),(\phi u)(t),(\psi u)(t)) = 0.
\tag{18}
$$

Let $X = C[0,1]$ be equipped with the norm $\|u\| = \max_{0 \leq t \leq 1}|u(t)|$. Then X is a Banach space. Define the cone $P \subset X$ by $P = \{u \in X, u(t) \geq 0 \text{ for } t \in [0,1]\}$.

In order to prove that the problem has a positive solution, we define an operator T_n on P by the formula

$$
(T_n u)(t) = \int_0^1 G(t,s) f_n(s,u_n(s),(\phi u_n)(s),(\psi u_n)(s)) ds.
\tag{19}
$$

The properties of the operator T_n are given in the following lemma.

Lemma 3.1. *Let* (H_1) *and* (H_2) *hold. Then* $T_n : P \to P$ *and* T_n *is a completely continuous operator.*

Proof Let $u \in P$ and let $\mu(t) = f_n(t,u_n(t),(\phi u_n)(t),(\psi u_n)(t))$ for a.e. $t \in [0,1]$. Then $\mu \in L^1[0,1]$ because $f_n \in Car([0,1] \times B_*)$, and μ is positive. Thus $T_n : P \to P$.

In order to prove that T_n is a continuous operator. Let $\{u_n\} \subset P$ be a convergent sequence and let $\lim_{m\to\infty} \|u_m - u\|_* = 0$. Then $u \in P$ and $\|u_m\|_* \leq S$ for $m \in \mathbf{N}$, where S is a positive constant. Keeping in mind that $f_n \in Car([0,1] \times B_*)$, we have

$$
\begin{aligned}
\lim_{m\to\infty} f_n(t,u_m(t),(\phi u_m)(t),(\psi u_m)(t)) \\
= f_n(t,u(t),(\phi u)(t),(\psi u)(t)) \quad \text{for} \quad a.e.t \in [0,1].
\end{aligned}
\tag{20}
$$

Since by (16) and (17),

$$
0 < f_n(t,u_m(t),(\phi u_m)(t),(\psi u_m)(t)) \leq \eta(t)(q(\frac{1}{n})+p(S)+p(1)+w(S)+v(S)),
\tag{21}
$$

the Lebesgue dominated convergence theorem gives

$$
\lim_{m\to\infty} \int_0^1 |f_n(t,u_m(t),(\phi u_m)(t),(\psi u_m)(t)) - f_n(t,u(t),(\phi u)(t),(\psi u)(t))| dt = 0.
\tag{22}
$$

Now we deduce from (22) Lemma 2.3

$$
\begin{aligned}
|(T_n u_m)(t) - (T_n u)(t)| \leq E \int_0^1 |f_n(s,u_m(s),(\phi u_m)(s),(\psi u_m)(s)) \\
- f_n(s,u(s),(\phi u)(s),(\psi u)(s))| ds
\end{aligned}
\tag{23}
$$

that $\lim_{m\to\infty}\|T_nu_m - T_nu\|_* = 0$, which proves that T_n is a continuous operator. Finally, let $\Omega \subset P$ be bounded in X and let $\|u\|_* \leq L$ for all $u \in \Omega$, where L is a positive constant. In view of $f_n \in Car([0,1] \times B_*)$, there exists $v \in L^1[0,1]$ such that

$$0 < f_n(t,u(t),(\phi u)(t),(\psi u)(t)) \leq v(t) \quad \text{for} \quad a.e.\, t \in [0,1] \quad \text{and all} \quad u \in \Omega. \quad (24)$$

Then

$$|(T_nu)(t)| \leq E \int_0^1 f_n(s,u(s),(\phi u)(s),(\psi u)(s))ds \leq E\|v\|_L$$

for $t \in [0,1]$ and $u \in \Omega$. Hence $\|T_nu\| \leq E\|v\|_L$ for $u \in \Omega$, and so $T_n(\Omega)$ is bounded in X. Let $0 \leq t_1 < t_2 \leq 1$. Then

$$\begin{aligned}
&|(T_nu)(t_2) - (T_nu)(t_1)|\\
&= \frac{1}{\Gamma(\alpha)}|\int_0^{t_1}(t_1 - s)^{\alpha-1}f_n(s,u(s),(\phi u)(s),(\psi u)(s))ds\\
&\quad - \int_0^{t_2}(t_2 - s)^{\alpha-1}f_n(s,u(s),(\phi u)(s),(\psi u)(s))ds|\\
&= \frac{1}{\Gamma(\alpha)}|\int_0^{t_1}[(t_1 - s)^{\alpha-1} - (t_2 - s)^{\alpha-1}]f_n(s,u(s),(\phi u)(s),(\psi u)(s))\\
&\quad - \int_{t_1}^{t_2}(t_2 - s)^{\alpha-1}f_n(s,u(s),(\phi u)(s),(\psi u)(s))|\\
&\leq \frac{1}{\Gamma(\alpha)}\left(\int_0^{t_1}\left((t_1 - s)^{\alpha-1} - (t_2 - s)^{\alpha-1}\right)v(s)ds + (t_2 - t_1)^{\alpha-1}\int_{t_1}^{t_2}v(s)ds\right).
\end{aligned}$$
$$(25)$$

Let us choose an arbitrary $\varepsilon > 0$. Since the function $t^{\alpha-1}$ is uniformly continuous on $[0,1]$ and $|t - s|^{\alpha-1}$ on $[0,1] \times [0,1]$, there exists $\delta > 0$ such that for each $0 \leq t_1 < t_2 \leq 1, t_2 - t_1 < \delta, 0 \leq s \leq t_1$, we have $(t_2 - t_1)^{\alpha-1} < \varepsilon, 0 < (t_1 - s)^{\alpha-1} - (t_2 - s)^{\alpha-1} < \varepsilon$. Then, for $u \in \Omega$ and $0 \leq t_1 < t_2 \leq 1, t_2 - t_1 < \min\{\delta, \sqrt[\alpha-1]{\varepsilon}\}$, we conclude from inequality (25) that the inequality

$$|(T_nu)(t_2) - (T_nu)(t_1)| < \frac{\varepsilon}{\Gamma(\alpha)}\|v\|_L$$

holds. Hence the sets of functions $T_n(\Omega)$ is bounded in $C[0,1]$ and equicontinuous on $[0,1]$. Consequently, $T_n(\Omega)$ is relatively compact in X by the Arzelà–Ascoli theorem. We have proved that T_n is a completely continuous operator. $\square$

The next result follows immediately from Lemma 2.1.

Lemma 3.2. *Let* (H_1) *and* (H_2) *hold. Then any fixed point of the operator* T_n *is a solution of problem* (3) *and* (4).

Lemma 3.3. *Let* Y *be a Banach space and* $P \subset Y$ *be a cone in* Y. *Let* Ω_1, Ω_2 *be bounded open balls of* Y *centered at the origin with* $\overline{\Omega_1} \subset \Omega_2$. *Suppose that* $T :$ $P \cap (\overline{\Omega_2} \setminus \Omega_1) \to P$ *is a completely continuous operator such that*

$$\|Ax\| \geq \|x\| \quad for \quad x \in P \cap \partial \Omega_1, \quad \|Ax\| \leq \|x\| \quad for \quad x \in P \cap \partial \Omega_2$$

hold. Then A *has a fixed point in* $P \cap (\overline{\Omega_2} \setminus \Omega_1)$.

In the following, we will give the existence result for the regular problem (18) and (2).

Lemma 3.4. *Let* (H$_1$) *and* (H$_2$) *hold. Then problem* (18) *and* (2) *has a solution.*

Proof By Lemmas 3.1 and 3.2, $T_n : P \to P$ is completely continuous and u is a solution of problem (18) and (2) if u solves the operator equation $u = T_n u$. In order to apply Lemma 3.3, we separate the proof into two steps:

Step 1. Let

$$\Omega_1 = \{u \in X : \|u\|_* < K\},$$

where K is as in (H$_2$). It follows from Lemma 2.3 and from the definition of T_n that $(T_n u)(t) \geq K(1 - t^{\alpha-1})$ for $t \in [0,1]$ and $u \in P$, and consequently,

$$\|T_n u\|_* \geq \|u\|_*, \quad for \quad u \in P \cap \partial \Omega_1. \tag{26}$$

Step 2. Inequality (16) and Lemma 2.3 imply that for $u \in P$,

$$|(T_n u)(t)| \leq \int_0^1 \eta(s)(q(\frac{1}{n}) + p(u(s)) + p(1) + w((\phi u)(s)) + v((\psi u)(s)))ds$$

$$\leq (q(\frac{1}{n}) + p(\|u\|) + p(1) + w(\|\phi u\|) + v(\|\psi u\|))\|\eta\|_L. \tag{27}$$

Hence, for $u \in P$, the inequality

$$\|T_n u\|_* \leq (q(\frac{1}{n}) + p(\|u\|) + p(1) + w(\|\phi u\|) + v(\|\psi u\|))\|\eta\|_L \tag{28}$$

is fulfilled. Since $\lim_{x \to \infty} \frac{p(x) + w(x) + v(x)}{v} = 0$ by (H$_2$), there exists $Q > 0$ such that

$$(q(\frac{1}{n}) + p(\|u\|) + p(1) + w(\|\phi u\|) + v(\|\psi u\|))\|\eta\|_L \leq S. \tag{29}$$

Let

$$\Omega_2 = \{u \in X : \|u\|_* < S\}.$$

Then

$$\|T_n u\|_* \le \|u\|_*, \quad \text{for} \quad u \in P \cap \Omega_2. \tag{30}$$

Applying Lemma 3.3, we conclude from (26) and (30) that T_n has a fixed point in $P \cap (\overline{\Omega_2} \setminus \Omega_1)$. Consequently, problem (18) and (2) has a solution by Lemma 3.2.

□

Lemma 3.5. *Let* (H_1) *and* (H_2) *hold. Let* u_n *be a solution of problem* (18) *and* (2). *Then the sequence* $\{u_n\}$ *is relatively compact in* $C[0,1]$.

Proof Denote $\mu(t) = f_n(t, u_n(t), (\phi u_n)(t), (\psi u_n)(t))$. We note that

$$u_n(t) = \int_0^1 G(t,s) f_n(s, u_n(s), (\phi u_n)(s), (\psi u_n)(s)) \mathrm{d}s, \quad t \in [0,1], \quad n \in \mathbf{N}. \tag{31}$$

It follows from Lemma 2.6 that

$$u_n(t) \ge K(1 - t^{\alpha-1}) \quad \text{for} \quad t \in [0,1], \quad n \in \mathbf{N}. \tag{32}$$

Therefore,

$$f_n(t, u_n(t), (\phi u_n)(t), (\psi u_n)(t)) \le \eta(t)(q(K(1 - t^{\alpha-1}))$$
$$+ p(u_n(t)) + p(1) + w((\phi u_n)(t)) + v((\psi u_n)(t))) \tag{33}$$

for a.e. $t \in [0,1]$ and $n \in \mathbf{N}$. Now, by (31) and Lemma 2.3,

$$u_n(t) \le M + (p(\|u_n\|) + p(1) + w(\|\phi u_n\|) + v(\|\psi u_n\|))\|\eta\|_L, \tag{34}$$

for $t \in [0,1]$ and $n \in \mathbf{N}$, where

$$M = \int_0^1 \eta(t) q(K(1 - t^{\alpha-1})) \mathrm{d}t < \infty. \tag{35}$$

In particular,

$$\|u_n\|_* \le M + (p(\|u_n\|_*) + p(1) + w(\|\phi u_n\|_*) + v(\|\psi u_n\|_*))\|\eta\|_L, \tag{36}$$

for $n \in \mathbf{N}$. Since $\lim_{x \to \infty} \frac{p(x) + w(x) + v(x)}{x} = 0$, there exists $S > 0$ such that

$$M + (p(x) + p(1) + w(x) + v(x))\|\eta\|_L < x \quad \text{for} \quad x \ge Q. \tag{37}$$

Therefore

$$\|u_n\|_* < Q \quad \text{for} \quad n \in \mathbf{N}. \tag{38}$$

Hence the sequence $\{u_n\}$ is bounded in $C[0,1]$.

Now, we prove that $\{u_n\}$ is equicontinuous on $[0,1]$. Let $0 \le t_1 < t_2 \le 1$. Then

$$|u_n(t_2) - u_n(t_1)| = |\int_0^1 (G(t_2,s) - G(t_1,s)) f_n(s,u(s),(\phi u)(s),(\psi u)(s)) ds|$$

$$= \frac{1}{\Gamma(\alpha)} |\int_0^{t_1} (t_1 - s)^{\alpha-1}$$

$$- \int_0^{t_2} (t_2 - s)^{\alpha-1} f_n(s,u_n(s),(\phi u_n)(s),(\psi u_n)(s)) ds|$$

$$= \frac{1}{\Gamma(\alpha)} |\int_0^{t_1} [(t_1 - s)^{\alpha-1} - (t_2 - s)^{\alpha-1}] f_n(s,u(s),(\phi u)(s),(\psi u)(s))$$

$$- \int_{t_1}^{t_2} (t_2 - s)^{\alpha-1} f_n(s,u_n(s),(\phi u_n)(s),(\psi u_n)(s))|$$

$$\le \frac{1}{\Gamma(\alpha)} \left(\int_0^{t_1} \left((t_1 - s)^{\alpha-1} - (t_2 - s)^{\alpha-1}\right) v(s) ds \right.$$

$$\left. + (t_2 - t_1)^{\alpha-1} \int_{t_1}^{t_2} v(s) ds \right).$$

$$(39)$$

We proceed analogously to the proof of Lemma 3.1. Let us choose an arbitrary $\varepsilon > 0$. Then there exists $\delta_0 > 0$ such that for each $0 \le t_1 < t_2 \le 1$, $t_2 - t_1 < \delta_0$ and $0 \le s \le t_1$, we have $t_2^{\alpha-1} - t_1^{\alpha-1} < \varepsilon$, $(t_2 - s)^{\alpha-1} - (t_1 - s)^{\alpha-1} < \varepsilon$. Let $0 < \delta < \min\{\delta_0, \sqrt[\alpha-1]{\varepsilon}\}$. Now, using the inequality

$$0 < f_n(t, u_n(t), (\phi u_n)(t), (\psi u_n)(t)) \le \eta(t)(q(K(1 - t^{\alpha-1})) + p(S) + p(1) + w(S) + v(S)$$

$$(40)$$

for a.e. $t \in [0,1]$ and all $n \in \mathbf{N}$, we conclude from (39) that for $0 \le t_1 < t_2 \le 1$, $t_2 - t_1 < \delta$, and $n \in \mathbf{N}$, the following inequalities are fulfilled:

$$|u(t_2) - u(t_1)| \le \frac{\varepsilon}{\Gamma(\alpha)} \left(\int_0^1 (1 - t)^{\alpha-1} \eta(t)(q(K(1 - t^{\alpha-1})) + p(S) + p(1) + w(S) \right.$$

$$+ v(S)) dt + \int_0^{t_2} \eta(t)(q(K(1 - t^{\alpha-1})) + p(S) + p(1) + w(S) + v(S)) dt$$

$$< \frac{2\varepsilon}{\Gamma(\alpha)} \int_0^1 \eta(t)(q(K(1 - t^{\alpha-1})) + p(S) + p(1) + w(S) + v(S)) dt.$$

$$(4)$$

As a result, $\{u_n\}$ is equicontinuous on $[0,1]$. Hence, $\{u_n\}$ is relatively compact $C[0,1]$ by the Arzela–Ascoli theorem. The proof is complete.

4   The Existence of Positive Solutions

Theorem 4.1. *Let* (H$_1$) *and* (H$_2$) *hold. Then BVP* (1) *and* (2) *has a positive solution u and*

$$u(t) \geq K(1 - t^{\alpha-1}) \quad \text{for} \quad t \in [0,1]. \tag{42}$$

Proof Lemmas 3.4 and 3.5 guarantee that BVP (18) and (2) has a solution u_n satisfying (32) and $\{u_n\}$ is relatively compact in $C[0,1]$. Hence there exist $u \in X$ and a subsequence $\{u_{k_n}\}$ of $\{u_n\}$ such that $\lim_{n\to\infty} u_{k_n} = u$ in X. Consequently, $u \in P$, u satisfies (42), and

$$\lim_{n\to\infty} f_{k_n}(t, u_{k_n}(t), (\phi u_{k_n})(t), (\psi u_{k_n})(t))$$

$$= f(t, u(t), (\phi u)(t), (\psi u)(t)) \text{ for } a.e.\ t \in [0,1]. \tag{43}$$

Since $\{u_n\}$ fulfills (38), where Q is a positive constant, it follows from inequality (38) and from Lemma 2.3 that the inequality

$$0 \leq G(t,s) f_{k_n}(s, u_{k_n}(s), (\phi u_{k_n})(s), (\psi u_{k_n})(s))$$

$$\leq \eta(s)(q(K(1 - s^{\alpha-1})) + p(S) + p(1) + w(S) + v(S)) \tag{44}$$

holds for a.e. $s \in [0,1]$ and all $t \in [0,1]$, $n \in \mathbf{N}$. Hence

$$\lim_{n\to\infty} \int_0^1 G(t,s) f_{k_n}(s, u_{k_n}(s), (\phi u_{k_n})(s), (\psi u_{k_n})(s)) \mathrm{d}s$$

$$= \int_0^1 G(t,s) f(s, u(s), (\phi u)(s), (\psi u)(s)) \mathrm{d}s \tag{45}$$

for $t \in [0,1]$ by the Lebesgue dominated convergence theorem. Now, passing to the limit as $n \to \infty$ in

$$u_{k_n}(t) = \int_0^1 G(t,s) f_{k_n}(s, u_{k_n}(s), (\phi u_{k_n})(s), (\psi u_{k_n})(s)) \mathrm{d}s, \tag{46}$$

we have

$$u(t) = \int_0^1 G(t,s) f(s, u(s), (\phi u)(s), (\psi u)(s)) \mathrm{d}s \quad \text{for} \quad t \in [0,1]. \tag{47}$$

Consequently, u is a positive solution of problem (1) and (2) by Lemma 2.3, the proof is complete. □

5 Example

In this section, we will give an example to illustrate our main results.

Example 5.1.

$$\begin{cases} D^{\alpha}u(t) + f(t, u(t), (\phi u)(t), (\psi u)(t))) = 0, & 0 < t < 1, \\ u(0) = 0, \quad u(1) = \int_0^1 g(s)u(s)\,ds, \end{cases} \tag{48}$$

where $f(t,x,y,z) = \frac{1}{(1-t)^{2/3}}(\frac{1}{x^{\lambda}} + x^{\mu} + y^{\nu} + z^{\omega})$, $\mu < 1$, $0 < \nu < 1$, $0 < \omega < 1$. Clearly, f satisfies all the conditions in Theorem 4.1. So (48) has a positive solution u.

Acknowledgements This work was supported by the NNSF (No: 11101385).

References

1. Agarwal, R.P., O'Regan, D.: Svatoslav StaněkPositive solutions for Dirichlet problems of singular nonlinear fractional differential equations. J. Math. Anal. Appl. **371**, 57–68 (2010)
2. Agarwal, R.P., Zhou, Y., He, Y.: Existence of fractional neutral functional differential equations. Comput. Math. Appl. **59**, 1095–1100 (2010)
3. Ahmada, B., Sivasundaram, S.: Existence results for nonlinear impulsive hybrid boundary value problems involving fractional differential equations. Nonlinear Anal. Hybrid Syst. **3**, 251–258 (2009)
4. Ahmad, B., Sivasundaram, S.: On four-point nonlocal boundary value problems of nonlinear integro-differential equations of fractional order. Appl. Math. Comput. **217**, 480–487 (2010)
5. Bai, Z., Lü, H.: Positive solutions for boundary value problem of nonlinear fractional differential equation. J. Math. Anal. Appl. **311**, 495–505 (2005)
6. Benchohra, M., Berhoun, F.: Impulsive fractional differential equations with variable times. Comput. Math. Appl. **59**, 1245–1252 (2010)
7. Guo, B., Han, Y., Xin, J.: Existence of the global smooth solution to the period boundary value problem of fractional nonlinear Schrödinger equation. Appl. Math. Comput. **204**, 468–47 (2008)
8. Rehman, M., Khan, R.A.: Existence and uniqueness of solutions for multi-point boundary value problems for fractional differential equations. Appl. Math. Lett. **23**, 1038–1044 (2010)
9. Tatar, N.: On a boundary controller of fractional type. Nonlinear Anal. **72**, 3209–3215 (2010)
10. Zhong, W., Lin, W.: Nonlocal and multiple-point boundary value problem for fraction differential equations. Appl. Math. Comput. **59**, 1345–1351 (2010)
11. Zhou, Y., Jiao, F., Li, J.: Existence and uniqueness for fractional neutral differential equation with infinite delay. Nonlinear Anal. **71**, 3249–3256 (2009)

Functional Aspects of the Hardy Inequality: Appearance of a Hidden Energy

J.L. Vázquez and N.B. Zographopoulos

Abstract We obtain new insights into the Hardy inequality and the evolution problem associated to it. Surprisingly, the connection of the energy of the new formulation with the standard Hardy functional is nontrivial, due to the presence of a Hardy singularity energy. This corresponds to a loss for the total energy. The problem arises when the equation is posed in a bounded domain.

We also consider an equivalent problem with inverse square potential on an exterior domain. The extra energy term is then present as an effect that comes from infinity, a kind of hidden energy. In this case, in an unexpected way, this term is additive to the total energy, and it may even constitute the main part of it.

1 Introduction

In this paper we present some results of [24]; we contribute new results on the Hardy inequality posed in a bounded domain or in an exterior domain of $\mathbb{R}^N$ and on the corresponding parabolic evolution. The motivation came from a functional difficulty we found in the work [26], where the following singular evolution problem was studied:

J.L. Vázquez (✉)
Departamento de Matemáticas, Universidad Autónoma de Madrid, 28049, Madrid, Spain
e-mail: juanluis.vazquez@uam.es

N.B. Zographopoulos
Department of Mathematics and Engineering Sciences, Hellenic Army Academy,
16673, Athens, Greece
e-mail: nzograp@gmail.com

Pinelas et al. (eds.), *Differential and Difference Equations with Applications*, Springer
Proceedings in Mathematics & Statistics 47, DOI 10.1007/978-1-4614-7333-6_60,
© Springer Science+Business Media New York 2013

$$\begin{cases} u_t = \Delta u + c_* \dfrac{u}{|x|^2}, \; x \in \Omega, \; t > 0, \\ u(x,0) = u_0(x), \; \text{for } x \in \Omega, \\ u(x,t) = 0 \text{ in } \partial\Omega, \; t > 0. \end{cases} \tag{1}$$

with critical coefficient $c_* = (N-2)^2/4$. The space dimension is $N \geq 3$ and Ω is a bounded domain in $\mathbb{R}^N$ containing 0, or $\Omega = \mathbb{R}^N$. More precisely, the authors in [26] studied the well posedness and described the asymptotic behavior of (1). Moreover, they obtained improved Hardy inequalities and completed the study of the spectrum of the associated eigenvalue problem. This problem is closely connected with the Hardy inequality:

$$\int_\Omega |\nabla u|^2 \, dx > \left(\frac{N-2}{2} \right)^2 \int_\Omega \frac{u^2}{|x|^2} \, dx, \tag{2}$$

which is well known to hold for any $u \in C_0^\infty(\Omega)$. For Hardy-type inequalities and related topics, we refer to [7, 13, 21–23]. Due to this connection, c_*, which is the best constant in the inequality, is also critical for the basic theory of the evolution equation. Indeed, the usual variational theory applies to the subcritical cases: $u_t = \Delta u + c u/|x|^2$ with $c < c_*$, using the standard space $H_0^1(\Omega)$, and a global in time solution is then produced. On the other hand, there are no positive solutions of the equation for $c > c_*$ (instantaneous blowup), [4,9,20]. In the critical case we still get existence, but the functional framework changes; this case serves as an example of interesting functional analysis and more complex evolution.

In order to analyze the behavior of the solutions of Problem (1) in [26], the Hardy functional

$$I_\Omega[\phi] := \int_\Omega |\nabla\phi|^2 \, dx - \left(\frac{N-2}{2} \right)^2 \int_\Omega \frac{\phi^2}{|x|^2} \, dx \tag{3}$$

is considered as the Dirichlet form naturally associated to the equation. This form is positive and different lower bounds have been obtained; for Hardy and Hardy-type inequalities we refer to [1, 2, 6, 8, 12, 14–19, 26] and the references therein. Note that the expression is finite for $u \in H_0^1(\Omega)$, but it can also be finite as an improper integral for other functions having a strong singularity at $x = 0$, due to cancelation between the two terms. To take this possibility into account, the Hilbert space H was introduced in [26] as the completion of the $C_0^\infty(\Omega)$ functions under the norm

$$\|\phi\|_{H(\Omega)}^2 = I_\Omega[\phi], \quad \phi \in C_0^\infty(\Omega). \tag{4}$$

According to Sect. 5 of [26], this space allows us to define in a natural way a self-adjoint extension of the differential operator $L(u) := -\Delta u - c_* u/|x|^2$ (the Friedreich extension) and then to use standard theory to generate a semigroup and describe the solutions using the spectral analysis. The study of the spectrum leads to an associated elliptic eigenvalue problem, the solution of which turns out to be a classical problem in separation of variables.

1.1   Problem with the Singularities

The separation of variable analysis produces some singular solutions. In particular, the maximal singularity (corresponding to the first mode of separation of variables) behaves like $|x|^{-(N-2)/2}$ near $x = 0$, and this function is not in $H_0^1(\Omega)$. Now, this solution must belong to the space H associated to the quadratic form, hence, the conclusion $H \neq H_0^1(\Omega)$. We recall that this is a peculiar phenomenon of the equation with critical exponent $c_* = (N-2)^2/4$. For values of $c < c_*$, the maximal singularity is still in $H_0^1(\Omega)$.

However, we have realized that with the proposed definition of H, there exists a problem with the solutions of the evolution problem having the maximal singularity. The verification is quite simple in the case where $\Omega = B_1$, the unit ball in $\mathbb{R}^N$ centered at the origin. Then, the minimization problem

$$\min_{u \in H} \frac{||u||_H^2}{||u||_{L^2}^2} \tag{5}$$

has as a solution the function $e_1(r) = r^{-(N-2)/2} J_0(z_{0,1} r)$, $r = |x|$, J_0 is the Bessel function with $J_0(0) = 1$, up to normalization, and $z_{0,1}$ denotes the first zero of J_0. This function plays a big role in the asymptotic behavior of general solutions of Problem (1). The minimum value of (5) is $\mu_1 = z_{0,1}^2$. Moreover, the quantity $I_{B_1}(e_1)$ is well defined as a principal value. Assuming that

$$||e_1||_H^2 = I_{B_1}(e_1), \tag{6}$$

from the definition of H, for any $\varepsilon > 0$, we should find a C_0^∞-function ϕ, such that $|e_1 - \phi||_H^2 < \varepsilon$. However, we may prove that $||e_1 - \phi||_H^2 \geq c > 0$, for any C_0^∞-function ϕ, which is a contradiction. Thus, we see that e_1 fails to be in H, since it cannot be approximated by C_0^∞-functions and this will happen for every function with the maximal singularity.

Therefore, under the assumption (6), the space H seems not to be correctly defined in [26] to apply the rest of the theory, since there exists a problem in dealing with very singular behavior near $x = 0$ that is not covered by approximation with infinitely smooth functions. Actually, this was our first impression.

.2   New Results

The examination of the difficulty shows that the proposed norm I_Ω is too detailed near the singularity and produces a topology that is too fine to allow the convergence of $\phi_n \in C_c^\infty(\Omega)$ to e_1. By means of a transformation already proposed in [8], we obtain a more suitable norm N that is equivalent to $I_\Omega^{1/2}$ on $C_0(\Omega)$ but is gross enough near the singular point. In this way, we are able to define a possibly larger closure, which we call $\mathcal{H}$, that contains all the functions needed for constructing the evolution.

We proceed next to reexamine the above-mentioned difficulty. We will show that the spaces H and $\mathscr{H}$ are indeed the same. What is different is the norm that was implicitly assumed to be acting in H for solutions that do not necessarily vanish at $x = 0$, which in principle seemed to be $I_\Omega^{1/2}$ taken in the sense of principal value. When both terms of I_Ω become infinite, the correct definition of the norm is a particular limit that we call the *cutoff limit*. This is explained in Sect. 2 where we examine the connection of the new norm with the Hardy functional; the difference is characterized in terms of a certain value, the *Hardy singularity energy* (HS energy for short) that we precisely define. We think that the existence of the two different norms that coincide on $C_c^1(\Omega \setminus \{0\})$ is quite interesting and was unexpected for us.

2. In Sect. 3 we discuss a result which has its own interest: the Critical Caffarelli–Kohn–Nirenberg Inequalities, in a bounded domain. It was well known (see [11]) that these inequalities are related to the Hardy inequality with $c < c_*$. The critical case is as expected related with the Hardy inequality with $c = c_*$. The proper functional setting that we had to consider for Problem (1) leads us naturally to these critical inequalities. We also give the connection of this new space with the Sobolev space $D^{1,2}(\mathbb{R}^N)$, with the use of a proper transformation.

3. In Sect. 4 we explore the existence of an analogue of the Hardy singularity energy for problems posed in exterior domains. The Kelvin transform suggests that the most natural problem to study is the following:

$$\begin{cases} |y|^{-4} w_t(y,t) = \Delta w(y,t) + c_* \dfrac{w(y,t)}{|y|^2}, & y \in B_\delta^c, \ t > 0, \\ w(y,0) = w_0(y), & \text{for } y \in B_\delta^c, \\ w(y,t) = 0 & \text{for } |y| = \delta, \ t > 0, \end{cases} \tag{7}$$

where $c_* = (N-2)^2/4$ is the critical coefficient, $B_\delta^c = \mathbb{R}^N \setminus B_\delta(0)$ is the standard exterior domain, and $\delta > 0$. Without loss of generality we take $\delta = 1$.

Problem (7) has the striking property that the Hardy functional posed in the exterior domain is not necessarily a positive quantity; we will show that for function which vanish at $\partial B_\delta^c(0)$ and behaving at infinity like $|y|^{-(N-2)/2}$, it may be negative We avoid the difficulty by basing our existence theory on the unitary equivalence vi the Kelvin transform. Results concerning subcritical potentials and/or Hardy-typ inequalities, in the case of unbounded domains, may be found in [2, 5, 14, 16, 17].

The novel feature of the bounded domain, namely, the HS energy term, does exis also in the case of the exterior domain, but it appears at infinity. Besides, there is big difference with the bounded domain case since in this case the new energy is n only additive to the total energy involved in the evolution; it may even represent th main part of it. The energy term at infinity looks to us like a "hidden" energy. Th seems to be the first study of an evolution problem with such curious properties an exterior domain.

Some further results on Hardy-type inequalities and the related hidden energi are contained in the forthcoming work [25].

2 Proper Functional Setting: Bounded Domain Case

We start the study by analyzing the case of a bounded domain $\Omega \subset \mathbb{R}^N$, $N \geq 3$.

2.1 Transformation and Definition of Spaces

The way we follow to address the difficulty mentioned in the introduction and to properly pose Problem (1) is to introduce a more convenient variable by means of the formula

$$u(x) = |x|^{-(N-2)/2} v(x). \tag{8}$$

We will write the transformation as $u = \mathscr{T}(v)$. Clearly, this is an isometry from the space $X = L^2(\Omega)$ into the space $\tilde{X} = L^2(d\mu, \Omega)$, $d\mu = |x|^{2-N}dx$. Many arguments of [26] were also based on transformation (8), which was first used in [8] and then in many papers concerning results on Hardy's inequalities. The great advantage of this formula is that it simplifies $I_\Omega(u)$, at least for smooth functions, into

$$I_1(v) := \int_\Omega |x|^{-(N-2)} |\nabla v|^2 \, dx. \tag{9}$$

It is easy checked that $I_\Omega(u) = I_1(v)$ for functions $u \in C_0^\infty(\Omega)$. However, the equivalence fails for functions with a singularity like $|x|^{-(N-2)/2}$ at the origin, as we have hinted before and will explain below in detail. Our proposal is to use this formulation for the definition of the new space, $\mathscr{H}$. An important observation is that when $u(x,t)$ is a solution of equation (1), then v satisfies the following associated equation: $v_t = |x|^{N-2} \nabla \cdot \left(|x|^{-(N-2)} \nabla v \right)$, with clear equivalence for $x \neq 0$. This last form gives the clue to the proper variational formulation to be followed here. First, the space associated to this equation through the quadratic form (9) is defined as the weighted space $\mathscr{H} = W_0^{1,2}(d\mu, \Omega)$, which is the completion of the $C_0^\infty(\Omega)$ functions under the norm

$$\|v\|_{\mathscr{H}}^2 = \int_\Omega |x|^{-(N-2)} |\nabla v|^2 \, dx. \tag{10}$$

Following the usual procedure of the Calculus of Variations, we take an appropriate base space which is $\tilde{X} = L^2(d\mu, \Omega)$, and then the quadratic form (9) has as form domain the subspace $\mathscr{H}$ where $I_1(v)$ is finite. Then, it can be proved that $L(v) = -|x|^{N-2} \nabla \cdot \left(|x|^{-(N-2)} \nabla v \right)$ is a positive self-adjoint operator in the space $D(L) = \{v \in \mathscr{H} : L(v) \in \tilde{X}\}$. It is also known that $D(L^{1/2}) = \mathscr{H}$. Hence, the variational approach works for v. See further analysis below.

We translate these results to the original framework. $\mathscr{H}$ is defined as the isometric space of $\mathscr{H} = W_0^{1,2}(|x|^{-(N-2)}dx, \Omega)$ under the transformation $\mathscr{T}$ given by (8). In other words, $\mathscr{H}$ is defined as the completion of the set

$$\left\{ u = |x|^{-\frac{N-2}{2}} v, \quad v \in C_0^\infty(\Omega) \right\} = \mathscr{T}(C_0^\infty(\Omega)),$$

under the norm $N(u) = \|u\|_{\mathscr{H}}$ defined by

$$\|u\|_{\mathscr{H}}^2 = \int_\Omega |x|^{-(N-2)} |\nabla(|x|^{\frac{N-2}{2}} u)|^2 dx. \tag{11}$$

2.2 Connection of Space $\mathscr{H}$ with the Hardy Functional

By Hardy functional we refer to $I_\Omega(u)$ defined in (3) with the integral defined in the sense of principal value around the origin when both separate integrals diverge. Denote by B_ε the ball centered at the origin with radius ε and by B_ε^c its complement in Ω. Assume now that $u \in \mathscr{H}$, so that $v = |x|^{(N-2)/2} u \in \mathscr{H}$. Then, we have that

$$I_{B_\varepsilon^c}[u] = \int_{B_\varepsilon^c} |\nabla u|^2 dx - \left(\frac{N-2}{2} \right)^2 \int_{B_\varepsilon^c} \frac{u^2}{|x|^2} dx. \tag{12}$$

By change of variables and integration by parts the following remarkable formula is obtained:

$$I_{B_\varepsilon^c}[u] = \int_{B_\varepsilon^c} |\nabla v|^2 |x|^{2-N} dx + \frac{N-2}{2} \varepsilon^{-(N-1)} \int_{S_\varepsilon} v^2 dS, \tag{13}$$

where dS is the surface measure. Next, we denote by Λ_ε the quantity:

$$\Lambda_\varepsilon(u) = \frac{N-2}{2} \varepsilon^{-(N-1)} \int_{S_\varepsilon} v^2 dS = \frac{N-2}{2} \varepsilon^{-1} \int_{S_\varepsilon} u^2 dS$$

that represents a kind of *Hardy energy at the singularity*. It is clear that

$$\lim_{\varepsilon \to 0} \int_{B_\varepsilon^c} |\nabla v|^2 |x|^{2-N} dx = \|v\|_{\mathscr{H}}^2.$$

In order to take the limit $\varepsilon \to 0$, in (12) we distinguish the following cases:

- If $u \in H_0^1(\Omega)$, then $u \in \mathscr{H}$ and we have $\Lambda(u) := \lim_{\varepsilon \to 0} \Lambda_\varepsilon(u) = 0$; thus, the limit as $\varepsilon \to 0$, in (12), gives the well-known formula $I_\Omega[u] = \|v\|_{\mathscr{H}}^2 := N^2(u)$, which holds for any $u \in H_0^1(\Omega)$. Note that the converse is not true: if $\Lambda(u) = 0$, it does not imply that $u \in H_0^1(\Omega)$. For example, take a function u such that v behaves zero like $(-\log|x|)^{-1/2}$.
- If $v \in \mathscr{H}$ is such that $\lim_{|x| \to 0} v^2(x) = v^2(0)$ exists as a real positive number, then $u \in \mathscr{H}$ but $u \notin H_0^1(\Omega)$. In this case $\Lambda(u) = \frac{N(N-2)}{2} \omega_N v^2(0)$, where ω_N denotes the Lebesgue measure of the unit ball in $\mathbb{R}^N$. $\Lambda(u)$ is then a well-defined positi

number and (12) implies that $I_\Omega[u] = ||v||^2_{\tilde{\mathcal{H}}} + \Lambda(u)$. We note that this is the case of e_1 and the case of the minimizer of the improved Hardy–Sobolev inequality; see [27], in the radial case. As it will be clear, this is the case for the minimizers of

$$\min_{u \in H} \frac{||u||^2_H}{||u||^q_{L^q}}, \qquad 1 \le q < \frac{2N}{N-2}. \tag{14}$$

- If $v \in \tilde{\mathcal{H}}$ is such that v at zero is bounded but the $\lim_{x \to 0} v^2(x)$ does not exist, i. e., v oscillates near zero. For example, let $v \sim \sin((-\log|x|)^a)$, $|x| \to 0$. Then, v belongs in $\tilde{\mathcal{H}}$ if $0 < a < 1/2$, so $u = |x|^{-(N-2)}v \in \mathcal{H}$. In this case, the limit $L(u)$ does not exist, since it oscillates, and from (12) we have that the same happens to the Hardy functional, in the sense that

$$\lim_{\varepsilon \to 0} \left(I_{B^c_\varepsilon}[u] - \Lambda_\varepsilon(u) \right) = ||v||^2_{\tilde{\mathcal{H}}}. \tag{15}$$

- If $v \in \tilde{\mathcal{H}}$ is such that $\lim_{x \to 0} v^2(x) = \infty$. For example, let $v \sim (-\log|x|)^a$, $|x| \to 0$. Then, v belongs in $\tilde{\mathcal{H}}$ if $0 < a < 1/2$, so $u = |x|^{-(N-2)}v \in \mathcal{H}$. It is clear that $\Lambda(u) = \infty$, and from (12) we have that the same happens to the Hardy functional, in the sense that (15) holds.

Note that in all cases Λ_ε is a nonnegative quantity, for every $\varepsilon > 0$ and so is $I_{B^c_\varepsilon}[u]$. As a consequence, we obtain a generalized form of the Hardy inequality valid in the limiting case of (15), when the Hardy functional is not defined or is infinite. We do not know if there is any physical meaning for the singularity energy we have found. It looks like an energy defect at the singularity.

2.3   The Spaces $\mathcal{H}$ and H Are the Same

We recall that H was introduced as the completion of the $C^\infty_0(\Omega)$ functions under the norm $I_\Omega^{1/2}$. The proof of $\mathcal{H} = H$ relies on showing that the set $C^\infty_0(\Omega\backslash\{0\})$ is a dense set in both spaces and on observing that the two norms coincide on that subset. The following lemma follows from [11, Lemma 2.1], which studies the subcritical case.

Lemma 1. *Holds that* $C^\infty_0(\Omega\backslash\{0\})$ *is a dense set in* H.

Next we will prove that the $C^\infty_0(\Omega\backslash\{0\})$-functions are also dense in $\mathcal{H}$ and hence in $\mathcal{H}$. The special cutoff functions that are dense in $\mathcal{H}$ are the ones that allow to prove that $\{0\}$ has zero capacity in two space dimensions.

Lemma 2. *The* $C^\infty_0(\Omega\backslash\{0\})$-*functions are dense in* $\mathcal{H}$.

From Lemmas 1 and 2 we have that the spaces H and $\mathcal{H}$ may both be defined the closure of $C^\infty_0(\Omega\backslash\{0\})$-functions, with respect to (4) and (11), respectively.

However, for such functions these two norms are equal, i. e., $||w||^2_{\mathscr{H}} = I_\Omega(w)$, for any $w \in C_0^\infty(\Omega \backslash \{0\})$. Thus,

Proposition 1. *The spaces H and $\mathscr{H}$ coincide.*

For the space $\tilde{\mathscr{H}}$, which is defined by (10), we have that

Lemma 3. *The space $\tilde{\mathscr{H}}$ contains all the functions that satisfy $v|_{\partial\Omega} = 0$ and $||v||_{\tilde{\mathscr{H}}} < \infty$.*

2.4 On the Norm of H

Let us examine further the definition of the norm N that will be considered for the space $H = \mathscr{H}$. We know that the norms $I_\Omega^{1/2}$ and N coincide on functions $H_0^1(\Omega)$ and also that $I_\Omega(u)$ is larger than $N^2(u) = ||u||^2_H$ when they differ. More precisely,

$$||u||^2_H = \lim_{\varepsilon \to 0} \left(I_{B_\varepsilon^c}[u] - \Lambda_\varepsilon(u) \right). \tag{16}$$

Now, if for any $u \in H$, we consider a sequence of cutoff approximations $u_\varepsilon(x) = \rho_\varepsilon(x)u(x)$ with ρ_ε as in Lemma (2), then $||u_\varepsilon||_H = I_\Omega(u_\varepsilon)^{1/2}$ and $u_\varepsilon \to u$ in H. The limit value $||u||^2_H = \lim_{\varepsilon \to 0} ||u_\varepsilon||^2_H$ is what we call the *cutoff value of the Hardy functional* and produces the correct norm on H.

As a conclusion, the space H as it is defined in Vazquez–Zuazua [26] is a well-defined space, as the completion of $C_0^\infty(\Omega)$-functions with respect to the norm $||\phi||^2_H = I_\Omega(\phi)$, $\phi \in C_0^\infty(\Omega)$. In this space there exist "bad functions," such that $I_\Omega(u)$ defined as an improper integral does not coincide with the limit of the sequence of cutoff approximations. Even more, it can happen that the principal value of the integrals in I_Ω is not well defined, either oscillating or infinite. For example, let u behave at the origin as $|x|^{-(N-2)/2}$; for this function, the quantity $I_\Omega(u)$ is well defined, but its norm in H is not $I_\Omega(u)$, but it is equal to $I_\Omega(u) - \Lambda(u)$. For the normalized first eigenfunction e_1, the norm is not $I_\Omega^{1/2}(e_1)$, but

$$||e_1||^2_H = I_\Omega(e_1) - \Lambda(e_1) = I_\Omega(e_1) - \frac{N(N-2)}{2}\omega_N = \mu_1.$$

As a result the minimization problem (5) and the following one $\min_{u \in H} I_\Omega(u)/||u||^2_L$ are not the same, since the first admits a minimizer while the other does not.

2.5 Application to the Evolution Problem

We now justify that the results described in [26] for the solutions of Problem (hold in the space $\mathscr{H}$ defined in (11), with Ω a bounded domain of $\mathbb{R}^N$, $N \geq 3$. is clear that this space is a Hilbert space and, as stated in [26, Theorem 2.2], $\mathscr{H}$

imbedded continuously in the Sobolev space $W^{1,q}(\Omega)$, $1 \leq q < 2$. Thus, the compact imbedding $\mathscr{H} \hookrightarrow L^p(\Omega)$, $1 \leq p < \frac{2N}{N-2}$ holds. Moreover, we may justify that all the results concerning the spectrum of the related eigenvalue problem given in [26] hold for $\mathscr{H}$.

The weak formulation (or the energy equation) of (1) is the following:

$$\frac{1}{2} \int u_t^2 = -\|u\|_H^2 = -\lim_{\varepsilon \to 0} \left(I_{B_\varepsilon^c}[u] - \Lambda_\varepsilon(u) \right)$$

for every $u \in H$. The space H is really the energetic space.

3 Further Properties of the Spaces

We observe that the above imbeddings give the following corollary, which completes the results obtained in [10] (see also [11]) concerning the Caffarelli–Kohn–Nirenberg Inequalities, in a bounded domain, in the limiting case where $a = \frac{N-2}{2}$.

Corollary 1 (Critical Caffarelli–Kohn–Nirenberg Inequalities). *Assume that v_n is a bounded sequence in $\mathscr{H}$. Then $u_n = |x|^{-(N-2)/2} v_n$ is a bounded sequence in $\mathscr{H}$. The compact imbeddings imply that up to some subsequence, u_n converges in $L^q(\Omega)$ to some u. Thus, we obtain the compact imbeddings*

$$\tilde{\mathscr{H}} \hookrightarrow L^q(|x|^{-q(N-2)/2} dx, \Omega), \quad \text{for any} \quad 1 \leq q < \frac{2N}{N-2}. \tag{17}$$

Then, for every $0 \leq s \leq \frac{N-2}{2} q$, we further obtain the compact imbeddings

$$\tilde{\mathscr{H}} \hookrightarrow L^q(|x|^{-s} dx, \Omega), \quad \text{for any} \quad 1 \leq q < \frac{2N}{N-2}. \tag{18}$$

Remark 1. In (17) it is clear that q cannot reach $\frac{2N}{N-2}$. For this value of q, the best that we can have are improved Hardy–Sobolev inequalities; see [3, 15, 27] and the references therein.

In addition, we can relate these spaces, in the radial case, with the space $\mathscr{D}^{1,2}(\mathbb{R}^N)$. If we denote by $\mathscr{H}_r(\Omega)$, $\tilde{\mathscr{H}}_r$, and $\mathscr{D}_r^{1,2}(\mathbb{R}^N)$ the subspaces of $\mathscr{H}$, $\tilde{\mathscr{H}}$, and $\mathscr{D}^{1,2}(\mathbb{R}^N)$, respectively, which consist of radial functions, we have that

Proposition 2. *For some function $v \in \tilde{\mathscr{H}}_r(B_R)$ we set*

$$v(|x|) = w(t), \quad t = \left(-\log\left(\frac{|x|}{R} \right) \right)^{-\frac{1}{N-2}}. \tag{19}$$

Then, $v \in \tilde{\mathscr{H}}_r(B_R)$ if and only if $w \in \mathscr{D}_r^{1,2}(\mathbb{R}^N)$ and

$$||v||_{\tilde{\mathscr{H}}_r(B_R)} = (N-2)^{-1}||w||_{D_r^{1,2}(\mathbb{R}^N)}. \tag{20}$$

Observe that (20) is independent of the radius R, and in the case where $N = 3$, the norm in $\tilde{\mathscr{H}}_r(B_R)$ coincides with the norm in $\mathscr{D}_r^{1,2}(\mathbb{R}^N)$. Similarly we may argue for the space $\mathscr{H}_r$. Transformation (19) was used in [27]. For a discussion concerning the construction of such transformations for Hardy-type inequalities, we refer to [25].

3.1 Nonexistence of H_0^1-Minimizers

The above transformations provide us with an extra argument concerning the nonexistence of H_0^1-minimizers. We will prove that these minimizers belong to H, they do not belong to H_0^1, and their behavior at the origin is exactly $|x|^{-(N-2)/2}$.

Proposition 3. *The minimizers of (14) cannot exist in $H_0^1(\Omega)$ and behave at the origin like $|x|^{-(N-2)/2}$.*

As mentioned before, the case of e_1 and the minimizer of the improved Hardy–Sobolev inequality (in the radial case) behave at the origin like $|x|^{-(N-2)/2}$. Thus, the Hardy functional for these functions is a well-defined positive number, although it does not represent their H-norm. These functions do not belong to the "worst" cases, where I_Ω is not well defined or is infinite. As in the case of e_1, we emphasize the fact that the minimization problem (14) and the following $\min_{u \in H} I_\Omega(u)/||u||_{L^q}^q$ are not the same.

4 The Case of the Exterior Domain

We consider Problem (7) describing the evolution (up to some weight) of the Hardy potential in an exterior domain. We may fix $\delta = 1$. Our arguments will be based on the unitary equivalence with the previous problem posed on a ball. For that we use the Kelvin transform in the form $u(x) = |y|^{N-2}w(y)$, $y = x/|x|^2$. These formula transform solutions $u(x,t)$ of Problem (1), i.e.,

$$\begin{cases} u_t(x,t) = \Delta u(x,t) + c_* \dfrac{u(x,t)}{|x|^2}, \ x \in B_1(0), \ t > 0, \\ u(x,0) = u_0(x), \ \text{for } x \in B_1(0), \\ u(x,t) = 0 \ \text{in } \partial B_1(0), \ t > 0, \end{cases} \tag{21}$$

defined in the unit sphere $B_1(0)$ into solutions $w(y,t)$ of Problem (7) posed in $B_1^c(0)$ its complement in $\mathbb{R}^N$. We will write the transformation as $u = \mathscr{K}(w)$.

4.1 Basic Properties

We will address the questions of existence of solutions for Problem (7) by means of this equivalence that will also be used as a clue to the proper variational formulation that will be followed. As we saw, the functional space which corresponds to (21) is H with norm given in (16). Our proposal is to use this formulation for the definition of the new space, $\mathcal{W}$. The space $\mathcal{W}$ is defined as the isometric space of H under the Kelvin transformation. In other words, $\mathcal{W}$ is defined as the completion of the set

$$\left\{ w(y) = |y|^{-N+2} u\left(\frac{y}{|y|^2}\right), \quad u \in C_0^\infty(B_1(0)), \quad |y| \geq 1 \right\},$$

under the norm $\|w\|_{\mathcal{W}}$ defined by $\|w\|_{\mathcal{W}}^2 = \lim_{\varepsilon \to 0} \left(I_{B_1(0) \setminus B_\varepsilon}[u] - \Lambda_\varepsilon(u)\right) u = \mathcal{K}(w)$.
The first eigenpair of the corresponding eigenvalue problem is $\mu_1 = z_{0,1}^2$, $\tilde{e}_1 = |y|^{-(N-2)/2} J_0\left(\frac{z_{0,1}}{|y|}\right)$. The well posedness of (7) in the space $\mathcal{W}$ is understood through the unitary equivalence with H. The existence, uniqueness, and stabilization results of Problem (21) apply for Problem (7).

4.2 Hardy Functional

Next, we investigate the connection of the space $\mathcal{W}$ with the Hardy functional, $I_{B_1^c(0)}$, defined as

$$I_{B_1^c(0)}[\phi] = \int_{\mathbb{R}^N \setminus B_1(0)} |\nabla \phi|^2 \, dx - \left(\frac{N-2}{2}\right)^2 \int_{\mathbb{R}^N \setminus B_1(0)} \frac{\phi^2}{|x|^2} \, dx, \tag{22}$$

which is positive for any compactly supported $\phi \in C^\infty(B_1^c(0))$ that vanishes on the boundary. We denote by I_ε and by $I_{1/\varepsilon}$ the Hardy functional defined on $B_1(0) \setminus B_\varepsilon$ and $B_{1/\varepsilon}(0) \setminus B_1(0)$, respectively.

Lemma 4. We have the following fundamental relation:

$$I_\varepsilon[u] = I_{1/\varepsilon}[w] + 2\Lambda_{1/\varepsilon}(w), \tag{23}$$

where

$$\Lambda_{1/\varepsilon}(w) = \frac{N-2}{2} \varepsilon \int_{\partial B_{1/\varepsilon}(0)} w^2 \, dS.$$

Moreover, it is clear that if $u = \mathcal{K}(w)$, then $\Lambda_\varepsilon(u) = \Lambda_{1/\varepsilon}(w)$.

When we apply these results to functions in the class $\mathcal{W}$ (by density), we are able to give the following unexpected definition of the norm of $\mathcal{W}$,

$$||w||_{\mathcal{W}}^2 = \lim_{\varepsilon \to 0} \left(I_{1/\varepsilon}[w] + \Lambda_{1/\varepsilon}(w) \right). \tag{24}$$

So, the weak formulation (or the energy equation) of (7) translates into

$$\frac{1}{2} \int w_t^2 = -||w||_{\mathcal{W}}^2 = -\lim_{\varepsilon \to 0} \left(I_{1/\varepsilon}[w] + \Lambda_{1/\varepsilon}(w) \right),$$

for every $w \in \mathcal{W}$.

4.3   Conclusions and Remarks

1. We have shown that a correcting term also appears as in the energy analysis of the problem posed in the exterior domain. Actually, the correcting term has the same absolute value as the Hardy singularity energy considered in the problem in a bounded domain, but now it represents a kind of energy at infinity. However, there is a big difference from the bounded domain case since in this case the singular energy acts in an additive way to the usual Hardy integral.
2. Moreover, this new term may be the main part of the total energy, since $I_{1/\varepsilon}$ may be also a negative quantity: we do the calculations for the normalized $\tilde{e}_1$ and we get that $I_{\mathbb{R}^N \setminus B_1(0)}(\tilde{e}_1) = 5.76 - \frac{N(N-2)}{2} \omega_N$, which is negative for $N = 3$.
3. As a conclusion, we can say that Λ is a hidden energy that "comes" from infinity and is not only a gain of the total energy but it may represent the main part of the total energy.

Acknowledgements The first author was supported by Spanish Project MTM2008-06326-C02-01 The second author was partially supported by P.E.B.E. of NTUA. The work partly done during a visit of the second author to Univ. Auónoma de Madrid was supported by the same project.

References

1. Adimurthi, Chaudhuri, N., Ramaswamy M.: An improved HardySobolev inequality and it application. Proc. Amer. Math. Soc. **130**, 489–505 (2002)
2. Adimurthi Esteban, M.J.: An improved Hardy–Sobolev inequality in $W^{1,p}$ and its applicatio to Schrödinger operators. Nonlinear differ. Equ. Appl. **12**, 243–263 (2005)
3. Adimurthi, Filippas, S., Tertikas, A.: On the best constant of Hardy–Sobolev inequalitie Nonlinear Anal. **70**, 2826–2833 (2009)
4. Baras, P., Goldstein, J.A.: The heat equation with a singular potential. Trans. Am. Math. So **284**(1), 121–139 (1984)
5. Barbatis, G., Filippas S., Tertikas A.: Critical heat kernel estimates for Schrödinger operato via Hardy–Sobolev inequalities. J. Funct. Anal. **208**, 1–30 (2004)
6. Blanchet, A., Bonforte, M., Dolbeault, J., Grillo, G., Vzquez, J.-L.: Hardy-Poincaré inequaliti and applications to nonlinear diffusions. C. R. Acad. Sci. Paris **344**, 431–436 (2007)
7. Brezis H., Marcus, M.: Hardy's inequality revisited. Ann. Sc. Norm. Pisa **25**, 217–237 (199

8. Brezis, H., Vázquez, J.L.: Blowup solutions of some nonlinear elliptic problems. Rev. Mat. Univ. Complutense Madrid **10**, 443–469 (1997)

9. Cabré, X., Martel, Y.: Existence versus explosion instantané pour des equations de lachaleur linéaires avec potentiel singulier. C. R. Acad. Sci. Paris **329**, 973–978 (1999)

10. Caffarelli, L., Kohn R., Nirenberg, L.: First order interpolation inequalities with weights. Compos. Math. **53**(3), 259–275 (1984)

11. Catrina F., Wang, Z.-Q.: On the Caffarelli–Kohn–Nirenberg inequalities: Sharp constants, existence (and nonexistence) and symmetry of extremal functions. Commun. Pure Appl. Math. **LIV**, 229–258 (2001)

12. Cianchi A., Ferone, A.: Hardy inequalities with non-standard remainder terms. Ann. I. H. Poincaré AN **25**, 889–906 (2008)

13. Davies, E.B.: A review of Hardy inequalities. Oper. Theory Adv. Appl. **110**, 55–67 (1999)

14. Ekholm T., Frank, R.L.: On lieb-thirring inequalities for Schrödinger operators with virtual Level. Commun. Math. Phys. **264**, 725–740 (2006)

15. Filippas S., Tertikas, A.: Optimizing improved Hardy inequalities. J. Funct. Analysis **192**, 186–233 (2002); Corrigendum, J. Funct. Analysis **255**, 2095 (2008)

16. Frank, R.L.: A simple proof of Hardy-Lieb-Thirring inequalities. Commun. Math. Phys. **290**, 789–800 (2009)

17. Frank, R.L., Lieb, E.H., Seiringer, R.: Hardy-Lieb-Thirring inequalities for fractional Schrödinger operators. J. Am. Math. Soc. **21**(4), 925–950 (2008)

18. Gazzola, F., Grunau, H.C., Mitidieri, E.: Hardy inequalities with optimal constants and remainder terms. Trans. Am. Math. Soc. **356**, 2149–2168 (2004)

19. Ghoussoub N., Moradifam, A.: Bessel potentials and optimal Hardy and Hardy-Rellich inequalities. Math. Ann. **349**, 1–57 (2011)

20. Goldstein J.A., Zhang, Q.S.: Linear parabolic equations with strong singular potentials. Trans. Am. Math. Soc. **355**, 197–211 (2003)

21. Kufner, A., Maligranda, L., Persson, L.-E.: The Hardy Inequality: About its History and Some Related Results. Vydavatelsky' Servis, Plzen (2007).

22. Maz'ja, V.G.: Sobolev Spaces. Springer, Berlin (1985)

23. Opic, B., Kufner, A.: Hardy type inequalities. Pitman Res. Notes Math. **219**, (1990)

24. Vázquez, J.L., Zographopoulos, N.B.: Functional aspects of the Hardy inequality. Appearance of a hidden energy, arxiv:1102.5661

25. Vázquez, J.L., Zographopoulos, N.B.: Functional aspects of Hardy type inequalities, in preparation.

26. Vázquez, J.L., Zuazua, E.: The Hardy inequality and the asymptotic behaviour of the heat equation with an inverse-square potential. J. Funct. Anal. **173**, 103–153 (2000)

27. Zographopoulos, N.B.: Existence of extremal functions for a Hardy–Sobolev inequality., J. Funct. Anal. **259**, 308–314 (2010)

Printed in the United States
By Bookmasters